ASSESSMENT OF TECHNOLOGIES FOR Improving Light-Duty Vehicle Fuel Economy—2025–2035

Committee on Assessment of Technologies for Improving Fuel Economy of Light-Duty Vehicles—Phase 3

Board on Energy and Environmental Systems

Division on Engineering and Physical Sciences

A Consensus Study Report of

The National Academies of

SCIENCES · ENGINEERING · MEDICINE

THE NATIONAL ACADEMIES PRESS

Washington, DC

www.nap.edu

THE NATIONAL ACADEMIES PRESS **500 Fifth Street, NW** **Washington, DC 20001**

This activity was supported by Award No. DTNH2217H00028 of the U.S. Department of Transportation and National Highway Traffic Safety Administration. Any opinions, findings, conclusions, or recommendations expressed in this publication do not necessarily reflect the views of any organization or agency that provided support for the project.

International Standard Book Number-13: 978-0-309-37122-3
International Standard Book Number-10: 0-309-37122-8
Digital Object Identifier: https://doi.org/10.17226/26092
Library of Congress Control Number: 2021951949

Additional copies of this publication are available from the National Academies Press, 500 Fifth Street, NW, Keck 360, Washington, DC 20001; (800) 624-6242 or (202) 334-3313; http://www.nap.edu.

Printed in the United States of America

Suggested citation: National Academies of Sciences, Engineering, and Medicine. 2021. *Assessment of Technologies for Improving Light-Duty Vehicle Fuel Economy—2025–2035*. Washington, DC: The National Academies Press. https://doi.org/10.17226/26092.

The National Academies of

SCIENCES • ENGINEERING • MEDICINE

The **National Academy of Sciences** was established in 1863 by an Act of Congress, signed by President Lincoln, as a private, nongovernmental institution to advise the nation on issues related to science and technology. Members are elected by their peers for outstanding contributions to research. Dr. Marcia McNutt is president.

The **National Academy of Engineering** was established in 1964 under the charter of the National Academy of Sciences to bring the practices of engineering to advising the nation. Members are elected by their peers for extraordinary contributions to engineering. Dr. John L. Anderson is president.

The **National Academy of Medicine** (formerly the Institute of Medicine) was established in 1970 under the charter of the National Academy of Sciences to advise the nation on medical and health issues. Members are elected by their peers for distinguished contributions to medicine and health. Dr. Victor J. Dzau is president.

The three Academies work together as the **National Academies of Sciences, Engineering, and Medicine** to provide independent, objective analysis and advice to the nation and conduct other activities to solve complex problems and inform public policy decisions. The National Academies also encourage education and research, recognize outstanding contributions to knowledge, and increase public understanding in matters of science, engineering, and medicine.

Learn more about the National Academies of Sciences, Engineering, and Medicine at **www.nationalacademies.org**.

The National Academies of
SCIENCES • ENGINEERING • MEDICINE

Consensus Study Reports published by the National Academies of Sciences, Engineering, and Medicine document the evidence-based consensus on the study's statement of task by an authoring committee of experts. Reports typically include findings, conclusions, and recommendations based on information gathered by the committee and the committee's deliberations. Each report has been subjected to a rigorous and independent peer-review process and it represents the position of the National Academies on the statement of task.

Proceedings published by the National Academies of Sciences, Engineering, and Medicine chronicle the presentations and discussions at a workshop, symposium, or other event convened by the National Academies. The statements and opinions contained in proceedings are those of the participants and are not endorsed by other participants, the planning committee, or the National Academies.

For information about other products and activities of the National Academies, please visit www.nationalacademies.org/about/whatwedo.

COMMITTEE ON ASSESSMENT OF TECHNOLOGIES FOR IMPROVING FUEL ECONOMY OF LIGHT-DUTY VEHICLES—PHASE 3

GARY MARCHANT, Arizona State University, *Chair*
CARLA BAILO, Center for Automotive Research
RODICA BARANESCU, NAE,[1] University of Illinois, Chicago (retired) (resigned September 2020)
NADY BOULES, NB Motors, LLC
DAVID L. GREENE, University of Tennessee, Knoxville (resigned March 2021)
DANIEL KAPP, D.R. Kapp Consulting, LLC
ULRICH KRANZ, Canoo
THERESE LANGER, American Council for an Energy-Efficient Economy
ZHENHONG LIN, Oak Ridge National Laboratory
JOSHUA LINN, University of Maryland, College Park
NIC LUTSEY, International Council on Clean Transportation
JOANN MILLIKEN, Independent Consultant, Alexandria, Virginia
RANDA RADWAN, Highway Safety Research Center, University of North Carolina, Chapel Hill
ANNA STEFANOPOULOU, University of Michigan and Automotive Research Center
DEIDRE STRAND, Wildcat Discovery Technologies
KATE WHITEFOOT, Carnegie Mellon University

Staff

ELIZABETH ZEITLER, Associate Director, Board on Energy and Environmental Systems (BEES), *Study Director*
REBECCA DEBOER, Research Assistant, BEES
BRENT HEARD, Associate Program Officer, BEES (beginning January 2020)
K. JOHN HOLMES, Director/Scholar, BEES
MICHAELA KERXHALLI-KLEINFIELD, Research Associate, BEES
KASIA KORNECKI, Associate Program Officer, BEES (beginning February 2020)
KATHERINE KORTUM, Senior Program Officer, Transportation Research Board
HEATHER LOZOWSKI, Senior Financial Business Partner, BEES
BEN WENDER, Senior Program Officer, BEES (until December 2019)
CATHERINE WISE, Associate Program Officer, BEES (beginning June 2020)

NOTE: See Appendix B, Disclosure of Conflicts of Interest.

[1]Member, National Academy of Engineering.

BOARD ON ENERGY AND ENVIRONMENTAL SYSTEMS

JARED COHON, NAE,[1] Carnegie Mellon University, *Chair*
VICKY BAILEY, Anderson Stratton Enterprises
CARLA BAILO, Center for Automotive Research
W. TERRY BOSTON, NAE, GridLiance GP, LLC, and Grid Protection Alliance
DEEPAKRAJ DIVAN, NAE, Georgia Institute of Technology
MARCIUS EXTAVOUR, XPRIZE
KELLY SIMS GALLAGHER, Tufts University
TJ GLAUTHIER, TJG Energy Associates, LLC
NAT GOLDHABER, Claremont Creek Ventures
DENISE GRAY, LG Chem Michigan, Inc.
JOHN KASSAKIAN, NAE, Massachusetts Institute of Technology
BARBARA KATES-GARNICK, Tufts University
DOROTHY ROBYN, Boston University
JOSÉ SANTIESTEBAN, NAE, ExxonMobil Research and Engineering Company
ALEXANDER SLOCUM, NAE, Massachusetts Institute of Technology
JOHN WALL, NAE, Cummins, Inc. (retired)
ROBERT WEISENMILLER, California Energy Commission (former)

Staff

K. JOHN HOLMES, Director/Scholar
ELIZABETH ZEITLER, Associate Director
HEATHER LOZOWSKI, Senior Financial Business Partner
BEN WENDER, Senior Program Officer (until December 2019)
BRENT HEARD, Associate Program Officer (beginning January 2020)
KASIA KORNECKI, Associate Program Officer (beginning February 2020)
CATHERINE WISE, Associate Program Officer (beginning June 2020)
MICHAELA KERXHALLI-KLEINFIELD, Research Associate
REBECCA DeBOER, Research Assistant

[1]Member, National Academy of Engineering.

Preface

Passenger car and truck manufacturers have faced corporate average fuel economy standards since 1978 and greenhouse gas emissions standards since 2012, governed by several statutes and specified in regulations from the U.S. Department of Transportation's National Highway Traffic Safety Administration (NHTSA) and the U.S. Environmental Protection Agency (EPA). Over this period, vehicle efficiency technology has advanced dramatically, including improvements to internal combustion engine powertrains; introductions of efficient hybrid, electric, and fuel cell vehicles; improvements to vehicle aerodynamics and mass reduction technologies; and introduction of limited vehicle automation. NHTSA and EPA have increasingly incorporated technology analysis into estimate costs and benefits of fuel economy and greenhouse gas standards. Beginning in 2007, Congress requested that the National Academies of Sciences, Engineering, and Medicine undertake periodic review of technologies for fuel economy standards. Most recently, NHTSA contracted with the National Academies to form the Committee on Assessment of Technologies for Improving Fuel Economy of Light-Duty Vehicles—Phase 3 to update the requested technology, consumer behavior, and policy analysis of vehicle efficiency technologies for 2025–2035.

The committee was asked to assess technologies for improving the fuel economy of light-duty vehicles in 2025–2035 and to provide updated estimates of the potential cost, fuel economy improvements, and barriers to deployment of these technologies. The committee was asked to consider internal combustion engine, electric, and fuel cell propulsion systems; non-powertrain technologies; the structure of the fuel economy regulations related to new technologies; shifts in personal transportation and vehicle ownership models; and consumer behavior associated with new efficiency technologies.

The committee comprised a wide array of backgrounds and sought input from agency officials, vehicle manufacturers, equipment suppliers, consultants, nongovernmental organizations, academicians, and many other experts. In addition to regular committee meetings, committee members held webinars on several critical topics, spoke in public sessions with experts in state and federal government, and conducted numerous information-gathering site visits to automobile manufacturers and suppliers. The committee put great effort into thorough preparation for these meetings, asked probing questions, and requested follow-up information in order to understand the perspectives of the many stakeholders. In addition, the committee commissioned a material substitution and mass reduction study from the Center for Automotive Research in order to better understand the opportunities for these advances. I greatly appreciate the considerable time and effort contributed by the committee's individual members throughout our information-gathering process, report writing, and deliberations, and especially for persevering through the challenges presented by the COVID-19 pandemic during the important final stages of completing our report.

The committee operated under the auspices of the National Academies Board on Energy and Environmental Systems, in collaboration with the Transportation Research Board. I would like to recognize the study staff for organizing and planning meetings, and assisting with information gathering and report development. The efforts of our hardworking and knowledgeable study director Elizabeth Zeitler, ably assisted by her National Academies colleagues Rebecca DeBoer, Michaela Kerxhalli-Kleinfield, Brent Heard, Kasia Kornecki, Catherine Wise, K. John Holmes, and Katherine Kortum, were critical to the committee's delivery of its report. I would also like to recognize Ben Wender and Janki Patel for their early input. Thanks are also due to the many experts and presenters, too numerous to name individually, who contributed to the committee's data-gathering process. Their contributions were invaluable and are listed in Appendix C.

Gary Marchant,
Chair, Committee on Assessment of Technologies
for Improving Fuel Economy of Light-Duty Vehicles—Phase 3

Acknowledgment of Reviewers

This Consensus Study Report was reviewed in draft form by individuals chosen for their diverse perspectives and technical expertise. The purpose of this independent review is to provide candid and critical comments that will assist the National Academies of Sciences, Engineering, and Medicine in making each published report as sound as possible and to ensure that it meets the institutional standards for quality, objectivity, evidence, and responsiveness to the study charge. The review comments and draft manuscript remain confidential to protect the integrity of the deliberative process.

We thank the following individuals for their review of this report:

GEORGE CRABTREE, NAS,[1] Argonne National Laboratory Joint Center for Energy Storage Research
PATRICK DAVIS, Strategic Marketing Innovations
DANIEL GASPAR, Pacific Northwest National Laboratory
CHRIS GEARHART, National Renewable Energy Laboratory
KENNETH GILLINGHAM, Yale University
ROY GOUDY, Nissan Technical Center North America
CHRIS HENDRICKSON, NAE,[2] Carnegie Mellon University
ASHLEY HORVAT, Greenlots
JEREMY MICHALEK, Carnegie Mellon University
MARGE OGE, U.S. Environmental Protection Agency (retired)
GREG PANNONE, IHS Markit
HUEI PENG, University of Michigan
GIORGIO RIZZONI, The Ohio State University
GARY W. ROGERS, Roush Enterprises
KRISTIN SLANINA, TrueCar
GUI-JIA SU, Oak Ridge National Laboratory
RICHARD YEN, Altair

Although the reviewers listed above provided many constructive comments and suggestions, they were not asked to endorse the conclusions or recommendations of this report nor did they see the final draft before its release. The review of this report was overseen by Susan Hanson, Clark University, and Andrew Brown Jr., Diamond Consulting, Engineering, and Management Services. They were responsible for making certain that an independent examination of this report was carried out in accordance with the standards of the National Academies and that all review comments were carefully considered. Responsibility for the final content rests entirely with the authoring committee and the National Academies.

[1]Member, National Academy of Sciences.
[2]Member, National Academy of Engineering.

Contents

Summary

The period from 2025 to 2035 could bring the most fundamental transformation in the 100-plus year history of the automobile. Battery electric vehicle (BEV) costs are likely to fall and reach parity with internal combustion engine vehicles (ICEVs). New generations of fuel cell vehicles will be produced. Connected and automated vehicle technologies will become more common, including likely deployment of some fully automated vehicles. These new categories of vehicles will for the first time assume a major portion of new vehicle sales, while ICEVs with improved powertrain, design, and aerodynamics will continue to be an important part of new vehicle sales and fuel economy improvement. An important driver of greater vehicle fuel economy will be the growing national priority to reduce greenhouse gas (GHG) emissions. These developments will impact automaker options for vehicle efficiency and bring about changes to consumer behavior and vehicle system services, including dealerships, vehicle service and repair, fueling and charging infrastructure, and transportation planning.

This report of the National Academies of Sciences, Engineering, and Medicine Committee on Assessment of Technologies for Improving Fuel Economy of Light-Duty Vehicles—Phase 3 addresses the potential for internal combustion engine, hybrid, battery electric, fuel cell, non-powertrain, and connected and automated vehicle technologies to contribute to efficiency in 2025–2035. It explores consumer and manufacturer responses to these technologies, and the regulatory aspects of fuel efficiency technologies. The report's messages are summarized here and developed in greater detail in the body of the report, with findings and recommendations and technology cost and effectiveness estimates. Specifically, Chapters 1–3 provide historical, regulatory, and technical context for vehicle fuel economy up to 2025. Chapters 4–10 discuss vehicle and fuel technologies. Chapters 11 and 12 discuss consumer and regulatory aspects of fuel efficiency. Chapter 13 synthesizes the previous chapters' content to make overall findings and recommendations about the future of light-duty vehicle fuel efficiency in 2025–2035, and advise Congress, the U.S. Department of Transportation (DOT), and the U.S. Environmental Protection Agency (EPA) as they move forward under existing or future mandates for vehicle efficiency. Significantly, the committee finds that the current statutory authority and regulatory structure for fuel economy is rapidly becoming outdated in legal, scientific, policy, technological, and global leadership perspectives, and should be updated before 2025–2035 to reflect national goals for transportation efficiency and emissions.

CONTEXT

Fuel economy requirements were first legislated in 1975 and have periodically increased with congressional action and regulations promulgated by DOT. During this time, the vehicle population, miles driven, and average vehicle performance have increased. Many efficiency technologies have achieved greater than 25% penetration by model year (MY) 2017, including variable valve timing, gasoline direct injection, 6-speed or greater transmissions, and improved tire rolling resistance. MY 2017 vehicles also showed greater than 15% penetration of variable valve lift, turbocharging, continuously variable transmissions, stop start, and 10% improvement in both aerodynamics and mass reduction, relative to the regulatory baseline. Alternative fuel vehicles have been developed, commercialized, improved in functionality, and proliferated in model availability. To advance from 2017 to 2025, automakers may pursue different pathways for efficiency improvements, but the least cost paths may include reductions in road loads such as rolling resistance, aerodynamic drag, and mass, as well as engine technologies like application of Miller and Atkinson cycles along with cooled exhaust gas recirculation (EGR) and transmission technologies such as 8-, 9-, and 10-speed transmissions.

INTERNAL COMBUSTION ENGINE-BASED POWERTRAIN TECHNOLOGIES

The future of efficiency for ICE-based powertrains in 2025–2035 will continue to focus on improving engine efficiency and reducing use of inefficient operating modes through engine technologies, complementary transmission technologies, and electrical assistance via hybridization.

In 2025–2035, thermal efficiency of turbocharged, downsized engines will improve by applying the Miller cycle to increase compression ratios. Potential enabling technologies include variable valve lift, variable compression ratio, cooled EGR, variable geometry turbine turbocharging, electric intake cam phasing, and increased fuel injection pressure. Further technologies to reduce pumping losses will include cylinder deactivation on 4- and 3-cylinder engines. These advances have the potential to achieve peak thermal efficiencies up to 40%.

Naturally aspirated engines of 2–2.5 L can readily utilize a simpler form of cooled EGR, cylinder deactivation, and variable valve lift. These engines are complementary to and used in strong hybrid applications as well, where lower engine performance demand can result in lower cost and very high peak thermal efficiency via the Atkinson cycle. Mostly in large sport utility vehicles (SUVs) and pickups, naturally aspirated V8 and V6 engines continue to fill some niches of high performance at lower cost, but they will continue to be substituted with downsized/boosted alternatives and/or hybridization.

Electric hybridization represents the ultimate efficiency approach for gasoline-fueled vehicles. In 2025–2035, there is likely to be expansion of 12 V stop-start systems, 48 V mild hybrids, and powersplit and P2 strong hybrids, and implementation of series strong hybrids and additional P2 offerings in larger vehicles. The electric components of the hybrid powertrain will have improved materials for higher efficiency and lower cost, including improved motor magnet materials, silicon carbide or gallium nitride power electronics, and battery cathodes with higher nickel content. The internal combustion engine can achieve higher efficiency when specifically developed to take advantage of hybrid synergies.

Transmissions in 2025 will typically have at least 6 speeds but will continue to transition to 8–10 speeds in 2025–2035, and some manufacturers will continue to utilize advanced continuously variable transmissions. Increased ratio span and more discreet operating points are highly complementary to the engine technologies described above. Integration of electrification and transmission will be a key development.

BATTERY AND FUEL CELL ELECTRIC VEHICLES

Automakers have developed electrified powertrain systems with zero or ultra-low tailpipe emissions. Many automakers have sold BEVs and plug-in hybrid electric vehicles (PHEVs), with a market penetration of roughly 2% in 2019. Electrified powertrains include an electric drive (electric motor, inverter, and controller) and a battery or fuel cell. Most automakers have converged on brushless permanent magnet synchronous motors with rare-earth magnets over induction motors owing to their superior power density, torque, and overall efficiency. Wide-bandgap power-switching devices offer potential electric drive cost and performance improvements in 2025–2035.

Lithium-ion batteries will be the dominant battery chemistry in 2025–2035; much uncertainty remains regarding near-term commercial readiness of advanced battery concepts (e.g., solid-state batteries). Incremental engineering and manufacturing improvements to current chemistries will result in a roughly 7% annual cost reduction through 2030. Estimated pack-level costs are $90–$115/kWh by 2025 and $65–$80/kWh in 2030; thus, price parity with ICEVs is expected in 2025–2030. Reducing battery cost and improving charging infrastructure options may increase consumer appeal and adoption of BEVs. Improved charging infrastructure could expand possibilities for shorter-range BEVs, although there is currently a strong industry trend toward increasing EV range. Full-fuel-cycle BEV emissions will decrease with decreasing electric grid emissions.

Fuel cell electric vehicles (FCEVs) may become cost competitive with ICEVs and BEVs in 2025–2035, particularly in larger vehicles and vehicles with heavier use such as taxi fleets. Lack of hydrogen fueling infrastructure as well as high hydrogen costs are obstacles to FCEV adoption. Three automakers—Honda, Hyundai, and Toyota—have introduced light-duty FCEVs for sale or lease in California, Japan, and Germany, where government-industry partnerships are building hydrogen refueling networks. Research and development (R&D) efforts to reduce precious metal content in fuel cell assemblies and lower the cost of producing, delivering, and storing hydrogen are under way.

NON-POWERTRAIN TECHNOLOGIES

Improvements in non-powertrain technologies for road and accessory load reduction will increase vehicle efficiency in 2025–2035. Lightweighting via materials substitution and design optimization is the largest opportunity for road load reduction. In 2025–2035, material use is expected to shift away from mild- and high-strength steels and toward generation three steels and aluminum, with some contribution from magnesium and polymer composites. Design optimization will be important to offset weight added for electrification. Improvements in aerodynamic drag include the replacement of outside mirrors with cameras, pending safety approval. Consumer preference for crossover utility vehicles (CUVs) and SUVs, rather than sedans, and increased electrification will influence fleetwide aerodynamic performance. Reduced tire rolling resistance of 10% to possibly 30% from the current baseline will likely occur in 2025–2035 from new materials, design, and construction. Accessory load reductions for efficiency include air conditioning improvements, accessory electrification, low-drag-resistance brakes, and secondary axle disconnect.

Vehicle safety must be considered as technologies are implemented to improve fuel economy. In particular, the National Highway Traffic Safety Administration (NHTSA) should study (1) the potential changes in crash type and severity that could occur as a result of increased advanced driver assistance system (ADAS) implementation and (2) the potential changes in mass disparity that could occur in a fleet with increased penetration of electric vehicles, ADAS, CUVs, SUVs, and pickup trucks. Furthermore, the Federal Motor Vehicle Safety Standards for crashworthiness should consider crash compatibility with emphasis on differences in vehicle mass and design.

CONNECTED AND AUTOMATED VEHICLES

Connected and automated vehicles (CAVs) use sensing, control, and communication technologies to respond to external information and take increasing control of tasks previously handled by the driver. Automation levels[1] are defined by the amount of driver intervention versus automation system control. Automation is developed for safety, convenience, accessibility, productivity, and commerce and entertainment. If designed with efficiency in mind, automated driving could substantially improve fuel efficiency, thereby lowering fuel costs, increasing driving range for electric vehicles, and delivering societal benefits through reduced fuel use and emissions.

[1] Automation levels used in this report are those defined by the Society of Automotive Engineers, ranging from Level 1 (driver assistance, steering or acceleration) and Level 2 (partial automation, both steering and acceleration), to Level 5 (full automation, all driving tasks and all driving modes, no driver intervention).

Technologies to enable automation include radar, lidar, cameras, ultrasound, data, and mapping technology. Technologies to enable connectivity to other vehicles and infrastructure include short-range radio and cellular systems. Several systems for Level 2 automation are already commercially available, with more than 10% of new vehicles equipped with Level 2 technologies by early 2019. Level 4/5 vehicles are being piloted in U.S. cities. Fully automated vehicles are expected to be deployed in 2025–2035, especially in fleets.

Vehicle automation and connectivity can increase or decrease fuel consumption. Decreases in fuel consumption of individual vehicles come primarily from velocity optimization for "eco-driving," and powertrain efficiency optimization, particularly for hybrid vehicles. Individual automation technologies may result in fuel savings of more than 8% in some driving conditions, although they can also increase fuel use if not implemented for efficiency. Connectivity and automation together could increase fuel efficiency of individual ICEVs by as much as 9% over an urban drive cycle and up to 20% for a PHEV over a combined urban/highway cycle. Off-cycle credits should be available for CAV technologies only to the extent that they improve the fuel efficiency of the vehicle on which they are installed and use realistic assumptions of technology adoption on other vehicles or infrastructure. The agencies should consider energy-saving CAV technologies in setting fuel economy standard stringency.

*System effects, particularly for significant deployment of fully autonomous vehicles, may also have fuel consump*tion impacts. Some of these effects have relatively straightforward relationships with fuel consumption, including increased vehicle miles traveled (VMT), changes in congestion, and changes in vehicle size and weights. Some have more complex relationships with fuel consumption, including changes in vehicle ownership models, and the interaction of electrification and automation. Research to date indicates that autonomous vehicles at full penetration could plausibly produce a 40% reduction to a 70% increase in energy consumption, and thus policies will be required to ensure that these vehicles achieve net energy savings. NHTSA should consider how autonomous vehicle properties and usage differ from conventional vehicles and how this should be reflected in the stringency and structure of fuel economy standards. NHTSA should consider regulating fuel efficiency of autonomous vehicles for fleet use more stringently than personally owned vehicles; an all-electric requirement should be considered, at least for urban areas.

ENERGY AND EMISSIONS IMPACTS OF NON-PETROLEUM FUELS

In 2025–2035, emerging alternative fuels, such as electricity, hydrogen, and low-carbon synthetic fuels, are expected to see increasing use in the light-duty fleet. All of these alternative fuels have the ability to decrease the well-to-wheels and criteria emissions of vehicles, and electricity and hydrogen also have zero tailpipe emissions. Making electricity a low-carbon fuel on a well-to-wheels basis in all regions of the United States requires further decarbonization of the electricity grid. Low-carbon hydrogen requires additional R&D to decrease costs and enable scale-up. Low-carbon synthetic fuels can serve as drop-ins for gasoline and diesel, thereby providing an opportunity to decrease the well-to-wheels emissions of existing and future ICEVs and hybrid electric vehicles. To be considered low-carbon, these fuels must be synthesized using emissions-free energy sources and derived from low-carbon feedstocks. Low-carbon synthetic fuel commercialization requires further development of enabling technologies, including direct air capture, carbon dioxide electrolyzers, and biorefineries.

In addition to the required technological developments, regulatory changes might be necessary in the long term to account for the use of low-carbon fuels in the light-duty fleet. Depending on their specific long-term goals, NHTSA and EPA should evaluate whether full-fuel-cycle emissions are more appropriate metrics to use in setting standards. Such an approach would be particularly relevant in a fleet with high use of low-carbon synthetic fuels, which provide no benefit compared to conventional gasoline when only tailpipe emissions are considered.

CONSUMER ACCEPTANCE AND MARKET RESPONSE

When and how various fuel-saving technologies are incorporated into vehicles depends on multiple market factors—including consumer demand and willingness to pay for efficiency technologies, how manufacturers respond to the standards, and barriers to technology adoption.

Since inception of regulations, vehicle fuel economy and emission rates have improved. There have also been increases in vehicle weight and power, although manufacturers are producing ICEVs with higher fuel economy and less performance than they would have otherwise. Over this time, the share of sales of light trucks has increased, raising concerns over mass disparity in vehicles. The extent to which these shifts have affected overall consumer welfare remains an area for study.

Understanding consumer value changes with vehicle attributes is important for understanding the effects of the standards. Consumer valuation may be assessed through the classic framework of utility maximization and/or through the lens of behavioral economics. Under either framework, lack of consumer understanding and familiarity and risk aversion are key barriers to the adoption of novel technologies. These barriers affect the extent of consumer demand for zero-emission vehicles (ZEVs)[2] and will affect acceptance for technologies like CAVs. Additional study is required to better understand effective interventions to reduce these adoption barriers (e.g., education, incentives, supporting infrastructure availability). Purchase subsidies have been found to increase sales of PHEVs, BEVs, and FCEVs. To continue to decrease vehicle energy use and emissions, federal subsidies should be continued and changed to operate as point-of-sale rebates with income eligibility considered. Effectiveness of subsidies should be studied in meeting goals of equity, sales, and/or electric VMT.

REGULATORY STRUCTURE AND FLEXIBILITIES

Vehicle fuel economy regulation began under the Energy Policy and Conservation Act of 1975, with the most recent regulation being the 2020 Safer Affordable Fuel Efficient Vehicles Rule. U.S. automakers have typically expressed preference for consistent and predictable regulations, to harmonize across global markets and accommodate long product cycles.

There are discrepancies between measured and real-world vehicle fuel economy. Given improvements in onboard technology for measuring real-world performance, the agencies should collect data on vehicle fuel consumption and emissions and consider how to adjust future crediting with the standards. The current approach to adjusting fuel economy involves the use of off-cycle crediting[3] to augment the test cycle fuel economy measurements. EPA and NHTSA should require the documentation for off-cycle credits to be transparent and detailed, available for comment, and publicly reported. Emerging vehicle technologies require particular considerations regarding crediting, test cycle procedures, and accounting for their full fuel cycle environmental impacts. The standards allow for credit trading, which appears to have reduced overall manufacturer compliance costs, although the effects are made more difficult to evaluate by transparency challenges.

The U.S. fuel economy program exists in the context of an increasingly globalized vehicle market influenced by a number of national regulations. In this context, the 2025–2035 Corporate Average Fuel Economy (CAFE) standard should be set and designed to depend on and incentivize a significant market share of ZEVs.

EMERGENT FINDINGS AND RECOMMENDATIONS FOR CONTINUED REDUCTION IN ENERGY USE AND EMISSIONS OF LIGHT-DUTY VEHICLES

In Chapter 13, the committee makes recommendations for Congress, DOT, and EPA under the current legislative authority.

> **SUMMARY RECOMMENDATION 1.** ***Growing Role of ZEVs*:** The agencies should use all their delegated authority to drive the development and deployment of zero-emission vehicles (ZEVs), because they represent the long-term future of energy efficiency, petroleum reduction, and greenhouse gas emissions reduction in the light-duty vehicle fleet. Vehicle efficiency standards for 2035 should be set at a level consistent with market dominance of ZEVs at that time, unless consumer acceptance presents a barrier that cannot be overcome by public policy and private sector investment. At the same time, maximum feasible fuel economy of

[2]As used in this report, a ZEV has zero emissions at the tailpipe. When upstream emissions are considered, ZEVs do not generally have zero emissions on a life cycle basis.

[3]Off-cycle credits are aspects of vehicle efficiency and emissions regulations that adjust for efficiencies or emissions reductions that are not directly measured on vehicle test cycles.

petroleum-fueled vehicles should be pursued, under the National Highway Traffic Safety Administration's interpretation of its existing authority, and as a portion of the U.S. Environmental Protection Agency's combined stringency assessment. The pathway to zero emissions should be pursued in a technology-neutral manner.

SUMMARY RECOMMENDATION 2. *Purchase Subsidies*: The U.S. federal battery electric vehicle, plug-in hybrid electric vehicle, and fuel cell electric vehicle purchase subsidies should be continued until financial and psychological consumer barriers to purchasing such vehicles have been overcome. However, it should be changed to point-of-sale rebates to increase effectiveness and lower fiscal burdens. Income eligibility should be considered for both policy equity and effectiveness. Research organizations in partnership with federal agencies should conduct studies to optimize which type of vehicles and electric ranges should receive more or less subsidy, with considerations of equity and policy effectiveness in promoting zero-emission vehicle sales and/or electric vehicle miles traveled share.

SUMMARY RECOMMENDATION 3. *Charging Infrastructure*: The U.S. Department of Transportation, *the U.S. Environmental Protection Agency, and the U.S. Department of Energy should coordinate to facili*tate electric charging and hydrogen refueling infrastructure deployment with relevant stakeholders, including state and local government agencies, business associations, and entitites. Congress should appropriate funds for and the agencies should create a national public-private partnership to lead this coordinating effort. For plug-in electric vehicle (PEV) charging, this coordinated effort should explicitly incorporate corridor fast charging, public charging at public parking spaces, PEV readiness of new and renovated homes and communities, and PEV readiness of workplace parking. For fuel cell electric vehicles, this coordinated effort should include support of hydrogen fuel infrastructure for light-duty vehicle (LDV) users in conjunction with medium- and heavy-duty vehicles and industry users, and deployment of LDV hydrogen refueling stations.

SUMMARY RECOMMENDATION 4. *Agency Coordination of Different Authorities*: The efforts of the National Highway Traffic Safety Administration (NHTSA) and the U.S. Environmental Protection Agency (EPA) to coordinate their fuel economy and greenhouse gas (GHG) emission standards since 2010 have been beneficial and should be continued to the extent feasible. However, the separate agency standards may now diverge because of the growing availability and benefits from zero-emission vehicles (ZEVs) and the agencies' different statutory authorities. The EPA can and must consider the availability and benefits of ZEVs and more efficient petroleum-fueled vehicles in setting the most stringent feasible GHG emission standards. In order to remain binding and relevant, NHTSA's program must consider the fuel economy or energy efficiency benefits provided by alternative fuel vehicles such as battery electric vehicles and fuel cell electric vehicles in setting the stringency of its corporate average fuel economy standards, either by NHTSA's interpretation of existing statute or by Congress passing a new or amended statute.

SUMMARY RECOMMENDATION 5. *NHTSA ZEV Authority*: To fulfill its statutory mandate of obtaining the maximum feasible improvements in fuel economy, the National Highway Traffic Safety Administration should consider the fuel economy benefits of zero-emission vehicles (ZEVs) in setting future corporate average fuel economy (CAFE) standards. The simplest way to accomplish that would be for Congress to amend the statute to delete the prohibition (42 U.S.C. § 32902[h][1]) on considering the fuel economy of dedicated alternative-fueled vehicles in setting CAFE standards. If Congress does not act, the Secretary of Transportation should consider ZEVs in setting the CAFE standards by using the broad authority under the statute to set the standards as a function of one or more vehicle attributes related to fuel economy, and define the form of the mathematical function. For example, recognizing that the maximum feasible average fuel economy depends on the market share of gasoline and diesel vehicles relative to ZEVs, the Secretary could consider redefining the function used for setting the standards to account for the expected decreasing share of gasoline and diesel vehicles relative to ZEVs. One possible mechanism to do this could be setting the standard as a function of a second attribute in addition to footprint—for example, the expected market share of ZEVs in the total U.S. fleet of new light-duty vehicles—such that the standards increase as the share of ZEVs in the total U.S. fleet increases.

SUMMARY RECOMMENDATION 6. *Fulfilling EPA Mandate*: If the National Highway Traffic Safety Administration is unable to consider alternative-fuel vehicles, and in particular zero-emission vehicles (ZEVs), in its stringency analysis, then the U.S. Environmental Protection Agency should continue under its mandate with divergent, more stringent standards, based on the advancements in ZEVs.

SUMMARY RECOMMENDATION 7. *Life Cycle Emissions*: Congress should define long-term energy and emissions goals for the corporate average fuel economy (CAFE) and greenhouse gas (GHG) programs, and the National Highway Traffic Safety Administration (NHTSA) and the U.S. Environmental Protection Agency (EPA) should set regulations that put the United States on a path to meet those goals. Considering other regulatory systems that may be implemented as part of a national program to reduce energy use and emissions in the fuel, electricity, and manufacturing sectors, the light-duty vehicle CAFE and GHG programs may or may not need to address the full vehicle and fuel life cycle emissions and energy consumption. Any vehicle or fuel life cycle requirements within the NHTSA or EPA programs should be set with appropriate lead-time for manufacturers to revise their upcoming product plans.

SUMMARY RECOMMENDATION 8. *ZEV Upstream Emissions Accounting*: In the longer term, it makes sense to address the full-fuel-cycle emissions of all vehicles, including zero-emission vehicles (ZEVs), especially as ZEVs become a progressively larger portion of the light-duty vehicle fleet. The National Highway Traffic Safety Administration and the U.S. Environmental Protection Agency should undertake a study of how and when to implement a full-fuel-cycle approach, including consideration of the potential benefits and drawbacks of the current temporary exclusion of upstream emissions for compliance of ZEVs. Based on that study, the agencies should decide whether and when to adopt a different approach for accounting for upstream ZEV emissions for compliance.

SUMMARY RECOMMENDATION 9. *Safety*: Improved crash compatibility will reduce the adverse effect of mass and geometric disparity on crash safety for passengers of all vehicles and vulnerable roads users, including pedestrians. The National Highway Traffic Safety Administration should study mass disparity in 2025–2035, improve federal motor vehicle safety standards testing protocols for crash compatibility, and further develop testing or computer-aided engineering fleet modeling to simulate real-world crash interactions between new vehicle designs and with vulnerable users at different impact speeds and impact configurations.

SUMMARY RECOMMENDATION 10. *Autonomous Vehicle Efficiency Regulation*: The agencies should consider regulating autonomous vehicles for fleet use differently from personally owned vehicles. Maximum feasible standards for these vehicles could be substantially more stringent than standards for personally owned vehicles; an all-electric requirement should be considered. To achieve the fuel-savings potential of autonomous driving and avoid its unintended consequences, the U.S. Department of Transportation should consider actions to guide the effects of autonomous driving on the U.S. transportation system and make recommendations accordingly to other agencies and to Congress.

SUMMARY RECOMMENDATION 11. *Novel Technology Barriers*: Because consumer resistance to novel technology is a significant issue in market penetration and acceptance of new technologies, policy interventions beyond purchase subsidies may be needed to address these barriers. Such policies may include investment in charging and refueling infrastructure, or consumer education and exposure to the new technology and its benefits.

SUMMARY RECOMMENDATION 12. *In-Use Performance*: The agencies should implement a program that measures fuel consumption and greenhouse gas emissions from the light-duty vehicle fleet in use. The purpose of the in-use program should be to evaluate and improve the effectiveness of the corporate average fuel economy program, not for year-by-year enforcement against individual manufacturers. New data sources and telematic technologies makes such in-use monitoring feasible, but safeguards must be established to minimize privacy risks for vehicle owners and operators.

SUMMARY RECOMMENDATION 13. ***Driving Patterns and Emissions Certification*****:** The agencies (U.S. Department of Transportation, U.S. Environmental Protection Agency, U.S. Department of Energy) should conduct a study on how well current driving patterns and new vehicle technology impacts are reflected by current vehicle certification test cycles. The results of this study should then be used to propose new light-duty vehicle test cycles, or adoption of the current or a new weighting of the existing 5-cycle test. The study of driving patterns and emissions and resulting changes in the test cycle may make it possible to eliminate some off-cycle treatment of fuel efficiency technologies and evaluate the energy saving impacts of those that remain.

SUMMARY RECOMMENDATION 14. ***Off-Cycle Technologies*****:** The U.S. Environmental Protection Agency (EPA) and the National Highway Traffic Safety Administration should consider off-cycle technologies in setting the stringency of the standards. The agencies should approve off-cycle credits on an annual cycle, require automakers to clearly and transparently document the test procedures and data analysis used to evaluate off-cycle technologies, and produce a compiled report on proposed credits that is available for public comment. The agencies should track the adoption of off-cycle credits in the vehicle fleet at the model level and report these data to the public, for example through the EPA Trends Report.

SUMMARY RECOMMENDATION 15. ***CAV Efficiency Regulation*****:** In setting the level of the standards, the agencies should consider connected and automated vehicle (CAV) technologies that can save energy. Off-cycle credits should be available for CAV technologies only to the extent that they improve the fuel efficiency of the vehicle on which they are installed. Credits should be based on realistic assumptions, where needed, regarding technology adoption on other vehicles or infrastructure.

SUMMARY RECOMMENDATION 16. ***Car and Truck Standards*****:** The National Highway Traffic Safety Administration and the U.S. Environmental Protection Agency should commission an independent group to study the effectiveness and appropriateness of separate standards for passenger cars and light trucks.

FUTURE POLICY SCENARIOS FOR CONTINUED REDUCTION IN ENERGY USE AND EMISSIONS OF LIGHT-DUTY VEHICLES

The committee considered the future of fuel efficiency technology, consumers, market, and regulatory aspects in 2025–2035; how Congress should move forward to update the legislative mandates for vehicle efficiency; and how DOT and EPA should update and better integrate their respective regulatory structures given the committee's assessment of the technology future. The committee made the following findings and recommendations for the future legislative and regulatory structure of the CAFE program:

FINDING 13.1: The current statutory authority for the CAFE program is becoming increasingly outdated as a result of legal, scientific, policy, technological, global leadership, and economic developments and trends.

RECOMMENDATION 13.1: Given the end of the latest legislative specification for corporate average fuel economy (CAFE) in 2030, Congress should extend the CAFE program and, as part of that reauthorization, evaluate and update the statutory goals of the CAFE program. With those goals in mind, Congress should consider changes to the program structure and design, and its interaction with other related policies and regulations.

RECOMMENDATION 13.2:[4] The statutory authorization for the corporate average fuel economy (CAFE) program should be amended to expressly include climate change as a core objective of the program, along with

[4]In discussion with stakeholders after the pre-publication version of this report was released, it became apparent that this recommendation could be interpreted in a way that would have unintended effects that were not examined by the Committee. This footnote was added to this final publication of the report in order to clarify that in making this recommendation, the Committee did not examine and did not intend for any changes to statute to disrupt the regulatory structure that provides for state greenhouse gas emissions standards, state zero emission vehicle requirements, and greenhouse gas emissions standards promulgated by the U.S. Environmental Protection Agency.

existing objectives such as energy conservation. Specifically, the statutory considerations for setting CAFE standards in 49 U.S.C. § 32902(f) should be amended to include the goal of reducing greenhouse gas emissions.

FINDING 13.2: The continued existence of two partially overlapping programs, the CAFE program administered by NHTSA and the GHG emissions program administered by EPA, imposes some duplication and extra costs on government and industry, but these additional burdens can be mostly offset by coordinating the two programs. In addition, the continued existence of the two separate programs provides some benefits that outweigh the duplicated costs and burdens, including the consideration of different unique and relevant factors by each agency, and the benefits of having the two agencies check and backstop each other's activities.

RECOMMENDATION 13.3: Congress should reauthorize the continuation of the National Highway Traffic Safety Administration (NHTSA) corporate average fuel economy (CAFE) program, notwithstanding its practical overlap with the U.S. Environmental Protection Agency light-duty vehicle greenhouse gas program. Congress can minimize any disruption from having two programs by eliminating any obstacles to coordinating the two programs, such as by eliminating the current prohibition that prevents NHTSA from considering zero-emission vehicles and other dedicated alternative-fueled vehicles in setting CAFE standards.

FINDING 13.3: Many studies suggest that reaching an economy-wide deep decarbonization goal will require new vehicles to be zero-emissions. To comprehensively address climate change, a transition to ZEVs needs to be in concert with a full move to net-zero GHG fuels and electricity, and also net-zero vehicle manufacturing GHG emissions.

RECOMMENDATION 13.4: To provide vehicle manufacturers a longer-term target to assist planning their ongoing technology investments and pathways, Congress should set a goal that all new light-duty vehicles will have net-zero greenhouse gas emissions by a specific date that aligns with a national deep decarbonization goal, and includes interim goals. This target should be technology neutral, to allow each manufacturer to choose its compliance pathway and technology strategy.

RECOMMENDATION 13.5: The Executive Branch should create an interagency task force with the objective of coordinating and integrating government efforts to achieve a cleaner, safer, and fairer transportation and mobility system.

RECOMMENDATION 13.6: The federal interagency committee on new mobility, along with state and local policy makers, should consider rules or incentives to encourage future autonomous vehicles, especially in fleets, to use zero or near-zero emission technologies. Furthermore, the impact of any incentives should be evaluated for their ability to promote an overall reduction in vehicle miles traveled and increase in the use of transit and shared rides.

CAFE has historically been the bedrock of U.S. vehicle energy efficiency and climate policy, eventually joined by EPA vehicle and other GHG programs. It is now entering a time of major change, with new technologies enabling a pathway to zero emissions, and a future of a diversity of energy sources and modes of mobility. The committee expects that CAFE will continue to play an important role in the future if the recommendations in this report are adopted, and serve as an example for other energy and climate policies administered by government agencies in the United States and around the world.

1

Introduction

The period from 2025 to 2035 will be a time of pivotal change for fuel economy of light-duty vehicles (LDVs) in the United States. By the time this report is released in 2021, the United States will be approximately 15 years into the modern era of increasing fuel economy standards, tracing back to 2005 when the light-duty truck standards began to increase. In this report, the committee projects and estimates the fuel economy technology improvements that may be feasible in the next 15 years. Energy savings by LDVs over the past 15-year period have come primarily from improvements in internal combustion vehicles, with only minor impact from alternative fuel powertrains. There continue to be incremental improvements available in internal combustion engines (ICEs); however, the most dramatic improvements in fuel economy and reductions in greenhouse gas (GHG) emissions come from electric-ICE hybrids, and battery electric vehicles (BEVs) and fuel cell vehicles. In the next 15 years, and in particular in 2025–2035, the two central issues are (1) whether the United States will regulate LDVs to deeply reduce GHG emissions and (2) how much BEVs and other alternative-fueled vehicles can penetrate the new vehicle fleet in the United States. Of course, other factors will also determine fuel economy improvements in this next era through 2035, including the development of connected, autonomous, and shared vehicles; other regulatory programs at the international, national, state, and local levels; and consumer response to new vehicle technologies.

This introductory chapter begins with a brief summary of the status of fuel consumption, energy efficiency, and GHG emissions of LDVs on U.S. roads today, and then provides further detail on some of the relevant changes expected in the 2025–2035 period.

1.1 A SNAPSHOT OF TODAY'S LDV FLEET

Passenger vehicle, on-road travel is the primary means of transportation in the United States. In 2017, there were almost 251 million LDVs registered in the United States, such as sedans, crossovers, sport utility vehicles (SUVs), vans, and passenger trucks. They traveled a total of 2.88 trillion miles and consumed 129 billion gallons of fuel, resulting in 4.82 trillion miles of passenger travel (FHWA, 2017). In that same year, LDV energy consumption represented 17% of total national energy use. That energy is provided primarily by gasoline (89%), diesel (3%), ethanol (8%), and electricity (0.04%) (Davis and Boundy, 2020; EIA, 2020). Despite a pause in 2008–2011 during a national recession, vehicle miles traveled (VMT) have continued to increase year over year, as have other indicators of vehicle use, such as the number of registered vehicles and total consumption of fuel (Figure 1.1).

Operation of the LDV fleet provides great value to individuals and to the nation, but also has large environmental, human health, and other costs. Reducing these costs motivates improvements in vehicle fuel economy

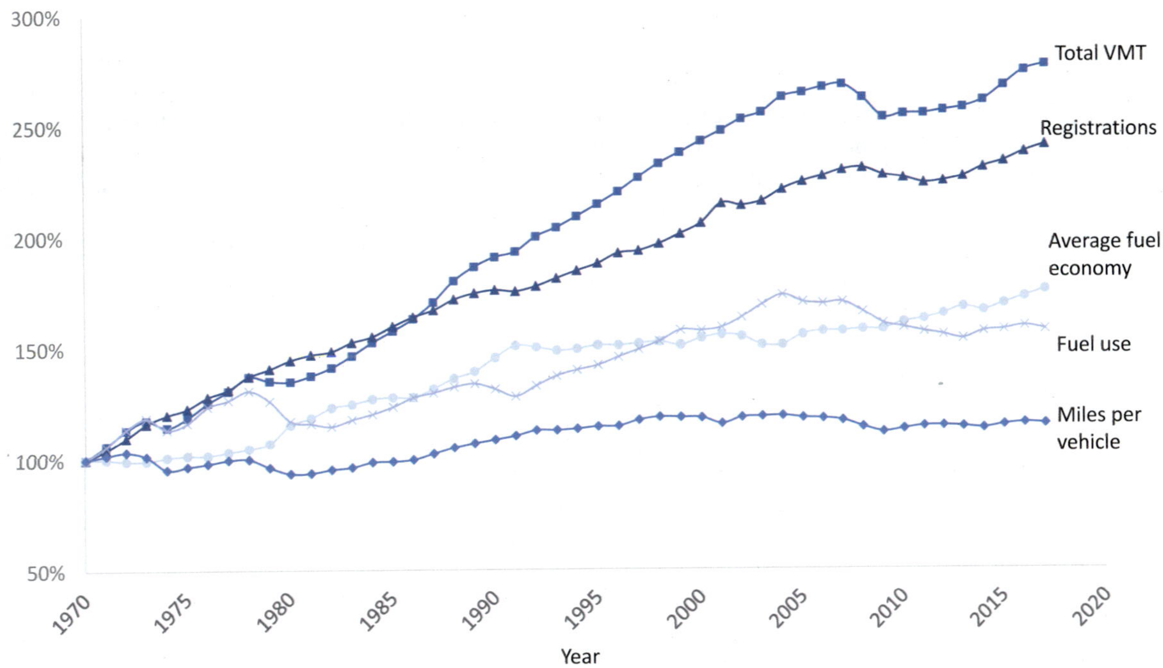

FIGURE 1.1 LDV transportation characteristics, including total VMT, vehicle registrations, average fuel economy, total fuel use, and miles per vehicle.
SOURCE: Committee generated using data from Davis and Boundy (2020).

and energy efficiency. The combustion of petroleum fuels to power LDVs produces 17% of U.S. GHG emissions (EPA, 2019), as well as a significant fraction of emissions of important air pollutants such as ozone precursors (volatile organic compounds and nitrogen oxides [NO_x]), carbon monoxide, sulfur oxides, and particulate matter, including black carbon. Use of petroleum fuels also exposes a major sector of the U.S. economy to the volatile world markets for gasoline and diesel, even with increased domestic production. Last, purchase of fuel is the largest operating expense to the user of an LDV, with consumers spending on average $2,109 on fuel, 2.7% of their income. To reduce these private and public costs from petroleum dependence, the U.S. government began requiring minimum fuel economy standards for passenger vehicles in 1978.[1] To meet these standards, automakers implemented technologies for fuel economy, ranging from engine and transmission improvements to vehicle design and lightweighting. The opportunities and costs of technologies for fuel economy to be implemented in the 2025–2035 vehicle fleet are the primary subject of this study, requested by the U.S. Department of Transportation (DOT) in response to a congressional mandate in the Energy Independence and Security Act of 2007.

1.2 A LOOK AT THE FUTURE

The future of LDV technologies is uncertain and likely disruptive, but there is opportunity for positive changes that will benefit vehicle users, vehicle owners, vehicle manufacturers, and the health of the planet and its people. The future fleets of LDVs in 2025–2035 will depend on technological availability (technology push); consumer acceptance and demand for new types of vehicles (technology pull); and regulatory, business, and economic factors. Key changes in technology push and pull, global market factors, and energy use implications for the future of LDVs are summarized below.

[1]Further discussion of the history of fuel economy regulation is found in Chapters 2 and 12.

1.2.1 Future of Technologies

Automakers and automotive industry suppliers have continuously improved technologies for LDVs, in response to consumer demand for vehicle features such as horsepower, comfort and convenience, carrying capacity, fuel economy, safety, and advanced technology. Safety, environmental, and other vehicle regulations have also driven technology development and implementation. Historically, major vehicle technology advancements have included improved engines and transmissions, emissions controls, introduction of air conditioning, introduction of seatbelts and airbags, and development of hybridized and fully electrified powertrains. Some of the most recent improvements include advanced engine technologies such as turbocharging and downsizing, 8- to 10-speed transmissions, optimized vehicle design and materials substitution, longer-range electric vehicles, and many safety and convenience features such as lane keeping and automatic braking. On the horizon, vehicle and travel system advances may include significant to total vehicle automation and connectivity, vehicle sharing in addition to personal vehicle ownership, improvements in cost and capabilities of electric vehicles and their infrastructures (including both battery and fuel cell vehicles), and implementation of low-carbon fuels. These technologies have been enabled by automotive-specific technology development, such as in mechanical and electrical engineering, but also by developments in other fields including consumer electronics, communications, control systems, and material science.

1.2.2 Future of Market in United States and Globally

Technology implementation is impacted not only by technology availability and cost but also by customer demand in the domestic and global vehicle market, and by regulatory policies. As the market grows for electric vehicles and autonomous vehicles—which offer a different ownership experience to the consumer—consumer expectations and demand may change. In 2018, 17.3 million new vehicles were sold in the United States (Jato, 2019b). Over the past 30 years, and especially in the past 10 years, U.S. customers have moved strikingly away from compact cars and sedans and into SUVs and crossover utility vehicles (CUVs) (Figure 1.2). In 2018, coupe,

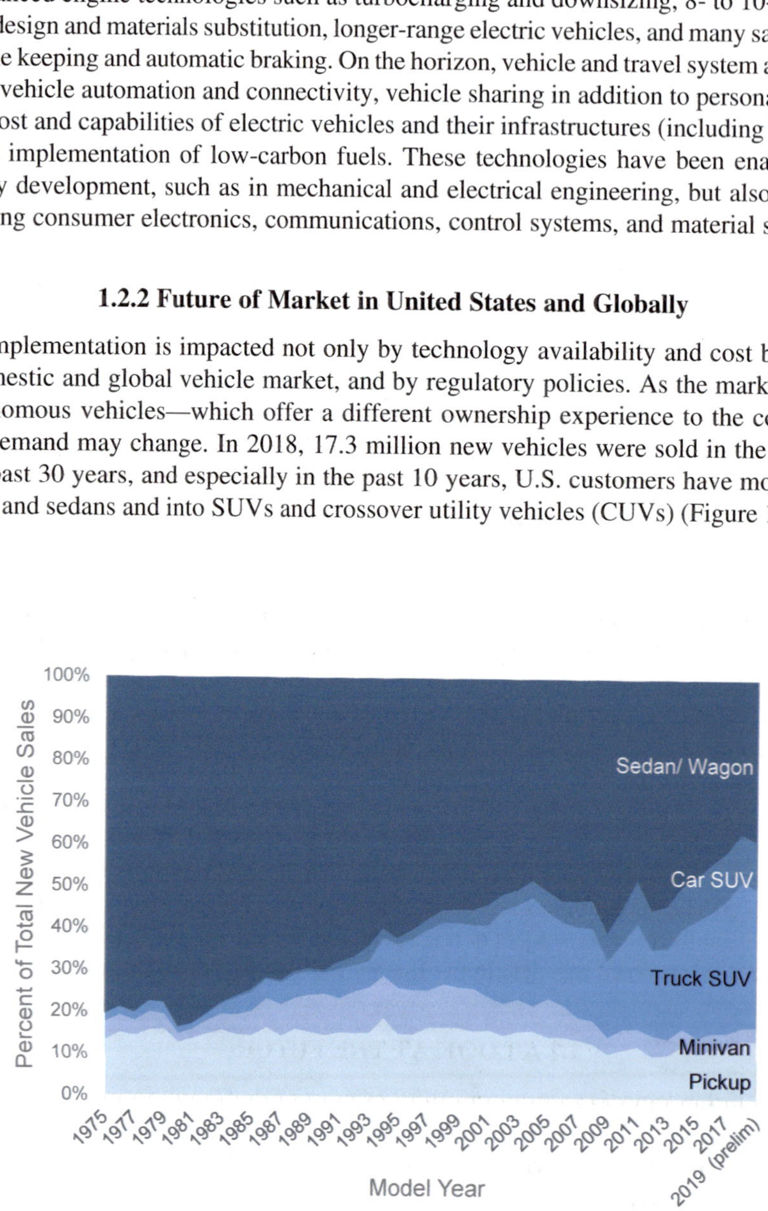

FIGURE 1.2 Vehicle classes over time, showing the reduction in market share of sedans/wagons and minivans, and the increase in both car and truck SUVs. The total share of vehicles classified as trucks (truck SUV, minivan, and pickup) was approximately 50% of vehicles in 2019, up from about 20% in 1975.
SOURCE: Committee generated using data from EPA (2020).

sedan, and wagon-type cars represented only 31% of LDV sales, with the remaining 69% being SUVs, CUVs (including those classified as cars), vans, and trucks. Consumers are purchasing these vehicles for the passenger room, ride height, ease of entry and exit from the vehicle, and cargo capacity. Consumers have also started moving toward greater purchases of alternative powertrain vehicles such as hybrids, BEVs and to a lesser extent, fuel cell vehicles.

All automakers selling in the U.S. market also sell vehicles in other countries, in a market of approximately 86 million new LDVs sold globally in 2018 (Jato, 2019a). Compared to the U.S. market, consumers in China, Japan, and Europe tend to purchase smaller vehicles and more sedans, although the shift to larger vehicles is also occurring globally. Fuel quality and price differs globally, impacting consumers' value for vehicle efficiency and preference of fuel type (diesel, gasoline, electricity). As the market for vehicle models and technologies is becoming increasingly globalized, automaker design decisions are responding to this global marketplace as well as national and regional environmental, health, and safety regulations.[2]

The combination of new types of vehicles, new models of vehicle ownership, and increasing globalization of the vehicle manufacturing industry driven by regulatory and market developments in several major markets around the world portend highly disruptive changes in the automobile industry over the next couple decades. Attempting to predict the timing and direction of these changes is difficult given the multiple factors that will affect future vehicle technologies and sales. Nevertheless, by carefully studying and integrating vehicle technology feasibility and costs; consumer expectations and shifts; and regulatory and market pressures at the state, national, and international levels, it is possible to project a series of reasonable scenarios for the future, which this report attempts to do.

1.2.3 Energy Use Implications

Because vehicle technologies can influence multiple desired vehicle attributes such as power, efficiency, convenience, and cost, automakers tune technology implementation to reflect the desired suite of vehicle attributes within the constraint of compliance with applicable regulatory requirements. In some cases, decisions are made to trade attributes off against one another, such as optimizing for performance[3] (power) versus fuel economy when turbocharging an engine or when implementing a lightweighted vehicle design. In other cases, there may be complementary benefits of technologies, such as safety features like automated cruise control or optimized engine controls that also yield fuel economy benefits. Some vehicle technologies may cause the total VMT to change, in addition to the per-mile change in fuel economy. For example, if automated and connected vehicle technologies become a significant part of the U.S. vehicle fleet, the changes in VMT may become even more important, as traveling by LDV becomes greatly easier, safer, more accessible, and more appealing to many travelers. Changes in VMT are important, as they impact total energy use of the light-duty fleet and hence total costs to individuals and to society.

1.3 LDV SYSTEM ENERGY USE

Vehicle energy consumption has significant costs (in fuel, energy, emissions, congestion, etc.) and benefits (movement of people and goods). Improving the energy efficiency of vehicles reduces fuel-based operating costs, as well as the emissions and other impacts associated with combustion onboard the vehicle or in the energy system used to power the vehicle (using liquid fuels, hydrogen, or electricity). Key considerations that influence system energy use and associated emissions and impacts include vehicle efficiency per mile, total vehicle use, and the life cycle energy and emissions of different vehicles and fuels. These considerations can be described under the following two aspects of vehicle energy and emissions impact: (1) rate-based performance standards versus total performance (e.g., grams per mile of emissions versus total emissions summed over the vehicles miles), and (2) vehicle-based versus full-fuel-cycle-based (including fuel production and transportation upstream emissions) versus full

[2]Further discussion of other national and regional automotive regulations is found in Chapters 3 and 12.

[3]In this report, performance refers to attributes related to engine and motor power such as vehicle horsepower and acceleration, and not to fuel economy or other desired attributes such as minimized noise, vibration, and harshness.

vehicle-life-cycle-based metrics (including all aspects of vehicle life cycle of full-fuel-cycle but also aspects such as vehicle manufacturing and end-of-life outcomes). Furthermore, these aspects can be applied to energy use, emissions, or petroleum consumption. Historically, LDV fuel efficiency in the United States has been regulated on a miles-traveled-per-volume-of-fuel basis, miles per gallon, with different minimum standards by vehicle class or footprint. Aspects of energy use and GHG emissions have been added to the regulatory structure over time.

When considering metrics for vehicle energy efficiency, some trade-offs include ease of measurement, control and attribution, and strength of the relationship between the metric and the goal of improved efficiency. In particular, measurability, control and attribution are important to addressing problems such as national-level economic, environmental or security costs or individual consumer costs and assigning them to a responsible party such as the automaker for vehicle fuel economy or the fuel/electricity producer for the off-vehicle portion of vehicle energy emissions. Also important is the choice of metric to prioritize the most relevant aspect of energy, emissions, or petroleum consumption for solving a given problem, such as improving U.S. energy security and reducing emissions leading to climate change. For example,

- *Fuel consumption per-mile metric is more easily measured and certified in vehicle testing, while a total* energy metric is more relevant to consumer costs and environmental, security, and other costs of nationwide GHG emissions.
- A measure of efficiency (or the related consumption) based on a metric other than liquid fuel volume, such as an energy or GHG metric, becomes more relevant as vehicles become increasingly efficient in using liquid fuels, as the type, source, and environmental impact of liquid fuels change, and as vehicles increasingly use non-liquid fuels like hydrogen and electricity.

As shown in Figure 1.3, depending on what metric you consider, different conclusions can be drawn about the performance of the transportation system. The above metrics expand the current definition of miles per gallon of

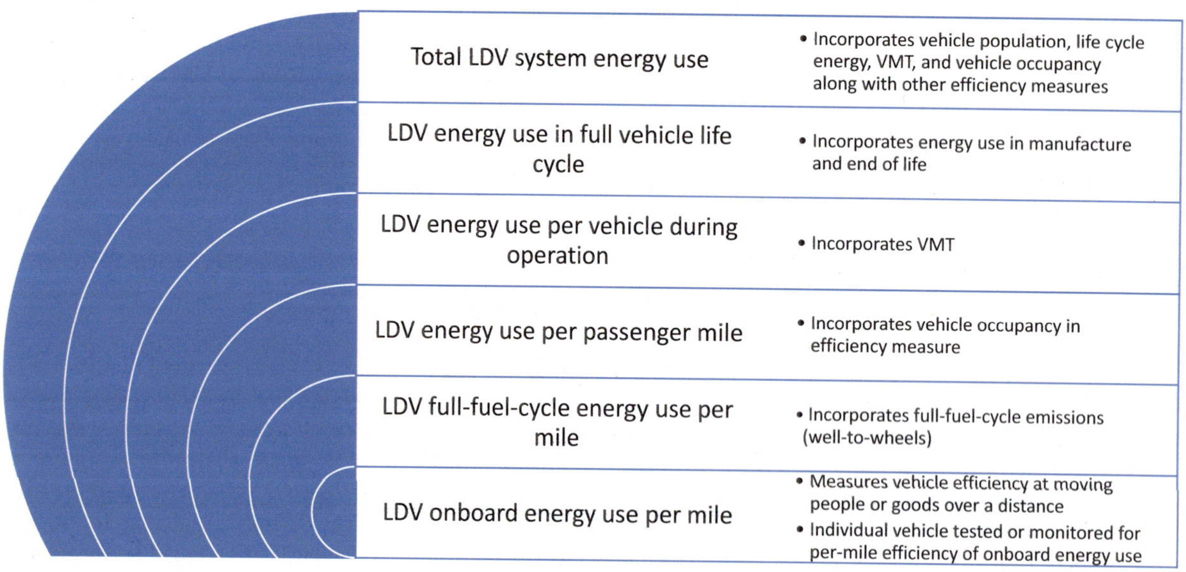

FIGURE 1.3 LDV system energy use can be measured as a per-vehicle, per-mile efficiency rate, or as total energy used per vehicle, or as total system energy use. Rates may include only onboard energy use, or incorporate full-fuel-cycle energy use and/or vehicle occupancy. Measures of total system energy use per vehicle build off the efficiency measures, further incorporating VMT, and may additionally include the full vehicle life cycle energy use of vehicle manufacture and end of life. Total system energy use incorporates the vehicle population along with the previous aspects. These same considerations can apply to fuel consumption or GHG emissions.

an individual vehicle at a point in time to include the vehicle's total travel, the transportation energy system's total consumption or emissions for all vehicles, and metrics that are not based on a liquid fuel. If per-mile efficiency is the primary metric, the system has become more efficient, increasing fuel economy 75% since 1970. If total system fuel use or associated energy use and GHG emissions are the relevant metrics, then total LDV system energy use has increased 57% owing to a small increase in per-vehicle VMT (+15%) and a large increase in vehicle population (+141%), even with more efficient vehicles (Davis and Boundy, 2020). In the period of 2025–2035, energy use aspects that may be relevant include fuel volume, energy use, and GHG emissions; per-mile impacts, per-vehicle impacts, and total impacts; and well-to-wheels and full vehicle life cycle analysis.

In 2025–2035, as the system boundary expands, it is likely that a wider variety of metrics will be relevant, including fuel consumption and related energy and GHG emissions per mile; total vehicle energy consumption or GHG emissions per year or lifetime; and total system energy consumption or GHG emissions per year or lifetime. This report will further discuss the appropriate vehicle energy system metrics in a later chapter, and report on vehicle efficiency using per mile metrics.

1.4 CONTEXT FOR FUEL ECONOMY IMPROVEMENTS

1.4.1 Key Changes for 2025–2035

Vehicle manufacturers are expected to make continued incremental changes in the fuel economy of vehicles powered by ICEs in the period from 2025 to 2035, and this report describes in detail the most significant changes that are expected in this time period. Significant changes are expected in the period from 2025 to 2035, including electric vehicles of various types approaching mainstream market attributes, as well as the deployment of new vehicle types—in particular, autonomous and connected vehicles. A driving factor in the total fuel economy of the U.S. LDV fleet in 2035 will be the success of these vehicles in gaining widespread acceptance and adoption across all new vehicle purchasers. Consumer perceptions and acceptance have always played an important role in the U.S. fuel economy program—for example, the current shift to greater numbers of crossovers and SUVs in the U.S. fleet is a reflection of consumer preferences, among other factors. Yet in the period 2025–2035, consumer expectations and behavior will play a much larger role than ever before in fuel economy, as the success of the new types of vehicles will depend not only on the cost, feasibility, and performance of the technology (technology push) but also on the acceptance of new types of vehicles by consumers that involve different modes of operation, refueling, and even ownership. In addition to consumer acceptance, other factors beyond the vehicle technology will also be crucial to the integration of autonomous and electrified vehicles, including the installation of appropriate recharging infrastructure, and transportation planning to allow such vehicles to thrive. Thus, this report necessarily goes beyond just vehicle technology to look at these other factors that will affect fuel economy of LDVs in the 2025–2035 period.

1.4.2 Pricing Fuels, Fuels Policy, and Fuel Energy Equivalency

Fuels have always played an important role in fuel economy and will play an even more important role going forward into the 2025–2035 period. Fuel prices affect consumer demand for more fuel efficient vehicles, which then influences manufacturer trade-offs between a variety of vehicle attributes. The rapid increase in natural gas and petroleum production in the United States beginning in 2009 has created increased supply of both commodities, helping to keep the price of gasoline for vehicles controlled. Yet, gasoline prices have historically fluctuated considerably in response to a number of domestic and international factors that are often unpredictable, so fuel costs are always somewhat of a wild card in projecting fuel economy trends in the future. One or more new liquid fuels may become more prevalent in the vehicle industry in the 2025–2035 period, including high-octane gasoline, low-carbon gasoline (e.g., California low-carbon fuel standard), and biofuels. Electricity and hydrogen used as fuels in LDVs create even more diversity in fuel costs, infrastructure, and propulsion technologies. Each of these fuel alternatives will have relevant fuel economy and GHG emission impacts, which are discussed in more detail elsewhere in this report.

1.4.3 Criteria Emissions Regulations

Since burning gasoline directly produces criteria air pollutant emissions and GHGs, criteria pollutants are directly tied to fuel economy and GHG emissions from ICEs. Criteria air pollutants are heavily regulated under the National Ambient Air Quality Standards (NAAQS) program, so revisions of criteria pollutant emission regulations often have implications for fuel economy. In some cases historically, the goals of increasing fuel efficiency and controlling criteria pollutants have been in tension. An example is the trade-off between optimizing NO_x emission control and maximizing fuel economy by adjusting the air-to-fuel ratio for the combustion process. In other cases, the interaction between fuel economy improvements and criteria pollutant reductions is synergistic rather than antagonistic. For example, shifting from an ICE to a BEV will generally reduce both fuel consumption and criteria pollutant emissions, with the level of benefits determined by the source used to generate the electricity used to charge such vehicles. Whether the interaction is synergistic or antagonistic, NAAQS have important implications for how fuel economy standards are achieved by automakers.[4]

1.4.4 Infrastructure—Highway Speed Limits, HOV Lanes, and Congestion Pricing

Infrastructure investment and regulations have always affected fuel consumption. Because new vehicle types such as autonomous and connected vehicles will likely present new models of vehicle ownership and use, infrastructure will be particularly important in the upcoming years in impacting fuel economy. Recharging or hydrogen fueling infrastructure, discussed in more detail later in this report,[5] will obviously be critical to the deployment of BEVs and fuel cell vehicles. Higher speeds consume more energy per mile traveled than traveling at lower speeds. Thus, speed limits will affect total fuel consumption and GHG emissions. High-occupancy vehicle (HOV) lanes also encourage more passengers per vehicle, which can reduce energy consumption. Providing HOV lanes for electric vehicles can also help incentivize such vehicles.[6] Last, a number of other regulatory and policy initiatives to reduce VMT will also affect overall fuel economy, such as ridesharing and carpooling programs, public transportation incentives, and urban planning initiatives that encourage less driving. New York City has recently decided to implement a congestion pricing system, and other cities are likely to follow suit in the 2025–2035 period; these initiatives will also reduce vehicle use and thus fuel consumption.

1.5 STATEMENT OF TASK

This report is organized to introduce the emissions, energy, and fuel consumption aspects of the LDV vehicle fleet today and into the future (Chapters 2 and 3); discuss vehicle technology packages likely to be prevalent in the model year 2025–2035 new vehicles; and discuss technology fuel consumption and costs (Chapters 4–10), as well as aspects of infrastructure and fuels related to those technologies. The report describes the consumer and regulatory aspects of fuel economy technologies (Chapters 10–12). Findings are made throughout the report, and recommendations are made on vehicle technology and regulatory matters. The overarching report findings and recommendations are highlighted in the final chapter (Chapter 13) of the report, as well as considerations for Congress, DOT, and the U.S. Environmental Protection Agency as they move forward under existing or future mandates for vehicle efficiency. The committee's full statement of task is reproduced below:

> The committee that will be formed to carry out this study will continue the work of the National Academies for the U.S. Department of Transportation's National Highway Traffic Safety Administration (NHTSA) in the assessment of technologies for improving the fuel economy of light-duty vehicles. While the committee will need to consider the near-term deployment of fuel economy technologies, it is tasked with looking out into the 2025 to 2035 time frame to provide updated estimates of the potential cost, fuel economy improvements, and barriers to deployment of technologies. The committee will need to broadly consider the types of technologies that might emerge over this

[4]Further analysis of NAAQS and the impacts on fuel economy are discussed in the regulatory background in Chapter 3.

[5]See Chapter 5.

[6]Incentive programs for electric vehicles and other zero-emission vehicles are discussed in more detail in Chapters 11 and 12.

BOX 1.1
Impacts of COVID-19 on Automotive Markets 2025–2035

The COVID-19 pandemic has transformed societies and economies in the United States and globally during the final year of drafting this report. It is still an ongoing, evolving situation at the time of this report publication. More than 123 million people around the world have been infected by the SARS-CoV-2 virus, with more than 29 million infections and 542,000 deaths confirmed in the United States as of March 22, 2021 (Dong et al., 2020). The disease is easily transmissible from person to person, especially when individuals congregate and share airspace (CDC, 2020; NASEM, 2020). To combat the pandemic, individuals have been encouraged to significantly change their behavior, including frequent hand washing, wearing a mask in public, limiting gatherings with others outside their household, and maintaining distance from others when in public settings or gatherings.

Consumers' behavioral changes in response to COVID-19 have led to unprecedented changes in light-duty vehicle transportation, including less travel and commuting, reduced petroleum demand, and mode share shift away from public transport and toward private transportation. The number of personal trips decreased by about 40% nationwide in March and April 2020 and remained down by 20%–30% through October 2020 as compared to the previous year (BTS, 2020). During the early pandemic, half of all workers in the United States worked from home, as compared to about 10% before the pandemic; some people moved to distant locations for telework. There were more deaths per miles traveled despite fewer vehicles on the road, owing to more risky driving behavior (Blanco, 2020). Vehicles are being used in new ways, including for "pandemic safe" socializing, such as camping, drive-through activities, and drive-in theaters and performances.

The pandemic has also impacted vehicle manufacturers. Early in the pandemic, automakers and suppliers had to shut down manufacturing across many countries at varying times, owing to government restrictions, risks to workers, and supply-chain disruptions. Return to work has experienced these same concerns (Koenig, 2020). Some automakers produced material for pandemic response, such as personal protective equipment and ventilators. Additionally, automakers and dealers have been affected by changes in consumer demand. Automotive sales remain down 20%–25% for 2020 versus 2019 in the United States, and the trend of increasing digital aspects of vehicle sales is accelerating (Madhok, 2020). Consumers have expressed greater interest in using passenger vehicles over other forms of transportation (Furcher et al., 2020). Some ride-hailing services and mobility-as-a-service pilots suspended service during the pandemic, and since resuming service, ridership remains down, especially for shared services.

The pandemic is not yet over, especially in the United States, and more behavioral and market changes will be observed, although it is unclear which will last. Some of these recent changes will likely end after the health emergency or the economic downturn pass; some may become long-term behavioral, societal, or market changes; and some may become periodic responses to future pandemics or flu season. For example, trips that were deferred or delayed because they require in-person service, like haircuts or surgeries, may return when the health emergency passes. Similarly, trips related to long-distance leisure travel may return as the health and economic emergencies pass. On the other hand, some activities that were previously undertaken in person may be permanently reduced as people adopt the convenience of going online for shopping and delivery, or meetings and appointments. The long-lasting impact of mode change shifts on personal vehicle sales and use are unclear, especially for advanced vehicles. For example, suspension of services and lack of demand for ride-hailing, in particular vehicle sharing, may influence automaker investment in car sharing and mobility as a service. However, the economic toll of the pandemic may make the total cost of ownership benefits more salient after the infection danger passes. Automaker investments in new technologies, especially battery electric vehicles and connected and automated vehicle (CAVs), may slow if there is insufficient investment capital, but CAVs in particular may have increased investment if they become desirable as more individuals choose to travel long distances in personal automobiles rather than planes, trains, or buses.

The pandemic's long-term impact on automaker investment, vehicle technology development, and consumer demand and vehicle use is uncertain, but it could influence the energy efficiency, petroleum use, and emissions for individual vehicles as well as the overall transportation system.

time period and their impacts on fuel consumption. It will also consider shifts in the personal transportation and vehicle ownership models and how such shifts might impact vehicle technologies. The committee will build on the assessments completed in earlier National Academies reports, including the first two phases of this series of studies, Assessment of Fuel Economy Technologies for Improving Light-Duty Vehicle Fuel Economy (2011) and Costs, Effectiveness, and Deployment of Fuel Economy Technologies for Light-Duty Vehicles (2015). It will reflect on developments since these reports were issued and investigate any new technologies and trends in consumer behaviors that may become important by 2035. In particular, the committee will:

1. Examine the costs (direct and indirect), fuel economy improvements, and potential implementation timing for high-volume production of technologies for internal combustion engine powertrains.
2. Examine the costs (direct and indirect), fuel economy improvements, and potential implementation timing for high-volume production of electric powertrain technologies. The committee shall include an examination of the cost, performance, durability, usable battery capacity and other issues related to critical components, including batteries, ultracapacitors, and power electronics and auxiliary vehicle systems such as heating and cooling. The committee will also address transition issues associated with meeting the infrastructure needs for such powertrains.
3. Examine the costs (direct and indirect), fuel economy improvements, and potential implementation timing for high-volume production of non-powertrain technologies including mass reduction, aerodynamics, low rolling resistance tires, and vehicle accessories. For mass reduction, the committee shall consider opportunities for a range of baseline vehicle materials, including steel, high-strength steel, mixed metal, aluminum, polymers, composites, and others. The committee shall include an examination of methodologies for cost assessment of mass reduction, including equipment and retooling costs, manufacturing issues, supply chain requirements, and implications for durability, safety, and reparability.
4. Consider the current and possible future role of flexibilities in the CAFE program on the introduction of new technologies, including credit trading, treatment of alternative fuel vehicles, off-cycle provisions, and flexibilities for small volume manufacturers.
5. Assess how shifts in personal transportation and vehicle ownership models might evolve out to 2035, how these changes could impact fuel economy-related vehicle technologies and operation, and how these changes might impact vehicle scrappage and vehicle miles traveled. Scenarios might be used to bound this task.
6. Examine consumer behavior issues associated with new fuel efficiency technologies, including acceptance of any utility or performance impacts and cost of new technologies. This could include considerations of consumers' willingness to pay for improvements in fuel economy and other vehicle attributes.
7. Write a final report documenting the committee's conclusions and recommendations.

1.6 REFERENCES

Blanco, S. 2020. "2019 U.S. Traffic Deaths Lowest Since 2014, but 2020 Numbers Aren't Looking Good." Car and Driver. October 1. https://www.caranddriver.com/news/a34240145/2019-2020-traffic-deaths-coronavirus/.

BTS (Bureau of Transportation Statistics). 2020. "Trips by Distance." Updated November 9, 2020. https://data.bts.gov/Research-and-Statistics/Trips-by-Distance/w96p-f2qv.

CDC (Centers for Disease Control and Prevention). 2020. "What You Should Know about COVID-19." Department of Health and Human Services. June 1.

Davis, S.C., and R.G. Boundy. 2020. *Transportation Energy Data Book: Edition 38*. Oak Ridge National Laboratory. https://TEDB.ORNL.GOV.

Dong, E., H. Du, and L. Gardner. 2020. An interactive web-based dashboard to track COVID-19 in real time. *The Lancet Infectious Diseases* 20(5):533–534. https://doi.org/10.1016/S1473-3099(20)30120-1.

EIA (Energy Information Administration). 2020. "April 2020 Monthly Energy Review." DOE/EIA-0035(2020/4). Monthly Energy Review. Washington, DC: Energy Information Administration. https://www.eia.gov/totalenergy/data/monthly/archive/00352004.pdf.

EPA (U.S. Environmental Protection Agency). 2019. *Inventory of U.S. Greenhouse Gas Emissions and Sinks 1990–2017*. EPA-430-R-19-001. April.

EPA. 2020. *The 2019 EPA Automotive Trends Report: Greenhouse Gas Emissions, Fuel Economy, and Technology since 1975*. EPA-420-R-19-002. March.

FHWA (Federal Highway Administration). 2017. "Annual Vehicle Distance Traveled in Miles and Related Data—2017 (1) by Highway Category and Vehicle Type." In Highway Statistics 2017. https://www.fhwa.dot.gov/policyinformation/statistics/2017/vm1.cfm.

Furcher, T., B. Grühn, I. Huber, and A. Tschiesner. 2020. "How Consumers' Behavior in Car Buying and Mobility Is Changing amid COVID-19." McKinsey & Company. September 22. https://www.mckinsey.com/business-functions/marketing-and-sales/our-insights/how-consumers-behavior-in-car-buying-and-mobility-changes-amid-covid-19.

Jato. 2019a. "Global Car Market Remains Stable During 2018, as Continuous Demand for SUVs Offsets Decline in Sales of Compact Cars and MPVs." February 21. https://www.jato.com/usa/global-car-market-remains-stable-during-2018-as-continuous-demand-for-suvs-offsets-decline-in-sales-of-compact-cars-and-mpvs/.

Jato. 2019b. "U.S. New Vehicle Sales Saw a Slight Increase in 2018 as SUVs Continue to See Market Share Growth." February 27. https://www.jato.com/usa/u-s-new-vehicle-sales-saw-a-slight-increase-in-2018-as-suvs-continue-to-see-market-share-growth/.

Koenig, B. 2020. "Auto Industry Gets Back Into Gear Following COVID-19 Shutdown." SME. August 17. https://www.sme.org/technologies/articles/2020/september/auto-industry-gets-back-into-gear-following-covid-19-shutdown/.

Madhok, A. 2020. "Weekly Update: COVID-19 Impact on Global Automotive Industry." Counterpoint Research [Insights]. September 15. https://www.counterpointresearch.com/weekly-updates-covid-19-impact-global-automotive-industry/.

NASEM (National Academies of Sciences, Engineering, and Medicine). 2020. *Airborne Transmission of SARS-CoV-2: Proceedings of a Workshop—in Brief.* Washington, DC: The National Academies Press. https://doi.org/10.17226/25958.

NRC (National Research Council). 2015. *Cost, Effectiveness, and Deployment of Fuel Economy Technologies for Light-Duty Vehicles*. Washington, DC: The National Academies Press. https://doi.org/10.17226/21744.

2

Fuel Economy, Greenhouse Gas Emissions, and Vehicle Efficiency Background

Consumers look for many vehicle attributes including improved vehicle fuel economy and energy efficiency to reduce their fueling costs, time spent refueling, and environmental impact. Government leaders in the United States have required increasing fuel economy and reduced greenhouse gas (GHG) emissions to reach national goals of energy security, improved consumer value, and reduced emissions of GHGs and other tailpipe pollutants. Automakers have responded to the consumer and regulatory impetus by improving energy efficiency of technologies for vehicle operation. More efficient technologies developed by automakers and automotive suppliers provide competitive advantage with consumers and meet regulatory requirements for fuel economy and emissions ratings of whole vehicles, but also enable improvement of other vehicle attributes valued by consumers, such as power, acceleration time, towing, and other capabilities. These improvements are occurring in technologies for petroleum-fueled vehicles and also for vehicles fueled by a diversity of energy carriers, including biofuels, blended fuels, hydrogen, and electricity. New technologies, implemented for fuel efficiency or to improve other vehicle attributes, often cost more than the technologies they replace. This has implications for vehicle cost, price, sales, consumer value, and the costs of fuel economy and emissions standards.

This chapter briefly describes the history of energy efficiency in light-duty vehicles (LDVs), including technical and regulatory aspects. It provides context for current fleet performance described in Chapter 3 and the technologies (Chapters 4–8) and policies for 2025–2035 described and recommended in the remainder of the report.

2.1 TECHNOLOGY PRINCIPLES AFFECTING VEHICLE EFFICIENCY

This report is tasked with informing the National Highway Traffic Safety Administration (NHTSA), Congress, and the public on fuel economy technology potential relevant in 2025–2035. To understand technology value and interactions, it is important to understand the physical principles controlling the efficiency of vehicle movement.

The movement of vehicles, and their drivers, passengers, and goods, requires input of energy. A portion of the energy input is converted to the desired output, travel of the vehicle mass, and a portion instead goes into energy loss pathways, such as aerodynamic, rolling resistance, and pumping losses. The forces impeding vehicle motion can be written as follows:

$$F = ma + R_a + R_{rr},$$

where ma is the inertial force, R_a is the aerodynamic resistance, and R_{rr} is the rolling resistance.

Reduction of energy loss pathways improves the efficiency of vehicle movement. Efficiency of vehicle movement can also be improved through reduction in vehicle mass and operational factors for more efficient travel of passengers and drivers.

2.1.1 Mass

If all other factors are equal, vehicles with less mass are more fuel efficient. Lighter vehicles require less power to overcome inertial force impeding vehicle motion than heavier vehicles, which have greater inertia. Consequently, mass reduction is an effective way of increasing vehicle fuel economy. Lightweighting is accomplished through vehicle design changes and use of lighter materials, such as aluminum, high-strength steel, and advanced composites.

2.1.2 Aerodynamics

Reductions in fuel consumption can be achieved by decreasing vehicle aerodynamic drag. Vehicle aerodynamic drag is the product of the drag coefficient, the frontal area, air density, and speed squared.[1] The drag coefficient measures the force of air resistance on the vehicle. A lower drag coefficient indicates that the vehicle will have less aerodynamic drag. The average passenger vehicle has a drag coefficient between 0.302 and 0.395 (EPA/NHTSA/CARB, 2016). Because aerodynamic drag increases with the square of speed, its effects increase at higher speeds. At low speeds, aerodynamic drag accounts for about one-fourth of fuel burned. At high speeds, more than half of fuel can be used to overcome aerodynamic drag. External vehicle add-ons (e.g., roof storage units) can degrade aerodynamic efficiency. Optimizing vehicle shape by streamlining and with active and passive design add-ons can reduce the drag coefficient, and hence the energy losses in aerodynamic drag.

2.1.3 Rolling Resistance

In addition to aerodynamic drag and inertial force owing to mass, tire rolling resistance is one of the many forces that must be overcome in order for a vehicle to move. Rolling resistance is the product of the repeated deformation of a tire during rotation resulting in energy loss.

$$F = C_{rr}N,$$

where F is the force of rolling resistance, N is the normal force, and C_{rr} is the rolling resistant coefficient. A lower rolling resistance coefficient indicates less energy loss. In addition to pavement conditions, tire pressure, vehicle mass, and vehicle brake type affect rolling resistance. Tire rolling resistance can be decreased through tread design and composition, inflation, stiffening sidewalls, and a smaller tire footprint.

2.1.4 Engine Thermal Efficiency

Spark ignition (Otto Cycle) engines, the most common type in the U.S. fleet, typically convert only about one-third of the total fuel energy consumed to indicated work done on the piston (indicated thermal efficiency). Of the remaining energy, approximately one-third is lost as heat rejected to the coolant and another one-third is lost as exhaust enthalpy. Of that indicated work at conditions representative of the Federal Test Procedure fuel economy test cycle, brake work delivered at the crankshaft (brake thermal efficiency) is further reduced by up to 40% owing to losses attributable to intake and exhaust pumping (5%), mechanical friction (8%), and engine driven accessory drive requirements (1%). In discussing technologies and approaches to improve the fuel efficiency

[1]The force required to overcome drag is represented by the product of the drag coefficient, C_d; the frontal area, A; and the square of speed, V^2. The formula is $F = \frac{1}{2}C_d AV^2$.

of an internal combustion engine (and thereby reduce its carbon dioxide [CO_2] emissions), it is convenient to categorize them as improving efficiency through the following:

1. Thermodynamic factors such as combustion timing, compression ratio, working fluid properties, or heat loss reduction.
2. Reducing losses to mechanical rubbing friction via design, surface treatments, lubricants, downsized/turbocharged engines (fewer number of cylinders).
3. Reducing engine pumping losses with technologies such as variable valve timing and lift, cylinder deactivation, and engine downsizing.
4. Additionally, accessory loads can be made more efficient and/or converted to electrically driven.

2.1.5 Electrical System Losses

Electrical loads in LDVs come primarily from accessories for safety, entertainment, and comfort, and electric drive components. Electrical accessories such as lights, electric power steering, air conditioning, power windows, seats, and door locks, seat and steering wheel warmers, windshield wipers, navigation systems, and entertainment systems require power. Energy losses from accessories can account for up to 2% of total fuel use in a typical vehicle. Driver assist systems for safety, comfort, and convenience are a growing source of electric loads and can be in the tens of percents of total fuel use (EPA, n.d.).

In electric and hybrid vehicles, energy losses occur primarily as resistive loses in the battery and other electronics, and motor losses when a permanent magnet motor is spinning but not powering the vehicle or generating charge on the battery. Resistive losses transform some of the stored energy into internal heat instead of external power, thereby lowering the energy efficiency. Internal resistance varies by material.

2.1.6 Operational Factors

Operational factors affect the efficiency of individual vehicle use and also of total energy use of the transportation system. Total energy use impacts will be discussed in later chapters. For an individual vehicle, operational factors include choice of vehicle for a trip, driver behaviors like speeding and idling, fueling choices such as use of premium fuel or charging of plug-in hybrid electric vehicles (PHEVs), efficient trip planning and routing, and vehicle maintenance.

2.1.7 Total Vehicle System Energy Flows

Figure 2.1 shows a schematic of system energy flows in a hybrid vehicle. Table 2.1 shows the related breakdown of energy losses and resulting power to the wheels for internal combustion engine vehicles (ICEVs), hybrid electric vehicles (HEVs), and battery electric vehicles (BEVs). The system losses are areas where vehicle efficiency can be improved.

2.2 FUEL CONSUMPTION, GHG EMISSIONS, AND ENERGY USE

As introduced in Chapter 1, vehicle energy consumption, fuel consumption, and GHG emissions are related metrics. At base is vehicle energy use. As shown in Table 2.1, different vehicles use different amounts of energy to provide the same power to the wheels. Energy consumption is directly related to fuel consumption for both liquid petroleum fuels and various forms of non-petroleum fuel, although the metric used varies (gallons of gasoline or diesel, kWh of electricity, and kg of hydrogen). GHG emissions are also related to vehicle energy consumption, although the fuel used to provide the energy determines the relationship. GHG emissions related to vehicle energy consumption occur not just on board the vehicle but also upstream in the fuel generation and transport to the vehicle. Chapter 10 discusses the detailed considerations around a well-to-wheels evaluation of vehicle energy use and GHG emissions.

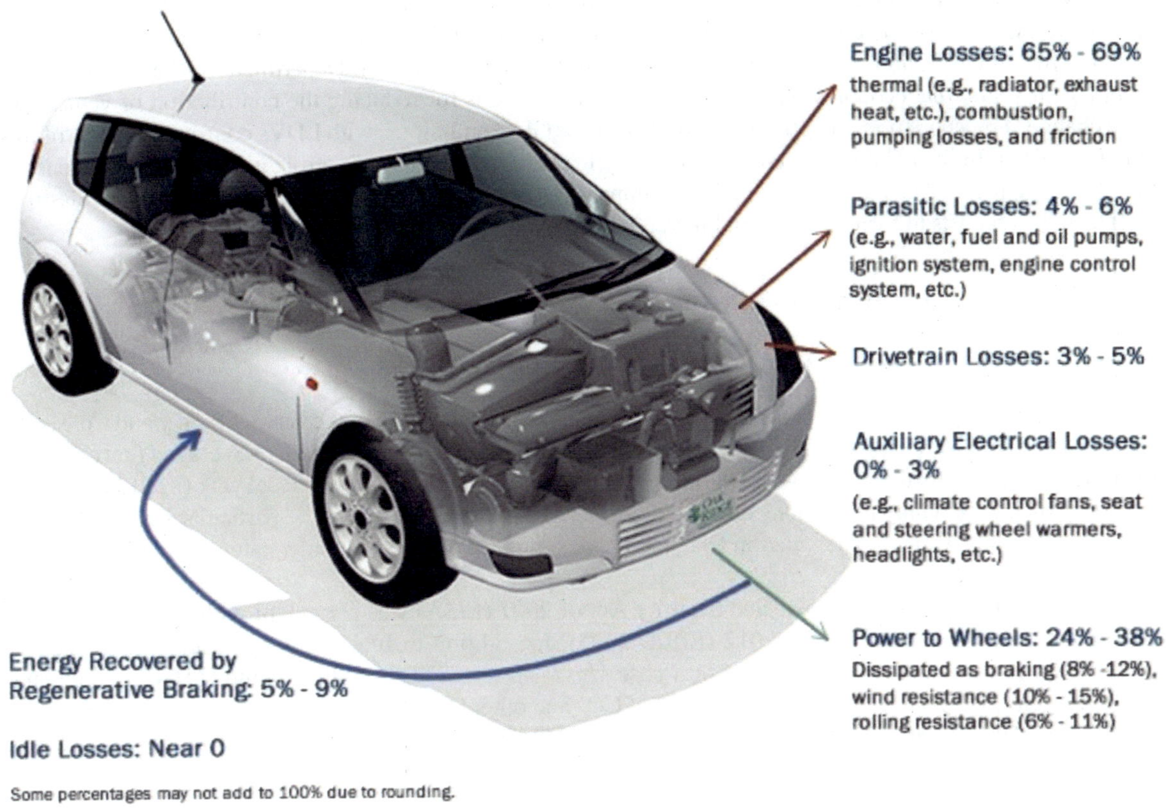

FIGURE 2.1 Total system energy flows for an illustrative HEV.
SOURCE: EPA (n.d.).

TABLE 2.1 Total Onboard Energy Flows as Estimated for ICEVs, HEVs, and BEVs

Energy Flow	ICEV % Energy Loss	HEV % Energy Loss	BEV % Energy Loss
Engine losses	68–72	65–69	N/A
Battery charging and electric drive system losses	N/A	N/A	20
Parasitic losses	4–6	4–6	N/A
Drivetrain losses	5–6	3–5	N/A
Auxiliary electrical losses	0–2	0–3	0–4
Idle losses	3	0	0
Regenerative braking recovery	N/A	+ 5–9	+ 17
Power to wheels	+ 16–25	+ 24–38	+ 86–90

NOTE: Some percentages may not add to 100% owing to rounding and variable ranges.
SOURCE: EPA (n.d.).

Energy efficiency is motivated by reduced use of resources. Concerns on the use of liquid fuels and efficiency were brought into U.S. public consciousness largely following the 1973 oil crisis, which resulted in shortages of automotive fuel at gas stations and saw sharp increases in the price of oil (U.S. Department of State, n.d.). However, in recent years, fuel economy has come into focus as a means for reducing the contribution of vehicles to climate change. Transportation comprised 29% of 2017 U.S. GHG emissions, with LDVs contributing 59% of this total (EPA, 2019). In the LDV sector, energy efficiency leads to reduced use of petroleum fuels and other energy carriers/sources, reduced GHG emissions, reduced impacts associated with energy use such as criteria emissions, as well as reduced costs to the consumer and increased protection from price volatility.

2.3 TECHNICAL, REGULATORY, AND STATUTORY HISTORY

2.3.1 Vehicle Efficiency Regulatory History

Vehicle efficiency has been an explicit government goal since the passage of the Energy Policy and Conservation Act of 1975 (EPCA, 1975). EPCA assigned authority for regulating manufacturer fleet-averaged fuel economy of LDVs to the U.S. Department of Transportation (DOT)'s NHTSA beginning with model year (MY) 1978. Since the enacting of fuel economy regulations under EPCA, there have been changes in the particular structure of fuel economy regulations, which are detailed in Chapter 12. A summary of fuel economy statutes and regulations is shown in Table 2.2.

Notably, the Energy Independence and Security Act of 2007 (EISA) was passed, providing a new mandate for fuel efficiency beginning for MY 2012 (EISA, 2007). In addition to fuel economy regulations mandated by EISA and EPCA, in 2007 the Supreme Court case *Massachusetts v. EPA* obligated the U.S. Environmental Protection Agency (EPA) to determine if emissions of CO_2 and other GHGs from motor vehicles were required to be regulated under Section 202 of the Clean Air Act (United States Code, 1990). Under these two new mandates, a national program of fuel economy and GHG regulations was implemented. NHTSA, EPA, and the California Air Resources Board produced a single set of requirements for MY 2012–2016, and a second national program for MY 2017–2025.

The most recent set of regulations is The Safer Affordable Fuel-Efficient (SAFE) Vehicles Rule, setting fuel economy and GHG emissions standards over 2021–2026. Under this new regulation, fuel economy and emissions standards increase by 1.5% each year through MY 2026, compared with 5% per year under the previously proposed standard, resulting in a projected 40.4 miles per gallon (MPG) required fuel economy in MY 2026, compared with 46.7 MPG projected under the previous regulations (NHTSA/EPA, 2020). Fuel economy standards as set by NHTSA can be complied with by achieving the appropriate weighted average of tested fuel economy, by paying a fine, and through various credit mechanisms, described further in Chapter 12 (EPA/NHTSA, 2012). Fuel economy standards, achieved fuel economy, and resulting vehicle energy efficiency improvements over time are shown in Figure 2.2.

NHTSA is required by EISA to set maximum feasible fuel economy standards through 2030, but for no more than 5 years. This means that standards beginning in MY 2027 can at most cover MY 2027–2031. This report is tasked with informing NHTSA, Congress, other federal agencies, and the public about fuel economy technologies, consumer behavior, and policy issues pertinent to the 2025–2035 time frame, and so can inform the upcoming standards.

2.3.2 Vehicle Technology History

Vehicle technology development has occurred in response to consumer demands for fuel economy and other vehicle attributes, as well as regulatory stringency drivers. Technologies have advanced to better use fuel in engines and motors, to better use the power output of those engines and motors, and to reduce the power required to move the vehicle. To better use fuel, engines have been improved to accomplish more complete combustion, and to transform more of that combustion energy into mechanical energy output of the engine. Similarly, motors have been improved to more efficiently convert electrical energy to mechanical energy. Transmissions and other aspects of the drivetrain have been improved to more effectively convert mechanical energy out of the motor or

TABLE 2.2 Statutes Governing Fuel Economy and GHG Standards for LDVs, Including Relevant Mandate by Year Enacted and Active Status, and Additional Regulatory Aspects by Model Year Implemented

Statute	Regulation(s)
Clean Air Act (1970, updated 1990) The CAA gives the EPA the authority to regulate air pollutants harmful to humans. Originally passed in 1970, the CAA has been amended several times since to include newly recognized pollutants and was the focus of a 2007 Supreme Court case establishing if GHGs fall under the CAA (42 U.S.C. 85 §7521–§7554; 42 U.S.C. 85 §7581–§7590). **Status: Active** • Gives EPA authority to set standards for any air pollutant that "may reasonably be anticipated to endanger public health." • Note regarding CO_2 and other GHGs from vehicles: Not explicitly included in CAA until 2007 (Ex. Ord. 13432); DOT, the U.S. Department of Energy, and EPA mandated to regulate to the extent determined to be practical.	• Average Fuel Economy Standards for Light Trucks MY 2008–2011 • Light-Duty Vehicle Greenhouse Gas Emission Standards and Corporate Average Fuel Economy Standards; Final Rule (2010, MY 2012–2016) • 2017 and Later Model Year Light-Duty Vehicle Greenhouse Gas Emissions and Corporate Average Fuel Economy Standards (MY 2017–2025) • The Safer Affordable Fuel-Efficient (SAFE) Vehicles Rule for Model Years 2021–2026 Passenger Cars and Light Trucks
Energy Independence and Security Act (2007) EISA enacts the three provisions of the corporate average fuel economy (CAFE) standards, the Renewable Fuel Standard, and lighting and appliance efficiency standards for the goal of reducing U.S. dependence on oil. This was the first statutory increase in fuel economy (FE) standards since EPCA in 1975. **Status: Active** DOT given authority to set FE standards through 2030. • Combined FE for MY 2020 of 35 MPG for total passenger and non-passenger fleet (Sec. 102). • FE standards through MY 2030 should be based on "maximum feasible FE standard." • FE regulations must be issued by DOT for at least 1 but not more than 5 MYs.	• Revisions and Additions to Motor Vehicle Fuel Economy Label • Average Fuel Economy Standards for Light Trucks MY 2008–2011 • Light-Duty Vehicle Greenhouse Gas Emission Standards and Corporate Average Fuel Economy Standards; Final Rule (2010, MY 2012–2016) • 2017 and Later Model Year Light-Duty Vehicle Greenhouse Gas Emissions and Corporate Average Fuel Economy Standards (MY 2017–2025) • SAFE Rule (2020)
Energy Policy and Conservation Act (1975) EPCA established a federal program to set energy targets for consumer products. It gives NHTSA the authority to set fuel economy standards for the purpose of reducing energy and oil consumption and led to the first CAFE standards in 1975 (EPCA, 1975). See Title III, Part A. Automotive Fuel Economy **Status: Inactive**	• Average Fuel Economy Standards for Light Trucks MY 2008–2011 • Light-Duty Vehicle Greenhouse Gas Emission Standards and Corporate Average Fuel Economy Standards; Final Rule (2010, MY 2012–2016) • 2017 and Later Model Year Light-Duty Vehicle Greenhouse Gas Emissions and Corporate Average Fuel Economy Standards (MY 2017–2025) • The Safer Affordable Fuel-Efficient (SAFE) Vehicles Rule for Model Years 2021–2026 Passenger Cars and Light Trucks *Note:* Only regulations after MY 2008 are included.
Alternative Motor Fuels Act of 1988 AMFA encourages the production of dual-fueled vehicles or those entirely using a fuel other than petroleum by providing credits toward the calculation of CAFE performance, with the goal of energy independence. It specifies this incentive program will last until 2004, at which point the DOT could choose to extend the program an additional four MYs (AMFA, 1988). **Status: Inactive (at least with regards to CAFE)** • Mandate for dual-fuel CAFE credit expired in MY 2008.	• Report to Congress: Effects of the Alternative Motor Fuels Act CAFE Incentives Policy (2002) • Automotive Fuel Economy Manufacturing Incentives for Alternative Fueled Vehicles (2004) (Extends incentives through MY 2008)
Energy Policy Act of 2005 The Energy Policy Act of 2005 calls for the development of grant programs, demonstration and testing initiatives, and tax incentives that promote alternative fuels and advanced vehicles production and use. EPAct 2005 also amends existing regulations, including fuel economy testing procedures and EPAct 1992 requirements for federal, state, and alternative fuel provider fleets. (EPAct, 2005). See Title VII: Vehicles and Fuels **Status: No active FE/GHG standards**	

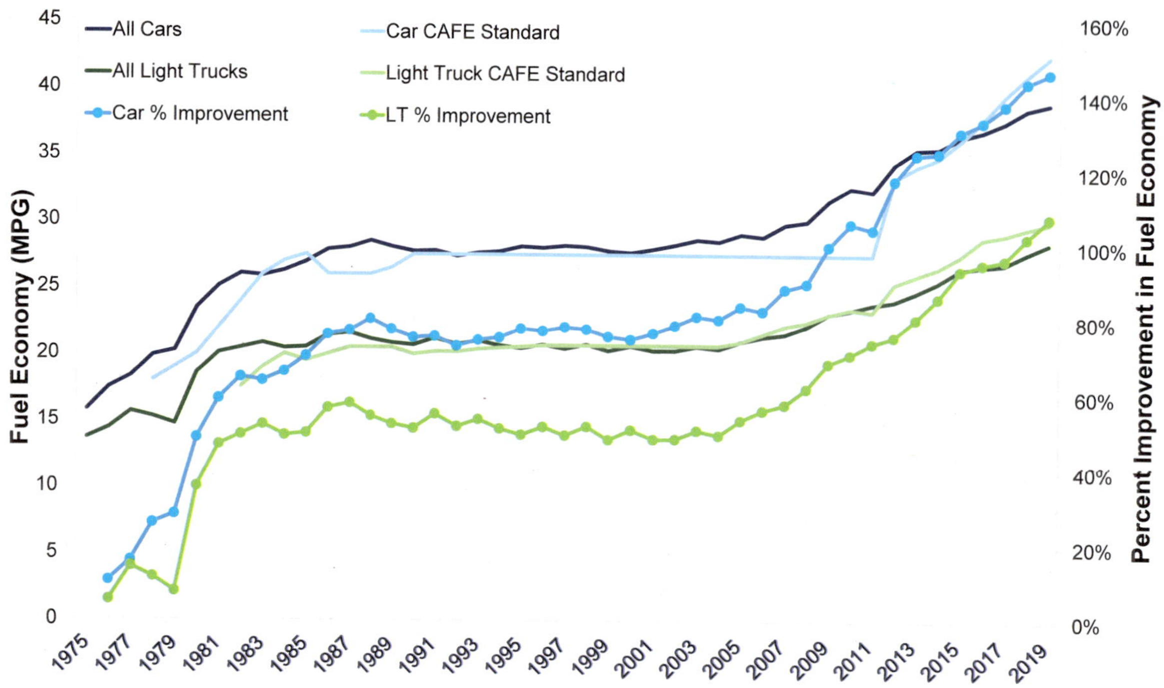

FIGURE 2.2 U.S. fleet fuel economy standard, achieved fuel economy in MPG, and vehicle energy efficiency as percent improvement in fuel economy from 1975 to 2018.
SOURCE: Committee generated using data from EPA (2019).

engine to movement in the wheels. Road load has been reduced through mass reduction, improved aerodynamics, and reduced rolling resistance. New propulsion systems that can recover braking energy and eliminate idling, and have inherently more efficient energy use, have been implemented (stop start, hybrids, and electric vehicles [EVs]). Figure 2.3 shows some of the vehicle technologies that were uncommon or nonexistent in 1975 that have been implemented to improve efficiency and have become commonplace.

While these efficiency improvements have been implemented to reduce fuel consumption, horsepower and other desired vehicle attributes have continued to increase. Figure 2.4 illustrates the energy improvement and reduced fuel consumption available as engines have been improved from 1975 to 2020.

2.3.3 Other Vehicle Regulations That Impact Vehicle Efficiency

2.3.3.1 Vehicle Safety Regulations

Federal Motor Vehicle Safety Standards issued by NHTSA reflect congressional laws pertaining to vehicle safety. Safety and fuel economy have potential trade-offs and synergies. One notable example is that reducing the mass of materials on the vehicle can present improvements in fuel economy; however, historically, there has been concern about the passenger safety implications of reducing vehicle mass, or "lightweighting." The precise relationship between vehicle mass and safety risk is complex, however, with the potential for vehicle material selections to be made that improve fuel economy through mass reductions but without necessarily decreasing safety. Recent technological developments such as advanced driver assistance systems can enhance both safety and fuel economy through assisting drivers with breaking, acceleration, and/or steering (NHTSA/EPA, 2020), although the impact of these devices is still under study. Further overall discussion of the relationship between vehicle mass and safety considerations is presented in Chapter 7. Safety considerations regarding specific vehicle types (e.g., BEVs, fuel cell electric vehicles [FCEVs], connected and automated vehicles) are presented in Chapters 5, 6, and 8.

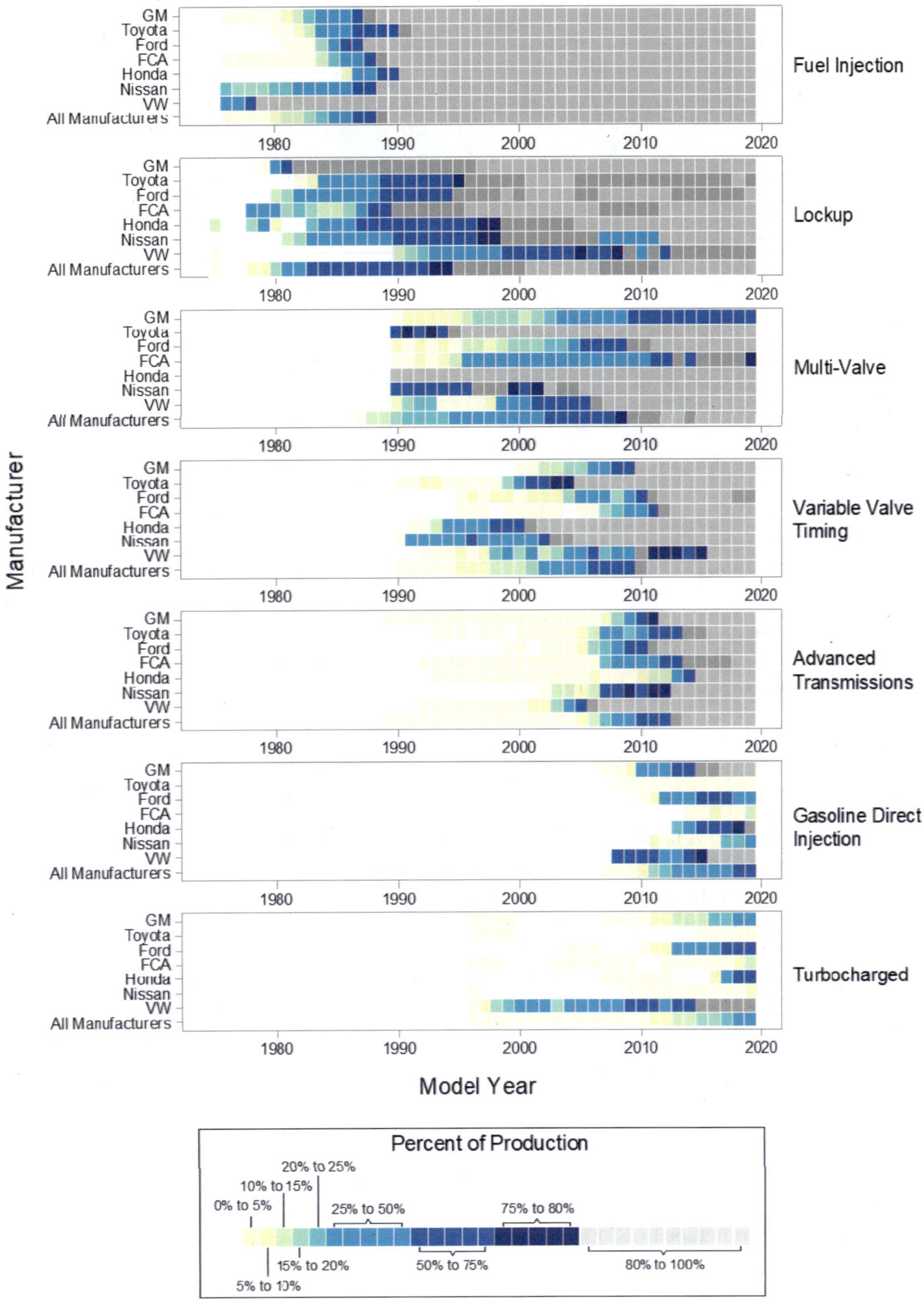

FIGURE 2.3 Implementation of fuel injection, lockup, multivalve engines, variable valve timing, advanced transmissions, gasoline direct injection, and turbocharging by major manufacturer, showing the time it takes for a technology to be implemented in a large percentage of a manufacturer's fleet, and the variability in implementation of technologies across different manufacturers.
SOURCE: EPA (2020).

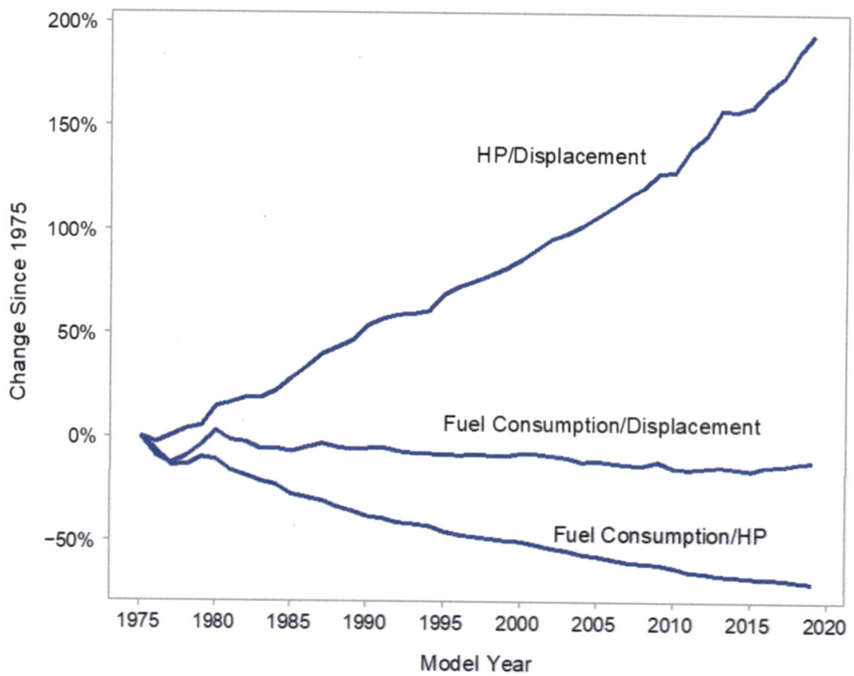

FIGURE 2.4 Measures of impact of improved gasoline engine technologies, including constant improvement of close to 200% in horsepower (hp) per displacement, slightly falling fuel consumption per displacement, and a more than 50% decrease in fuel consumption per horsepower over the fleet since 1975.
SOURCE: EPA (2020).

2.3.3.2 Criteria Pollutants

Under the Clean Air Act, EPA began regulating nitrogen oxide emissions from LDVs, with the 1990 amendment expanding standards to non-methane organic gases, carbon monoxide, particulate matter, and formaldehyde (EPA, 2015c). EPA considers vehicle fuel economy as part of its GHG rules (EPA, 2015b), as reducing the quantity of fuel combustion reduces the vehicle emissions released. In addition to climate change-related benefits, reductions in vehicle emissions present benefits to public health, with exposure to criteria pollutants associated with negative health outcomes (Utell et al., 1994). Switching vehicles to non-petroleum sources of energy, such as electricity, has the potential to provide reductions in criteria pollutants where the vehicle is operating, and also at the source of electricity emissions, if non-combustion processes are used, or if emissions reductions are implemented at the power plant.

2.3.3.3 Vehicle Efficiency Information and Labeling

EPA fuel economy labeling has been on vehicles since 1974, with the most recent label design (2012) providing ratings on vehicle smog and GHG emissions and estimates of how much the consumer will save on fuel in the next 5-year period (compared with an average new vehicle), among other information (EPA, 2015a). Fuel economy labels provide information to inform consumers' purchases. However, the way consumers perceive and experience value from fuel economy is complex, as detailed in Chapter 11.

2.3.3.4 State and Local LDV Efficiency and Emissions Regulations

Several states and localities also have requirements for vehicle fuel economy, energy use, or GHG emissions. California's Zero-Emission Vehicle (ZEV) program is a prominent example, establishing a credit system to require automakers to produce a certain number of vehicles that do not necessarily emit directly at the tailpipe (e.g., PHEVs, BEVs, FCEVs). Estimates show that this credit system may lead to 8% of 2025 new vehicle sales in California being ZEVs, and these ZEV regulations have been adoption by nine other states (CARB, 2020). In September 2020, Governor Newsom signed an executive order requiring 100% of new passenger cars and light trucks purchased in the state of California to be zero-emissions by 2035, although the regulations that will enforce meeting this goal have not yet been developed. The current California ZEV program is discussed further in Chapter 12.

2.3.3.5 Incentives for Different Fuels and Powertrains

Governments (international and U.S. national, state, and local governments) have provided incentives for the adoption of different powertrain technologies and use of different fuel types. Such incentives have included purchase subsidies and operational incentives. One example of purchase subsidies for powertrains is the U.S. federal government plug-in electric vehicle tax credit, which provides consumers with a tax rebate ranging from $2,500 to $7,500 depending on vehicle specifications (AFDC, 2020a). A discussion of the economic considerations pertaining to incentives is presented in Chapter 11. Incentives and requirements for the use of alternative fuels are also present, notably, the Renewable Fuel Standard (RFS) and the California Low Carbon Fuels Standard (LCFS). The RFS program sets volume requirements for the presence of different renewable fuels in the fuel mix sold in the United States (AFDC, 2020b). The California LCFS requires yearly decreases in the carbon emissions intensity of gasoline, diesel, and their substitutes. This is accomplished through decreasing life cycle GHG emissions standards to be met by fuel providers, with tradeable credits between under- and over-emitting providers. Further discussion of alternative fuels' GHG and energy use impacts are found in Chapter 10, and discussion of incentivizing fuels through regulation is presented in Chapter 12.

2.4 TEST CYCLE AND REAL-WORLD FUEL ECONOMY

Manufacturer compliance with the fuel economy standard is based on an individual vehicle model's performance of a two-cycle test procedure under controlled conditions on a vehicle dynamometer (49 U.S.C. § 32904[c]). A driver completes one of two speed versus time traces on a dynamometer while the vehicle's energy consumption (FCEVs, EVs) or emissions (HEVs, ICEVs) or a combination (PHEVs) are collected to determine its energy consumption, fuel consumption, and/or GHG emissions. One time trace is meant to mimic a city driving cycle and one a highway driving cycle, and they are weighted at 55% for the city cycle and 45% for the highway cycle. These tests are typically conducted by auto manufacturers, with occasional compliance checks by the EPA.

The two-cycle test overestimates the efficiency of vehicles relative to what drivers experience in typical driving. Fuel economy estimates for customer communication and labeling purposes for new LDVs are determined based on five standardized fuel economy test cycles conducted in a highly controlled laboratory environment. The five tests are the city cycle, highway cycle, a high-speed cycle, a cycle including air conditioning, and a cold temperature cycle, each of which are meant to simulate specific real-world driving conditions. Off-cycle credits can be awarded to technologies that deliver real-world fuel economy benefits or decrease emissions but are insufficiently counted on the official test cycle. Other driving conditions excluded from the tests that may reduce fuel economy, such as wind, low tire pressure, rough roads, hills, snow, or ice, are accounted for in the fuel economy labels by decreasing the measured fuel economy by an adjustment factor; however, this adjustment is not applied to CAFE compliance measurements.

The fuel economy label values, which include correction factors and are better estimates of real-world fuel economy, are about 20% lower on average than the fuel economy calculated using the test cycles. This discrepancy is particularly important because compliance with CAFE standards is based solely on the test cycles, with no adjustment for real-world conditions that are not accounted for in the cycles. This gap between test cycle and real-world fuel economy could adversely affect advances in real-world fuel economy in two key ways. It could incentivize manufacturers to design their vehicles to minimize fuel consumption based on the specific parameters of the test cycles, rather than real-world driving conditions. Second, this system does not reward manufacturers for non-powertrain technologies that reduce fuel consumption in real-world conditions but do not impact the test cycle results.

Up-to-date real-world fuel economy data are critical to evaluating the performance of vehicles, of the standards, and of the test cycles as compliance measures, but the United States currently has no database of or method to collect such data. The data could be collected through sampling of vehicle emissions, through collection of information stored in the vehicle's onboard diagnostic unit, or through remote sensing methods.

2.5 REFERENCES

AFDC (Alternative Fuels Data Center). 2020a. "Qualified Plug-In Electric Vehicle Tax Credit." https://afdc.energy.gov/laws/409.

AFDC. 2020b. "Renewable Fuel Standard." https://afdc.energy.gov/laws/RFS.html.

AMFA (*Alternative Motor Fuels Act of 1988.*) 1988. *United States Code*. Vol. 102. https://www.govinfo.gov/content/pkg/STATUTE-102/pdf/STATUTE-102-Pg2441.pdf.

CARB (California Air Resources Board). 2020. "Zero-Emission Vehicle Program." https://ww2.arb.ca.gov/our-work/programs/zero-emission-vehicle-program/about.

EISA (Energy Independence and Security Act). 2007. United States Code. Vol. 121.

EPA (U.S. Environmental Protection Agency). n.d. "Where the Energy Goes." https://www.fueleconomy.gov/feg/atv-hev.shtml.

EPA. 2015a. "Learn About the Fuel Economy Label." Collections and Lists. September 3. https://www.epa.gov/greenvehicles/learn-about-fuel-economy-label.

EPA. 2015b. "Carbon Pollution from Transportation." Overviews and Factsheets. September 10. https://www.epa.gov/transportation-air-pollution-and-climate-change/carbon-pollution-transportation.

EPA. 2015c. "Light Duty Vehicle Emissions." Overviews and Factsheets. October 13. https://www.epa.gov/greenvehicles/light-duty-vehicle-emissions.

EPA. 2019. *Inventory of U.S. Greenhouse Gas Emissions and Sinks 1990–2017*. EPA-430-R-19-001. April.

EPA. 2020. *The 2019 EPA Automotive Trends Report: Greenhouse Gas Emissions, Fuel Economy, and Technology Since 1975*. EPA-420-R-19-002. March.

EPA/NHTSA (U.S. Environmental Protection Agency and National Highway Traffic Safety Administration). 2012. "2017 and Later Model Year Light-Duty Vehicle Greenhouse Gas Emissions and Corporate Average Fuel Economy Standards." Federal Register 77(99).

EPA/NHTSA/CARB (U.S. Environmental Protection Agency, National Highway Traffic Safety Administration, and California Air Resources Board). 2016. *Draft Technical Assessment Report: Midterm Evaluation of Light-Duty Vehicle Greenhouse Gas Emission Standards and Corporate Average Fuel Economy Standards for Model Years 2022–2025*. EPA-420-D-16-901. https://www.nhtsa.gov/staticfiles/rulemaking/pdf/cafe/Draft-TAR-Final-Executive-Summary.pdf.

EPAct (Energy Policy Act of 2005). 2005. United States Code. Vol. 119.

EPCA (Energy Policy and Conservation Act). 1975. United States Code. Vol. 89.

NHTSA/EPA (National Highway Traffic Safety Administration and U.S. Environmental Protection Agency). 2020. *The Safer Affordable Fuel-Efficient (SAFE) Vehicles Rule for 2021–2026 Passenger Cars and Light Trucks*. April 30. https://www.govinfo.gov/content/pkg/FR-2020-04-30/pdf/2020-06967.pdf.

U.S. Department of State. n.d. "Oil Embargo, 1973–1974." Office of the Historian. https://history.state.gov/milestones/1969-1976/oil-embargo.

United States Code. 1990. Clean Air Act Amendments, Pub. L. No. 101–549 (1990). https://www.govinfo.gov/content/pkg/STATUTE-104/pdf/STATUTE-104-Pg2399.pdf.

Utell, M.J., J. Warren, and R.F. Sawyer. 1994. Public health risks from motor vehicle emissions. *Annual Review of Public Health* 15:157–178. https://doi.org/10.1146/annurev.pu.15.050194.001105.

3

2025 Baseline of Vehicles

3.1 COMPARATIVE BENCHMARKS FOR 2016–2026 VEHICLES

This chapter summarizes applicable developments related to the evolution of vehicle efficiency and carbon dioxide (CO_2)-reduction technology in the near-term time frame and examines the current fleet and makes estimates of technology penetration in 2020–2025 using the most recent and comprehensive baseline databases for 2016–2018 new vehicles and regulatory analyses through 2026. There are several applicable regulation levels and multiple technology pathways that are relevant to the amount and types of vehicle technologies that will be deployed to increase efficiency through 2025 and 2026. This chapter describes the committee's assumptions for its evaluation of benchmark technologies, efficiency, and CO_2 emission levels by vehicle class and therefore helps create a link between 2018–2020 regulatory developments and the committee's chapters on 2025–2035 technologies. In addition, the chapter provides related global regulatory context for continued automotive industry investments in vehicle efficiency and electric vehicles.

3.2 BASELINE VEHICLE CLASSES

The latest complete detailed data set, for which all the critical vehicle attributes (e.g., make, model, engine, transmission, emissions, fuel economy, size, mass, vehicle class, sales, application of efficiency technology) are included, is the model year (MY) 2017 data set used in the March 2020 regulatory analysis (NHTSA/EPA, 2020). The vehicle emission levels, efficiency, technologies applied, vehicle classes, and other characteristics from this 2017 data set are analyzed and applied in this analysis.

Throughout this report, two types of vehicle breakdowns, regulatory and vehicle classes, are used, as shown in Table 3.1. For regulatory purposes, the light-duty vehicle (LDV) fleet is fundamentally split into passenger cars and light trucks, each of which gets its own set of regulatory targets for each MY. For the committee's discussion of the relative efficiency, emission levels, and technologies throughout the report, representative vehicle classes are established. The analysis applies five major vehicle classes as also shown in Table 3.1, based on the vehicle body types used by the regulatory agencies in their analyses. The classes broadly cover the market and distinguish how different efficiency technologies may be applied in each class and have different cost and effectiveness values.

TABLE 3.1 Vehicle Classes and Sales-Weighted Attributes for MY 2017

	Class	Sales	Percent of Total Sales	Percent Light Trucks	Test Fuel Economy (MPG)	Test CO_2 Emissions (g/mi)	Label Fuel Economy (MPG)
Regulatory category	Passenger car	8,955,057	53%	0%	37.6	237	28.9
	Light trucks	8,054,950	47%	100%	26.9	330	20.7
Vehicle class	Small car	4,393,901	26%	0%	42.2	211	32.5
	Medium car	2,102,788	12%	0%	33.4	266	25.7
	Crossover	4,565,184	27%	50%	33.3	267	25.6
	Sport utility vehicle	3,889,793	23%	95%	26.6	334	20.5
	Pickup	2,058,341	12%	100%	23.6	376	18.2
Total		17,010,007	100%	47%	31.6	281	24.4

NOTE: CO_2 = carbon dioxide; MPG = miles per gallon.
SOURCE: NHTSA/EPA (2020).

The small car class includes subcompact and compact vehicles, and the medium car class includes midsize and large cars. The crossover class includes small car platform-based sport utility vehicles (SUVs) and hatchbacks; the SUV class includes the larger SUVs and minivans; and the pickup class covers those pickups that fall within the LDV classification. The smaller classes tend to be lighter and deliver higher fuel economy in miles per gallon (MPG). These classes are used in the sections below to assess trends going forward for technology adoption for 2025 and beyond vehicles. The test cycle efficiency in MPG is shown, along with the corresponding grams of CO_2 per mile (g CO_2/mi), based on the assumption of 8,887 grams of CO_2 per gallon of gasoline. A simple estimate of consumer label fuel economy is shown, based on a simple 23% reduction from the test cycle CO_2 level. As indicated in the most recent trends reports (EPA, 2020), the general trend is toward a higher share of LDVs in the crossover class.

Figure 3.1 summarizes the relative uptake of efficiency technologies in MY 2017 vehicles, as represented in the National Highway Traffic Safety Administration (NHTSA) 2017 reference fleet. As shown, technologies that can improve efficiency and performance in the engine, transmission, hybrid, electric, and road load areas have seen penetration across the five classes. Many of these technologies had much lower shares in MY 2008 vehicles, the baseline from which the 2012–2016 CO_2 and corporate average fuel economy (CAFE) regulations were developed. For example, of all MY 2008 vehicles sold, 2% had gasoline direct injection, 3% had turbocharging, 10% had 7-or-greater transmission gears, and 0% had non-hybrid stop-start. Since the 2016 database of MY 2017 vehicles used in Figure 3.1 was developed, several technologies have seen further increases, for example, to 31% turbocharging and 28% stop-start in MY 2018 (EPA, 2019).

3.3 FUTURE YEAR CO_2 REDUCTION AND INCREASED EFFICIENCY TO 2025

To understand the likely efficiency technology packages of vehicles in the new vehicle fleet around 2025, the committee reviewed the required stringency and projected least-cost paths to reach the original 2021–2025 5% and newly revised 2021–2026 1.5% yearly increases in fuel economy, as assessed by the U.S. Environmental Protection Agency (EPA) and NHTSA in their regulatory documents. The committee considers these to represent a reasonable approximation of possible futures for the auto industry to globally deploy technologies. Another benchmark is the 3.7% annual reduction in CO_2 emissions agreed to by the California Air Resources Board and several automakers (CARB, 2019).

MY 2017 vehicle models are shown in comparison to vehicle footprint-indexed target lines of the regulatory standards for 2012, 2017, and 2020 fuel economy (top panels) and CO_2 emission (bottom panels) in Figure 3.2. The figure displays regulatory target lines for the original Obama administration 2025 standards and newly adopted

Area	Technology	Passenger Cars	Light Trucks	Light-duty Vehicles
Engine	Variable valve timing	62%	79%	70%
	Variable valve lift	22%	19%	20%
	Gasoline direct injection	22%	37%	29%
	Turbocharging	30%	18%	24%
	Cylinder deactivation	3%	22%	12%
	High compression ratio	3%	1%	2%
Transmission	6-speed or less	53%	51%	52%
	7- or 8-speed	10%	24%	17%
	9- or 10-speed	3%	9%	6%
	Continuously variable	28%	14%	21%
Hybrid	Start-stop	14%	20%	17%
	Mild hybrid	0%	0%	0%
	Strong hybrid	3%	1%	2%
Electric	Plug-in hybrid electric	1%	0%	1%
	Battery electric	1%	0%	1%
	Fuel cell electric	0%	0%	0%
Road load reduction	Mass reduction (10% or more)	20%	14%	17%
	Tire rolling resistance reduction (10% or more)	43%	55%	49%
	Aerodynamic reduction (10% or more)	29%	15%	23%

FIGURE 3.1 Percent of MY 2017 passenger cars, light trucks, and all LDVs with various efficiency technologies, as represented in the NHTSA 2017 reference fleet. NOTE: Technologies are defined as in NHTSA's CAFE model, and some technologies, such as variable valve timing or gasoline direct injection, will also be present as a part of other technologies, such as turbocharging and variable valve lift.
SOURCE: NHTSA/EPA (2020).

2026 standards. The figure includes passenger car standards and MY 2017 vehicles on the left and light truck standards and models on the right. The sales-weighted average for 2017 vehicles is also shown. As indicated the individual vehicle models are scattered across a wide range of vehicle footprints, fuel economy, and CO_2 emissions. The footprint-indexed target lines are designed to ensure that vehicles across different sizes see efficiency improvements from additional technology, rather than to promote the shift toward smaller vehicles, to comply with the standards.

3.4 MY 2020 VEHICLES WITH LOWEST CO_2 EMISSIONS

To help assess the deployment of automotive technologies, this section analyzes the top-performing low-CO_2-emission models, excluding hybrids, as compared to newly adopted Trump administration 1.5% and original Obama administration 5% per year improvements through 2026. Although there are several dozen hybrid models that compare even more favorably than the conventional non-hybrid models illustrated below, this section's analysis focuses on determining the lowest-CO_2 vehicles using conventional engine and transmission technologies, as these show the more dominant, mainstream, and low-cost technologies.

As a first step, the lowest-CO_2 MY 2020 vehicles in two major vehicle classes, midsize cars and crossover light trucks, were identified as compared to the footprint-indexed CO_2 standards. Figure 3.3 shows the MY 2012–2021

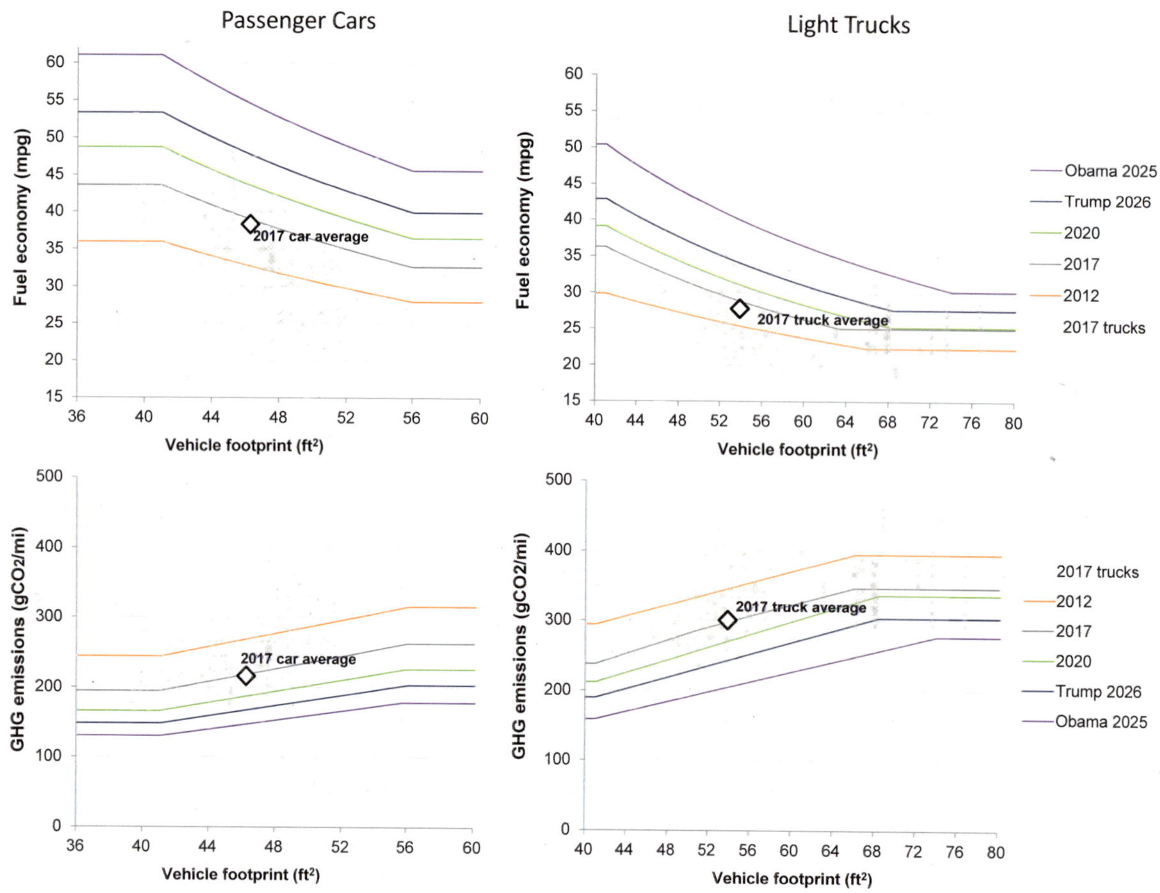

FIGURE 3.2 Fuel economy (top) and GHG emission (bottom) target versus footprint curves for MY 2012–2025 for passenger cars (left) and light trucks (right). The fuel economy or GHG emissions and vehicle footprint for the sales-weighted average of all MY 2017 vehicle models is plotted as a diamond.
SOURCE: Committee generated based on NHTSA/EPA (2020) and EPA (2021).

footprint indexed standards and future 1.5% (Trump administration) and 5% (Obama administration) improvements per year from 2021 on, and the lowest-CO_2 non-hybrid medium car and crossover light truck models from MY 2020. The car models are the Honda Accord, Nissan Altima, and Toyota Camry. Note that the California-automaker 3.7% per year CO_2 reduction benchmark would be approximately equivalent to the 2025 Obama administration line, but for 1 year later (i.e., 2026). The crossover models are all-wheel-drive versions of the Ford Escape, Honda CR-V, and Toyota RAV4. Further details on the vehicle specifications are summarized below. Also shown in Figure 3.3 are the CO_2 emission levels for the comparable 2010 versions of the models to show how the models' test cycle CO_2 emissions have declined. As shown, the MY 2010 models' test cycle CO_2 emissions typically matched the 2012–2014 footprint-indexed greenhouse gas (GHG) targets. The 2020 models' test cycle CO_2 emissions approximately match the 2020 footprint-indexed GHG targets, before credits outside the test cycle are factored into their CO_2 levels.

Figure 3.3 also shows the 2020 models, including the assumed use of applicable technology credits that can be expected to be widely deployed in the 2025 time frame. The available technology credits include air conditioning credits and off-cycle credits. For historical context, when averaged over all MY 2016 vehicles, passenger cars had 9 grams per mile (g/mi) air conditioning and 2 g/mi off-cycle credits; light trucks had 11 g/mi for air conditioning and 4 g/mi off-cycle credits. To provide an applicable comparison for how the MY 2020 vehicle compares

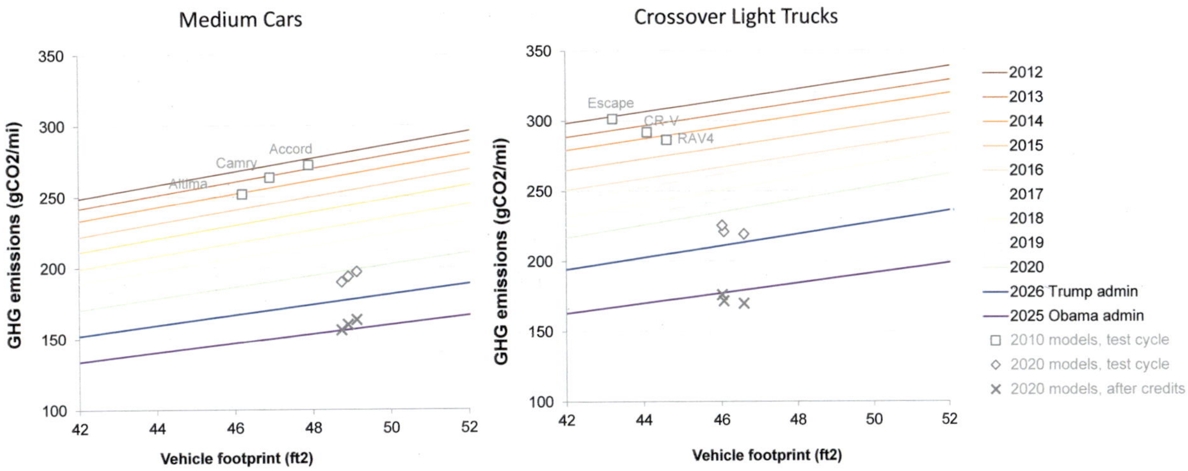

FIGURE 3.3 GHG emission targets versus footprint for MY 2012–2025 for medium cars (left) and crossover light trucks (right), with points representing GHG emissions and footprint of selected high-volume, low-CO_2 emission car and crossover models for MY 2010 and 2020. Squares and diamonds represent the test cycle vehicle emissions and footprints in MY 2010 and MY 2020, respectively. Cross marks represent MY 2020 emissions and footprint values, where the emissions are adjusted with estimated GHG credit values.
SOURCE: Committee generated based on NHTSA/EPA (2020) and EPA (2021).

against the MY 2025 standards, more complete usage of the technology credits is assumed in the figure for air conditioning (18.8 g/mi for cars, 24.4 for light trucks) and off-cycle credits (17 g/mi for cars, 24 for light trucks). Off-cycle credits from a predefined technology menu are limited to an average of 15 g/mi (CARB, 2019), and cars typically get lower, and light trucks higher, credit values (EPA, 2020). Additional off-cycle technologies outside the menu, and therefore without such menu restrictions, are increasingly being granted credits (EPA, 2020; Lutsey and Isenstadt, 2018). The effect of these two types of credits is that the lowest-CO_2 2020 medium cars and crossovers approximately match the original 5% per year 2024–2025 footprint-indexed CO_2 target curves.

To further understand the technology trends involved with the Figure 3.3 analysis, Table 3.2 shows the detailed vehicle specifications for the low-CO_2 car and crossover models depicted. From 2010 to 2020, the six different models each saw CO_2 emissions reduce by 22% to 29%, while also getting larger by 2% to 7%, while also increasing power by 1% to 13%. These trends are also being seen in the wider fleet over this time period. The vehicle models each had efficiency technologies added in the vehicle redesign and refresh cycles that occurred at different points within the 2010–2020 time period. As shown in MY 2020, several leading low-CO_2 models have variable valve timing and/or lift, turbocharging, direct injection, and advanced transmissions (8-speed or continuously variable). Technologies like cylinder deactivation and stop-start are also deployed on some models. The overall fleet trends for the most recent complete fleet-wide database on adoption of these technologies is shown above in Figure 3.1, which similarly shows that many efficiency technologies have only been deployed in a small percentage of new models.

This analysis illustrates a few key aspects of the functioning of the 2012–2025 CO_2 standards. As shown in Table 3.2, different technology pathways are being deployed on different vehicle models by different companies. For example, some of the lowest CO_2 models in 2020 (as compared to their CO_2 standard lines) have turbocharging, stop-start, cylinder deactivation—while others do not. Considering the leading models of Table 3.2 and the fleet-wide trends of Figure 3.1 together, it appears likely that the various efficiency technologies could see much greater deployment by 2025 if the incrementally more stringent 2021–2025 CO_2 standards had remained in place. Another key finding is that companies are deploying technologies that enable the fleet to meet the increasingly stringent standards while simultaneously delivering increased acceleration performance and increased size for passenger and cargo capacity. Although this analysis is based on selected

TABLE 3.2 Vehicle Model Attributes for Selected 2010 and 2020 Vehicle Models

			Vehicle Attributes			Change from MY 2010 to 2020			
Vehicle Class	Year	Model	Footprint (ft^2)	Test Cycle CO_2 (g/mi)	Power (hp)	Footprint (ft^2)	Test Cycle CO_2 (g/mi)	Power (hp)	Efficiency Technologies
Car	2010	Camry (2.5 L)	46.9	263.6	179	—	—	—	6-speed transmission
		Accord (2.4 L)	47.9	272.3	190	—	—	—	Variable valve timing/lift, 5-speed
		Altima (2.5 L)	46.1	251.9	175	—	—	—	Variable valve timing, continuously variable transmission
	2020	Camry (2.5 L)	48.7	189.7	203	4%	–28%	13%	Variable valve timing, direct injection, high compression ratio, cooled exhaust gas recirculation, 8-speed, road load reduction
		Accord (1.5 L)	48.9	193.6	192	2%	–29%	1%	Variable valve timing/lift, turbocharging, direct injection, continuously variable transmission
		Altima (2.5 L)	49.1	197.0	188	7%	–22%	7%	Variable valve timing, direct injection, continuously variable transmission
Crossover light truck	2010	CR-V (2.4 L)	44.1	292.0	180	—	—	—	Variable valve timing/lift, 5-speed
		Escape (2.5 L)	43.2	301.5	175	—	—	—	Variable valve timing, 6-speed
		RAV4 (2.5 L)	44.7	303.4	166	—	—	—	Variable valve timing, 4-speed
	2020	CR-V (1.5 L)	46.1	221.0	190	4%	–24%	6%	Variable valve timing/lift, turbocharging, direct injection, continuously variable transmission
		Escape (1.5L)	46.0	225.4	180	7%	–25%	3%	Variable valve timing, turbocharging, direct injection, cylinder deactivation, 8-speed
		RAV4 (2.5 L)	46.6	219.2	203	4%	–28%	22%	Variable valve timing, direct injection, high compression ratio, cooled exhaust gas recirculation, 8-speed, road load reduction, stop-start

SOURCE: Committee generated using MY 2010 and 2020 vehicle data from EPA (2021) and German (2018).

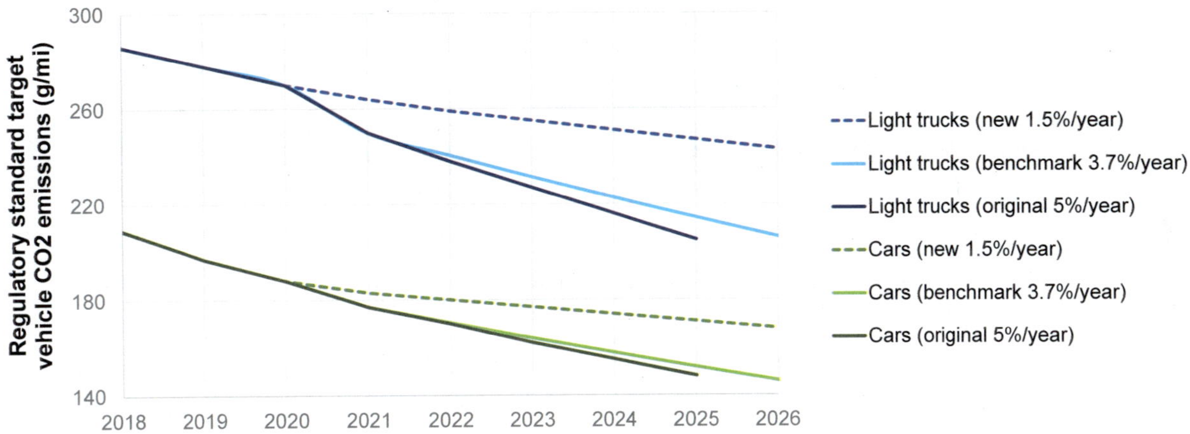

FIGURE 3.4 Car and light truck regulatory target CO_2 emissions for MY 2018–2026 with original standards (5%/year), new 2020 rollback (1.5%/year), and compromise benchmark (3.7%/year).
SOURCE: Committee generated based on NHTSA/EPA (2020) and CARB (2019).

models, it illustrates in a detailed model-by-model manner what is largely happening on a fleet-wide basis (see EPA, 2019).

FINDING 3.1: The latest complete data set of vehicles released as part of the 2020 National Highway Traffic Safety Administration regulatory analysis (based on the MY 2017 fleet) showed considerable penetration of previously identified fuel-efficient technologies across all five vehicle classes under study. There is also ongoing growth of newly developed fuel-efficient technologies by manufacturers up through the current 2020 models. There is roughly a 20% gap between the average fuel economy of the 2017 fleet and the original 2025 standard, when estimated manufacturer use of credits is taken into account.

3.5 BENCHMARK FOR MY 2025 AND MY 2026

Figure 3.4 shows the original 2012-adopted and the newly revised March 2020 standards through 2026. The original 2012-adopted standards maintain approximately 5% per year CO_2 emission reductions for 2020 through 2025. The March 2020 standards would increase fuel economy by 1.5% per year from 2020 through 2026. The announced framework terms by California and four automakers, labeled as the "benchmark" values for cars and light trucks in Figure 3.4, provides a path between the original and rolled-back standards. The benchmark emission levels include lower annual CO_2 reduction targets to 3.7% per year from 2021 through 2026 and additional flexibilities that reduce the required test-cycle CO_2 reductions, as assessed below.

3.6 BENCHMARK FOR MY 2025

Benchmark vehicle emission levels for 2025 are defined here by the committee to provide an approximation for technology packages that are likely to be commonplace around 2025, the first year of this committee's analysis. Benchmark efficiency and emission levels are approximated based on the terms of the July 2019 deal between California and four automakers, including incorporation of existing trends for the usage of technology credits. Most of the evaluated technologies assessed below have CO_2-reduction effectiveness, measured as a percentage improvement on the standard regulatory test cycles. Because there are also a variety of technology credits (related to air conditioning, off-cycle, and electric vehicle accounting) that affect how much test-cycle vehicle improvement is ultimately needed, an estimate of the potential impact of non-test-cycle technologies toward compliance is included.

TABLE 3.3 Regulatory and Test-Cycle CO_2 Emissions for Benchmark 2025 Emission Levels

	Regulation CO_2 Emissions			Technology Credit Assumptions for 2025			Combustion Vehicle Test Cycle CO_2 Emission Level		
	2017[a] (g/mi)	2025 Target (g/mi)	Change, 2017–2025	Air Conditioning[b] (g/mi)	Off-Cycle (g/mi)	Electric Vehicle[c] Share	2017[a] (g/mi)	2025 Target (g/mi)	Change, 2017–2025
Car	223	154	–31%	19	17	6.5%	240	198	–18%
Light truck	306	213	–30%	24	24	2.5%	330	271	–18%
Average	262	182	–31%	21	20	4.6%	283	233	–18%

[a] MY 2017 vehicles have 10 g/mi air conditioning and 3 g/mi off-cycle credits for cars; 17 g/mi air conditioning and 7 g/mi off-cycle credits for light trucks.

[b] Air conditioning credits include 6.0 g/mi for efficiency and 13.8 g/mi for refrigerant leakage for cars; 7.2 g/mi for efficiency and 17.2 g/mi for refrigerant leakage for light trucks.

[c] Electric vehicle shares in 2025 from regulatory agency March 2020 central case analysis. Electric vehicles, based on regulatory provisions, are multipliers and counted as 0 g/mi for battery electric vehicles (and to extent they are powered by electricity for plug-in hybrid electric vehicles).

Table 3.3 outlines a set of assumptions that translate overall regulatory CO_2-emission requirements—incorporating deployment of technology credits and electric vehicles—to the test-cycle combustion vehicle CO_2 improvements to meet the benchmark 3.7% annual reduction emission levels by 2025. As the agencies have assumed, the maximum allowable air conditioning credit use is included at 21 g CO_2/mi. In addition, following regulatory developments and automaker trends, off-cycle credit usage is estimated to increase to 20 g CO_2/mi by 2026. The framework terms of the July 2019 agreement by California and four automakers include the expanded use of off-cycle credits of up to 15 g/mi from the off-cycle menu and additional permits beyond the menu (CARB, 2019). The March 2020 standards also streamline processes for technology credits, and recent trends and analysis support the feasibility of automakers achieving these credits (EPA, 2019; Lutsey and Isenstadt, 2018). Electric vehicles, based on the Advanced Vehicle Technology credit provisions, are counted as 0 g/mi and with applicable multipliers that vary by technology and model year. The national share of new vehicles that are plug-in electric vehicles is assumed to be 5% in MY 2025. Consistent with recent trends, this electric share is above the regulatory requirements but is well below many automakers' public announcements (as discussed more below). After accounting for the various technology credits, more than half of benchmark 2025 CO_2-emission reduction (18%, versus the overall 31% compared to 2017) would come from test-cycle vehicle efficiency improvements.

Table 3.4 summarizes the test-cycle CO_2 emission levels, test-cycle fuel economy, and consumer label fuel economy levels estimated for MY 2017 and the 2025 benchmark for each of the five vehicle classes. The table shows values only for the combustion vehicles (i.e., the 99% in 2017 and assumed 95% in 2025 of vehicle sales that

TABLE 3.4 Combustion Vehicles' 2017 and Benchmark 2025 CO_2 and Fuel Economy by Class

	MY 2017			MY 2025			2017–2025 Change	
Class	Test CO_2 (g/mi)	Test MPG	Label MPG	Test CO_2 (g/mi)	Test MPG	Label MPG	CO_2 Emissions	Fuel Economy
Small car	212	42	32	194	46	35	–9%	10%
Medium car	274	32	25	206	43	33	–25%	33%
Crossover	269	33	25	220	40	31	–18%	22%
Sport utility vehicle	336	26	20	265	34	26	–21%	27%
Pickup	376	24	18	312	29	22	–18%	21%
Total	283	31	24	233	38	29	–18%	22%

NOTE: CO_2 = carbon dioxide; MPG = miles per gallon.
SOURCE: Committee analysis of data sets from NHTSA/EPA (2020).

are not plug-in electric). Table 3.4 accounts for the use of off-cycle, air conditioning, and electric vehicle crediting as in Table 3.3. As one example from the table, the average crossover vehicle would see its consumer label fuel economy improve from 25 MPG in 2016 to 31 MPG in 2026, for a 22% fuel economy increase, approximately the same as the fleet average increase across all five classes. The annualized test-cycle CO_2 reduction would be 2.4%, and the annualized fuel economy increase would be 2.5%, for all combustion vehicles over 2017–2025.

3.7 TECHNOLOGY PACKAGES IN 2025

The above analysis provides the fleet-level and class subdivided CO_2 emissions and fuel economy context that is expected to impact the technology packages that will be commonplace in 2025. Examining the regulatory analysis associated with the last several proposed and adopted regulations provides examples of efficiency packages, including technology pathways with turbocharging, naturally aspirated engines, and hybridization, and their approximate costs.

Technology packages for two of the five vehicle classes, medium cars and SUVs, are summarized. Similar packages are available in the other vehicle classes. In each case, technology packages that are expected to play prominently in reaching the CO_2 and efficiency benchmarks above are shown, including with agency estimates of the overall CO_2 emission-reduction benefit and technology cost. Technology package details are based on the EPA (2016, 2017a–d) and NHTSA/EPA (2018, 2020) regulatory files.

For cars, as indicated above, 2017 models typically have variable valve timing (62% of 2017 sales) and 6-speed transmissions (53%). There are many other technologies (e.g., turbocharging 30%, direct injection 22%, variable valve lift 22%, stop-start 14%, and road load technologies) that are also increasing in share in new vehicles. The regulatory agency analyses of potential future-year standards confirms that many of the lower-percentage technologies in Figure 3.1 can be expected to comprise a larger share of new vehicles in 2025.

The above benchmark analysis showing high fuel economy, mass market vehicles indicates that combustion cars could reduce CO_2 emissions by 18% from 2017 to 2025 (see Table 3.2). The agencies' technology evaluations, as illustrated in Table 3.5 for medium cars, provide an estimation of the technologies that can help meet those emission-reduction levels. Technology packages with greater and lesser emission reduction (ranging from 17% to 21%) are shown. Various companies have deployed some of the technologies already to varying degrees, and often many of the technologies are implemented together during vehicle redesign or new powertrain development cycles. Also, as indicated, the data are taken from NHTSA and EPA analyses to show differences in how they assembled and evaluated the technology effectiveness and cost of the packages. Based on these packages and the 18% CO_2-reduction 2025 benchmark, a typical cost increase for medium cars from 2017 to 2025 could be approximately $800 per vehicle. For context, $800 is approximately 2% of the average cost of a reference medium car from MY 2017.

TABLE 3.5 Technology Package CO_2-Reduction and Associated Cost for Medium Car Class to Obtain a 17% to 21% Reduction in Test-Cycle Fuel Economy from 2017 to 2025

Analysis	Test-Cycle CO_2 and Fuel Consumption Reduction	Incremental Cost per Vehicle (2018$)	Technology Package to Meet the Required Improvement in CO_2 Emissions and Fuel Consumption
EPA	17.3%	$567	Engine friction reduction, low rolling resistance tires 20%, improved accessories 2, aerodynamic 20%, mass reduction 10%
NHTSA	19.1%	$798	Rolling resistance 20%, 10-speed transmission, aerodynamic 15%, mass reduction 7.5%, variable valve lift
NHTSA	20.8%	$1,119	Rolling resistance 20%, 10-speed transmission, aerodynamic 15%, mass reduction 7.5%, variable valve lift, direct injection
EPA	21.0%	$788	Engine friction reduction, low rolling resistance tires 20%, improved accessories 2, aerodynamic 20%, mass reduction 10%, 10-speed transmission

SOURCE: NHTSA rulemaking analysis from original and amended standards (NHTSA/EPA, 2020).

TABLE 3.6 Technology Package CO_2-Reduction and Associated Cost for SUV Class to Obtain a 21%–23% Reduction in Test-Cycle Fuel Economy from 2017 to 2025

Analysis	Test-Cycle CO_2 and Fuel Consumption Reduction	Incremental Cost per Vehicle (2018)	Technology Package to Meet the Required Improvement in CO_2 Emissions and Fuel Consumption
EPA	20.9%	$995	Engine friction reduction, low rolling resistance tires 20%, improved accessories 2, aerodynamic 20%, mass reduction 10%, 10-speed transmission, direct injection, turbocharger (18 bar)
EPA	22.5%	$1,143	Engine friction reduction, low rolling resistance tires 20%, improved accessories 2, aerodynamic 20%, mass reduction 10%, 10-speed transmission, direct injection, turbocharger (18 bar), variable valve lift
NHTSA	22.9%	$1,308	Rolling resistance 20%, 8-speed transmission, aerodynamic 15%, mass reduction 5%, variable valve lift, turbocharging (18 bar); added to original vehicle with variable valve timing, direct injection

SOURCE: NHTSA rulemaking analysis from original and amended standards (NHTSA/EPA, 2020).

For the SUV class, baseline 2017 models typically have variable valve timing (79% of 2017 sales) and 6-speed transmissions (51%). There are many other technologies (e.g., direct injection 37%, cylinder deactivation 22%, start stop 20%, variable valve lift 19%, turbocharging 18%, and road load technologies) that also have significant shares. The agency analyses of potential future-year standards indicate that many of the lower-percentage technologies in Figure 3.1 can be expected to comprise a larger share of new vehicles in 2025.

The above benchmark analysis indicates that combustion vehicles in the SUV class could reduce CO_2 emissions by 21% from 2017 to 2025 (see Table 3.4). Table 3.6 shows agencies' technology evaluations of technology packages around that level of emission-reduction levels. Packages with a similar emission-reduction of 21% to 23% are shown, as some companies have deployed more or less of the technologies already, and often many of the technologies are implemented together during vehicle redesign or new powertrain development cycles. Based on these packages and the 21% CO_2-reduction 2025 benchmark, a typical cost increase for SUVs from 2017 to 2026 could be approximately $1,000–$1,300 per vehicle. For context, $1,000–$1,300 is approximately 2%–3% of the average cost of a reference SUV from MY 2017.

Although there are uncertainties about the precise technologies that will be deployed, their costs, their consumer interest, and their ultimate technology uptake through 2026, the regulatory agencies' modeling of the expected uptake provides a reasonable range of potential outcomes for the U.S. LDV market. These agencies' projected percentages for the above-discussed technologies are summarized in Box 3.1. The committee notes that the agencies' projected market shares are not a forgone conclusion nor do they offer a precise baseline upon which the report's estimates are based. Notably, the auto industry and consumer decisions are based on complex factors beyond minimizing the regulatory cost of compliance. For example, many technologies have mutual benefits (e.g., turbocharging with moderate engine downsizing increases acceleration and efficiency) and trade-offs (e.g., engine downsizing alone increases efficiency and reduces acceleration). Technologies also may align differently with automakers branding or market positioning, such as branding for eco-friendliness, high power, aerodynamics, or innovative technology. Technology effects can have different market acceptance and profitability implications, which are not incorporated in regulatory cost minimization analysis, that influence their deployment.

3.8 INTERNATIONAL MARKET AND REGULATIONS

There are broader global considerations for automakers' technology deployment decisions. Many companies are actively developing global vehicle platforms that can more rapidly deploy engine or transmission technologies at higher annual volumes across continents. This could increase the likelihood that technologies that are being widely deployed in Europe and Asia are also deployed in the United States, even if the technologies are beyond what is minimally needed for compliance. In addition to creating global platforms to reduce engineering and supply chain costs, automakers are making technology decisions and investments that go far beyond regulatory compliance for 2025–2026 (Lutsey, 2018).

BOX 3.1
Regulatory Agency Estimation of 2025 Technology Adoption

The preceding analysis provides an assessment of the near-term vehicle efficiency improvements and associated technologies that can be expected in the MY 2025 time frame. Figure 3.1.1 shows the actual MY 2017 technology (as in Figure 3.1, above) and compares with the agency estimates for MY 2025. The estimated 2025 technology deployment is provided as a range to include the March 2020 agency analysis of the original 5% per year 2021–2025 standards and the newly revised downward 1.5% per year 2021–2026 standards. The 5% per year case requires roughly 10% greater deployment of powertrain technologies like direct injection, turbocharging, cylinder deactivation, high compression ratio, and 9–10 speed transmissions. Another difference in the two cases is that the 5% per year scenario requires more electrification of various types (4.6% plug-in, 9.4% strong hybrid, 4.3% mild hybrid) compared to the 1.5% per year case (3.1% plug-in, 2.5% strong hybrid, 0.1% mild hybrid). The committee considers the range to represent a reasonable approximation of possible futures for the auto industry to efficiency technologies in the United States through 2025.

Area	Technology	MY 2017	Estimated MY 2025
Engine	Variable valve timing	70%	91% - 95%
	Variable valve lift	20%	77% - 85%
	Gasoline direct injection	29%	64% - 73%
	Turbocharging	24%	34% - 45%
	Cylinder deactivation	12%	28% - 38%
	High compression ratio	2%	19% - 28%
Transmission	6-speed or less	52%	3% - 4%
	7- or 8-speed	17%	43% - 60%
	9- or 10-speed	6%	9% - 18%
	Continuously variable	21%	21% - 22%
Hybrid	Start-stop	17%	15% - 17%
	Mild hybrid	0.01%	0.1% - 4.4%
	Strong hybrid	2.2%	2.5% - 9.4%
Electric	Plug-in hybrid electric	0.7%	3.1% - 4.6%
	Battery electric	0.6%	2.5% - 4.0%
	Fuel cell electric	0.02%	0.02%
Road load reduction	Mass reduction (10% or more)	17%	30% - 69%
	Tire rolling resistance reduction (10% or more)	49%	100%
	Aerodynamic reduction (10% or more)	23%	67% - 94%

FIGURE 3.1.1 Percent of MY 2017 and 2025 vehicles with efficiency technologies.
NOTE: Technologies are defined as in the NHTSA's CAFE model, and some technologies, such as variable valve timing or gasoline direct injection will also be present as a part of other technologies, such as turbocharging and variable valve lift.
SOURCE: NHTSA/EPA (2020).

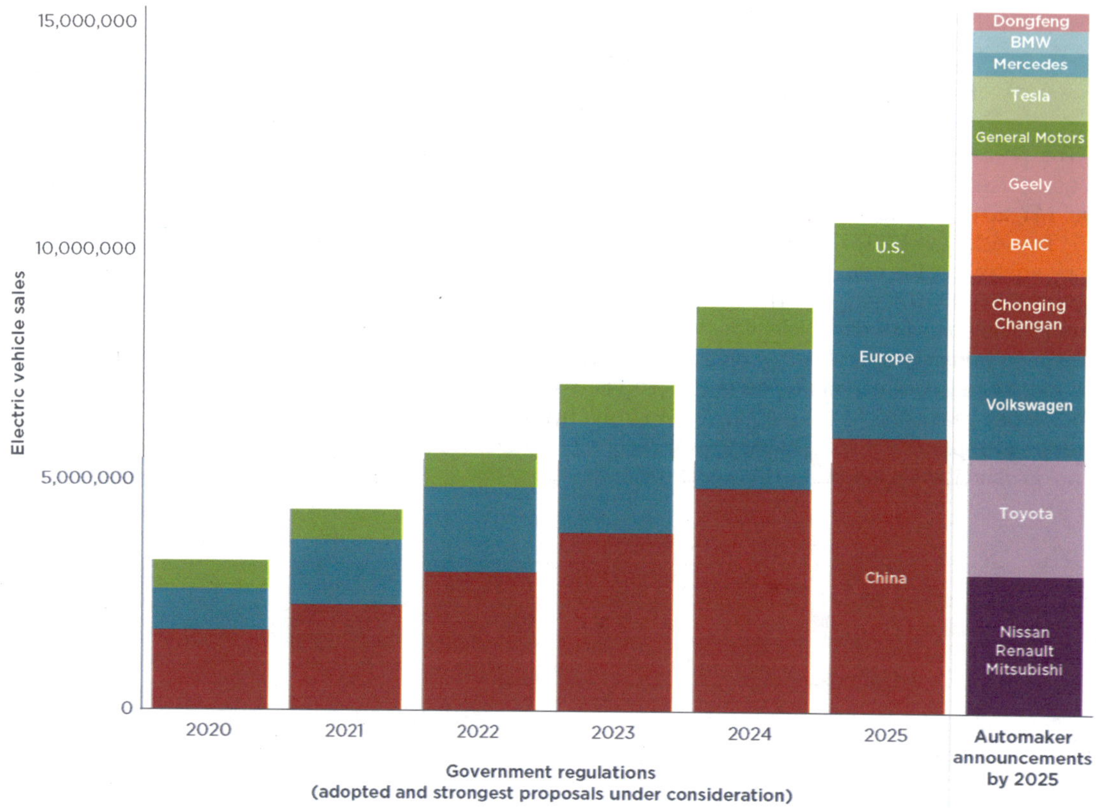

FIGURE 3.5 Estimated government regulations in 2020–2025 and 2025 automaker targets for electric vehicles.
SOURCE: Lutsey (2018).

This is especially important in the case of advanced technologies—for example, with electric vehicle technologies, about which many automakers have announced long-term global technology strategies.

In light of broader policy, market, and technology developments, the global automotive industry is supplementing its combustion efficiency investments with major plans for high-volume electric vehicle production. Many manufacturers have announced that they will greatly increase their electric vehicle deployment within the next 5 years in the United States and elsewhere. Tallying the company announcements indicates that automotive investments surpassing $300 billion are under way, amounting to more than 15 million new plug-in electric vehicle sales annually, by 2025 (Lienert and Chan, 2019; Lienert et al., 2019; Lutsey, 2018). As shown in Figure 3.5, the electric vehicle requirements would be 50% greater than the regulatory requirements in China, Europe, and North America. International markets and regulatory aspects, and their influence on the U.S. vehicle fleet, are discussed further in Chapters 11 and 12, respectively.

3.9 REFERENCES

CARB (California Air Resources Board). 2019. "California and Major Automakers Reach Groundbreaking Framework Agreement on Clean Emission Standards." https://ww2.arb.ca.gov/newscalifornia-and-major-automakers-reach-groundbreaking-framework-agreement-clean-emission.

EPA (U.S. Environmental Protection Agency). 2016. "Proposed Determination on the Appropriateness of the Model Year 2022–2025 Light-Duty Vehicle Greenhouse Gas Emissions Standards Under the Midterm Evaluation." https://www.federalregister.gov/documents/2016/12/06/2016-29255/proposed-determination-on-the-appropriateness-of-the-model-year-2022-2025-light-duty-vehicle.

EPA. 2017a. "Final Determination on the Appropriateness of the Model Year 2022–2025 Light-Duty Vehicle Greenhouse Gas Emissions Standards Under the Midterm Evaluation." https://www.epa.gov/sites/production/files/2017-01/documents/420r17001.pdf.

EPA. 2017b. "EPA Technical Projects to Inform the Midterm Evaluation." https://www.epa.gov/regulations-emissions-vehicles-and-engines/midterm-evaluation-light-duty-vehicle-greenhouse-gas-ghg#technical-projects.

EPA. 2017c. "EPA Publications Informing the Midterm Evaluation." https://www.epa.gov/regulations-emissions-vehicles-and-engines/midterm-evaluation-light-duty-vehicle-greenhouse-gas-ghg#publication.

EPA. 2017d. "Optimization Model for Reducing Emissions of Greenhouse Gases from Automobiles [OMEGA]." (Version v1.4.56.) https://www.epa.gov/regulations-emissions-vehicles-and-engines/optimization-model-reducing-emissions-greenhouse-gases.

EPA. 2020. *The 2019 EPA Automotive Trends Report: Greenhouse Gas Emissions, Fuel Economy, and Technology since 1975.* EPA-420-R-19-002. March.

EPA. 2021. "Download Fuel Economy Data." https://www.fueleconomy.gov/feg/download.shtml.

German, J. 2018. "How Things Work: OMEGA Modeling Case Study Based on the 2018 Toyota Camry." International Council on Clean Transportation. https://theicct.org/publications/how-things-work-omega-modeling-case-study-based-2018-toyota-camry.

Lienert, P., and C. Chan. 2019. "Investment in Electric Vehicles." Reuters. September 1. https://graphics.reuters.com/AUTOS-INVESTMENT-ELECTRIC/010081ZB3HZ/INVESTMENT.jpg.

Lienert, P., N. Shirouzu, and E. Taylor. 2019. "Exclusive: VW, China Spearhead $300 Billion Global Drive to Electrify Cars." Reuters. January 10. https://www.reuters.com/article/us-autoshow-detroit-electric-exclusive/exclusive-vw-china-spearhead-300-billion-global-drive-to-electrify-cars-idUSKCN1P40G6.

Lutsey, N. 2018. *Modernizing Vehicle Regulations for Electrification.* International Council on Clean Transportation. https://theicct.org/publications/modernizing-regulations-electrification.

Lutsey, N., and I. Isenstadt. 2018. *How Will Off-Cycle Credits Impact U.S. 2025 Efficiency Standards?* International Council on Clean Transportation. https://theicct.org/publications/US-2025-off-cycle.

NHTSA/EPA (National Highway Traffic Safety Administration and U.S. Environmental Protection Agency). 2018. "2018 NPRM for Model Years 2021–2026 Passenger Cars and Light Trucks. Compliance and Effects Modeling System." https://www.nhtsa.gov/corporate-average-fuel-economy/compliance-and-effects-modeling-system#compliance-and-effects-modeling-system-downloads.

NHTSA/EPA. 2020. "2020 Final Rule for Model Years 2021–2026 Passenger Cars and Light Trucks. Compliance and Effects Modeling System." https://www.nhtsa.gov/corporate-average-fuel-economy/compliance-and-effects-modeling-system#compliance-and-effects-modeling-system-downloads.

4

Internal Combustion Engine-Based Powertrain Technologies

The spark-ignition internal combustion engine (ICE) fueled by gasoline is by far the dominant form of propulsion in the current U.S. fleet, with 96% of model year (MY) 2019 vehicles containing a spark-ignition ICE engine, and 91% powered exclusively by a spark-ignition ICE (EPA, 2020a). Even with increasing levels of electrification, the spark-ignition engine will continue to play a significant role in ICE-only powertrains as well as in hybrid and plug-in hybrid powered vehicle powertrains in 2025–2035. This chapter discusses the technological development of conventional powertrains, specifically ICEs and transmissions and including those developed for integration into hybrid electric applications of all levels. The electric components and batteries used in hybrid powertrain systems are addressed in Chapter 5.

In 2025–2035, automakers will pursue a variety of powertrain options to improve fuel economy. Therefore, rather than focusing solely on individual technologies and their fuel economy benefits and costs, conventional and hybrid powertrain technologies are described in the context of "pathways" representing significant trends in engine technology and development. This pathway approach allows technologies to be evaluated, including their potential contribution in specific system-level applications. For example, the benefit of cylinder deactivation is different for a downsized/boosted 4-cylinder engine than for a large displacement, naturally aspirated 8-cylinder engine and might therefore be prioritized differently for different applications. Additionally, manufacturers will pursue multiple strategies to satisfy customer requirements in different vehicle classes and carlines. For example, in the midsize crossover segment alone, the largest and fastest growing vehicle class, U.S. market MY 2020 powertrains include diesel, hybrid, plug-in hybrid, naturally aspirated ICE, downsized/boosted ICE, and battery electric vehicles (BEVs). The efficiency approach pursued by automakers will likely be a mix of some or all of these pathways in addition to non-powertrain technologies. Last, because the impacts of these technologies differ by their implementation, engine technology efficiency must be assessed within the context of powertrain systems.

4.1 DOWNSIZED/BOOSTED ICE PATHWAY

Downsized/boosted engines are ICEs where the swept volume (displacement) of the engine has been reduced, while vehicle performance is maintained by pressure charging the intake air using a turbocharger or supercharger. Such engines represented 34% of the market in 2019 (EPA, 2020a). Their current technology status and opportunities for energy improvement and the committee's estimated costs and capabilities of future turbocharged/boosted technologies in 2025–2035 are described below.

4.1.1 Current Technology in Downsized/Boosted ICE Pathway

In 2015, the National Research Council (NRC, 2015) described the next steps toward turbocharged/downsized engines as turbocharging and downsizing Level 1 (33% downsized), stoichiometric direct injection, variable valve timing (VVT), dual cam phasing, and idle stop-start. In MY 2019, engines with these technologies are available from most manufacturers and found in all light-duty vehicle (LDV) segments (EPA, 2020a). In particular, turbocharged engines with gasoline direct injection enabling engine displacement reduction have achieved a market penetration rate of 24% in MY 2017, as noted in Chapter 3, Table 3.2, and are likely to be a predominant engine type in 2025.

4.1.2 Efficiency Aspects Improved by Downsizing and Boosting

Automotive engines have traditionally been oversized for regular operation so that they can meet the peak performance demands of a given vehicle application. Such engines were naturally aspirated and operated at far less than peak engine efficiency when operating on the relatively lightly loaded fuel economy drive cycles. To improve performance and engine efficiency, pressure-charged engines were developed that can generate higher torque levels at lower relative engine speeds. Such engines can offer comparable vehicle performance from a smaller engine. These smaller displacement engines, often with fewer cylinders, run in a more efficient region of the engine's speed and load map during normal operation owing to lower throttling and frictional losses. On the other hand, boosted engines are more prone to knocking owing to the higher density air/fuel charge and therefore tend to be tuned with a lower compression ratio (CR) than modern naturally aspirated engines with gasoline direct injection. The result is a lower peak efficiency for downsized/boosted engines than that of a naturally aspirated engine, but more time spent operating in the higher efficiency region.

Initially, the downsized/boosted engine offerings were largely based on the suite of technologies applied to already existing engine platforms. Manufacturers have further improved upon the basic concept through optimized engine architectural design and technologies. For instance, the bore/stroke ratio has been optimized for downsized/boosted engines. The first downsized/boosted engines were designed like older naturally aspirated engines, as square or over-square (bore size greater than or equal to stroke dimension), which, for naturally aspirated engines, allows for improved breathing (larger valve sizes) and revving capability to generate power at higher engine speeds. Purpose-designed downsized/boosted engines, on the other hand, tend to be under square (bore less than stroke) similar to diesels, enabling a more compact package with greater structural rigidity under the higher pressure operating conditions of turbocharged engines. Lowering the bore/stroke ratio has several additional benefits for downsized/boosted engines. It yields a lower surface-to-volume ratio in the combustion chamber and therefore improves thermal efficiency via reduced heat losses. A smaller bore is less prone to knocking—the ultimate constraint on boosted engines—owing to the shorter flame travel distance. Higher piston speeds can enhance charge turbulence and combustion speeds. Furthermore, the lower operating speeds of downsized/boosted engines reduce the frictional penalty associated with a longer stroke. This same engine design trend is observed in newer naturally aspirated 4-cylinder engines that focus on efficiency, including Atkinson and hybrid applications. Reduced displacement engines may benefit from reduction in cylinder count. As an engine moves toward a lower displacement on the same configuration (e.g., 2.0 L to 1.5 L 4-cylinder), the mechanical friction does not necessarily scale proportionately, and friction mean effective pressure can become a higher percentage of total work on the smaller engine. There are also efficiency losses from the individual cylinder displacement getting smaller (with approximately 500 cc being optimum). To counter these effects and better address the needs for smaller engines, a number of 3-cylinder engines are being introduced in displacements up to 1.5 L, such as in the 2020 Ford Escape.

4.1.3 2025–2035 Downsized/Boosted ICE Pathways

The 2025–2035 period will see further opportunities to improve the efficiency of downsized/boosted engines. Technology evaluations presented in the 2015 NRC report and previous National Highway Traffic Safety Administration (NHTSA) and U.S. Environmental Protection Agency (EPA) studies focused primarily on additional engine downsizing opportunities enabled through higher engine operating pressures (and torque) utilizing exhaust gas recirculation (EGR) to control knock. The improvements evaluated were two technology bundles that further downsized a baseline engine beyond the Level 1 turbocharging system described above that is likely to be common in the 2025 fleet:

- *Level 2:* 24 bar brake mean effective pressure (BMEP)/50% downsize utilizing variable geometry turbocharger with and without cooled EGR.
- *Level 3:* 27 bar BMEP/56% downsize utilizing two-stage turbocharging with low and high pressure cooled EGR.

Concerns for implementations of Level 2 and Level 3 downsized/boosted engines discussed in the NRC and NHTSA/EPA evaluations were the ability of EGR to control knock at full load and thereby enable BMEP levels higher than 25 bar, and low-speed preignition, especially on U.S. conventional fuel, which is relatively low octane. These remain significant issues for implementation of highly downsized engines. Furthermore, the dual loop EGR system with supporting two-stage boosting as described in Level 3 is extremely complex and expensive. The resulting reductions in fuel consumption would be limited since they derive primarily from incremental pumping loss reductions and are possibly offset by CR reduction owing to the concerns noted above. An extremely downsized engine operating at very high BMEP levels will be forced to retard combustion to avoid knock and to enrich the air-to-fuel ratio to protect hardware from excessive exhaust gas temperatures at high loads, which will increase emissions.

The benefits of further downsizing, as well as the corresponding constraints, depend largely on the octane level of the dominant fuel in use. In the United States, regular fuel has a research octane number (RON) of 91, also reported at the pump as 87 anti-knock index. Automakers have been proposing for some time to increase the octane level of regular fuel in the United States, which could be achieved via the ongoing increase in ethanol blend levels associated with the Renewable Fuel Standard. U.S. consumers have historically been unwilling to comply with premium fuel requirements, so that is not a viable option. An increase in nominal octane rating of U.S. fuel similar to that in Europe (95 RON) might allow for an increase of approximately 1 CR, which is a common adjustment between otherwise similar North American and European engine offerings. That might translate to an approximate 2% efficiency opportunity (NRC, 2015, Fig 2.12). However, those benefits would largely only apply to new models and could not be design implemented until the fuel is widely available. Additionally, even in Europe where higher octane is standard, engine developments do not indicate a trend toward aggressive further downsizing. This likely stems from a shift to emphasize real-world driving fuel economy and emissions, and test cycle performance on the new worldwide harmonized light-duty test procedure and real driving emissions cycles, which are more heavily loaded. Since there appears to be no activity to change U.S. octane levels going forward, the technology look-ahead in this report will concentrate on the current EPA Tier 3 fuel with 91 RON and low sulfur.

While there does not appear to be a trend toward more aggressive engine downsizing, there have been significant improvements made to the engine offerings in this category. Ongoing evolutionary developments in boosting, fuel delivery systems, and combustion control have allowed for better knock avoidance and improved efficiency. In addition to the basic engine architectural optimizations described above, manufacturers have incorporated numerous features to lower engine friction, such as variable capacity oil pumps and piston/bore coatings. Most new gasoline turbocharged direct injection (GTDI) engines incorporate exhaust manifolds that are cast integrally with the cylinder head and can therefore be cooled. In order to reduce friction as well as enable variable lift technologies and cylinder deactivation, valve train architectures have migrated from direct acting mechanical bucket type (low cost, high-rev capable) to roller finger follower type. Numerous engine entries now use so-called dual fuel injection systems employing a combination of port and direct injection. These systems offer combustion advantages and lower cold start particulate matter emissions in the face of future stringent particulate matter emissions standards.

To evaluate the potential for further improvements to downsized/turbocharged engines in 2025–2035, the numerous newer engine offerings in this category were canvassed along with the technologies and design approaches being applied to them (see below). The fundamentals behind the technologies and how they might combine going forward are discussed. A recent EPA benchmarking evaluation of a 2016 Honda 1.5 L L15B7 GTDI engine is an example of a downsized/boosted engine that showed significant improvement over previously benchmarked engines of this type. The engine demonstrated brake thermal efficiency levels of 37%–38%, consistent

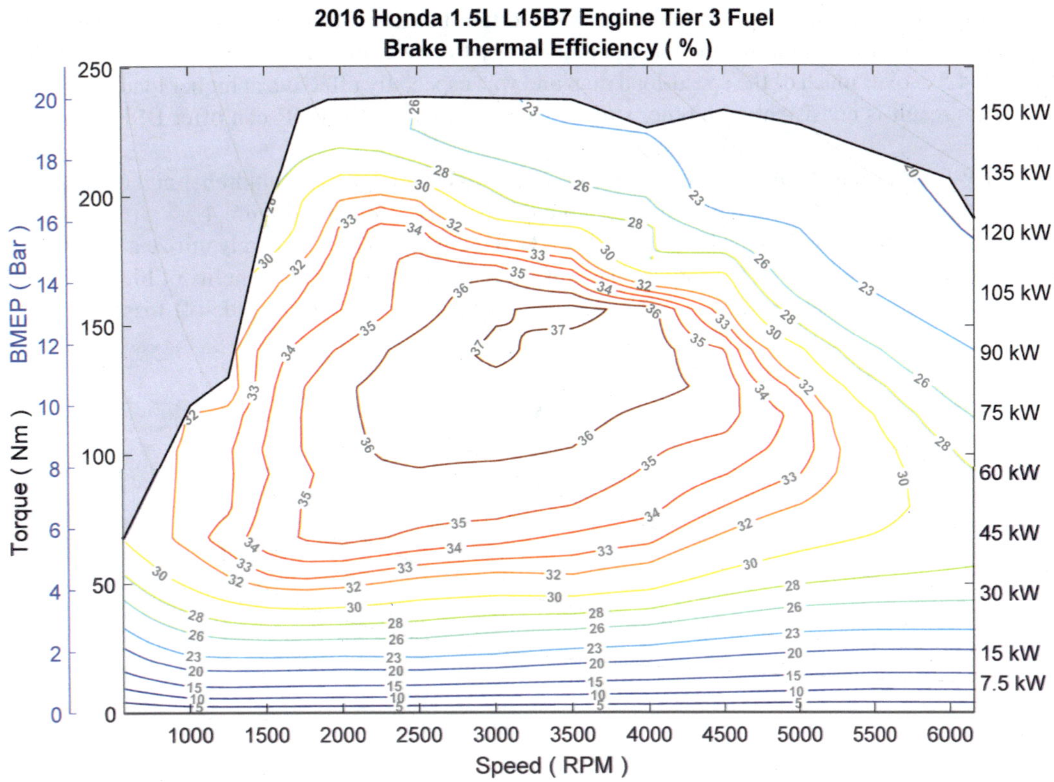

FIGURE 4.1 Engine torque versus speed map for the 2016 Honda 1.5 L L15B7 turbocharged engine, showing areas of low fuel consumption and high efficiency near the center of the map (contours labeled with percent brake thermal efficiency), and low efficiency under low load, and at high speed and high torque.
SOURCE: EPA (2020e).

with Honda's published results with a relatively nominal set of technologies (Figure 4.1). As previously described, these engines operate in the more efficient areas of the engine speed/load map to derive their benefit. However, the absolute levels of best brake-specific fuel consumption (BSFC) can be worse than naturally aspirated engines, primarily owing to their lower CR to prevent knock. Level 1 downsized/boosted engines tend to have CRs in the 10.0–10.5:1 range, lower than those in modern naturally aspirated engines with gasoline direct injection, which are typically at least 12:1. Based on data from Heywood (1988), this sort of CR gap could account for roughly a 3% reduction in BSFC (Heywood, 1988; NRC, 2015). Therefore, the greatest improvement opportunity for downsized/boosted engines will come from improvements in thermal efficiency as opposed to further downsizing at extremely high load operation, with CR as an area of focus.

Low-pressure loop-cooled exhaust gas recirculation (LP-EGR) has also been studied extensively as a technology with the ability to enhance the efficiency of a GTDI engine in numerous ways. EGR lowers peak combustion temperatures, thereby lowering heat losses to the coolant. It also improves thermal efficiency by increasing the specific heat ratio of the working fluid, less effectively than lean operation, but without the same emission constraints. Cooled EGR can also serve as an effective knock mitigant, potentially enabling a higher CR. Last, it can provide some incremental pumping loss improvement in lightly loaded conditions. LP-EGR systems place higher demands on the range of authority of the boosting system and initially required costly two-stage turbochargers. However, the recent development of variable nozzle type turbochargers for gasoline applications offers a more cost effective solution.

A 2018 EPA study (Conway et al., 2018) investigated the effects of LP-EGR on a 1.6 L GTDI engine to determine whether cooled EGR alone could enable an increase in CR from the nominal 10.5:1 up to 12.0:1. This bench-based study met the additional boosting requirements with a surrogate secondary device (electric supercharger) and increased ignition system energy to ignite the more dilute mixtures. In this case, under idealized boosting conditions and without accounting for the incremental energy to power the supercharger, the LP-EGR improved BSFC roughly 4.5% over much of the speed-load map and was especially effective at higher loads owing to knock mitigation. This result is consistent with other studies showing that cooled EGR can offer BSFC improvements of at least 3%.

Beyond LP-EGR as a technology to improve efficiency of GTDI engines, Stuhldreher and colleagues (2018) identify and describe several technologies for boosted engines, as illustrated in Figure 4.2.

Conway et al. (2018) also examined whether a 1.6 L GTDI engine could effectively utilize a 12:1 CR leveraging the knock mitigation benefits of EGR alone. The results showed the expected benefits of higher CR at lighter loads including most fuel economy drive cycles, but the study concluded that the trade-off losses in BSFC were

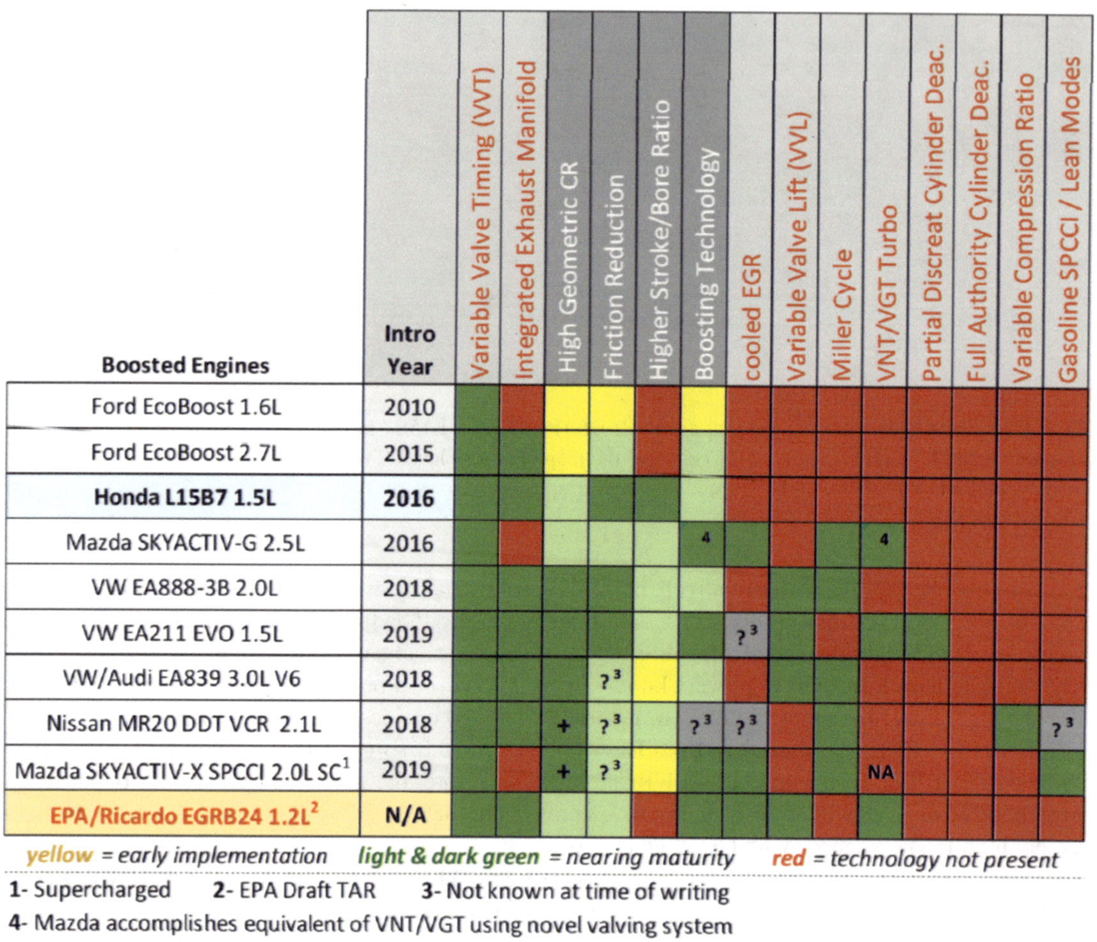

FIGURE 4.2 Example production boosted engines (except the EPA/Ricardo engine), with implemented efficiency technology solutions in green and yellow. Unimplemented efficiency technologies are shown in red. Superscripted numbers, question marks, and plus signs provide further context for the implementation designations, and are detailed in Appendix C of the figure's source. The technology frontier, primarily the technologies in red, is examined in the discussion below.
NOTE: SPCCI = Spark Controlled Compression Ignition; TAR = Technical Assessment Report; VGT = Variable-geometry turbochargers; VNT = variable-nozzle turbines.
SOURCE: Republished with permission of SAE International, from Stuhldreher et al., 2018, Benchmarking a 2016 Honda Civic 1.5-Liter L15B7 turbocharged engine and evaluating the future efficiency potential of turbocharged engines, *SAE International Journal of Engines* 11(6); permission conveyed through Copyright Clearance Center, Inc.

TABLE 4.1 Examples of Advanced Downsized, Boosted Engine Technologies Available in the MY 2020 Fleet

Manufacturer/Engine Design	Technology Approach	Models Incorporating Technology
Audi 2.0 L TFSI	Miller-like cycle enabled by an extremely short intake cam duration with intake phase and two position lift control (using both early and late intake valve control depending on engine operating conditions) and dual fuel (port and direct injection).	Audi TT, TT S, S1, S3, A3, A4, A5, A6, Q2, Q3, Q5; SEAT León Cupra, Alhambra, Ateca, Cupra Ateca; Škoda Superb, Kodiaq; Volkswagen Golf GTi, Polo GTi, T-Roc, Atlas, Passat, Arteon, CC, Beetle, Tiguan, Sharan.
Nissan 2.0 L VC-Turbo	Fully variable CR engine (ranging from 8.0–14.0:1) that also includes a "wide range" turbocharger with electric waste gate for boost control capability, electric control of intake cam phaser, and dual port and direct injection.	INFINITI QX50 crossover, Altima SR and Platinum front-wheel drive grades.
Volkswagen 1.5 L TSI evo (96 KW version)	Process based on Miller cycle with 12.5:1 geometric CR, in which an electrically variable geometry turbocharger provides the expanded boost requirements of this approach. The engine also uses a higher pressure direct injection fuel system (350 bar) and employs cylinder deactivation.	Golf and Golf Variant in Comfortline, Highline, and SOUND specifications in Europe. Not currently available in the United States.

NOTE: TFSI = turbo fuel stratified injection; TSI = turbocharged stratified injected; VC = variable compression.
SOURCE: Committee generated using information from Green Car Congress (2017), Lisle (2019), and Sessions (2018).

significant and unacceptable at higher loads. Therefore, the paper recommended further study of the potential for higher CR employing Miller cycle operation. The multiple engine offerings now in the market and considerable research activity utilizing this concept suggest that industry has reached the same conclusion.

The Miller cycle increases geometric CR to take advantage of the efficiency benefits of increased expansion while limiting compression (and knock). This is accomplished via adjustments to the timing of intake valve closing (typically late but can also be early). Manufacturers and engine developers are working on methods to aggressively leverage the benefits of the Miller cycle by using various complementary engine technologies to augment or manage trade-offs. Some of those technologies are listed in Table 4.1.

In summary, applying the Miller cycle concept in various ways to improve thermal efficiency of downsized/boosted engines seems to be a significant focus going forward into 2025–2035. The resulting benefits and corresponding costs will depend on the extent of technologies incorporated and the system-level effectiveness. Those enabling technologies already included in various applications are electric intake cam phasing, variable intake lift control, variable geometry turbocharging, cooled EGR, ignition and fuel system enhancements (also for particulate control), and variable CR. Because the Miller cycle can impose some constraints on performance, these implementations to date have occurred on engines of somewhat higher displacements (2.0 L) or, in the case of the Volkswagen (VW) 1.5 L EVO, on a power de-rated version in comparison to the non-Miller variant.

A focus on thermal efficiency and real-world emissions will constrain more aggressive engine downsizing. However, several technology approaches are being applied to further reduce throttling losses and allow these already downsized engines to operate in more efficient regions of the speed-load map. These include variable intake valve lift technologies (both two position and fully variable), cylinder deactivation, transmissions with additional discrete gear steps, and continuously variable transmissions (CVT). For example, Hyundai has recently introduced an all-new 1.5 L GTDI engine that features a first ever continuously variable valve duration technology. Cylinder deactivation can yield improved fuel economy in this engine category on 4-cylinder variants, and even the aforementioned Ford Dragon 1.5 L 3-cylinder engine deactivates 1 cylinder under certain conditions to operate on 2 cylinders. Additionally, the development of transmissions with additional discrete gear steps (8, 9, 10 speeds) or CVT can also enable the engine to operate closer to its peak efficiency regions; for instance, the subject Honda 1.5 L L15B7 is mated to a CVT.

Downsized/boosted engines utilizing the added expansion ratio of Miller cycle (geometric CR of 12:1 or greater) combined with aggressive use of cooled EGR and the necessary supporting technologies as listed have the potential to achieve brake thermal efficiency levels of 40% while meeting all other driving and emissions requirements (see Table 4.2). Much of the current development activity to further improve the efficiency of these

TABLE 4.2 Six Example Miller Cycle Packages for Downsized/Boosted ICE Vehicles in MY 2025–2035, with Associated Technology Cost Projections and Effectiveness Values

Baseline Engine	Example FE Technology Advance Above Baseline	Technology Cost[a]			Technology Effectiveness (Fuel Consumption Reduction Relative to Baseline Engine for the Technology as Part of the Described Bundle)
		MY 2025	MY 2030	MY 2035	
Example: Honda L15B7 1.5 L GTDI	Cooled Low-Pressure EGR (CEGR1)	$250	$230	$230	2%–3%
• Includes NHTSA technologies: VVT, stoichiometric gas direct injection (SGDI), TURBO1	Variable Geometry Turbocharging (VGT)	$55	$50	$50	None, enabler for technology bundle
• Includes technologies from Stuhldreher et al. (2018): integrated exhaust manifold, friction reduction, higher stroke/bore ratio, boosting technology including electric wastegate	Miller Cycle Implementation including 12:1 CR[b]	$0 w/ enabling technology present	$0 w/ enabling technology present	$0 w/ enabling technology present	2%–3%
Example: Hyundai-Kia Smartstream 1.5 L Turbo GDI Engine					
• Includes NHTSA technologies: SGDI, TURBO1, cooled LP EGR (CEGR1)	VGT	$55	$50	$50	None, enabler for technology bundle
• Includes technologies from Stuhldreher et al. (2018): integrated exhaust manifold, boosting technology (twin scroll and electric wastegate), friction reduction, higher stroke/bore ratio • Includes other technologies: continuously variable valve duration, 350 psi direct injection	Miller Cycle[b]	$0 w/ enabling technology present	$0 w/ enabling technology present	$0 w/ enabling technology present	2%–3%
Example: VW EA211 EVO 1.5 L GTDI Engine					
• Includes NHTSA technologies: VVT, SGDI, TURBO1, VTG, DEAC (cylinder deactivation of 2 cylinders)	CEGR1	$250	$230	$230	2%–3%
• Includes technologies from Stuhldreher et al. (2018): high compression ratio, Miller cycle, high stroke/bore ratio, friction reduction • Includes other technologies: 350 psi direct injection	Variable Valve Lift (VVL)	$210	$205	$200	1%–2%

Example: Ford 1.5 L 3-Cylinder GTDI Engine • Includes NHTSA technologies: VVT, SGDI, TURBO1, DEAC (1 cylinder) • Includes technologies from Stuhldreher et al. (2018): integrated exhaust manifold, low friction (3 cylinders) • Includes other technologies: dual port-fuel/direct-injection (PF/DI) fuel injection, higher BSFC from larger cylinder size relative to 4-cylinder	VVL[c]	$125	$125	$125	1%–2%
	CEGR1	$250	$230	$230	2%–3%
	Miller Cycle[b]	$0 w/ enabling technology present	$0 w/ enabling technology present	$0 w/ enabling technology present	2%–3%
Example: Audi 2.0 L TFSI GTDI Engine • Includes NHTSA technologies: VVT, VVL (2-position), SGDI, TURBO1 • Includes technologies from Stuhldreher et al. (2018): Integrated exhaust manifold, boosting technology (electric wastegate), higher stroke/bore ratio, high compression ratio (11.65:1), Miller cycle, friction reduction • Includes other technologies: dual PFI/DI fuel injection	CEGR1	$250	$230	$230	2%–3%
	VGT	$55	$50	$50	None, enabler for technology bundle
Example: Nissan 2.0 L VC-Turbo Engine (Variable Compression Ratio) • Includes NHTSA technologies: VVT, SGDI, TURBO1, variable compression ratio (VCR) (8.0–14.0:1) • Includes technologies from Stuhldreher et al. (2018): Integrated exhaust manifold, boosting technology ("wide range turbocharger," electric wastegate), high compression ratio, Miler cycle • Includes other technologies: dual PFI/DI fuel injection, electric intake variable cam timing (VCT) control	CEGR1	$250	$230	$230	2%–3%
	VGT	$55	$50	$50	None, enabler for technology bundle

[a]All costs are for I4 engines in 2018$; projections use learning curves from NHTSA/EPA (2020); costs rounded to the nearest $5.

[b]Miller cycle engine with 12:1 CR is covered in the associated technology packages; cost is function of the building blocks. NHTSA/EPA (2020); Stuhldreher et al. (2018).

[c]Costs adjusted for deletion of DEAC.

NOTE: The baseline engines represent advanced downsized/boosted engines in 2020, which may represent typical downsized/boosted engines in 2025. The example fuel economy technology advancement above the baseline illustrates some "next step" technologies that could be implemented in 2025–2035. Technology cost and effectiveness are for the individual added technologies, as they contribute to the efficiency of the technology bundle.

engines and manage their challenges involves leveraging the capabilities of electrified assistance via hybridization of various levels, which is discussed in Section 4.4.

FINDING 4.1: Downsized/boosted engines continue to increase market penetration and, along with improved efficiency naturally aspirated engines, have become the common engine types in 2020, as previously predicted for this time period. Manufacturers have also further optimized new engine designs to complement these technology trends, including the emergence of 3-cylinder engines.

FINDING 4.2: Efficiency improvements to current downsized/boosted engines through the application of additional technology is possible and ongoing. Many of the technologies that will contribute to the next generation of downsized/boosted engines are already present in various forms in the 2020 fleet. These include, for example, the application of Miller cycle (or Atkinson cycle in the naturally aspirated case) to allow for higher expansion ratios and therefore greater thermal efficiency, cooled EGR, friction reduction technologies, and the application of cylinder deactivation to already downsized 4- and even 3-cylinder engines.

FINDING 4.3: By 2025, non-electrified internal combustion engines could implement technologies such as currently represented in 2020 by the advanced downsized/boosted engines described in Tables 4.1 and 4.2, which would offer improvements in efficiency of up to 5% over a current baseline downsized/boosted engine. For the period of 2025–2035, those same non-electrified engines could deploy additional technologies as outlined in Table 4.2 for a potential further efficiency improvement of up to 5%. Improvements beyond these levels for 2025–2035 will likely require some degree of hybridization.

4.2 NATURALLY ASPIRATED ICE PATHWAY

Naturally aspirated ICEs do not employ any sort of air pressure charging to assist performance. The multiple approaches of synergizing these engines with electric hybridization will be covered under that dedicated pathway (Section 4.4). In the current U.S. market, this type of engine typically falls into two very different vehicle and customer usage categories, which will be discussed separately to better assess and prioritize applicable FE technologies. The first are larger displacement engines, that is, V8s, that also need to provide performance—load carrying, towing, and so on—and fall primarily in the full-size sport utility vehicle (SUV) and pickup truck vehicle classes. The second category comprises I4 engines with significant fuel economy technology, which are positioned as an alternative to downsized/boosted engine systems.

Larger displacement naturally aspirated engines, mostly V8s with 5.0 L–5.7 L displacement, continue to play a significant role in the U.S. market, primarily in the large SUV and pickup truck vehicle classes. Typically, these engines are optional offerings to provide additional performance in the vehicles, which are utilized for load carrying, towing, and so on. Engines in this category are therefore oversized and highly throttled on more lightly loaded duty cycles, including fuel economy test cycles. Compared to other engine types, they will derive greater benefit from technologies that reduce pumping losses, such as cylinder deactivation. Offerings from General Motors (GM) and Fiat Chrysler are 2 valve per cylinder overhead valve (single cam in block) configurations, which allow for cost effective application of cylinder deactivation and also employ dual equal camshaft phasing for fuel efficiency. GM has also added "Dynamic Skip Fire," a control strategy developed by Tula Technologies that further expands the capability and benefits of cylinder deactivation. The GM 5.3 L engine has also incorporated gasoline direct injection and stop-start, and the CR with direct injection is 11.0:1. Ford's 5.0 L engine in the F150 is a 32 V dual overhead camshaft configuration, and utilizes dual cam phasing and dual fuel injection (direct and port fuel injection points) with stop start. Fiat Chrysler offers a 48 V mild hybrid option as their Ram pickup truck fuel economy leader. All of these vehicles are now also migrating 8-, 9-, and 10-speed transmissions into their offerings. Looking to the future, these large, performance-oriented V8 engines will likely be limited in volume, and/or hybridized, or replaced by smaller, boosted alternatives to achieve more power with less displacement.

In the midsize car and crossover segments, some manufacturers continue to offer larger naturally aspirated 4-cylinder engines in the 2.5 L displacement range. Several of the best fuel economy performing MY 2020 midsize and crossover vehicles identified in Chapter 3 (Toyota Camry, Nissan Altima, Toyota RAV4) fall into this category.

Engine technologies commonly utilized in this pathway include dual variable cam phasing, homogeneous gasoline direct injection, and stop-start. In this pathway, gasoline direct injection is primarily used to improve thermal efficiency via increased CR, and there are examples in the U.S. market of up to 13:1 CR on 91 RON fuel. The engine architectural benefits described in the downsized/boosted pathway (e.g., lower bore/stroke; see Section 4.1.3) offer similar efficiency benefits in naturally aspirated engines. Being sized to meet peak performance requirements, naturally aspirated engines are more throttled than downsized/boosted engines on lighter duty test cycles and can therefore derive greater benefit from technologies that reduce pumping loss. There are various applications utilizing cylinder deactivation, cam profile switching, and variable valve lift in this category. The Toyota 2.5 L Toyota New Global Architecture engine used in the MY 2020 Camry and RAV4 is one example of a naturally aspirated large I4 that has achieved high levels of efficiency (Figure 4.3 and Table 4.3). With a low bore/stroke ratio and corresponding high tumble combustion system with direct and port fuel injection, the engine achieves a CR of 13:1 (14:1 in Atkinson cycle form for hybrid applications). It reaches a peak thermal efficiency of 40% by using EGR (cooled internal to the cylinder head) along with a suite of actions to lower friction.

Similar to downsized/boosted systems, transmission technology can be complementary in naturally aspirated powertrains and allow the engine to operate closer to peak efficiency. The MY 2020 Toyota Camry and Nissan Altima referenced previously use 8-speed and CVT transmissions, respectively. Naturally aspirated engines will be constrained in comparison to boosted engines in both specific performance potential and their ability to incorporate the benefits of dilute combustion via EGR. Many strong hybrid powertrains currently use highly efficient naturally aspirated engines utilizing the Atkinson cycle made possible by hybrid electric synergies described in Section 4.5.2. Implementation will likely continue as these engines are lower in cost than boosted variants so can help defray hybridization cost. Therefore, as hybrid penetration grows, so will these engine types.

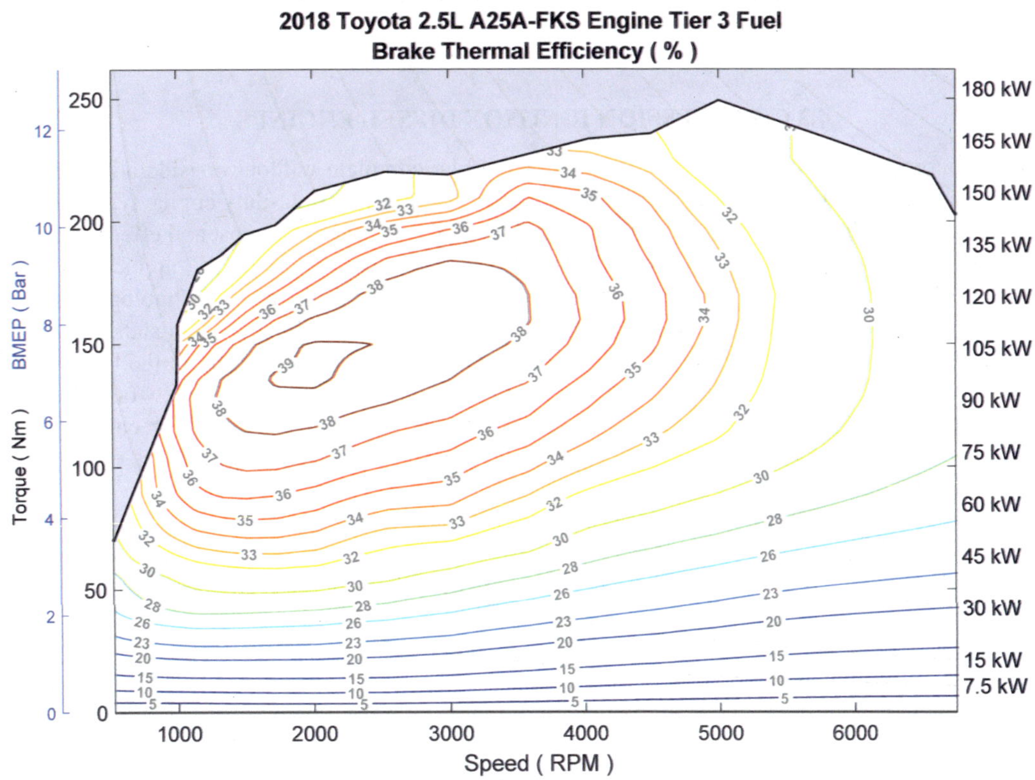

FIGURE 4.3 Engine torque versus speed map for the 2018 Toyota 2.5 L A25A-FKS naturally aspirated engine, showing areas of low fuel consumption and high efficiency near the center of the map (contours labeled with percent brake thermal efficiency), and low efficiency under low load, and at high speed and high torque.
SOURCE: EPA (2020b).

TABLE 4.3 One Example Package for Naturally Aspirated, Unhybridized ICE Vehicles in MY 2025–2035, with Associated Technology Cost Projections and Effectiveness Values

Baseline Engine	Example FE Technology Advance Above Baseline	Technology Cost[a]			Technology Effectiveness (Fuel Consumption Reduction for the Technology as Part of the Bundle)
		MY 2025	MY 2030	MY 2035	
Example: Toyota 2.5 L New Global Architecture Naturally Aspirated Engine • Includes NHTSA Technologies: VVT, SGDI, HCR1 (13:1 non-hybrid electric vehicle [HEV]) • Includes EPA Technologies: integrated exhaust manifold (with EGR cooling function), higher stroke/bore ratio, friction reduction • Includes NASEM technologies: Dual PFI/DI fuel injection, electric intake VCT control, Cooled EGR	Cylinder Deactivation (DEAC)	$110	$105	$100	2%–4%

[a]All costs for I4 engine in 2018$; costs reduced by 1%/year from MY 2020 value reported in NRC (2015) and rounded to the nearest $5.
NOTE: The baseline engine represents an advanced naturally aspirated, unhybridized engine in 2020, which may represent typical naturally aspirated engines in 2025. The example fuel economy technology advancement above the baseline illustrates a "next step" technology that could be implemented in 2025–2035. Technology cost and effectiveness are for the individual added technologies above the baseline engine, as they contribute to the efficiency of the technology package.

4.3 COMPRESSION IGNITION DIESEL ENGINES

No fuel economy assessment of ICE technologies would be complete without consideration of the diesel engine, given that it has the highest thermodynamic cycle efficiency of any light-duty engine type (NRC, 2015). The 2015 NRC report dedicated an entire chapter to diesel technologies, including cost and effectiveness projections relative to spark ignition engines. The technical content in that report is largely still applicable—including the fundamentals behind diesel's fuel efficiency advantage, the potential of future diesel technologies, and perhaps most importantly the challenges and cost associated with meeting Tier 3 criteria emission standards. However, since the time of that publication, the penetration outlook for diesel engines, especially in the U.S. market, has been affected by several very significant developments. This section provides a brief review of the technology, a qualitative update on the cost and effectiveness relative to current and future spark ignition engines, and rationale for why diesel engines will not be included as a significant contributor to the efficiency of the LDV fleet in 2025–2035.

The diesel engine's efficiency advantage over a spark ignition gasoline engine stems from three factors:

1. The higher CR of compression ignition engines relative to spark ignition engines, which as previously discussed provides a thermodynamic expansion advantage;
2. The operation of diesel engines on an overall lean mixture (excess air), which is more thermodynamically efficient based on a higher ratio of specific heats; and
3. The ability, based on that lean operation, to control torque with the amount of fuel injected and therefore operate without throttling and corresponding losses.

In addition, diesel fuel has a higher energy content than gasoline, giving another roughly 11% fuel economy advantage incremental to gasoline when measured on a volumetric basis (miles/gallon). At the same time, however, diesel fuel has a higher carbon density so has greater carbon dioxide (CO_2) emissions per gallon of fuel burned.

Thus, shifting the regulatory focus from fuel consumption to greenhouse gas (GHG) emissions would mitigate some of the advantage of the diesel pathway.

The technical challenge for the diesel engine is the ability to meet stringent criteria emission standards, especially nitrogen oxides (NO_x) and particulates, and to manage the cost associated with that challenge while maintaining a compelling fuel efficiency advantage versus improving spark ignition engines. This is particularly true as it relates to California Low-Emission Vehicle (LEV) III and U.S. Federal Tier 3 criteria emissions standards. Lean air-to-fuel ratio operation precludes the use of the three-way catalysts to control tailpipe NO_x. The technologies already deployed on modern diesel engines to control emissions at the engine feed gas level include the following: higher fuel injection pressures (in some cases with in-cylinder combustion pressure sensing), high levels of cooled EGR to limit NO_x forming peak combustion temperatures, and two-stage and variable turbine geometry (VTG) boosting systems to maintain performance while supplying the required EGR with the associated cooling systems. In the emissions aftertreatment system, technology options include diesel particulate filter, diesel oxidation catalysts for carbon monoxide and hydrocarbons, NO_x storage catalyst and selective catalytic reduction (requiring the consumer to maintain an additive, typically urea), and the corresponding onboard diagnostic system and sensor set. While used individually or in combinations in 2020, in 2025–2035 virtually all of these countermeasures would likely be necessary to achieve the aforementioned U.S. standards. Another strategy to address criteria emissions is to actively limit the fuel economy and performance advantage of the diesel engine in order to control emissions, especially off-cycle emissions.

For modeling purposes, NHTSA has characterized the potential of future improvements in diesel efficiency as a single technology bundle labeled "Advanced Diesel" (EPA/NHTSA, 2012), which includes benefits from engine downsizing and downspeeding, friction reduction, and combustion improvements. Downsizing with corresponding high BMEP operation is not likely to occur owing to real-world and off-cycle emissions concerns. Downspeeding, which per the 2015 NRC report requires additional transmission ratio span, is likely still viable given the penetration of 8-, 9-, and 10-speed transmissions. Friction reduction is also credible given that diesel CRs are trending downward as a result of increasingly stringent emissions standards. Combustion improvements, such as low temperature combustion concepts, could potentially be more emissions driven. In the meantime, while the cost of diesel technology will remain relatively high, the fuel efficiency advantage relative to spark ignition engines will erode as those engines incorporate technologies to approach diesel-like efficiencies. As outlined in Section 4.6 below, these include reduced throttling technologies such as downsizing with boosting, higher CR with direct fuel injection, Miller/Atkinson cycle application, and dilute mixtures using EGR in lieu of lean. Combined with the current and likely future higher price of diesel fuel relative to gasoline, a consumer payback equation based on fuel cost savings will be difficult.

By far the biggest impact on diesel as a future technology pathway has come from the aftermath of a U.S. diesel emissions cheating scandal (EPA, 2015), which shifted automaker priorities, negatively affected consumer views of diesel, and virtually eliminated the light-duty U.S. market. Another diesel emissions scandal recently came to light (EPA, 2020c) and may have related consequences. In addition, the broader scrutiny brought to diesel worldwide with regard to off-cycle and real-world emission effects both at a policy maker and consumer level has led to significant diesel share erosion, even in Europe where diesel represented roughly 50% of light-duty markets in some countries. This loss of diesel contribution has had a negative effect on the EU CO_2 compliance progress and driven incremental recovery actions. EPA and NHTSA did not include diesel in their possible cost-effective compliance path for the original 2017–2025 corporate average fuel economy (CAFE) rule for 2025 (EPA/NHTSA, 2012) even prior to these events, so the 2025 starting point for this study is not affected.

Despite their limited expected penetration in the U.S. light-duty fleet, diesel engines will continue to play a significant role in the U.S. commercial truck segment. They also still represent a large global share of engines such that the technologies required to meet the aforementioned challenges will continue to be developed. Diesel can offer high fuel efficiency at heavy load conditions (e.g., towing) relative to other technologies, along with compelling performance. These properties could make them attractive into the future to some customers of full-size pickup trucks and large SUVs focused on those attributes and willing to pay for them. However, consistent with feedback from manufacturers, diesel will be considered a specialty niche and will not be developed as a significant pathway to U.S. fuel economy improvement for the purposes of this 2025–2035 report.

4.4 TRANSMISSION PATHWAY

Manufacturers have continued to develop automatic transmissions and CVTs for the U.S. market to improve their efficiency and customer performance as well as to complement the technology trends of ICEs described in this chapter. Manual transmissions have all but left the U.S. light-duty market except in sports performance categories and therefore will not be discussed. The 2015 NRC report provides a still applicable description of transmission fundamentals and a comparison of the various types of automatic transmissions, along with the fuel efficiency elements by which they can provide improvements. In fact, the findings from that report with respect to transmissions are still largely applicable today. Herein, the focus will be on developments since 2015, including updates on previously forecasted trends and a look to the future for 2025–2035.

As of MY 2016, 54.9% of the U.S. market comprised 6-speed planetary automatics, which at that point had undergone about 10 years of refinement since their first widespread deployment in the mid-2000s (EPA, 2020a). Evolutionary improvements to planetary automatic transmissions between their introduction and current implementations included internal efficiency actions, such as variable displacement pumps, lower friction fluids, clutch materials, bearings, and seals. Improved automatic transmissions controls can allow for more fuel efficient aggressive transmission shift schedules and torque convertor lock-up strategies. However, they must be managed to avoid customer concerns with drivability; performance; and noise, vibration, and harshness (NVH). The trend toward downsized/boosted engines with fewer cylinders and consequent higher torque amplitudes at lower frequencies present additional potential NVH concerns that limit lugging and/or require enhanced torque convertor damping.

Eight-, nine-, and ten-speed planetary automatic transmissions have continued a steady rollout in replacement of six-speeds and, as of MY 2019, have achieved a U.S. market penetration of 44% (EPA, 2020a). As stated in the 2015 NRC report, upgrading to those higher transmission speeds offers fuel economy improvements in the 2%–3% range, depending on the maturity of the 6-speed (or in some cases > 6 speeds) in the reference case (NRC, 2015). (It should be noted, however, that NHTSA provided higher estimates at that time.) In addition to providing high-value fuel economy improvement based on the understood cost, an increase in the number of transmission speeds can also be a marketable customer feature. The additional ratio span with the potential for lower "launch" gears can be synergistic with the trend toward engine downsizing to support low-speed performance (turbo lag zones). Furthermore, with more steps, the planetary automatics can approach the ability of a CVT to allow the engine to operate in regions of best fuel consumption. However, the incorporation of engine technologies that greatly expand the speed-load regions of low brake specific fuel consumption can minimize the benefits of higher transmission speeds. Many research efforts aim to develop new transmission architectures to deliver these benefits and maintain or improve the transmission spin losses, while incorporating additional gears, clutches, and so on.

The penetration of CVT transmissions is also increasing in the United States, primarily owing to implementation in high-volume offerings from Japanese manufacturers (e.g., Nissan, Honda, Toyota, and Subaru). Additionally, Toyota and Ford Powersplit (PS) hybrid models utilize CVT-type technology. As described in the 2015 NRC report, CVTs derive their fuel economy benefit by allowing the engine to operate at its most efficient speed and load for a given power demand; however, they have higher losses than a conventional automatic. To decrease these losses, the CVTs in today's market have taken several actions. Since major power losses occur with the hydraulic pump and belt, the pumps can be variable displacement or even an on-demand electric pump employed by Toyota to enable stop-start functionality. The Toyota CVT has incorporated a launch gear to initially accelerate the vehicle and provide an overall increase in ratio span while reducing the CVT pulley ratio. CVTs have historically experienced customer acceptance concerns associated with the NVH associated with their typically higher engine speeds and "droning" sound as well as absence of traditional shifting sensations. Manufacturers have addressed these concerns with control strategies that program in fixed-ratio set points and even offer paddle shifters to provide U.S. customers the aesthetic of a traditional automatic. CVTs will be particularly effective in a powertrain system using a high efficiency or Atkinson cycle naturally aspirated engine of higher displacement by keeping the engine in its most efficient operating points, reducing the benefit of additional engine technologies such as cylinder deactivation.

Dual clutch transmissions (DCTs) have one clutch for the odd gears and one for the even gears and can operate manually or automatically, depending on configuration. Having two clutches allows the vehicle to maintain torque from engine to the wheels while the next gear is being set up. DCTs have a higher fuel efficiency potential relative to a planetary automatic owing to the lower spin losses of their layshaft (manual transmission) architecture and, in the case of a dry clutch DCT, no hydraulic losses. Despite its avoidance of hydraulic losses, dry clutch is torque limited and more difficult to manage. On the other hand, wet clutch versions require a pump and, if needing excessive clutch slip to address customer concerns, may achieve less than the theoretical FE. Additionally, while lower in cost on a teardown basis, the actual cost of sophisticated clutch and controls at relatively low volume make them cost more than a conventional clutch and controls. DCTs originally developed in Europe, where the transmission manufacturing infrastructure was based on manual transmissions and importantly so was the customer expectation. In the early to mid-2000s, EPA, NHTSA, and automakers thought DCTs would develop into a significant trend in the U.S. market. However, attempts by some automakers to introduce this technology to the U.S. market were met with significant customer acceptance issues—for instance, customers accustomed to a torque convertor-based automatic transmission performance seem to have concerns with a start-up clutch, mostly at lower speeds. Therefore, some automakers have since transitioned away from DCT, and other automakers scrapped introduction plans prior to launch. Even some European luxury brands that did not experience the same problems are moving back toward conventional automatics. Beyond the consumer concerns, the appeal of the DCT for fuel economy has also diminished. As described above, the planetary automatics evolving to 8-, 9-, and 10-speeds have improved so as to close the gap without the same level of risk. By 2020, DCTs are anticipated to make up only less than 5% of the U.S. market and are not expected to grow measurably. The 2015 NRC report recommended that NHTSA and EPA reflect this lowered expectation in their analyses.

In summary, automatic transmissions, most notably 8-, 9-, and 10-speed planetary systems have continued to expand in the U.S. market and offer fuel economy improvements of about 2–3%, incremental to an advanced 6-speed. Further improvements are possible but will likely be of diminishing value. The various transmission types will continue, as the differences between them do not indicate any clear winning technology, and they continue to become more similar in terms of fuel economy benefit and cost. Their fuel economy benefits can also be interdependent with engine technology, and therefore the engine and transmission must be viewed as a system. Historically, automatic transmissions have shown potential for customer dissatisfaction, and any push to improve their efficiency must be managed with this risk. Perhaps the most significant future trend in transmission development will be the integration of electrification. In addition to the PS CVT hybrid transmission, numerous parallel (P2) strong and mild hybrid configurations incorporate the electric motor into the transmission. There are also examples of hybrids utilizing planetary automatics, such as the Ford rear-wheel drive (RWD) 10-speed, and notably some utilizing DCTs, which is a potential pathway for those transmissions to grow.

FINDING 4.4: Transmission technology continues to play a role in improving fuel efficiency. The U.S. fleet is in the midst of a transition from advanced 6-speed planetary automatic transmissions to 8-, 9-, or 10-speeds, which will likely expand to much of the fleet while some manufacturers will continue to develop advanced continuously variable transmissions. Transmission contributions to vehicle level efficiency are highly interdependent with engine technology. While further gains from additional gear steps beyond 10-speed are not seen as likely, the future development of transmissions may well focus on the integration of electrification.

4.5 HYBRIDIZED POWERTRAIN PATHWAY

Hybrid powertrain systems are forms of electrified powertrains that contain ICEs and electrical machines and can operate on liquid fuel only or, for plug-in hybrid vehicles, with the added option of operating on electric energy charged from the power grid. The extent of electrification of the powertrain varies, as does the resulting capability for efficiency improvements. As described later in this section, the incorporation of electrification to liquid-fueled vehicles represents the greatest opportunity for improvements in fuel efficiency and reduction in CO_2 emissions. Hybrid electric vehicles (HEVs) derive their primary benefit from the capability of the electrical machine, along with sufficient storage capacity of the battery, to capture and store energy under deceleration (regenerative braking). In addition, they can allow for reduced idling time and a more efficient and/or downsized ICE, taking advantage of the available electric motor propulsion assistance. Chapter 4 of the 2015 NRC report provided descriptions, terminology, architectural definitions, and a still applicable primer on how hybrid powertrains work. Table 4.4 gives a summary of levels of hybridization in LDVs, noting typical technologies, system voltage, and efficiency capabilities.

By far, the most fuel-efficient gasoline-only vehicles in the U.S. fleet are strong hybrids. There are offerings from numerous manufacturers in every segment under study. Despite wide availability of models, current sales volumes remain relatively low nationwide. The low sales can be attributed in part to the added cost to the manufacturer

TABLE 4.4 Description and Energy Improvement Aspects of Different Levels of Hybridization

Hybridization Level	Illustrative Technology Description	System Voltage (V)	% Idling Reduced	% Brake Energy Recovered	% Fuel Consumption Reduction Relative to a Conventional Vehicle	Electric Assist or Electric Operation Capability?
Stop-Start	Robust starter Controls for customer comfort and NVH	12	~100%	0%	2%–3%[a]	Assist? No. Operation? No.
Mild Hybrid, BISG architecture	BISG Motor, 10–15 kW 0.5–1.0 kWh battery Optimized ICE	48		50%[b]	6%–10%[c]	Assist? Yes. Operation? Some implementations allow short, low power driving on electric motor.
Strong Hybrid	≥ 1 motor, 30–90 kW Parallel P2, Powersplit, or Series architectures 1.0–.08 kWh battery Optimized ICE	200–300		70%–80%[b]	~20%–35%[d]	Assist? Yes. Operation? PS: In low load operation. P2: In low load operation. Series: Yes.
PHEV	≥ 1 motor, 60–100 kW Power- or energy-optimized battery, depending on EV range Charger for fueling with grid electricity Optimized ICE	400		70%–80%	Varies by the ratio of miles traveled fueled with electricity or gasoline. Fueling with gasoline is similar efficiency to a hybrid; fueling with electricity is similar to a BEV.	Assist? Yes. Operation? Yes, supplemented with ICE drive in some applications.

[a]As reported in EPA/NHTSA (2012), NRC (2015), and NHTSA/EPA (2020); relative to baseline engine.
[b]Lee et al. (2018).
[c]Relative to conventional vehicle.
[d]Relative to conventional vehicle of same model. Varies with hybrid architecture and vehicle class.
NOTE: BISG = belt integrated starter generator; PHEV = plug-in hybrid electric vehicle. Percent of idling reduced and brake energy recovered are illustrative, as the amount varies by hybrid implementation and under different use conditions.

associated with hybridization, particularly for strong hybrids. In the past, there has also been a strong correlation between hybrid sales and fuel prices, and fuel prices have been low relative to historic standards.

Strong hybrids represent the greatest potential to improve the fuel efficiency of gasoline-only powered vehicles. Even some of the 2020 vehicles that were previously mentioned as high performers in their base ICE form (Toyota Camry, RAV4, Ford Escape) offer strong hybrid options that further improve FE ratings by at least 33%. In seeking to improve vehicle efficiency, fuel economy, and GHG emissions, HEVs will have several advantages and disadvantages relative to conventional ICEs and BEVs/fuel cell electric vehicles (FCEVs). In using electrification to improve efficiency, hybrid powertrains currently have lower absolute cost relative to BEVs and FCEVs, but also a lower ultimate fuel efficiency or CO_2 reduction potential. Hybrid powertrains require less battery energy capacity than BEVs or plug-in hybrid electric vehicles (PHEVs), and do not require a fuel cell and hydrogen storage system like FCEVs. Unlike BEVs and FCEVs, HEVs do not depend on enhancements in the electric or hydrogen fueling infrastructure and will not require any modification to existing consumer behavior. As the costs of batteries come down, BEVs will reduce in cost such that they become less expensive for the manufacturer to produce than HEVs. HEV potential in the market will depend on customer requirements and manufacturer responses to fuel economy regulations as applied to non-zero-emission vehicles.

4.5.1 Hybrid Architectures

This section will cover 48 V mild hybrid electric vehicles (MHEVs) and 200 V or greater strong hybrids as they pertain to the U.S. market. While both mild and strong hybrids appear in numerous configurations, there does not seem to be any ongoing trend toward intermediate voltage levels. Stop-start systems operating at 12 V (micro hybrid) have become common in ICE vehicles and will not be discussed in depth. However, some manufacturers have shared customer satisfaction concerns with stop-start implementations, which can often be ameliorated by higher voltage electrical starting assistance provided by hybrids. Mild and strong hybrids contain a larger battery, electronics capable of operating above 12 V, and one or more electric machines, such as motors and generators, amongst other components required to utilize electric energy in storing braking energy and assisting or providing propulsion. Schematic diagrams of P0 mild hybrid, P2 mild or strong hybrid, PS strong hybrid, and series strong hybrid architectures are shown in Figure 4.4. The position of the electrical machine is often important to

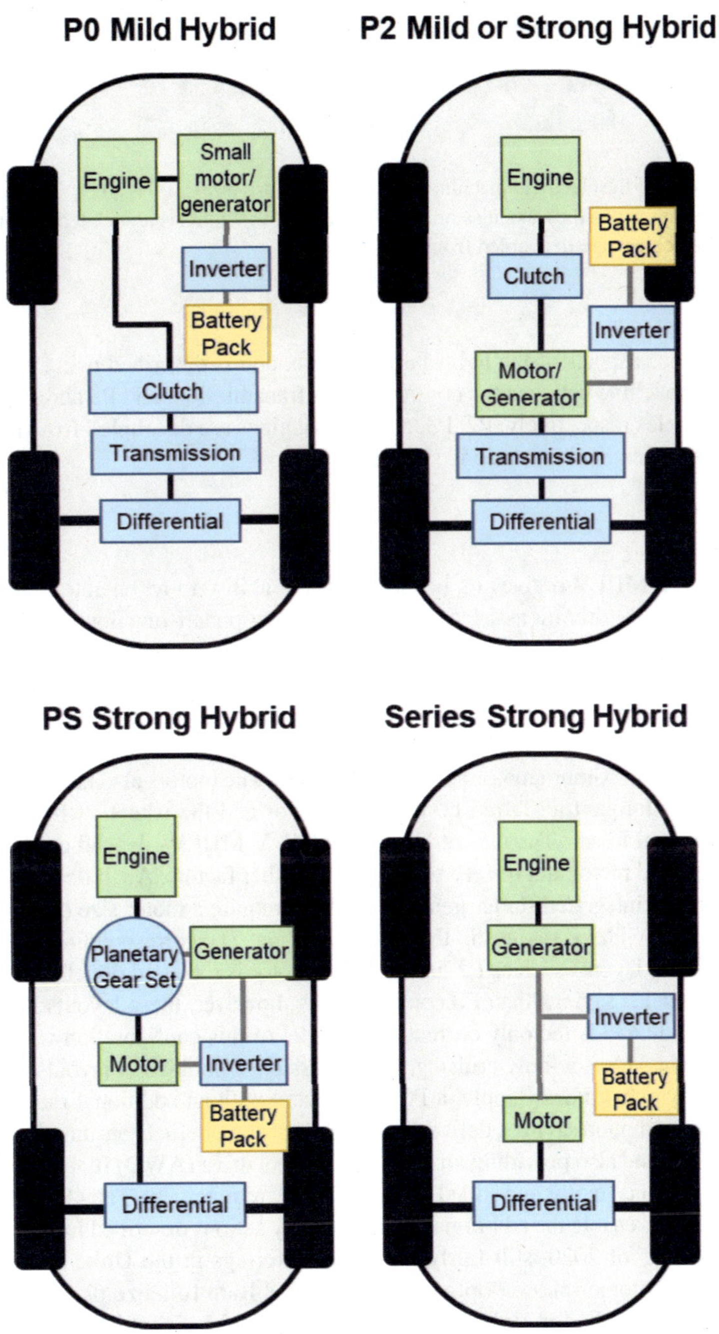

FIGURE 4.4 Basic mild and strong hybrid architectures including BISG P0, P2, PS, and Series.
SOURCE: Committee generated, adapted from Phillips (2018) and NRC (2015).

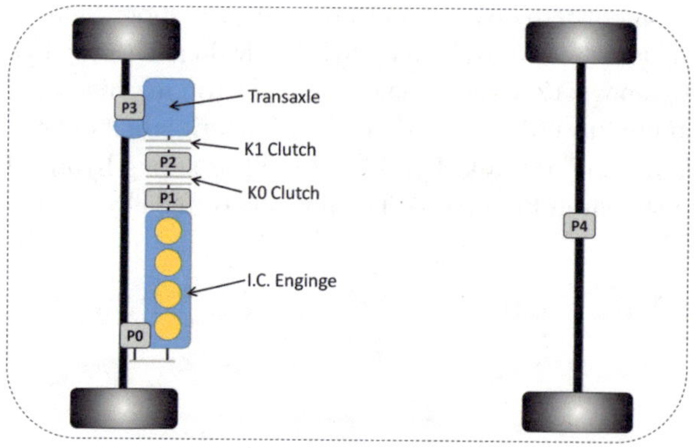

FIGURE 4.5 Different positions of the electrical machine within a hybrid drivetrain. P0 is a belt-integrated electrical machine, P2 for a machine between the engine and the transmission, and P3 and P4 for machines associated with the front and rear axles respectively. P2, P3, and P4 machines are decoupled from the engine.
SOURCE: Lee et al. (2018).

the efficiency and performance aspects of the hybrid architecture, and is described using P0 for a belt-integrated electrical machine, P2 for a machine between the engine and the transmission, and P3 and P4 for machines associated with the front and rear axles respectively. P2, P3, and P4 machines are decoupled from the engine. A diagram of the P0–P4 locations of the electrical machine is shown in Figure 4.5.

4.5.1.1 48 V MHEV

Like all hybrids, the 48 V MHEV derives its benefit from the ability to regenerate braking energy, augment engine performance with electrical driving assistance, and enhance stop-start functionality. The 48 V MHEV can offer a significant improvement in fuel efficiency at a relatively low absolute cost compared to strong HEV or BEV and can also be fairly straightforward to implement, with most current versions using a belt driven starter-generator incorporated into the engine's accessory drive. Along with their lower absolute costs than strong hybrids, 48 V MHEVs have relatively limited benefits. Since most applications to date use a belt-driven machine in a P0 configuration, the belt transfer capability limits the maximum motor and braking torque. The motors also have regeneration constraints at low speed. Additionally, all torque transferred between the motor and the wheels, whether driving or braking, is exposed to the ICE's frictional losses. The realized benefits of 48 V MHEVs depend on the application, driving cycle, battery state of charge, and motor and battery sizing, among other factors. Analytical studies have shown that P0 configurations, such as a belt integrated starter generator (BISG) require a motor size of at least 10 kW and offer no further benefit above 20 kW, with FE savings in the 6%–10% range. The corresponding optimal battery sizing is in the 0.5–1.0 kWh range. Other layouts such as P2, in which the motor is mounted directly to the output side of the crankshaft, can provide greater FE savings than P0 configurations; however, these layouts have higher integration costs. The Mercedes-Benz CLK 450 is the only current U.S. model of this configuration with a 16 kW motor and 1 kWh battery capable of limited electric-only cruising. Other possible combination layouts have been studied and shown to offer even greater benefits. For example, a P0 + P4 layout with an additional motor at the rear axle can potentially double the fuel consumption savings, delivering > 10% reduction in fuel consumption by maximizing 48 V regenerative braking capability and also providing an electric all-wheel drive (AWD) function. However, this layout comes with the cost of the additional motor and added battery capacity requirement. The 48 V MHEV provides a sufficient degree of electrification to enable the additional ICE efficiency actions described in Section 4.2 of this report.

BISG 48 V MHEVs are, as of 2020, still fairly limited in offerings in the United States in comparison to strong hybrid models. It is a customer-chosen option on V6 and V8 Ram full-size pickups and is also available in models from Mercedes-Benz, Audi, and Volvo. There appears to be significant interest in 48 V MHEVs in the European market, particularly for premium models, as automakers attempt to quickly recover from the fleetwide CO_2 deterioration resulting from erosion of the diesel market share and shift from sedans toward crossover utility vehicles (CUVs)/SUVs. The MHEV is also effective when added to a diesel powertrain, which is a larger share in global markets. By taking advantage of the 48 V electrical architecture, manufacturers can incorporate additional customer features, such as improved comfort and driver assist technologies. Those companies extensively implementing 48 V in Europe will likely utilize the technology in their U.S. market, as is being seen in current models. When automakers report corporate electrification goals, the 48 V mild hybrid is often part of the percent of vehicles electrified in corporate messaging. Similar to stop-start systems, 48 V BISG hybridization could represent an evolutionary step in ICE powertrains because, while not capable of delivering the fuel efficiency benefit of strong hybridization, it can be "added-on" relatively simply and at lower cost.

4.5.1.2 Strong Hybrid

Strong hybrids use a larger motor/generator and battery at higher voltage levels than mild hybrids and, as a result, can achieve much greater levels of braking regeneration and engine support, often including some degree of electric-only driving. Strong hybrids are much more expensive to implement than mild hybrids but in turn provide a far greater fuel economy improvement (see Table 4.4). The three primary strong hybrid architectures, PS, P2, and series, are shown in Figure 4.4. PS hybrids incorporate a planetary gear set that connects the motor, generator, and engine. Power from the engine is transferred to both the battery (via the generator) and the wheels, with the exact ratio of the split optimized to achieve maximal efficiency. P2 hybrids add a motor and battery in P2 to a conventional vehicle architecture, which includes a clutch between the motor/generator and engine. In this configuration, the engine and battery both provide power to the wheels. Both PS and P2 hybrids recover energy during braking with the motor operating as a generator. To implement a series hybrid architecture, the motor and power battery must be sized to provide all of the vehicle propulsion, as there is no combined driving assist directly from the ICE to the wheels, as in the P2 and PS hybrids. All power delivered to the wheels comes from the electric motor, and therefore the vehicle has the instant torque performance characteristic of a BEV. Given its limited operating requirements as a generator only, the ICE can therefore be made to be extremely efficient and/or low cost. Advantages and disadvantages of the various hybrid architectures are discussed below in the context of specific vehicle offerings.

Despite a relatively challenging market demand, manufacturers have continued to expand the number of strong hybrid offerings across more vehicle segments in recent years. The types of hybrid architectures in the market have also proliferated to adapt to different customer needs in larger vehicle classes and to allow for performance orientations, AWD, towing, and so on. Manufacturers with a legacy in strong hybrid continue to be bullish about their role in 2025–2035. Other manufacturers state that they will focus strictly on BEVs as the end game. In either case, strong hybrid represents the ultimate efficiency potential of vehicles relying on petroleum fuel as the sole source of energy. As described in Section 4.4.2 below, future developments in ICE efficiency often focus on the synergies of the engine operating in a hybrid powertrain system. However, given that strong hybrids already enjoy a significant FE advantage over base ICE vehicles, manufacturers may not be compelled to add significant technology to their engines to achieve further efficiency improvements.

An overview of the current U.S. market can provide a look at the diversity of hybrid choices, in both the various technical approaches used by different manufacturers as well as their FE potential. In the small car segment alone, there are offerings from Hyundai (Ioniq), Toyota (Prius), and Honda (Insight) that all achieve greater than 50 miles per gallon (MPG) in label fuel economy and greater than 70 MPG in compliance fuel economy. Three distinct hybrid configurations are common: P2 integrated into a DCT, PS, and Honda 2-motor (Figure 4.6). Also important to consider is whether strong hybrids still provide a compelling fuel economy improvement when compared to an unhybridized vehicle that has taken advantage of many of the available FE technologies in that category. Chapter 3 highlighted several high performing MY 2020 vehicles that approach the original 2025 CO_2 standards (with off-cycle credits) by incorporating extensive FE technology. Two of those, the Toyota Camry and Ford Escape, also offer hybrid variants that provide a roughly 40% improvement in FE even above the highly performing conventional model (Table 4.5). With strong hybridization making its way into luxury and larger SUV and crossover vehicle classes, manufacturers may configure those powertrains to enhance performance as well as FE, which can provide a different marketing position for those products. Table 4.5 shows comparisons of hybrid and non-hybrid vehicles in the same carline.

The predominant hybrid architecture in the U.S. market to date has been the PS used by Toyota and Ford and built on volume success of the Prius. PS hybrids typically have higher efficiency and slightly lower cost than

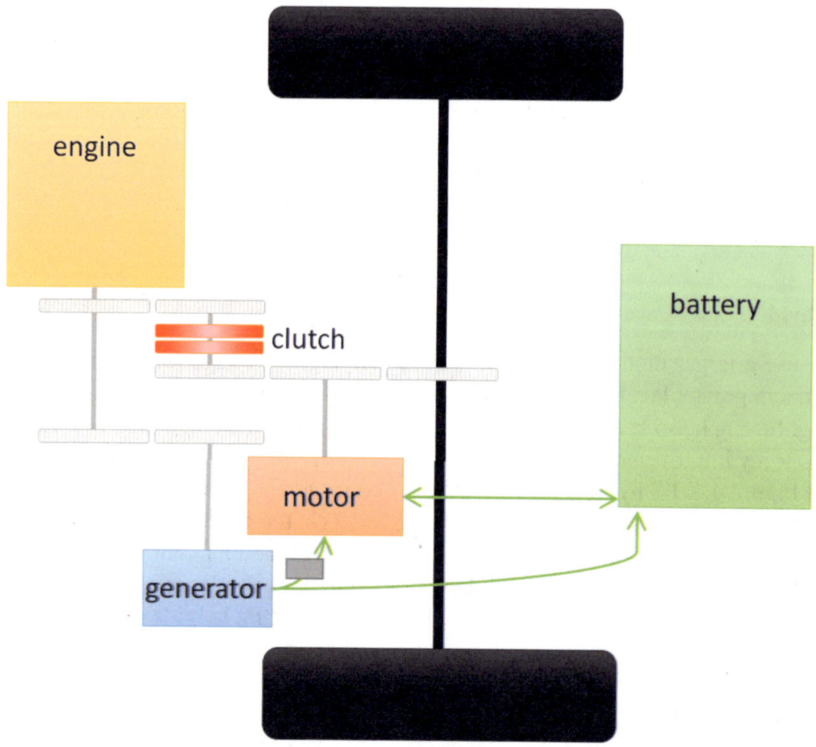

FIGURE 4.6 A diagram of a basic Honda 2-motor strong hybrid architecture, where green lines represent electrical connections and gray represents mechanical connections through gears. A small gray box between motor and generator indicates location of the inverter/converter.
SOURCE: Committee generated, adapted from Sherman (2013) and Wheel Network (2018).

P2 hybrids; however, the PS architecture has some known disadvantages in off-cycle driving modes, torque capacity, high power operation (e.g., towing), AWD compatibility, and reverse (Kapadia et al., 2017). P2 variants have historically experienced more success in Europe, due in part to the orientation of the transmission manufacturing infrastructure toward manual transmissions like DCTs, which can be easily integrated into the P2 architecture. However, in order to support the demands of larger vehicles and trucks, more P2 variants are coming to the U.S. market offering improved capabilities, for example AWD and RWD architecture useful for towing, and still showing significant FE improvements over conventional ICE vehicles. One example is the Ford Explorer 10-speed, which provides a RWD hybrid architecture that could also apply to trucks. Another is the Acura MDX, which combines a 3.0 L V6 engine with a 47 hp motor integrated into a 7-speed DCT and two 36 hp motors in the rear axle for AWD and torque vectoring. Offerings of series hybrids are more limited. Nissan's ePower hybrid architecture will be the first pure series HEV in the U.S. market and is being offered based on its success in the Japanese market. In the ePower hybrid configuration, all vehicle energy is produced by the engine acting as a generator, charging the battery, which in turn drives the electric motor. However, since all engine energy is exposed to double conversion before reaching the wheels, this hybrid configuration has an efficiency disadvantage compared to other architectures.

Alongside the proliferation of HEV offerings, automakers are offering configurations of many strong hybrid powertrains as plug-in hybrids with the addition of an onboard charger and an energy-type battery[1] offering varying levels of electric-only range. Depending on the all-electric range of the PHEV, the battery and motor may be sized more like an HEV (shorter electric range, fewer operating conditions) or more like a BEV (longer electric

[1]The distinction between energy-type and power-type batteries is described later in this report in Section 5.3.3.9.

TABLE 4.5 Select Light-Duty Passenger Vehicle Models with Hybrid and Non-Hybridized Options, Demonstrating Fuel Economy Improvements from Hybridization

Class	Year, Vehicle, Engine Details	Valve Info	Fuel Injection	Power (hp)	Curb Weight (lb)	Footprint (ft^2)	MY 2019 Sales Volume	FE (MPG)	% FE Improvement (Hybrid Versus Conventional)	% Change in Fuel Consumption (Hybrid Versus Conventional)
Crossover	2020 ESCAPE FWD 1.5 L TC	Intake/Exhaust, Hydraulic Actuated VCT	GDI	181	3298	46.01	Conventional: 241,387	41.166	42%	–30%
	2020 ESCAPE FWD 2.5 L NA HEV	Hydraulic Actuated VCT	MFI	200	3534		2020 is the first MY of the Escape Hybrid	58.344		
	2020 ESCAPE AWD 1.5 L TC	Intake/Exhaust, Hydraulic Actuated VCT	GDI	181	3474			39.431	45%	–31%
	2020 ESCAPE AWD 2.5 L NA HEV	Hydraulic Actuated VCT	MFI	200	3668			57.042		
Medium car	2020 CAMRY LE FWD 2.5 L NA	Intake and exhaust	GDI	203	3296	48.46	Conventional: 336,978	43.451	65%	–39%
	2020 CAMRY LE FWD 2.5 L NA HEV			208	3472		Hybrid: 26,043	71.781		
	2020 CAMRY XLE FWD 2.5 L NA			203	3391			42.669	47%	–32%
	2020 CAMRY XLE FWD 2.5 L NA HEV			208	3572			62.929		

NOTE: FWD = front-wheel drive; MFI = multipoint fuel injection; NA = naturally aspirated; TC = turbocharged. Switching to the hybrid version can also mean changes in valves, fuel injection, horsepower, weight, and footprint of the vehicle, all of which affect its final fuel economy.
SOURCE: Committee generated using information from EPA (2020d), Ford (2020), Toyota (2020a), Toyota (2020b), Toyota Newsroom (2020), and Tulumba (2020).

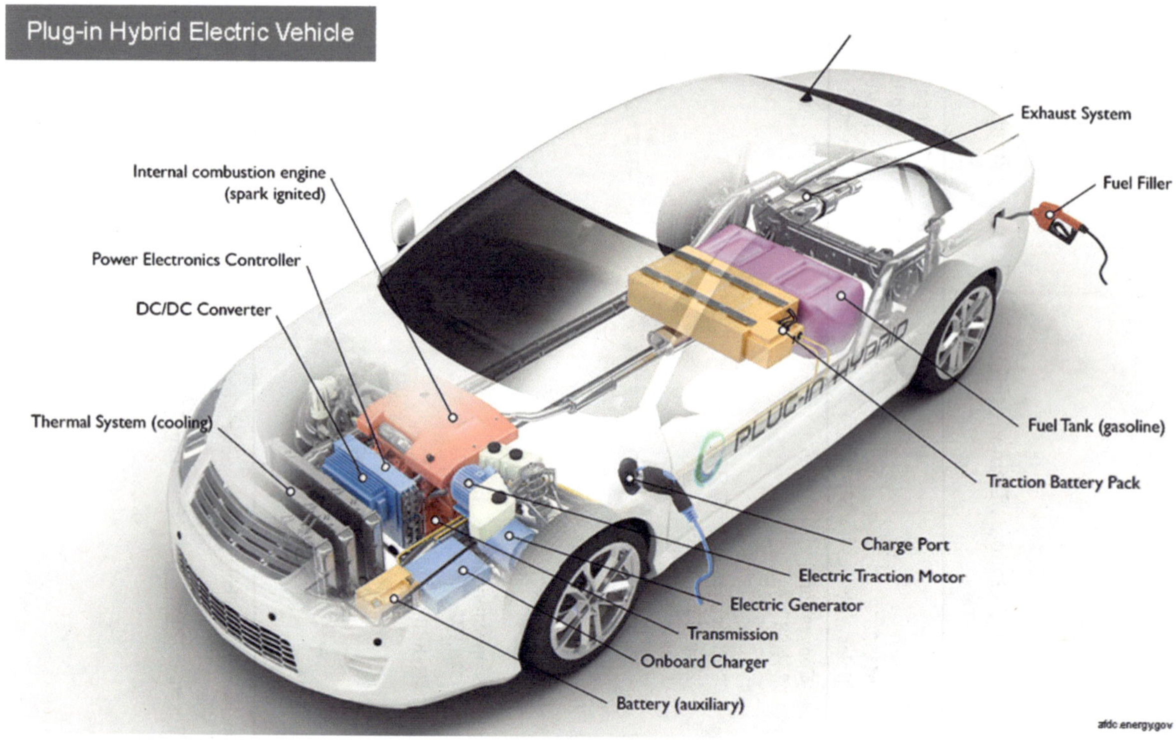

FIGURE 4.7 Diagram of a PHEV, showing key components for the combustion and electric aspects of the propulsion system. SOURCE: DOE Alternative Fuels Data Center (n.d.).

range, greater operating conditions). Figure 4.7 depicts a generic plug-in hybrid architecture. Manufacturers have differing views on the role of PHEVs in the future, which are influenced by the treatment of policy in the global markets they participate in.

4.5.2 Efficiency Opportunities for the ICE Implemented in Hybrid Architectures

This section addresses the potential to improve the efficiency of the ICE in the context of a hybrid powertrain system. Beyond the aforementioned regenerative braking, the available electric power can complement the ICE in a strong hybrid propulsion system in several ways. First, in addition to the alternator/generator function, accessories such as water pump and air conditioning can be electrified, thereby eliminating the need to power a belt-driven accessory drive using the engine. Electrifying these accessories also reduces the requirement of the ICE to provide high torque at low engine speeds where the engine is most prone to knocking, thus opening a wider range of CRs. Second, depending on the level of electric assistance available, modifications to the ICE, such as downsizing and/or improving efficiency while optimizing cost, can help meet efficiency and performance goals while offsetting the cost of the hybrid electric components. The hybrid propulsion system also can augment peak vehicle power needs reducing the engine's high-speed horsepower requirement. Additionally, lightly loaded operating areas of low engine efficiency with high throttling losses can be minimized through electric-only driving or load leveling (to charge battery under these modes). Figure 4.8 illustrates these opportunities with a modified engine map that indicates regions where the motor can replace or supplement the engine.

To date, automakers have offered the required engine performance of hybrids through the use of relatively low cost, naturally aspirated engines that achieve high efficiency by aggressively utilizing the Atkinson cycle with geometric CR up to 14:1. However, looking ahead, there are still significant opportunities to improve the overall powertrain system efficiency by further improving the ICE in a hybrid context. For example, turbocharging in

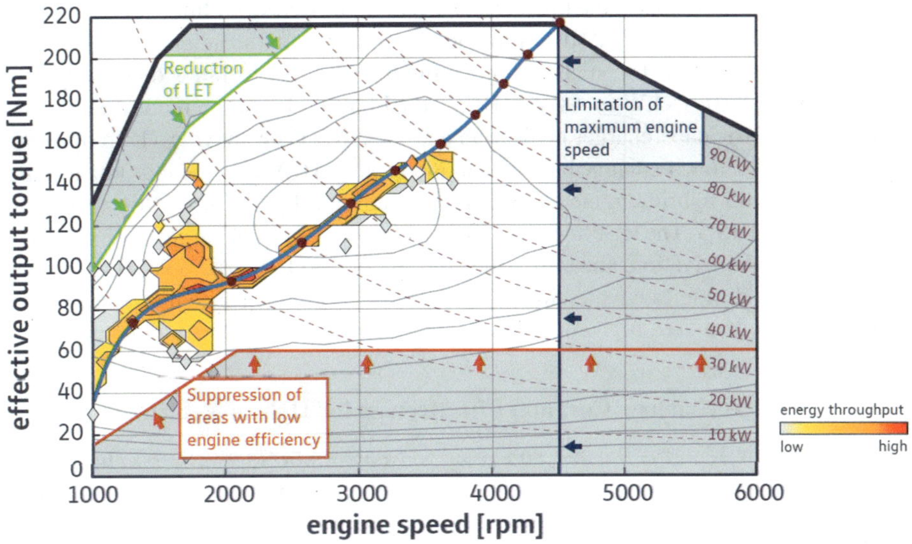

FIGURE 4.8 Example engine map of the VW 1.5 L TSI evo engine. The areas covered in gray boxes illustrate some areas of torque versus speed that no longer have to be provided by the engine alone. For example, motor-driven propulsion can replace low load operation at the lower engine torques, and the motor can supplement higher load operation at high speed and/or low speed high torque. Additionally, not pictured for this engine, in a series hybrid configuration, the motor provides all vehicle speed and torque, and therefore the engine's operation can be highly tuned to operate at its most efficient point, as a generator. SOURCE: Brannys (2019).

combination with hybridization can allow for engine downsizing and potentially enable the use of new 3-cylinder engines with displacements as low as 1.0 L. Significantly, however, the future boosted ICE in a hybrid is being approached differently than strictly enabling downsizing. As described in Section 4.1, the Miller cycle provides opportunities for significant improvements in thermal efficiency. In a hybrid with its limited engine speed/load topology requirement, the Miller cycle can be implemented at lower cost than for an unhybridized ICE, or the technologies can be tailored to achieve even lower BSFC given the more limited operating regime. The ultimate expression of the implementation of Miller cycle engines in HEVs is in a series hybrid, where the engine, with no mechanical coupling to the wheels, can operate strictly at its peak efficiency for any given power requirement. The PS hybrid architecture can approach this level with its CVT-like functionality. An engine operating in a series hybrid application that demonstrated up to 10% improvement in BSFC over a state-of-the-art downsized/boosted Miller cycle engine meeting all conventional application requirements was reported by VW (Brannys, 2019). The paper notes, however, that the technical enablers employed in this engine concept (15:1 CR, external cooled EGR to suppress knock, reoptimized VTG turbocharger based on EGR flow requirements, reduced intake cam lift, and a passive prechamber ignition required to ignite the dilute mixture, and more) can only work to this dramatic extent in a series hybrid context. Numerous studies by engine developers have described the ultimate potential of the ICE to occur as a so-called dedicated hybrid engine with brake thermal efficiency potential approaching 45%. However, there still is room for significant improvement through optimization of the total engine and hybrid systems. Some of technology possibilities include the following:

- Boosting systems (which can include electrically assisted devices) do not have the same low-end torque and peak power requirements. They can therefore be cost-optimized and/or designed for peak efficiency over a limited operating range. They can also be focused to support flow of EGR where applicable.
- Cooled low-pressure EGR systems can be optimized to improve fuel efficiency in the dominant speed/load regions (as opposed to being used for high load knock mitigation).
- Engine designs that do not have the same peak power requirements can be designed for lower speed (lower friction) with lower intake cam lifts and optimized combustion systems (high charge motion).

ICE improvements are also possible at the lower voltage end of the hybrid architecture spectrum. As described in the overview to Section 4.5, the 48 V MHEV provides its primary fuel efficiency improvement through its capacity to regenerate braking energy. It can also improve the functionality of idle stop-start systems. The added electrical launch assistance of 10–15 kW in 48 V MHEVs could allow for some modest engine downsizing. In some demonstrations, an electric supercharger made viable at 48 V has been shown to help overcome the vehicle launch/turbo lag concerns of a much smaller boosted engine in an MHEV application. For instance, Volvo has recently introduced a new mild hybrid engine architecture including a family of 4-cylinder gasoline and diesel engines that integrate a 48 V BISG. The gasoline 48 V BISG versions are turbocharged with external low pressure cooled EGR, utilize cylinder deactivation, and utilize the aforementioned E-charger in a performance variant. This is just one example of the technologies covered in the "conventional" engine space now starting to integrate with "electrification." Others include diesel, high compression, and downsized/boosted engine technologies.

4.5.3 Technology Cost and Effectiveness for Hybrid Vehicles

Tables 4.6 and 4.7 below report estimated and projected cost and effectiveness values for representative PS and P2 strong hybrids of example vehicle classes. Cost estimates are provided for key components in 2025, 2030, and 2035. The engine and transmission component and system changes are those employed in the studied examples to convert to the respective hybrid systems. Engine and transmission component costs were estimated for the hybrid relative to the ICE from NHTSA/EPA values in the Safer Affordable Fuel Efficient (SAFE) rule with reductions for learning of 1%/year. The motor, generator, and power electronics systems were scoped based on component power, torque, and weight to establish a 2020 baseline technology. Costs were estimated using baseline component costs estimates for 2020 from suppliers, automakers, industry consultants, and literature. Cost reductions for 2025–2035 are projected based on the introduction, scale up, and learning of new technologies like cerium rare earth magnets in motors and gallium nitride in power electronics and amount to about 20% for motor/generator and 35%–45% for power electronics. Motor, generator, and power electronic assumptions are detailed in the footnotes of Tables 4.6 and 4.7. Battery technology was scoped based on Li-ion power-type batteries and 2020 values for battery capacity. Battery costs were estimated based on the assumed battery chemistry and were reduced over time using a learning rate of 5%/year. Beyond the aforementioned motor, generator, power electronics, and battery, other necessary components for electrification/hybridization may include an electronic control unit upgrade; a high-voltage harness; regenerative brakes;[2] alternating current (AC) modifications; a direct current (DC)/DC converter; and battery monitoring, safety, and thermal management systems. These costs were estimated using data from a presentation to the committee on hybrid costs (Duleep, 2020), with cost reductions based on a learning rate of 1%/year. Hybrid costs have significant associated uncertainty, in part owing to differences in how different automakers implement their hybrid systems. While Tables 4.6 and 4.7 report single cost estimates for each year for an example conversion of an ICE vehicle to an HEV, in reality there will be a range of costs, given uncertainties in future technology penetration and differences in costs and technology implementation for various automakers. For example, using input from suppliers and automakers, as well as the committee's own expertise, the total cost to convert a conventional midsize car or CUV to a P2 or PS hybrid ranges from about $2,000–$3,000 in MY 2020. Notably, these costs are significantly lower than the costs of hybridization reported in the 2020 SAFE Rule, which are approximately $3,500–$6,700 for MY 2020 and $2,900–$5,700 in MY 2025 for small cars to medium-size SUVs depending on hybrid architecture (NHTSA/EPA, 2020).[3]

[2]The regenerative braking function is added to capture vehicle energy during braking and convert it to useful electric power. In hybrid vehicles, where an electric motor exists in some fashion to provide partial propulsion power, the motor can be controlled in a generation mode during braking to capture part of the vehicle energy during braking. Required equipment for regenerative braking includes added controls and sensors, and a vacuum pump for the existing hydraulic system.

[3]Approximate range from summing the costs in the SAFE rule for battery, transmission, and non-battery electrification components of P2 and PS hybrids and learning the costs from MY 2017 to MY 2020 or MY 2025. Battery costs for P2 not provided in SAFE Rule but assumed to be the same as PS battery costs for a given vehicle class. Reported range is small car P2 to medium SUV PS and includes both performance and non-performance vehicles.

TABLE 4.6 Projected Costs and Effectiveness of Representative PS Hybrid Technology Packages, 2025–2035

Example Vehicle	Components	Technology Cost (2018$)[a] MY 2025	MY 2030	MY 2035	Technology Effectiveness (Percent Change in Fuel Consumption)[b]
Medium Car Naturally Aspirated to PS Hybrid	Engine Modifications[c]	Total: $55	Total: $50	Total: $50	Medium Car Hybridization Effectiveness Average: –42% Range: –32% to –47%
	• Electric water pump	$55	$50	$50	
	• Increased compression ratio	No change	No change	No change	
	8-speed AT to eCVT[d]	–$435	–$410	–$390	
	Motor[e]	$320	$290	$260	
	Generator[e]	$140	$125	$115	
	Battery (1.0 kWh, Li-ion)[f]	$550	$425	$330	
	Inverter/power electronics[g]	$490	$440	$310	
	Battery Monitoring, Safety, and Thermal Management Systems[h]	$330	$315	$300	
	ECU Upgrade[h]	$45	$40	$40	
	High-Voltage Harness[h]	$130	$125	$120	
	Regenerative Brakes[h]	$170	$165	$155	
	AC Modifications[h]	$170	$165	$155	
	DC/DC converter (1.1 kWh)[i]	$90	$90	$90	
	Total	$2,055	$1,820	$1,535	
CUV Turbocharged, Downsized (TCDS) to PS Hybrid	Engine Modifications[c]	Total: –$340	Total: –$325	Total: –$305	CUV/SUV Hybridization Effectiveness Average: –34% Range: –26% to –39%
	• 3-cylinder to 4-cylinder	$95	$90	$85	
	• GDI to MPI	–$165	–$155	–$145	
	• TCDS deletion	–$295	–$280	–$265	
	• DCP to intake only VCT	–$30	–$30	–$30	
	• Electric water pump	$55	$50	$50	
	8-speed AT to eCVT[d]	–$435	–$410	–$390	
	Motor[e]	$375	$340	$305	
	Generator[e]	$100	$90	$80	
	Battery (1.1 kWh, Li-ion)[f]	$605	$470	$360	
	Inverter/PE[g]	$515	$470	$330	
	Battery Monitoring, Safety, and Thermal Management Systems[h]	$330	$315	$300	
	ECU Upgrade[h]	$45	$40	$40	
	High-Voltage Harness[h]	$130	$125	$120	
	Regenerative Brakes[h]	$170	$165	$155	
	AC Modifications[h]	$170	$165	$155	
	DC-DC converter (1.1 kWh)[i]	$90	$90	$90	
	Total	$1,755	$1,535	$1,240	

[a]Vehicle specifications and costs are general estimates for the vehicle class and do not represent specific vehicles; all costs rounded to the nearest $5.

[b]Effectiveness calculated using combined, unadjusted fuel economy values for MY 2020 vehicles from DOE/EPA Fuel Economy Guide Dataset (EPA, 2020d).

[c]Costs based on MY 2020 values reported in NRC (2015), updated to 2018$ and reduced 1%/year.

[d]Cost of eCVT (Duleep, 2020) with savings for removing 8-speed automatic transmission (Duleep, 2020; NRC, 2015). All costs in 2018$ and reduced 1%/year from base year (MY 2011 for eCVT and removing 6-speed AT; MY 2025 for going from 8-speed AT to 6-speed AT) through 2035.

[e]Assumes motor component cost reductions of 10% every 5 years, starting from MY 2020. Assumes introduction of cerium rare earth magnets in 2030.

[f]Assumes $550/kWh 2025 cost, with 5%/year cost reductions through 2035.

[g]Inverter cost assumptions in 2025: uses silicon carbide switches, 25% cost reduction from 2020; in 2030: uses gallium nitride on silicon devices, 10% cost reduction from 2025, reduction in device losses decreases cooling costs by 10%; in 2035: uses gallium nitride on silicon devices at high switching frequencies, reduces filtering and cooling needs by 75% compared to low frequency switching. Controller costs decrease by 25% from 2020 to 2025 and 2025 to 2030, based on normal electronic cost reductions.

[h]1%/year cost reduction from 2011 estimate obtained from Duleep (2020).

[i]Cost estimate from Duleep (2020); assumes $80/kW.

NOTE: AT = automatic transmission; DCP = dual cam phasing; ECU = engine control unit; MPI = multipoint injection.

TABLE 4.7 Projected Costs and Effectiveness of Representative P2 Hybrid Technology Packages, 2025–2035

		Technology Cost (2018$)[a]			Technology Effectiveness (Percent Change in Fuel Consumption)[b]
Example Vehicle	Components	MY 2025	MY 2030	MY 2035	
Medium Car Naturally Aspirated to P2 Hybrid	Engine Modifications[c]	Total: $55	Total: $50	Total: $50	Medium Car Hybridization Effectiveness Average: –39% Range: –33% to –44%
	• Balance shaft deletion	May be possible	May be possible	May be possible	
	• Electric water pump	$55	$50	$50	
	Transmission Modifications	Total: –$5	Total: –$5	Total: –$5	
	8-speed AT to 6-speed AT[d]	–$55	–$50	–$50	
	Electric transmission pump[e]	$50	$45	$45	
	Motor[f]	$240	$215	$195	
	Battery (1.5 kWh, Li-ion)[g]	$825	$640	$495	
	Inverter/power electronics[h]	$315	$280	$175	
	Battery Monitoring, Safety, and Thermal Management Systems[i]	$330	$315	$300	
	ECU Upgrade[i]	$45	$40	$40	
	High-Voltage Harness[i]	$130	$125	$120	
	Regenerative Brakes[i]	$170	$165	$155	
	AC Modifications[i]	$170	$165	$155	
	DC-DC converter (1.1 kWh)[j]	$90	$90	$90	
	Total	$2,365	$2,080	$1,770	
SUV Turbocharged, Downsized (TCDS) to P2 Hybrid	Engine Modifications[c]	Total: $370	Total: $350	Total: $335	SUV Hybridization Effectiveness Average: –10% Range: –6.5% to –14%
	• I4 to V6	$470	$445	$425	
	• Dual VVT on 2 more cylinders	$40	$40	$35	
	• GDI on 2 more cylinders	$100	$95	$90	
	• TCDS deletion	–$295	–$280	–$265	
	• Electric water pump	$55	$50	$50	
	Transmission Modifications	Total: $50	Total: $45	Total: $45	
	10-speed AT	No change	No change	No change	
	Electric transmission pump[e]	$50	$45	$45	
	Motor[f]	$485	$435	$390	
	Battery (1.5 kWh, Li-ion)[g]	$825	$640	$495	
	Inverter/PE[h]	$300	$265	$165	
	Battery Monitoring, Safety, and Thermal Management Systems[i]	$330	$315	$300	
	ECU Upgrade[i]	$45	$40	$40	
	High-Voltage Harness[i]	$130	$125	$120	
	Regenerative Brakes[i]	$170	$165	$155	
	AC Modifications[i]	$170	$165	$155	
	DC-DC converter (1.1 kWh)[j]	$90	$90	$90	
	Total	$2-965	$2,635	$2,290	

[a]Vehicle specifications and costs are general estimates for the vehicle class and do not represent specific vehicles; all costs rounded to the nearest $5.

[b]Effectiveness calculated using combined, unadjusted fuel economy values for MY 2020 vehicles from DOE/EPA Fuel Economy Guide Dataset (EPA, 2020d).

[c]Costs based on MY 2020 values reported in NRC (2015), updated to 2018$ and reduced 1%/year.

[d]MY 2020 costs from NRC (2015), updated to 2018$ and reduced by 1%/year.

[e]Estimated at $50 in MY 2025 and reduced 1%/year through 2035.

[f]Motor component cost reductions of 10% every 5 years, starting from MY 2020. Assumes introduction of cerium rare earth magnets in 2030.

[g]Assumes $550/kWh 2025 cost, with 5%/year cost reductions through 2035.

[h]Inverter cost assumptions in 2025: uses silicon carbide switches, 25% cost reduction from 2020; in 2030: uses gallium nitride on silicon devices, 10% cost reduction from 2025, reduction in device losses decreases cooling costs by 10%; in 2035: uses gallium nitride on silicon devices at high switching frequencies, reduces filtering and cooling needs by 75% compared to low frequency switching. Controller costs decrease by 25% from 2020 to 2025 and 2025 to 2030, based on normal electronic cost reductions.

[i]1%/year cost reduction from 2011 estimate obtained from Duleep (2020).

[j]Cost estimate from Duleep (2020); assumes $80/kW.

Fuel economy improvements provided by strong hybrids appear to be similar within a given vehicle class and hybrid architecture. The effectiveness values reported in Tables 4.6 and 4.7 are the range and average fuel consumption reduction for 2020 vehicles of the given class where both hybrid and conventional models are produced. In 2025–2035, the effectiveness of PS and P2 hybrids with respect to their ICE counterpart will likely remain constant, or could improve slightly if engine efficiency improvements are increasingly targeted to engines in hybrid systems, as discussed above.

FINDING 4.5: Strong hybrids represent the maximum fuel efficiency possible in vehicles powered only by gasoline, both in the current fleet and into the future. This is owing to their ability to maximize braking energy recuperation and to augment the operation of the engine. Strong hybrid offerings in the U.S. market span all vehicle classes and average 35%–40% reduction in fuel consumption in midsize and crossover vehicles compared to similar conventional powertrain vehicles. More cost information is needed, but the committee estimates that midsize and crossover strong hybrids have an incremental cost of around $2,000–$3,000 above a conventional vehicle in 2020, with expected decreases in the cost increments in 2025–2035. Many of the future developments of the internal combustion engine itself are focusing on the added efficiency potential of the engine when integrated into a hybrid system.

FINDING 4.6: Mild hybrids, defined here as 48 V, represent a viable pathway to realize some of the same CO_2 reduction benefits of hybridization described for strong hybrids, but at a lower overall cost. Most current production mild hybrid vehicles have employed a belt driven machine in the P0 location, as it is the lowest cost and most straightforward to implement, but has some inherent efficiency limitations. Other mild hybrid architectures such as 48 V P2 can achieve greater efficiency gains but at a higher implementation cost than the belt driven P0 architecture. Analytical studies have shown the maximum efficiency potential of mild hybrid vehicles from multiple motor architectures such as P0+P3 or P0+P4, but more cost data are needed.

RECOMMENDATION 4.1: The National Highway Traffic Safety Administration should update its teardown and full system simulation for mild hybrid, and both PS and P2 strong hybrid vehicles, which are now and will continue to be the highest efficiency vehicles containing an internal combustion engine.

4.6 ADVANCED COMBUSTION TECHNOLOGIES

For many years, the industry has endeavored to bridge the thermal efficiency gap between spark ignition and compression ignition engines that results from the diesel engine's higher CR and higher ratio of specific heats (dilute mixtures from excess air and/or EGR). Spark ignition engines are constrained to operate at a stoichiometric air to fuel ratio to accommodate the functional requirements of the three-way catalyst necessary to meet emissions standards. Recirculated exhaust gas has therefore been commonly used in this manner, but it is not as effective as air owing to its lower specific heat ratio. Homogeneous charge compression ignition (HCCI) has long been recognized as the ultimate solution to this challenge. In this combustion concept, a premixed (homogeneous) charge of fuel and air is compressed until it auto-ignites throughout the chamber volume rather than through a traveling flame front. The theoretical advantages of the HCCI concept with respect to improved efficiency are as follows:

- Higher CR than conventional spark-ignition engines (to promote auto-ignition).
- Lean or dilute operation with air providing a higher specific heat ratio for improved thermal efficiency.
- Low temperature combustion and rapid heat release that reduce heat losses. The low peak combustion temperatures also dramatically reduce the production of NO_x feedgas emissions, mitigating dependence on the three-way catalyst.
- Lower pumping losses owing to dilute operation relative to a throttled spark-ignition engine.

Finding 2.7 of the 2015 NRC report stated that while lean HCCI had the potential to improve fuel consumption by up to 5%, many challenges remained before it could be implemented. Highlighted were issues associated with limited engine load range for operation (making it somewhat incompatible with the trend toward downsizing) and difficulty controlling mode switching. The finding also cited a DOE-funded project indicating that the constraint of super ultra-low emissions vehicle emissions would eliminate the fuel consumption benefits of HCCI. The finding concluded that HCCI technology would not likely have an impact by 2025 but with further development may contribute by 2030, or after the full benefits of downsized/turbocharged engines had been realized.

In the development of this report, manufacturers were asked again about the potential of lean operation and, if anything, were more skeptical than in 2015. This skepticism can largely be attributed to the even more stringent criteria emission constraints and focus on real-world emissions looking ahead, as well as the added cost associated with enabling hardware, lean NO_x aftertreatment, and so on. Nonetheless, there are ongoing developments to maximize the efficiency of the ICE, particularly in the context of downsized, turbocharged engines. Much of that work seems to focus on Lambda=1 operation using Miller cycle and high levels of EGR as the diluent. Some of these efforts include the development of advanced ignition systems and/or the utilization of prechamber combustion (both active and passive) to ignite the dilute mixtures. There is also significant work focused on developing engines in hybrid applications (including series hybrid) where the engine's function and operating requirements can be more oriented toward efficiency.

One exception to this is Mazda's recent introduction of their SkyActiv-X technology applied to a 2.0 L engine in Japan and Europe. This technology represents the first production application of gasoline compression ignition, in this case using assistance from a spark event, dubbed spark-controlled compression ignition (SPCCI). Over the years, multiple concepts have been developed as compromise positions between spark ignition, compression ignition, and pure HCCI as described above, with SPCCI being one of them. In the Mazda 2.0 L SkyActiv-X, three distinct combustion modes are utilized: lean air to fuel ratio SPCCI at light to moderate loads, lean exhaust gas to fuel ratio using cooled EGR at higher loads, and conventional spark ignition at full load. The technology is offered only in combination with Mazda's mild hybrid system. It employs an extremely high CR (16.3:1) to optimize compression ignition, along with swirl control valves, high pressure fuel injection, in-cylinder pressure transducers (to provide combustion feedback control), and a supercharger (to supply needed excess air) in support of the lean SPCCI system. An external cooled EGR system is also included for the lean ratio of EGR gas plus air to fuel combustion mode region. While the SkyActiv-X technology was originally announced for U.S. introduction, recent reports indicate a deferral, based on Mazda's assessment that the U.S. customer is currently not willing to trade off performance or pay a premium for fuel economy technology.

FINDING 4.7: It is reasonable to assume that if compelled by regulation and/or competition, manufacturers will continue to deploy more of the fuel-efficient technologies utilized in 2020 high fuel economy internal combustion engine vehicle models. However, these technologies are not equally applicable or affordable across all vehicle segments. For example, small cars may be more cost constrained or some vehicles will place higher customer priority on other attributes such as performance and towing. In addition, while there is certainly further internal combustion engine improvement potential, the cumulative gains can become costly in terms of benefit to cost and challenging in terms of emission constraints or other attribute trade-offs. Therefore, some manufacturers may choose to transition to alternative pathways such as electrification.

4.7 REFERENCES

Brannys, S. 2019. "Maximum Efficiency Concept of a 1.5l TSI evo for Future Hybrid Powertrains." Volkswagen AG, New Gasoline Engines II, Aachen Colloquium.

Conway, G., D. Robertson, C. Chadwell, J. McDonald, J. Kargul, D. Barba, and M. Stuhldreher. 2018. "Evaluation of Emerging Technologies on a 1.6 L Turbocharged GDI Engine." SAE Technical Paper 2018-01-1423. https://doi.org/10.4271/2018-01-1423.

DOE (U.S. Department of Energy) Alternative Fuels Data Center. n.d. "Alternative Fuels Data Center: How Do Plug-In Hybrid Electric Cars Work?" https://afdc.energy.gov/vehicles/how-do-plug-in-hybrid-electric-cars-work.

Duleep, G. 2020. "Cost and Benefit of Power-Split Hybrid Designs." Presentation to the Committee on Assessment of Technologies for Improving Fuel Economy of Light-Duty Vehicles—Phase 3. October 30.

EPA (U.S. Environmental Protection Agency). 2015. "VW Notice of Violation, Clean Air Act," September 18. https://www.epa.gov/sites/production/files/2015-10/documents/vw-nov-caa-09-18-15.pdf.

EPA. 2020a. *2019 EPA Automotive Trends Report*. EPA-420-R-20-006. March. Washington, DC. https://www.epa.gov/sites/production/files/2020-03/documents/420r20006.pdf.

EPA. 2020b. *2018 Toyota 2.5L A25A-FKS Engine Tier 3 Fuel—ALPHA Map Package*. Version 2020-07. Ann Arbor, MI: US EPA, National Vehicle and Fuel Emissions Laboratory, National Center for Advanced Technology.

EPA. 2020c. "Tampered Diesel Pickup Trucks: A Review of Aggregated Evidence from EPA Civil Enforcement Investigations." November 20. https://www.epa.gov/sites/production/files/2021-01/documents/epaaedletterreportontampereddieselpickups.pdf.

EPA. 2020d. *2020 Fuel Economy Guide—2020 Datafile*. August 21. https://www.fueleconomy.gov/feg/download.shtml.

EPA. 2020e. 2016 Honda 1.5L L15B7 Engine Tier 3 Fuel—ALPHA map Package. Version 2018-05. Ann Arbor MI: US EPA National Vehicle and Fuel Emissions Laboratory, National Center for Advanced Technology.

EPA/NHTSA (U.S. Environmental Protection Agency and National Highway Traffic Safety Administration). 2012. *Joint Technical Support Document: Final Rulemaking for the 2017–2025 Light-Duty Vehicle Greenhouse Gas Emission Standards and Corporate Average Fuel Economy Standards*. EPA-420-R-12-901. http://www.epa.gov/otaq/climate/documents/420r12901.pdf.

Ford. 2020. "2020 Escape." Ford. https://www.ford.com/suvs-crossovers/escape/models/escape-se/.

Green Car Congress. 2017. "VW Offering New Variant of Millerized 1.5 TSI on Golf and Golf Variant with Coasting Function." August 11. https://www.greencarcongress.com/2017/08/20170811-vw.html.

Heywood, J.B. 1988. *Internal Combustion Engine Fundamentals*. McGraw-Hill Series in Mechanical Engineering. New York: McGraw-Hill.

Kapadia, J., D. Kok, M. Jennings, M. Kuang, B. Masterson, R. Isaacs, and A. Dona. 2017. Powersplit or parallel—selecting the right hybrid architecture. *SAE International Journal of Alternative Powertrains* 6(1):68–76. https://doi.org/10.4271/2017-01-1154.

Lee, S., J. Cherry, M. Safoutin, A. Neam, J. McDonald, and K. Newman. 2018. "Modeling and Controls Development of 48 V Mild Hybrid Electric Vehicles." SAE Technical Paper 2018-01-0413. doi:10.4271/2018-01-0413.

Lisle, G. 2019. "'International Engine of the Year' Awards: Audi's 2.0 TFSI Engine Wins in Its Class." Audi MediaCenter. May 22. https://www.audi-mediacenter.com/en/press-releases/international-engine-of-the-year-awards-audis-20-tfsi-engine-wins-in-its-class-11682.

NHTSA/EPA (National Highway Traffic Safety Administration and U.S. Environmental Protection Agency). 2020. *Final Regulatory Impact Analysis: The Safer Affordable Fuel-Efficient (SAFE) Vehicles Rule for 2021–2026 Passenger Cars and Light Trucks*. March. https://www.nhtsa.gov/sites/nhtsa.dot.gov/files/documents/final_safe_fria_web_version_200330.pdf.

NRC (National Research Council). 2015. *Cost, Effectiveness, and Deployment of Fuel Economy Technologies for Light-Duty Vehicles*. Washington, DC: The National Academies Press.

Phillips, J. 2018. "Hybrid Vehicles: The WORST of Both Worlds?" ElectronicDesign. June 12. https://www.electronicdesign.com/markets/automotive/article/21806623/hybrid-vehicles-the-worst-of-both-worlds.

Sessions, R. 2018. "Nissan's New VCR Engine: Worth the Effort?" *SAE Automotive Engineering*. https://www.sae.org/news/2018/05/2019-infiniti-qx-50-w-vc-turbo-review.

Sherman, D. 2013. "Explaining the Honda Accord's Shrewdly Designed New Hybrid System." Car and Driver. October 1. https://www.caranddriver.com/features/a15114066/explaining-the-honda-accords-shrewdly-designed-new-hybrid-system-tech-dept.

Stuhldreher, M., J. Kargul, D. Barba, J. McDonald, S. Bohac, P. Dekraker, and A. Moskalik. 2018. Benchmarking a 2016 Honda Civic 1.5-Liter L15B7 turbocharged engine and evaluating the future efficiency potential of turbocharged engines. *SAE International Journal of Engines* 11(6):1273–1305. https://doi.org/10.4271/2018-01-0319.

Toyota. 2020a. "2020 Camry Full Specs." Toyota. https://www.toyota.com/camry/2020/features/mpg/2532/2540/2514.

Toyota. 2020b. "2020 Camry Hybrid Full Specs." Toyota. https://www.toyota.com/camry/2020/features/mpg/2559/2561/2560.

Toyota Newsroom. 2020. "Toyota Motor North America Reports December 2019, Year-End Sales." January 3. https://pressroom.toyota.com/toyota-motor-north-america-reports-december-2019-year-end-sales/.

Tulumba, C. 2020. "2019 US Vehicle Sales Figures by Model." GoodCarBadCar. https://www.goodcarbadcar.net/2019-us-vehicle-sales-figures-by-model/.

Wheel Network. 2018. *2019 Honda Insight Two-Motor Hybrid Explained*. Youtube Video. June 27. https://www.youtube.com/watch?v=IeV9EHmxOM8.

5

Battery Electric Vehicles

5.1 INTRODUCTION

Motivated by global environmental pressures calling for reduced tailpipe emissions and reduced dependence on petroleum as a source of energy for ground transportation, most automakers have been working for decades on the development of electrified powertrain systems with zero or ultra-low tailpipe emissions. Recent advances in electric drive technologies and battery technologies have made it possible for vehicle manufacturers to commercially deploy battery electric vehicles (BEVs). Globally, electric vehicle (EV) growth surpassed 7 million sales from 2010 through 2019, which was about twice as fast as initial hybrid vehicle growth from 2000 through 2009 (Cui et al., 2020). Currently, market penetration of plug-in electric vehicles (PEVs) has been limited to about 2% in the United States through 2019. To make further progress and move from early adopters to mainstream consumers, EVs will need to overcome the barriers of limited model availability, relatively high cost compared to conventional vehicles, relative convenience of charging versus gasoline refueling, and consumer awareness.

The assumption throughout the chapter is that vehicle electrification improves fuel economy (e.g., in hybrid electric vehicles [HEVs] and plug-in hybrid electric vehicles [PHEVs]), and/or eliminates the use of petroleum-based fuels (e.g., BEVs). If full fuel cycle emissions per mile are considered, the assumptions are more complex and depend upon the upstream emissions of the charging electricity source. When and where electricity is generated with low carbon sources, emissions per mile are significantly reduced relative to an internal combustion engine vehicle (ICEV). However, when and where electric systems depend upon high emitting generation facilities, the emission benefits are reduced. In 2025–2035, the committee anticipates that the U.S. grid will continue to work toward net-zero emissions, which will drive a decrease in total emissions for electrified vehicles. Life cycle emissions from EVs are summarized in Box 5.1, with additional charging and fuel aspects discussed in Section 5.4 and in Chapter 10.

At the core of all electrified powertrains is the electric drive consisting of an electric motor, an inverter, and an electronic controller and, of course, the battery. The electric drive is also critical in HEVs, discussed in Chapter 4, and fuel cell vehicles, which are the subject of Chapter 6. A key objective of this chapter is to explore technologies impacting the size, weight, efficiency, and cost of the electric propulsion system components for 2025–2035. While battery technology is still advancing on multiple fronts to enhance performance and reduce cost (from battery chemistry, to packaging and manufacturing), electric drive technology is relatively mature and has been greatly optimized over the years to achieve the current impressive performance (power and torque densities and efficiency). There are, however, several opportunities in both the motor and power electronics areas that appear

BOX 5.1
Overall Battery Electric Vehicle Emissions

The emissions implications of the shift to BEVs include upstream fuel and vehicle manufacturing processes. EVs have similar vehicle assembly-related emissions, except that they differ by manufacturing batteries and electric powertrains in place of the engine, transmission, and exhaust systems. Instead of the upstream extraction, refining, and distribution of petroleum-based fuel, EVs have electricity-related emissions from the primary energy extraction and use of electric power. See Chapter 10 for more information about the generation and use of electricity as a low-carbon vehicle fuel.

Figure 5.1.1 shows average U.S. life cycle carbon dioxide (CO_2) emissions for conventional, hybrid, and electric cars and crossovers in 2018. The figure includes the average U.S. light-duty conventional vehicle (29 miles per gallon [MPG] car, 26 MPG crossover), a typical efficient hybrid (52 MPG car, 40 MPG crossover), and a representative electric vehicle (0.28 kilowatt hour per mile [kWh/mile] car, 0.33 kWh/mile crossover). In all cases, average upstream fuel-level and vehicle-level emissions are assumed. The average EV emissions assume a 70 kWh capacity battery pack with manufacturing emissions of 75 kilograms CO_2 per kWh, and average U.S. 2018 electricity emissions of 449 grams CO_2 per kWh for electric vehicle charging. The result of these average U.S. assumptions is that EVs have approximately 54% lower CO_2 than average U.S. vehicles, and 26% to 31% lower than hybrids, within the same vehicle class. EVs on a California grid, reflective of decarbonization trends, have 70% lower CO_2 emissions than the U.S. average.

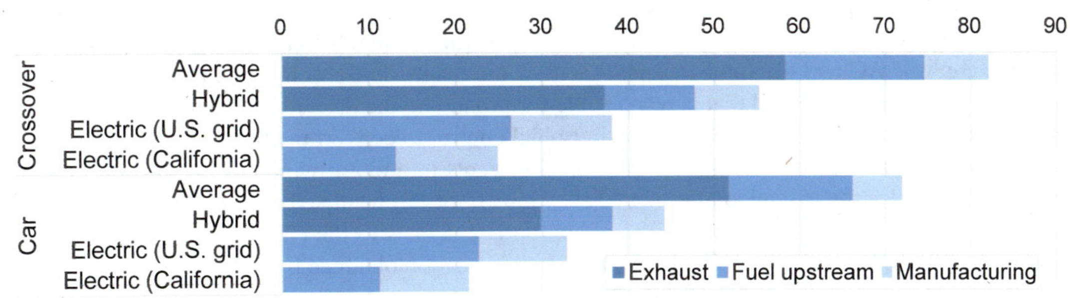

FIGURE 5.1.1 Lifetime life cycle emissions of CO_2 for of conventional and electric vehicles in 2018, assuming 14 year vehicle lifetimes driving 12,000 miles per year, reported in metric tons.

There are many complexities with such analyses. For example, the average grid has experienced declining CO_2 emissions at about 3%/year over 2005–2018, so actual per-mile EV CO_2 emissions decline as vehicles age. Most U.S. EVs are in lower-carbon northeast and west coast electric regions, also making emissions lower. EVs can be charged where there is grid capacity (e.g., excess capacity overnight can bring higher fossil emissions, or excess daytime solar can result in lower emissions) depending on electric utility energy sources and customer programs. In some regions, EVs are powered more by fossil sources on the grid, resulting in higher emissions; however, those situations are decreasing (especially coal generation) as the electric power system evolves. Yet, including average upstream energy sources for vehicle-level and fuel-level effects, BEVs generally deliver average carbon emission benefits over the most efficient combustion vehicles (although certain factors can limit emission benefits; see Holland et al., 2016 and Yuksel et al., 2016). The committee expects the trend of decreasing emissions from BEVs to continue.

promising for reducing the electric drive cost and weight and further enhancing drive efficiency, which would ultimately translate into increased electric range and energy savings. Section 5.2 reviews the state of the art in electric drive technologies and explores the potential impact of new opportunities.

The cost of battery technology will be a key determinant for BEVs to reach cost parity with combustion vehicles within the next decade. Section 5.3 explores the myriad options for automotive battery materials and cell packing, and assesses their relative cost, efficiency, and in the case of beyond-lithium technologies, possible deployment timelines. The section also describes battery management systems (BMS), thermal effects on battery lifetime, and safety principles. Battery performance, life cycle, and real-world battery usage are also described. Approaches to overcome current limitations, improve performance, improve customer acceptance, and reduce cost are discussed within the battery section as well. After summarizing cost reduction opportunities in each technology section, overall vehicle cost estimates that are expected to be realized in 2025–2035 are provided.

5.2 THE ELECTRIC DRIVE

Several electric drive technologies, including brush and brushless direct current (DC) and alternating current (AC) motors, have been investigated over the years for vehicle propulsion. However, thanks to its high efficiency and power density (attributes critical for achieving desirable range in electrified vehicles), the propulsion drive of choice used by most major automakers has been the brushless permanent magnet synchronous motor (PMSM) with rare-earth (RE) magnets (NdFeB) (Figure 5.1). General Motors (GM), Ford, Toyota, Nissan, Tesla, and Honda have used such motors for almost all electrified vehicles produced today.

The PMSM consists of a stationary part (stator) fitted with 3-phase copper windings placed in its slotted structure and a rotating member (rotor) fitted with permanent magnets assembled around its peripheral. The stator windings carry 3-phase AC and the rotor magnets produce the magnetic field. It is the interaction between the stator currents and the magnetic field that is responsible for producing the desired propulsion torque.

Most automakers use a 3-phase inverter with sinusoidal control to convert the battery's DC voltage to alternating 3-phase voltage, and then drive 3-phase sinusoidal currents into motor windings, as shown in Figure 5.2. The inverter uses six electronic semiconductor switches mostly of the insulated-gate bipolar transistor (IGBT) type. The role of the electronic controller is to send appropriate signals to the electronic switches to switch the currents on and off at the appropriate timing in response to information obtained by current sensors. This controls the current level and shape (sinusoidal) to the demanded level.

Permanent magnets come in various magnetic strength levels (measured by their maximum energy product) based on their material composition, as shown in Figure 5.3. NdFeB, an alloy of neodymium, iron, and boron, is the strongest and most widely used RE magnet. Strong magnets produce higher magnetic field, hence requiring

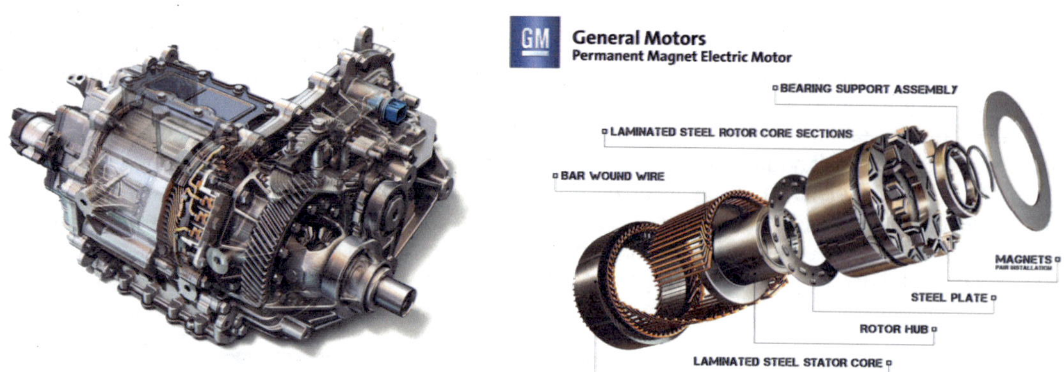

(a) Motor and gears assembled

(b) Motor parts

FIGURE 5.1 Many automakers' motor of choice, brushless PMSM.
SOURCE: (a) Chevrolet Pressroom (2016); (b) Chevrolet Pressroom (2011).

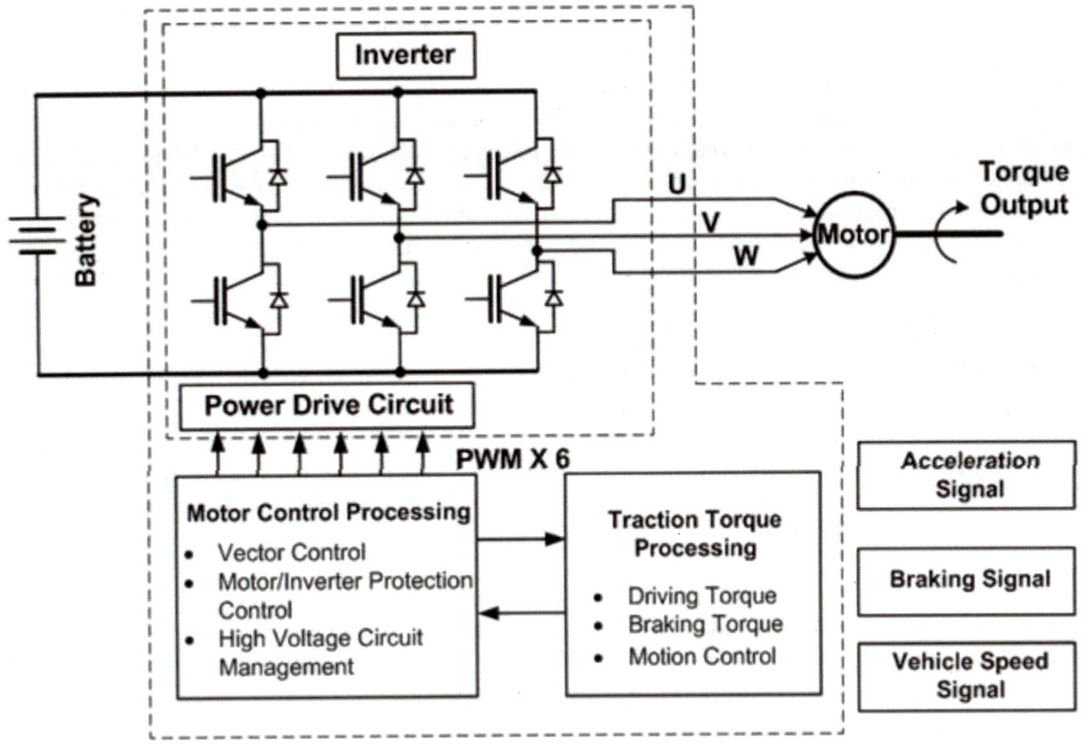

FIGURE 5.2 Brushless PMSM—power and control electronics.
SOURCE: Rajashekara (2013).

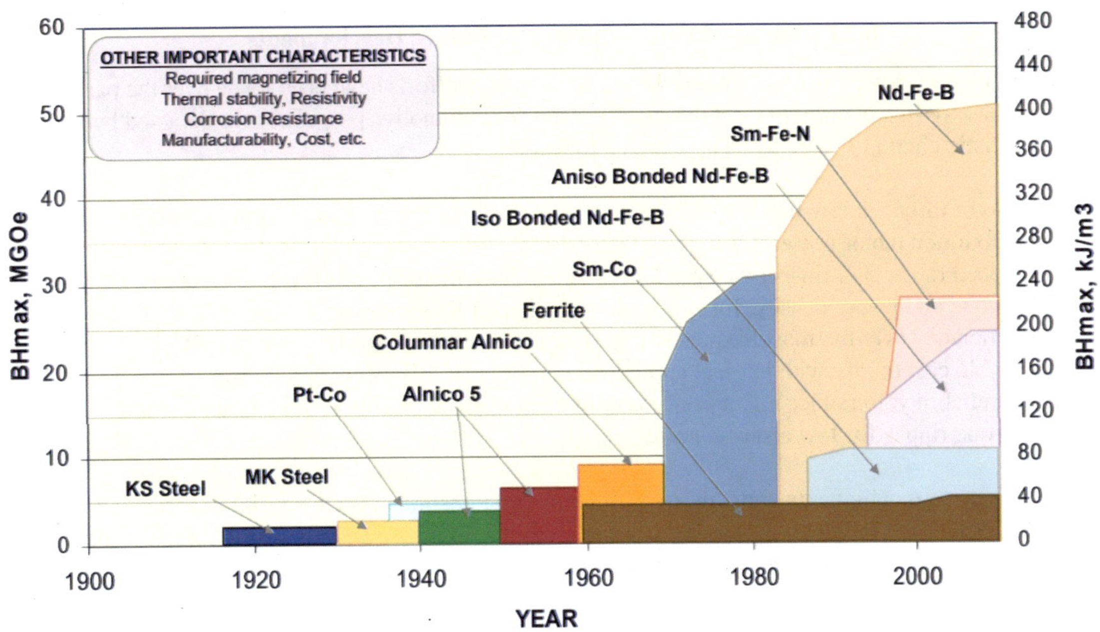

FIGURE 5.3 History of improvements in magnet strength.
SOURCE: Constantinides (2011).

less motor current for a given torque. This results in less ohmic loss in the motor, and thus higher drive efficiency and power density.

However, the above analysis and situation has been disrupted by the unstable cost of RE magnets owing to Chinese dominance of neodymium magnet production based on their control of much of the world's sources of RE mines (Vekasi, 2019; China Power Team, 2020). The prices jumped from $80/kg in 2010 to $460/kg in 2011. Owing to deep concerns about long-term availability of these materials, many users started to look for alternatives. There has been an intensive research effort focusing on developing technologies aimed at the reduction or elimination of RE magnets in motors (Ames National Laboratory, 2012). The effort to eliminate RE magnets explored two possibilities (Buress, 2016):

- Maintaining the brushless PMSM motor type but developing advanced high-energy non-RE magnets (AlNiCo [Ley, 2016], ferrite, or dysprosium-free RE magnets) to replace the NdFeB magnets having the super-expensive dysprosium content currently used; and
- Reconsidering other non-permanent magnet motor types (e.g., induction motors, switched reluctance motors, synchronous reluctance motors, and wound field excited motors) but incorporating innovative structures/assemblies, and effective thermal and noise management techniques as lower cost alternatives (El-Refaie, 2016; Ludois, 2015; Omekanda, 2013).

It should be understood, however, that moves to replace RE magnets with lower-energy magnets would lead to lower motor efficiency and/or increase motor size and weight, which would have a negative impact on energy consumption and range depending on the approach taken. In the meantime, the price of NdFeB-RE magnets has come back down to reasonable levels of $50–$60/kg in 2020, encouraging automakers to stay the course of brushless motors equipped with RE magnets. The new developments resulting from the above-mentioned research could be revisited and pursued for commercial implementation if RE magnet supply channels are disturbed again (Sekulich, 2020).

The following sections provide a summary of the current status of motors and power electronics, as well as research efforts and ongoing innovation in the field that could have an impact on electrified vehicle energy consumption and electric-only range.

5.2.1 Motors—Current Status and Future Developments

Electric motor technology is a mature one; however, intensive efforts have been made over the past decade to optimize the motor design in order to meet the specific needs of automotive propulsion, as depicted by the torque-speed characteristic chart in Figure 5.4. These are as follows:

- High motor torque at low motor speeds, for adequate vehicle acceleration and hill climbing;
- High maximum motor power, for high-speed cruising;
- Wide speed range, 3–4 times base speed, at the maximum motor power level, for cruising performance;
- High torque and power density, for low motor weight and longer range;
- High efficiency over the most frequently used range of operation, for longer E-range;
- Reasonable cost (parity with internal combustion engines [ICE]), for affordability;
- Higher reliability, to reduce maintenance cost in view of the exposure to road G-forces; and
- Low torque ripple, for low acoustic noise.

To achieve these requirements, most recent production systems incorporate special materials and advanced manufacturing techniques into their motor designs. For example, to minimize iron loss (hysteresis and eddy current loss) at high speeds (high frequencies) and high torque (high flux density) motors use special thin (0.25 mm) electrical steel laminations featuring high flux carrying capability and low loss-factor to improve efficiency (Thanh and Min-Fu, 2017). Also, to achieve the high torque without excessive stator heating, flat wire conductors are being used instead of the traditional round wires for the stator windings. These provide for a higher stator slot fill and thereby minimize ohmic loss and maximize efficiency. Furthermore, a new hairpin winding manufacturing

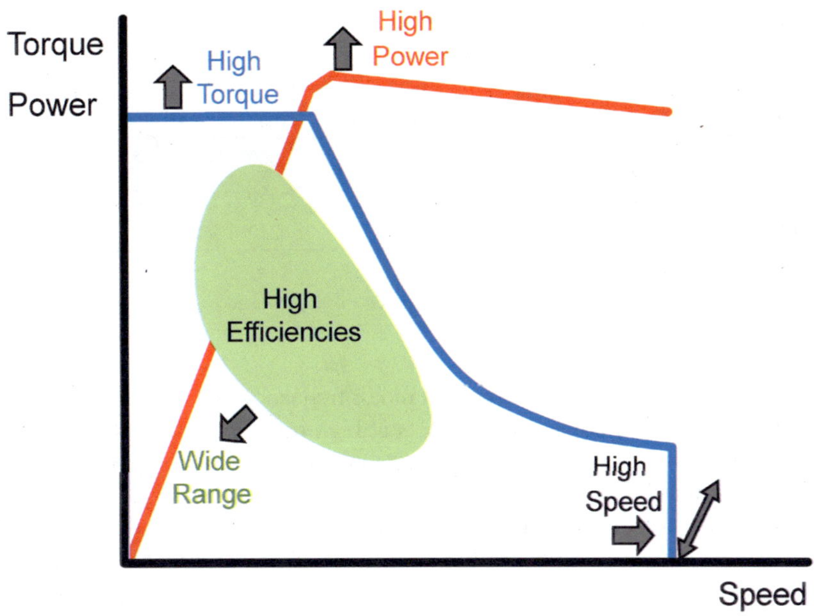

FIGURE 5.4 Visual depiction of required motor torque-speed and power-speed characteristics for automotive propulsion applications (see bulleted list above).

technique is used to form the wire in a way that minimizes the size of the windings overhang and connections and further minimizes the motor size, weight, and material cost and maximizes efficiency. See Figure 5.5 for a comparison between the traditional winding and the hairpin approaches.

Also, to achieve a desirable high motor speed (to minimize motor size, which is roughly inversely proportional to motor speed) while maintaining a low total motor cost, a single-stage gearbox with a gear ratio of about 7:1 to 10:1 is being used by most automakers. Increasing motor speed would result in unacceptable levels of the gear

(a) Traditional stranded winding

(b) Hairpin winding

FIGURE 5.5 Stator winding.
SOURCE: (a) iStock; (b) © 2020 IEEE. Reprinted, with permission, from Liu et al., 2020, Experimental investigation on oil spray cooling with hairpin windings, *IEEE Transactions on Industrial Electronics* 67(9):7343–7353, doi: 10.1109/TIE.2019.2942563.

TABLE 5.1 Propulsion Motor Performance Status Summary—Motor Only

Application	Power (kW)	Gear Ratio	Motor Only Weight[a] (kg)	Power Density[a] (kW/kg)	Max Motor Torque (Nm)	Torque Density[a] (Nm/kg)	Motor Only Cost[a] ($)	Specific Cost[a] ($/kW)
GM Bolt	150	7.05	43	3.5	360	8.4	714	4.8
Tesla Model 3 Rear	188	9.03	45	4.2	380	8.4	750	4.0
BMW i3	125	9.7	31	4.0	250	8.1	496	4.0

[a] Estimated.
SOURCE: Committee-generated data, partially based on motor weight and cost data presented by Munro and Associates to the committee on September 24, 2019.

audible noise at high gear ratios as well as an increase in motor frequency and iron loss, which would impact the efficiency negatively. Furthermore, to protect the magnets at high rotor speeds, a rotor cross-section with deep slots is used to house the RE magnets and provide adequate support and robustness against centrifugal forces. This construction is typically known as buried or interior magnet construction.

Examples for the performance of some of the brushless PMSMs used in BEVs deployed in the market today are given in Table 5.1. This represents the state of the art in propulsion motor technology to date. While all use RE permanent magnets in their design, the BMW i3 motor is designed to have an improved performance at higher speed using a special rotor construction which contributes to an additional torque component (reluctance torque). This type of construction is known as hybrid PM-reluctance motor. Comparing the three vehicles in Table 5.1 reveals that the motor power density and specific cost, which are based on the motor weight and cost without the gearbox, improve with increasing the gear ratio. Adding the gearbox weight and cost, which increase with the gear ratio, will offset this improvement but still show cost improvement, as shown in Table 5.2.

For high-performance vehicles, the use of two motors simultaneously helps achieve the desirable 0–60 miles per hour acceleration performance. For example, the Tesla Model 3 has a PMSM for the rear axle and an induction motor for the front. Using the combination of induction motor and PMSM, as opposed to using the same type motor for front and rear, results in an improved overall efficiency by relying on the induction motor at high speeds and turning the PMSM off. The magnetic field and its associated loss in an induction motor can be easily controlled, unlike in PMSMs. Mechanically disconnecting the PMSM via a clutch in the two-motor system may yield further efficiency improvements by avoiding the substantial magnetic losses associated with the permanent magnet's constant magnetic field at high speeds. Of course, there are trade-offs between clutch weight and cost versus efficiency gain, which need to be evaluated. Current motor research points to several areas that could potentially impact future propulsion motor performance and cost in the 2025–2035 time frame:

1. *New magnet material* (ARPA-E, 2015): Ames Laboratory is developing a new class of permanent magnets based on the more commonly available element cerium to replace the scarcest and most expensive RE element, dysprosium, which is used in today's RE magnets for high temperature stability (dysprosium comprises ~3%–6% by weight of NdFeB magnets). Cerium is four times more abundant and

TABLE 5.2 Propulsion Motor Performance Status Summary—Motor with Gearbox

Application	Power (kW)	Gear Ratio	Motor + Gear Weight[a] (kg)	Power Density[a] (kW/kg)	Max Motor Torque (Nm)	Output Torque Density[a] (Nm/kg)	Motor + Gear Cost[a] ($)	Specific Cost[a] ($/kW)
GM Bolt	150	7.05	59	2.5	360	43.0	895	6.0
Tesla Model 3 Rear	188	9.03	71	2.6	380	48.3	1044	5.6
BMW i3	125	9.7	49	2.5	250	49.5	703	5.6

[a] Estimated.
SOURCE: Committee generated data, partially based on motor weight and cost data presented by Munro and Associates to the committee on September 24, 2019.

TABLE 5.3 Potential Impact of Future Motor Technologies on Various Vehicle Classes

BEV 300 Motor Technologies	Vehicle Class (Power, Torque)	Technology Cost by Class	Motor Total Weight, Cycle Efficiency
• Cerium magnets	Small (110 kW, 142 Nm)	$ 531	43 kg, 90.5%
• Higher gear ratio (14:1)	Medium (180 kW, 233 Nm)	$ 868	67 kg, 91%
	Crossover (150 kW, 194 Nm)	$ 724	57 kg, 90.7%
	SUV (220 kW, 285 Nm)	$ 1,061	81 kg, 91.2%
	Truck (250 kW, 324 Nm)	$ 1,206	92 kg, 91.5%

significantly less expensive than dysprosium. The result is a cost-effective cerium alloy of neodymium, iron, and boron co-doped with cerium and cobalt, with properties that are competitive with traditional sintered magnets containing dysprosium. With magnet cost representing roughly 20% of motor cost (approx. $150 at $75/kg), reducing magnet cost results in a substantial motor cost reduction if RE magnet prices climb to the levels seen in 2011 ($480/kg). Toyota has also announced development of a neodymium-reduced (50%), heat-resistant cerium magnet, stating that it will likely be utilized in power steering applications in the first half of the 2020s, and in propulsion motor applications within the next 10 years (Toyota USA Newsroom, 2018). It is estimated that this technology could reduce the magnet cost by approximately 30%.

2. *Higher motor speed:* Many of the automakers are actively developing high-speed motors. However, because of the negative impact on the gearbox weight and cost, it is not clear what would be the optimum motor speed/gear ratio. To illustrate the point, doubling the speed of the GM Bolt motor from its current 8,800 rpm to 17,600 rpm for the same output power would result in a motor with half the active length, weight, and active material cost. While the gear weight and cost are expected to increase, doubling the speed could result in a total (motor + gear) weight increase of approximately 1 kg but a total cost reduction of approximately $240. This decrease in cost is largely owing to a decrease in required RE magnet material, which is by far the most expensive part of the motor. Some of the challenges with this approach include noise, as well as reliability issues stemming from increasing the gear ratio. One should also keep in mind that operating at higher speed and reduced motor size will also result in a decreased cooling surface, which should be taken into account in sizing the motor cooling system for proper thermal management of the motor.

Table 5.3 provides a summary of estimated potential cost and effectiveness impact of the above technologies by 2025 on the various vehicle classes. The following assumptions were made:

1. New cerium-based magnet material to reduce magnet cost by 30% from current prices.
2. New gearing with a higher gear ratio of 14:1 instead of the 9:1 assumed in current systems.

From Table 5.2 (current) and Table 5.3 (future) for a medium-size vehicle (Tesla Model 3 rear), one can conclude that there is a potential for weight and cost reduction of approximately 5% and 16%, respectively.

5.2.2 Power and Control Electronics—Current Status and Future Developments

Inverter and controller technologies are also relatively mature, thanks to the industry's sustained efforts aimed at increasing their performance and efficiency while reducing their size and cost (Zhao, 2016a). These efforts include the following:

1. *Design optimization of the silicon semiconductor IGBT switches:* Design for minimum conduction and switching losses, which translate to high inverter efficiency.
2. *System integration:* An example of an effective system integration is the collaborative effort among GM, Oak Ridge National Laboratory, the National Renewable Energy Laboratory, and suppliers, resulting in

TABLE 5.4 Power Density and Cost of Current Inverter Topologies

Application	Power (kW)	Inverter Weight[a] (kg)	Power Density[a] (kW/kg)	Inverter Cost[a] ($)	Specific Cost[a] ($/kW)
17 GM Bolt	150	10.0	15.0	700	4.7
Tesla Model 3	188	5.5	34.1	800	3.5
BMW i3	125	19.0	6.6	1100	8.8

[a] Estimated.

NOTE: BMW i3 system has all electronics including the battery charger integrated with the drive motor, resulting in higher numbers compared with the other listed examples.

SOURCE: Committee generated data, partially based on motor weight and cost data presented to the committee on September 24, 2019 by Munro and Associates.

achieving new higher levels for efficiency and power-density while maintaining a capability for scalability in their next-generation inverter. This was achieved by an innovative packaging in a design, which integrates active components and reduces/eliminates supporting components.

3. *High-performance control:* The use of advanced high-performance control techniques, such as deadbeat direct torque control with loss observer, further reduces drive loss and enhances drive efficiency. Additional known areas of advanced control focus on the following:
   - Sensor (observers) reduction or elimination, with significant cost implications.
   - Acoustic and electromagnetic noise reduction.
   - Improved reliability (fault tolerance, diagnostics and prognostics).

Examples for the performance of some of the propulsion inverters used in electrified vehicles deployed in the market today are given in Table 5.4. The power electronics in these three vehicles are not listed here together for the purpose of comparison, as they are adapting different integration philosophies in their execution, but rather as a representation of the state of the art of propulsion power electronics. So, while the Tesla Model 3 inverter (power stage, filtering, and controller) is integrated with the motor drive and tapping into its cooling system, the Chevrolet Bolt inverter and controller are housed with the DC/DC converter and power distribution cabling and all necessary cooling lines in a separate box (listed weight and cost does not include DC/DC converter and distribution). The BMW i3 system has all electronics including the battery charger integrated with the drive motor, resulting in higher numbers compared with the other listed examples. The direct connection between the power electronics and electric motor claimed to be responsible for reducing the overall weight of the drivetrain by about 1.5 kg owing to reduced cabling length (Green Car Congress, 2013).

While most automakers still use IGBT power switching devices, including the Chevrolet Bolt and BMW i3, the Tesla Model 3 inverter uses the new, more expensive but more efficient silicon carbide (SiC) devices. SiC devices belong to a new category of power switching devices, known as wide-bandgap (WBG) devices, which have been evolving in recent years and might emerge as an impactful technology for electrified vehicles in the 2025–2035 time frame. There are two types of materials used in WBG device construction: (1) SiC and (2) gallium nitride (GaN). They have the capability to operate at higher voltages (> 600 volts), temperatures (> 200°C), and frequencies (> 1 MHz), and exhibit a 100-fold lower on-resistance (Figure 5.6)—compared with Si-based devices such as the IGBTs currently being used in automotive inverters.

The higher switching speeds (10 times faster) of WBG devices lead to very low switching loss, which along with their low on-resistance (low conduction loss) could eliminate up to 90% of the loss in power-electronic devices. This could result in very high inverter and converter efficiencies (typically on the order of 99% compared to 96% for the Si-based devices). With less energy expended as heat, and the capability to operate at higher temperature, WBG devices require less cooling and smaller heat sinks. This could result in an overall reduced system size, weight, and material cost. Furthermore, with WBG-based devices operating at higher frequencies, smaller inductors and capacitors can be used in power circuits. The inductance and capacitance scale down in proportion to the frequency: a 10-fold increase in frequency produces a 10-fold decrease in the capacitance and inductance. This can result in a substantial decrease in the weight, volume, and cost of typically large and heavy passive components.

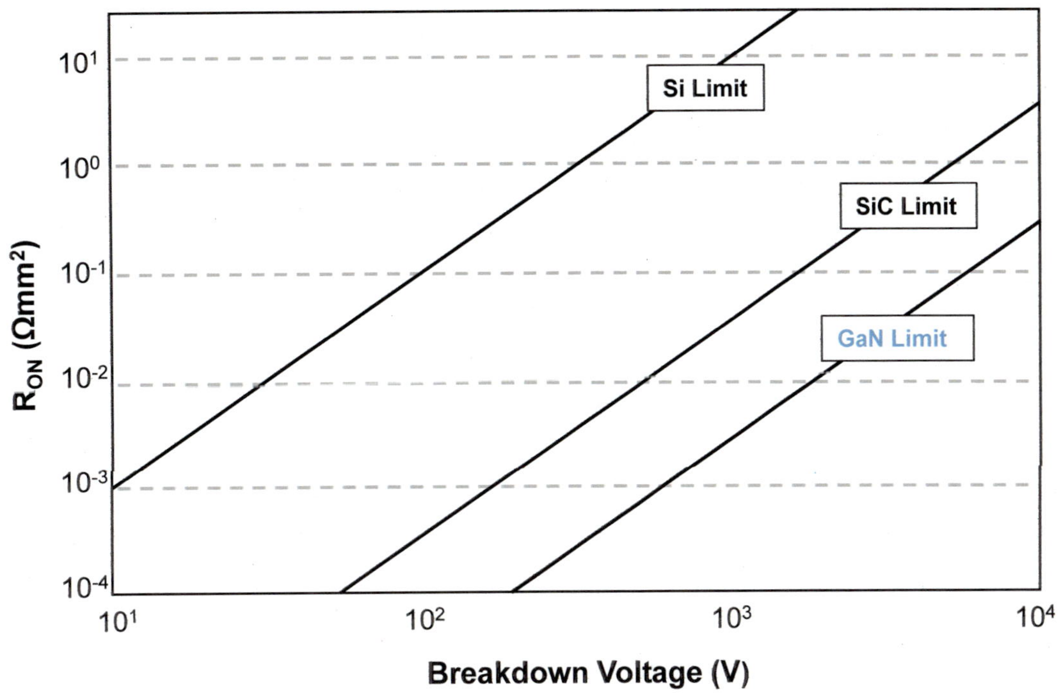

FIGURE 5.6 Characteristic comparison of Si, SiC, and GaN devices on-resistance and breakdown voltage.
SOURCE: Strydom et al. (2017).

On the other hand, increasing switching frequency may impact the motor iron loss (eddy and hysteresis), which should be a consideration in the motor design and its material selection.

As such, WBG devices have become a focus of current research and are expected to come to fruition in the time frame 2025–2035. Research organizations of automakers and suppliers are active in research to understand the ultra-fast switching of WBG devices and are developing high-frequency circuitry and high-temperature components necessary to sustain and take advantage of WBG devices. Some of these research areas include WBG device characterization, as well as evaluating converter and inverter technologies. Inverter efficiencies of more than 99% have been achieved in a GM program (Jaksic, 2019). It should be noted that currently the cost of WBG devices is higher than silicon devices, but they are expected to eventually be competitive as manufacturing capabilities (e.g., yield, wafer size, etc.) improve and their market grows.

GaN offers some advantages versus SiC. In addition to its lower on-resistance (low conduction loss; see Figure 5.6), there is evidence that GaN also exhibits lower switching loss at high frequencies (Figure 5.7). It should be noted that while both SiC and GaN technologies still need further improvements (PowerAmerica, 2018), SiC-based devices are further ahead in their development than GaN devices, as they were the subject of years of targeted RD&D for aerospace applications, which could afford the high cost of SiC. Most of the WBG device investigations to date have used SiC devices in their experimental builds simply because of availability. Another advantage of lateral GaN devices is that a thin layer of active GaN can be grown on silicon, a cheap substrate. Therefore, GaN on Si devices present a potential cost advantage compared to SiC. However, the advancement of GaN devices faces several challenges that must be resolved first before their broad implementation:

1. The difference in thermal coefficient of expansion between GaN and Si in GaN on Si devices causes issues at high temperatures which may limit their usage at these temperatures. This led researchers to explore GaN on SiC substrates, both having a similar coefficient of expansion. GaN on SiC is, however, more expensive than GaN on Si and comes close to the cost of the more mature SiC technology.

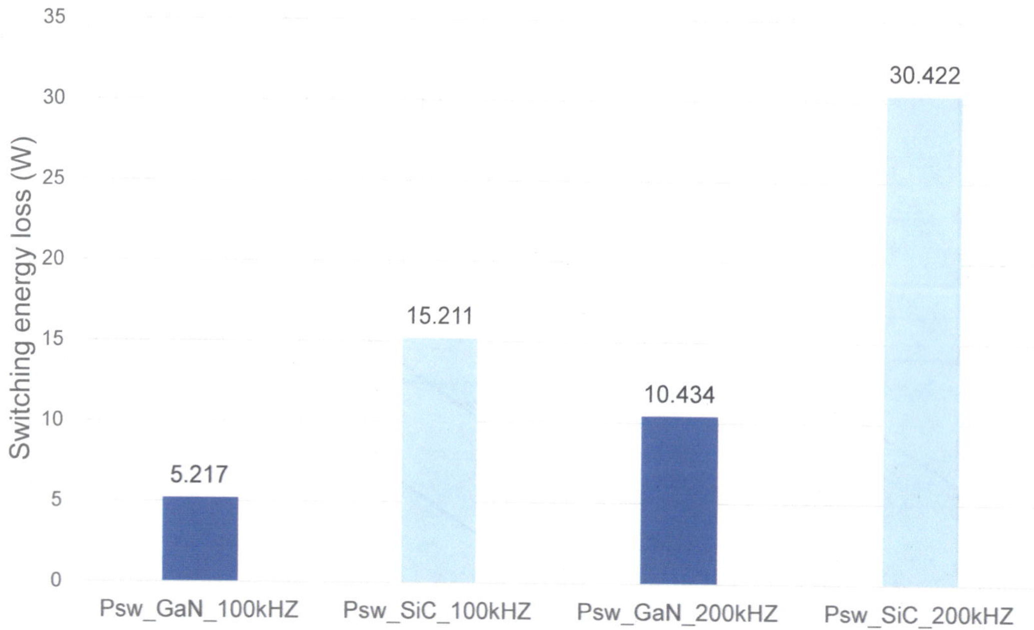

FIGURE 5.7 Switching loss comparison between GaN and SiC MOSFET.
SOURCE: Modified from Xu and Chen (2017).

2. Designing a GaN-based device that can withstand high breakdown voltage is a challenge. More established GaN devices utilize a lateral device architecture where the current flow is constrained to a thin section of GaN material. However, higher power applications (e.g., EVs) require higher breakdown voltages and thus more material, making these lateral devices unattractive (Chowdhury and Mishra, 2013). Significantly larger chip sizes would be needed to accommodate this higher breakdown voltage which poses manufacturing challenges. Therefore, researchers are redesigning devices to allow current to pass through the bulk of the GaN material via vertical device architectures. Technical developments needed to realize vertical devices include the production of high-quality GaN substrates and development of reliable selective-area doping processes to control current flow within the device (Hu et al., 2018). Both of these areas are currently priorities for ongoing Advanced Research Projects Agency–Energy (ARPA-E) programs. The ARPA-E Strategies for Wide-Bandgap, Inexpensive Transistors for Controlling High-Efficiency Systems (SWITCHES) program, started in 2013, funds numerous projects to improve the processing of GaN vertical devices and GaN substrates for applications including automobiles. The Power Nitride Doping Innovation Offers Devices Enabling (PNDIODES) program is an extension of SWITCHES focusing specifically on developing selective-area doping processes for GaN power electronics.

While research toward resolving the issues associated with GaN continues, the debate among proponents of GaN versus SiC fills the literature (Boutros et al., 2012; Patterson, 2015; Allan, 2017; Ferdowsi et al., 2017; Green Car Congress, 2017; Guerra, 2017; Slovick, 2017; Transphorm Inc., 2017; Wolfspeed, 2017; Els, 2018; Li, 2018; Business Wire, 2019; Davis, 2019; Semiconductor Today, 2019; Arrow Electronics, 2020; Bakeroot, 2020; Schweber, 2020).

As with motors, an attempt is made here to estimate the potential cost, weight, and efficiency of propulsion inverters assuming that the above discussed technologies have matured for commercial implementation by 2025. These estimates are summarized in Table 5.5 below. The following relatively conservative assumptions are made:

- The baseline for the estimates is today's Tesla Model 3 inverter, using SiC devices and a high degree of integration as described above; see Table 5.4.

TABLE 5.5 Potential Impact of Future Inverter Technologies on Various Vehicle Classes

BEV 300 Inverter Technologies	Vehicle Class (Power)	Technology Cost by Class	Inverter Weight, Efficiency
• GaN-based power switching devices	Small (110 kW)	$ 334	2.3 kg, 98.5%
• High frequency switching (100 kHz)	Medium (180 kW)	$ 471	3.8 kg, 99%
	Crossover (150 kW)	$ 412	3.2 kg, 98%
	SUV (220 kW)	$ 550	4.7 kg, 99%
	Truck (250 kW)	$ 609	5.3 kg, 99%

- The cost of GaN power switching devices is 25% lower than today's SiC; this decrease in cost includes the effects of resolving manufacturing issues and increasing production volume.
- The reduced conduction and switching loss (at high switching frequency) will lead to reducing cooling needs by 75%.
- Switching at higher frequency (100 kilohertz) will result in reduced filtering components size, weight, and cost by 75%, particularly for DC/DC converters.
- Natural electronics cost reduction trajectory leads to 25% controller cost reduction.
- Inverter cost includes power stage, cooling and mechanical assembly, filtering, and electronic controller only. It does not include power distribution, DC/DC converter, or charging electronics.

5.2.3 Findings and Recommendation for Motors and Power Electronics

FINDING 5.1: The majority of automakers have converged on using permanent magnet synchronous motors with rare earth magnets as the drive motor for electrified vehicles owing to their superior efficiency, torque, and power density. Although permanent magnet synchronous motors are more costly (approx. 50%–70%) than induction motors, the efficiency gain is important for reducing the costs of the powertrain as a whole.

FINDING 5.2: The industry has converged on the use of a single-stage gearbox for electric propulsion systems, with a gear ratio between 7:1 and 10:1. Increasing the gear ratio to 14:1 in a medium-size vehicle (Tesla Model 3 rear), for example, could potentially lead to a weight and cost reduction of approximately 5% (4 kg) and 16% ($176), respectively. While the cost saving is considerable, the weight reduction is small and would contribute to only an insignificant range increase (< 1 mile).

FINDING 5.3: While the majority of the automakers are still using insulated-gate bipolar transistor (IGBT) power-switching devices in their power electronic circuitry, some are considering the use of wide-bandgap (WBG) devices in their next-generation propulsion systems, owing to their lower loss (only 10% of IGBTs). This could result in boosting inverter and converter efficiencies to 99% (from 96%), while reducing the size and weight of the cooling system components by ~75%. The efficiency gain translates to adding roughly 9–10 miles to a vehicle with a 300-mile range.

FINDING 5.4: There are two types of wide-bandgap devices: silicon carbide (SiC) and gallium nitride (GaN). Most automakers are focusing on SiC owing to its widespread availability. Given the inherent cost advantage of GaN on Si devices compared with SiC devices, GaN on Si could ultimately become the most cost effective among these two competing technologies, provided improvements in GaN device architectures lead to usable performance.

RECOMMENDATION 5.1: The U.S. Department of Energy should continue funding research on advancing gallium nitride on silicon wide-bandgap device technology to help expedite its readiness for the automotive market and advance the practical utilization of its efficient high switching frequency capability.

5.3 BATTERIES FOR EVS

5.3.1 Basic Principles

Today's EV technology is based primarily on lithium-ion batteries. While the Toyota Prius established a significant market for HEVs using nickel metal hydride batteries, newer Prius models are based on lithium ion as well. All PHEVs and BEVs utilize lithium-ion batteries; to date, lithium ion is the only chemistry that can supply the necessary energy and power density for automotive performance. Lithium-ion batteries are a form of chemical energy storage in which a lithium containing cathode is used in conjunction with a lithium accepting anode, between which lithium ions shuttle back and forth during charge and discharge cycles (Figure 5.8).

The amount of energy stored in the battery is proportional to the voltage differential between the anode and the cathode and the amount of lithium ions that can be moved back and forth. Both parameters are dependent upon the specific active materials used within the battery. Other inactive components within the battery, such as separator, electrolyte, and current collectors, are necessary for the electrochemical cell to operate, but decrease the cell-level energy density on an energy per unit weight or volume basis.

5.3.2 Today's Performance

Battery performance determines key attributes of vehicle performance. Key metrics of the battery include energy and power density (both gravimetric and volumetric), lifetime, safety, and cost. The cell energy density is the determining factor for driving range and will depend upon the active materials used within the cell, which define the cell voltage and capacity, as well as the inactive materials, which add weight and volume to the battery. The amount of energy stored is proportional to the amount of lithium ions shuttled back and forth in the cell.

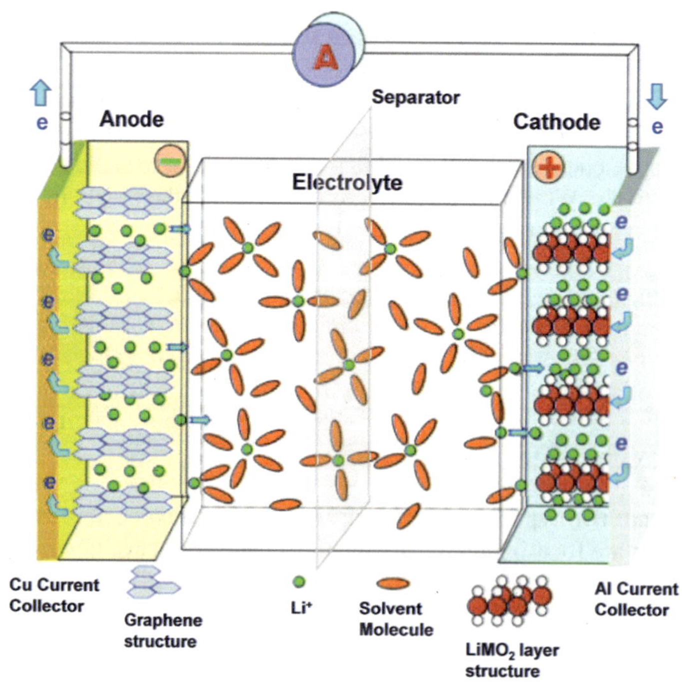

FIGURE 5.8 Schematic of lithium-ion battery.
SOURCE: Reprinted with permission from K. Xu, 2004, Nonaqueous liquid electrolytes for lithium-based rechargeable batteries, *Chemical Reviews* 104(10):4303–4418, copyright 2004 American Chemical Society.

Therefore, the energy density will depend upon the amount of cathode in the cell and the amount of anode required to store the lithium from the cathode. In general, cathodes that contain more usable lithium per unit volume and anodes that can hold more lithium will result in higher energy density.

Power density affects the rate at which the battery can be charged or discharged and plays a large role in automotive performance. The power performance of the cell depends upon the inherent kinetic properties of the active materials (lithium-ion transport properties in the electrolyte and interface layers), and the physical characteristics (thickness, porosity, tortuosity) of the anode, cathode, and separator. Kinetic properties of the materials are temperature dependent and can limit low-temperature performance of the cells. The power (or rate) performance of the active materials is dependent upon the state of charge (SOC), as both the ionic and electronic conductivity of active materials are a function of SOC. Poor conductivity at low states of charge limits the depth of discharge (DOD) at which the cell can be used. Thus, not only do power characteristics of the cell affect driving parameters such as acceleration and charge acceptance during braking but they can also affect the driving range owing to limitations on DOD of the cell.

Cell lifetime can be defined as the time at which the cell capacity falls below a predetermined value (typically 80% of initial capacity), or a cell resistance at which a predetermined capacity cannot be achieved on charge or discharge at a specific rate. These effects may be observed after the battery has undergone hundreds or thousands of cycles or has spent significant amounts of calendar time at high temperature. Failure mechanisms that occur upon cycling or calendar storage will depend upon the specific use case of the battery. For example, BEVs use a wide SOC of the cell over several thousand cycles resulting in true capacity loss and impedance growth. HEVs use a relatively narrow SOC for hundreds of thousands of cycles, with resistance growth being a major issue. Catastrophic failure of the cell can also occur but is more closely linked to cell safety considerations.

The performance of cells across all key metrics will depend upon the application for which they are designed. Table 5.6 summarizes the energy density of some commercial cells used for BEVs.

Safety is a key consideration for all automotive applications and must be considered whenever large amounts of energy are stored in small volumes. Battery safety will depend upon the specific types and amounts of active materials used within the cell, as well as the properties of the inactive components. For example, thin separators which prevent the anode from touching the cathode in a physical short are desirable to improve energy density, but thin separators are also more susceptible to punctures during use, resulting in potential safety hazards. High-quality manufacturing processes are required to eliminate flaws causing internal cell shorting that can lead to a fire. Last, engineering of battery modules and packs with good thermal management can prevent a series of events within the cell from causing thermal runaway and fire. Further discussion of this is given in Section 5.3.5 on thermal management.

5.3.3 Materials and Limitations

Most commercial automotive batteries contain a cathode intercalation material with a graphite-based anode, as well as a separator and an electrolyte. This section will discuss the many different material options for battery

TABLE 5.6 Examples of Energy Densities for Automotive Cells

Vehicle	Type	Format	Specific Energy (Wh/kg)	Energy Density (Wh/L)
Tesla Model 3	BEV	Cylindrical 21700	250	721[a]
Nissan Leaf	BEV	Pouch 33Ah	224	460[b]
BMW i3	BEV	Prismatic 94Ah	174	352[c]
Chevy Bolt	BEV	Pouch 60 Ah	237	444[d]

[a] Field (2019).
[b] Lima (2018).
[c] Kane (2018).
[d] Bower (2019).

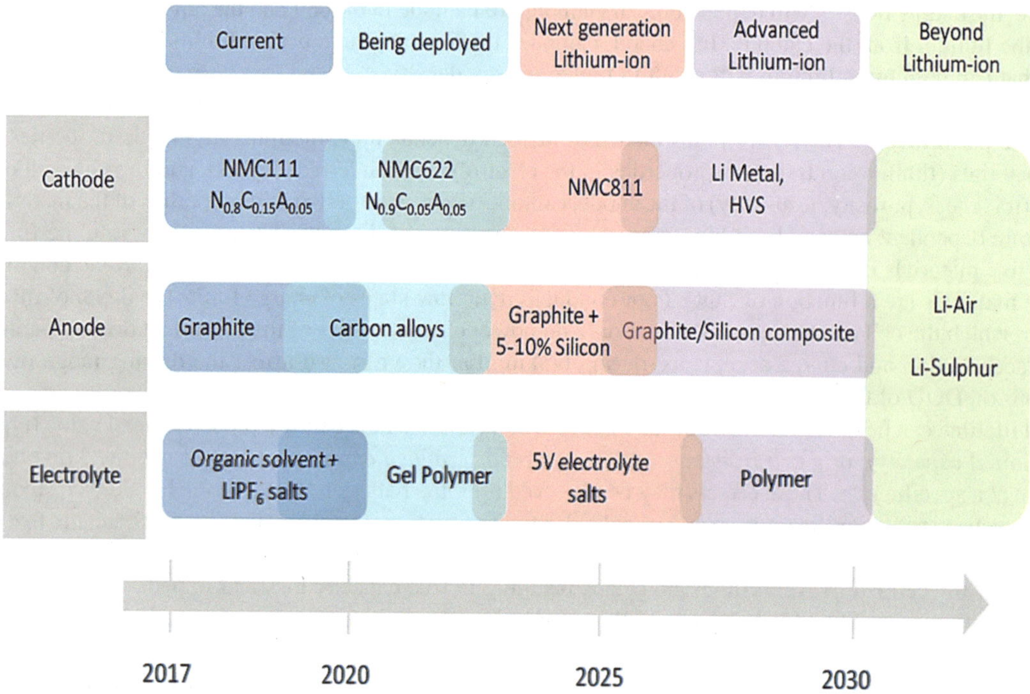

FIGURE 5.9 Uncertain timeline for beyond lithium-ion technologies.
NOTE: HVS = high voltage spinel. The diagram shows the likely beginning of commercialization of a given technology.
SOURCE: Mihet-Popa and Saponara (2018).

components, along with the advantages and disadvantages of each. A summary of the uncertain timeline of battery evolution for each of these components is shown in Figure 5.9. A key focus of the industry is to move toward cheaper cathode materials that include less cobalt.

5.3.3.1 Cathode Materials

The composition of the cathode relates to the energy density of the battery. Commercial cathodes used in lithium-ion batteries are generally intercalation materials, wherein lithium ions can move into (intercalate) and out of (deintercalate) the structure without major phase transitions. Intercalation structures consist of transition metal cations (e.g., Ni, Mn, Co, and Fe) and oxygen or phosphate anions.

The most commonly used cathode materials are layered oxides of nickel, manganese, and cobalt (NMCs), such as $LiNi_{0.33}Mn_{0.33}Co_{0.33}O_2$ (NMC111). These consist of two-dimensional "layers" of transition metals, with lithium ions contained between the layers. The lithium ions can move into or out of the layers with modest changes in the layer spacing of the structure. There are, however, limits to the amount of lithium that can be removed from the structures. At higher voltage, larger quantities of lithium are removed, and phase changes in the material can start to occur (transition metals tend to move into the lithium layer and cause structural rearrangements). These rearrangements are sometimes irreversible and prevent lithium from reintercalating into the structure causing the energy density of the battery to deteriorate. In addition, the phase transitions can be accompanied by a loss of oxygen in the structure causing release of reactive oxide/oxygen to the organic electrolyte, which is a safety concern.

Other structured materials used in lithium-ion batteries include transition metal phosphates (olivines) such as lithium iron phosphate ($LiFePO_4$) or lithium manganese phosphate ($LiMnPO_4$). These materials provide one-dimensional lithium transport through tunnel like channels in the crystal structure. Olivine materials are advantaged over cathodes in that nearly all the lithium can be removed from the structure without irreversible phase changes or release of oxygen; this structural stability results in long cycle life for these materials. Vehicle applications requiring lower energy densities, such as stop-start or MHEV, can effectively use $LiFePO_4$. Recently, Tesla and CATL announced a "cell to pack" technology that uses low-cost $LiFePO_4$ chemistry as a cathode (Manthey, 2020). Owing to the inherent safety of $LiFePO_4$, cells can be placed directly in packs without the secondary control of using modules within the packs. Elimination of the modules not only reduces cost but also increases system-level energy density owing to lower weight and volume. Although near theoretical capacities can be achieved with olivine materials, they are disadvantaged in energy density owing to the relatively low weight percent of lithium contained in the materials.

Most automakers are expanding the use of higher nickel containing NMCs to improve energy densities. These materials, such as $LiNi_{0.6}Mn_{0.2}Co_{0.2}O_2$ (NMC622) or $LiNi_{0.8}Mn_{0.1}Co_{0.1}O_2$ (NMC811), take advantage of nickel's 2-electron redox chemistry, increase the amount of lithium that can be cycled in and out of the material, and thus increase the specific capacity and energy density of the cell. NMC811 is primed to potentially be the fastest growing chemistry: its use increased from 1% in 2018 to 12% in early 2020 in China (Statista, 2020a). NMC811 is being deployed by BMW, GM, Nio, and Volkswagen, among other automakers, and suppliers include LG Chem and CATL, a testament to how quickly NMC activity is advancing (LeVine, 2020). Tesla cells also use a high nickel material, $LiNi_{0.8}Co_{0.1}Al_{0.1}O_2$ (nickel cobalt aluminum, NCA), to achieve high energy density. While these materials improve energy density and use a lower amount of expensive and problematic cobalt, they suffer from poorer stability as nickel tends to migrate into the lithium layer more readily than other elements.

For the next several years, automotive battery suppliers and automakers are pushing toward higher nickel materials operated at higher voltage to improve energy density. While NMC111 and NMC532 were common around 2015, NMC622 is the most common cathode chemistry in 2019, and NMC811 has entered commercial vehicle models. Yet, challenges remain regarding material stability to ensure that a target lifetime and safety performance can be met. Solutions for increased stability include the following:

- Doping small amounts of multivalent cations (e.g., Al^{3+}, Si^{4+}, Ti^{4+}, Zr^{4+}, Ta^{5+}) into the crystal structure to stabilize the layered material as more lithium is removed, preventing irreversible phase changes and increasing material stability (Weigel et al., 2019).
- Coating the surface of the cathode particles, which also serves to stabilize the reactive materials at the surface.
- New electrolytes that form passivation layers on the high energy cathode materials can extend lifetime and improve safety.

However, all these approaches increase the cost of the cathode material and thus the overall cost of the lithium-ion cell. Several studies demonstrate the potential of incremental and next-generation NMC technologies in particular to increase cell performance and deliver greater gravimetric (Wh/kg) and volumetric (Wh/L) energy density (Wentker et al., 2019).

5.3.3.2 Anode Materials

Improvements in anode materials, specifically graphite, are focused on fast charge requirements. BEVs need to compete with ICEVs in total travel time for long-distance driving. For travel beyond the range of the BEV, extended recharge times make these vehicles less attractive for consumers. Lithium-ion batteries using graphite intercalation anode materials suffer from lithium plating during charge at high current densities. Plating of lithium metal results in reduced battery lifetimes and safety concerns. The current density limitation of graphite involves both the diffusion rate of lithium within the graphite and the rate of transport across the solid electrolyte interphase (SEI), which is formed owing to reduction of electrolyte on the surface of the anode.

Today's commercial anodes used in automotive cells are primarily graphite-based. Graphite's layered structure allows lithium-ion intercalation and deintercalation similar to what occurs in the layered oxide cathodes. Different types of graphite may be used including natural or artificial graphite. Both types have similar specific capacities and performance profiles, but artificial graphite tends to be at least twice as expensive.

Battery performance will be affected by the graphite particle size, morphology, and functional groups, and there are various advantages and disadvantages to using different graphite or carbon materials. For example, amorphous hard carbon anodes exhibit superior lifetime and safety, whereas artificial graphite exhibits higher energy density. Meanwhile, natural graphite is the least cost prohibitive. Any given cell design will have to factor in these anode material trade-offs. Different types of graphites are available in the marketplace. The cheapest material is natural flake graphite, which can provide good electrode density and lower cost cells. However, the material cannot provide good rate or power performance owing to its flake morphology. Natural graphite can be spheroidized and carbon coated to improve the rate performance, while increasing costs. The process also yields a high degree of graphitization, translating to high specific capacity. Artificial or synthetic graphite is more expensive than natural graphite but has much higher purity which leads to long cycle life. The artificial graphite can be produced in a variety of particle sizes and morphologies with good rate performance. Amorphous carbons (e.g., hard carbon, soft carbon) are used for more specialized applications and are generally not widely used in automotive cells.

A key attribute of graphite is the surface functionality. As the graphite is lithiated during battery charge, the potential of the lithiated carbon drops to very low potential—at which the organic electrolyte is not reductively stable. As reduction of the electrolyte occurs, the reaction products precipitate onto the graphite surface forming an SEI layer. The composition of the SEI is dependent upon the specific electrolyte formulation, the graphite surface, the age of the battery, and many other factors. Without formation of the protective SEI, the lithiated graphite will continue to reduce the bulk electrolyte eventually leading to total consumption of electrolyte. Owing to the complexity of studying the SEI, it is difficult to predict which electrolytes and graphites work best together, and optimum electrolytes must be developed for specific anode materials.

Further improvements in energy density require new anode materials to replace graphite. The most promising material is silicon, which can exist in a variety of forms including silicon oxides, silicon alloys, nano-Si/graphite composites, and silicon nanowires, among others.[1] While silicon-based anodes have very high specific capacities, the density of the lithium silicon alloy is very dependent upon lithium content. To date, electrodes with high silicon content (> ~ 8%) have not been demonstrated to have cycle life adequate for automotive applications. Key challenges facing use of silicon in anodes include low first cycle efficiency (owing to formation of irreversible phases), varying quality and consistency of starting material options, and manufacturing challenges associated with prelithiation and nanoparticle dispersion. In addition, the SEI formed on silicon anodes is not as robust as that formed on silicon. This can lead to shorter calendar life of the silicon-based cells.

Several studies demonstrate the potential of incremental and next-generation NMC technologies in particular to greatly increase cell performance and deliver greater specific cell energy (watt-hours per kilogram [Wh/kg] cathode or cell material), cell energy density watt-hours per liter (Wh/L), and cost (dollars per kilowatt-hour [$/kWh]). Combining improved cathodes with silicon containing anodes can substantially improve cell-level energy densities. Figure 5.10 shows how higher-nickel and lithium or manganese-rich NMC batteries can deliver 30% to 75% Wh/kg improvement over baseline NMC611 technology that has been the most prevalent BEV technology in the 2019 market.

5.3.3.3 Separators

The separator provides a physical barrier between the anode and the cathode to prevent shorting. Automotive separators must have stringent quality control to ensure pinholes and tears are not present in the membranes. High performance separators consist of a polymer layer or layers coated with inorganic particles, such as Al_2O_3. These inorganic coatings can improve overall safety of the battery in case of a thermal event. If the temperature of the battery gets high enough such that the polymer layer in the separator melts, the inorganic particles will physically separate the anode from the cathode.

In order to achieve the highest possible energy density, separators should be as thin as possible. A thinner separator takes up less space in the cell, resulting in a smaller cell for a given capacity. However, thinner separators are

[1]Silicon forms alloys with lithium, rather than intercalating lithium ions, and has a theoretical capacity of 3579 mAh/g.

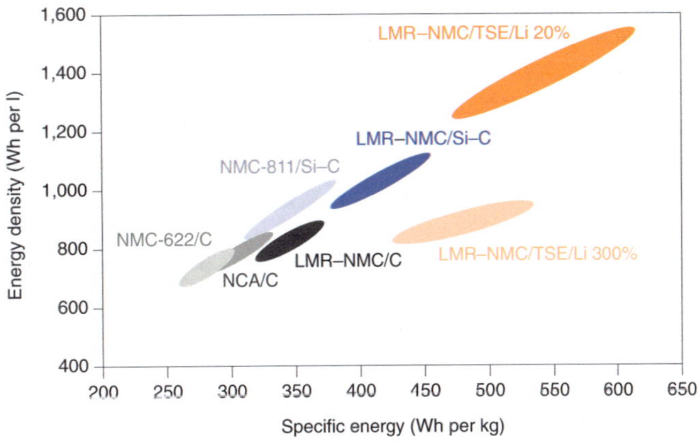

FIGURE 5.10 Potential material combinations for improved lithium-ion performance from advanced cathodes and anodes. NOTE: C = carbon (graphite); Li (20%) = lithium anode with 20% excess lithium relative to cathode; Li (300%) = lithium anode with 300% excess lithium relative to cathode; LMR-NMC = lithium manganese rich NMC; NCA = lithium nickel cobalt aluminum oxide; NMC622 = LiNi0.6Mn0.2Co0.2O2; NMC811 = LiNi0.8Mn0.1Co0.1O2; NMC = lithium nickel manganese cobalt oxide; Si = silicon; Si-C = silicon-carbon composite; TSE = thiophosphate-based solid electrolyte (e.g., Li7P3S11). SOURCE: Reprinted by permission from Springer Nature: Schmuch et al. Performance and cost materials for lithium-based rechargeable automotive batteries, *Nature Energy* 3:267–278, © 2018.

more prone to puncture during use or tear during cell manufacturing, so this trade-off must be managed. Table 5.7 shows key characteristics of separators used for automotive applications.

Key improvements in traditional separator technology involves development of robust, thin, low-cost, high-temperature materials to prevent catastrophic failure in the event of a thermal event.

5.3.3.4 Electrolyte

The electrolyte provides the medium by which lithium ions can move between the anode and the cathode. In addition, it infiltrates the electrodes enabling lithium ions to move into and out of bulk electrolyte. Electrolytes are complex formulations of solvents, salts, and additives. A high dielectric constant solvent, such as ethylene carbonate (EC), is required to solubilize the lithium salt. Most high dielectric solvents have viscosities that are too high to allow fast lithium transport. Therefore, solvents such as EC are diluted with other low viscosity solvents. Typically, linear carbonates (e.g., dimethyl carbonate or ethyl methyl carbonate) are used as low viscosity diluents.

TABLE 5.7 Important Separator Properties for Automotive Applications

Property	Typical Values	Comments
Thickness	10–40 microns	Trend is thinner to improve cell energy density, but need to balance with safety
Air permeability (Gurley value)	< 1,000 sec	Reflects porosity and pore structure for a given thickness
Porosity	35%–50%	Higher porosity yields better power performance, but need to balance with safety
Shrinkage	< 3%	Minimize shrinkage at elevated temperatures for safety
Tensile strength	Variable	Needs to withstand battery manufacturing process
Puncture strength	Variable	Needs to withstand puncture from lithium dendrites or sharp particulates as the cell is under some pressure

TABLE 5.8 Common Electrolyte Components

Component	Examples	Function	Comment
High dielectric constant solvent	Ethylene carbonate, propylene carbonate	Solvates Li^+	High viscosity detrimental to rate, power, and low temperature performance; participates in SEI formation
Low viscosity solvent	Ethyl methyl carbonate, diethyl carbonate, dimethyl carbonate	Lowers viscosity	Volatile, flammable solvents detrimental to safety
Salt	$LiPF_6$, LiFSI, $LiBF_4$	Provides anion for Li^+	Expensive, corrosive, moisture sensitive
SEI additives	Vinylene carbonate, fluorinated ethylene carbonate	Anode SEI stabilizer	Adds cost
Cathode active additives	1,3-propane sultone, nitriles	Cathode passivation stabilizer	Regulatory concerns, adds cost

A lithium-ion salt (or salts) is added to the formulation as a source of anions required to complex the lithium cations. Almost all commercial electrolyte formulations use lithium hexafluorophosphate ($LiPF_6$) as the primary salt. As the lithium ions approach the electrode for intercalation, the solvation sphere and/or anion interaction must be such that the cation can be released to enter the active material. No other salt performs as well as $LiPF_6$. One of the important functions of $LiPF_6$ is passivation of the aluminum current collector, without which corrosion will occur. $LiPF_6$ also plays an important role in the composition of the SEI layer on the anode. However, $LiPF_6$ has deficiencies in that it is expensive, reacts with water, and has poor thermal stability. In the presence of water or at temperatures above about 60°C, $LiPF_6$ generates acidic species such as hydrogen fluoride (HF), which is very detrimental to battery performance and poses safety hazards. The use of $LiPF_6$ as an electrolyte salt requires stringent (and costly) manufacturing processes to keep moisture out of the battery and also requires good thermal management of batteries when in use.

Last, additives are essential for long life of lithium-ion batteries. Solvents, salts, and additives participate in SEI formation on the anode—but additives can enhance the stability and conductivity of the SEI such that good power performance over many cycles can be achieved. As higher nickel cathode materials are more reactive at the upper voltage cutoff, additives are also required to improve the oxidative stability of the organic electrolytes, resulting in passivation layers at both the anode SEI and cathode interface (cathode electrolyte interphase). Common electrolyte components are listed in Table 5.8.

Electrolyte development offers many different approaches to battery improvement. New additives to promote more robust SEI layers on the anode can enable longer cycle life, better low temperature power, lower resistance at high temperatures, and better safety. High voltage additives can stabilize high energy cathodes by forming passivation layers. Other types of additives can scavenge harmful species such as HF. New solvents are being studied to yield less flammable or non-flammable electrolytes, which could contribute to better safety. Several large efforts by companies such as Air Products and Honeywell to develop alternatives to $LiPF_6$ have thus far been unsuccessful, but lower cost and more stable alternatives to $LiPF_6$ should be a research target.

5.3.3.5 Cell Component Cost Reduction

To achieve widespread adoption, BEVs need to approach cost parity with ICEVs. Current battery costs have been a significant barrier to lowering the cost of EVs across more vehicle segments. As shown in Figure 5.11, 70% of the battery cost is owing to the material costs, with the remainder being factors such as manufacturing labor, R&D, and overhead. Efforts to reduce overall BEV costs must focus on reducing the cell cost—which translates to use of cheaper higher energy density materials and more efficient manufacturing methods. The individual cells must be packaged in modules and packs, which further adds to the battery cost. Automotive companies are looking across the value chain from materials through the pack assembly to reduce total system costs.

As shown in Figure 5.11, the materials in the cell account for about 70% of the product cost of the cell. Of the material cost, the cathode accounts for the largest fraction of the cost, typically accounting for 40%–45% of

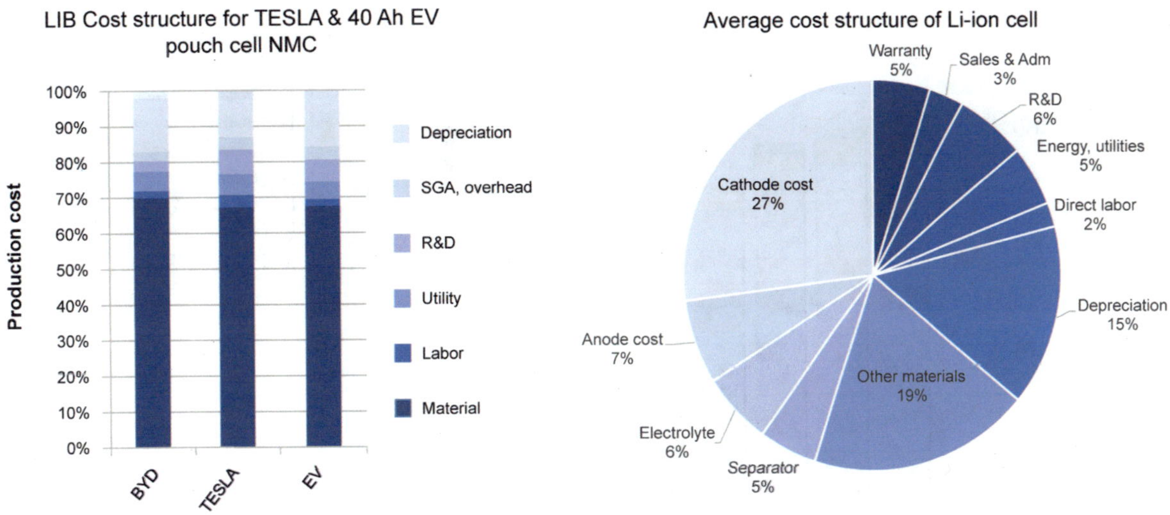

FIGURE 5.11 Cost structure for lithium-ion cells, assuming an average mix of cylindrical, prismatic, and laminate cells.
SOURCE: Pillot (2019).
NOTE: LIB = lithium ion battery; SGA = selling, general, and administrative.

material cost, followed by the anode and separator with 10%–15% each (Wentker et al., 2019). Other components include the aluminum and copper current collectors, tabs, cases, and packaging materials.

The constituent raw materials in the cathodes account for approximately 50% of the cathode cost. Figure 5.12a shows results of calculations for total material costs (cell level) per kWh for varying cathode compositions (Wentker et al., 2019). A shift from today's NMC532//graphite to a high nickel (NMC811 or NCA) can reduce materials costs from \$80 to near \$70/kWh, primarily owing to the improved energy density of the higher nickel materials and reduced cost owing to minimization of cobalt content. Figure 5.12b shows the sensitivity of various cathode costs to base cobalt market price.

In addition to using lower cost materials, the absolute costs of materials have decreased over time, as shown in Figure 5.13. From 2011 to 2017, a decrease of 5%–10% in the raw materials cost was observed. This decrease mirrored a reduction in the constituent metal prices over that time frame and so may not be sustainable. However, no cost reduction owing to process improvements in cathode powder manufacture was observed.

5.3.3.6 Cell Design

Cell performance is strongly affected by the material components, but the cell design can be equally important. The cell design encompasses the design of the cathode and anode electrodes—including the active material content, the loading of material onto the current collector, and the porosity of the electrodes. A cell design for higher power applications such as HEVs will have thinner, more porous electrodes. This design allows for ample flow of electrolyte containing lithium ions into and out of the electrodes so that power and rate performance is not limited by bulk mass transport of lithium to the surface of the active materials. Each anode and cathode requires a current collector, and a separator is required between the two. Cells containing thinner, less dense electrodes will have a relatively higher weight percent and volume of these inactive components.

Cells designed for high energy will, therefore, tend to have thick, dense electrodes such that the weight and volume percent of inactive components will decrease. However, these types of electrodes can be limited in terms of their rate and power performance. Specially optimized electrolytes with lower viscosity or thinner SEI layers can help to overcome this problem. Thin current collectors are also beneficial in improving overall energy density but can cause problems in the electrode manufacturing process with film breakage or curvature as the electrode dries.

Another important parameter in cell design is the ratio of capacity of the anode to the cathode. In general anode layers are designed to have slightly higher capacity (5%–10%) than the cathode. This ensures that the anode can

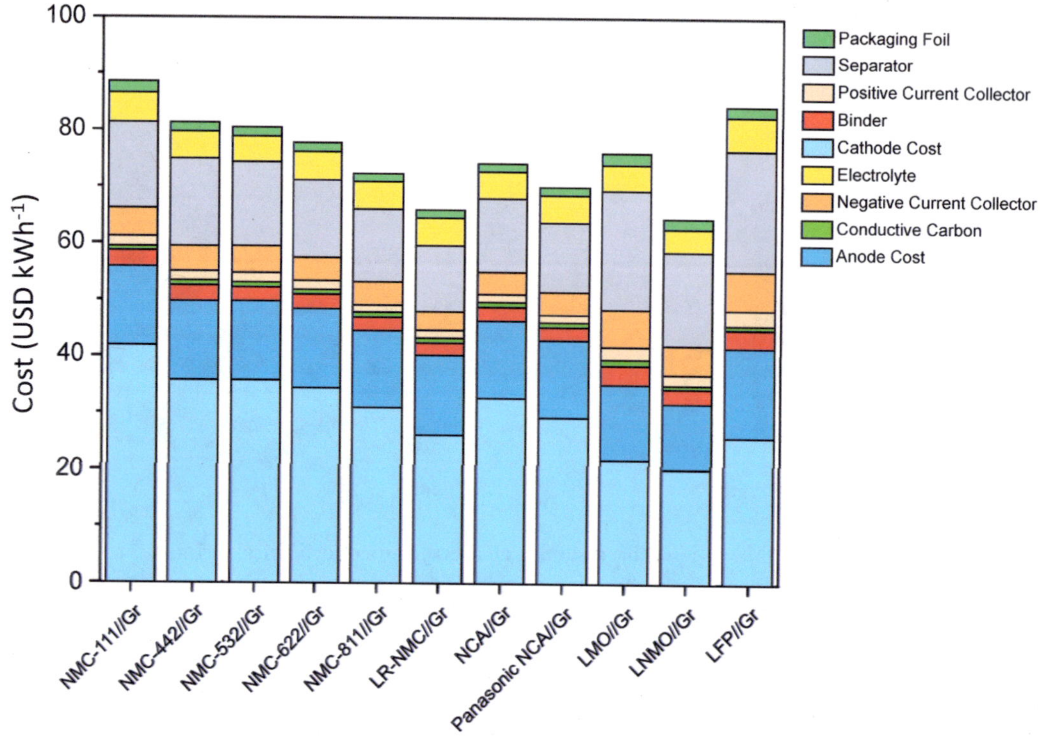

(a)

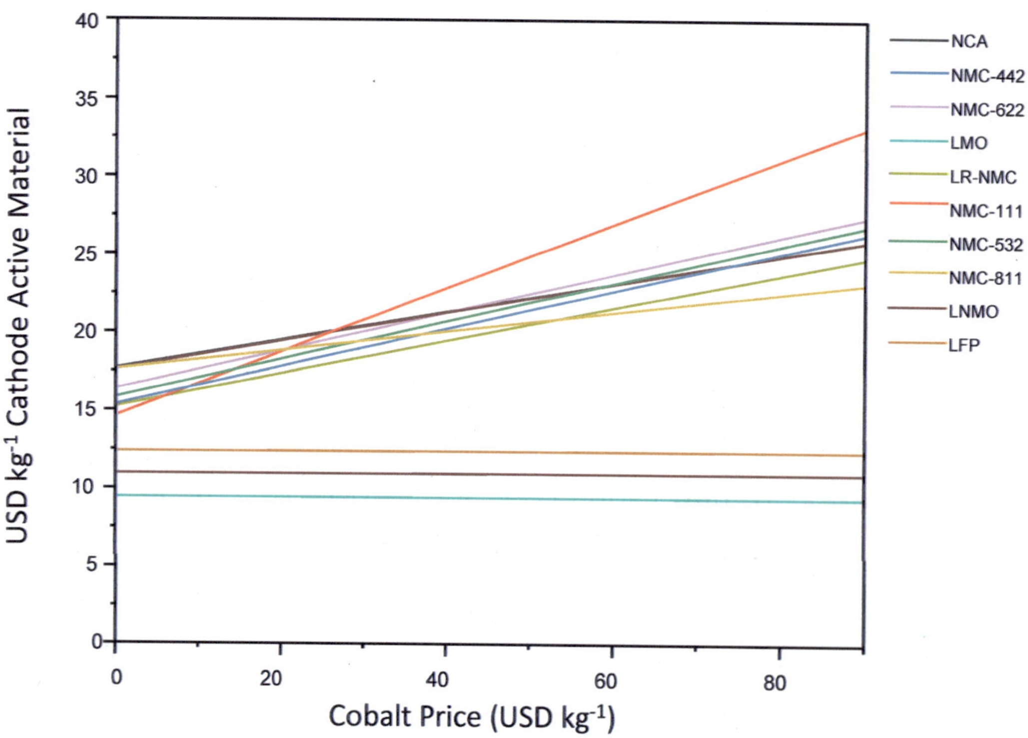

(b)

FIGURE 5.12 (a) Raw materials costs per kWh for varying cathode compositions; (b) Cathode cost with respect to cobalt market price.
SOURCE: Wentker et al. (2019).

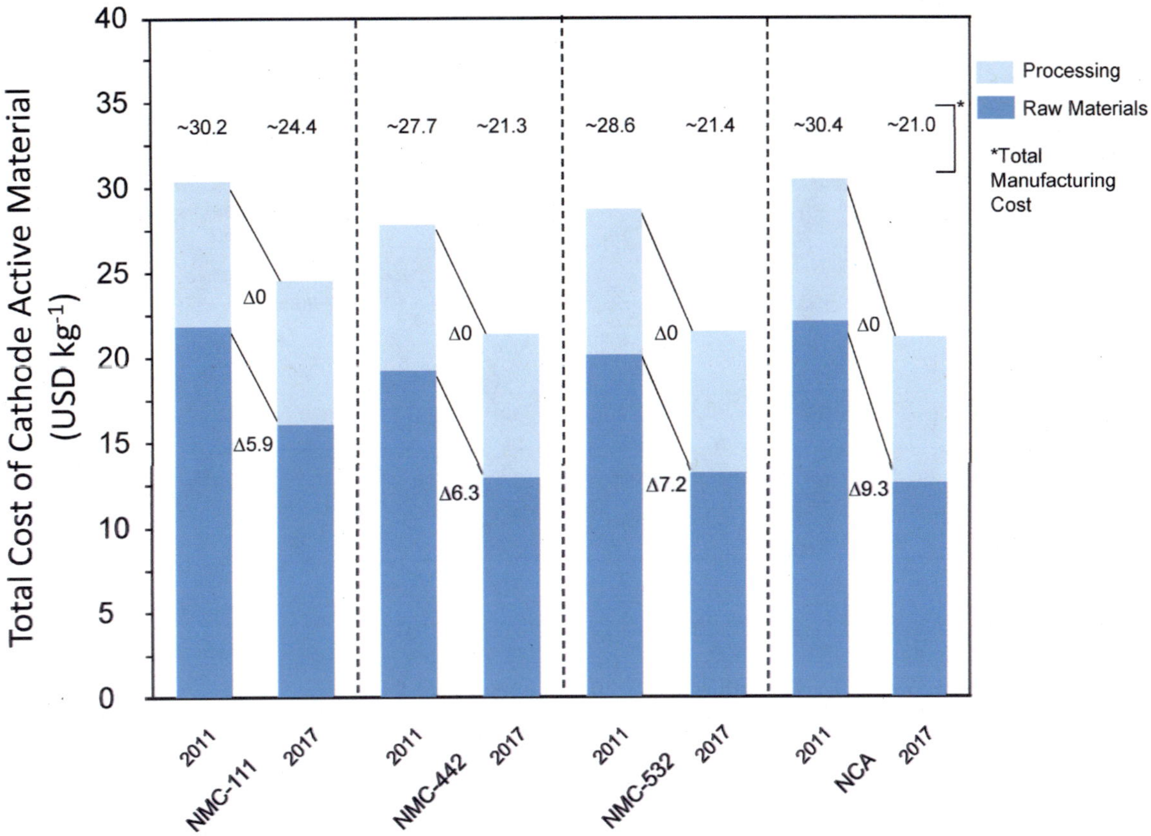

FIGURE 5.13 Total cost changes in cathode active materials for NMC-111, NMC-442, NMC-532, and NCA between 2011 and 2017.
SOURCE: Wentker et al. (2019).

always intercalate all of the lithium ions coming over from the cathode. If the anode cannot accommodate all of the lithium ions, lithium metal plating can occur on the anode, which results in capacity fade and safety concerns. Higher ratios of anode to cathode (> 1.10) capacity provides better insurance against such events, but the excess anode takes up space and adds weight to the cell with no energy density benefit. In addition, excess anode results in more SEI formed, which consumes lithium and lowers energy density. Another safety factor built into most cells is extra anode area relative to the cathode. In other words, the anode is slightly larger than the cathode. Again, this takes up extra space and adds weight to the cell.

Increasing electrode thickness reduces the volume and weight of inactive materials in a given cell size. In addition to improving cell-level energy density, this also reduces cell costs, as shown in Figure 5.14. Improvements in technology of cathode coating for designed electrodes can enable cost reduction while maintaining performance.

5.3.3.7 Manufacturing Processes

Battery manufacturing constitutes approximately 30%–50% of battery costs, depending upon the location of manufacture and scale. The process is capital intensive and consists of multiple complex operations, an example of which is shown in Figure 5.15.

In addition to the high-cost equipment, the battery manufacturing process is energy intensive. Large furnaces are required to evaporate the solvents from the coated electrodes. Owing to the sensitivity of the cell chemistry to moisture, the cell assembly must be performed in a dry room, which incurs large energy costs.

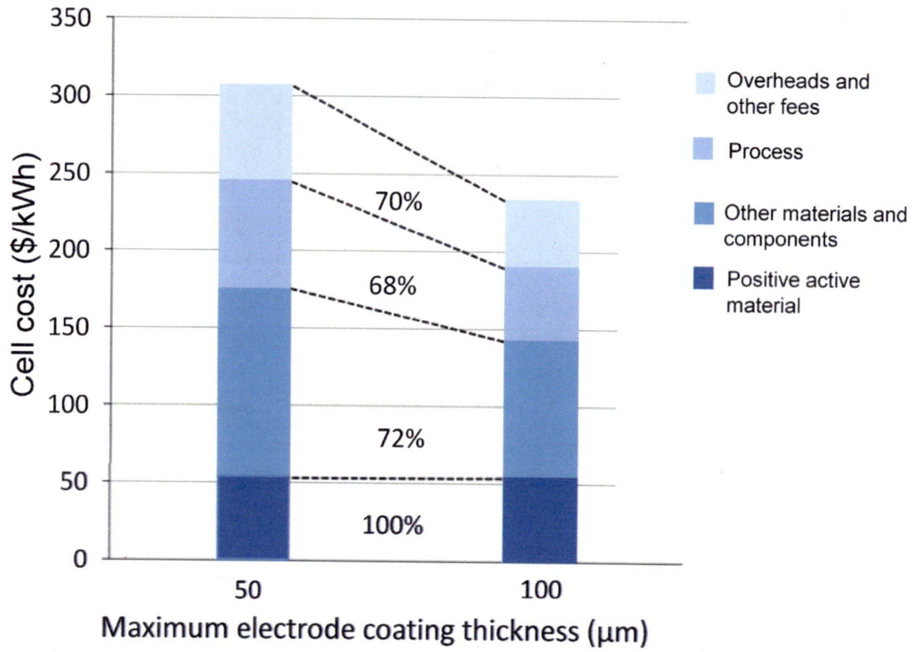

FIGURE 5.14 Cost breakdown for an NMC cell with two different electrode thicknesses.
SOURCE: Patry et al. (2015).

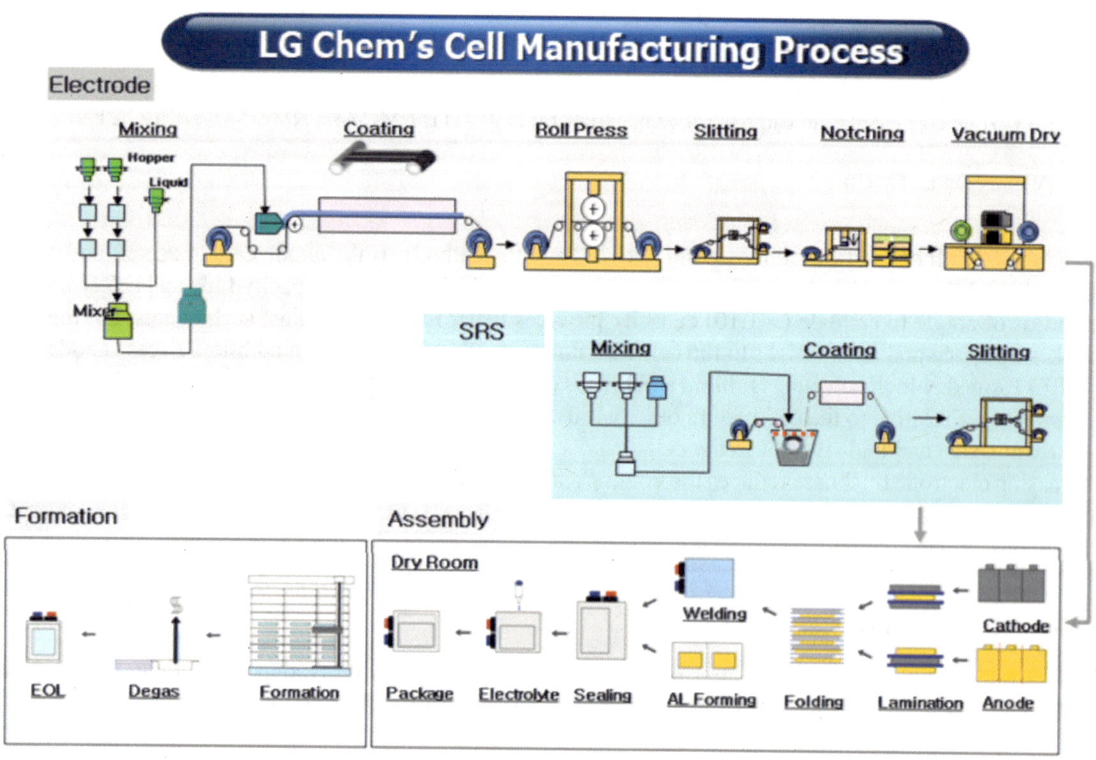

FIGURE 5.15 Schematic of pouch cell battery manufacturing process.
NOTE: SRS = safety reinforced separator.
SOURCE: Koo (2012).

Beyond increasing volume, there are opportunities to reduce the cost of battery manufacturing. Coating thicker and wider electrodes reduces energy costs to dry the solvent. Currently, anodes are coated from aqueous slurries, but cathodes still use an organic solvent, N-methyl-2-pyrrolidone, which requires safety equipment and must be recycled from the drying furnace. Elimination of these organic solvents would reduce processing costs for electrode manufacturing.

The other large cost factor is the dry room manufacturing. At this point, there are not many technical approaches that eliminate the need for a dry room. However, materials that are less sensitive to moisture would be advantageous in cost reduction. Last, the formation process requires expensive equipment and holds up inventory. Formation process for some products can take as long as one week. Ex situ SEI chemistries that would eliminate the need for slow formation cycles could shorten this time and reduce the cost of formation equipment. At current time, there are no viable technologies for liquid electrolyte cells that address this problem.

5.3.3.8 Cell and Pack Cost Reduction

As growth of vehicle electrification has occurred, costs have come down owing to cell-level and pack-level improvements, and they are expected to decrease even further as volume increases. For example, GM announced that LG Chem cells cost $145/kWh total energy in 2019 but will be reduced to $100/kWh by 2021–2022 (Cole, 2015; Gardner, 2017). Similarly, Volkswagen reported that its battery cell costs were around €100/kWh ($108/kWh) in 2018, and battery system costs are reducing to below €100/kWh by 2020 (Witter, 2018). As battery costs continue to come down, various studies suggest cell costs will be 73% to 84% of the total battery pack cost with higher production volume in the 2025–2030 time frame (Anderman, 2017; Pillot, 2019; UBS, 2017). While those companies have focused on nickel manganese cobalt (NMC) technology, Tesla has similarly approached the same $100/kWh cell-level cost in 2020 with NCA technology, which has lower amounts of expensive cobalt (P3, 2020). These announcements underscore how quickly battery costs are declining as automakers and the suppliers move to higher volume and lower cost materials.[2]

Decreases in cell costs owing to material changes, process changes, and volume translate to decreases in pack costs (Wentker et al., 2019). However, it is noted that material costs and battery costs have reduced to below the numbers shown. For example, cobalt prices in 2019–2020 have consistently been about half of the 2017–2018 prices applied in that study. As previously shown, cell cost is decreased as cobalt content in the cathode is minimized. The sensitivity of the cell cost to the constituent metal pricing can be translated to pack costs as shown in Figure 5.16, which gives an example for a low cobalt NCA/graphite cell. As manufacturing scale increases, the overall production costs drop. The effect on the total cost will depend upon the fraction of the cost that is owing to materials versus process. Therefore, Figure 5.16 shows the effect on costs for scenarios where the materials account for 60% to 80% of the total pack cost (Wentker et al., 2019).

5.3.3.9 BEV Versus HEV Cell Technologies

Much of the previous technology descriptions focused on improved energy density, which is most relevant for BEVs. Cells for HEVs are designed more for power than energy, as their primary role is to support a downsized engine when higher power is required and to run auxiliary equipment in a stop-start fashion. The HEV is not plugged in to recharge, so the battery must be able to capture energy lost during braking. This requires fast charge acceptance to capture a maximum amount of energy. A larger amount of regenerative braking energy that can be captured and stored results in more energy that can be used to augment the engine and ultimately better fuel efficiency. Thus, the electrodes in HEV batteries are thinner and less dense than BEV cells. The cells are also operated over a narrower SOC than BEV cells, which enables the long cycle life (hundreds of thousands) required for HEVs.

Today's lithium-ion battery chemistries can meet the performance requirements for HEVs, with some differences along the hybrid spectrum, as illustrated in Figure 5.17. SOC conditions for batteries used in various electrified

[2]As of 2019, five battery suppliers delivered batteries to supply at least 200,000 EVs per year (Sharpe et al., 2020).

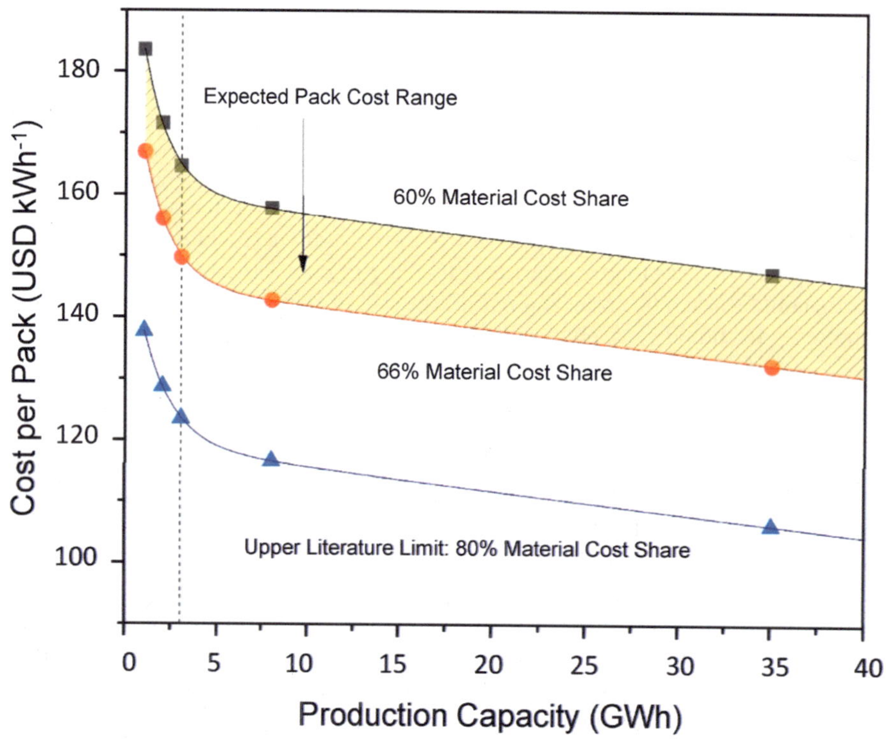

FIGURE 5.16 Changes in cost per cell pack as a function of production volume.
SOURCE: Wentker et al. (2019).

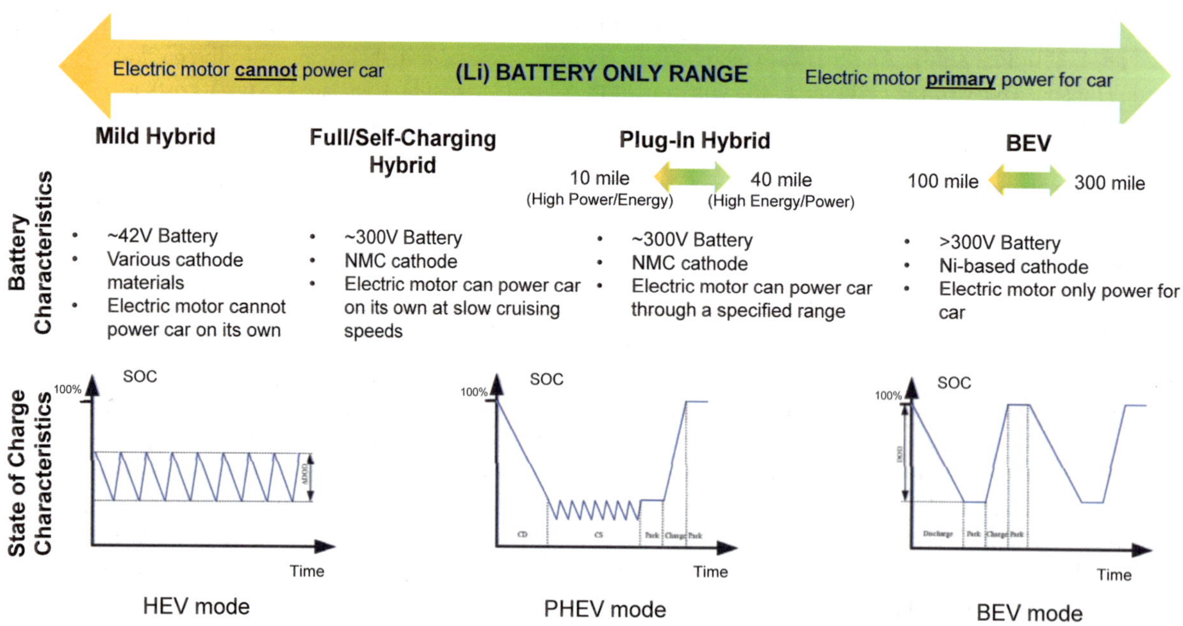

FIGURE 5.17 Summary of battery differences along the spectrum of mild hybrid to BEV. HEV, PHEV, and BEV batteries vary dramatically in their size and SOC characteristics.
SOURCE: Committee generated using images reprinted from Han et al., A review on the key issues of the lithium-ion battery degradation among the whole life cycle, *eTransportation* 1: 100005, © 2019 with permission from Elsevier.

vehicles are also shown in this figure. An in-depth discussion of SOC implications will follow in Section 5.3.4; however, considering SOC in the context of the different battery requirements between HEVs and BEVs points to very different power and life cycle considerations. Material cost reductions for batteries can still play an important role in overall cost reductions across the mild hybrid to BEV spectrum, but increases in production volume—the key approach for cost reduction in BEVs—can be leveraged in cells for HEVs and PHEVs as well.

5.3.3.10 Next-Generation Technologies

Current trends in more traditional battery materials rely on small incremental improvements toward higher nickel, higher voltage cathodes, and silicon containing anodes. As indicated by the references cited above, greater increases in specific energy (e.g., above about 400 Wh/kg) and cost reductions (e.g., below about $60/kWh) will likely need to originate from next-generation technologies that go well beyond the lithium-ion technologies that are relatively well known in 2020. These future technologies are often referred to as "beyond lithium" technologies and encompass varied approaches and chemistries. The time frame for solving key technical challenges for these next-generation chemistries is unclear.

Lithium Metal Anodes

Today's anodes serve as hosts to take up lithium shuttling from the cathode, with carbon or silicon having theoretical capacities of 370 or 3579 mAh/g, respectively. From an energy density perspective, lithium is also an ideal anode as it is 100% active material with a specific capacity of 3844 mAh/g. The use of lithium anodes is under development but has many challenges that are summarized in Figure 5.18.

Plating and de-plating of lithium is not uniform. The lithium tends to grow dendrites, which are needle-like structures. The dendrites can puncture or grow thorough separators, resulting in a battery short, which can precipitate a safety event. Even if dendrites can be prevented, the plated lithium tends to be low density high surface area material. High surface area lithium is very reactive. The reactivity with liquid electrolyte results in rapid capacity fade and consumption of electrolyte and lithium. In addition, puncture or other damage to the battery can expose high surface area lithium to the atmosphere, which will result in a fire. Last, the lower density plated lithium causes relatively large dimensional changes which can exert large forces on the structure of the battery module or pack. Thus, additional space needs to be incorporated into the design to accommodate these dimensional changes—which negatively affects volumetric energy density. These technical challenges need to be overcome while using the least amount of excess lithium possible. In order to realize the maximum energy density benefit of a lithium metal anode, no lithium would be theoretically built into the anode. A copper current collector would

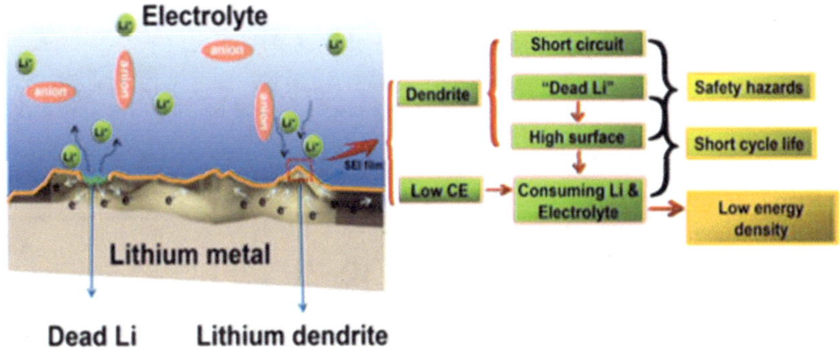

FIGURE 5.18 Challenges with the use of lithium metal anodes.
SOURCE: Reprinted from Wu et al., Perspectives for restraining harsh lithium dendrite growth: Towards robust lithium metal anodes, *Energy Storage Materials* 15: 148–170, © 2018, with permission from Elsevier.

be built into the cell. The plating and stripping of lithium would be performed on lithium solely coming from the cathode added to the cell. Realistically, this is not possible, as the issues with non-uniform lithium plating and lithium consumption owing to electrolyte reactivity prevent such a cell from cycling very long as there is no excess lithium. Therefore, a factor of twice the lithium in the cathode is targeted to keep energy density high yet achieve stable cycling. At the current time, there are no commercial suppliers of low-cost thin lithium foils to meet this target.

Owing to the technical challenges of safety and life of lithium metal batteries as well as the commercial challenges of low-cost lithium electrodes, it is not anticipated that these will have any significant penetration into automotive markets before 2035.

Solid-State Electrolytes

As previously described, today's organic electrolytes are volatile and flammable. Significant increases in safety can be achieved by replacing these liquids with solid-state materials. Safer materials may allow a reduction in system-level thermal management, allowing for improved system-level energy density and reduced costs. In addition, solid electrolytes may enable safer use of lithium metal anodes by mitigating growth and penetration of dendrites—which ultimately results in energy density improvements. There are, however, many challenges to the development and implementation of solid-state electrolytes into batteries.

First, solid electrolytes need to transport lithium ions similar to liquid electrolytes over a temperature range appropriate for automotive applications, −30°C to 60°C. The lithium-ion conductivity of various families of solid ion conductors is shown in Figure 5.19. As shown, some solid electrolytes have inherent lithium-ion conductivities equal to or better than typical liquid electrolytes.

However, lithium transport also needs to occur between the solid electrolyte and the active material particles. While liquids generally wet the particles to provide a cohesive interface even during expansion and contraction of

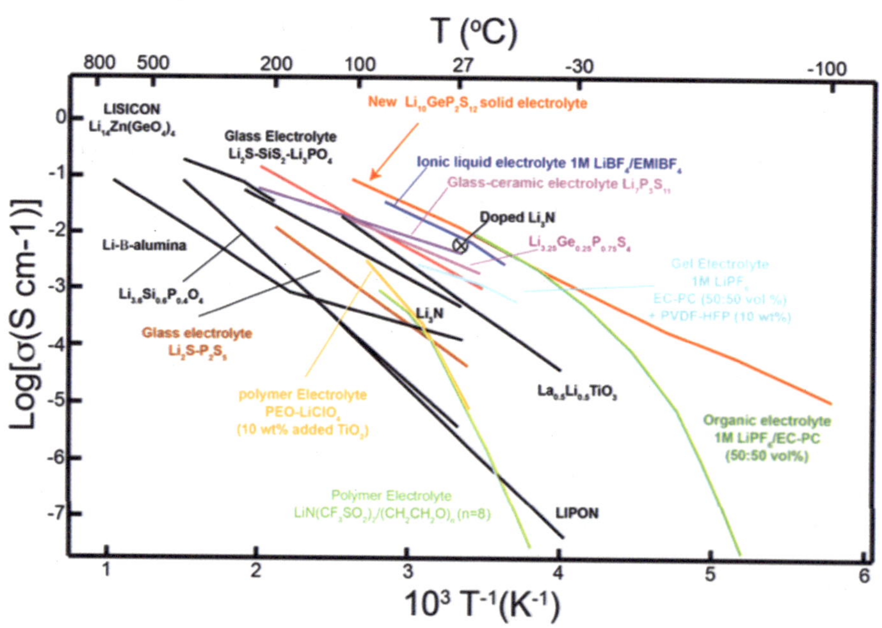

FIGURE 5.19 Ionic conductivities of solid electrolytes compared to liquid electrolytes. While solid electrolytes have inherent safety benefits, their conductivities are more than an order of magnitude lower than liquid electrolytes.
SOURCE: Reprinted by permission from Springer Nature: Kamaya et al. A lithium superionic conductor, *Nature Materials* 10(9): 682–686, © 2011.

the active material particles, the solid-to-solid interface is not as robust. The following considerations are important for engineering cells with solid electrolyte:

- Large amounts of solid electrolyte may need to be added to achieve adequate lithium transport within the electrodes, which decreases gravimetric energy density.
- High stack pressures are often required to minimize impedance between the electrode layer and the solid electrolyte layer within the cells. These pressures require heavier and more costly structures to contain the cells.
- Solid electrolyte needs to be chemically and electrochemically stable on the surfaces of the electrodes. Some of the most highly conductive solid electrolytes, such as sulfur-containing materials, are not stable at typical cathode potentials in advanced lithium-ion batteries. The cathodes need to be coated with thin layers of $LiNbO_3$ or other materials, which add cost and complexity to the active material manufacturing process. Some of these materials, such as the sulfides, are also not stable on lithium metal anodes.
- Lithium metal anodes are used to improve cell energy density, and the solid electrolyte needs to be wetted by the lithium metal in order to minimize formation of high surface area lithium and lithium dendrites. Ideally, the shear modulus of the solid electrolyte should be a factor of eight higher than that of lithium metal to avoid puncture by dendrites. Experimentally, even very hard ceramic materials suffer from dendrite penetration owing to growth along grain boundaries.
- Manufacturing processes need to be modified. In the case of the highly conducting sulfides, the materials release toxic and explosive H_2S when exposed to moisture. While lithium-ion battery manufacturing is done in a dry room, the release of even small amounts of H_2S is problematic. Other materials, such as ceramics, are hard, brittle materials that require significant engineering to fit into conventional roll-to-roll manufacturing processes.

While lithium metal anodes are not expected to be have significant use in automotive cells before 2035, solid-state electrolytes can be used with conventional anodes such as graphite or silicon. Although the energy density advantage of a solid-state cell is not realized with conventional anodes, the elimination of organic liquid electrolytes can be a safety advantage. Integration of a high conductivity solid electrolyte into conventional lithium-ion electrodes has been demonstrated and may be commercially relevant by 2030. In terms of cost, as described in previous sections, cell costs are dominated by cathode costs, which would not change. Some cost savings might be realized with the minor components: while a solid electrolyte would still likely cost more than today's liquid electrolyte, a solid electrolyte would not require a separator. However, processing and manufacture costs to integrate a solid electrolyte would probably be higher than the addition of a liquid electrolyte.

Lithium-Sulfur Batteries

Lithium-sulfur batteries use lithium metal as an anode and low-cost, high-capacity sulfur as a cathode. As such, they are subject to all the technical challenges previously listed for lithium metal anodes. Since the sulfur cathode is not typically prelithiated, all the lithium in the cell must come from the anode—so a thin lithium foil is required. The advantages of a sulfur cathode are multiple, as shown in Table 5.9—high capacity, high availability, and low cost (Zhao, 2016a). However, the relatively low voltage and practical approaches necessary to achieve good cycle life negate some of these advantages.

The cell is built in the charged state, and on first discharge lithium ions move from the anode to the sulfur cathode. The sulfur is reduced at the cathode, and the S-S bonds in the sulfur break. Ultimately, a series of polysulfides, S_n^{x-}, are created. Complete reduction of sulfur results in the formation of Li_2S in the cathode.

In addition to the challenges of using lithium metal in the cell, lithium-sulfur cells have additional technical hurdles. First, the sulfur cathode is not electronically conducting, which is required for a rechargeable battery. This is typically managed by embedding the sulfur into an electronically conductive carbon type matrix. Even with good dispersion of sulfur in the conductive matrix, less than 50% of the sulfur can be typically utilized. Between the addition of the carbon matrix and the poor utilization of sulfur, a theoretical capacity of more than 1000 mAh/g

TABLE 5.9 Redox Properties of Various Lithium Cells

Properties	$LiCoO_2$	$LiNiO_2$	$LiMn_2O_4$	$LiFePO_4$	Sulfur
Redox couple	Co^{4+}/Co^{3+}	Ni^{4+}/Ni^{3+}	Mn^{4+}/Mn^{3+}	Fe^{3+}/Fe^{2+}	$S/S_n^{x-}/S^{2-}$
Voltage (V)	3.6	4	3.9	3.5	2.1
Specific capacity (mAh g^{-1})[a]	274	274	148	170	1675
Discharge capacity (mAh g^{-1})[b]	145	160	105	155	400
Environmental friendliness	Poor	Fair	Good	Good	Good
Availability	Low	Fair	High	High	High
Cost	High	Fair	Low	Low	Very low

[a]Theoretical.
[b]Practical.
SOURCE: Fan et al. (2018); licensed under CC BY-NC-ND 4.0, https://creativecommons.org/licenses/by-nc-nd/4.0.

becomes a practical capacity of a few hundred (Figure 5.20). Since energy density depends upon both the specific capacity and the cell voltage, the net result is that it is difficult to demonstrate significant improvements in practical energy density over traditional lithium ion.

Another significant challenge for lithium-sulfur cells is achieving long cycle life owing to soluble species formed at the cathode. While Li_2S is completely insoluble, some of the intermediate higher order species are soluble in the electrolyte. These dissolved species can migrate to the anode, where they are reduced to lower order and precipitate. This results in loss of active material at the cathode and formation of high resistance layers on the anode—both of which are very detrimental to lithium-sulfur cycle life.

While large improvements in lithium-sulfur technology have been observed in the past few years, significant improvements are still required. Commercial cells are available with stated energy densities of 450 Wh/kg, but cycle life of these cells is only a few hundred cycles. Owing to the low densities of both sulfur and lithium metal, the volumetric energy densities of these cells are lower than those of today's lithium-ion batteries. Owing to the technical challenges of safety and life of lithium-sulfur batteries as well as the commercial challenges of low-cost lithium electrodes, it is not anticipated that these will have any significant penetration into automotive markets before 2035.

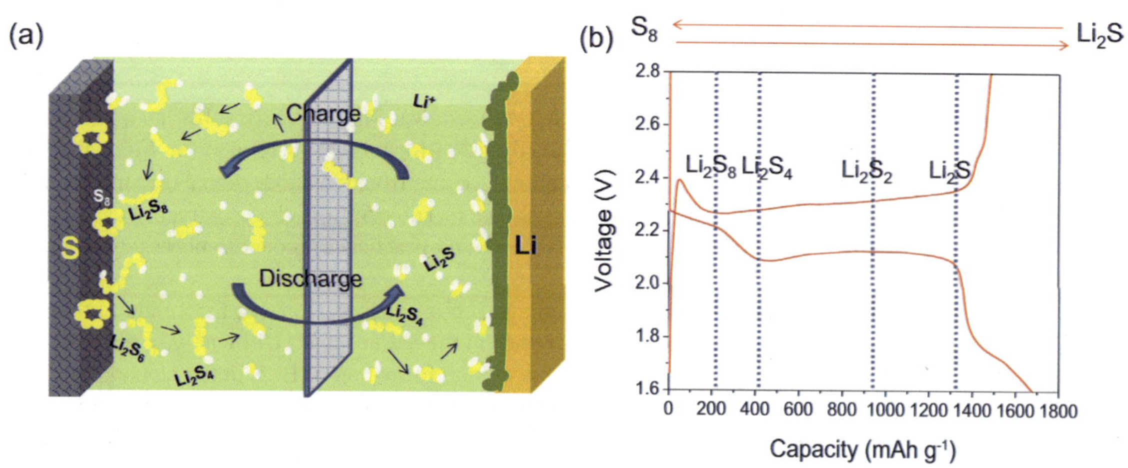

FIGURE 5.20 The electrochemistry occurring in a lithium-sulfur cell, showing cell structure (a) and lithium as well as sulfur speciation at various capacities and voltages (b).
SOURCE: Fan et al. (2018); licensed under CC BY-NC-ND 4.0, https://creativecommons.org/licenses/by-nc-nd/4.0.

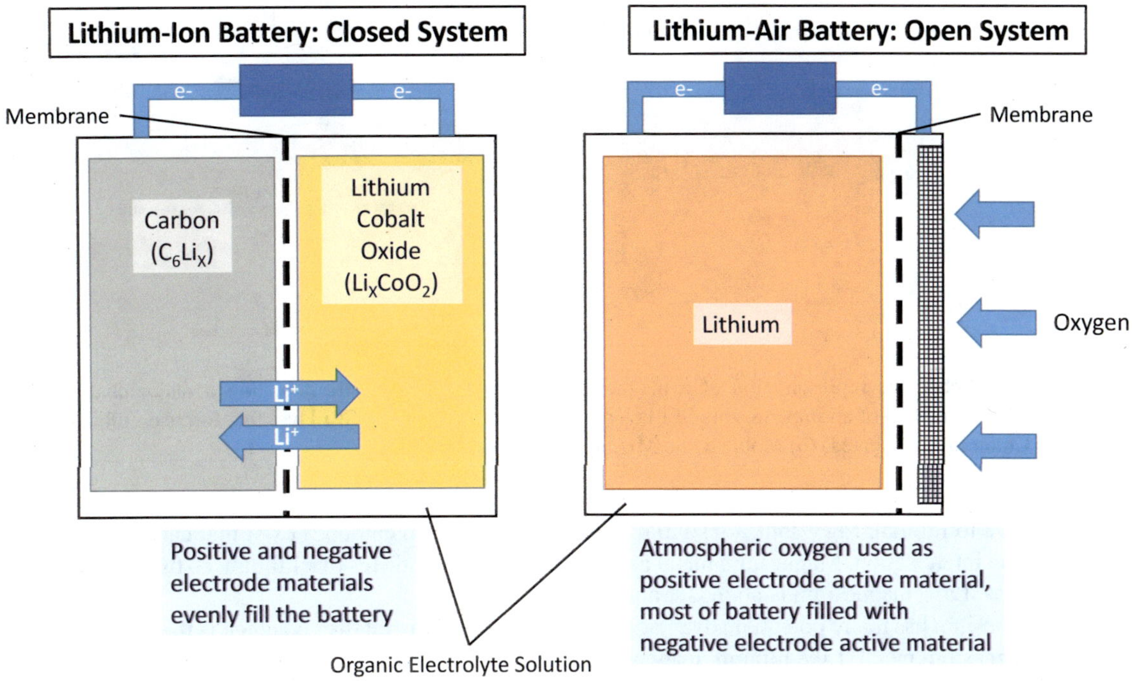

FIGURE 5.21 Schematic comparison of a traditional lithium-ion battery with a lithium-air battery.
SOURCE: Committee generated using NTT (2016) and NTT (2020).

Lithium-Air Batteries

Theoretically, lithium-air batteries have tremendous potential to improve cell-level energy density and cost, as no cathode material is required. As shown in Figure 5.21, oxygen from the environment serves as the active material. A traditional lithium-ion battery is a closed system where the cathode takes up a substantial amount of space. In an open system with oxygen coming from the environment, the battery would mainly consist of just the lithium anode.

In theory, lithium-air architectures would present a substantial energy density improvement. Practically, lithium-air batteries require significant development. On the cathode side, a cheap, efficient oxygen reduction catalyst is required to achieve high reversible capacities. The catalyst needs to be incorporated into some type of structure, which takes up space in the cell—diminishing the energy density advantage. The structures that contain the catalyst must be porous to allow transport of oxygen through the system, but those pores can become blocked by insoluble reduction products of oxygen, such as Li_2O. The anode in these cells is lithium metal, which cannot be exposed to moisture or CO_2 (which would result in formation of insoluble LiOH or Li_2CO_3). The cell needs to contain a membrane through which O_2 can rapidly transport but that blocks CO_2 and moisture. Current prototypes of lithium-air cells are frequently operated under enhanced oxygen environments to achieve high power performance.

In addition, the lithium metal anode is subject to all the performance and safety issues previously addressed. Owing to the technical challenges of lithium-air cells, it is not anticipated that these will have any significant penetration into automotive markets by mid-century.

Magnesium Batteries

Magnesium batteries continue to be of interest to the industry owing to multiple advantages over lithium-ion batteries. Magnesium batteries consist of a cathode that can intercalate/deintercalate magnesium ions, a separator, electrolyte, and a magnesium metal anode (Figure 5.22a). Because magnesium is multivalent (Mg^{2+} versus Li^+), the movement of a magnesium ion from anode to cathode translates to two electrons—meaning greater storage

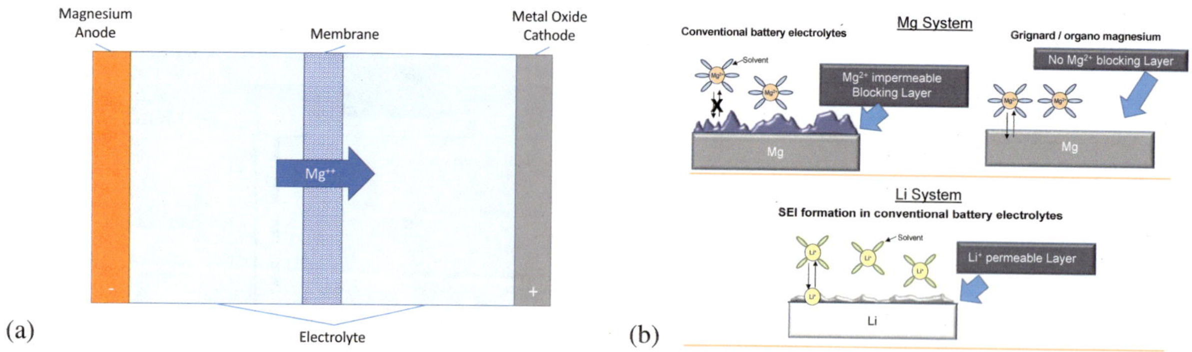

FIGURE 5.22 (a) Schematic representation of a magnesium battery. (b) Schematic depicting a major challenge with magnesium anodes: formation of an impermeable SEI layer. This is in contrast to the SEI layer that forms on lithium anodes. SOURCE: (a) Committee generated; (b) Mohtadi and Mizuno (2014).

of energy relative to lithium. Mg^{2+} ions are similar in size to Li^+ ions, so cathodes exist that can fit the ions into their structure via intercalation. Magnesium metal anodes do not form dendrites like lithium, so they have potential safety advantages. Last, magnesium is abundant and low cost.

While magnesium has many potential advantages to lithium, a significant disadvantage is the reaction of electrolyte with magnesium metal. Like lithium, magnesium metal is very reactive with an organic electrolyte. However, on lithium metal, organic electrolytes are reduced during the formation cycle to form the SEI. The relatively stable SEI allows lithium-ion transport into and out of the anode but electrically insulates the remaining electrolyte from further reaction with the anode. On magnesium metal, the electrolyte reacts and forms an SEI. However, the resulting SEI does not allow magnesium ion transport, as shown in Figure 5.22b. Owing to the complexity of the electrolyte development and further needs to improve cathodes for these systems, it is unlikely that magnesium batteries will have any significant penetration into automotive markets before 2035.

5.3.4 Battery Management Systems

The BMS is the combination of hardware and software responsible for ensuring reliable and safe operation by estimating the remaining usable battery capacity, and health of the battery. The BMS relies heavily on estimation of SOC and state of health (SOH). Together, these estimations act like the ubiquitous fuel gauge in a conventional vehicle, telling the driver how much range remains on the vehicle. To ease range-anxiety, modern PHEVs and BEVs have complex estimation algorithms for the SOC and SOH to translate the remaining battery energy to miles based on recognizing driving and terrain patterns.

The performance and longevity of EV battery packs relies on constraining their operation so that current, SOC, and temperature are regulated within prescribed limits. Enforcement of constraints is achieved by the communication of BMS algorithms with the vehicle electronic control unit to limit the power input (charging) or the power being drawn from the battery pack. Enforcing these limits can cause power denials affecting vehicle torque generation, braking, and charging time. These power limits are encompassed in the real-time estimation of the pack state of power (SOP) based on lithium concentration, and temperature- and age-dependent internal impedance.

The BMS also balances all the cells in a pack—a function that is critical to the safe and efficient function of a BEV. Cell balancing is necessary to extract the maximum energy from the pack, as the cell with the lowest capacity or extremum in SOC will limit the total cyclable energy (charging and discharging must be terminated when any cell reaches its limits). Over time, the fraction of stored charge relative to the total capacity in each cell will begin to drift owing to differences in the temperature-dependent self-discharge rate and the rate of capacity loss.

The BMS also monitors sparse temperature measurements sampled strategically from key cells in a pack. The measurements are typically complemented by the estimation of internal cell temperatures since they might be several degrees higher during high-power operation than the sensed values. The battery SOP limits account for the estimated highest internal temperature and the thermal gradients developed in the pack depending on cell and pack geometric features, packaging, and cooling from the vehicle thermal management system. The recognition of abnormal battery system conditions and fault detection are also performed in the BMS to assess the conditions against calibrated thresholds and issue messaging to the vehicle owner for condition-based maintenance.

Ultimately, the BMS needs to be able to estimate critical battery states and rely on an accurate prediction of complex electrochemical, thermal, and mechanical phenomena, typically through the combined use of models and measurements. Under normal operation, the BMS enforces safety through set limits that restrict internal states of individual cells inside the pack, with the capability of preventing harmful operation that would result in lithium plating, metal dissolution, or particle cracking. These limits are typically calibrated in the laboratory using a lengthy and comprehensive set of experiments at the beginning of life (BOL) of a cell or pack that probe aging and other harmful mechanisms, building safety margins for the entire life of the pack. Most BMS algorithms are currently reactive and enforce voltage and temperature limits based on short predictions—a combination of data-driven and physics-based models with various levels of fidelity (from equivalent-circuit models to first-principle models). More recently, model-free BMS are being developed based on data-driven predictions (Attia et al., 2020). Critically, BMS research is still addressing the following:

- Computation of first-principles models that can run microcontrollers in real time and provide predictive capability of inner physical states (Dubarry and Beck, 2020).
- Identification of physics-based model parameters to reflect the real battery age using onboard measurements under real-world use. This endeavor is much more difficult than off-line model tuning using lab experiments owing to limited, sparse, and noisy real-world data (Dubarry et al., 2020).
- Machine learning based on aggregation of a plurality of onboard sensing and real-world use features to inform (predict) long-term use. Onboard prediction is much more difficult than prediction under full DOD and repeated cycling conditions (Sulzer et al., 2020; Severson et al., 2019). Despite these difficulties, data collection across academic laboratories and from all EVs across manufacturers and environments worldwide may advance the recognition of features, clustering, and data analytics for the prediction of battery life (Che et al., 2020; Aykol et al., 2020).
- Adaptation of the BMS to slow down aging by adjusting fast charging protocols with optimized pulses if packs are used aggressively (Choe et al., 2013), or stretching their utilization (power, energy, and range limits) if the packs are gently used (Lam et al., 2020).

These efforts involving data, models, and algorithms are considered highly proprietary, but much fundamental and precompetitive research remains to be done that would benefit all battery chemistries, form factors, and applications. Concentrated efforts could leverage the U.S. Department of Energy (DOE) results from a multiyear program called CAEBAT: Computer-Aided Engineering for Electric Drive Vehicle Batteries that developed multiscale multidomain models and the software integration for the design of cells and packs. Such a limited effort was championed and coordinated by the ARPA-E project AMPED for advancing models, algorithms, integrated sensing, data, and power electronics. Data analytics and machine-learning with physics-informed features will enable accurate estimations and predictions of battery SOC, SOP, and SOH in real-world vehicle-use. Onboard or telemetric data collection of battery signals (voltage, current, and temperature) along with driving patterns could provide customized battery life prediction based on the battery's past history and likely future use patterns. This kind of customized onboard prediction can be a key advance in driver convenience, manufacturer warranty management, battery design, planning for end of life, second use or recycling, and EV policy making.

5.3.4.1 State of Charge

Battery SOC describes the remaining battery capacity and, therefore, remaining driving range. Since battery behavior is affected by several factors such as operating temperature, current direction, and history, battery SOC is a function of these factors. Battery SOC is defined as the ratio of available capacity to the nominal capacity. Many studies have been conducted to accurately estimate battery SOC, the earliest ones from the National Aeronautics and Space Administration estimating astronaut backpack range (Pop et al., 2005) and automotive applications (Verbrugge and Tate, 2004; Plett, 2004). Today's estimation methods can be grouped into the following categories:

- *Coulomb counting:* Coulomb counting relies on the integration of the current drawn from and supplied to a battery over time. Unknown initial SOC is one of the main problems and is circumvented with voltage inversion after a rest. Sensor accuracy is also important since accumulated errors can lead to a drift in the estimated battery SOC.
- *Voltage inversion:* Battery SOC can be estimated by using voltage measurement, which is referred to as voltage-inversion method since this method utilizes the one-to-one relationship between voltage and battery SOC. The relationship can be implemented or programmed using a look-up-table, piecewise linear function, or mathematical function (Pop et al., 2005). Including temperature and c-rate dependency for SOC correction makes this estimation process more complicated and more expensive than coulomb counting.
- *Combination of coulomb counting and voltage inversion:* Considering the deficiencies of coulomb-counting-based and voltage-inversion-based SOC estimation, some early implementations of onboard SOC estimation algorithms attempted to combine both methods. The need for heuristics tuning was made redundant with the adoption of modern model-based estimation.
- *Model-based closed-loop SOC estimation:* In model-based estimation, the output error injection (measured voltage) and model-based prediction (predicted voltage) assuming an SOC and comparing with the actual measured voltage; the error between measured and estimated voltage can then be used to reduce the error in the assumed SOC from a closed-loop estimation as a means to combine the coulomb counting and the voltage inversion in a systematic way. This is depicted in Figure 5.23.

Model-based closed-loop estimation has been adopted as the most widely used method for battery SOC estimation. The representative models used for estimation are nonlinear extended Kalman filters, as first presented by Plett (2004) using equivalent circuit models of battery cell behavior. Thereafter, a slew of other model-based techniques—including extended Kalman filters for electrochemical models (Plett, 2004; Xia et al., 2014), sigma-point/unscented Kalman filters (Ji et al., 2013; Hu and Yurkovich, 2012), particle filters (He et al., 2014; Hannan et al., 2017), sliding mode observers (Wang et al., 2017), and their variants—have been proposed. Wang et al. (2018) published a comprehensive review of model-based methods for SOC estimation.

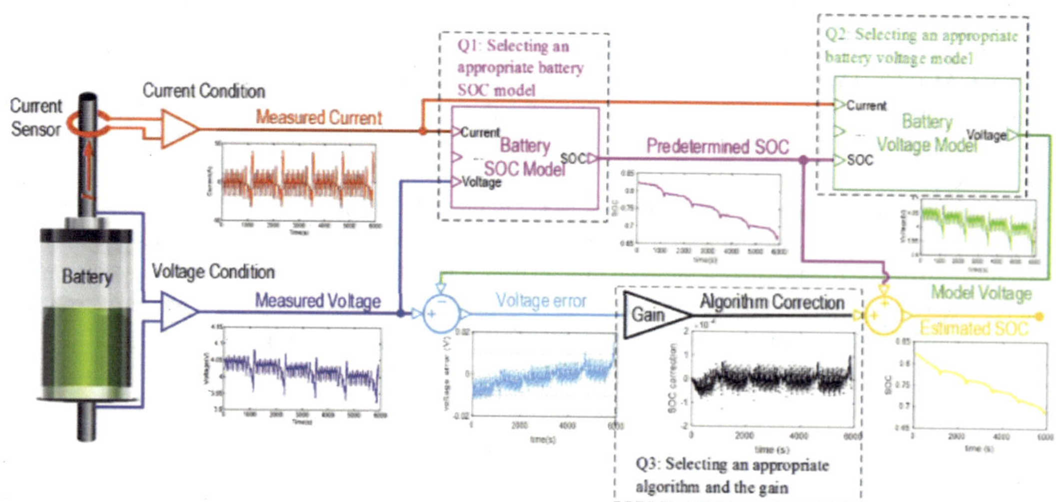

FIGURE 5.23 Schematic of an SOC estimation method in the form of a flow chart that integrates measurable outputs with models and algorithmic corrections.
SOURCE: Reprinted from Zheng et al., Investigating the error sources of the online state of charge estimation methods for lithium-ion batteries in electric vehicles, *Journal of Power Sources* 377: 161–188, © 2018 with permission from Elsevier.

Nevertheless, model-based estimation methods suffer from two issues in practice: (1) the need for knowledge of model parameters (Lin, 2017), which are often difficult to obtain and subject to change over time; and (2) limited fidelity even for the complicated electrochemical model (Speltino et al., 2009). Therefore, in recent years, data-driven and machine learning approaches for battery state estimation have received increasing attention. However, the data-driven approaches need to be handled with caution. The drawbacks include the need for massive training data, data labeling, and specialized tests or measurements to extract features; overfitting under biased or noisy data; and the heavy computational load required for training. Last, the resulting model states may not represent any physical state of the battery, making their interpretation difficult.

5.3.4.2 State of Power

Battery SOP, or power capability, refers to the constant power that can be safely drawn from or provided to the battery over a finite window of time (Verbrugge and Koch, 2006; Wang et al., 2011). Information on the battery SOP is useful when making decisions for optimal power split in the core of hybrid powertrain systems (Lin et al., 2003). Battery SOP estimation is also important for battery thermal management (Kim et al., 2013, 2014) and charging limitations where thermal constraint as well as electrical constraints are considered, as shown in Figure 5.24.

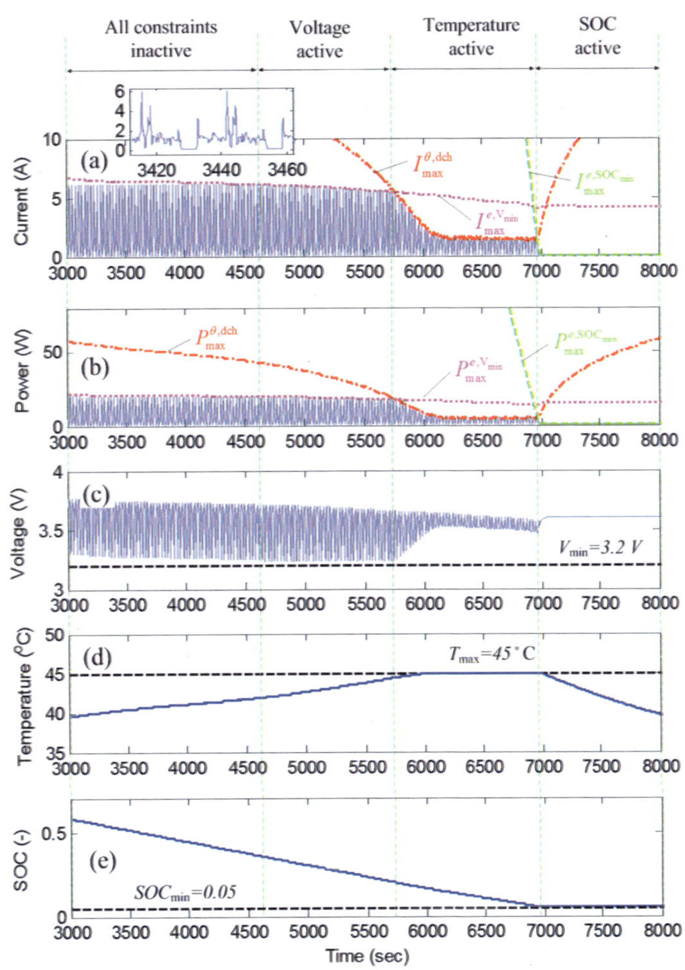

FIGURE 5.24 Battery SOP estimation during battery operations at 30°C ambient temperature with natural convection (6 W/m^2/K): (a) current, (b) power, (c) terminal voltage, (d) temperature, and (e) SOC.

SOURCE: Republished with permission of the American Society of Mechanical Engineers, from Kim et al., 2014, "Maximum Power Estimation of Lithium-Ion Batteries Accounting for Thermal and Electrical Constraints," ASME 2013 Dynamic Systems and Control Conference, Paper DSCC2013-3935, https://doi.org/10.1115/DSCC2013-3935; permission conveyed through Copyright Clearance Center, Inc.

It is expected that future SOC and SOP algorithms will impose constraints associated with internal stresses that can fracture the electrode particles and consume the lithium inventory and hence cause capacity loss or stress rates that can cause layer delamination and impedance increase (Lin, 2018).

5.3.4.3 State of Health

Batteries degrade over time so that a battery that is several years old and has been through many charge cycles will not hold a charge as well as a brand-new one. The degree of battery degradation is quantified by impedance increase and capacity loss; both of these measures decrease the power capability and energy availability, and consequently the vehicle range. SOH estimation seeks to identify the capacity fade or impedance increase in an aged pack. The impedance increase is particularly important in HEVs, since batteries act as power buffers for the ICEs or the fuel cells on board. Capacity fade is more relevant to BEVs that can travel long enough distances to reach the stored energy limit. The pack degradation depends on the degradation of the weakest (lowest capacity and highest resistance) cell in a series string, making cell-to-cell balancing an important BMS functionality. Cell-to-cell variability is caused by non-uniform temperature distributions owing to cooling and manufacturing tolerances.

The factors that affect battery degradation and cause aging are SOC, current, and temperature. These factors influence the cell's potential and the rate of change in potential that each electrode experiences, potentially resulting in internal particle stress, cracking, and more SEI build-up from phase transitions in the electrode material. Fast charging or low temperature are aggravating conditions that can lead to lithium plating. Beyond the loss of cyclable lithium, lithium plating can cause internal shorts and lead to thermal runaways. High SOC also coincides with high cathodic overpotential, which could cause dissolution of the cathode electrode metal oxide. High currents are also damaging for Li-ion batteries, especially the ones with thick electrodes (high-capacity cells), because they cause internal overpotential gradients owing to limitations in electrode and electrolyte diffusivity.

The most accurate method for estimating the actual battery capacity on board a vehicle is to fully charge and then fully discharge while counting (integrating) the current drawn, also known as coulomb counting. Electronic devices are fully charged and discharged more often than most EVs. Assuming the median driving distance, most large battery packs do not utilize more than 20% of their stored energy, which begs for other methods that will provide accurate estimation of the remaining capacity on board the vehicle.

Most methods rely on comparing the measured voltage versus coulomb counting versus the BOL open circuit voltage and inferring the capacity loss. Researchers have shown that capacity loss can be estimated based on identifiable peaks and plateaus in incremental capacity analysis and differential voltage curves (Mendoza et al., 2017; Lin, 2018; Zhao, 2016b), which do not require complete charge and discharge. The degradation can be identified by matching the changes of the aged open circuit voltage curve to various degradation modes. But all these methods also require operation until a certain DOD—associated with electrode phase transitions—is reached.

Without such data, the estimation suffers from the loss of accuracy; hence, the reliability of the estimation results has to be questioned. The estimation uncertainty of the SOH at shallow depths of discharge is shown to be inaccurate by up to 30% (Lee et al., 2020b), posing significant concern on the ability to predict automotive battery end of life. Additional measurements, such as the cell expansion as the cell charges, allow SOH derivation under limited DODs (Mohtat et al., 2019), raising the possibility of reducing the range anxiety with the additional cost of pack sensors. Another possible option is that the BMS can prompt the user to fully discharge occasionally, followed by a slow-charging protocol, to enable a more accurate SOH estimation.

Life Cycle Prediction

Life cycle models have been extensively researched and published for both physics-based and empirical modeling approaches using laboratory data. Physics-based models employ an electrochemical model with a side reaction sub-model to capture the mechanism of battery degradation (Kleinet al., 2013; Schwunk et al., 2013; Tanim et al., 2015), such as the SEI growth and lithium deposition. These models can predict the capacity loss and resistance growth over time driven by inputs, such as current and temperature. The empirical approach, however, uses a heuristic formula obtained from historic data fitting. In this approach, SOH metrics (e.g., capacity loss and resistance growth) are expressed as a function of time and impacting factors (e.g., current magnitude, temperature, DOD, etc.).

These open-loop prediction methods are generally convenient to implement in practice and provide prognostic tools for chemistry selection and component sizing. They rely on a rich data set for fitting the model, but they have to be used in combination with an estimation of the actual degradation state for higher accuracy predictions of future aging. It is worth noting that recent machine learning techniques (Severson et al., 2019) predict the life of a cell based on how certain features changed between the first and the 100th cycle. This technique's predictive ability is limited, however, by the repetition of the duty cycle.

Testing Battery Degradation and Real-World Use

Battery capacity degradation is considered a barrier for market penetration of BEVs. Battery life is commonly measured by the number of charge-discharge cycles before the battery capacity is degraded to 80% of its original capacity. However, the most common testing method is based on an accelerated test with deep discharge and full recharge cycles. This method is reasonable for technology benchmarking but does not represent real-world end-use factors and therefore is inadequate for informing consumers and BEV makers—for example, on total cost of ownership, range anxiety over vehicle lifetime, BEV range design, and battery warranty offering. Real-world driving involves shallow discharges and micro-cycling that is very different from the degradation testing done in the laboratory under accelerating stress conditions.

Some laboratory-based battery testing studies show that battery capacity degradation progresses more slowly with smaller SOC windows during charge-discharge cycles (Omar et al., 2014). It is also reported that the BEV leasing company Tesloop has its Tesla Model S vehicles driven more than 400,000 miles without significant battery capacity degradation (Tesloop, 2018), although it is not yet clear if smaller SOC windows are the explanation. Recent Nissan battery life data suggest that the battery itself may last more than 10 years beyond the life of the vehicle (Loveday, 2019).

Indeed, Figure 5.25 shows the life cycle for various DOD versus the 100% DOD that typically is exercised in laboratory tests. How conditions affect the SOC and DOD, both on average and within a cycling window, are critical for assessing battery life; thus, they contribute to a combination of onboard cell diagnostics for estimating the battery SOH. Together with a (off-line trained) prognostic model, these measures can provide life prediction based on the intended or learned duty cycle as shown in Figure 5.25.

If real-world battery lifetime can be even longer than previously understood with rich and reliable data from real-world BEV operation, it could have implications on battery R&D priority, battery warranty provision, consumer confidence and acceptance of BEVs, and the role of electrification in fuel economy policies.

Life prediction also guides battery sizing (Samad et al., 2018), warranty, and resale value, as discussed below. Other economic and environmental assessments that guide decisions for repurposing batteries to grid application or recycling automotive packs depend on accurate estimations of the battery life. Some of these considerations are discussed in Box 5.2.

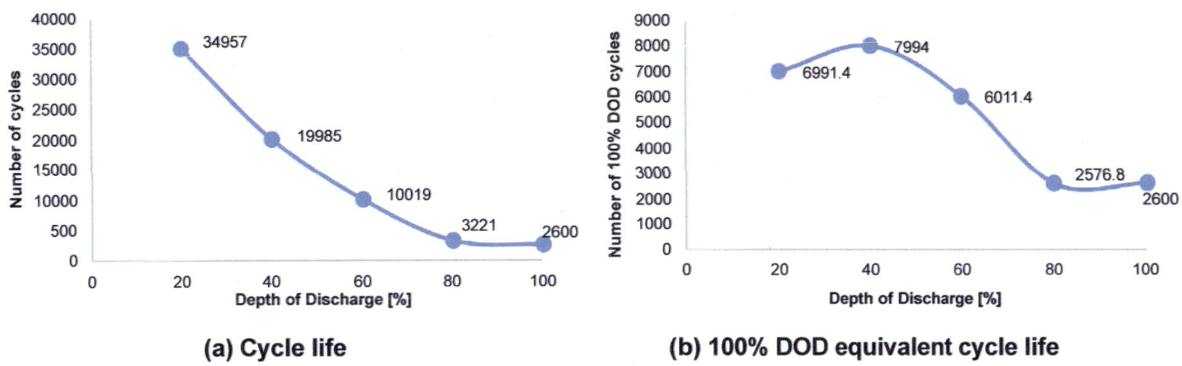

FIGURE 5.25 (a) Cycle life and (b) influence of DOD in cycle life.
SOURCE: Reprinted from Han et al., A review on the key issues of the lithium-ion battery degradation among the whole life cycle, *eTransportation* 1: 100005, © 2019 with permission from Elsevier.

BOX 5.2
The Battery Electric Vehicle Ecosystem

Large-scale deployment of EVs will affect processes throughout the vehicle ecosystem and present new practical issues that will require specific infrastructure to address safety and recycling of battery materials, and require consideration of the life cycle costs of batteries. The introduction of millions of Li-ion batteries—the expected dominant EV battery type for the foreseeable future—presents safety challenges for manufacturing and distribution, everyday use, and end-of-life processes.

Manufacturing

High production levels of BEVs will require manufacturing large numbers of batteries, which presents specific challenges including the following:

- Strain on the already unstable raw mineral supply chain for cobalt, a common cathode material, will increase as demand for Li-ion batteries rises.
- Increased demand for lithium presents concerns about its global availability.
- Large distances between battery manufacturing sites and automaker locations could present logistical challenges and create a bottleneck in the supply chain.

Commercial Use

Commercial use of BEVs at scale will require diverse stakeholders to employ modifications or precautions relevant to their services, such as the following:

- The vehicle maintenance industry will require both repair technicians and aftermarket parts manufacturers to be highly trained in BEV design and specific safety concerns.
- Transportation of vehicles will require additional care to avoid damage to the battery pack. Tow truck drivers need to be familiar with BEV design and safety concerns.
- Emergency responders will need to quickly identify BEVs and be trained to handle the unique hazards of BEVs, such as how to safely quench a Li-ion battery fire.
- Insurance providers will need to adapt policies to account for risks that are unique to or increased for BEVs.
- Roadside assistance services will need to prepare for emergency remote charging of Li-ion batteries and towing to the nearest charging stations as an increasing number of vehicles become stranded on roadways owing to depleted batteries.
- Electricity providers will need to prepare for increased demand on the grid owing to BEV charging.
- Auto dealers will need to install chargers, introducing safety and cost issues.

End of Life, Safety, and Disposal

Last, BEVs introduce unique end-of-life considerations compared to their ICE counterparts, particularly when millions of vehicles are retired annually, such as the following:

- Car storage at end of life will require vehicle disposal sites to implement the necessary precautions for storing BEVs prior to battery removal, such as avoiding water.
- Battery reuse and recycling will require either identifying a market for retired EV batteries with degraded performance or establishing supply chains to recover raw materials from retired batteries.
- Battery disposal supply chains will need to prepare for higher volumes to safely dispose of toxic battery components that cannot be recycled.

Many automakers use active cooling and heating systems to help keep the battery at a healthy temperature and limit exposure to damaging conditions. Some also advise owners against fully charging or discharging the battery, since storing batteries at 100% or discharging them to 0% tends to cause fast degradation. Tesla owners, for example, may control their maximum charge, while they are advised to a "daily" charging to 80%–90% unless they know they will need the full range and to plugging in as soon as possible whenever the battery energy is very low.

5.3.4.4 Accidents, Faults, and State of Safety

Trends in Li-ion battery technology feature a continuous increase in battery energy density, which also increases the severity of battery failures. But for comparison, it is important to note that the heat of combustion for Li-ion cells (E_B(Wh) = 0.14 M_B(g)) is an order of magnitude smaller than the one associated with similar mass of gasoline (E_G(Wh) = 12.8 M_G(g)). Given the considerably longer range reached by 1 kg of gasoline than 1 kg of battery, the fire heat release should be compared for similar vehicle range as it is shown in Figure 5.26.

To maintain safety, the BMS must prevent each individual cell in the pack from overcharging, discharging, or reaching too high a temperature to minimize failures associated with harmful operation. Some failures, however, involve mechanical abuse or other unanticipated events that can lead to battery internal short circuit and self-heating (Feng et al., 2018). At elevated temperatures, exothermic battery side reactions will become active, starting with the decomposition of the SEI layer. This leads to gas evolution that further leads to cell swelling and potentially cell rupture and gas venting. The resulting hazards include toxic off-gassing, smoke, fire, and even an explosion if combustible gases accumulate (Nedjalkov et al., 2016; Abada et al., 2016).

Regarding the risk of electrochemical failure, the 2017 National Highway Traffic Safety Administration report concludes that the propensity and severity of fires and explosions from the accidental ignition of flammable electrolytic solvents used in Li-ion battery systems are anticipated to be somewhat comparable to or perhaps slightly less than those for gasoline or diesel vehicular fuels (Stephens et al., 2017). The overall consequences for Li-ion batteries are expected to be less because of the much smaller amounts of flammable solvent released in a catastrophic failure situation. Recent studies, however, have measured the gas emission from lithium-ion batteries during malfunction for different fault scenarios, showing a large variety of species with mostly toxic to highly toxic properties (Nedjalkov et al., 2016). Although statistically rare, battery fires pose specific hazards and safety challenges that must be understood (Bravo-Diaz et al., 2020).

Detection of a thermal runaway event inside a battery pack should be made immediately after the fault to avoid further damage. From a regulation perspective, the proposed global technical regulation No. 20 on EV safety also requires that in an event of thermal runaway, the vehicle shall provide an advance warning indication to allow egress of 5 minutes prior to hazardous conditions inside the passenger compartment (NTSB, 2020). This requirement is deemed to be satisfied if the thermal propagation does not lead to a hazardous situation for the vehicle occupants.

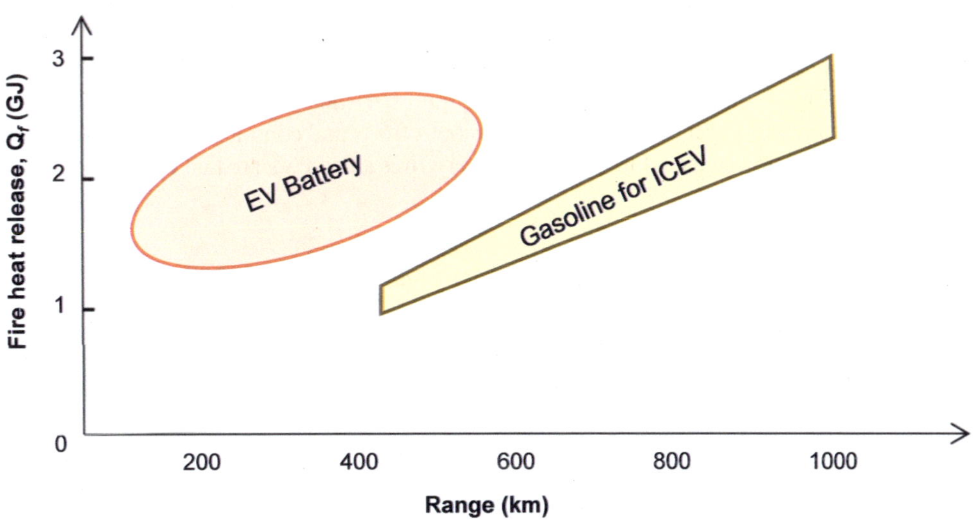

FIGURE 5.26 Fire heat release for EV and gasoline vehicles by range.
SOURCE: Reprinted by permission from Springer Nature: Sun et al., 2020, A review of battery fires in electric vehicles, *Fire Technology* 56(4): 1361–1410, © 2020.

Conventional methods of battery thermal runaway detection are usually based on voltage and surface temperature measurements. These methods work well for a single cell but are difficult to apply in large scale battery packs used in automotive and grid storage applications. Moreover, in EV battery packs, the batteries are connected in parallel. For example, the Tesla Model S battery packs come with 74 cells in parallel. A large number of parallel-connected batteries will suppress the battery fault voltage signal. Because the other healthy cells in parallel will continue to supply the nominal voltage, the pack voltage with a single cell at fault will be very similar to that of a healthy battery pack, making the fault detection using voltage alone challenging.

Note here that overcharging (discharging) protection for cells in parallel connections requires additional hardware and design measures that for most small parallel arrays is accomplished by internal disconnects, such as positive temperature coefficient devices embedded in cells that limit current at the cell level. Although these internal cell measures exist, cell rupture and leakage can still occur; thus, gas detection methods are also used to identify cell failure events. The composition of battery vent-gas during a thermal runaway event includes CO_2, CO, H_2, and volatile organic compounds. The gas venting is considered the precursor of thermal runaway events (Cai et al., 2019a). Although the gas venting will not necessarily lead to thermal runaway every time, such faults should be identified to prevent escalating the fault (Cai et al., 2019b). Ongoing research is investigating the target gas species that (1) appears with good consistency in all vent tests, (2) is present in first venting event, and (3) is measurable by an onboard sensor.

The additional sensors and the redundancy that would be required for the detection of the onset of a fault could increase the cost of the pack and reduce the pack energy density, but it is important to systematically address fault detection as EVs start aging and becoming an important portion of the 2035 automotive landscape (Sulzer et al., 2020).

From 2013 through late summer 2020, there have been 17 EV fires in the United States. For Tesla alone, there have been 19 EV fire incidents globally. Among these, 8 accidents are related to collision, 5 accidents are related to charging, 3 accidents happened while the vehicle was parked, and 3 accidents happened during operation. Given the concerns and publicity that EV fires attracted, Tesla has included a fire-safety comparison in their 2019 impact report showing 1 EV fire incident per 175 million miles, whereas there is 1 ICEV fire per 19 million miles. The conclusion drawn was that EVs are approximately 10 times safer than their counterparts based on linear interpolation of the statistics. Notably, the U.S. average age of vehicles on the road is 12 years, whereas the EV average population age cannot be larger than 4 years (Statista, 2020b). EVs seem to be statistically at least two times safer than ICEVs when the population age is taken into account. While these safety statistics are encouraging, it is worth noting that many fire departments in the United States do not yet have protocols in place to respond to EV-specific incidents, and battery-related fires are currently being probed by safety regulators in the United States and abroad (Levin, 2020; Foldy, 2020; IDTechEx, 2020).

Recent recalls of Hyundai's Kona EV and Ford Kuga PHEV along with an investigation on Chevy Bolt EV are drawing public attention, despite the fact that there is still no statistical evidence that EVs are less safe than ICEVs (Riley, 2020). All manufacturers are investing enormous efforts and continue improving their cells, chemistry, pack design, and battery management toward zero battery fires since EVs are that emerging technology that all manufacturers wish to accelerate.

5.3.5 Thermal Management

Lithium-ion batteries' cycle life or capacity is considerably affected by operating temperature owing to irreversible chemical reactions. Thermal management systems monitor and control the battery pack temperature at certain locations. Cooling systems are also designed to maintain a uniform temperature distribution in the battery packs to avoid battery aging via hot spots.

5.3.5.1 Temperature Sensing and Estimation

The temperature of a battery cell or a pack can be directly measured using thermocouples. Despite this simple principle, there are two major concerns on temperature monitoring: (1) whether the measured temperature can be representative of the whole cell, and (2) how many sensors are required to monitor all the batteries inside a pack.

For a relatively small cell (e.g., 18650 type), the Biot number[3] of the battery cell is small, suggesting that the heat transfer at the surface is much smaller than the internal heat transfer by conduction. Hence, no significant temperature gradient inside the cell is expected, meaning that measured temperature can be considered as cell temperature. However, as reported in literature (Forgez et al., 2010; Lin et al., 2013; Kim et al., 2014) for a relatively large cylindrical battery (e.g., 26650 type), core temperature can be considerably higher than surface temperature.

Cost constraints do not allow the temperature measurements of all the cells in a pack limiting the onboard monitoring and state awareness (Lin, 2014). Thermal nonuniformity inside battery cell and across packs requires temperature estimators utilizing control-oriented battery thermal models such as lumped parameter model (Forgez et al., 2010; Lin et al., 2013; Debert et al., 2013) and reduced order model (Kim et al., 2013). The combination of estimators and optimal deployment of temperature sensors under frugal sensing can provide the real-time information to guide and adapt computationally compact thermal models that can propagate the sparse temperature sensing to the entire pack and inform the battery management and power limits even under aged cell conditions (Lin, 2018).

Beyond thermal management under normal conditions, temperature monitoring is extremely important for detecting and diagnosing faults and irregularities like fast degradation through increased cell resistance (Lin et al., 2013) or malfunctioning of the cooling system (Kim et al., 2013). Temperature sensing for safety measures against thermal runaway, however, is too slow, and other sensing requirements will increase cost but dramatically improve mitigation actions, improve first responders' response, and reduce the impact of EV accidents. Comprehensive analysis and improvements in sensing will be required as EV market penetration increases and as a transition occurs to more reactive and energy dense material (e.g., nickel) and away from cobalt. Many committees are currently working on safety standards (e.g., SAE International, 2014). Additional efforts are needed in workforce development and consumer education for the operation, transportation, storage, and disposal of EVs.

5.3.5.2 Cooling Systems

Thermal management for heating and cooling of lithium-ion batteries relies on the existence of auxiliary systems using a medium such as fluid (air or liquid), solid, or phase change material. The thermostatic control, one of the heuristic or rule-based control techniques, has been widely employed for temperature regulation. The basic idea of the thermostatic control is that cooling is applied when the measured temperature exceeds a target maximum temperature. The cooling control (fan-off or pump-off) is turned off when the measured temperature falls below another temperature threshold. Model predictive control can provide optimal and smooth cooling or heating if a good preview of low load or high load demands are predicted accurately (Zhu et al., 2018).

Fluids such as air and liquid are commonly used as a heat transfer medium for thermal management. Air is in direct contact with modules for heat transfer whereas liquid can be in either direct or indirect contact with modules—for example, submerging modules in a dielectric fluid or placing a heat sink plate between the modules. It can be found that there is a trade-off between the heat rejection and power consumption for cooling. Moreover, these performances are influenced by design factors such as the number of cells in series and parallel in the cooling circuit, along with the size of the gap between cells and arrangements. Parallel cooling is advantageous to minimize cell-to-cell variations of temperature, and temperature nonuniformity, which is important to minimize localized degradation in the module or pack. Toyota Prius and Ford Fusion use a parallel cooling for thermal management by supplying the conditioned air from the cabin. Moreover, a parametric study has been conducted to optimize design parameters for both air and liquid cooling systems (Park and Jung, 2013), wherein it was reported that a liquid cooling system consumes much less power compared with an air-cooling system owing to better thermal properties of liquid and high heat exchange efficiency of the radiator. Nevertheless, air cooling involves simple design, lower cost, easier maintenance, and shorter warm-up period over liquid cooling. Mahamud and Park (2011) proposed a method using reciprocating air flow to achieve uniform temperature

[3]Biot number (Bi) is the ratio of heat convected to the surroundings to heat conducted to the surface.

distribution across a battery pack. It was shown that a reciprocating flow method can reduce the cell temperature difference of the battery system by about 4°C (72% reduction) and the maximum cell temperature by 1.5°C compared to the unidirectional flow case.

Phase change material can be an alternative option for heat transfer material. This option is advantageous in terms of efficiency since parasitic power consumption can be removed by using the latent heat of phase change (solid to liquid) at constant temperature or the melting point. Even though this method is beneficial to obtain the minimum temperature distribution in a battery module or pack, there are several concerns associated with the increase of volume and weight of the system owing to phase change material (Bandhauer et al., 2011). Moreover, warm-up of the battery in cold temperature would be difficult owing to the low thermal conductivity of phase change material.

Most existing or planned thermal auxiliary systems depend on a combination of fluid and solid media for an effective heat conduction from the cells to a cooling system.

5.3.5.3 Cold Weather Packages for Light-Duty EVs

Light-duty EVs are reported to have lower range in cold weather owing to lower powertrain efficiency and higher energy consumption for cabin heating. To mitigate this issue, some automakers include the so-called cold weather package into their standard model or as an additional option for customers in cold weather regions. Additionally, battery capacity and charging speed decrease in cold weather. Reduction in range from battery capacity decrease and resulting reduction in trips that can be completed is compounded by the reduced capability to extend range via charging. Slow EV charging in cold weather may be especially unacceptable to drivers as a long wait in a cold car may be unpleasant or dangerous. Even with the availability of optional cold-weather upgrades, automakers want their vehicles to be sellable and drivable in all climates in the United States. The impact of weather on range and charging may mean that longer electric ranges are more cost-effective in cold regions.

In 2018, GM had a survey for its Chevy Bolt and Volt EV owners to inquire about their preference on the different Cold Weather Packages with suggested prices (Malone, 2018). The Cold Weather Package usually consists of a dedicated battery heater, a heat pump system, and multimode thermal management. A dedicated battery heater can warm up the battery pack indirectly by heating the liquid coolant and running the warmer liquid coolant through the pack. A smaller liquid cooler volume typically through a shorter circuit is utilized for more effectiveness and speed of response. Warming up the battery enables higher regenerative braking and faster charging that is typically reduced at low temperatures owing to the slow diffusion and associated propensity of lithium to plate instead of intercalating in the anode material (graphite). In both cases, the vehicle range improves. The heater can be powered by the charger but the fastest way to heat up the pack is to discharge it and generate heat through internal Joule heating (I^2R) (Vlahinos and Pesaran, 2002). Using the battery discharge current to power the heater can double the effectiveness and speed up the time to reach a bulk temperature threshold (Mohan et al., 2019). Once this temperature threshold is achieved, the battery can be charged quickly and safely from the charger or high currents from regenerative breaking.

Many automakers, including Tesla, suggest immediately plugging the car into the charger to maintain a room temperature and avoid the lengthy warm-up process. Trickle heating is, however, energy consuming owing to the ambient heat dissipation and wasteful if applied during long periods of parking or storage. Again, discharging some of the energy stored in the battery is the fastest and most efficient way to warm up and condition the battery before charging it up.

Replacing a resistive heating system with a heat pump in an EV can reduce the energy consumption for cabin heating, which is critical in cold weather. A heat pump is more effective than a resistive heating system in that it extracts the heat from outside instead of generating heat itself (Hu et al., 2020). A multimode thermal management system can have different system settings to optimize for different temperatures. For example, Hyundai Kona model year (MY) 2019 (Nisewanger, 2018) includes three coolant loop modes, for mild, hot, and cold temperatures. For other EV models, Tesla Model Y includes a heat pump and a battery heater as a standard while the Kia Soul has both a heat pump and battery heater as additional options.

5.3.6 Battery Pack Cost

Reflecting the battery innovations discussed above, battery cost estimates are drawn from various battery cost reports as well as information from various automakers and other experts. The battery pack cost is a critical factor for the future prospects of EV adoption, owing to it being a high fraction of the EV cost. Average battery pack prices have declined by at least 80% within 10 years: DOE indicates that battery pack prices declined from more than $1,000 to $197/kWh from 2008 to 2018 (Simmons and Chalk, 2019; DOE, 2019a), and an industry survey indicates that the cost decline was $1,160 to $176/kWh, from 2010 to 2018 (Goldie-Scot, 2019). Technological improvements in cell chemistry, cell design, pack design, and manufacturing have resulted in higher energy density and lower costs than predicted: to provide context for how these cost reductions have greatly exceeded earlier projections, the National Research Council (NRC, 2013) estimated that average battery costs would still be $200–$250/kWh in 2030 and reach $175–$200/kWh by 2050.

Table 5.10 summarizes recent applicable technical studies that quantify EV battery pack costs for continued advances in battery technology and volume. These studies are state of the art in terms of offering transparent bottom-up engineering and manufacturing analysis with specificity on lithium-ion battery chemistries and production volume and include details on battery pack production (e.g., material, cell, and pack costs; cost versus production volume; bottom-up cost engineering approach). The table also includes automaker statements from Volkswagen, GM, and Tesla.

The battery pack costs from these studies and automaker announcements pertain to 2018 through 2030. They generally assume production volumes of 100,000 EV battery packs per year for 2020 and 500,000 units per year for 2025. For context, there were five battery suppliers in 2018 that supplied batteries for at least 200,000 EVs (Lutsey et al., 2018). Several of the estimates indicate that costs will decline to $120 to $135/kWh by 2025. The research studies are corroborated by several industry statements. Tesla (2018), reaching higher volume more quickly than others, indicated that it will reach $100/kWh much sooner, and Berckmans et al. (2017) found that even greater

TABLE 5.10 Technical Reports and Automaker Statements on EV Battery Pack Cost

Type	Report	Battery Specifications and Cost Elements Included
Technical reports	Ahmed et al. (2018)	Pouch NMC 6,2,2-graphite, production volume-based; includes total cost to automaker for material, process, overhead, depreciation, warranty.
	Anderman (2017)	Cylindrical 21700, NCA 83,13,4, production volume-based; includes cost of material, capital, pack integration, labor, overhead, depreciation, R&D, administration, warranty, profit.
	Anderman (2018)	Pouch NMC 8,1,1-graphite, production volume-based; includes cost of materials, capital, pack integration, labor, overhead, depreciation, R&D, administration, warranty, profit.
	Berckmans et al. (2017)	Pouch NMC 6,2,2-graphite anode, production volume-based; includes material, process, labor, overhead, depreciation, profit.
		Pouch NMC 6,2,2-silicon alloy anode, production volume-based; includes material, process, labor, overhead, depreciation, profit.
	UBS (2017)	Pouch NMC 6,2,2-graphite, production volume-based; includes material, process, labor, overhead, depreciation, profit.
Automaker statements	Davies (2017)	Volkswagen statement. Associated with planned production volume of 100,000 per year by 2020 for I.D. series.
	Lienert and White (2018)	GM statement related to Chevrolet Bolt (NMC 6,2,2); associated time frame for production volume has not been stated.
	Tesla (2018)	Tesla statement related to Model 3 production volume of 500,000 with Panasonic battery production (cylindrical 21700, NCA 83,13,4) in Nevada by 2020.

NOTE: NCA = nickel cobalt aluminum; NMC = nickel manganese cobalt oxide (numbers refer to the proportion of each element).

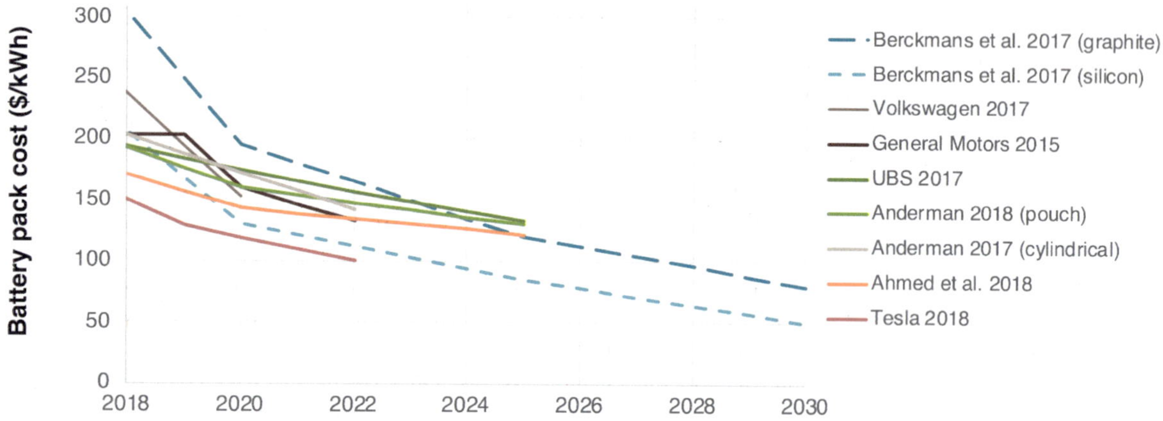

FIGURE 5.27 EV battery pack costs from technical studies and automaker statements.
SOURCE: Modified from Lutsey and Nicholas (2019a).

battery cost declines will occur. An industry survey by Bloomberg New Energy Finance (BloombergNEF) projects a volume-weighted average battery cost reduction from $176/kWh in 2018 to $62/kWh in 2030 (Goldie-Scot, 2019). In addition, Tesla's October 2020 battery analysis indicates the viability of high-volume production battery pack costs reaching approximately $50/kWh in the 2025–2030 time frame (P3, 2020).

Figure 5.27 shows findings from the studies cited above in Table 5.10 to illustrate the likely range of battery pack costs for 2020 to 2030. Several estimates indicate that battery pack costs will decline to $130–$160/kWh by 2020 to 2022, and then to $120–$135/kWh by 2025. However, Tesla states it will reach $100/kWh by 2022, associated with its NCA-based battery pack technology and based on its earlier high-production volume. Berckmans et al. (2017) find that even greater battery cost declines can be achieved with NMC cathode batteries, if the anode can transition from the 2018-dominant graphite to a silicon alloy while overcoming cycle-life issues.

The estimates of overall vehicle prices in the chapter summary (Section 5.5) are based on these references and a baseline 2018 average battery cost of $128/kWh at the cell level, and $176/kWh at the pack level (for a 45-kWh battery pack). The assumption for projected 2030 costs is that incremental improvements to current Li-ion technology will result in ~7% cost reduction per year. This level of cost reduction includes battery improvements in lithium-ion technology as described above, pack manufacturing improvements, and a shift to high-production volume (battery suppliers serving 500,000 EVs per year). Although battery costs are reduced by 7% per year from 2018 through 2030, the precise cell and pack costs will differ by battery pack size. Based on these assumptions, high-volume battery production is expected to decrease battery pack costs to $90–$115/kWh by 2025 and $65–$80/kWh by 2030. To apply battery cost estimates for the vehicle-level costs in the summary section, we apply a decreasing pack-to-cell cost ratio with increasing pack capacity. Our pack-to-cell cost ratio ranges from 1.54 for a 16 kWh pack down to 1.20 for 112 kWh and larger packs, based on Safoutin et al. (2018). This means larger battery packs (e.g., for a 300-mile range sport utility vehicle [SUV]) have lower per-kilowatt-hour pack costs, compared to smaller packs.

5.3.7 Findings and Recommendations for Battery Technologies for EVs

FINDING 5.5: Innovation in materials, components, and packaging is required for improvements in cost and energy density of EV batteries. Lithium-ion batteries, thanks to incremental improvements in materials, supplier competition, and production scale, will be the dominant battery technology in 2025–2035. Further improvements toward "beyond lithium ion" technologies all require breakthroughs that are not guaranteed. Advances in solid-state electrolytes may become commercially relevant after 2030 and could enable the use of lithium metal anodes, which have an extremely high theoretical specific capacity of 3860 mAh/g.

FINDING 5.6: EV batteries are generally composed of cells, modules, and a pack. A cluster of cells makes a module, and a cluster of modules makes a pack. While material improvements (Finding 5.5) are related to the cell-level, pack-level improvements such as wiring and balancing, cooling, and BMS advances will be critical (to meet projected costs) as well.

FINDING 5.7: Battery degradation affects ownership costs and consumer acceptance of plug-in electric vehicles. Relative to laboratory-based estimation, battery degradation under real-world driving is less understood. It is challenging to leverage laboratory aging data to predict the remaining useful life under real-world driving and charging patterns that involve mixed operating conditions, various temperatures, powers, depth of discharge, and voltage.

RECOMMENDATION 5.2: Evaluation of battery life in distance and calendar years should be conducted with real-world driving in mind. The U.S. Department of Energy Vehicle Technologies Office (VTO), in collaboration with the private sector, should investigate real-world battery life for battery EVs and plug-in hybrid electric vehicles using simulation and testing. Results from these studies will inform VTO's battery research investments and target setting, impact flexible design of the new corporate average fuel economy standards, and potentially lead to greater consumer acceptance and faster market penetration of plug-in electric vehicles.

FINDING 5.8: With an increasing emphasis on improving battery energy density, and extending range and performance, some automakers are currently considering new chemistries and new battery management systems to address safety concerns, which could result in increased battery costs.

FINDING 5.9: Although EV fires have attracted attention and raised safety concerns among some consumers, statistical comparison of fire incidents (between battery electric vehicles and internal combustion engine vehicles of all ages) suggests that EVs may be less prone to fire incidents than conventional vehicles. However, the slow and hidden evolution of an internal battery fault is still an issue that needs to be addressed.

RECOMMENDATION 5.3: Efforts by automakers to pack more energy density into EV batteries and speed up charging, motivated by both consumer demand and policy incentives, should be matched with efforts to understand, detect, mitigate, and manage fire risks. The National Highway Traffic Safety Administration in collaboration with the U.S. Department of Energy and automakers will need to lead an effort on advancing battery electric vehicle system safety, and preventing and responding to battery failure.

RECOMMENDATION 5.4: Comparison of fire safety between battery electric vehicles (BEVs) and internal combustion engine vehicles (ICEVs) needs to be further studied, especially between BEVs and ICEVs at similar ages, and between BEVs with different chemistries, battery management system designs, vehicle classes, charging rates, and electric ranges, and even over time between BEV product generations.

5.4 ELECTRIC CHARGING INFRASTRUCTURE

A charging infrastructure for PEVs can be viewed as a technology to reduce fuel consumption of on-road light-duty vehicles. Rather than directly improving fuel efficiency, as with ICE efficiency technologies, a better charging infrastructure allows more charging events, by existing PEVs or new PEVs. Better charging infrastructure permits more electricity to contribute to the total energy needed to power total vehicle miles traveled (VMT), thus reducing the average per-mile fuel consumption of on-road light-duty vehicles. Better charging infrastructure also alleviates range anxiety and encourages adoption of BEVs, resulting in more vehicle miles being powered by electricity. A charging infrastructure consists of a network of chargers, as simple as a regular 120 V power outlet or as sophisticated as a high-power charging station, at different locations to allow PEVs to be safely recharged with grid electricity and to extend vehicle driving range. Dedicated electric charging stations serve a similar role

TABLE 5.11 EV Charging Infrastructure Specifications in the United States

Charging Level	Input Voltage (V AC)	Typical Power (kW AC)	Electric Vehicle Range per Charging Hour (miles)	Location
Level 1	120	1.2–1.4	3–4	Primarily home, some workplace
Level 2	208–240	3.3–6.6	10–20	Home, workplace, and public
DC fast	400–1000	50 or more	150–1000	Public, frequently intercity

NOTE: AC = alternating current; DC = direct current; kW = kilowatt; V = volt.
SOURCE: Nicholas (2019).

to gasoline stations, but the overall charging infrastructure has an advantage in that the ubiquity of the electric grid and charging capability at different power levels allows electric chargers to be installed at private homes, residential communities, workplaces, customer parking areas of business entities or any institutions, or dedicated parking garages. This diversity of charging location, power and time options, and associated accessibility has profound implications with respect to technical attributes, costs, impact on VMT electrification and PEV adoption, the infrastructure need, and usage behavior. In general, when the charging infrastructure becomes more powerful, affordable, available, and convenient, more PEVs will be adopted and more charging events will occur, resulting in more VMT powered by electricity and fewer by petroleum energy. Business models for charging infrastructure are not yet clear or well understood. It is clear, however, that the improvement of charging infrastructure requires investment. The key questions are when, where, how many, and with what technologies to invest and deploy more chargers to achieve the most cost-effective energy impact.

5.4.1 Electric Charging Infrastructure Technologies and Costs

5.4.1.1 AC Level 1, AC Level 2, and DC Fast Chargers

For a given electric range, longer available charging time at some locations (e.g., at-home overnight charging) allows less-expensive, low-power charging. Other locations or travel contexts make high-power charging necessary (e.g., stopping for recharges during long-distance or urgent trips, or complete dependency on public charging owing to lack of access to home charging). As a result, a wide range of charging technologies, largely differentiated by charging power, have been developed and are being deployed. In this chapter a charger is defined as the electric device necessary to connect a PEV with the existing power grid to achieve safe charging. It can be in the form of a charging cable connecting the vehicle directly to a 120 V power outlet, or an electric vehicle supply equipment (EVSE) connection, or a wireless charging pad. The three most common types of charging technologies—AC Level 1, AC Level 2, and DC fast chargers[4]—are the focus of this report, as they are expected to dominate the charging infrastructure in 2025–2035 (Engel et al., 2018b; Nicholas, 2019). Their voltage, power, distance extended per charging hour, and typical application locations are summarized in Table 5.11. Note that even with the most powerful DC fast charging, the charging speed is still far lower than gasoline refueling at about 250 miles per minute.[5] Other charging technologies, including static or dynamic wireless charging, extreme fast charging (xFC), and battery swapping, are not expected to be mature enough or widely deployed to have significant impacts during 2025–2035.

A key distinction between Level 1 or 2 AC and DC fast charging is the location of the AC-DC converter. A battery can only be charged directly with DC electricity. Most PEVs are equipped with an onboard converter (called a charger in Figure 5.28) to accept 120 V–240 V AC electricity from the grid and convert it to DC electricity to

[4]The schematics of these charging technologies can be found in Smith and Castellano (2015).

[5]Assuming 10 gallons per minute of maximum dispensing flow rate according to 40 CFR 80.22 through 80.33 and the average fuel economy of 25 MPG for MY 2019 new vehicles (EPA, 2020).

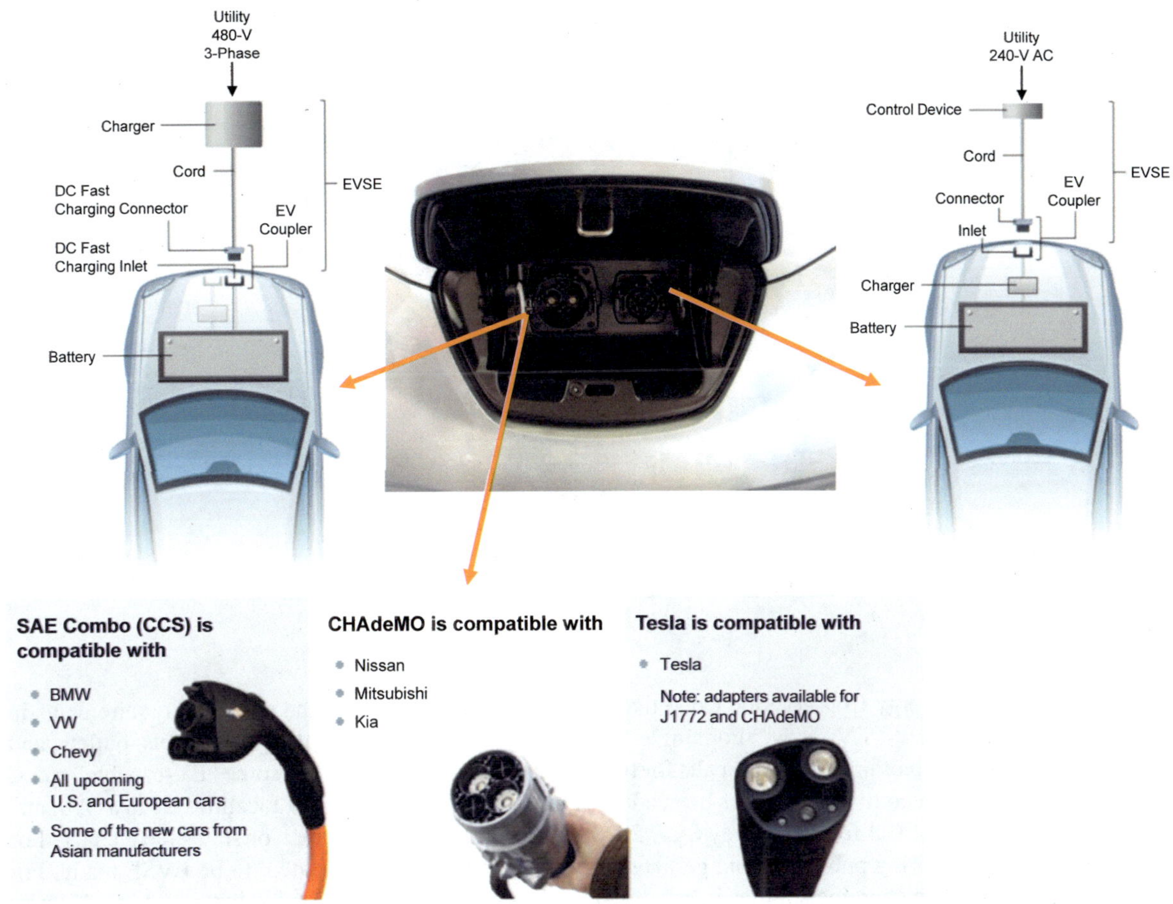

FIGURE 5.28 Typical AC and DC inlets on a BEV, charging system illustration, and DC fast-charging interface standards. SOURCE: Clean Cities (2012).

charge the battery. In contrast, DC fast chargers have built-in converters and directly provide DC electricity that bypasses the onboard converter and feeds into the battery. Many PEV models are equipped with two charging inlets—one for AC and one for DC, as shown in Figure 5.28. In terms of connector design, most PEV models in the United States adopt the SAE 1772 for Level 1 and 2 AC charging. For DC fast charging, there currently are three types of connectors—SAE Combined Charging System (CCS), CHAdeMO, and Tesla—shown in Figure 5.28. The SAE CCS combines single-phase AC, three-phase AC, and DC high-speed charging in a single set of charger, connector, and inlet system, making it compatible with PEV models in both the United States and Europe. Adaptors are available to connect otherwise incompatible EVSE. The lack of a uniform charger interface standard means a need for multiple charger networks and has implications on infrastructure costs and user convenience. Efforts are ongoing to harmonize these different standards (e.g., Buckley, 2011; ANL, 2020; Wolbertus et al., 2020; EPRI, 2019).

The differences among Level 1 AC, Level 2 AC, and DC fast charging affect location availability and required hardware upgrades. In a sense, the electric charging infrastructure is already ubiquitous for Level 1 AC, as PEVs can charge directly with a regular 120 V power plug. Level 2 AC requires a 240 V power plug, which is available with commercial buildings and in residential units, usually for electric dryers but also in garages outfitted for PEV readiness. Installing a 240 V plug and the associated wiring in a home or public garage is much cheaper during new construction than in retrofitting an existing building or garage. The International Code Council has approved

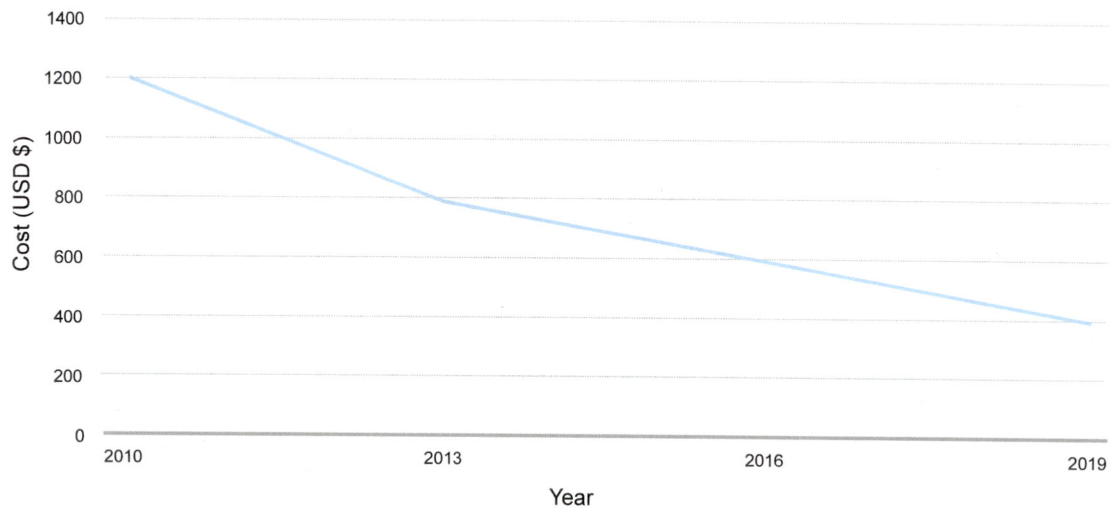

FIGURE 5.29 Cost reduction in residential Level 2 chargers, 2010 to 2019.
SOURCE: RMI, from Nelder and Rogers (2019).

the 2018 International Energy Conservation Code that recommends installation of the electrical requirements in all new homes to make them "EV-ready." For single-family homes, that means installing the panels, outlets, and conduits. For multifamily buildings, the code calls for two "EV-ready" parking spots and some "EV-capable" spots. Such an EV-ready home recommendation has been adopted by an increasing number of localities, such as Atlanta, Denver, and some parts of California (Energy Star, 2020; SWEEP, 2020). Some cities or states have enacted or proposed legislation requiring public parking garages, if to be constructed or renovated, to be EVSE ready. For example, "Electric vehicle charging stations in open parking lots and parking garages," a law passed in 2013 by the City Council in New York, requires that a minimum of 20% of parking spaces in new-construction open lots (or older lots being upgraded) be readied for EV charging (New York City Council, 2013). Specifically, the law requires the spaces to be embedded with at least 1-inch conduit that can support later installation of an EVSE to an electric supply panel with ≥ 3.1 kilowatts of capacity. An EVSE is usually needed to connect the PEV's Type 2 inlet with the 240 V power plug or source. DC fast charging stations require 480 V power sources that are available typically only in industrial or commercial settings. Installation of DC fast charging stations may vary in cost by tens of thousands of dollars depending on readiness of the site (Smith and Castellano, 2015).

More powerful chargers are generally more expensive owing to the higher manufacturing costs for charger components with higher amperage ratings. The cost of owning and installing chargers can be largely categorized in the equipment capital cost, the installation cost, and the operation and maintenance cost. Networked chargers can be connected to the internet and monitored and controlled remotely, including implementation of pricing mechanisms and charging location discovery for customers. Table 5.12 shows charger hardware costs for various charging options in 2019. For context, the cost of a commercial-use Level 2 charger in 2011 was estimated to be $1,875–$4,500 (in 2009$, or $2,234–$5,362 in 2019$) (NRC, 2013), indicating only a small reduction in cost from 2011 to 2019. Meanwhile, the cost of a home-use Level 2 charger has decreased by 67% from 2010 to 2019 (Nelder and Rogers, 2019). Despite a clear decline in the hardware costs of Level 2 residential chargers over the past decade (Figure 5.29), cost reduction seems slower for commercial-use chargers based on the above literature, for reasons unknown.

Installation costs estimated by the International Council on Clean Transportation (ICCT) are shown in Table 5.13 for workplace and public Level 2 chargers and in Table 5.14 for DC fast chargers. The data in both tables demonstrate that the per-charger installation cost can decrease significantly with increasing the number of chargers per site. Charger lifespan, a key parameter for calculating total cost of ownership, is not well understood.

TABLE 5.12 Cost Ranges for Chargers and Infrastructure Components

Cost Element	Lowest Cost	Highest Cost
Level 2 residential charger	$380 (2.9 kW)	$689 (7.7 kW)
Level 2 commercial charger	$2,500 (7.7 kW)	$4,900 (16.8 kW); outlier: $7,210 (14.4 kW)
DCFC (50 kW)	$20,000	$35,800
DCFC (150 kW)	$75,600	$100,000
DCFC (350 kW)	$128,000	$150,000
Transformer (150–300 kVa)	$35,000	$53,000
Transformer (500–750 kVa)	$44,000	$69,600
Transformer (1000+ kVa)	$66,000	$173,000
Data contracts	$84/year/charger	$240/year/charger
Network contracts	$200/year/charger	$250/year/charger
Credit card reader	$325	$1,000
Cable cost	$1,500	$3,500

NOTE: DCFC = direct-current fast chargers; kVA = kilovoltampere.
SOURCE: RMI (2019).

Operation and maintenance costs are typically assumed as a percentage of the equipment capital cost, ranging from 1%–7% (Serradilla et al., 2017), and also need further studying.

Besides the costs of equipment, installation, and maintenance, the costs of charging infrastructure are also affected by electricity cost, which in turn is affected by time of use. The total cost of the charging infrastructure system also depends on its scale, the number of PEVs, and the total electric vehicle miles traveled (eVMT). The total infrastructure cost can be divided by the total electricity consumption to estimate a levelized cost (in $ per kWh) that can be easily compared to electricity prices. Including home, workplace, and public charging equipment costs and electricity prices, the levelized cost of charging has been estimated at $0.15/kWh for light-duty BEVs (assuming a charging mix of 13% residential Level 1, 68% residential Level 2, 14% workplace or public Level 2, and 5% DC fast charging) and $0.14/kWh for light-duty PHEVs (assuming a charging mix of 40% residential Level 1, 41% residential Level 2, and 19% workplace or public Level 2) on average in the United States.

5.4.1.2 Extreme Fast Charging

One major technology direction is xFC, typically aimed at 400 kilowatt (kW) or higher of charging power, which is equivalent to extending about 200 miles of driving in less than 10 minutes (DOE, 2017). xFC could achieve similar convenience as gasoline refueling. This goal is natural but perhaps unnecessary, since most of

TABLE 5.13 Installation Costs per Level 2 Public and Workplace Charger, by Chargers per Site, California Only

		1 Charger per Site	2 Chargers per Site	3–5 Chargers per Site	6+ Chargers per Site
California	Labor	$2,471	$1,786	$1,491	$1,747
	Materials	$1,235	$958	$1,014	$908
	Permit	$283	$172	$110	$65
	Tax	$156	$121	$128	$115
	Total	$4,148	$3,039	$2,745	$2,837

SOURCE: Nicholas (2019).

TABLE 5.14 Installation Costs per DC Fast Charger by Power Level and Chargers per Site

Power Level (kW)	Chargers per Site	Labor	Materials	Permit	Taxes	Total
50	1	$19,200	$26,000	$200	$106	$45,506
	2	$15,200	$20,800	$150	$85	$36,235
	3–5	$11,200	$15,600	$100	$64	$26,964
	6–50	$7,200	$10,400	$50	$42	$17,692
150	1	$20,160	$27,300	$210	$111	$47,781
	2	$15,960	$21,840	$158	$89	$38,047
	3–5	$11,760	$16,380	$105	$67	$28,312
	6–20	$7,560	$10,920	$53	$45	$18,577
350	1	$27,840	$37,700	$290	$154	$65,984
	2	$22,040	$30,160	$218	$123	$52,541
	3–5	$16,240	$22,620	$145	$92	$39,097
	6–10	$10,440	$15,080	$73	$62	$25,654

SOURCE: Nicholas (2019).

the eVMT will be powered by electricity from home, workplace, or secondary[6] fast (6–60 kW) public charging. In these charging events, the cost per unit of time for the traveler is lower than that for gasoline refueling, where the traveler needs to dedicate all attention to the refueling activity. The total time value associated with electricity charging and gasoline refueling may be a better metric of comparison. From this perspective, it is worth questioning how many PEV drivers will need 400 kW of charging power and for those who will, how often. Occasional urgent needs for fast charging may not require a full charge of 200 or more miles. If the need is for a quick extension of, for example, 50 miles in order to arrive at home or another charger location with ample dwell time, the required power for 10 minutes of charging is only 90 kW (assuming 0.3 kWh/mile of electricity rate for driving). Enabling more miles to be extended is certainly valuable, but the value of additional miles charged diminishes—that is, the next additional 50 miles charged is less valuable than the first 50 miles charged. However, the presence of xFC may be perceived as an assurance for usability of the BEV at any time, critical at least for some buyers, and thus motivates BEV adoption.

While xFC can be valuable for PEV operation, it may or may not be cost effective, depending on the additional cost required for technology improvements. After all, profitable business models for 100 kW DC fast chargers are yet to be proved, not to mention for 400 kW xFC, which are presumably more expensive than the current DC fast charging systems. Two types of xFC systems are being pursued, with important cost implications. One is an AC-connected system, where a low-frequency step-down transformer interfaces between the medium-voltage distribution network and a three-phase AC bus, providing 250–480 line-to-line voltage to each charger at the charging station. Such systems use mature technologies but are bulky, complicated, and expensive, especially in public places where the step-down transformer has not been built. Another technology direction under development, xFC chargers based on DC-connected systems (using a single front-end AC/DC converter rather than one with each charger; see Figure 5.30), could potentially reduce costs and footprints and increase system efficiency, especially with the use of a solid-state transformer. xFC infrastructure costs are not well understood. The equipment cost per xFC charger was estimated to be $245,000 based on early prototypes (Francfort et al., 2017), about 4–10 times the cost of current DC fast chargers, but it is expected to reach similar cost levels of the current DC fast charging systems (Borlaug et al., 2020).

[6]As opposed to charging as the main trip purpose. When the driver has a main trip purpose such as eating lunch or shopping, the concurrent charging event is viewed as a secondary trip purpose.

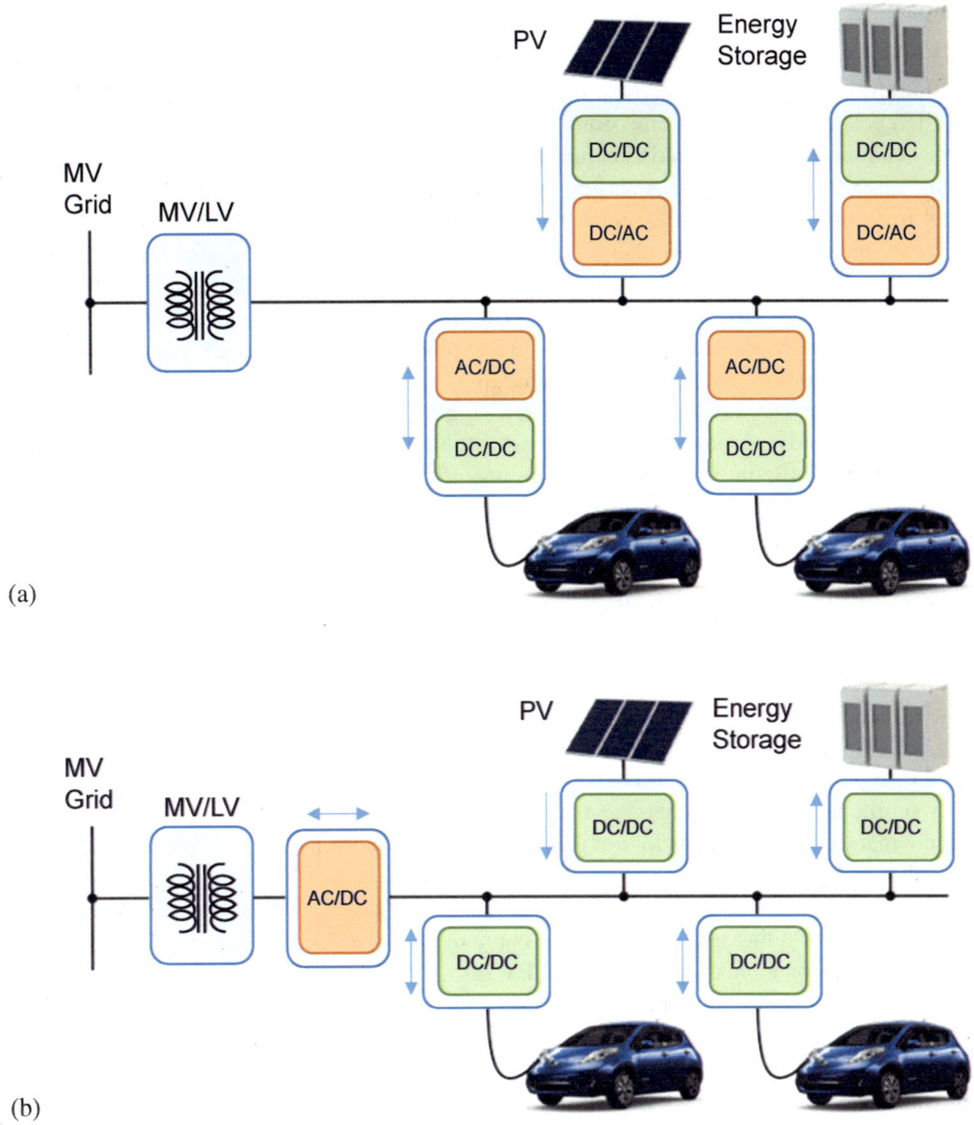

FIGURE 5.30 Configurations for xFC stations. (a) AC-connected system. (b) DC-connected system.
NOTE: LV = low voltage (mains voltage); MV = medium voltage (distribution voltage, 2–35 kV); PV = photovoltaic.
SOURCE: Reprinted with permission from Tu et al., 2019, Extreme fast charging of electric vehicles: A technology overview, *IEEE Transactions on Transportation Electrification* 5(4):861–878, https://doi.org/10.1109/TTE.2019.2958709; © 2019 IEEE.

In addition to cost uncertainty, there are technical challenges to xFC chargers, such as the impact of high power demand on the distribution system, stress and aging of distribution transformers, integration of multiple EVs with random charging demand while maintaining grid stability, communication between the xFC charging station and the existing grid system, lack of fast acting protective devices, lack of standards for xFC practices, adverse effect on battery degradation, higher requirements and costs for vehicle electric components owing to high current, and compatibility with non-xFC-capable PEVs. Solutions are being studied and developed, and include distribution system upgrades, smart charging techniques, and onsite storage (Nicholas, 2019).

5.4.1.3 Grid Impact

The electricity demand from PEVs will likely represent a small fraction of the total electricity demand in the 2025–2035 time frame. Incremental increase in generation capacity will be needed. From the perspective of PEVs reducing life cycle greenhouse gas emissions, it is critical that that added generation capacity comes from renewable sources and the existing generation capacity transitions to low-carbon technologies.

The main challenge will be for the transmission and distribution systems, especially for the residential feeder circuits and local transformers, to meet the power demand from a large number of PEVs charging at the same time, each expecting the rated power of the charger, altogether in the same grid subsystem, and even concurrently with peak demand from other electricity end-users (Clement-Nyns et al., 2010; Dharmakeerthi et al., 2014; Engel et al., 2018a; Nicholas and Hall, 2018). Simply put, the grid impact of PEVs is not an energy or kWh issue, but a peak power or kW issue. For the same feeder-circuit of 150 homes with 25% local EV penetration, for example, the local peak power load could increase by 30% if charging activities are not intervened. Public fast charging can be highly volatile. Unmanaged charging demand at a single DC fast charger could easily exceed the peak-load capacity of a residential feeder-circuit transformer. Multiple DC fast chargers together could easily require peak power capacity beyond the megawatt level. Three potential strategies for addressing these challenges are grid upgrades, demand management and off-peak charging, and onsite storage. These solutions only address the technical challenges of the power system from supply to demand as a whole. There are some challenges related to transactions and pricing between different entities of the system. For example, demand charge can be costly for fast charging, especially xFC during peak hours. In theory, grid upgrades could increase demand charge as they add to the upstream cost. Onsite storage could reduce demand charge but add to end-user (chargers) capital costs.

5.4.2 Electric Charging Behavior

Even when a charger is present, PEV drivers may or may not plug in; the probability of charging therefore depends not only on the availability of charging networks but also on the temporal and spatial nature of travel activities. There are many possible reasons for not needing a recharge when a charger is present, such as sufficient SOC, proximity to home, high fees, and energy cost. BEVs and PHEVs are distinct in this regard, as the decision to recharge when a charger is available is an economic issue (energy cost) for PHEV users but an operation feasibility issue for BEV users. Understanding charging behavior is crucial for assessing the cost-effectiveness of charging infrastructure improvements on increasing eVMT and PEV sales. Overall, charging activities could increase when the charging infrastructure becomes more powerful, affordable, available, and convenient.

Charging behavior may be reflected in charging activities. An Idaho National Laboratory study of the first mass-market PEVs, Nissan Leafs and Chevy Volts, showed that early Chevy Volt drivers powered more than 70% of their miles using electricity (utility factors [UF] of 0.72 and 0.75 for different model years with slightly different ranges; Smart et al., 2014). A more recent study exploring PHEV UF for various ranges of PHEVs in California found similar results for the Volt (35/38 mi range, 0.67 UF; 53 mi range, 0.68 UF) and lower UF for the plug-in Prius (11 mi range, 0.18 UF), C_{max} (20 mi range, 0.41 UF), and Fusion (20 mi range, 0.35 UF) (Raghavan and Tal, 2020). The study also found that longer range vehicles charged farther from home. In the United States, "Public charge infrastructure usage was low (1.4 events per week) but was also very location dependent, some of the public charge stations had between 7 and 11 charge sessions a day. Public fast charging was used more intensively (7.2 charges a week)" (Raghavan and Tal, 2020). Utilization of the charging infrastructure grows with EV adoption (Wolbertus et al., 2016; Lee and Clark, 2018).

These observed charging activities do not reveal the underlying reasons of charging behavior and thus do not show to what extent charging activities, and the eVMT they enable, can be further motivated. BEVs have to be recharged to meet minimum travel needs, but more frequent charging of BEVs to accomplish more trips increases BEV utility for the owners. Charging frequency has to increase if the BEV range is shorter or long-distance trips are more frequent. Charging frequency can, but does not have to, increase if charging opportunities are more available. The downside is the increased physical and mental effort owing to more charging events (see discussion of charging convenience later), while the benefit includes feasibility of more affordable, shorter range BEVs or better battery life owing to reduced numbers of deep charge-discharge cycles. A charging behavior survey by University of California, Davis, shows that, contradictory to the common assumption, BEV drivers do not plug in their vehicles every night, especially when free public charging motivates substitution for home charging (Nicholas et al., 2017). It is important to recognize the possibility that charging behavior may change significantly as a result of expansion of the charging infrastructure, behavior adaption, and consumer prioritization of cost and time in diverse contexts. More surveys and behavior analysis are needed.

In contrast to BEVs, PHEVs do not need to be recharged to be used. Surveys of early Chevrolet Volt adopters show that they are diligent in plugging in at home (Hardman et al., 2016), but it is unclear if that is owing to early adopter enthusiasm for advanced technology or environmental stewardship, or a conscious decision to save energy cost. In a PHEV, the cost saving is about 1.4 cents per mile when electricity replaces gasoline (assuming 45 miles per gallon [MPG], $2 per gallon, 0.3 kWh/mile, and $0.1/kWh). This translates into $0.56 of savings for each full recharge of 40 miles. If the effort of plugging in is perceived to be significant (e.g., pulling a charging cable in a disorganized garage or on a hot day), this cost saving may not be enough to motivate the charging action. This possibility seems to be reflected by the finding by University of California, Davis, that drivers of the Prius PHEV (13 miles of charge depleting range; discontinued) rarely recharged their vehicles. Their purchases were found to be motivated by obtaining the HOV lane access tag. Similarly, several studies indicated that PHEVs without home chargers in Shanghai were rarely plugged in and were purchased to avoid the high license fee of a gasoline-only vehicle (Ou et al., 2020; SHEVDC, 2017). When chargers are not widely available and each recharge saves so little money, consumers may lose the early adopter enthusiasm over time and skip recharging, especially when the gasoline price is low.

Charging convenience, shaped by technology, location, and management of chargers, affects charging behavior, PEV adoption, and business models. Technologies being developed for increased convenience include wireless charging and robotic charging. For example, DOE and Oak Ridge National Laboratory worked with industry to develop a 20 kW, 90% efficiency system (DOE, 2016), and Tesla has developed robotic charging stations (Yvkoff, 2019). Such technologies could also enable the use of autonomous EVs with autonomous charging. Convenience encompasses many factors that allow charging to be easy to use and worry-free for the consumer, including availability of chargers at desirable locations, charger station maintenance and functionality, ubiquity of chargers to prevent worry about finding a charger when traveling an unfamiliar route, redundancy of chargers to avoid charging congestion, and streamlined user experience enabled by connectivity and no-payment (free charging or free member use). Locations of chargers need to match the vehicle dwell locations and dwell time to maximize charging opportunities. Kontou et al. (2019) uses travel data to estimate the relationship between charger availability and charging opportunities and suggests that charger locations should be optimized to increase the probability of parked vehicles encountering chargers.

Business models for public EV charging are still uncertain, even 5 years after the publication of a previous National Academies report on EV deployment that extensively explored the EV charging business models (TRB and NRC, 2015). Public chargers are most often owned by a site (a shopping mall, hotel, apartment complex, etc.) or by a third party (e.g., a charging provider, electric utility, or an automaker). As noted above, charger utilization has been variable and mostly low for public chargers. Low utilization does not necessarily indicate a lack of utility for the charging facility owner, although it does indicate that business models based on high charging utilization and throughput may not be operative. Some chargers have been placed intentionally by networks of charging providers to provide range confidence or strategic range extension. Other existing charging siting may not be motivated to encourage charger use but rather may be motivated to encourage EV drivers to patronize a nearby business, or customers to perceive a business commitment to sustainability. The needs for public charging may still be nascent, with small volumes of BEVs on the road, and with unclear motivations for public charger placement, low utilization is not indicative of lack of future need for public charging.

Additionally, improved business models for public charging could improve the customer experience. User-friendly transaction interfaces could improve charging convenience. In theory, the combination of wireless charging, vehicle-to-charger connectivity, and autonomous payment (e.g., free charging or membership-type business models) could maximize the charging convenience during the transaction by making it essentially free of physical and mental efforts. Availability of charging infrastructure has been found to significantly affect adoption of PEVs, but the impact of charging convenience on PEV adoption and charging behavior requires further research.

Maness and Lin (2019) borrow insights from the marketing literature and use a modeling approach to demonstrate the impact of free charging on the adoption of PEVs. More research is needed on how consumers value more charging convenience with better technologies (e.g., wireless charging), better location (e.g., aligned with travel activities), and better management (e.g., free charging or membership unlimited charging).

5.4.3 Electric Charging Infrastructure Needs

Unlike biofuel- or hydrogen-powered vehicles, PEVs do not require a public charging infrastructure in order to be accepted by some consumers owing to the feasibility of home or workplace charging enabled by the existing electricity infrastructure. On the other hand, deployment of more (and more powerful) public chargers will increase feasibility and appeal of PEV products, and may be necessary for certain consumers including those who do not have access to a home charger or need public charging for long trips with a BEV as their primary vehicle. Therefore, the needs for the charging infrastructure can be estimated by understanding how and where PEVs are used and parked and where new PEV buyers are.

The total number of chargers needed has been estimated based on anticipated on-road PEV population and charging demands. However, the definition of "infrastructure needs" is still vague, as it is unclear whether these estimates represent the minimum, desirable, optimal, or ideal level of deployment for the given PEV population target. The shares of charging activities among home, work, and public chargers are not consistently assumed, possibly resulting in different estimates of needs for public chargers. According to the Alternative Fuels Data Center, as of September 20, 2020, there are 88,122 public charging outlets (at 27,129 charging locations) and 10,817 private charging outlets in the United States, excluding residential chargers (AFDC, 2020). This translates into about 0.06 charging outlets per PEVs on the road. A study by Edison Electric Institute found that 9.8 million charge ports are needed to support 18.7 million PEVs by 2030 (Cooper and Schefter, 2018; Figure 5.31). That equals 0.52 chargers per PEV overall, including about 0.12 non-home chargers per PEV, about double the current level. Based on this study, the International Code Council has adopted a new guideline that recommends all new homes in the United States to be EV-ready (Coren, 2020). According to ICCT, by 2025, about 2.2 million chargers (including

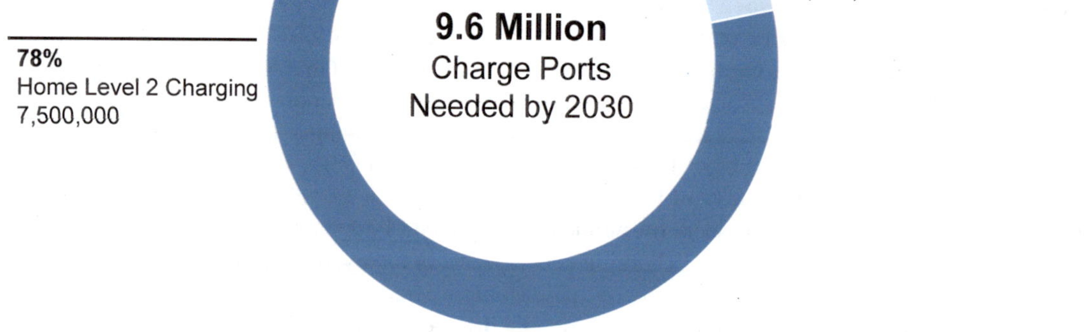

FIGURE 5.31 Projected EV charging infrastructure needs in 2030.
SOURCE: Cooper and Schefter (2018).

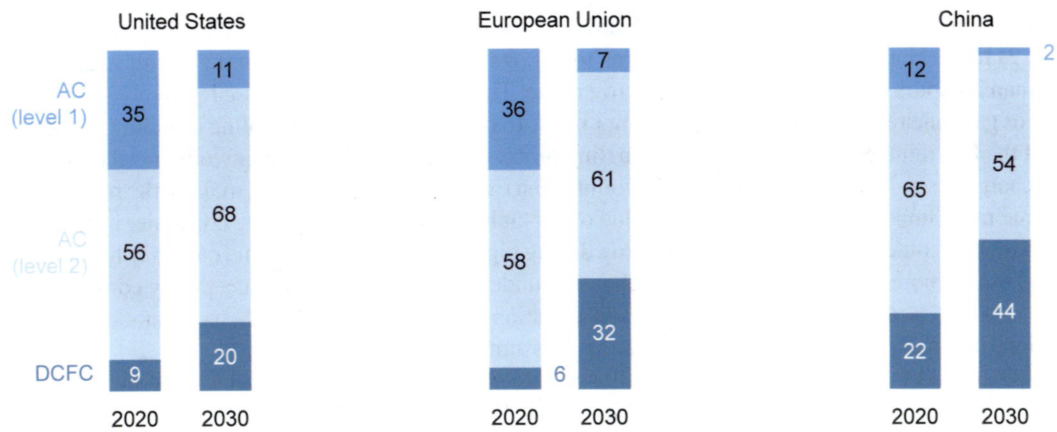

Figures may not sum to 100%, because of rounding.

FIGURE 5.32 Energy demand by charging technology in percent kilowatt hours (home-centered scenario).
SOURCE: Exhibit from Engel et al., 2018, "Charging Ahead: Electric-Vehicle Infrastructure Demand," August, McKinsey & Company, www.mckinsey.com. Copyright © 2021 McKinsey & Company. All rights reserved. Reprinted by permission.

2.1 million for home charging) are needed for 2.6 million PEVs in the 100 most populous U.S. metropolitan areas. That equates to about 0.85 chargers per PEV, much higher than Edison Electric Institute's estimate for 2030.

Home charging is widely viewed as most necessary, in both the number of chargers and the magnitude of electric load. Many studies have concluded that home charging is more cost-effective and important than workplace and public charging (Lin and Greene, 2011; Hardman et al., 2018; Lee et al., 2020a). Early EV charging activity data show that 82% of charging events were conducted at home (Smart and Schey, 2012). Edison Electric Institute estimates that by 2030, 78% of chargers will be home chargers. Similarly, McKinsey projects that by 2030 in the United States, about 80% of PEV electricity will come from home charging. These U.S. projections are much higher than the projected 68% and 60% for EU and China, respectively (Figure 5.32), possibly owing to the higher percentage of homes with garages in the United States. Long hours of home parking and the low time cost of home charging explain the importance of home charging, which makes inexpensive low-power charging feasible with residential distribution systems.

However, not every household or vehicle has easy access to home charging, as indicated by measures of dedicated home parking places, which can serve as a proxy for Level 1 AC charging availability. About two-thirds of all occupied housing units in the United States have garages or carports (Census Bureau, 2017), where 120 V outlets are usually available or inexpensive to add. More directly, more than half of new vehicle buyers in the United States park their vehicle within 25 feet of a 120 V power outlet (Axsen and Kurani, 2012). Although about 80% of households have off-street parking, only 56% of vehicles have a dedicated off-street parking space (with only 47% at an owned residence) (Traut et al., 2013). Thus, while much of the new vehicle market will have easy access to home charging, expanding the market to all vehicle buyers will require enabling PEV charging for those in dense cities or multiple-unit developments, where a higher proportion of drivers are unable to access dedicated home charging. According to several studies, a lack of home charging capability is perceived as one major barrier to PEV purchase (Ajanovic and Haas, 2016; Nilsson and Nykvist, 2015; Axsen and Kurani, 2013). Upgrades to Level 1 or 2 wiring to enable home charging can be expensive, depending on home circumstances, but are much cheaper if done during home construction or renovation. For those without dedicated parking spaces, available solutions include residential curbside Level 2 charging stations (Hall and Lutsey, 2017), on-street chargers (Grote et al., 2019), and residential driveway chargers (Traut et al., 2013). An alternative solution to lack of home charging access would be exclusive dependence on workplace or public chargers. However, reliance on non-home charging can be subject to high charging cost and inconvenience that could amount to significant vehicle operating cost over the vehicle lifetime (Ou et al., 2018).

Another important factor in evaluating the electric charging infrastructure needs is charging time. Level 1 (or even Level 2) home charging is usually regarded as too slow based on the calculation that a full recharge would take more than 25 hours even with a 100-mile electric range. This "tank refueling" reasoning likely stems from the experience of gasoline refueling, where the consumer needs to be attentive during gasoline refueling and thus loses the value of the few minutes of time during the refueling process. Similarly, if the driver would need to be attentive during the charging session, long charging hours would mean extremely high time cost, making the reduction of the charging time more important. However, if the value of per-unit time is low because the consumer is capturing the value of the time via other activities, such as sleeping during nighttime charging, the time cost may be significantly reduced and even ignored, making long hours of charging under these circumstances acceptably convenient. Then analysis should be based on whether the average and variation of travel distance can be accommodated by available charging time and power (Lin and Greene, 2011). Assuming 10 hours of night charging at 4 miles charged per hour, typical of a Level 1 connection, the 40 miles of extended range can cover about 75% of the travel days of U.S. drivers or the round-trip commuting travel of 80% of U.S. workers, based on the National Household Travel Survey (NHTS) 2009 data (Van Haaren, 2011). In fact, home charging with 110 V outlets for long-range Tesla EVs is a popular topic in the Tesla owner's forum (Yamauchi, n.d.), which suggests the practicability of Level 1 charging at least for some owners. A Level 2 charger in a home garage can extend the electric range by 150 miles per 10 hours of nighttime charging. This can cover about 95% of travel days of U.S. drivers and meet the round-trip commuting need of virtually everyone, based on the NHTS 2009 data (Nicholas et al., 2013).

Nonetheless, home charging alone cannot satisfy all travel needs and must be supplemented with workplace and public charging. Studies have suggested that a limited number of DC fast chargers is required to support intercity travel, regular charging demand by PEV owners without home or workplace charging, and urgent charging demand by those with access to other charging types. The needs for non-home chargers depend on home charging availability as well as BEV ranges. The marginal benefit of additional public chargers in enabling more electric miles has been found to decrease with longer BEV ranges. With Level 2 home charging, long-range BEVs depend less on non-home chargers. On the other hand, more non-home chargers can make the more affordable short-range BEVs more useful (Lin, 2014; Wenig et al., 2019; Peterson and Michalek, 2013; Lin and Greene, 2010). Non-home chargers may also provide the psychological benefit of "range confidence." For example, TEPCO showed that the addition of a second quick DC charger for their EV fleet in the Tokyo area led to significantly more use of EVs, without actual increased use of the second charger (Botsford and Sczczepanek, 2009). Such range confidence may, in the longer term, increase public charger utilization as BEVs are used by drivers even more extensively. The range confidence benefit may be important for understanding the value of some low-utilization public chargers and needs to be further studied. Other considerations that justify deploying more non-home chargers include BEV range reduction owing to cold weather and intercity or long-distance travel.

5.4.4 Impact on eVMT and PEV Adoption

When charging infrastructure becomes more available, powerful, and convenient to use, more VMT will be electrified. For PHEVs, a better charging infrastructure increases the UF—that is, the share of VMT powered by electricity. A case study in Massachusetts showed that the addition of workplace and public charging increased the UF of PHEVs 40%–88% from 70% with only home charging (Wood et al., 2017; Figure 5.33). For BEVs, a better charging infrastructure increases the daily effective driving range and usability, which reduces dependence on a backup vehicle for long trips and increases eVMT. For example, studies have shown that in areas where drivers have access to 50-kW or 120-kW fast charge stations, annual eVMT increased by more than 25%, even in cases where fast charging was used for only 1%–5% of total charging events (Howell et al., 2017; Keyser et al., 2017) (Figure 5.34).

Improving the availability, speed, and convenience of the charging infrastructure can accelerate PEV adoption, as charging infrastructure concerns have been found to be among the top reasons that consumers resist PEV purchase (Figure 5.35). The effect of charging infrastructure on PEV purchase depends on how consumers value additional improvement of charging infrastructure. Their willingness to pay for public charging has estimated to be similar in magnitude to the $7,500 tax credit (Greene et al., 2020).

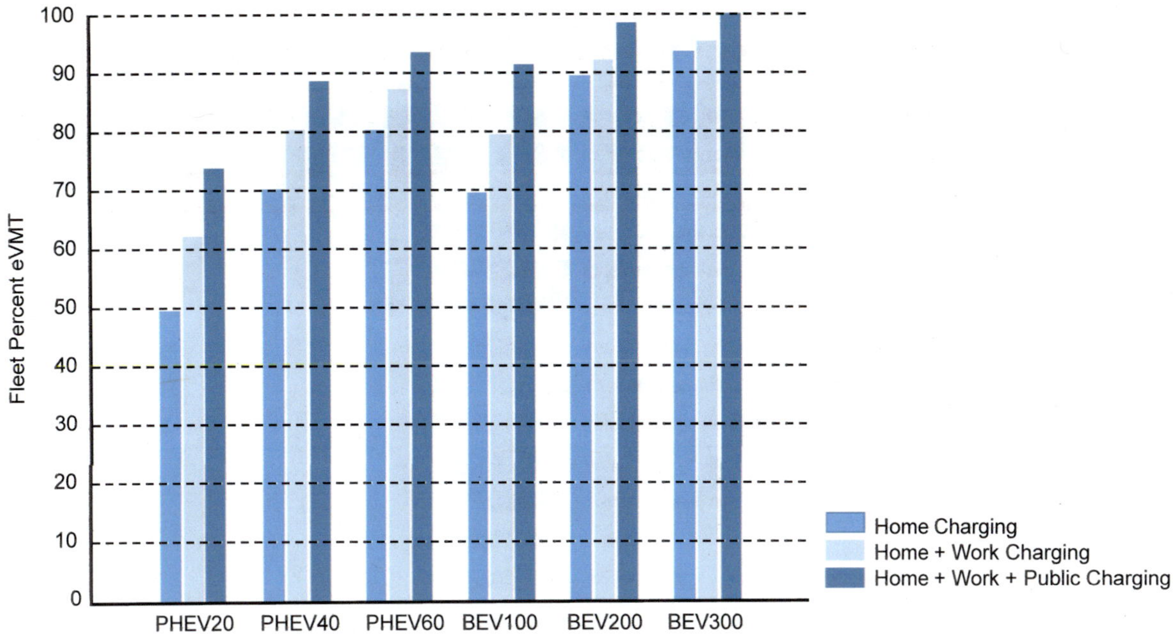

FIGURE 5.33 Fleet percent of eVMT by different vehicle types and type of charging.
SOURCE: Wood et al. (2017).

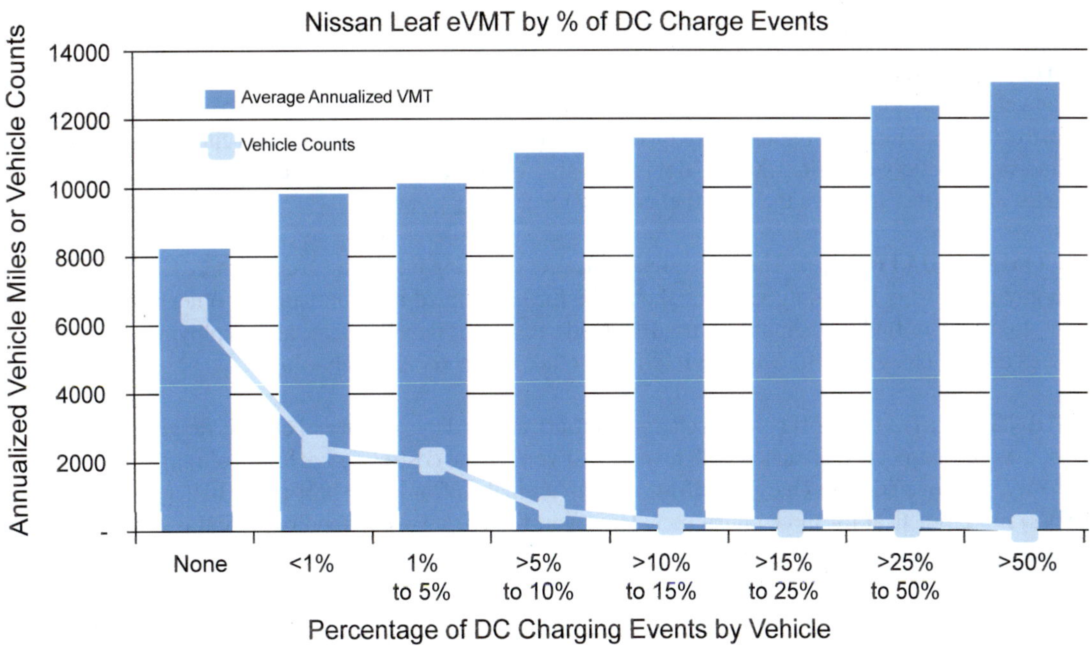

FIGURE 5.34 Better charging infrastructure leads to more eVMT.
SOURCE: Howell et al. (2017).

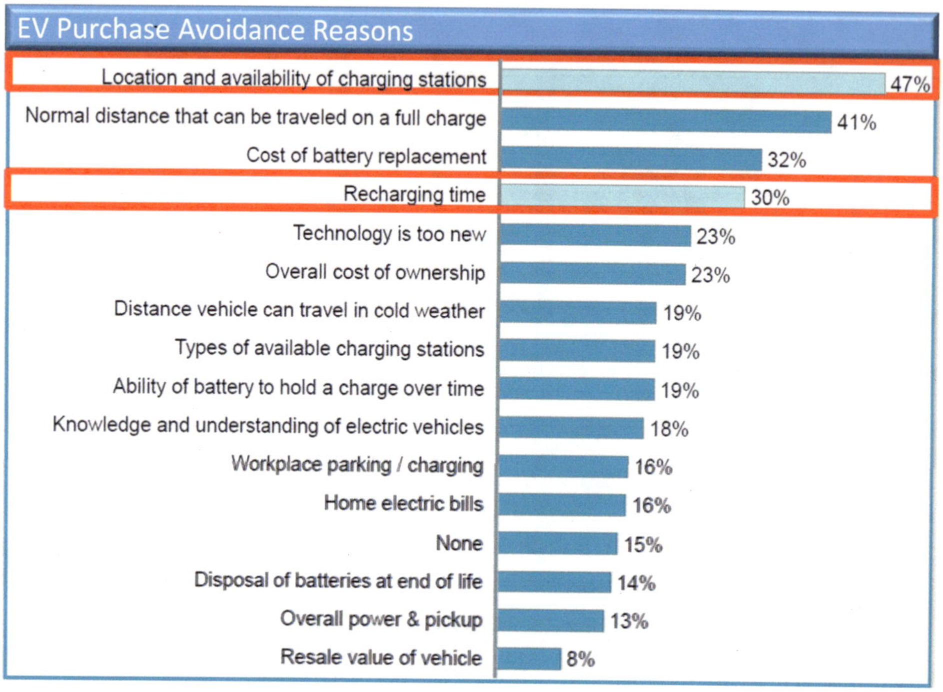

FIGURE 5.35 Charging infrastructure-related issues affecting EV purchase.
SOURCE: New Vehicle Experience Study, Strategic Vision (2017).

5.4.5 Findings and Recommendations on EV Infrastructure

FINDING 5.10: Better charging infrastructure increases electrified vehicle miles traveled, reduces petroleum-fueled vehicle miles traveled, and thus improves the average fuel economy of on-road vehicles. Better charging infrastructure can increase the usability of the electric portion of plug-in hybrid vehicle propulsion and of limited-range battery electric vehicles. Better charging infrastructure can also encourage adoption of plug-in vehicles and further improve the fleet fuel economy.

RECOMMENDATION 5.5: The cost-effectiveness of charging infrastructure for improving the overall fuel economy of existing light-duty vehicles should be further studied and compared to other vehicle efficiency technologies. The findings should be used to guide research, development, and deployment priorities and policies around charging infrastructure relative to other fuel economy technologies.

FINDING 5.11: The trade-off between charging infrastructure deployment and electric range has been studied, but consensus has not been reached with respect to diversity and cost-effective levels of ranges for consumers. In theory, improvements in charging infrastructure can make affordable short-range battery electric vehicles (BEVs) more practical for more consumers. However, there is a strong trend in the industry toward offering long-range BEV products.

RECOMMENDATION 5.6: The U.S. Department of Energy should further study the consumer preferences for electric range, including before-purchase stated preferences, revealed choices, and after-ownership opinions. Consumer value of range anxiety, range uncertainty, and charging availability should be further studied.

FINDING 5.12: Home charging is expected to dominate with respect to number of charge ports and capital costs. About 7.5 million out of 9.6 million charge ports, or ~80% by 2030, are expected to be in homes.

Expensive public fast chargers appear to offer both values of psychological assurance and practical utilization, but the extent and marginal effects are not well understood. Low utilization public chargers are common.

RECOMMENDATION 5.7: Owing to the high cost and low utilization of direct current fast chargers (at least as appears currently), the U.S. Department of Energy should investigate the consumer value, expected utilization, and business models of public charging for the purpose of guiding further deployment decisions.

FINDING 5.13: The most discussed charging technologies are Level 1, Level 2, and DC fast chargers. The equipment costs for at-home Level 2 chargers have come down significantly in recent years, but the installation cost varies greatly, mostly depending on awareness, EV attitude, and experience of the deployment decision maker. Installation cost is much cheaper for home builders who are aware and supportive of the EV trend and have experience installing the Level 2 wiring for new constructions. There has been slower progress in reducing equipment costs of Level 2 chargers in commercial settings.

RECOMMENDATION 5.8: *Government agencies, automakers, and utilities should proactively educate* charging infrastructure decision makers (builders, employers, business entities with large parking capacity, etc.) on the EV trends and low-cost opportunities for charger deployment.

FINDING 5.14: Preconstruction EV-ready guidelines have been adopted, indicating a good level of education and outreach to construction and planning stakeholders, but the real-world implementation or enforcement is not clear.

RECOMMENDATION 5.9: Actual impacts of preconstruction EV-ready guidelines for buildings should be tracked and analyzed to identify barriers to implementation.

FINDING 5.15: Charging behavior, in terms of likelihood of plugging in when a charger is available, has been studied, but strategies to make charging more convenient and increase charging events are less studied. Electric vehicle automakers are trying to make charging more convenient and available.

RECOMMENDATION 5.10: In conjunction with deployment of chargers, strategies for encouraging public use should also receive attention. Strategies can include convenience, user-friendliness, free charging or membership business models (as opposed to pay-per-use), and pricing, among others.

5.5 SUMMARY OF EV COSTS

This section takes input from the above sections and analyzes the overall vehicle costs and cost differences for propulsion and EVs. This section's analysis applies expected combustion vehicle cost increases from previous chapters with EV component and battery costs (Sections 5.2 and 5.3). The bottom-up vehicle cost framework is based on evaluation of EVs of differing electric range from two ICCT papers (Lutsey and Nicholas, 2019a and 2019b). These papers relied on a detailed engineering teardown assessment of the Chevrolet Bolt BEV with a 60-kWh battery pack, electric power output of 145 kW, and consumer label range of 238 miles (UBS, 2017) and updated battery cost data, per the above studies.

This bottom-up vehicle cost analysis offers the committee's best estimate of the relative technology costs over time, as EV costs decline, per the underlying technologies assessed below. This vehicle cost analysis is neither conservative nor optimistic. The committee recognizes that there is research indicating EV costs, especially for batteries, declining faster or slower than represented here. For example, among the battery cost studies cited, some indicate that batteries will remain at higher cost (e.g., Anderman, 2017; Ahmed et al., 2018; Anderman, 2018), while others indicate lower cost (Berckmans et al., 2017; P3, 2020) than those applied here through 2030. The committee's estimates, as analyzed here, rely on lithium-ion battery innovations (cathode, anode, and pack-level) that reduce the use of costly materials and increase battery plant-level production volume to at least 500,000 battery packs per year, which are consistent with industry developments. It is also emphasized that this analysis, by design, sought to inform on how EV price parity is ultimately not a point in time but a broad range of years over which different vehicles, across vehicle classes and electric ranges, approach conventional vehicle prices.

As defined in Chapter 4, the analysis includes representative vehicles in five classes: small car, medium car, crossover, SUV, and pickups. The comparable average conventional gasoline vehicle prices in MY 2016 were about $21,000 for small cars, $34,000 for medium cars, $28,000 for crossovers, $41,000 for SUVs, and $36,000 for pickups. For the various representative vehicles, powertrain components are scaled to vehicle power, vehicle-level manufacturing costs are scaled to the vehicle footprint, and indirect conventional vehicle costs are treated as a percentage of direct manufacturing costs.

The evaluated BEVs include electric ranges of 150–300 miles, and the PHEVs have electric range capabilities of 20–60 miles. The initial EV efficiency is based on existing MY 2018 models (DOE, 2019b). These efficiency values account for increased electricity use per mile for longer-range EVs owing to larger, heavier battery packs, as well as other attributes regarding the utility of vehicles (e.g., more crossovers have all-wheel drive; SUVs have four-wheel drive and higher towing capacity). EV efficiency is assumed to improve by 1% per year owing to vehicle-level (aerodynamic, tire, mass reduction) and electric powertrain improvements, as discussed above. These effects incrementally reduce the battery pack size, for a given vehicle class and range, over time.

Vehicle cost increases for increased efficiency improvements for conventional gasoline vehicles are applied. Per above, 2% per year fuel economy improvements are included—starting from consumer label values of approximately 34 MPG for the small car, 27 MPG for the medium car, 26 MPG for the crossover, 20 MPG for the SUV, and 18 MPG for the pickup in 2018. The associated incremental price increases amount to $500–$900 for cars and crossovers, and $900-$1,000 for SUVs and pickups for the expected efficiency increase by 2025. These combustion vehicle technologies are incorporated with a 0.35% annual price increase from 2018 on.

Figure 5.36 illustrates vehicle manufacturing costs, including conventional and BEV technology components for two of the five vehicle classes. As indicated, BEV costs in 2018 are substantially higher than conventional vehicle costs. The incremental cost for BEVs is at least $8,500 for the medium car (i.e., $36,800 BEV150 versus $28,300 conventional) to about $26,000 for the long-range SUV (i.e., $57,000 BEV300 versus $31,000 conventional). From 2018 to 2025, the absolute cost of each BEV is reduced by $9,000 (BEV150) to $13,000 (BEV300) for the medium cars, and from $14,000 (BEV150) to $19,000 (BEV300) for the SUVs.

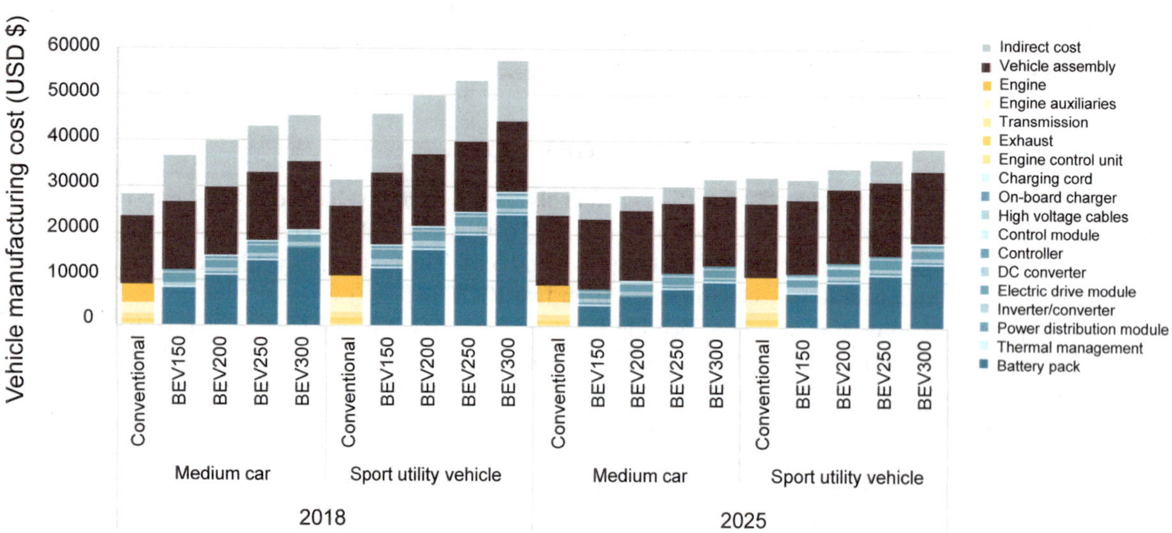

FIGURE 5.36 Vehicle technology costs for ICEs and BEVs for 2018 and 2025 for the medium car and SUV classes. The figure shows the level of detail for the cost analysis's engine-related components (yellow), electric components (blue), vehicle assembly costs (brown), and indirect costs (gray).
SOURCE: Lutsey and Nicholas (2019a).

TABLE 5.15 BEV Technology Cost Element Estimates by Vehicle Class and Year

Technology Detail and Cost	Year Vehicle Class	2025 BEV200	2030	2025 BEV300	2030
Battery capacity (kWh)	Small car	49	47	78	76
	Medium car	60	59	97	94
	Crossover	66	64	104	101
	SUV	95	92	149	144
	Pickup	105	101	167	162
Battery cost	Small car	$5,154	$3,475	$8,193	$5,524
	Medium car	$6,358	$4,282	$9,790	$6,659
	Crossover	$6,979	$4,699	$10,207	$6,996
	SUV	$9,662	$6,563	$13,791	$9,287
	Pickup	$10,215	$7,003	$15,462	$10,412
Motor cost	Small car	$548	$521	$548	$521
	Medium car	$896	$852	$896	$852
	Crossover	$747	$710	$747	$710
	SUV	$1,095	$1,041	$1,095	$1,041
	Pickup	$1,244	$1,183	$1,244	$1,183
Inverter cost	Small car	$414	$394	$414	$394
	Medium car	$578	$550	$578	$550
	Crossover	$508	$483	$508	$483
	SUV	$672	$639	$672	$639
	Pickup	$742	$706	$742	$706

NOTE: Power ratings by vehicle class in kilowatts: small car 110, large car 180, crossover 150, SUV 220, pickup truck 250.

Several assumptions for markups are applied to link the vehicle *cost* to the vehicle *price*. All vehicles include a 15% dealer markup for dealer incentives and marketing. Automaker profit margins of 3% for small cars, 7% for medium cars, 10% for crossovers, 15% for SUVs, and 10% for pickups are included based on discussions with industry experts and to match the bottom-up costs with available average vehicle prices for the five classes. These approximations across technology types ensure that EVs have the same dealer markup and automaker profit built in as assumed for conventional vehicles for each vehicle class over time.

Table 5.15 summarizes some of the critical technical and cost elements for the BEV cost estimations for 2025 and 2030; as there is higher uncertainty related to battery technology past 2030, rigorous cost estimates past this point are not attempted. The table shows the battery capacity, including incremental reduction for efficiency improvements for BEV200 and BEV300 technology packages across the five vehicle classes. The associated costs are shown for the battery pack, the motor, and the inverter, per the discussion above.

Figure 5.37 shows the changing vehicle technology prices from 2020 through 2030 for the five vehicle classes. Each segment includes the average conventional gasoline vehicle (gray line) with incrementally increasing prices for efficiency improvements. The figure shows the decreasing prices for the EVs of various ranges from the 20-mile PHEV (PHEV20) to the 300-mile BEV (BEV300).

As shown in Figure 5.37, reducing BEV prices results in their reaching price parity with conventional vehicles during the 2023–2030 time frame. The BEV150 vehicles achieve price parity soonest, crossing the conventional vehicle line by 2023–2026. The longer-range BEV300s achieve price parity 4–5 years later than the BEV150s in each case: this is primarily owing to longer-range BEVs having larger battery packs, thus adding substantial costs over the shorter-electric-range versions of the same vehicle type. To give a sense of this price difference for a prospective vehicle buyer in 2030, compared to the shorter-range BEV150 small car, a BEV300 would be priced $3,500 higher. Similarly, longer-range 300-mile crossovers would be priced $4,500 above the BEV150, and the 300-mile pickup would be priced $6,000 over the BEV150 version, by 2030.

PHEVs with 20 miles (PHEV20) to 60 miles (PHEV60) of electric range are also shown in Figure 5.37. The PHEV price differential versus conventional gasoline vehicles is reduced by 2030, but there are no price parity points with conventional vehicles in any vehicle class. The PHEV20 small car price differential with

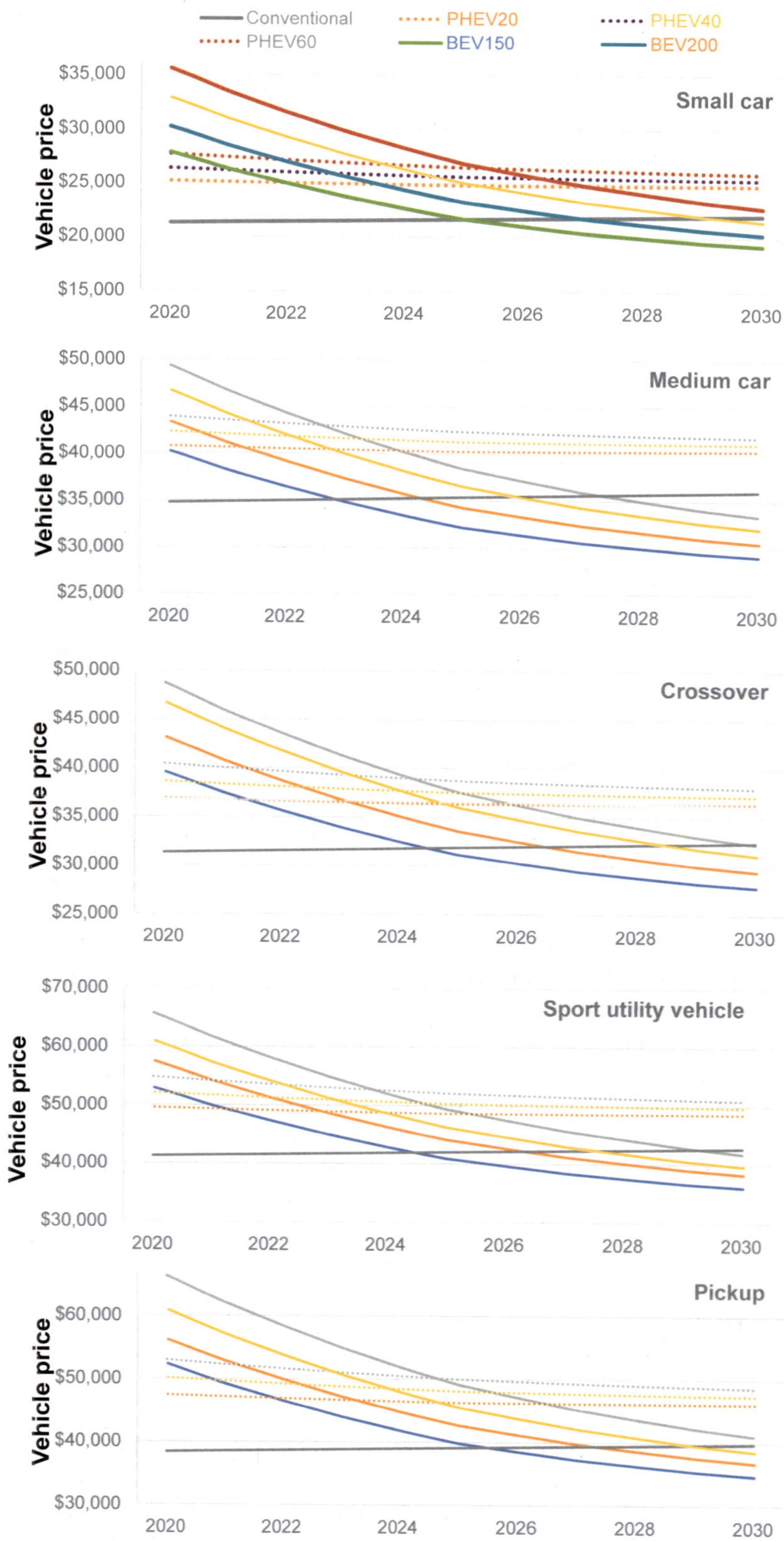

FIGURE 5.37 Price of conventional and electric vehicles for five classes for 2020–2030.

conventional vehicles declines from $4,500 in 2020 to $2,900 in 2030. For an example of a larger vehicle class and larger pack, the PHEV60 SUV cost differential drops from $13,000 in 2020 to $8,400 in 2030. PHEVs do not have a price parity point like BEVs because the battery pack—where there are large price reductions—is a much lower contributor to the PHEV price and because the PHEVs retain the combustion powertrain in addition to all the electric components.

The finding from this vehicle analysis is that one of the EV barriers, upfront vehicle cost, is likely to incrementally subside over 2025–2030, first for shorter-range EVs and later for longer-range vehicles. Incorporated in this figure's analysis are substantial lithium-ion battery improvements and battery producers increasing to higher production volume to supply an expanding global EV volume. As indicated through the course of this report, the upfront cost differential of EVs versus comparable conventional vehicles is one of several major EV barriers, which also include model availability, charging convenience, and consumer awareness. This multibarrier aspect underscores the importance of considering policy and benefits of EVs over a longer-term time frame while simultaneously promoting the uptake of advanced ICEVs, HEVs, and PHEVs that also contribute substantially to emission reductions.

5.5.1 Findings and Recommendation for EV Cost Summary

FINDING 5.16: Although battery pack costs are reduced by 7% per year from 2018 through 2030 with continued improvements to lithium-ion battery technologies and higher volume production, the precise cell and pack costs will differ by battery pack size. The key cost driver for EVs is the battery, which for high-volume battery production is expected to decrease to $90-$115/kWh by 2025 and $65-$80/kWh by 2030 at the pack level. The committee views breakthrough changes in "beyond lithium ion" materials technology, like solid-state batteries or lithium metal, as unlikely by 2030.

FINDING 5.17: Engineering (packing) solutions for lowering battery cost are also a promising solution for economic deployment of battery electric vehicles in the 2025–2030 time frame. However, owing to multiple players and various packing solutions, companies will be slow to converge on a common cell type (e.g., prismatic, pouch, etc.) owing to substantial overhead investments in manufacturing processes.

FINDING 5.18: Battery electric vehicles with increasing electric range are expected to reach first-cost parity with combustion vehicles during 2025–2030 for companies moving to high-production volume. Reducing battery cost in addition to meeting specifications for greater durability and rapid charging capabilities will widen their appeal.

RECOMMENDATION 5.11: Fuel economy standards should avoid overcrediting performance (e.g., range) of plug-in electric vehicles; overcrediting range may result in greater use of additional safety features, sensors, and algorithms, which in turn will increase cost. Plug-in electric vehicles with policy-motivated oversized batteries could slow market penetration. Policy makers should align any regulatory incentives with customer needs, in order to ensure that automaker decisions about battery electric vehicle range are based on customer demands rather than regulatory credit.

5.6 REFERENCES

Abada, S., G. Marlair, A. Lecocq, M. Petit, V. Sauvant-Moynot, and F. Huet. 2016. Safety focused modeling of lithium-ion batteries: A review. *Journal of Power Sources* 306:178–192.

AFDC (Alternative Fuels Data Center). 2020. "Alternative Fuels Data Center: Electric Vehicle Charging Station Locations." https://afdc.energy.gov/fuels/electricity_locations.html#/find/nearest?fuel=ELEC.

Ahmed, S., P. Nelson, N. Susarla, and D. Dees. 2018. "Automotive Battery Cost Using BatPaC." Presented at the IEA Workshop on Batteries for Electric Mobility, March 7, 2018, Paris. https://www.iea.org/media/Workshops/2018/Session2ShabbirAhmedANL.pdf.

Ajanovic, A., and R. Haas. 2016. Dissemination of electric vehicles in urban areas: Major factors for success. *Energy* 115: 1451–1458. http://dx.doi.org/10.1016/j.energy. 2016.05.040.

Allan, R. 2017. "SiC and GaN vs. IGBTs: The Imminent Tug of War for Supremacy," Power Electronics, Informa PLC. July 27. https://www.powerelectronics.com/automotive/sic-and-gan-vs-igbts-imminent-tug-war-supremacy.

Ames National Laboratory. 2012. "Beyond Rare Earth Magnets." *FY 2012 APEEM Annual Progress Report*, Section 3. https://www.energy.gov/sites/prod/files/2014/03/f10/vtpn08_rogers_ape_2011_o.pdf.

Anderman, M. 2017. *The Tesla Battery Report: Tesla Motors: Battery Technology, Analysis of the Gigafactory and Model 3, and the Automakers' Perspectives.* Total Battery Consulting. http://www.totalbatteryconsulting.com/industry-reports/Tesla-report/Extract-from-the-Tesla-Battery-Report.pdf.

Anderman, M. 2018. *The xEV Industry Insider Report.* Total Battery Consulting. June. https://totalbatteryconsulting.com/industry-reports/xEV-report/Extract-from-the-2018-xEV-Industry-Report.pdf.

ANL (Argonne National Laboratory). 2020. "EV-Smart Grid Interoperability Center." https://www.anl.gov/es/evsmart-grid-interoperability-center.

ARPA-E (Advanced Research Projects Agency–Energy). 2015. "Cerium-Based Magnets, Ames National Laboratory, Novel High Energy Permanent Magnet Without Critical Elements." https://arpa-e.energy.gov/?q=slick-sheet-project/cerium-based-magnets.

Arrow Electronics. 2020. "Silicon Carbide (SiC) vs. Gallium Nitride (GaN)." Arrow.com. January 22. https://www.arrow.com/en/research-and-events/articles/silicon-carbide-and-gallium-nitride-compared.

Attia, P.M., A. Grover, N. Jin, K.A. Severson, T.M. Markov, Y.-H. Liao, M.H. Chen, et al. 2020. Closed-loop optimization of fast-charging protocols for batteries with machine learning. *Nature* 578(7795):397–402. https://doi.org/10.1038/s41586-020-1994-5.

Axsen, J., and K.S. Kurani. 2012. Who can recharge a plug-in electric vehicle at home? *Transportation Research Part D: Transport and Environment* 17:349–353. http://dx.doi.org/10.1016/j.trd. 2012.03.001.

Axsen, J., and K.S. Kurani. 2013. Hybrid, plug-in hybrid, or electric-What do car buyers want? *Energy Policy* 61: 532–543. http://dx.doi.org/10.1016/j.enpol.2013.05.122.

Aykol, M., P. Herring, and A. Anapolsky. 2020. Machine learning for continuous innovation in battery technologies. *Nature Reviews Materials* 5(10):725–727. https://doi.org/10.1038/s41578-020-0216-y.

Bakeroot, B., and S. Decoutere. 2020. "New Power Electronics Will Make EVs More Efficient and Less Expensive." Wards Auto. March. https://www.wardsauto.com/industry-voices/new-power-electronics-will-make-evs-more-efficient-and-less-expensive.

Bandhauer, T.M., S. Garimella, and T.F. Fuller. 2011. A critical review of thermal issues in lithium-ion batteries. *Journal of the Electrochemical Society* 158(3):R1–R25.

Berckmans, G., M. Messagie, J. Smekens, N. Omar, L. Vanhaverbeke, and J. Van Mierlo. 2017. Cost projection of state of the art lithium-ion batteries for electric vehicles up to 2030. *Energies* 10(9):1314. http://www.mdpi.com/1996-1073/10/9/1314.

Borlaug, B., S. Salisbury, M. Gerdes, and M. Muratori. 2020. Levelized cost of charging electric vehicles in the United States. *Joule* 4(7):1470–1485. https://doi.org/10.1016/j.joule.2020.05.013.

Botsford, C., and A. Sczczepanek. 2009. "Fast Charging vs. Slow Charging: Pros and Cons for the New Age of Electric Vehicles." In EVS24 International Battery, Hybrid and Fuel Cell Electric Vehicle Symposium. Stavanger, Norway.

Boutros, K.S., R. Chu, and B. Hughes. 2012. "GaN Power Electronics for Automotive Applications." IEEE Energytech. May.

Bower, G. 2019. "Tesla Model 3 2170 Energy Density Compared To Bolt, Model S P100D." InsideEVs. February 7. https://insideevs.com/news/342679/tesla-model-3-2170-energy-density-compared-to-bolt-model-s-p100d/.

Bravo Diaz, L., X. He, Z. Hu, F. Restuccia, M. Marinescu, J.V. Barreras, Y. Patel, et al. 2020. Review—Meta-review of fire safety of lithium-ion batteries: Industry challenges and research contributions. *Journal of the Electrochemical Society* 167(9):090559. https://doi.org/10.1149/1945-7111/aba8b9.

Buckley, S. 2011. "EV Manufacturers Get Harmonized, Agree to Build a Universal Charging System." Engadget. https://www.engadget.com/2011-10-14-ev-manufacturers-get-harmonized-agree-to-build-a-universal-char.html.

Burress, T. 2016. "Advanced Electric Motor Research." Presentation at the DOE Annual Merit Review, June 7, Washington, DC. http://energy.gov/eere/vehicleBs/downloads/vehicle-technologies-office-merit-review-2016-advanced-electric-motor.

Business Wire. 2019. "Transphorm Ships Over Half a Million GaN Power Devices for Multi-Kilowatt Class Applications." Business Wire. November. https://www.businesswire.com/news/home/20191107005291/en/Transphorm-Ships-Million-GaN-Power-Devices-Multi-kilowatt.

Cai, T., A.G. Stefanopoulou, and J.B. Siegel. 2019a. Modeling Li-ion battery temperature and expansion force during the early stages of thermal runaway triggered by internal shorts. *Journal of the Electrochemical Society* 166(12):A2431.

Cai, T., A.G. Stefanopoulou, and J.B. Siegel. 2019b. Early detection for Li-ion batteries thermal runaway based on gas sensing. *ECS Transactions* 89(1):85.

Census Bureau. 2017. 2017 "American Housing Survey." https://www.census.gov/programs-surveys/ahs/data.2017.List_1739896299.html.

Chevrolet Pressroom. 2011. "Chevrolet Showcases Spark EV Electric Motor." Media.Gm.Com. October 26. https://media.gm.com/media/us/en/chevrolet/news.detail.html/content/Pages/news/us/en/2011/Oct/1026_spark_elec_mtr.html.

Chevrolet Pressroom. 2016. "Drive Unit and Battery at the Heart of Chevrolet Bolt EV." Media.Gm.Com. January 11. https://media.gm.com/media/us/en/chevrolet/home.detail.html/content/Pages/news/us/en/2016/Jan/naias/chevy/0111-bolt-du.html.

China Power Team. 2020. "Does China Pose a Threat to Global Rare Earth Supply Chains?" https://chinapower.csis.org/china-rare-earths/.

Choe, S.-Y., X. Li, and M. Xiao. 2013. Fast charging method based on estimation of ion concentrations using a reduced order of electrochemical thermal model for lithium ion polymer battery. *World Electric Vehicle Journal* 6(3):782–792. https://doi.org/10.3390/wevj6030782.

Chowdhury, S., and U.K. Mishra. 2013. Lateral and vertical transistors using the AlGaN/GaN heterostructure. *IEEE Transactions on Electron Devices* 60(10):3060–3066. https://doi.org/10.1109/TED.2013.2277893.

Clean Cities. 2012. *Plug-In Electric Vehicle Handbook for Public Charging Station Hosts*. Washington, DC: U.S. Department of Energy. http://www.afdc.energy.gov/pdfs/51227.pdf.

Clement-Nyns, K., E. Haesen, and J. Driesen. 2010. The impact of charging plug-in hybrid electric vehicles on a residential distribution grid. *IEEE Transactions on Power Systems* 25(1):371–380. https://doi.org/10.1109/TPWRS.2009.2036481.

Cole, J. 2015. "LG Chem 'Ticked Off' with GM For Disclosing $145/kWh Battery Cell Pricing—Video." InsideEVs. October 23. https://insideevs.com/news/327874/lg-chem-ticked-off-with-gm-for-disclosing-145-kwh-battery-cell-pricing-video/.

Constantinides, S. 2011. "Arnold Magnetics, DOE Vehicle Technologies Program, APEEM R&D FY12 Kickoff Meeting, ORNL, Nov. 2–4." https://www.arpa-e.energy.gov/sites/default/files/documents/files/Breakout_Session_Magnetics_LowRes.pdf.

Cooper, A., and K. Schefter. 2018. "Electric Vehicle Sales Forecast and the Charging Infrastructure Required Through 2030." Edison Electric Institute. https://www.edisonfoundation.net/-/media/Files/IEI/publications/IEI_EEI-EV-Forecast-Report_Nov2018.ashx.

Coren, M.J. 2020. "New US Building Codes Will Make Every Home Ready for Electric Cars." Quartz. January 9. https://qz.com/1781774/new-us-building-codes-require-plugs-for-electric-cars/.

Cui, H., D. Hall, and N. Lutsey. 2020. "Update on the Global Transition to Electric Vehicles through 2019." Briefing. International Council on Clean Transportation. https://theicct.org/sites/default/files/publications/update-global-EV-stats-sept2020-EN.pdf.

Davies, C. 2017. "VW I.D. EV Boast: We'll Hugely Undercut Tesla's Model 3 Says Exec." Slashgear. July 12. https://www.slashgear.com/vw-i-d-ev-boast-well-hugely-undercut-teslas-model-3-says-exec-17491688/.

Davis, S. 2019. "The Great Semi Debate: SiC or GaN?" Power Electronics, Informa PLC. February 15. https://www.powerelectronics.com/power-management/great-semi-debate-sic-or-gan.

Debert, M., G. Colin, G. Bloch, and Y. Chamaillard. 2013. An observer looks at the cell temperature in automotive battery packs. *Control Engineering Practice* 21(8):1035–1042.

Dharmakeerthi, C.H., N. Mithulananthan, and T.K. Saha. 2014. Impact of electric vehicle fast charging on power system voltage stability. *International Journal of Electrical Power and Energy Systems* 57(May):241–249. https://doi.org/10.1016/j.ijepes.2013.12.005.

DOE (U.S. Department of Energy). 2016. "Wireless Charging for Electric Vehicles." Video. June 16. https://www.energy.gov/eere/videos/wireless-charging-electric-vehicles.

DOE. 2017. *Enabling Fast Charging: A Technology Gap Assessment*. https://www.energy.gov/sites/prod/files/2017/10/f38/XFC%20Technology%20Gap%20Assessment%20Report_FINAL_10202017.pdf.

DOE. 2019a. "Vehicle Technologies Office's Research Plan to Reduce, Recycle, and Recover Critical Materials in Lithium-Ion Batteries." https://www.energy.gov/sites/prod/files/2019/07/f64/112306-battery-recycling-brochure-June-2019%202-web150.pdf.

DOE. 2019b. "Fuel Economy Dataset." https://www.fueleconomy.gov/feg/download.shtml.

Dubarry, M., G. Baure, and D. Anseán. 2020. Perspective on state-of-health determination in lithium-ion batteries. *Journal of Electrochemical Energy Conversion and Storage* 17(4):044701. https://doi.org/10.1115/1.4045008.

Dubarry, M., and D. Beck. 2020. Big data training data for artificial intelligence-based Li-ion diagnosis and prognosis. *Journal of Power Sources* 479(December):228806. https://doi.org/10.1016/j.jpowsour.2020.228806.

El-Refaie, A. 2016. "Alternative High-Performance Motors with Non-Rare Earth Materials." GE Global Research, 2016 DOE-VTO Annual Merit Review, June 7.

Els, P. 2018. "SiC Is Revolutionising EV/HEV Power Electronics, but Is the Packaging up to the Job?" Automotove IQ. August. https://www.automotive-iq.com/electrics-electronics/articles/sic-is-revolutionising-evhev-power-electronics-but-is-the-packaging-up-to-the-job.

Energy Star. 2020. *Building Electric Vehicle-Ready Homes*. https://www.energystar.gov/sites/default/files/asset/document/Building%20Electric%20Vehicle-Ready%20Homes_1.pdf.

Engel, H., R. Hensley, S. Knupfer, and S. Sahdev. 2018a. "How Electric Vehicles Could Change the Load Curve." McKinsey. August 8. https://www.mckinsey.com/industries/automotive-and-assembly/our-insights/the-potential-impact-of-electric-vehicles-on-global-energy-systems.

Engel, H., R. Hensley, S. Knupfer, and S. Sahdev. 2018b. "Charging Ahead: Electric-Vehicle Infrastructure Demand." McKinsey. August 8, updated October 2018. https://www.mckinsey.com/industries/automotive-and-assembly/our-insights/charging-ahead-electric-vehicle-infrastructure-demand.

EPA (U.S. Environmental Protection Agency). 2020. *2019 Automotive Trends Report*. https://www.epa.gov/automotive-trends/download-automotive-trends-report.

EPRI (Electric Power Research Institute). 2019. *Interoperability of Public Electric Vehicle Charging Infrastructure*. https://www.epri.com/research/products/3002017164.

Fan, X., W. Sun, F. Meng, A. Xing, and J. Liu. 2018. Advanced chemical strategies for lithium–sulfur batteries: A review. *Green Energy and Environment* 3(1):2–19. https://doi.org/10.1016/j.gee.2017.08.002.

Feng, X., M. Ouyang, X. Liu, L. Lu, Y. Xia, and X. He. 2018. Thermal runaway mechanism of lithium ion battery for electric vehicles: A review. *Energy Storage Materials* 10:246–267.

Ferdowsi, M., P. Shamsi, and B. Baddipadiga. 2017. "Gallium Nitride (GaN) Based High Frequency Inverter for Energy Storage Applications." InnoCit. https://eesat.sandia.gov/wp-content/uploads/2017/12/Mehdi_Ferdowsi.pdf.

Field, K. 2019. "Tesla Model 3 Battery Pack and Battery Cell Teardown Highlights Performance Improvements." CleanTechnica. January 29. https://cleantechnica.com/2019/01/28/tesla-model-3-battery-pack-cell-teardown-highlights-performance-improvements/.

Foldy, B. 2020. "Auto Makers Grapple with Battery-Fire Risks in Electric Vehicles." Wall Street Journal. October 19. https://www.wsj.com/articles/auto-makers-grapple-with-battery-fire-risks-in-electric-vehicles-11603099800.

Forgez, C., D. Vinh Do, G. Friedrich, M. Morcrette, and C. Delacourt. 2010. Thermal modeling of a cylindrical $LiFePO_4$/graphite lithium-ion battery. *Journal of Power Sources* 195(9):2961–2968.

Francfort, J., S. Salisbury, J. Smart, T. Garetson, and D. Karner. 2017. "Considerations for Corridor and Community DC Fast Charging Complex System Design." INL/EXT—17-40829, 1459664. Idaho National Laboratory. https://doi.org/10.2172/1459664.

Gardner, G. 2017. "GM: Cheaper Batteries Will Lead to Electric Car Profits." USA Today. November 17. https://www.usatoday.com/story/money/business/2017/11/17/gm-thinks-cheaper-batteries-make-electric-cars-more-profitable/869372001/.

Goldie-Scot, L. 2019. "A Behind the Scenes Take on Lithium-Ion Battery Prices." Bloomberg New Energy Finance. March 5. https://about.bnef.com/blog/behind-scenes-take-lithium-ion-battery-prices/.

Green Car Congress. 2013. "BMW Unveils the Production I3 in New York, London and Beijing; Efficiency, Dynamics and a Supporting Ecosystem of Services." https://www.greencarcongress.com/2013/07/i3-20130730.html.

Green Car Congress. 2017. "Mitsubishi Electric Develops World's Smallest SiC Inverter for HEVs." March 9. http://www.greencarcongress.com/2017/03/20170309-sic.html.

Greene, D.L., E. Kontou, B. Borlaug, A. Brooker, and M. Muratori. 2020. Public charging infrastructure for plug-in electric vehicles: What is it worth? *Transportation Research Part D: Transport and Environment* 78:102182. https://doi.org/10.1016/j.trd.2019.11.011.

Grote, M., J. Preston, T. Cherrett, and N. Tuck. 2019. Locating residential on-street electric vehicle charging infrastructure: A practical methodology. *Transportation Research Part D: Transport and Environment* 74:15–27. doi:10.1016/j.trd.2019.07.017.

Guerra, M. 2017. "SiC Power Modules Charge up Venturi Formula E Cars." Electronic Design. July. http://www.electronicdesign.com/power/sic-power-modules-charge-venturi-formula-e-cars.

Hall, D., and N. Lutsey. 2017. "Emerging Best Practices for Electric Vehicle Charging Infrastructure." White Paper. International Council on Clean Transportation.

Hall, D., and N. Lutsey. 2019. "Estimating the Infrastructure Needs and Costs for the Launch of Zero-Emission Trucks." White Paper. International Council on Clean Transportation. August.

Han, X., L. Lu, Y. Zheng, X. Feng, Z. Li, J. Li, and M. Ouyang. 2019. A review on the key issues of the lithium ion battery degradation among the whole life cycle. *eTransportation* 1(August):100005. https://doi.org/10.1016/j.etran.2019.100005.

Hannan, M.A., M. Lipu, A. Hussain, and A. Mohamed. 2017. A review of lithium-ion battery state of charge estimation and management system in electric vehicle applications: Challenges and recommendations. *Renewable and Sustainable Energy Reviews* 78:834–854.

Hardman, S., A. Jenn, G. Tal, J. Axsen, G. Beard, N. Daina, and P. Plötz. 2018. A review of consumer preferences of and interactions with electric vehicle charging infrastructure. *Transportation Research Part D: Transport and Environment* 62:508–523.

Hardman, S., E. Shiu, and R. Steinberger-Wilckens. 2016. Comparing high-end and low-end early adopters of battery electric vehicles. *Transportation Research Part A: Policy and Practice* 88:40–57. https://doi.org/10.1016/j.tra.2016.03.010.

He, W., N. Williard, C. Chen, and M. Pecht. 2014. State of charge estimation for li-ion batteries using neural network modeling and unscented Kalman filter-based error cancellation. *International Journal of Electrical Power and Energy Systems* 62:783–791.

Holland, S.P., E. T. Mansur, N. Z. Muller, and A. J. Yates. 2016. Are there environmental benefits from driving electric vehicles? The importance of local factors. *American Economic Review* 106(12):3700–3729.

Howell, D., S. Boyd, B. Cunningham, S. Gillard, L. Slezak, S. Ahmed, I. Bloom, et al. 2017. *Enabling Fast Charging: A Technology Gap Assessment*. October. https://www.energy.gov/sites/prod/files/2017/10/f38/XFC%20Technology%20Gap%20 Assessment%20Report_FINAL_10202017.pdf.

Hu, J., Y. Zhang, M. Sun, D. Piedra, N. Chowdhury, and T. Palacios. 2018. Materials and processing issues in vertical GaN power electronics. *Materials Science in Semiconductor Processing* 78:75–84. https://doi.org/10.1016/j.mssp.2017.09.033.

Hu, X., Y. Zheng, D.A. Howey, H. Perez, A. Foley, and M. Pecht. 2020. Battery warm-up methodologies at subzero temperatures for automotive applications: Recent advances and perspectives. *Progress in Energy and Combustion Science* 77:100806. https://doi.org/10.1016/j.pecs.2019.100806.

Hu, X., Y. Che, X. Lin, and S. Onori. 2021. Battery health prediction using fusion-based feature selection and machine learning. *IEEE Transactions on Transportation Electrification* 7(2):382–398. https://doi.org/10.1109/TTE.2020.3017090.

Hu, Y., and S. Yurkovich. 2012. Battery cell state-of-charge estimation using linear parameter varying system techniques. *Journal of Power Sources* 198:338–350.

IDTechEx. 2020. "China to Enforce Electric Vehicle Safety by 2021 Says IDTechEx." PR Newswire. May 15. https://www.prnewswire.com/news-releases/china-to-enforce-electric-vehicle-safety-by-2021-says-idtechex-301060053.html.

International Energy Conservation Code. 2018. CE217-19 Part II. https://newbuildings.org/wp-content/uploads/2019/05/CE217-P2.pdf.

Jaksic, M. 2019. "Integrated Wide Band Gap Power Module for Next Generation Plug-in Vehicles." General Motors, DE-EE0007285. June 12. https://www.energy.gov/sites/prod/files/2019/06/f63/elt082_jaksic_2019_o_5.2_11.21am.pdf.

Ji, Y., Y. Zhang, and C.Y. Wang. 2013. Li-ion cell operation at low temperatures. *Journal of the Electrochemical Society* 160:A636–A649.

Kamaya, N., K. Homma, Y. Yamakawa, M. Hirayama, R. Kanno, M. Yonemura, T. Kamiyama, et al. 2011. A lithium superionic conductor. *Nature Materials* 10(9):682–686. https://doi.org/10.1038/nmat3066.

Kane, M. 2018. "BMW I3 Samsung SDI 94 Ah Battery Rated For 524,000 Miles." InsideEVs. May 28. https://insideevs.com/news/338067/bmw-i3-samsung-sdi-94-ah-battery-rated-for-524000-miles/.

Keyser, M., A. Pesaran, Q. Li, S. Santhanagopalan, K. Smith, E. Wood, S. Ahmed, et al. 2017. Enabling fast charging—battery thermal considerations. *Journal of Power Sources* 367:228–236. https://doi.org/10.1016/j.jpowsour.2017.07.009.

Kim, Y., J.B. Siegel, and A.G. Stefanopoulou. 2013. "A Computationally Efficient Thermal Model of Cylindrical Battery Cells for the Estimation of Radially Distributed Temperatures." American Control Conference. Washington, DC.

Kim, Y., S. Mohan, J.B. Siegel, and A.G. Stefanopoulou. 2014. Maximum Power Estimation of Lithium-Ion Batteries Accounting for Thermal and Electrical Constraints. *Proceedings of the ASME 2013 Dynamic Systems and Control Conference*. V002T23A003. ASME. https://doi.org/10.1115/DSCC2013-3935.

Klein, R., N.A. Chaturvedi, J. Christensen, J. Ahmed, R. Findeisen, and A. Kojic. 2013. Electrochemical model based observer design for a lithium-ion battery. *IEEE Transactions on Control Systems Technology* 21:289–301.

Kontou, E., C. Liu, F. Xie, X. Wu, and Z. Lin. 2019. Understanding the linkage between electric vehicle charging network coverage and charging opportunity using GPS travel data. *Transportation Research Part C: Emerging Technologies* 98:1–13. https://doi.org/10.1016/j.trc.2018.11.008.

Koo, R. 2012. "Advanced Li-Ion Polymer Battery Cell Manufacturing Plant in USA." Presented at the 2012 DOE Hydrogen and Fuel Cells Program and Vehicle Technologies Program Annual Merit Review and Peer Evaluation Meeting, May 16. https://www.energy.gov/sites/prod/files/2014/03/f10/arravt001_es_koo_2012_p.pdf.

Lam, F., A. Allam, W.T. Joe, Y. Choi, and S. Onori. 2020. Offline multiobjective optimization for fast charging and reduced degradation in lithium-ion battery cells using electrochemical dynamics. *IEEE Control Systems Letters*: 1. https://doi.org/10.1109/LCSYS.2020.3046378.

Lee, H., and A. Clark. 2018. "Charging the Future: Challenges and Opportunities for Electric Vehicle Adoption." Belfer Center for Science and International Affairs, Cambridge, MA: Harvard University. August.

Lee, J.H., D. Chakraborty, S.J. Hardman, and G. Tal. 2020a. Exploring electric vehicle charging patterns: Mixed usage of charging infrastructure. *Transportation Research Part D: Transport and Environment* 79:102249.

Lee, S., P. Mohtat, J.B. Siegel, A.G. Stefanopoulou, J. Lee, and T. Lee. 2020b. Estimation error bound of battery electrode parameters with limited data window. *IEEE Transactions on Industrial Informatics* 16(5):3376–3386. doi:10.1109/TII.2019.2952066.

Levin, A. 2020. "Half of U.S. Fire Departments Are Ill-Prepared for EV Fires, NTSB Says." Automotive News. October 9. https://www.autonews.com/regulation-safety/half-us-fire-departments-are-ill-prepared-ev-fires-ntsb-says.

LeVine, M. 2020. "A New Battery Breakthrough That Could Save Electric Vehicles During a Recession." Medium Marker. March 24. https://marker.medium.com/a-new-battery-breakthrough-that-could-save-electric-vehicles-during-a-recession-c193ebdd3a5d.

Ley, J. 2016. "Unique Lanthanide-Free Motor Construction." Presentation at the DOE Annual Merit Review, June 7, Washington, DC. http://energy.gov/eere/vehicles/downloads/vehicle-technologies-office-merit-review-2016-unique-lanthide-free-motor.

Li, Z. 2018. "Where SiC Outperforms GaN." Power Electronic Tips, February 27. https://www.powerelectronictips.com/sic-outperforms-gan/.

Lienert, P., and J. White. 2018. "GM Races to Build a Formula for Profitable Electric Cars." Reuters. https://www.reuters.com/article/us-gm-electric-insight/gm-races-to-build-a-formula-for-profitable-electric-cars-idUSKBN1EY0GG.

Lima, P. 2018. "2018 Nissan Leaf Battery Real Specs." PushEVs. January 29. https://pushevs.com/2018/01/29/2018-nissan-leaf-battery-real-specs/.

Lin, C., H. Peng, J.W. Grizzle, and J. Kang. 2003. Power management strategy for a parallel hybrid electric truck. *IEEE Transactions on Control Systems Technology* 11(6):839–849. https://doi.org/10.1109/TCST.2003.815606.

Lin, X. 2017. Analytic analysis of the data-dependent estimation accuracy of battery equivalent circuit dynamics. *IEEE Control Systems Letters* 1(2):304–309. doi: 10.1109/LCSYS.2017.2715821.

Lin, X. 2018. Theoretical analysis of battery SOC estimation errors under sensor bias and variance. *IEEE Transactions on Industrial Electronics* 65:7138–7148.

Lin, X., H.E. Perez, J.B. Siegel, A.G. Stefanopoulou, Y. Li, R.D. Anderson, Y. Ding, et al., 2013. Online parameterization of lumped thermal dynamics in cylindrical lithium ion batteries for core temperature estimation and health monitoring. *IEEE Transactions on Control Systems Technology* 21(5):1745–1755. https://doi.org/10.1109/TCST.2012.2217143.

Lin, Z. 2014. Optimizing and diversifying electric vehicle driving range for U.S. drivers. *Transportation Science* 48(4):635–650. http://dx.doi.org/10.1287/trsc.2013.0516.

Lin, Z., and D.L. Greene. 2010. "Rethinking FCV/BEV Vehicle Range: A Consumer Value Trade-off Perspective." Presentation at the 25th World Battery, Hybrid and Fuel Cell Electric Vehicle Symposium and Exhibition, Shenzhen, China, November 5–9.

Lin, Z., and D.L. Greene. 2011. Promoting the market for plug-in hybrid and battery electric vehicles: Role of recharge availability. *Transportation Research Record* 2252(1):49–56. http://dx.doi.org/10.3141/2252-07.

Liu, C., Z. Xu, D. Gerada, J. Li, C. Gerada, Y.C. Chong, M. Popescu, et al. 2020. Experimental investigation on oil spray cooling with hairpin windings. *IEEE Transactions on Industrial Electronics* 67(9):7343–7353. doi: 10.1109/TIE.2019.2942563.

Loveday, S. 2019. "Nissan LEAF Batteries to Outlast Car by 10–12 Years." InsideEVs. May 24. https://insideevs.com/news/351314/nissan-leaf-battery-longevity/.

Ludois, D. 2015. "Brushless and Permanent Magnet Free Wound Field Synchronous Motors for EV Traction." Presentation at the DOE Annual Merit Review. June 9. Washington, DC. http://energy.gov/eere/vehicles/downloads/vehicle-technologies-office-merit-review-2015-brushless-and-permanent-magnet.

Lutsey, N., and M. Nicholas. 2019a. *Update on Electric Vehicle Costs in the United States through 2030.* International Council on Clean Transportation. https://www.theicct.org/publications/update-US-2030-electric-vehicle-cost.

Lutsey, N., and M. Nicholas. 2019b. *Electric Vehicle Costs and Consumer Benefits in Colorado in the 2020–2030 Time Frame.* International Council on Clean Transportation. https://theicct.org/publications/ev-costs-colorado-2020-2030.

Lutsey, N., D. Meszler, A. Isenstadt, J. German, and J. Miller. 2018. *Efficiency Technology and Cost Assessment for U.S. 2025–2030 Light-Duty Vehicles.* International Council on Clean Transportation. http://www.theicct.org/US-2030-technology-cost-assessment.

Mahamud, R., and C. Park. 2011. Reciprocating air flow for Li-ion battery thermal management to improve temperature uniformity. *Journal of Power Sources* 196(13):5685–96. https://doi.org/10.1016/j.jpowsour.2011.02.076.

Malone, W. 2018. "Chevy Bolt EV Survey Hints at Cold Weather Package with Heat Pump." InsideEVs. October 24. https://insideevs.com/news/340561/chevy-bolt-ev-survey-hints-at-cold-weather-package-with-heat-pump.

Maness, M., and Lin, Z. 2019. Free charging: Exploratory study of its impact on electric vehicle sales and energy. *Transportation Research Record* 2673(9):590–601.

Manthey, N. 2020. "CATL to Kick-off LFP Cell Supply for Tesla China Model 3." electrive.com. July. https://www.electrive.com/2020/07/20/catl-to-kick-off-lfp-cell-supply-for-tesla-china-model-3/.

Mendoza, S., J. Liu, P. Mishra, and H. Fathy. 2017. On the relative contributions of bias and noise to lithium-ion battery state of charge estimation errors. *Journal of Energy Storage* 11:86–92.

Mihet-Popa, L., and S. Saponara. 2018. Toward green vehicles digitalization for the next generation of connected and electrified transport systems. *Energies* 11(11):3124. https://doi.org/10.3390/en11113124.

Mohan, S., J.B. Siegel, A.G. Stefanopoulou, and R. Vasudevan. 2019. An energy-optimal warm-up strategy for Li-ion batteries and its approximations. *IEEE Transactions on Control Systems Technology* 27(3):1165–1180. https://doi.org/10.1109/TCST.2017.2785833.

Mohtadi, R., and F. Mizuno. 2014. Magnesium batteries: Current state of the art, issues and future perspectives. *Beilstein Journal of Nanotechnology* 5(1):1291–1311. https://doi.org/10.3762/bjnano.5.143.

Mohtat, P., S. Lee, J.B. Siegel, and A.G. Stefanopoulou. 2019. Towards better estimability of electrode-specific state of health: Decoding the cell expansion. *Journal of Power Sources* 427:101–111, https://doi.org/10.1016/j.jpowsour.2019.03.104.

Nedjalkov, A., J. Meyer, M. Köhring, A. Doering, M. Angelmahr, S. Dahle, A. Sander, et al. 2016. Toxic gas emissions from damaged lithium ion batteries—analysis and safety enhancement solution. *Batteries* 2(1):5. https://doi.org/10.3390/batteries2010005.

Nelder, C., and E. Rogers. 2019. "Reducing EV Charging Infrastructure Costs." Rocky Mountain Institute. https://rmi.org/insight/reducing-ev-charging-infrastructure-costs/.

New York City Council. 2013. *Local Law 130 of 2013.* City Council Int. No.1176-A of 2013. https://www1.nyc.gov/assets/buildings/local_laws/ll130of2013.pdf.

Nicholas, M. 2019. *Estimating Electric Vehicle Charging Infrastructure Costs Across Major U.S. Metropolitan Areas*. International Council on Clean Transportation. https://theicct.org/publications/charging-cost-US.

Nicholas, M., and D. Hall. 2018. "Lessons Learned on Early Electric Vehicle Fast-Charging Deployments." International Council on Clean Transportation. July. https://theicct.org/sites/default/files/publications/ZEV_fast_charging_white_paper_final.pdf.

Nicholas, M., G. Tal, and J. Woodjack. 2013. "California Statewide Charging Assessment Model for Plug-In Electric Vehicles: Learning from Statewide Travel Surveys." International Council on Clean Transportation. http://www.its.ucdavis.edu/wp-content/themes/ucdavis/pubs/download_pdf.php?id=1832.

Nicholas, M., G. Tal, and J. Woodjack. 2017. "Lessons from In-Use Fast Charging Data: Why Are Drivers Staying Close to Home?" Research Report. UC Davis Institute of Transportation Studies, January. https://itspubs.ucdavis.edu/download_pdf.php?id=2699.

Nilsson, M., and B. Nykvist. 2015. Governing the electric vehicle transition—near term interventions to support a green energy economy. *Applied Energy* 179:1360–1371. http://dx.doi.org/10.1016/j.apenergy.2016.03.056.

Nisewanger, J. 2018. "Exclusive: Details on Hyundai's New Battery Thermal Management Design." Electric Revs. 2018. https://electricrevs.com/2018/12/20/exclusive-details-on-hyundais-new-battery-thermal-management-design.

NRC (National Research Council). 2013. *Transitions to Alternative Vehicles and Fuels*. Washington, DC: The National Academies Press.

NTSB (National Transportation Safety Board). 2020. "Highway Special Project HWY19SP002." https://data.ntsb.gov/Docket/?NTSBNumber=HWY19SP002.

NTT (Nippon Telegraph and Telephone Corporation). 2016. "Lithium-Air Secondary Battery Technology: NTT Device Technology Laboratories." https://www.ntt.co.jp/dtl/e/technology/sd_product-lithium-battery.html.

NTT. 2020. "Lithium-Air Secondary Battery Technology: NTT Device Technology Laboratories." https://www.ntt.co.jp/dtl/e/technology/sd_product-lithium-battery.html.

Omar, N., M.A. Monem, Y. Firouz, J. Salminen, J. Smekens, O. Hegazy, H. Gaulous, et al. 2014. Lithium iron phosphate based battery—assessment of the aging parameters and development of cycle life model. *Applied Energy* 113:1575–1585. https://doi.org/10.1016/j.apenergy.2013.09.003.

Omekanda, A. 2013. "Switched Reluctance Machines for EV and HEV Propulsion: State-of-the-art." 2013 IEEE Workshop on Electrical Machines Design Control and Diagnosis (WEMDCD). March 11–12. Paris, France.

Ou, S., Z. Lin, X. He, and S. Przesmitzki. 2018. Estimation of vehicle home parking availability in China and quantification of its potential impacts on plug-in electric vehicle ownership cost. *Transport Policy* 68:107–117. https://doi.org/10.1016/j.tranpol.2018.04.014.

Ou, S., Z. Lin, X. He, S. Przesmitzki, and J. Bouchard. 2020. Modeling charging infrastructure impact on the electric vehicle market in China. *Transportation Research Part D: Transport and Environment* 81:102248. https://doi.org/10.1016/j.trd.2020.102248.

P3. 2020. "Tesla Battery Day 2020—Technology Announcement Analysis." Electrive.net. https://www.electrive.net/wp-content/uploads/2020/09/200923_Tesla_Battery-Day_P3-Assessment-published.pdf.

Park, S., and D. Jung. 2013. Battery cell arrangement and heat transfer fluid effects on the parasitic power consumption and the cell temperature distribution in a hybrid electric vehicle. *Journal of Power Sources* 227 (April): 191–98. https://doi.org/10.1016/j.jpowsour.2012.11.039.

Patry, G., A. Romagny, S. Martinet, and D. Froelich. 2015. Cost modeling of lithium-ion battery cells for automotive applications. *Energy Science and Engineering* 3(1):71–82. https://doi.org/10.1002/ese3.47.

Patterson, G. 2015. Automotive opportunities for power GaN. *Power Electronics Europe* (4):20–23.

Peterson, S.B., and J.J. Michalek. 2013. Cost-effectiveness of plug-in hybrid electric vehicle battery capacity and charging infrastructure investment for reducing US gasoline consumption. *Energy Policy* 52:429–438.

Pillot, C. 2019. "The Rechargeable Battery Market and Main Trends 2011–2020." Presented at the Avicenne Energy, Paris, France, May 28. http://cdn.ceo.ca.s3.amazonaws.com/1em2t4r-02%20-%20Presentation%20Avicenne%20-%20Christophe%20Pillot%20-%2028%20Mai%202019.pdf.

Plett, G.L. 2004. Extended Kalman filtering for battery management systems of LiPB-Based HEV battery packs: Part 3. State and parameter estimation. *Journal of Power Sources* 134(2):277–292. https://doi.org/10.1016/j.jpowsour.2004.02.033.

Pop, V., H.J. Bergveld, P.H.L. Notten, and P.P.L. Regtien. 2005. State-of-the-art of battery state-of-charge determination. *Measurement Science and Technology* 16(12):R93–R110. https://doi.org/10.1088/0957-0233/16/12/R01.

PowerAmerica. 2018. "PowerAmerica Strategic Roadmap for Next Generation Wide Bandgap Power Electronics." White Paper. February. https://poweramericainstitute.org/wp-content/uploads/2018/02/PowerAmerica_Roadmap_Final-Public-Version-February-2018.pdf.

Raghavan, S.S., and G. Tal. 2020. Influence of user preferences on the revealed utility factor of plug-in hybrid electric vehicles. *World Electric Vehicle Journal* 11(1):6. https://doi.org/10.3390/wevj11010006.

Rajashekara, K. 2013. "Trends in Electric Propulsion—IEEE Transportation Electrification Community." IEEE Transportation Electrification Community Newsletter. October. https://tec.ieee.org/newsletter/october-2013/trends-in-electric-propulsion.

Riley, C. 2020. "Ford Delays Production of Plug-in Escape after SUV Fires in Europe." CNN. October 14. https://www.cnn.com/2020/10/14/business/ford-escape-hybrid-delay/index.html.

Robinius, M., J. Linβen, T. Grube, M. Reuβ, P. Stenzel, K. Syranidis, P. Kuckertz, et al. 2018. Comparative analysis of infrastructures: Hydrogen fueling and electric charging of vehicles. *Energie and Umwelt/Energy and Environment* 408.

SAE International. 2014. "Safety Standard for Electric and Hybrid Vehicle Propulsion Battery Systems Utilizing Lithium-based Rechargeable Cells." https://www.sae.org/standards/content/j2929/.

Safoutin, M., J. McDonald, and B. Ellies. 2018. Predicting the future manufacturing cost of batteries for plug-in vehicles for the U.S. Environmental Protection Agency (EPA) 2017–2025 light-duty greenhouse gas standards. *World Electric Vehicle Journal* 9(3):42. https://www.mdpi.com/2032-6653/9/3/42.

Samad, N.A., Y. Kim, and J.B. Siegel. 2018. On power denials and lost energy opportunities in downsizing battery packs in hybrid electric vehicles. *Journal of Energy Storage* 16:187–196. https://doi.org/10.1016/j.est.2018.01.013.

Schmuch, R., R. Wagner, G. Hörpel, T. Placke, and M. Winter. 2018. Performance and cost of materials for lithium-based rechargeable automotive batteries. *Nature Energy* 3(4):267–278. https://doi.org/10.1038/s41560-018-0107-2.

Schweber, B. 2020. "GaN Power Devices: Potential Benefits, and Keys to Successful Use." Mouser Electronics. https://www.mouser.com/applications/gan-power-devices/.

Schwunk, S., N. Armbruster, S. Straub, J. Kehl, and M. Vetter. 2013. Particle filter for state of charge and state of health estimation for lithium-iron phosphate batteries. *Journal of Power Sources* 239:705–710.

Sekulich, D. 2020. "Chinese Rare Earth Metals Surge in Price." Northern Miner. https://www.northernminer.com/news/chinese-rare-earth-metals-surge-in-price/1003825605/.

Semiconductor Today. 2019. "Transphorm's Gen III GaN Platform Earns Automotive Qualification." February 26. http://www.semiconductor-today.com/news_items/2019/feb/transphorm_260219.shtml.

Serradilla, J., J. Wardle, P. Blythe, and J. Gibbon. 2017. An evidence-based approach for investment in rapid-charging infrastructure. *Energy Policy* 106:514–524.

Severson, K.A., P.M. Attia, N. Jin, N. Perkins, B. Jiang, Z. Yang, M.H. Chen, et al. 2019. Data-driven prediction of battery cycle life before capacity degradation. *Nature Energy* 4(5):383–391. https://doi.org/10.1038/s41560-019-0356-8.

Sharpe, B., N. Lutsey, C. Smith, and C. Kim. 2020. *Power Play: Canada's Role in the Electric Vehicle Transition*. Retrieved from the International Council on Clean Transportation, https://theicct.org/publications/canada-zev-transition.

SHEVDC. 2017. *Shanghai New Energy Vehicle Market Characteristics and User Behavior Research Report (in Chinese).* SHEVDC.org. https://www.shevdc.org/report/original_report/1718.jhtml.

Simmons, D., and S. Chalk. 2019. Presentation to the National Academies Committee on Modernizing the U.S. Electricity System. March 4. https://sites.nationalacademies.org/cs/groups/depssite/documents/webpage/deps_191612.pdf.

Slovick, M. 2017. "Making the Jump to Wide Bandgap Power." Electronic Design. December. http://www.electronicdesign.com/analog/making-jump-wide-bandgap-power.

Smart, J., and S. Schey. 2012. Battery electric vehicle driving and charging behavior observed early in the EV Project. *SAE International Journal of Alternative Powertrains* 1(1):27–33. https://doi.org/10.4271/2012-01-0199.

Smart, J., T. Bradley, and S. Salisbury. 2014. Actual versus estimated utility factor of a large set of privately owned Chevrolet Volts. *SAE International Journal of Alternative Powertrains* 3(1):30–35. https://doi.org/10.4271/2014-01-1803.

Smith, M., and J. Castellano. 2015. *Costs Associated with Non-Residential Electric Vehicle Supply Equipment.* U.S. Department of Energy's (DOE) Clean Cities program. November. https://afdc.energy.gov/files/u/publication/evse_cost_report_2015.pdf.

Speltino, C., D. Di Domenico, G. Fiengo, and A.G. Stefanopoulou. 2009. "Comparison of Reduced Order Lithium-Ion Battery Models for Control Applications." Pp. 3276–3281 in *Proceedings of Joint 48th IEEE Conference on Decision and Control and 28th Chinese Control Conference.* October.

Statista. 2020a. "EV Batteries in China and Rest of World by Chemistry 2020." https://www.statista.com/statistics/964355/ev-batteries-elements-market-share-in-china-vs-rest-of-world/.

Statista. 2020b. "Age of US Automobiles and Trucks since 1990." https://www.statista.com/statistics/185198/age-of-us-automobiles-and-trucks-since-1990/.

Stephens, D., P. Shawcross, G. Stout, E. Sullivan, J. Saunders, S. Risser, and J. Sayre. 2017. *Lithium-Ion Battery Safety Issues for Electric and Plug-In Hybrid Vehicles.* Report No. DOT HS 812 418. Washington, DC: National Highway Traffic Safety Administration. https://www.nhtsa.gov/sites/nhtsa.dot.gov/files/documents/12848-lithiumionsafetyhybrids_101217-v3-tag.pdf.

Strategic Vision. 2017. "New Vehicle Experience Study." www.strategicvision.com/nves.

Strydom, J., D. Reusch, S. Colino, and A. Nakata. 2017. *Using Enhancement Mode GaN-on-Silicon Power FETs.* Application Note AN003. El Segundo, CA: Efficient Power Conversion. https://epc-co.com/epc/DesignSupport/ApplicationNotes/AN003-UsingEnhancementMode.aspx.

Sulzer, V., P. Mohtat, S. Lee, J.B. Siegel, and A.G. Stefanopoulou. 2020. "Promise and Challenges of a Data-Driven Approach for Battery Lifetime Prognostics." https://arxiv.org/abs/2010.07460.

Sun, P., R. Bisschop, H. Niu, and X. Huang. 2020. A review of battery fires in electric vehicles. *Fire Technology* 56(4):1361–1410. https://doi.org/10.1007/s10694-019-00944-3.

SWEEP (Southwest Energy Efficiency Project). 2020. "EV Infrastructure Building Codes: Adoption Toolkit." https://www.swenergy.org/transportation/electric-vehicles/building-codes#who.

Tanim, T.R., C.D. Rahn, and C.Y. Wang. 2015. State of charge estimation of a lithium ion cell based on a temperature dependent and electrolyte enhanced single particle model. *Energy* 80:731–739.

Tesla. 2018. "Tesla 2018 Shareholder Meeting." Online video. June 5. https://youtu.be/OIdjv6oHzio.

Tesloop. 2018. "Tesloop's Tesla Model S Surpasses 400,000 Miles (643,737 KM)." Tesloop. https://www.tesloop.com/blog/2018/7/16/tesloops-tesla-model-s-surpasses-400000-miles-643737-kilometers.

Thanh, A., and H. Min-Fu. 2017. Performance evaluation of interior permanent magnet motors using thin electrical steels. *IEEJ Journal of Industry Applications* 6(6):422–428. doi:10.1541/ieejjia.6.422.

Toyota USA Newsroom. 2018. "Toyota Develops New Magnet for Electric Motors Aiming to Reduce Use of Critical Rare-Earth Element by up to 50% | Corporate | Global Newsroom." Toyota Motor Corporation Official Global Website. February 20. https://global.toyota/en/newsroom/corporate/21139684.html.

Transphorm Inc. 2017. "Transphorm Announces First Automotive-Qualified GaN FETs, Second Generation Platform Passes AEC-Q101 Tests for High Power Automotive Applications." Company press release. March 17. https://www.transphormusa.com/en/news/transphorm-announces-first-automotive-qualified-gan-fets/.

Traut, E.J., T.C. Cherng, C. Hendrickson, and J.J. Michalek. 2013. US residential charging potential for electric vehicles. *Transportation Research Part D: Transport and Environment* 25:139–145.

TRB and NRC (Transportation Research Board and National Research Council). 2015. *Overcoming Barriers to Deployment of Plug-in Electric Vehicles.* Washington, DC: The National Academies Press.https://doi.org/10.17226/21725.

Tu, H., H. Feng, S. Srdic, and S. Lukic. 2019. Extreme fast charging of electric vehicles: A technology overview. *IEEE Transactions on Transportation Electrification* 5(4):861–878. https://doi.org/10.1109/TTE.2019.2958709.

UBS. 2017. "UBS Evidence Lab Electric Car Teardown: Disruption Ahead?" Q-Series newsletter. https://neo.ubs.com/shared/d1ZTxnvF2k/.

Van Haaren, R. 2011. *Assessment of Electric Cars' Range Requirements and Usage Patterns Based on Driving Behavior Recorded in the National Household Travel Survey of 2009.* Earth and Environmental Engineering Department, Columbia University, Fu Foundation School of Engineering and Applied Science, New York.

Vekasi, K. 2019. "China's Control of Rare Earth Metals." National Bureau of Asian Research (NBR). Pacific Energy Summit, August 13. https://www.nbr.org/publication/chinas-control-of-rare-earth-metals/.

Verbrugge, M., and B. Koch. 2006. Generalized recursive algorithm for adaptive multiparameter regression: Application to lead acid, nickel metal hydride, and lithium-ion batteries. *Journal of the Electrochemical Society* 153(1):A187. https://doi.org/10.1149/1.2128096.

Verbrugge, M., and E. Tate. 2004. Adaptive state of charge algorithm for nickel metal hydride batteries including hysteresis phenomena. *Journal of Power Sources* 126(1):236–249. https://doi.org/10.1016/j.jpowsour.2003.08.042.

Vlahinos, A., and A.A. Pesaran. 2002. "Energy Efficient Battery Heating in Cold Climates." Future Car Congress, Arlington, VA. SAE International.

Wang, S., M. Verbrugge, J.S. Wang, and P. Liu. 2011. Multi-parameter battery state estimator based on the adaptive and direct solution of the governing differential equations. *Journal of Power Sources* 196(20):8735–8741. https://doi.org/10.1016/j.jpowsour.2011.06.078.

Wang, T., S. Chen, H. Ren, and Y. Zhao. 2018. Model-based unscented Kalman filter observer design for lithium-ion battery state of charge estimation. *International Journal of Energy Research* 42(4):1603–1614. https://doi.org/10.1002/er.3954.

Wang, Y., H. Fang, L. Zhou, and T. Wada. 2017. Revisiting the state-of-charge estimation for lithium-ion batteries: A methodical investigation of the extended Kalman filter approach. *IEEE Control Systems* 37(4):73–96. https://doi.org/10.1109/MCS.2017.2696761.

Weigel, T., F. Schipper, E. Erickson, F. Susai, B. Markovsky, and D. Aurbach. 2019. Structural and electrochemical aspects of LiNi0.8Co0.1Mn0.1O2 cathode materials doped by various cations. *ACS Energy Letters* 4(2):508–516.

Wenig, J., M. Sodenkamp, and T. Staake. 2019. Battery versus infrastructure: Tradeoffs between battery capacity and charging infrastructure for plug-in hybrid electric vehicles. *Applied Energy* 255:113787.

Wentker, M., M. Greenwood, and J. Leker. 2019. A bottom-up approach to lithium-ion battery cost modeling with a focus on cathode active materials. *Energies* 12(3):504. https://doi.org/10.3390/en12030504.

Witter, F. 2018. "Shaping the Transformation Together." Volkswagen Group, investor meetings, Frankfurt and London, 29th–30th November 2018. Retrieved from https://www.volkswagenag. com/presence/investorrelation/publications/presentations/2018/11_november/2018_11_29-30_ Volkswagen_Group_Presentation_Investor_Meetings_Frankfurt_London.pdf.

Wolbertus, R., R. Van den Hoed, and S. Maase. 2016. Benchmarking charging infrastructure utilization. *World Electric Vehicle Journal* 8(4):754–771. https://doi.org/10.3390/wevj8040754.

Wolbertus, R., S. Jansen, and M. Kroesen. 2020. Stakeholders' perspectives on future electric vehicle charging infrastructure developments. *Futures* 123(October):102610. https://doi.org/10.1016/j.futures.2020.102610.

Wolfspeed. 2017. "Wolfspeed Introduces New SiC MOSFET for EV Drive Trains." Press release. January. https://www.wolfspeed.com/news/900v10mohm.

Wood, E., S. Raghavan, C. Rames, J. Eichman, and M. Melaina. 2017. "Regional Charging Infrastructure for Plug-In Electric Vehicles: A Case Study of Massachusetts." January. https://www.nrel.gov/docs/fy17osti/67436.pdf.

Wu, F., Y. Yuan, X. Cheng, Y. Bai, Y. Li, C. Wu, and Q. Zhang. 2018. Perspectives for restraining harsh lithium dendrite growth: Towards robust lithium metal anodes. *Energy Storage Materials* 15:148–170. https://doi.org/10.1016/j.ensm.2018.03.024.

Xia, B., C. Chen, Y. Tian, W. Sun, Z. Xu, and W. Zheng. 2014. A novel method for state of charge estimation of lithium-ion batteries using a nonlinear observer. *Journal of Power Sources* 270:359–366.

Xu, J., and D. Chen. 2017. "A Performance Comparison of GaN E-HEMTs Versus SiC MOSFETs in Power Switching Applications." *Bodo's Power Systems*. June.

Xu, K. 2004. Nonaqueous liquid electrolytes for lithium-based rechargeable batteries. *Chemical Reviews* 104(10):4303–4418. https://doi.org/10.1021/cr030203g.

Yamauchi, M. n.d. "Tesla Charging Speed on a 110 Volt Outlet." PluglessPower. https://www.pluglesspower.com/learn/can-tesla-charge-regular-110v-wall-outlet-technically-yes/.

Yuksel, T., M.-A. M. Tamayao, C. Hendrickson, I.M.L. Azevedo, and J.J. Michalek. 2016. Effect of regional grid mix, driving patterns and climate on the comparative carbon footprint of gasoline and plug-in electric vehicles in the United States. *Environmental Research Letters* 11(4):044007. https://doi.org/10.1088/1748-9326/11/4/044007.

Yvkoff, L. 2019. "Robotic EV Fast-Charging Stations Planned For San Francisco." Forbes. August 1. https://www.forbes.com/sites/lianeyvkoff/2019/08/01/robotic-ev-fast-charging-stations-planned-for-san-francisco/.

Zhao, X. 2016a. Lithium/sulfur secondary batteries: A review. *Journal of Electrochemical Science and Technology* 7(2):97–114. https://doi.org/10.5229/JECST.2016.7.2.97.

Zhao, Z. 2016b. "Next Generation Inverter." General Motors. DOE-VTO Annual Merit Review. June 7.

Zheng, Y., M. Ouyang, X. Han, L. Lu, and J. Li. 2018. Investigating the error sources of the online state of charge estimation methods for lithium-ion batteries in electric vehicles. *Journal of Power Sources* 377:161–188. https://doi.org/10.1016/j.jpowsour.2017.11.094.

Zhu, C., F. Lu, H. Zhang, J. Sun, and C.C. Mi. 2018. A real-time battery thermal management strategy for connected and automated hybrid electric vehicles (CAHEVs) based on iterative dynamic programming. *IEEE Transactions on Vehicular Technology* 67(9):8077–8084. https://doi.org/10.1109/TVT.2018.2844368.

6

Fuel Cell Electric Vehicles

6.1 BACKGROUND

Fuel cell electric vehicles (FCEVs) offer high efficiency, petroleum-free transportation, and zero tailpipe emissions just like battery electric vehicles (BEVs). Several automakers are planning to offer FCEVs and BEVs as complementary zero-emission vehicle (ZEV) technologies to fulfill different customer needs, with BEVs typically in smaller vehicle size classes with shorter driving ranges, and FCEVs in larger vehicle size classes with longer daily driving ranges and shorter refueling times.

FCEVs have an architecture similar to series hybrids, as shown in Figure 6.1, with the engine and generator replaced by a fuel cell. Most FCEVs use a hydrogen-powered fuel cell combined with a battery that stores energy generated from regenerative braking and provides supplemental power to the electric traction motor. The fuel cell and battery are sized to provide the most efficient combination of constant and peak power.

Although FCEVs are not in mass production currently, automakers have sold or leased more than 8,000 in the United States, mostly in California, where they are refueled at more than 40 hydrogen stations (CaFCP, 2020). In 2014, some automakers announced plans to introduce FCEVs in the Northeast United States beginning in 2016 (Toyota USA Newsroom, 2014); however, those plans have been delayed largely owing to the prohibition of hydrogen-powered vehicles in tunnels and on the lower deck of two-tier bridges in that region. Several studies conducted over the past 4 years have addressed the risks and implications of potential traffic incidents involving FCEVs in tunnels; these will be summarized later in this chapter.

The most significant hurdle to FCEV deployment is the lack of an extensive hydrogen infrastructure. The U.S. Department of Energy (DOE) recently launched the "H2@Scale" initiative to address the challenges associated with hydrogen infrastructure, and some industry-led efforts are also in place. Government-industry programs on FCEV deployment and hydrogen infrastructure development are generally much stronger in Asia and Europe, particularly in Japan and Germany, than in the United States. A key driver for these efforts is the potential use of hydrogen as a storage sink for the renewable energy system with versatile applications in transportation, heat for buildings, and feedstock for industry.

All automakers engaged in FCEV development are adding a focus on medium- and heavy-duty vehicle (MHDV) applications for fuel cell powertrains, while some are shifting their short-term fuel cell focus to MHDVs entirely and emphasizing BEVs for light-duty vehicle (LDV) applications. Fuel cells offer an alternative to batteries in difficult-to-electrify applications such as vehicles with heavy payloads or high vehicle miles traveled (VMT) that need lighter weight powertrains, longer driving ranges, and/or quicker refueling times. FCEVs are also well-suited for medium-duty applications such as delivery trucks, municipal vehicles, and other tethered fleets

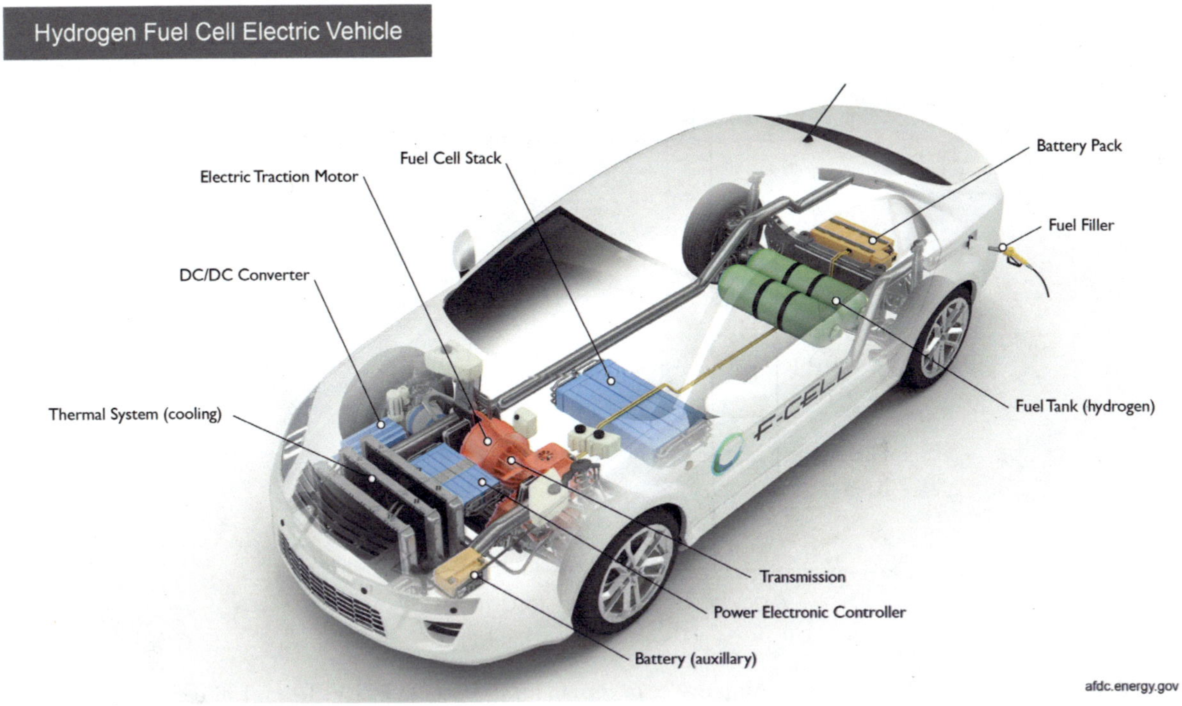

FIGURE 6.1 Schematic of an FCEV showing series hybrid configuration. Component placement will vary by vehicle manufacturer. In some FCEVs today, the fuel cell stack is placed under the hood, and three tanks are used to store hydrogen. SOURCE: AFDC (n.d.a).

that require fewer refueling locations or one centrally located refueling station. Fuel cells are also being tested as range extenders for BEVs in fleet applications.

Research and development (R&D) efforts, led by both government and industry worldwide, continue to drive down fuel cell technology costs and improve performance. This chapter provides basic information about fuel cells and today's commercial FCEVs, and describes the status of automotive fuel cell technology and current R&D activities aimed at improving the technology. A number of studies are described that estimate the current cost of automotive fuel cell systems based on state-of-the-art technology (not yet commercial) projected to high-volume production levels, as well as the cost of fuel cell technology in current commercial vehicles at today's manufacturing volumes. Studies are also presented that attempt to predict the future cost of hydrogen and fuel cell technologies based on technology improvements and the economies of scale anticipated through increased demand. The results of these studies, some of which are more optimistic than others, depend on the scenarios and assumptions used, of course, and key assumptions are identified. Several scenarios assume a significantly increased role for hydrogen as a zero-emission energy carrier in a carbon-constrained future; as such, there is a relatively high level of uncertainty in the projected time frames. In some cases, FCEVs are compared to BEVs to note similarities or differences in vehicle attributes and applications, cost and performance status, or projected timelines for development and deployment. This chapter also provides information on the status of hydrogen refueling infrastructure, plans to accelerate infrastructure development, and R&D efforts to improve hydrogen technologies. The chapter ends with findings and recommendations for automotive fuel cells and hydrogen refueling infrastructure.

6.2 FUEL CELL BASICS

Like batteries, fuel cells are composed of an anode, a cathode, and an electrolyte. Unlike batteries, fuel cells do not need to be periodically recharged—instead, they need to be refueled with hydrogen. While there are

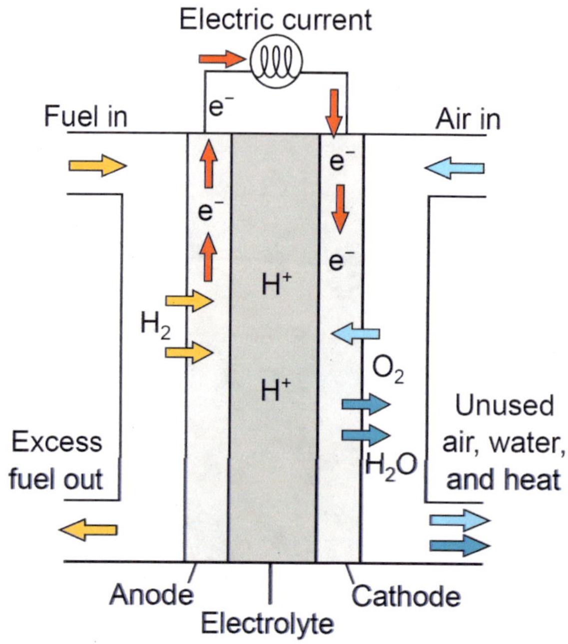

FIGURE 6.2 Schematic of a PEM fuel cell. Air provides oxygen to the cathode. In FCEVs today, hydrogen is stored in an onboard compressed hydrogen tank.
SOURCE: Mattuci (2015).

several types of fuel cells, the proton exchange membrane (PEM)—also sometimes called a polymer electrolyte membrane—is the fuel cell technology of choice for transportation applications owing to its low operational temperature, quick start-up, and high power density.

As shown in Figure 6.2, the PEM fuel cell works by passing hydrogen through the anode and oxygen (from air) through the cathode. At the anode site, hydrogen molecules are split into electrons and protons. The protons pass through the electrolyte membrane, while the electrons are forced through a circuit, generating an electric current and heat. At the cathode, the protons, electrons, and oxygen combine to produce water. In addition to electricity, hydrogen fuel cells produce only water and heat.

The maximum theoretical voltage of a single hydrogen/air fuel cell is 1.16 V (Thomas and Zalbowitz, 1999). However, in practice, the cell usually generates about 0.6 V to 0.9 V. The cells are stacked, or placed in series, to generate sufficient voltage to meet vehicle requirements. The key components of the PEM fuel cell stack (shown in Figure 6.3) are as follows:

- Membrane-electrode assembly (MEA), the "heart" of the fuel cell, comprised of catalyst/electrode and electrolyte/membrane (sometimes called the catalyst coated membrane or CCM), and gas diffusion layer (GDL). Platinum or platinum alloys are the catalysts typically used today,[1] and perfluorosulfonic acid (PFSA) ionomers, such as Nafion®, are commonly employed as the electrolyte.
- Other stack hardware required for electrical connections and/or insulation and the flow of fuel and air are current collectors and bipolar plates (BPs, or separator plates) with channels to distribute fuel and air. A variety of materials can be used for BPs, including metals, coated metals, graphite, or carbon composites.

The fuel cell stack is supported by balance-of-plant (BOP) components—pumps, sensors, heat exchanger, gaskets, compressor, and blower or humidifier—that manage the ancillary functions of injection and recirculation

[1] Automakers are working toward the same level of platinum as used in internal combustion engine catalytic converters.

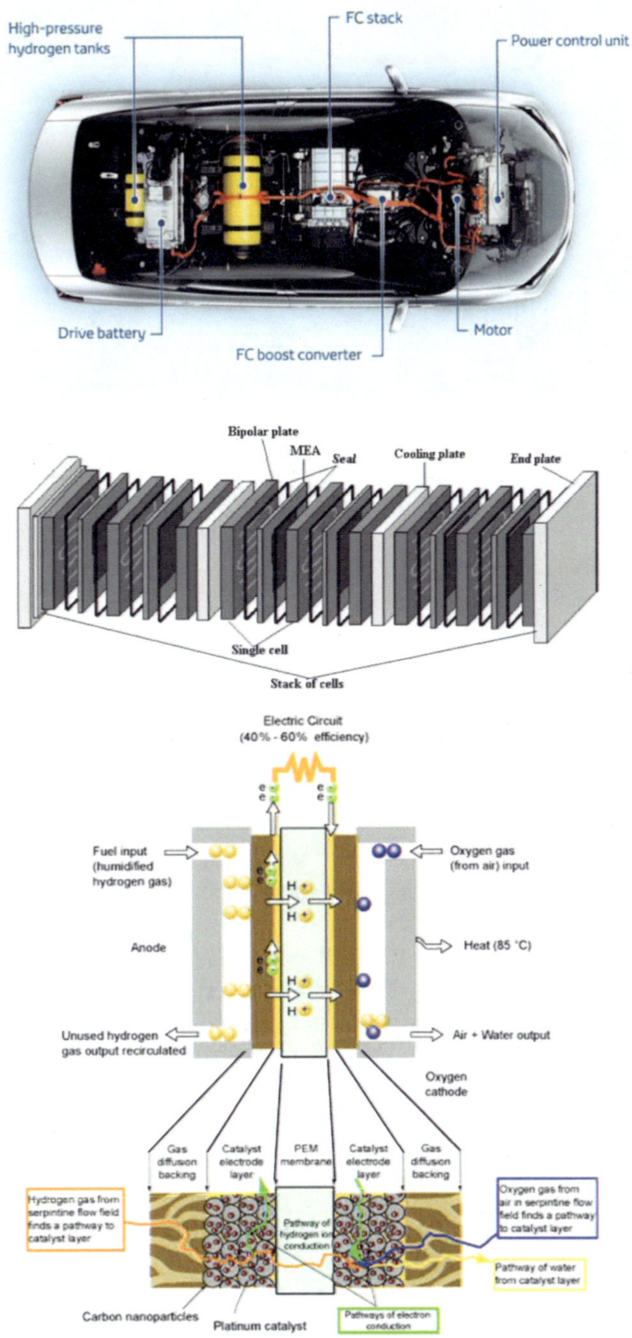

FIGURE 6.3 A schematic of the PEM fuel cell system, from the full vehicle powertrain assembly (elements of a Toyota FCEV), down to the exploded stack-level view, and finally a single cell, and the structure of a membrane electrode assembly. SOURCE: (top) Reprinted under CC BY-NC-ND 4.0 from M. Kendall, 2018, Fuel cell development for new energy vehicles (NEVs) and clean air in China, *Progress in Natural Science: Materials International* 28(2):113–120, https://doi.org/10.1016/j.pnsc.2018.03.001; (middle) Reprinted by permission from Springer Nature Customer Service Centre GmbH: [Springer Nature] PEM Fuel Cell Fundamentals by Yuan and Wang, in *PEM Fuel Cell Electrocatalysts and Catalyst Layers*, J. Zhang (ed.), © 2008; (bottom) Reprinted under CC BY 4.0 from X. Ren, X. Zhang, T. Teng, and C. Li, 2020, Research on estimation method of fuel cell health state based on lumped parameter model, *Energies* 13(23):6425, https://doi.org/10.3390/en13236425.

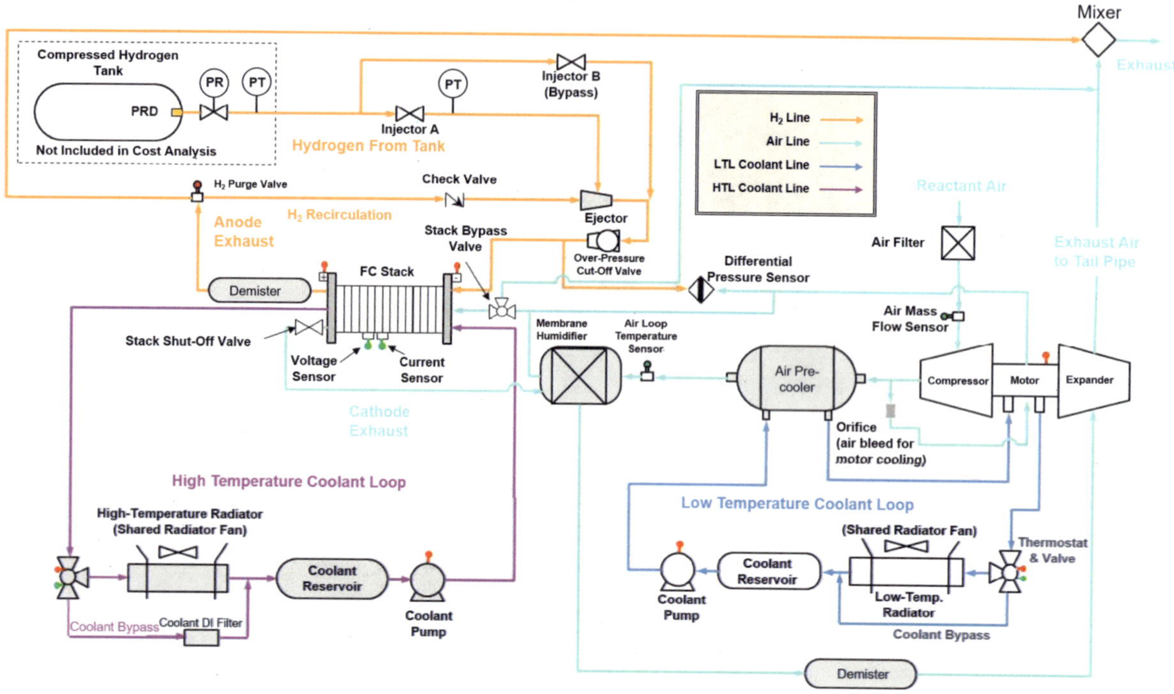

FIGURE 6.4 Basic flow schematic of main components in an automotive fuel cell power system, including the hydrogen storage system. Automakers' fuel cell system designs may vary somewhat.
SOURCE: James et al. (2018).

of hydrogen, air supply, and thermal and water management. The fuel cell stack together with the BOP components comprise the fuel cell system.

Other vehicle components that support the fuel cell system include the battery, electric motor, and power electronics (see Chapter 5), and the onboard hydrogen storage system. A generic flow schematic showing the basic components of an automotive fuel cell power system is shown in Figure 6.4. Automakers continue working on design improvements to simplify the system, improve performance, and reduce costs. For example, Toyota eliminated the external humidifier in its Mirai FCEV by modifying the fuel cell stack structure and operating conditions to use water generated at the cathode to humidify the anode MEA—a so-called self-humidifying stack design (Green Car Congress, 2016).

6.3 FCEV CURRENT STATUS AND PLANNED DEVELOPMENTS

6.3.1 FCEVs Today

While some automakers have shifted their fuel cell development efforts entirely to MHDV applications for the near term,[2] both automakers and hydrogen suppliers have expressed that increased focus on MHDV FCEVs will help build up a refueling infrastructure that will make hydrogen more available and less costly, facilitating more widespread deployment of light-duty FCEVs in the future. Three automakers—Honda, Hyundai, and Toyota—have introduced light-duty FCEVs for sale or lease in places where government-industry partnerships are building a network of hydrogen refueling stations, namely California, parts of Europe, South Korea, and Japan. In some

[2]For example, Daimler recently announced it will end light-duty FCEV development and phase out production of its F-Cell vehicle (Automotive News Europe, 2020) to focus on MHDVs (Daimler Truck AG, 2020).

TABLE 6.1 A Summary of Commercial Light-Duty FCEVs in the United States

	Honda Clarity	Hyundai Nexo Limited	Hyundai Nexo Blue	Toyota Mirai
	Vehicle Information			
Model Year	2020	2020	2020	2020
Vehicle Class	Medium	CUV	Small SUV	Medium
Horsepower	174	161	161	151
0–60 (sec)	8	8	8	9
Range (miles)	360	354	380	312
Fuel Economy (mi per kg H_2) comb/city/hwy	66 67 66	56 53 58	60 64 66	66 65 66
Fuel Economy (MPGE) comb/city/hwy	68 68 67	57 59 54	61 65 58	67 67 67
Cost	$379/mo lease	$62,185 MSRP	$58,735 MSRP	$58,550 MSRP
Fuel Cell System Warranty	8 yrs/100,000 mi	10 yrs/100,000 mi	10 yrs/100,000 mi	8 yrs/100,000 mi
Incentives	3 yrs/$15,000 complimentary fuel	3 yrs/$13,000 complimentary fuel	3 yrs/$13,000 complimentary fuel	3 yrs/$15,000 complimentary fuel
U.S. Availability	California (lease only)	California	California	California and Hawaii
	Powertrain Components			
Fuel Cell System Max Power	103 kW	95 kW	95 kW	114 kW
Battery	346 V Li Ion	240 V Li Ion	240 V Li Ion	245 V NiMH
Motor	130 kW Permanent Magnet AC Synchronous	120 kW Permanent Magnet AC Synchronous	120 kW Permanent Magnet AC Synchronous	112 kW Permanent Magnet AC Synchronous

NOTE: CUV: crossover utility vehicles; SUV: sport utility vehicle.
SOURCE: Photos are from FCHEA (n.d.). Data are from fueleconomy.gov and automaker/vehicle websites.

cases, the automakers themselves, in addition to energy companies and gas suppliers, have invested in hydrogen refueling stations to support the introduction of FCEVs.

Table 6.1 provides information on the powertrain components and other characteristics of the 2020 FCEVs currently available—Honda Clarity, Hyundai Nexo, and Toyota Mirai. These vehicles demonstrate several improvements over first-generation FCEVs: increased efficiency and power density, reduced size, and increased driving range. Figure 6.5 provides an example of fuel cell powertrain improvements, showing data for Hyundai's Nexo over its predecessor, the lease-only Tucson (Seredynski, 2018). In addition to improving the fuel cell stack materials and design, Toyota was able to reduce the size and weight of the fuel cell system from its previous FCEV model by employing a direct current (DC)-DC boost converter to step up the voltage from the fuel cell and increase the voltage of the motor (Green Car Congress, 2015). Many automakers that are focusing on MHDV applications are working on a variety of designs, including delivery trucks, municipal vehicles, and long-haul trucks.[3] Rather than

[3]DHL (Plug Power, 2019), FedEx (Galbach, 2020), and UPS are testing fuel cells to extend the range of their battery-electric delivery vans and/or trucks (Luth, 2019). The UPS trucks use a 45 kWh battery with a 32 kW fuel cell that continuously charges the battery, extending the vehicle's range from around 60 miles to 125 miles (UPS, 2017). Toyota is demonstrating fuel cells in heavy-duty freight handling trucks at the Port of Los Angeles and Long Beach (Toyota USA Newsroom, 2019a). Hyundai has announced plans to begin testing fuel cell trucks in Switzerland in September 2020. The truck is powered by a 190 kW fuel cell drive system using two 95 kW stacks; seven onboard compressed hydrogen tanks will provide enough fuel for an estimated 400 km (~ 248 mile) range (Hampel, 2020).

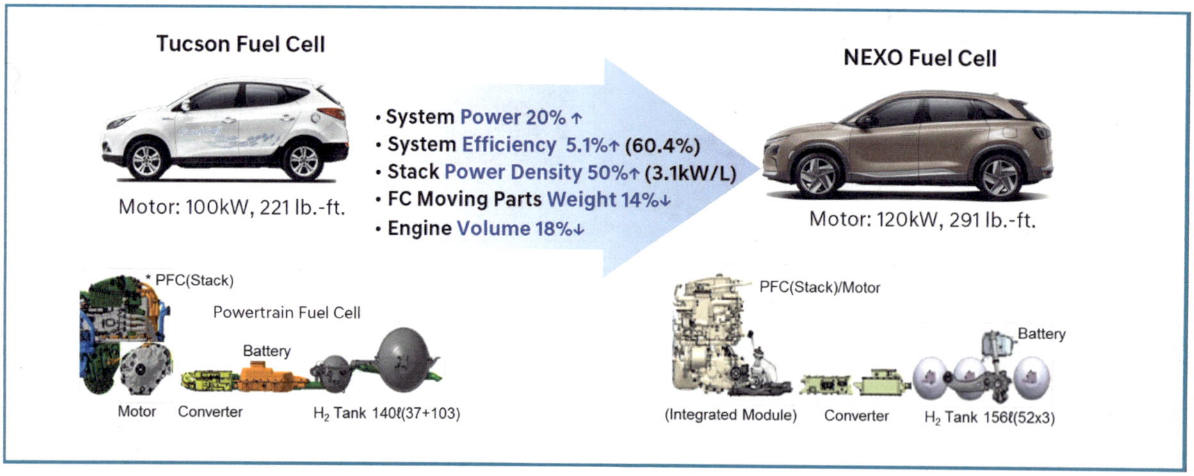

FIGURE 6.5 Comparison of 2019 Hyundai Nexo FCEV and predecessor Hyundai Tucson FCEV.
SOURCE: Republished with permission of SAE International, from P. Seredynski, 2018, 2019 Hyundai Nexo: Fuel-Cell Refinement, SUV Luxury, SAE International Online, November 13; permission conveyed through Copyright Clearance Center, Inc.

curtailment of wind and solar in times of resource excess, electrolysis can take advantage of excess electricity supply and make green hydrogen a cost-effective form of energy storage.[4]

6.3.1.1 FCEV Performance and Cost

Consumer Reports, Car and Driver, and MotorTrend describe driving performance and road handling in the Clarity, Mirai, and Nexo as similar to that of a BEV and typical of front-wheel drive vehicles—smooth, quiet, and responsive, although with a lower acceleration of 0 to 60 mph in 8–9 seconds (Consumer Reports, n.d.; MotorTrend, 2017; Car and Driver, 2019). Toyota's move to rear-wheel drive in the 2021 model, together with aerodynamic enhancements, is expected to improve the Mirai's acceleration, road handling, and range (Toyota USA Newsroom, 2019b).

Improvements in FCEV fuel economy and cost depend on technology progress, particularly reducing the size and weight of the fuel cell and hydrogen storage systems and increasing the efficiency of the fuel cell system. Technology progress is being driven by the automakers and the fuel cell industry as well as other R&D efforts conducted largely in the United States, Europe, Japan, Korea, and China. Through U.S. DRIVE,[5] the DOE has set technical and cost targets in collaboration with industry, periodically updating and revising the targets and their time frames based on technology progress and available R&D funding. For example, vehicle simulation studies conducted by Argonne National Laboratory (ANL) (using the Autonomie model[6]) project that, while improvements in batteries, energy management, and lightweighting will help, fuel cell system improvements are needed to significantly increase vehicle fuel efficiency, and that achieving DOE fuel cell targets can lead to fuel savings of about 40% by 2030 on the U.S. Environmental Protection Agency (EPA) combined driving cycle (Kim et al., 2016) when compared to the 2015 reference case technology (model year [MY] 2020 FCEV). Materials and com-

[4]The start-up company Nikola has announced its plans to build fuel cell pickup trucks and semi-trucks, including a fleet for Anheuser Busch to be delivered by 2025, and plans to build renewable hydrogen refueling stations to support the delivery fleet (O'Dell, 2018). The company plans to use excess wind and solar energy that would otherwise be curtailed to make hydrogen via electrolysis.

[5]U.S. DRIVE (Driving Research and Innovation for Vehicle efficiency and Energy sustainability) is a voluntary, non-binding, and nonlegal partnership among DOE; USCAR, representing Chrysler Group LLC, Ford Motor Company, and General Motors; Tesla Motors; five energy companies—BP America, Chevron Corporation, Phillips 66 Company, ExxonMobil Corporation, and Shell Oil Products US; two utilities—Southern California Edison and DTE Energy; and the Electric Power Research Institute.

[6]See http://www.autonomie.net.

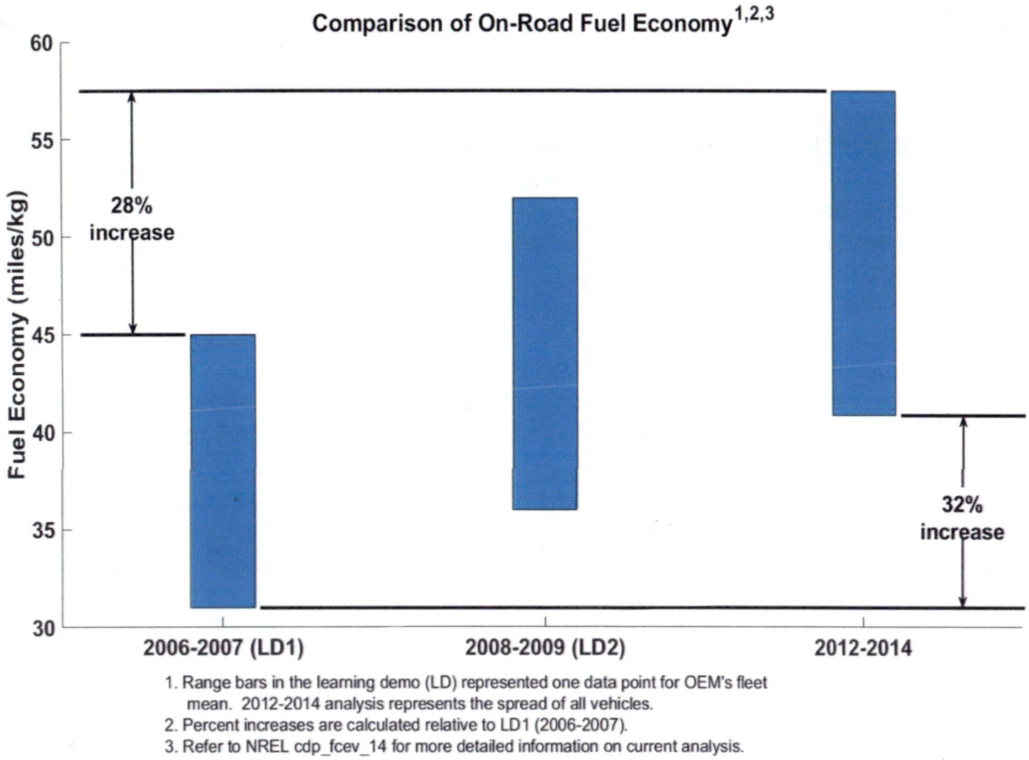

FIGURE 6.6 FCEV on-road fuel economy trends for precommercial FCEVs.
NOTE: Today's commercial vehicles have even higher fuel economies. For comparison, EPA combined fuel economy ratings for today's production FCEVs are considerably higher: 68, 61, and 67 mi/kg for the Clarity, Nexo, and Mirai, respectively.
SOURCE: NREL (2016).

ponent R&D efforts focused on reducing fuel cell system size, weight, and cost will also drive increases in fuel cell efficiency. Improvements in fuel cell materials may enable modifications to BOP and other vehicle components, leading to reduced vehicle size and weight. For example, a more efficient fuel cell stack may require less cooling and a smaller radiator, leading to less drag and greater fuel economy.

Figure 6.6 shows on-road fuel economy trends for a precommercial FCEV fleet monitored through the National Renewable Energy Laboratory (NREL) FCEV Learning Demonstration (Wipke et al., 2012). It does not include commercial FCEVs on the road today. The data indicated an approximately 30% increase in average on-road fuel economy, calculated from on-road fuel cell stack current, for the Learning Demonstration fleet from 2006 through 2014 vehicles, ranging from 31–45 miles per kilogram (mi/kg) hydrogen for Gen 1 vehicles and 36–52 mi/kg hydrogen for Gen 2. Today's EPA combined fuel economy ratings for FCEVs are considerably higher: 68, 61, and 67 mi/kg for the Clarity, Nexo, and Mirai, respectively (EPA, 2019).

ANL Autonomie vehicle simulation studies, also based on fuel cell technologies meeting established targets, project that by 2030 the FCEV total cost of ownership (TCO) will decrease to 43 cents per mile, comparable to conventional internal combustion engine (ICE) vehicles (Vijayagopal et al., 2017). Deloitte China has revealed similar TCO trajectories for U.S., European, and Chinese markets (Deloitte China, 2020). Another study using Autonomie looked at the TCO of BEVs and FCEVs in different LDV classes from 2020 through 2040, and the fraction of vehicle owners in those classes, to project potential market sizes (Morrison et al., 2018). The study, which assumed that an affordable hydrogen refueling infrastructure will be available, projected that FCEVs may have a cost advantage over BEVs for larger vehicles like passenger vans and sport utility vehicles (SUVs), while BEVs may have a cost advantage for smaller vehicles like compacts and midsize sedans. A key factor in the Morrison et al. (2018) study is mass compounding—as the capacity of the powertrain increases, the mass of the glider and other vehicle components increases, such that a greater fraction of that capacity is eventually used to move the mass of the powertrain rather than the mass of vehicle, passengers, and cargo, leading to a nonlinear relationship between vehicle cost and range. Mass compounding in FCEVs has less of an impact than in BEVs, especially in heavier vehicles, because the fuel cell powertrain has a higher energy density.

Another recent analysis by the McKinsey Center for Future Mobility for the Hydrogen Council also suggests that FCEVs will likely be the lower cost option for decarbonization of heavy-duty trucks, long-distance buses, and large passenger vehicles with long ranges, achieving cost parity with BEVs in those applications in the 2030 time frame (Hydrogen Council, 2020). The analysis for passenger vehicles shown in Figure 6.7 suggests that FCEVs will be more cost competitive in vehicles with heavier use and longer-range requirements, such as large passenger cars, SUVs, and taxi fleets. The study also points out that some drivers of smaller vehicles may be willing to pay for the increased flexibility provided by the longer range and quicker refueling of the FCEVs.

Other studies are less optimistic about FCEVs achieving cost parity in the 2030 time frame. To assess the status and expected future cost and performance of automotive PEM fuel cells, in 2017 Carnegie Mellon University (CMU) led an expert elicitation assessment of fuel cell system cost, stack durability, and stack power density under DOE's high-volume production scenario of 500,000 units per year (Whiston et al., 2019). The study included 39 experts from academia, government, and industry, who assessed the median 2017 automotive cost to be $75 per kilowatt (kW), stack durability to be 4,000 hours, and stack power density to be 2.5 kilowatts per liter (kW/L). For comparison, DOE cited the 2017 status to be $53/kW, 3,900 hours, and 3.0 kW/L, respectively. (It should be noted that Toyota and Honda reported a stack power density of 3.1 kW/L, more in line with the DOE status.) The experts

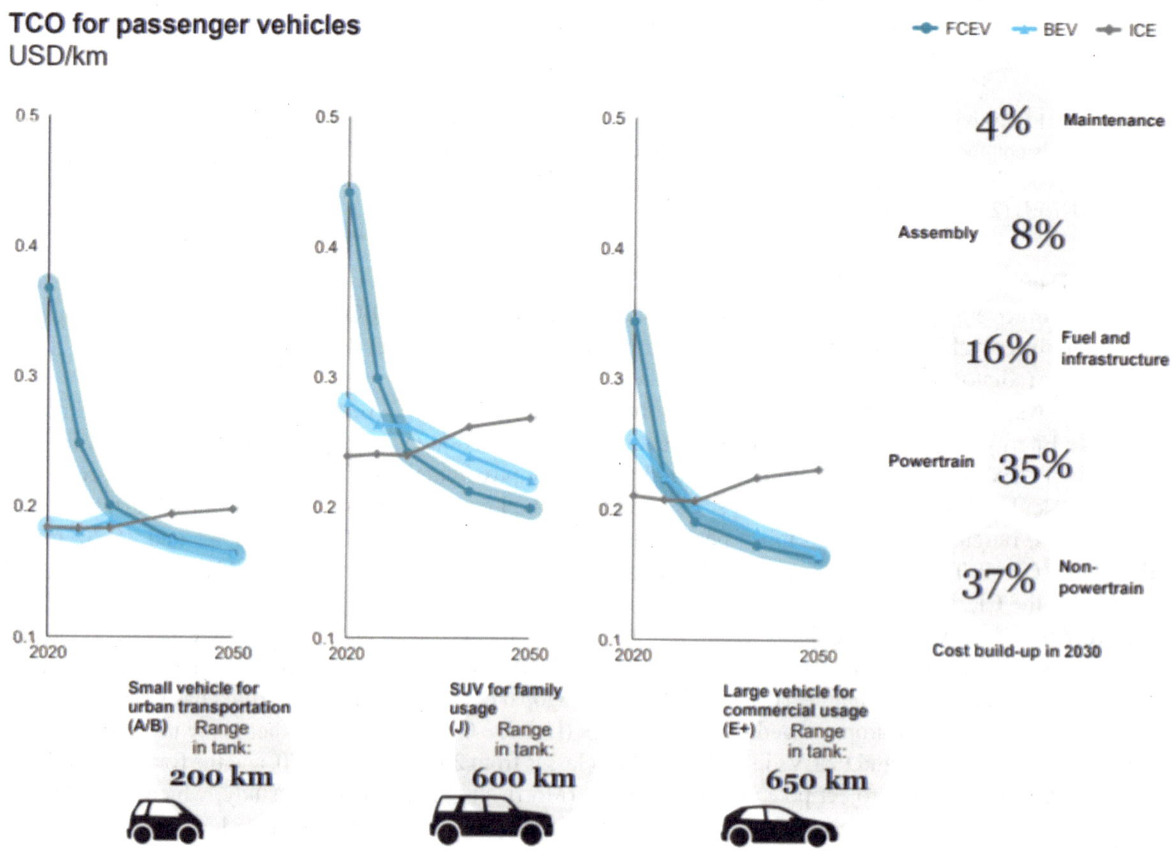

FIGURE 6.7 TCO projections for ICE, BEV, and FCEV passenger vehicles.
SOURCE: Hydrogen Council (2020).

in the CMU study ranged widely in their assessments—from $40 to $500/kW for cost, from 1,200 to 12,000 hours for durability, and from 0.5 to 4 kW/L for power density—demonstrating the difficulty in assessing performance and cost of technologies still under development. When asked to project into the future, many experts expected that DOE's ultimate targets of $30/kW and 8,000 hours durability would be achieved by 2050, and 3 kW/L power density by 2035. The study identified high platinum group metal (PGM) loading as the most significant barrier to reducing fuel cell cost, followed by membranes and BPs. The experts also noted the uncertainty of reaching production volumes of 500,000 units/year in the near term, citing the learning required to manufacture at that scale. As described later in this chapter, DOE's fuel cell R&D activities are heavily focused on reducing or eliminating PGM content and developing improved membranes. DOE supports BP R&D to a lesser extent, and support for fuel cell manufacturing R&D has been relatively low or nonexistent.

As FCEVs advance, and high-volume manufacturing capability is developed, the availability of real-world data will enable more certainty in the status and projections for fuel cell technology. To that end, Hyundai announced in February 2020 that it will provide DOE with five Nexo FCEVs (Hyundai Motor Group Newsroom, 2020), enabling DOE to collect, analyze, and publish data regarding fuel cell performance, durability, and reliability.

6.3.1.2 FCEV Energy Management

Fuel cell voltage is dependent on operating conditions such as temperature, pressure, applied load, and fuel/oxidant flow rates. The standard measure of performance for a fuel cell is the polarization curve (Figure 6.8), which indicates the cell voltage behavior as a function of current density (load).

The actual open circuit voltage of a fuel cell is lower than the theoretical value owing to fuel crossover through the electrolyte and internal currents. Other types of losses that cause voltage drops are as follows:

- Activation polarization, which dominates at low current densities, is owing to the voltage overpotential (typically, 0.1–0.2 V) required to overcome the activation energy of the electrochemical reaction on the catalyst surface and is largely driven by the slow kinetics of the oxygen reduction reaction (ORR).

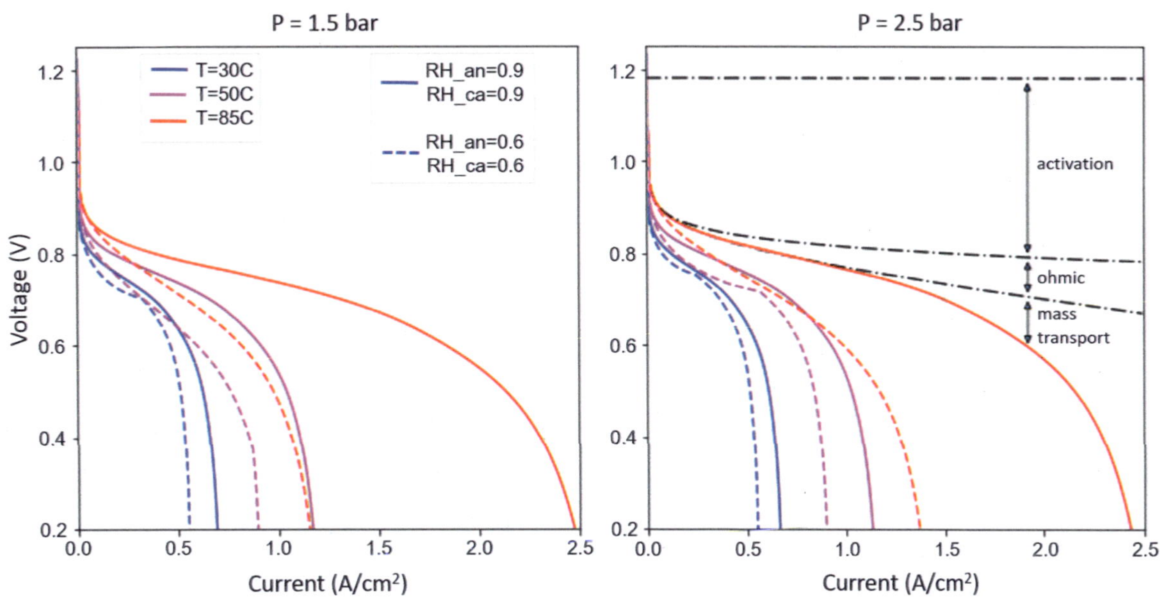

FIGURE 6.8 PEM fuel cell polarization curve at 1.5 atm (left) and 2.5 atm (right) and at varying temperatures and relative humidities.

SOURCE: Reprinted by permission from Springer Nature Customer Service Centre GmbH: Springer London, *Encyclopedia of Systems and Control* (Fuel Cell Vehicle Optimization and Control, Figueroa-Santos and Stefanopoulou), 2021.

- Ohmic polarization, which dominates at moderate power densities, is owing to ionic and electronic resistance in the fuel cell components—electrolyte, electrodes, and so on.
- Mass transport losses (also known as concentration polarization), which dominates at high current densities, is owing to mass transport losses from the decrease in reactant concentration at the surface of the electrodes as fuel is used.

As shown in Figure 6.8, decreases in fuel cell voltage depend on cell temperature, pressure, and relative humidity (RH) (Figueroa-Santos and Stefanopoulou, 2021). Within a three fuel cell system, these operating conditions are managed by BOP or auxiliary components—that is, compressor (or blower), humidifier, heat exchanger, and so on.

The total voltage of a fuel cell stack (V_{st}) is the product of the single cell voltage (V_{fc}) and the number of cells (N_{fc}) in series in the stack ($V_{st} = V_{fc}N_{fc}$). Fuel cell stack power (P_{st}) is the product of the fuel cell voltage (V_{fc}), fuel cell current (I_{fc}), and the number of cells (N_{fc}) in series in the stack ($P_{st} = N_{fc}V_{fc}I_{fc}$). The net power of the fuel cell system ($P_{fc,\ net}$) is the stack power (P_{st}) minus the power required to operate the BOP components (P_{aux}). Fuel cell stack efficiency (η) is the net power of the fuel cell stack ($P_{fc,\ net}$) divided by the energy value of hydrogen (E_{H2}) *consumed* ($\eta = P_{fc,\ net}/E_{H2}$). *Thus, minimizing BOP power requirements and hydrogen consumption increases the* efficiency of the fuel cell system. As described later in this chapter, R&D activities are focused on new materials to enable fuel cell system operation at lower RH, higher temperature, and lower pressure, with the aim of simplifying BOP requirements, increasing fuel cell system efficiency, and reducing costs. At the vehicle level, control strategies are critical to managing BOP operation to optimize fuel cell vehicle performance and minimize fuel cell and battery degradation (Figueroa-Santos and Stefanopoulou, 2021).

FCEV control strategies must manage several trade-offs to minimize voltage losses and optimize performance, while also keeping manufacturing and operating costs affordable. Hydrogen and air must be maintained at a certain stoichiometric ratio to ensure fuel cell efficiency. High hydrogen and air flow rates provide higher stack power density and efficiency but lower net power from the fuel cell system owing to the higher power consumption of the BOP components. High reactant flow control, if not fully humidified, may cause dehydration and subsequent degradation of the MEA. The fuel cell generates water at the cathode that helps self-humidify the MEA at the end of the cathode channel, but cannot support the dry channel entry unless there is external air flow humidification. The anode self-humidification is achieved via water diffusion to the anode side through the membrane and cross-flow configuration. The temperature of the fuel cell also needs to be controlled, typically 60°C–80°C—too high can cause MEA dehydration, shrinkage, pinholes, and cracks. The humidity of the reactants needs to be controlled to keep the membrane hydrated and enable water distribution that avoids dehydration or flooding, both of which increase ohmic voltage losses.

Fuel cell power is regulated using DC/DC converters, as the voltage changes in response to the load. A converter controls the power split and can be used to avoid abrupt transients or changes in fuel cell power demand to avoid fuel cell degradation. To obtain the highest energy efficiency (and minimize hydrogen consumption), the fuel cell system under dynamic load must be operated close to the maximum efficiency point during most of the fuel cell operation. This is typically accomplished by using full-fuel-cell system optimization or an extremum-seeking controller (Bizon, 2017; Zhou et al., 2017). PEM fuel cell efficiency increases to a maximum around 60% in the low to medium range of fuel cell power (typically, near 20% of peak power)—that is, at part load where most driving takes place—and then drops in the high-power region of fuel cell operation.

Fuel cells operate most effectively at constant load. When the load (current) on a fuel cell is changed, the heat and water balance change and it takes time for the fuel cell to reach a new equilibrium point. These changes can lead to catalyst and MEA degradation and reduced fuel cell durability. Hence, FCEVs are typically hybridized with a battery to improve the system durability and powertrain lifetime by reducing the fuel cell's exposure to transients and high current spikes and to repeated startup and shutdown cycles.

FCEV energy management strategies are important to provide the performance characteristics that drivers demand and also to optimize the durability of the fuel cell system and hybrid battery. Over the past 10 years, original equipment manufacturers (OEMs) and others have ramped up studies on the effects of transients and stop-start sequences on fuel cell performance and durability, examining stack failure modes owing to mechanical and chemical membrane degradation, voltage loss during operation, and corrosion of catalysts and support materials, with the aim of developing control strategies to mitigate these effects (Eberle et al., 2012). For example, in its HydroGen4 demonstration fuel cell vehicle, General Motors observed that dynamic loads caused humidity transients from 10% RH to greater than 100% (liquid water), leading to significant expansion and contraction of the membrane. Such repeated membrane structural changes create mechanical stresses that can cause microscopic cracks and lead to crossover of reactants, deteriorating performance, and eventually failure of the fuel cell system. Studies also show that voltage cycling leads to a loss of the electrochemically active surface area (ECSA) of the platinum catalyst, which is correlated to an increase in activation overpotential (Ahluwalia et al., 2020a). The ECSA loss increases exponentially at the upper potential limit and with temperature, and linearly with humidity and dwell time at the upper potential limit (Kneer, 2019). Degradation mechanisms and failure modes occurring during idle and stop-start cycles have also been examined (Eberle et al., 2012).

The results of fuel cell transient studies are used to design operating strategies to minimize catalyst and membrane degradation and increase fuel cell lifetime under load cycling, and during idle, start-up, and shutdown. Fuel cell and battery operations can be controlled to limit high potentials on the fuel cell, with the fuel cell system serving primarily as a low dynamic power source and the battery (or ultracapacitor) providing quick response needs. However, the power split must also consider the battery state of charge to prevent overdischarging or overcharging the battery. FCEV batteries must be capable of accommodating increased current and storing energy generated by the fuel cell. Control strategies that provide the optimal power split between the fuel cell and battery are needed to enable optimized performance and reliability in FCEVs, and to minimize cost.

Several review articles have surveyed many different FCEV control strategies that have been proposed, and described their pros and cons. Assessments address characteristics such as ease of implementation, computational complexity/cost, and responsiveness to real-time driving conditions (Yue et al., 2019). Dijoux et al. (2017) describe the state of the art in fault-tolerant control strategies, in which fault diagnostics are used to trigger corrective control actions. Hames et al. (2018) identified the most common control strategies to be peaking power source strategy, operating mode control strategy, fuzzy logic control strategy, and equivalent consumption minimization strategy (ECMS), stating a preference for ECMS owing to its simplicity and its ability to minimize hydrogen consumption and enable high-level FCEV performance (Hames et al., 2018; Kaya and Hames, 2019). In addition to ECMS, Figueroa-Santos and Stefanopoulou (2021) include descriptions of rule-based control strategies, dynamic programming, Pontryagin's minimum principle, model predictive control, and machine learning. Song et al. (2018) point out that a single energy management strategy cannot adequately address the complexity of real-world driving conditions and propose a multimode control strategy, including one based on pattern recognition.

Control strategies used in today's commercial FCEVs are proprietary; therefore, it is unclear which of the strategies is most common. General Motors described the successful implementation of individual control actions that significantly improved fuel cell stack durability in its precommercial HydroGen4, including standby mode (i.e., turning off the fuel cell system), voltage-suppression and oxygen-depletion, hydrogen injection during long off-times, and an automated stack recovery procedure (Eberle and von Helmolt, 2010). Toyota has also described control methods such as reducing air compressor power at low loads to increase fuel economy in the 2017 Mirai (Hasegawa et al., 2016). To better understand FCEV operation and assist development of energy management strategies, ANL performed a technology assessment of the 2017 Toyota Mirai (see Figure 6.9), correlating fuel cell system parameters and operation to outputs on varying drive cycles and over a wide range of temperatures (Lohse-Busch et al., 2018).

In collaboration with General Motors, ANL's modeling and analysis efforts are also providing input for design of FCEV control strategies by examining thermal and water management issues, design-point and part-load operation, efficiencies, and fuel economies (Ahluwalia et al., 2020b). Their recent efforts have identified compressor-expander operating conditions and coolant exit temperatures that will limit ECSA loss to levels that enable the U.S. DRIVE LDV fuel cell electrode durability of 8,000 hours to be achieved. The fact that many automakers are adding a focus on development of medium- and heavy-duty FCEVs, which have a durability target of 30,000 hours (Adams, 2020), suggests confidence in developing energy management strategies to enable significant improvements in fuel cell durability for mobile applications in 2025–2035.

To decrease the cost of FCEVs, the sizing and selection of the BOP components will need to be considered along with mitigating fuel cell degradation and reducing auxiliary losses (Wu et al., 2020). Additional cost savings will come from sizing the fuel cell, hydrogen storage tank, and battery for various drive cycles (Sundström and

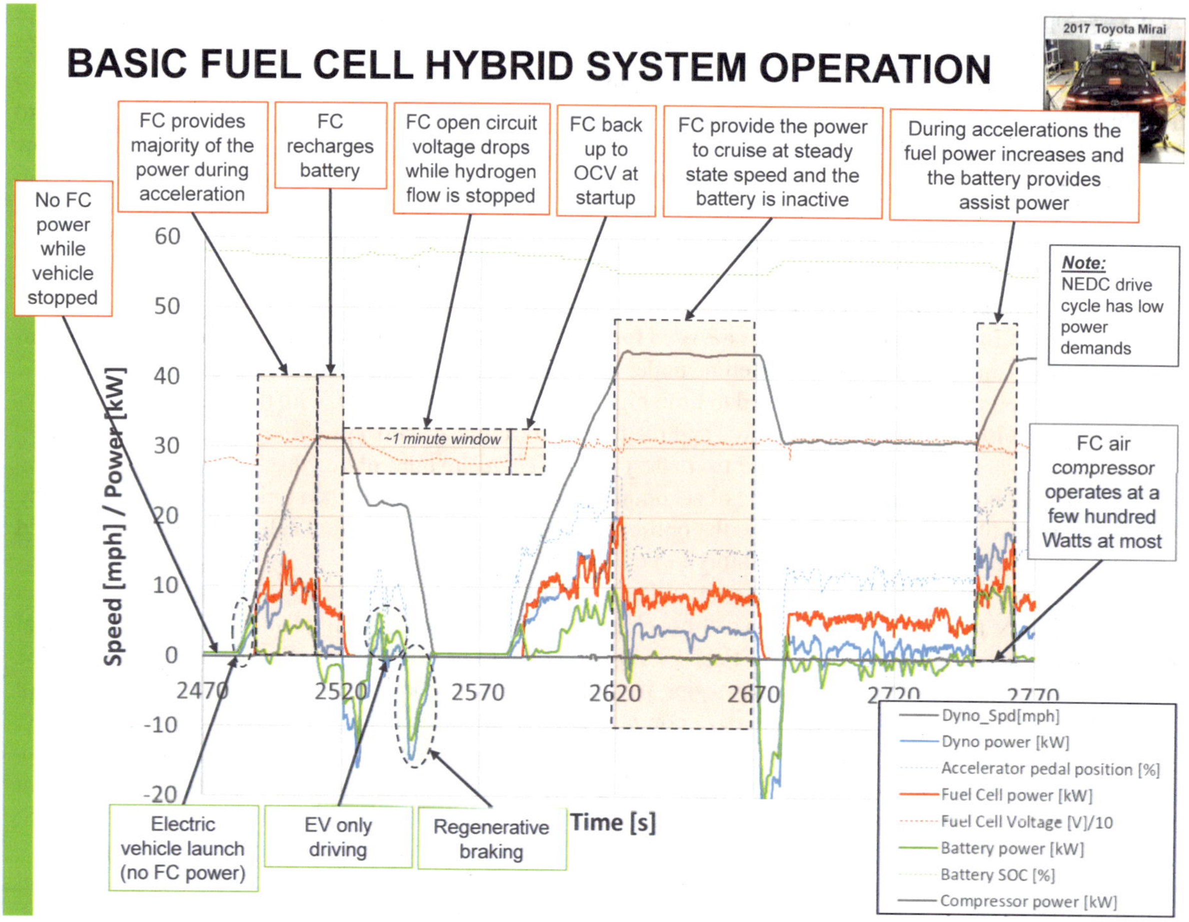

FIGURE 6.9 Dynamometer testing of 2017 Toyota Mirai showing fuel cell, battery, and compressor operation. An example of results from ANL's technology assessment of Mirai operations.
NOTE: OCV is open circuit voltage—a figure of merit for fuel cells, defined as the maximum operating voltage of the fuel cell, which occurs when no current is flowing (i.e., when no load is applied).
SOURCE: Lohse-Busch et al. (2018).

Stefanopoulou, 2007; Jiang et al., 2019). Current studies on optimizing components to minimize cost of ownership for fuel cell-powered trucks offer insights as well (Sim et al., 2019). Connectivity and automation will allow even higher efficiency gains for these advanced powertrains (Kim et al., 2020).

Plug-In Fuel Cell Vehicles

Plug-in fuel cell vehicles (PFCVs) have been explored as an approach to combine the best features of BEVs and FCEVs and mitigate the shortcomings. PFCVs have a moderately sized battery to provide some all-electric range, and a hydrogen fuel cell system (including hydrogen tank) smaller than that in a pure FCEV to extend the vehicle's driving range and enable quick refueling. PFCVs can recharge with electricity from the grid and/or refuel with hydrogen from a refueling station. The advantages over a gasoline plug-in hybrid electric vehicle (PHEV) are increased efficiency and zero tailpipe emissions. Studies suggest that a combination of low power fuel cells and high energy batteries is optimal in terms of manufacturing cost and environmental benefits (Fox et al., 2012; Lane et al., 2017). A 2017 case study based in California determined that PFCVs would require significantly fewer hydrogen refueling stations than FCEVs and put less strain on the electric grid than BEVs (Lane, 2017). The study determined that PFCVs with 40 miles electric-only range provided the highest efficiency of any alternative vehicle, the lowest well-to-wheels (WTW) GHG emissions, and the lowest infrastructure costs if limited to Level 1 charging.

There are no PFCVs commercially available today. Ford developed one of the first in 2008, the HySeries PFCV, which had a fuel cell system that was about 60% of full size (Ford Edge HySeries—Ford, 2008). In 2014, Audi announced its sporty A7 h-tron quattro concept car, powered by a hydrogen fuel cell and an 8.8 kWh Li-ion battery pack—the same pack used in the Audi A3 Sportback e-tron plug-in hybrid (Edelstein, 2014; Rügheimer, 2014). The battery, which could be recharged from the grid or through regenerative braking, provided up to 31 miles of electric-only range, and four hydrogen storage tanks provided 310 miles of vehicle range. Audi recently announced that it was increasing investment in FCEVs, with pilot production planned in 2021 and larger scale production in the late 2020s; however, the company provided no details on the types of vehicles (Goodwin, 2019).

Daimler developed the Mercedes-Benz GLC F-CELL in 2013—a PFCV with a 13.5 kWh battery (9.3 kWh net) and a fuel cell stack that was about 30% smaller than that in the Mercedes B Class F-CELL, their pure FCEV (Green Car Congress, 2018a,b). A 7.2 kW onboard charger enabled full recharging from a standard, residential power socket in around 1.5 hours. Several GLC F-CELL vehicles were sold for promotional purposes; however, Daimler canceled plans to begin leasing the F-CELL when it shifted focus to fuel cells for MHDVs in 2020. In fact, using fuel cells as range extenders in battery-powered MHDVs has gained significant traction (Sturgess, 2017). Toyota, UPS, FedEx, and others are currently testing fuel cell as range extenders in drayage and delivery trucks, for example (Hanlin, 2019). Swedish company myFC is developing scalable, modular systems that combine batteries, hydrogen fuel cells, and "power balancing technology" for plug-in hybrid applications (Lawrence, 2020).

A 2013 study suggests that PFCVs could be particularly competitive during the near term when hydrogen availability and, to some extent, recharging availability are low by providing drivers with two options for refueling their vehicles: extended range relative to BEVs, and reduced energy costs compared to pure FCEVs (less hydrogen is required) and ICE PHEVs (due primarily to increased vehicle efficiency and partly to the hydrogen subsidy) (Lin et al., 2013). As discussed in FCEV energy management strategies, optimizing the sizing of the fuel cell, hydrogen storage tank, and battery is key to maximizing PFCV performance and durability, and minimizing cost of ownership.

6.3.1.3 FCEV Safety

Safety concerns around FCEVs are related to the use of hydrogen generally, and high-pressure hydrogen in particular, especially in the event of a collision. In the United States, FCEVs are required to pass the same Federal Motor Vehicle Safety Standards (FMVSS) crash tests as ICE vehicles. Global Technical Regulation (GTR) No. 13, an agreement among Japan, Europe, and North America, sets the safety requirements for the integrity of onboard compressed and liquid hydrogen storage systems, including tests for pressure cycling, burst, permeation, and bonfire that are more stringent than the FMVSS No. 304 requirements for compressed natural gas tanks. In fact, the Hyundai Nexo FCEV earned a Top Safety Pick+ award from the Insurance Institute for Highway Safety (IIHS-HLDI, 2019).

Phase I of GTR 13, established in 2013, specifies that each participating country will use its existing national crash tests (GlobalAutoRegs, n.d.). GTR 13 Phase II, expected to be finalized by the end of 2020, will harmonize FCEV crash test requirements internationally with the goal of creating global standards (United Nations Economic and Social Council, 2017).

As mentioned previously, some automakers and hydrogen suppliers had announced plans to introduce FCEVs and hydrogen stations in Northeast states in 2016. However, as of this writing FCEVs remain prohibited from the tunnels and lower tier of double-deck bridges in Massachusetts, New York, and New Jersey (Port Authority of NY and NJ, 2016; State of New Jersey, 2019; Massachusetts Department of Transportation, 2019).[7] In 2017, Sandia National Laboratories completed an FCEV tunnel safety study to address a lack of data on this topic and to

[7]An alternative fuel vehicle powered by propane or natural gas may use Port Authority of NY and NJ tunnels and the lower level of the George Washington Bridge if the vehicle conforms to applicable federal regulations and industry standards, displays required markings to identify its alternative fuel system, and has a fuel capacity that does not exceed 150 pounds (AFDC, n.d.b.).

determine the risks and implications of traffic incidents in tunnels involving hydrogen fuel cell vehicles (La Fleur et al., 2017). The study included a risk analysis that examined a number of different tunnel configurations and crash scenarios and determined that hydrogen-fueled FCEVs are unlikely to pose additional hazards relative to other LDVs. In most scenarios examined, hydrogen is not released or does not ignite. The study determined that, in scenarios where hydrogen does ignite, the most likely result is a jet flame from the release of hydrogen through the thermally activated pressure relief device owing to the heat from an accident-related hydrocarbon fire. Where assumptions had to be made, the most conservative assumptions were used to ensure that the worst cases were analyzed. For example, a six-fold overestimate of hydrogen release was used in the models to increase the heat released by the jet flame and the height of the flame; hence, the study noted, observed temperatures should be lower than those predicted by the models. The analysis determined that the jet flame could cause localized concrete spalling where it hits the tunnel ceiling, which is not expected to occur with ventilation. With or without ventilation, the structural epoxy and steel structure of the tunnel would not be compromised.

To address follow-up questions regarding FCEV safety in tunnels and to assist highway tunnel officials, Sandia National Laboratories recently published two additional safety reports—one providing a comprehensive overview of studies related to the safety of alternative fuel vehicles (AFVs) within tunnels (La Fleur et al., 2020), and one providing the same for FCEVs (Glover et al., 2020). The reports also identify knowledge gaps to guide future safety research efforts and to enable a complete hazard analysis and recommendations for the safe use of AFVs in tunnels. The European Union's HyTunnel-CS project is expected to address some of the research gaps for FCEVs in tunnels and confined spaces, with the goal of enabling hydrogen vehicles entering underground environments to present no more risk than fossil-fueled vehicles. One task in the HyTunnel-CS project will develop engineering solutions to prevent and mitigate accidents involving hydrogen releases (HyTunnel-CS, 2019).

In 2019, the American Institute of Chemical Engineers, in collaboration with DOE, launched the Center for Hydrogen Safety (CHS) to promote hydrogen safety and best practices worldwide. The CHS provides information and tools on the safety aspects of hydrogen and fuel cell technologies and resources to those designing, approving, or using hydrogen systems and facilities, or responding to hydrogen incidents.

6.3.2 FCEV Plans

Hyundai and Toyota have been the most active in the commercialization of light-duty FCEVs, with both automakers recently stating their plans to ramp up production. Hyundai has announced plans to add production capacity for 500,000 FCEVs per year by 2030 (Hyundai USA, n.d.). Toyota recently announced that it is increasing FCEV production capacity to 30,000 vehicles per year for worldwide sales (Eisenstein, 2020). In 2019, the company announced that it will release its second generation Mirai for MY 2021, stating that the vehicle will have 30% more driving range (~400 miles) owing to increased hydrogen storage capacity and enhanced fuel cell performance (Toyota USA Newsroom, 2019b). For the 2021 Mirai, Toyota is switching from a front-wheel drive platform to rear-wheel drive and introducing a sleeker design for improved aerodynamics. In 2019, Toyota released almost 24,000 patents, royalty-free, to help accelerate the deployment of FCEVs (Toyota USA Newsroom, 2019a).

Other automakers are planning later FCEV deployments. BMW, for example, announced that it is looking to 2025–2030 for introducing commercial FCEVs, basing the decision on projections from other automakers that fuel cell cost will be equivalent to that of conventional technology in that time frame (Crosse, 2020). At the 2019 Frankfurt Motor Show, BMW unveiled its i Hydrogen NEXT SUV, which uses a Toyota-based fuel cell powertrain, and announced that the SUV will enter limited production in 2022, with 2025 the earliest target year for offering the vehicle to customers (BMW Group, 2019).

Figure 6.10 provides estimated commercialization timelines for FCEVs and other hydrogen applications based on a survey conducted by McKinsey and Company and analyses conducted by the Hydrogen Council.[8] As discussed throughout this chapter, improvements in the cost and performance of fuel cell and hydrogen technologies are needed for widespread commercialization.

[8]The Hydrogen Council is a group of chief executive officers leading global businesses in energy, transportation, and related industries with significant investments in the development of hydrogen and fuel cell systems and markets.

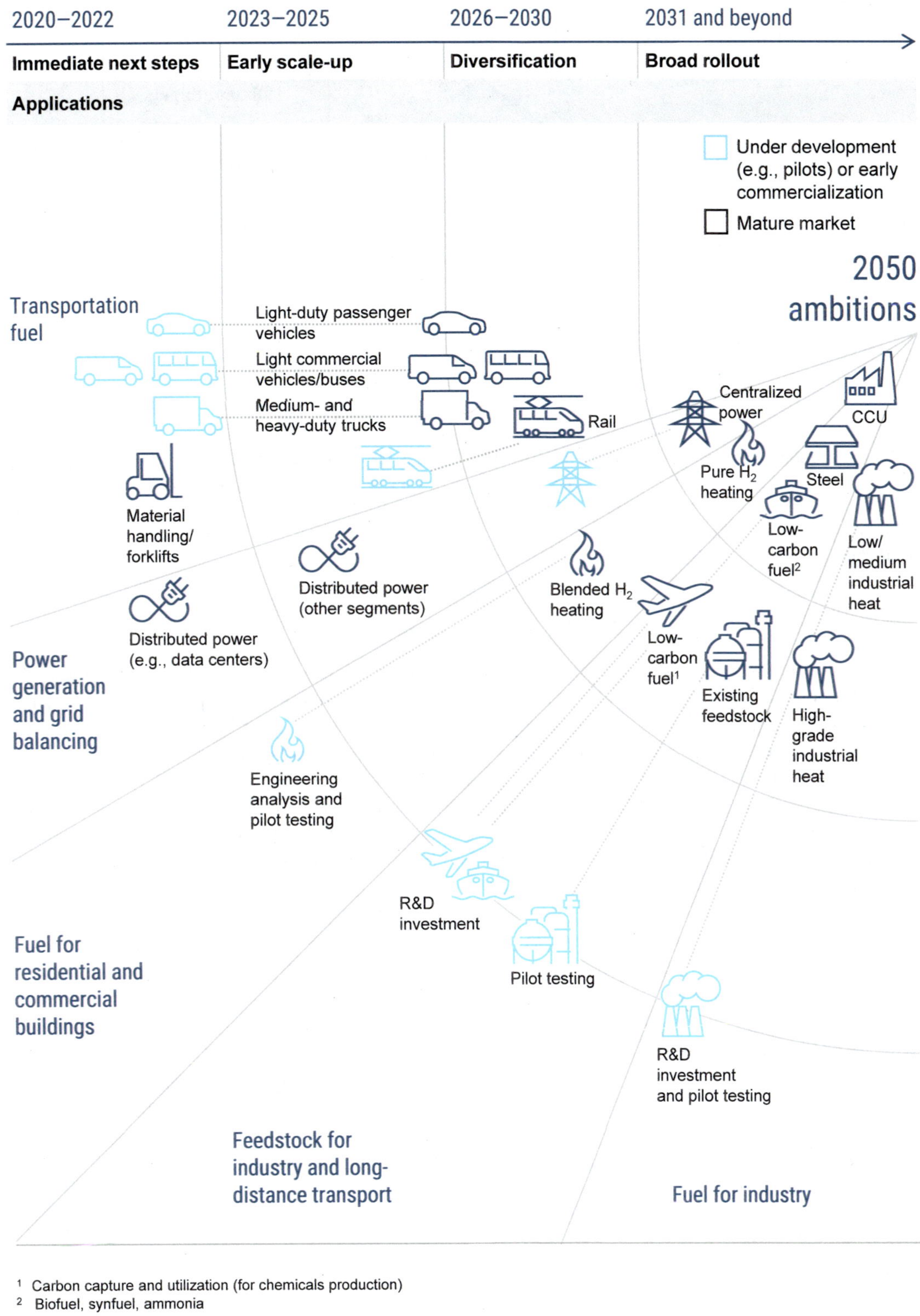

FIGURE 6.10 Development timeline for hydrogen and fuel cell systems in transportation and other applications.
SOURCE: FCHEA (2020).

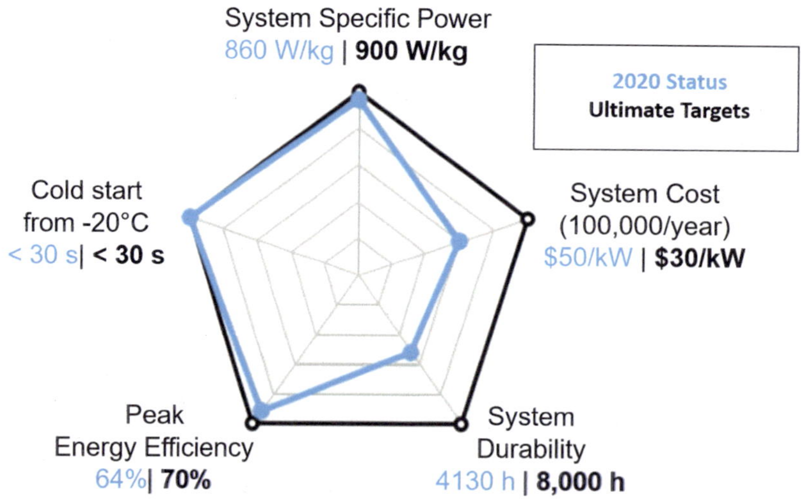

FIGURE 6.11 Automotive fuel cell power system status (blue) versus targets (black). Cost status is for a modeled system when manufactured at a volume of 500,000 units per year. While not included in the chart, the volumetric density target is 850 watts per liter; however, an accurate estimate of current automotive fuel cell system volume is not available owing to a lack of public information.
SOURCE: Padgett and Kleen (2020).

6.4 FCEV TECHNOLOGY R&D

For widespread deployment of FCEVs, further reductions in the costs of fuel cell and hydrogen storage systems are needed, while increasing fuel cell durability and maintaining or improving system performance. Key metrics for automotive fuel cell systems include energy efficiency, power density (volumetric) and specific power (gravimetric), durability, and cost. High specific power and power density are important for transportation applications, to minimize the weight and volume of the power system, respectively, as well as the cost. There is significant worldwide investment in fuel cell and hydrogen R&D, most notably in the United States, Europe (Germany in particular), Japan, Korea, and China. In many cases, researchers look to the U.S. DOE technology targets, established in collaboration with industry, to guide development efforts (U.S. DRIVE, 2017a).

The status of automotive fuel cell systems relative to the ultimate DOE targets is shown in the spider chart in Figure 6.11. As indicated, improvements are needed in fuel cell system peak efficiency and power density, and particularly in cost and durability (U.S. DRIVE, 2017a).

Cost. The DOE cost target for automotive fuel cell systems is $30 per kW_{net} at an annual production capacity of 500,000 units. In a design for manufacture and assembly (DFMA) analysis conducted for DOE, the modeled cost of an 80 kW_{net} power automotive PEM fuel cell system based on next-generation laboratory technology is projected to be $45 per kW_{net} at a volume of 500,000 units per year (James et al., 2018). This projected cost is based on an analysis of state-of-the-art components that have been developed and demonstrated at a laboratory scale through the DOE Hydrogen and Fuel Cell Technologies Office,[9] and reflects a 67% decrease since 2006 (see Figure 6.12). The latest cost reductions have come primarily from (1) use of a platinum-cobalt catalyst on high surface area carbon that led to increased power density, (2) reduction in platinum loading on the cathode, and (3) an improved BP stamping process.

This DFMA cost analysis is meant to provide a benchmark for informing early-stage R&D efforts focused on reducing fuel cell materials and manufacturing costs. Because long-term durability data for automotive fuel cells is lacking, this cost analysis is for a model system meeting beginning-of-life performance requirements. These cost estimates are also for technologies in the pipeline but not yet commercial; therefore, they do not take into account

[9]The DOE Fuel Cell Technologies Office recently changed its name to the Hydrogen and Fuel Cells Technologies Office.

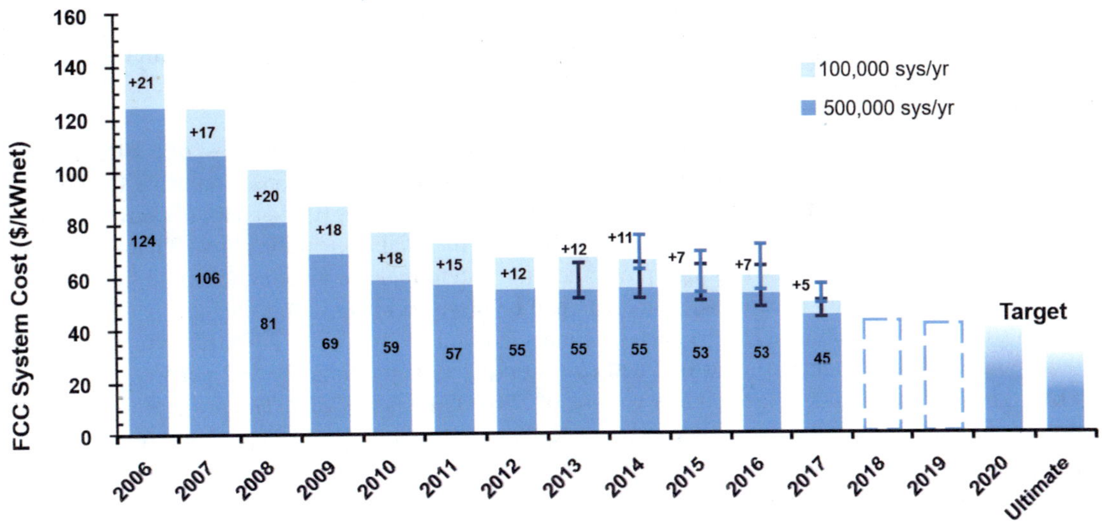

FIGURE 6.12 Modeled cost of an 80-kW_{net} PEM fuel cell system based on projection to high-volume manufacturing (100,000 and 500,000 units per year). Error bars represent the 90% confidence interval from a stochastic uncertainty analysis and reflect manufacturing uncertainty in the modeled system.
SOURCE: James et al. (2018).

some of the strategies used by automakers to ensure sufficient fuel cell durability to meet powertrain warranties in commercial FCEVs today. This will be discussed in more detail in the next section on durability.

A breakdown of PEM fuel cell stack cost, shown in Figure 6.13, indicates that the catalyst is the largest cost component at both low and high volumes. This drives the continued focus on reducing or eliminating the use of platinum catalysts. For PGM-based catalysts, research is focused on both decreasing PGM loading and increasing

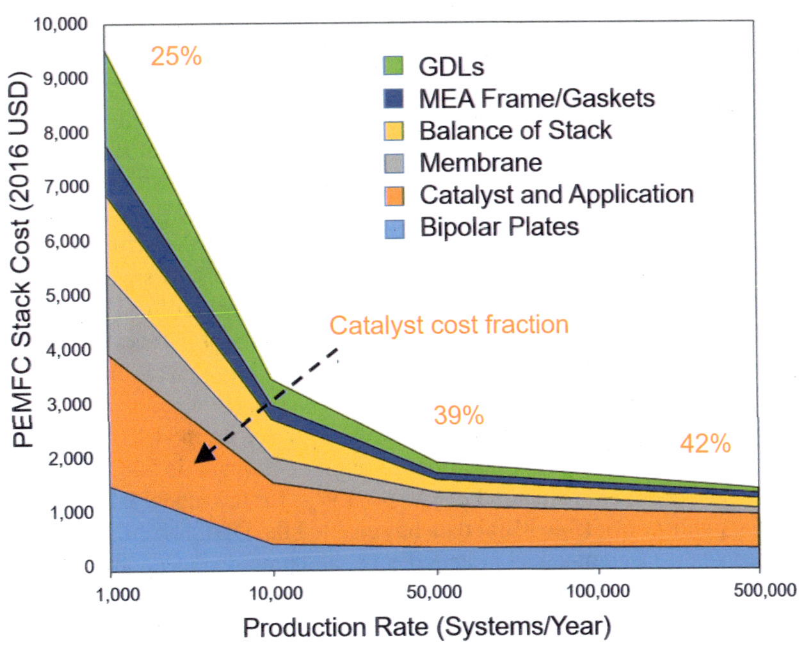

FIGURE 6.13 Breakdown of the projected fuel cell stack cost at 1,000 to 500,000 systems per year (2017 analysis).
SOURCE: Papageorgopoulos (2019).

MEA power density to reduce material costs. Current state-of-the-art MEAs with very low PGM loadings experience a reduction in performance when operating at high power. Commercial fuel cells are expected to use PGM-based catalysts in 2025–2035; however, some experts believe that a transition to PGM-free catalysts is needed for FCEV cost competitiveness in the longer term.

The balance of plant (BOP) system is projected to represent more than 60% of the cost of a 2025 automotive fuel cell system at a production rate of 500,000 systems per year, with the air loop responsible for 50% of the BOP cost. This is largely owing to the cost of the compressor-expander-motor (CEM) unit, as current automotive fuel cell systems operate at around 2.5 bar. With current fuel cell materials, operation above 2 bar is necessary to achieve high fuel cell stack efficiency and high power density, and to manage membrane humidification requirements. BOP issues and alternative membrane materials will be discussed later in this chapter.

According to the aforementioned DFMA analysis conducted for DOE, expected advances in materials, design, and manufacturing could reduce projected fuel cell system costs to approximately \$42 per kW_{net} in 2020 and \$37 per kW_{net} in 2025, when manufactured at a volume of 500,000 units per year. The projected economies-of-scale impact for a 2025 automotive fuel cell system ranges from approximately \$155 per kW_{net} to \$65 per kW_{net} *to \$35 per* kW_{net} *for annual production rates of 1,000, 10,000, and 1 million, respectively.*

Durability and its impact on cost. DOE originally set a target of 5,000 hours durability for automotive fuel cell systems, which corresponds to an expected lifetime of 150,000 miles driven within a particular range of speeds. In 2016, based on industry feedback, the ultimate durability target was increased to 8,000 hours to enable 150,000 miles for typical drivers on lower average speed drive cycles, such as city driving (Wilson et al., 2016). There is no publicly available fuel cell durability data from automakers, and it is extremely difficult to obtain long-term durability data from laboratory fuel systems. Therefore, DOE relies on precommercial vehicles being tested in NREL's demonstration program. The maximum number of operating hours recorded for a single FCEV in an NREL demonstration was 5,600 hours, recorded in 2015 (Kurtz et al., 2016; Kurtz et al., 2017). The warranty period currently provided by automakers for the fuel cell components in commercial FCEVs for MY 2020 is 100,000 miles (Table 6.1).

Fuel cell durability more than doubled versus precommercial FCEVs over the 9-year period from 2006 to 2015. While the interim target of 5,000 hours durability seems within reach, another doubling of current durability is required to reach the 8,000-hour target. In today's commercial FCEVs, automakers employ approaches to ensure that fuel cell durability is sufficient to enable competitive powertrain warranties, such as high loadings of PGM catalysts, corrosion-resistant BP materials, and system control strategies that reduce fuel cell degradation—all of which add cost to the fuel cell system. Therefore, the cost of fuel cell systems in commercial vehicles today is higher than the DFMA-modeled cost estimates for state-of-the-art laboratory systems described in the preceding section.[10] To address this issue, DOE used publicly available data for the Toyota Mirai to estimate the cost of fuel cell technology in today's commercial FCEVs at current commercial production volumes. The DFMA analysis estimated the Mirai's fuel cell system cost to be \$165 per kW_{net} at a manufacturing volume of 3,000 systems per year. Using materials and performance data for state-of-the-art laboratory fuel cell systems in the DFMA cost analysis yields an estimate of \$113 per kW_{net} for the model system at a production volume of 3,000 per year—a difference of \$52 per kW_{net}, almost 50%. The higher cost estimate for the Mirai fuel cell system is attributed to higher platimum loading, use of titanium rather than stainless steel BPs, additional balance of stack components, and higher cost components for its larger size—allowing for improved system durability and end-of-life performance.

DOE subseqently developed a durability-adjusted cost estimate for its model fuel cell system produced at higher manufacturing volumes. The current durability-adjusted cost estimate is \$68 per kW_{net} at a manufacturing volume of 500,000 sytems per year, compared to \$46 per kW_{net} for the model laboratory system (Kleen and Padgett, 2021). Use of the publicly available Mirai data has enabled the DOE baseline system design and DFMA cost model to be validated against a commercial system design, and provided an approach to account for fuel cell durabilities below the target value. Realizing the DOE model system cost estimate is dependent on scaling today's

[10]The cost estimate is also lower owing to the delay between laboratory demonstration and commercial deployment of state-of-the-art technology.

laboratory materials and components, successfully incorporating them into vehicles, and achieving the fuel cell durability target.

While durability requirements are greater for MHDVs, particularly those that carry heavy loads, these vehicles can tolerate higher costs than LDVs, which allows for durability-enhancing approaches such as higher PGM loadings and oversized fuel cell stacks. System control strategies will also be employed to minimize fuel cell degradation, while longer-term R&D is focused on MEAs with improved durability. Much has been learned from fuel cell buses, which have routinely exceeded 20,000 hours of operation in the United States, and close to 30,000 hours in some cases, without major repairs or replacement of the fuel cell stack (Eudy and Post, 2018).

6.4.1 Fuel Cell Materials and Component Development

The two major technical challenges for automotive fuel cells—cost and durability—are strongly interrelated. PEM fuel cells with higher platinum loadings on the electrodes have longer lifetimes but higher cost than those with lower loadings. Reducing a membrane's thickness decreases MEA cost but also makes it more prone to mechanical degradation (e.g., cracks, pinholes, fatigue). Replacing state-of-the-art materials with alternatives—PGM-free catalysts and membranes other than PFSA—often results in poor performance and reduced lifetimes under the harsh conditions of the PEM fuel cell. A significant amount of research is devoted to understanding fuel cell degradation to guide materials development and system design. Research directed at durability improvements is intertwined with research aimed at improving performance and lowering cost of catalysts, electrodes, membranes, and MEA fabrication.

Within the United States, DOE supports several R&D efforts aimed at reducing the cost of PEM fuel cells while increasing durability and maintaining or improving performance. These include two consortia led by DOE national laboratories:[11]

- Fuel Cell Consortium for Performance and Durability (FC-PAD) focuses on improving performance and durability, while simultaneously reducing cost.
- Electrocatalysis Consortium (ElectroCat) focuses on development of PGM-free catalysts.

The following sections summarize these and other efforts to develop improved materials and components for automotive fuel cells.

6.4.1.1 Electrodes

Today's automotive fuel cells use platinum-based catalysts, primarily owing to their relatively high activity for the hydrogen oxidation reaction (HOR) on the anode and for the ORR on the cathode. Despite a 50% reduction in the platinum content of PEM fuel cells over the past decade (U.S. DRIVE, 2017a), platinum-based catalysts remain the single highest cost contributor to the fuel cell stack (Thompson and Papageorgopoulos, 2019). Information on platinum loadings in commercial FCEVs is not publicly available. However, James et al. (2018) estimate the 2017 Toyota Mirai to have a total platinum (Pt) loading of 0.365 milligrams per centimeter squared (mg/cm^2), which for a 114 kW stack is estimated at 40 g Pt per FCEV, or 0.350 grams per kilowatt (g/kW). The U.S. DRIVE target, which is thought to be loosely based on the amount of platinum in today's internal combustion engine vehicle (ICEV) catalytic converters, is <0.1 g/kW, or <10g Pt for a 100-kW fuel cell system. State-of-the-art laboratory MEAs have demonstrated Pt loadings of 0.125 g/kW (Kongkanand, 2017). To achieve the cost reduction projected for DOE's 2025 fuel cell system ($37 per kW$_{net}$), a total platinum loading of 0.088 g/kW was used in the cost model. For comparison, the U.S. DRIVE/DOE 2025 target is <0.10 g/kW. For an 80 kW$_{net}$ fuel cell system, the reduction from 0.125 g/kW Pt in state-of-the-art laboratory MEAs to 0.088 g/kW corresponds to a reduction in stack platinum cost of $551 to $334, or $6.80 per kW to $4.18 per kW. R&D efforts continue to focus on further reducing PGM content by designing catalysts with high and stable platinum dispersion and modifying electrode structures to prevent the performance losses that occur at ultra-low platinum loadings (Kongkanand and Mathias, 2016).

[11] See https://www.energy.gov/eere/fuelcells/hydrogen-and-fuel-cell-technologies-office-consortia.

The kinetics of the HOR on the Pt anode are very fast; even low Pt loadings have little negative impact on anode performance (Holton and Stevenson, 2013). Studies indicate that a Pt loading as low as 0.025 mg/cm^2 is possible without losing performance from HOR kinetics, consistent with the DOE 2020 anode Pt target (Banham and Ye, 2017). The ORR at the cathode has much slower kinetics owing to its more complicated mechanism and therefore requires higher Pt loading (DOE's 2020 target is <0.10 mg/cm^2). The cathode also operates in a more corrosive environment and is subject to flooding from the water produced there, making catalyst stability as important as catalyst activity. Incomplete oxygen reduction at the cathode can produce significant amounts of hydrogen peroxide, which causes oxidative degradation of the membrane. Thus, there is significantly more research focused on cathode improvements to lower the cost of PEM fuel cells and increase their power density, efficiency, and durability. Essential catalyst characteristics for high-performance PEM fuel cell electrodes include the following (Holton and Stevenson, 2013):

- *High activity:* The ability to adsorb the reactant strongly enough to facilitate the reaction but not so strongly that the catalyst becomes obstructed by the reactant or products.
- *High selectivity:* The ability to make the desired product and minimize the production of undesirable intermediates or side products.
- *Good stability:* The ability to perform and endure in the operating environment of the fuel cell—acidic conditions; oxidants; reactive radicals; temperature, pressure, and voltage changes.
- *Tolerance to impurities:* The ability to resist poisoning by impurities in the air/fuel stream or materials.

Current ORR research is focused on platinum alloy and non-PGM catalysts that have slightly lower oxygen binding energies than pure Pt—strong enough to drive cleavage of the O=O bond but weak enough to release reaction intermediates and products (Wang et al., 2019). The most common approaches used to improve Pt activity are alloying with one or more other metals, layering Pt on or just below the surface of another metal, the core-shell method in which a "core" of lower cost metal is coated with Pt, and alloying Pt followed by dealloying to produce a Pt lattice structure that retains some of the properties of the alloy structure (Holton and Stevenson, 2013).

Platinum alloy catalysts that show promise include ordered PtM intermetallics, with M=Co being the most promising first row transition metal. Some PtM catalysts have been incorporated into MEAs exceeding the DOE 2020 activity target of 0.44 amps per milligram Pt and demonstrating encouraging durability (< 40 mV voltage loss after 30,000 cycles at 0.6–0.95 V) (Gröger et al., 2015; Wang et al., 2019). Research is also focused on advanced carbon supports that enable increased dispersion of PtM nanoparticles and stronger metal-support interactions to prevent particle migration, reduce carbon corrosion, and increase durability. Other approaches include putting ionomers in the catalyst layers to improve catalyst utilization, using thinner GDLs (~100–150 microns) and larger-pore microporous layers to improve water and gas transport, and improving electrode flow-field structures (Shinozaki et al., 2011; Nakagaki, 2015). Further advances in both catalysts and electrodes are needed to achieve high power density at ultra-low Pt loading.

One indication of the complex nature of catalyst development is seen in the trade-off between selectivity and stability. While binary and ternary Pt alloys supported on high surface area carbon have higher selectivity toward hydrogen peroxide formation in the ORR reaction than unalloyed Pt, they generally do not lead to membrane degradation (Sethuraman et al., 2009). This is owing to their increased stability, which limits migration of Pt ions into the membrane.

PGM-free ORR catalysts that demonstrate equivalent performance and durability to platinum-based catalysts are considered to be the longer-term (beyond 2035) and higher-risk approach. All non-PGM catalysts under development need significant improvements in both activity and stability to be viable in automotive fuel cells. To design improved catalysts and electrodes, research is focused on developing a greater understanding of the role of different metals in promoting catalytic activity, as well as the role of the surrounding ligand structure and morphology. Approaches include macrocyclic compounds $Co\text{-}N_4$ (e.g., CoTMPP, TMPP = tetramethoxyphenyl porphyrin) and $Fe\text{-}N_4$ (e.g., FeTPP, TPP = tetraphenyl porphyrin), heat-treated macrocyclic compounds, heat-treated transition metal-nitrogen-carbon (M-N-C), and atomically dispersed and nitrogen coordinated metal sites (Fe, Co, Mn) (Wang et al., 2019).

It is difficult to project timelines for successful development of new PEM fuel cell catalysts capable of meeting the demands of automotive drive cycles. In addition to the materials development challenges, maintaining performance of catalysts and electrodes when incorporated into MEAs presents an additional challenge for both

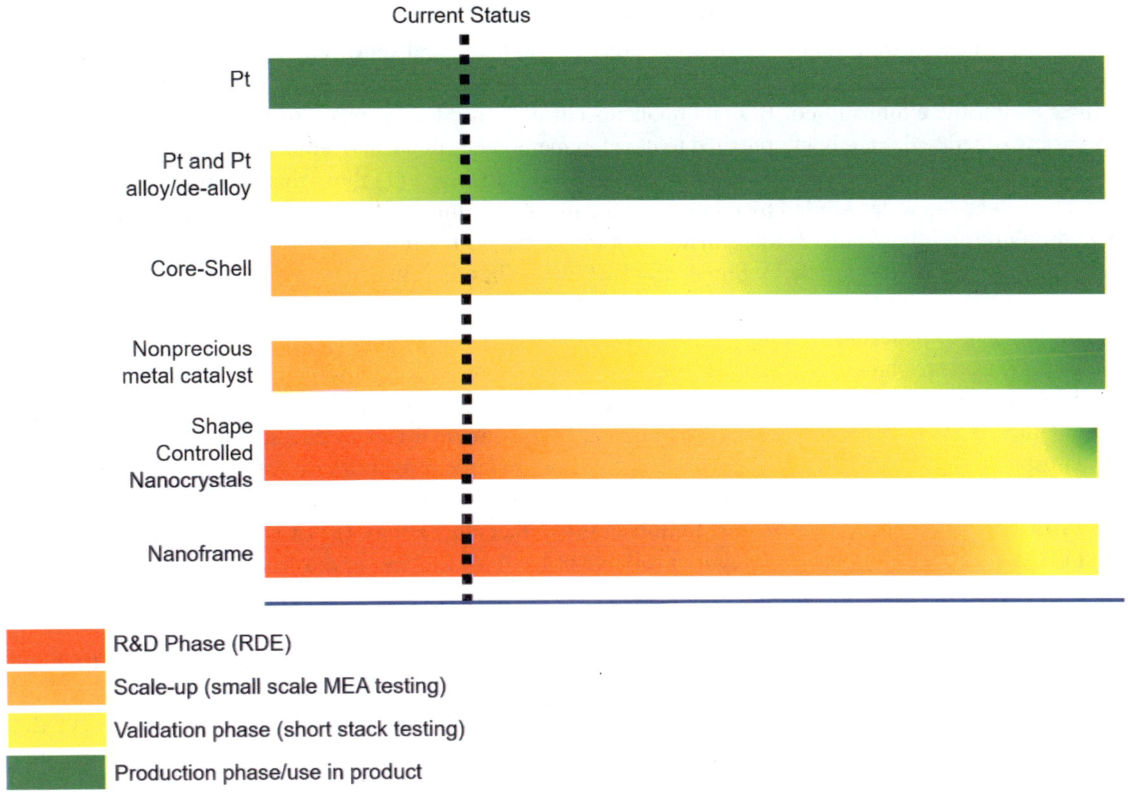

FIGURE 6.14 Relative development timelines for ORR catalysts.
SOURCE: Reprinted with permission from D. Banham and S. Ye, 2017, Current status and future development of catalyst materials and catalyst layers for proton exchange membrane fuel cells: an industrial perspective, *ACS Energy Letters* 2(3):629–638, © 2017 American Chemical Society.

low- and no-PGM materials as membrane-electrode interface interactions come into play. Figure 6.14 presents a relative timeline for ORR in six categories: (1) Pt/C, (2) Pt and Pt alloy/dealloy, (3) core-shell, (4) nonprecious metal catalysts (PGM-free), (5) shape-controlled nanocrystals, and (6) nanoframes (Banham and Ye, 2017).

Some PGM-free catalysts have demonstrated sufficient performance for use in backup power and/or portable power fuel cell applications. The first commercial PGM-free fuel cell, a 30 W stack for emergency or backup power, was announced in 2017 (Fuel Cells Bulletin, 2017; Banham et al., 2019). However, these applications have considerably lower performance and durability requirements than automotive applications. In 2025–2035, automotive fuel cells are likely to see a gradual lowering of Pt content, leading to reduced FCEV cost; however, current PGM-free catalysts are far from meeting automotive performance targets. PGM-free catalysts are unlikely to be in commercial FCEVs in that time frame, and their success beyond that is uncertain.

6.4.1.2 Membranes

Traditional PEM fuel cells use perfluorinated polyethylene membranes that when hydrated swell and form hydrophilic (water-filled) proton-conducting channels and hydrophobic backbones that allow for proton transport. Nafion®, which was developed by DuPont in the 1960s, is still the state-of-the-art membrane. It demonstrates high proton conductivity and good mechanical and chemical stability when operated below 90°C and at RH greater than 40%. Operating pressure is typically around 2.5 bar, as this simplifies humidification and water management in addition to enabling higher power density.

At temperatures higher than 90°C, the membrane can become dehydrated, which leads to decreased proton conductivity. Higher temperatures can also cause irreversible membrane degradation owing to Nafion's relatively low glass transition temperatures (110°C–135°C). However, if these issues could be resolved, higher-temperature fuel cells would benefit from improved reaction kinetics and decreased sensitivity to fuel impurities (e.g., CO), both of which enable reduced platinum catalyst loadings and higher efficiency owing to the production of useful waste heat and/or the elimination of BOP components currently needed for water management.

A variety of strategies are being pursued to develop membranes that can operate at low RH and temperatures up to 120°C. These strategies typically fall into two general categories: (1) those that still require water for conduction but reduce the water needed by controlling the membrane microstructure and/or increase water retention using hydrophilic additives, and (2) those that do not require water but provide conduction through an alternative mechanism. A recent review article by Sun et al. (2019) describes the following approaches:

- *Nafion-based composite membranes*: Adding hygroscopic inorganic or other fillers to Nafion® is designed to increase water retention, reduce reactant crossover, enhance proton mobility, and improve mechanical stability.
- *PBI-based composite membranes*: Thermoplastic polymers such as polybenzimidazole (PBI) have good chemical resistance, high oxidative stability, and good thermal and mechanical properties above 80°C. Phosphoric acid-doped PBI type membranes have shown the most promise for operation up to 200°C at ambient pressure and have been the most extensively studied.
- *PEEK-based composite membranes:* Sulfonated polyether ether ketone (SPEEK) is an attractive alternative to Nafion® because it is commercially available at low cost and has microstructure and morphology that enable superior water uptake and high protonic conductivities when filled with metal oxides, solid acids, metal-organic frameworks (MOFs), or carbon nanotubes.
- *Mixed electron-proton conducting composite membranes:* An alternative approach to the traditional fabrication of membrane-electrode assemblies is to mix or replace Nafion in the catalyst ink with an electron-conducting polymer such as polypyrrole or polyaniline, thus introducing electronic conductivity in parallel with protonic conductivity.

While some promising results have been achieved, a membrane that meets all of the U.S. DRIVE targets has not yet been developed. The primary challenge is developing a membrane that has sufficient conductivity at 120°C and lower RH while maintaining mechanical stability and durability during fuel cell operation. Other challenges include making an electrocatalytic layer that is compatible with both the membrane and the catalyst, and meeting the established cost target. It is unclear if successful high-temperature fuel cell membranes will be in commercial FCEVs in 2025–2035.

Anion Exchange Membranes

Another, longer-term approach to enabling higher temperature operation is to replace PEMs with anion exchange membranes (AEMs). AEMs, also called alkaline anion exchange membranes (AAEMs), conduct hydroxide anions (OH^-) rather than protons (H^+).

Both proton exchange membrane fuel cells (PEMFCs) and anion exchange membrane fuel cells (AEMFCs) produce water as a by-product. However, in contrast to PEMFC technology, in an AEMFC the hydroxide anion is transported from the cathode to the anode, opposite to the proton conduction direction in a PEMFC, and water is generated at the anode, while at the same time water is a reactant at the cathode. This distinctive water transport scenario, together with the alkaline medium, represent a unique feature of AEMFCs (You et al., 2019). Figure 6.15 shows a schematic drawing of transport in an AEMFC (Dekel, 2018).

The alkaline environment in AEM fuel cells provides several advantages over the acidic environment in PEM fuel cells, including the following:

- Enabling the use of lower cost non-PGM oxygen reduction catalysts owing to their improved stability in alkaline environments, and less expensive metal hardware.
- Enabling a wider selection of fuels for the fuel cell, as the electro-oxidation kinetics for many liquid fuels are improved in an alkaline environment. Liquid fuels like methanol or hydrazine could be used directly in the fuel cell, for example, or dimethyl ether (DME), a potentially carbon-neutral liquid fuel that can be produced from renewably sourced hydrogen and carbon dioxide (CO_2), and is non-toxic and easy to liquefy.

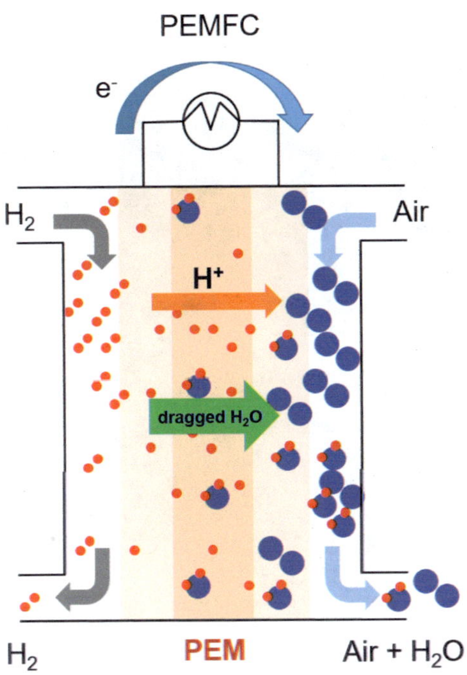

FIGURE 6.15 Schematic of PEMFC (left) compared to AEMFC (right).
SOURCE: Dekel (2018).

While much progress has been made, AEMFCs remain a significantly less mature technology than PEMFCs. Further development of alkaline membranes is needed, as well as integration of catalysts and membranes into high-performance MEAs. Current R&D efforts are focused on developing Pt-free AEMFCs, and understanding, enabling, and validating their long-term stability in fuel cell operation at high temperatures and with low water content. In 2018, reported performance data indicated that stable AEMFC operation was limited to less than 1,000 hours. Researchers cite chemical degradation of the cationic functional groups at low water content as the primary reason for durability limitations (Dekel et al., 2019). AEMFCs are also susceptible to carbonization from CO_2 in the air (Ul Hassan et al., 2020). Pathways are needed to minimize the impact of CO_2 on cell operation, including material, operational, and engineering solutions. Significant advances are needed before AEMFCs can be considered a viable alternative to PEMFCs; thus, it is unlikely that they will be in commercial FCEVs in 2025–2035.

6.4.1.3 Gas Diffusion Layers

The GDL in the PEM fuel cell is used for optimal distribution of reactants to the catalyst layer and for water management within the MEA (Tomas et al., 2017). The GDL can consist of a single layer or a double layer (GDL plus a microporous layer, MPL). The most commonly used GDL materials are carbon cloth and carbon paper. The GDL keeps the membrane humidified while also preventing flooding of the cathode by "wet-proofing" the GDL with hydrophobic poly(tetrafluoroethylene) (PTFE) to facilitate transport of water away from the active catalyst layer and to prevent the pores in the carbon cloth or paper from becoming clogged with water (see Figure 6.16). The MPL, which consists of carbon or graphite particles mixed with PTFE binder, provides improved electrical contact and facilitates water transport in and out of the diffusion layer, and enhances the chemical and mechanical stability of the catalyst layer and membrane. The GDL must have good electrical conductivity, high permeability for gases and liquids, and high chemical stability, and must be able to withstand the temperatures and compression forces of the fuel cell stack (Spiegel, 2018).

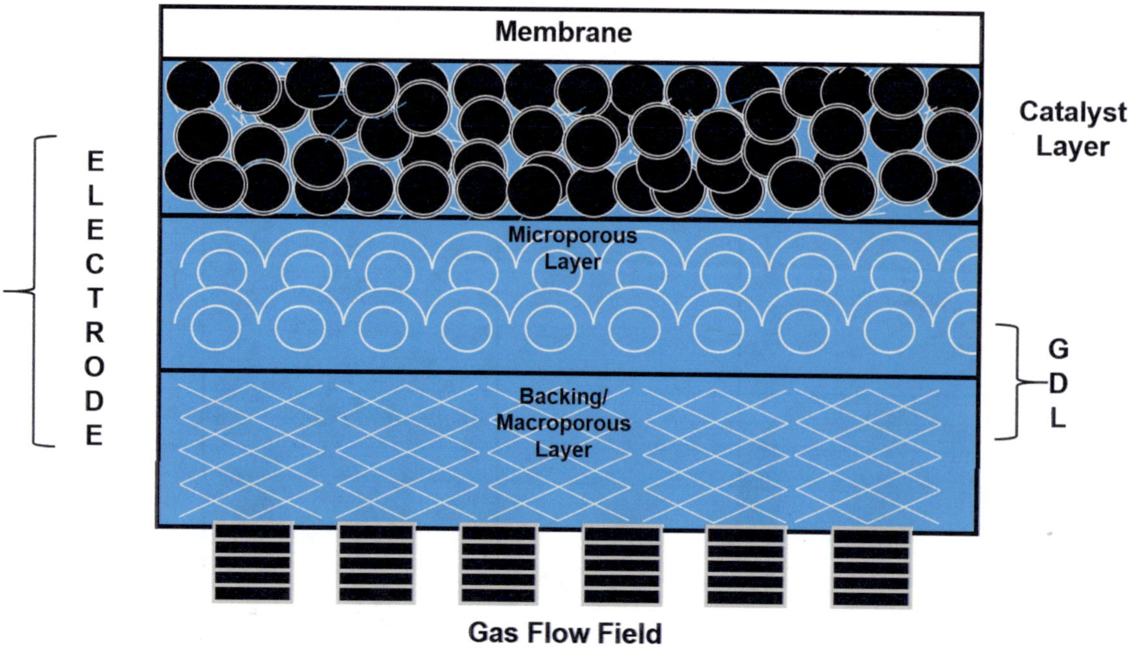

FIGURE 6.16 A 2D view of a gas diffusion electrode indicating the catalyst layer and GDL, comprising a backing layer and MPL.
SOURCE: Jayakumar et al. (2017).

Current GDL materials have long-term durability issues and complex manufacturing processes, which impact their cost. Research is focused on improving current materials and developing alternative materials and fabrication processes. Approaches include the following (Borup et al., 2019):

- Using lower cost carbon fibers (CFs);
- Using lower carbonization temperatures to reduce processing costs;
- Developing low-cost gas phase surface treatments to replace PTFE treatments;
- Developing super-hydrophobicity coatings to prevent water flooding and transport losses;
- Incorporating hydrophilic pathways separate from hydrophobic domains to provide pathways for water removal, including through laser patterning; and
- Incorporating porous metals—for example, sintered metal powders or fibers.

6.4.1.4 Bipolar Plates

BPs are a key component in PEM fuel cells, performing several essential functions. They connect each cell electrically, supply the reactant gases—hydrogen and oxygen (from air)—through flow channels, and remove heat and reaction by-products (water) from the cell. These functions require that BP materials have high electrical conductivity, high gas impermeability, good mechanical strength, and high corrosion resistance (Taherian, 2014). FCEVs contains hundreds of BPs (James et al., 2018).

Traditionally, BPs were fabricated from high-density graphite due its superior corrosion resistance, chemical stability, high thermal conductivity, and availability. However, graphite plates are costly to produce, bulky, and have mechanical properties that may make them unsuitable for fuel cell applications that require good structural durability against shock and vibration. For these reasons, many fuel cell manufacturers have moved away from graphite plates and use metal or composite plates instead.

Metal plates have higher electrical conductivity and lower gas permeability than composite plates, as well as higher strength, better shock resistance, and better high-volume manufacturability owing to their stampability (for flow fields).

Metal plates can be made thinner than composite plates—as thin as 1 mm—enabling smaller and lighter fuel cell stacks, and some, like stainless steel, are currently lower in cost. The downside to metal BPs is that they are more susceptible to corrosion than carbon plates, particularly in the acidic environment of the PEM fuel cell, which results in leaching of metal ions into the fuel cell membrane. This has been addressed by applying corrosion-resistant coatings to the metal surface, such as conductive polymer films, metal nitride/carbide films, noble metal films, or carbon. However, these additional processing steps increase the cost of these metal plate options.

Because of their robustness, metal BPs are the current choice of FCEV manufacturers (typically 316 stainless steel or titanium). Hyundai moved from carbon plates in its precommercial FCEV to unspecified metal plates in its first commercial FCEV, the Tucson, which the company began leasing to customers in 2014. Hyundai continues to use metal plates in its current commercial FCEV, the Nexo (Castillo, 2017). The Toyota Mirai fuel cell stack uses titanium BPs, which are more costly than stainless steel.

There is still some debate among fuel cell developers about whether carbon or metal BPs are the better choice for transportation fuel cell systems. Despite their shortcomings, fuel cell manufacturer Ballard Power Systems believes that polymer-based carbon composite plates are a better choice because of their higher corrosion resistance, higher durability, easier formability (which enables fabrication of thin plates and greater design flexibility), and the elimination of coating and welding (which enables lower cost manufacture) (Bach, 2019). It is unknown if coated metal plates will have the durability needed for vehicular applications (8,000 hours is the current U.S. DRIVE target for LDVs), and especially heavy-duty vehicles such as buses and trucks, which require lifetimes more than 20,000 hours. According to the U.S. DRIVE Fuel Cell Technical Team Roadmap, automotive fuel cell systems containing stainless steel BPs have demonstrated 4,130 hours durability (U.S. DRIVE, 2017a). According to Ballard, carbon plates used in fuel cell transit buses have reached more than 30,000 hours of operation without issues, and in material handling vehicles (forklifts) beyond 10,000 hours.

Aluminum is another BP material being explored for automotive fuel cell applications because it is relatively lightweight and inexpensive, conductive, high strength, and formable. The downside is aluminum's low corrosion and oxidation resistance. Current R&D efforts are aimed at aluminum-coated BPs fabricated through solid phase processing, which has been used to improve performance and lower cost in other applications (Ross, 2019).

The current cost of PEMFC BPs is well above the U.S. DRIVE 2025 cost target of \$2/kW for the finished plate at a production volume of 500,000 fuel cell stacks per year. DFMA analysis of the current cost of coated 316 stainless steel fabricated by progressive stamping is \$5.40/kW, excluding welding, for production volumes of 500,000 stacks/year (Huya-Kouadio et al., 2018). The cost of the plate material alone was estimated to be \$2.90/kW. Less expensive materials are needed as well as less expensive manufacturing processes (James, 2017). A comparison of the cost of metallic, carbon composite, and expanded graphite plate materials suggested that metal plates may be the lowest cost pathway, and that achieving the DOE 2020 target may be possible by using lower cost plate material, improving the manufacturing process, and increasing the power density of the fuel cell stack. However, a recent analysis of the cost to manufacture embossed flexible graphite BPs for LDV systems showed that they can be lower cost than metal plates and also meet the DOE 2020 target of \$3/kW for finished plates (James, 2019). The U.S. DRIVE 2025 cost target of \$2/kW will be a significant challenge with current commercial bipolar materials and processes.

6.4.1.5 BOP Components

The BOP in a PEM fuel cell system includes components associated with the air loop; humidifier and water recovery loop; coolant loops; fuel loop (excluding fuel storage); system controller; sensors; and miscellaneous items such as mounting frames, belly pan, and wiring and piping (James et al., 2018). As mentioned earlier, the BOP system is projected to represent more than 60% of the cost of 2025 automotive fuel cell systems at a production rate of 500,000 systems per year. In addition, parasitic power demands of BOP components result in a lower net system efficiency, largely owing to the air compression system, which typically consists of an integrated air compressor, exhaust gas expander, and an electronic CEM unit. The CEM unit is the highest cost component in the BOP system, projected to account for about 50% of the total BOP cost in 2025 automotive fuel cell systems. Current R&D efforts are focused on new and improved materials with potential to simplify system design, improve system efficiency, and reduce BOP cost.

Today's off-the-shelf air compressors were not designed for the operational characteristics of automotive fuel cell systems, which require compressors that are oil free, high pressure with low flow rate, high efficiency, and low weight and volume (Yu et al., 2015). Automakers have tested several types of compressors in FCEVs, including scroll and screw compressors, centrifugal turbocompressors, and Roots compressors (Kerviel et al., 2018). All have trade-offs. Centrifugal and Roots compressors are smaller and lower cost than screw and scroll compressors. The Toyota Mirai uses a Roots compressor, which has a higher power density than the centrifugal compressor but lower efficiency and pressure ratio, and pulsation characteristics that require a larger sound absorber (Fumihiro et al., 2015). Therefore, many FCEV manufacturers, including Honda, have adopted centrifugal compressors for FCEVs.

Current CEM R&D efforts are focused on different controller designs to increase CEM efficiencies from current levels of ~80% to >90%. Successful development of higher temperature, lower humidity membrane electrode assemblies is expected to lead to lower pressure fuel cell systems, simplifying air compression requirements and reducing cost in the longer term.

6.4.1.6 Fuel Cell Manufacturing R&D

The high cost of automotive fuel cell systems today is primarily owing to low production volumes. Current production processes are slow, expensive, and labor intensive; higher levels of automation are needed. High-volume manufacturing methods for fuel cell stack components are still evolving, and fuel cell stacks are assembled mostly in a manual process (Gurau et al., 2018). To achieve economies of scale, emerging fuel cell manufacturing processes must be optimized and scaled up to factory production volumes. However, as noted later in this chapter, the lack of a hydrogen fueling infrastructure has limited deployment of and demand for automotive fuel cell systems. Today's low demand, as well as uncertainty about future market volume, has limited industry investments in fuel cell manufacturing development (DOE HTAC MSC, 2014; Mayyas and Mann, 2019).

High-volume MEA manufacturers have implemented automation via continuous roll-to-roll processes, replacing the traditional spray coating process used for catalyst deposition in low-volume production. Roll-to-roll processes enable higher throughput, more uniform catalyst layers, and the ability to include infrared or optical systems for quality control. DOE-supported work at NREL has focused on improving inspection methods for in-line quality control using techniques such as infrared thermography. A Manufacturing Demonstration Facility at Oak Ridge National Laboratory is focused on additive manufacturing and low-cost CF.

Input from fuel cell manufacturers identified the following overarching needs for low-cost automated fuel cell stack assembly: simplifying component design of each component, reducing parts, and replacing materials not compatible with rapid serial production. Figure 6.17 shows an example of reducing parts by combining the GDL with the CCM. Gurau et al. (2018) reported that the most significant technological challenges to automated stack assembly are (1) the difficulty in aligning fuel cell components (BPs, MEAs, and gaskets) in the stack to avoid leaks of reactant gases, and (2) the diversity of fuel cell components that need to be handled by the robotic arm.

DOE has supported fuel cell manufacturing R&D at varying and relatively low levels of funding over the past decade (DOE, 2017). Efforts have been largely focused on fuel cell MEAs, BPs, and CFs/composites for hydrogen storage tanks. Moving forward, increased manufacturing R&D will be critical to achieving the economies of scale needed to reduce the cost of fuel cell and hydrogen technologies.

6.4.1.7 Onboard Hydrogen Storage

Because of the size and weight constraints of LDVs, high volumetric and gravimetric energy densities are important characteristics for LDV fuels. To be comparable to conventional gasoline vehicles, automakers have targeted driving ranges of 300 to 500 miles and vehicle refueling times of 3 to 5 minutes for FCEVs. Hydrogen's high gravimetric energy density (33 kilowatt-hour per kilogram [kWh/kg] based on lower heating value) provides an advantage over other fuels; however, hydrogen's very low volumetric energy density (~1 kilowatt-hour per liter [kWh/L] at 700 bar) is a distinct disadvantage.[12] See Figure 6.18 for a comparison of energy content in various fuels.

[12]For comparison, lower heating values for gasoline are 12 kWh/kg and 9 kWh/L.

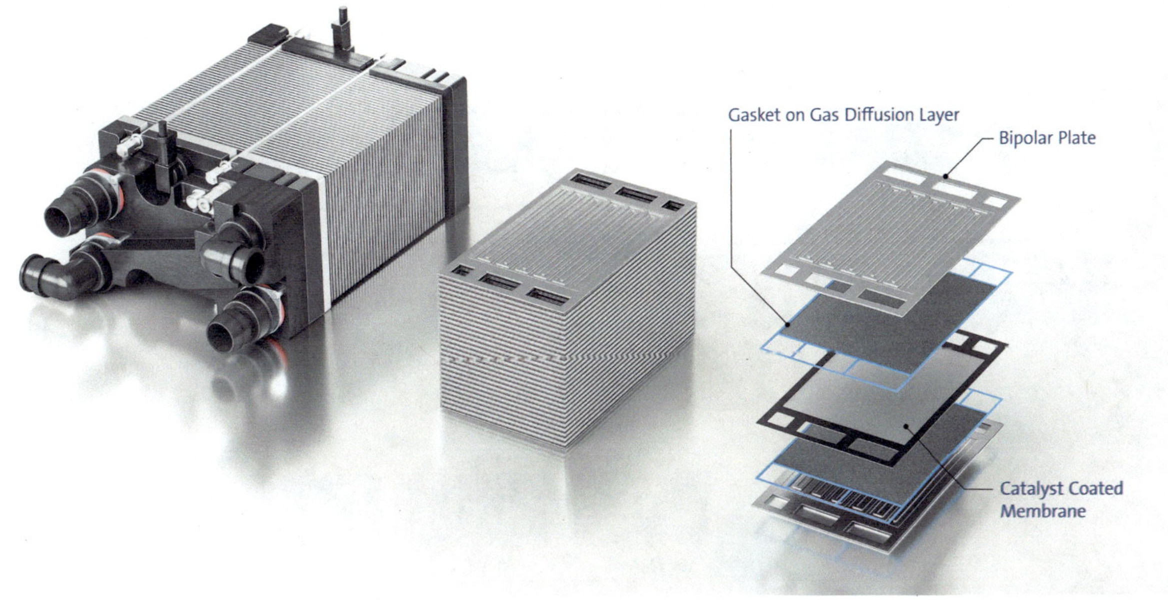

FIGURE 6.17 Fuel cell stack design with combined GDL and gasket.
SOURCE: Freudenberg Sealing Technologies (2018).

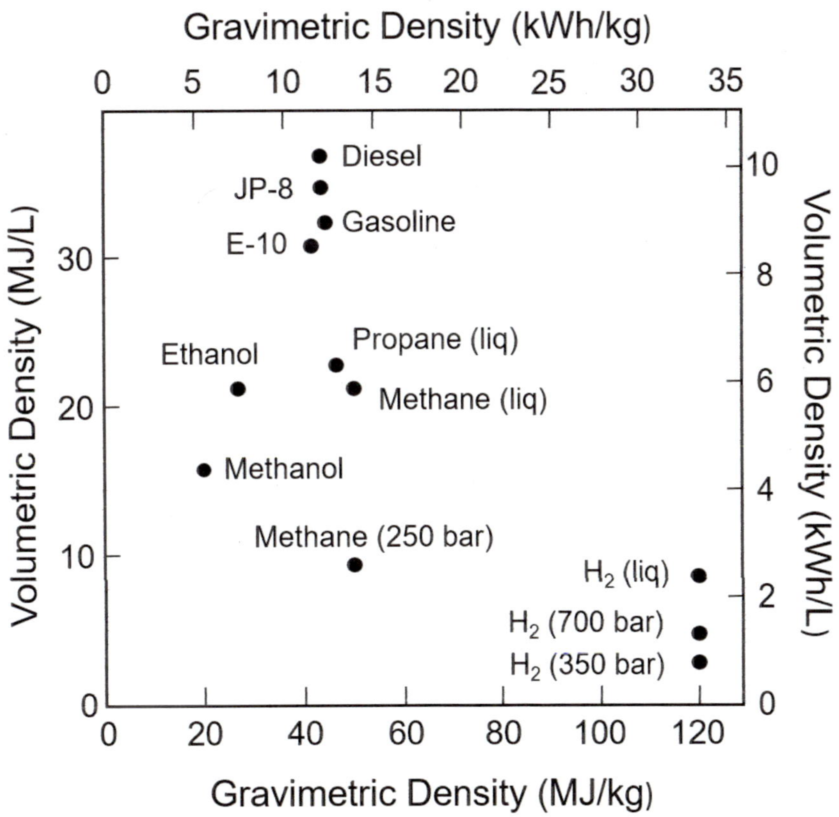

FIGURE 6.18 Comparison of the volumetric and gravimetric densities of various fuels.
SOURCE: U.S. DRIVE (2017c).

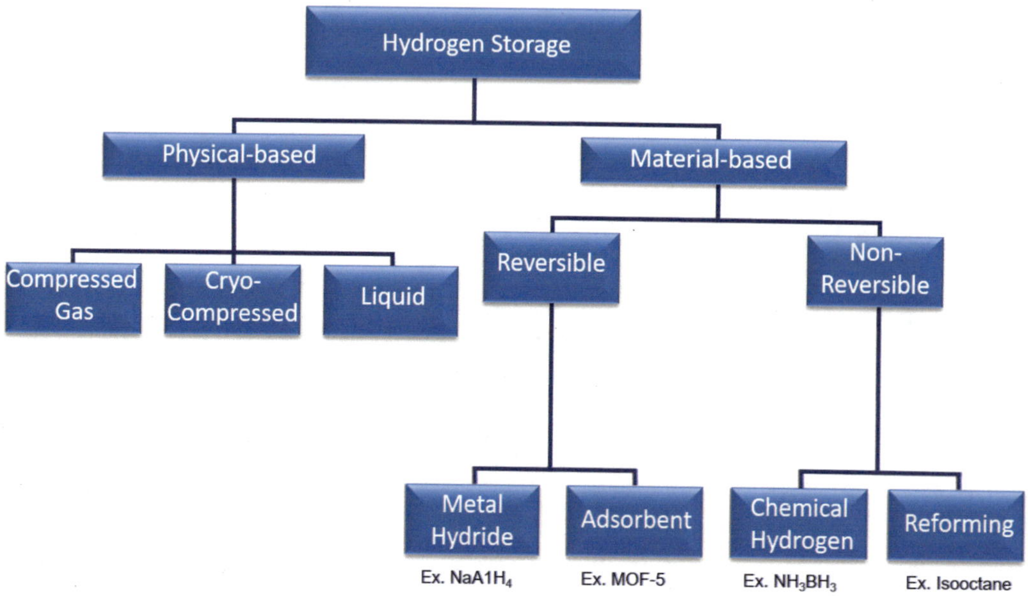

FIGURE 6.19 Hydrogen storage technologies under development.
SOURCE: U.S. DRIVE (2017c).

Conventional hydrogen storage methods, particularly high-pressure gas cylinders, are a well-established technology and the current method of choice for FCEVs; however, the tanks are bulky and expensive. Therefore, R&D activities are under way to lower the cost and reduce the volume of compressed hydrogen tanks, and to develop alternative methods of hydrogen storage to enable affordable, lightweight, and compact hydrogen storage systems for FCEVs. Alternative approaches include liquid, cryo-, or cold-compressed hydrogen; physisorption of hydrogen on materials with a high specific surface area; hydrogen intercalation in metals and hydrides; and chemical hydrogen storage methods.

Hydrogen storage approaches can be broadly characterized in two categories (see Figure 6.19):

1. Physical-based storage technologies, in which elemental hydrogen is stored as compressed hydrogen gas, cold- or cryo-compressed hydrogen, or liquid hydrogen. These storage technologies will dominate in 2025–2035, particularly gaseous and liquid storage.
2. Material-based (or solid-state) storage technologies, in which hydrogen is bound to other elements within materials—adsorbents, metal hydrides, or chemical hydrogen storage materials. These approaches are not likely to be in commercial FCEVs in 2025–2035.

The different hydrogen storage approaches and their volumetric densities are shown in Figure 6.20.

Besides energy density and cost, important performance characteristics for onboard hydrogen storage systems include the following:

- Operating pressure—pressure vessels must be reinforced with high-strength containment materials that impact system weight, volume, and cost.
- Operating temperature—temperature-dependent materials and systems require heat management equipment, which adds complexity and cost.
- The rate at which the system can release and curtail hydrogen upon demand in response to the vehicle's acceleration and braking.

The U.S. DRIVE Partnership has established technology-neutral targets for energy capacity, charging/discharging rates, durability/operability, dormancy, and cost of onboard hydrogen storage systems. Systems must also meet

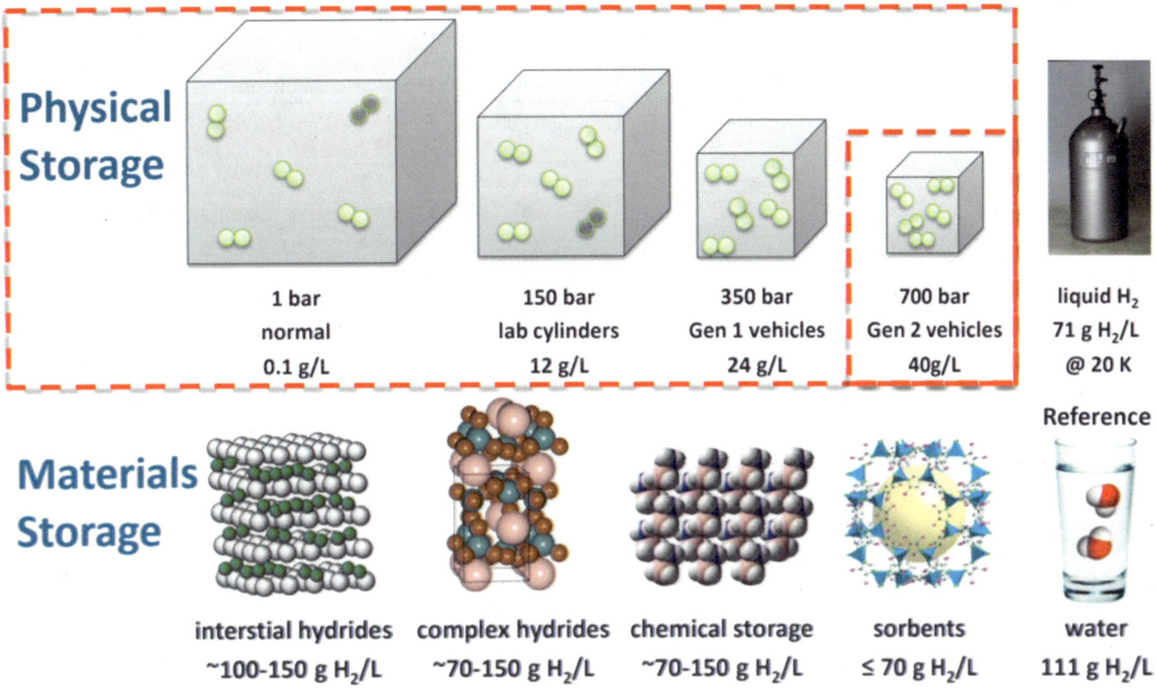

FIGURE 6.20 Comparison of volumetric energy densities for different physical and materials hydrogen storage approaches. SOURCE: Stetson (2015).

established standards for fuel quality (SAE J2719) and environmental health and safety (SAE J2579) (U.S. DRIVE, 2017b). Gravimetric and volumetric energy capacity targets for 2025 are as follows: 1.8 kWh/kg (5.5 wt%) and 1.3 kWh/L, respectively, at a cost of $9/kWh (U.S. DRIVE, 2017b). The ultimate targets are 2.2 kWh/kg (6.5 wt%), 1.7 kWh/L, and $8/kWh. Targets are for a complete system, including tank, material, valves, regulators, piping, mounting brackets, insulation, added cooling capacity, and all other BOP components. These targets are designed to enable greater than 300-mile range across the majority of the current LDV fleet. To that end, the targets exclude "unusable" energy—that is, any hydrogen left in the tank below minimum fuel cell system pressure, flow, and temperature requirements—and any energy/fuel used to extract the hydrogen from the storage medium. The latter may be the case for material-based storage approaches—for example, heating a metal hydride to release hydrogen (U.S. DRIVE, 2017b).

Table 6.2 provides a summary of different hydrogen storage methods. The following sections provide information on hydrogen storage systems in commercial FCEVs and other storage approaches under development.

Physical-Based Hydrogen Storage

Compressed hydrogen gas. Compressed hydrogen tanks are used in commercial FCEVs today and are likely to be the hydrogen storage technology used in 2025–2035 FCEVs. State-of-the-art tanks contain hydrogen gas at 350 or 700 bar in composite overwrapped pressure vessels (COPVs), which are constructed using CF reinforced polymers that are wrapped around metallic (Type-III) or polymeric (Type-IV) liners (typically, high-density polyethylene) (Gangloff, 2017); see Figure 6.21. Today's commercial FCEVs typically use 700-bar Type IV pressure vessels for onboard hydrogen storage (Yamashita et al., 2015).

Compressed gas has several advantages over other hydrogen storage approaches. The kinetics of compressed hydrogen are well-suited for mobile applications. Hydrogen flow is responsive to vehicle driving demands, increasing or decreasing in response to acceleration, and deceleration and braking. The materials involved—typically CF and nylon-6—are not toxic or environmentally harmful. The onboard hydrogen supply system

TABLE 6.2 Basic Hydrogen Storage Methods

Storage Method	ρ_m [mass%]	ρ_v [kg H_2 m^{-3}]	T [°C]	p [bar]	Phenomena and Remarks
High-pressure gas cylinders	13	< 40	RT	800	Compressed gas (molecular H_2) in light weight composite cylinders (tensile strength of the material is 2000 MPa)
Liquid hydrogen in cryogenic tanks	size dependent	70.8	–252	1	Liquid hydrogen (molecular H_2) continuous loss of a few % per day of hydrogen at RT
Adsorbed hydrogen	≈ 2	20	–80	100	Physiosorption (molecular H_2) on materials, e.g., carbon with a very large specific surface area, fully reversible
Absorbed on interstitial sites in a host metal	≈ 2	150	RT	1	Hydrogen (atomic H) intercalation in host metals, metallic hydrides working at RT are fully reversible
Complex compounds	*< 18*	*150*	*> 100*	*1*	*Complex compounds ($[AlH_4]^-$ or $[BH_4]^-$),* desorption at elevated temperature, adsorption at high pressures
Metals and complexes together with waters	< 40	> 150	RT	1	Chemical oxidation of metals with water and liberation of hydrogen, not directly reversible?

NOTE: The gravimetric density ρ_m, the volumetric density ρ_v, the working temperature T, and the pressure p are listed. RT stands for room temperature (25°C).
SOURCE: Züttel (2003).

operates at ambient temperature, so there is no need for thermal management equipment to store or release hydrogen onboard the vehicle.

The disadvantages of compressed hydrogen tanks are the large volume required and the high cost of the tank, primarily owing to the cost of CF. From an infrastructure standpoint, FCEV drivers, used to a liquid refueling infrastructure, will need to adapt to gaseous refueling equipment and processes, which may vary from station to station. Automakers are providing hydrogen refueling tips and other resources for early adopters.[13] Another consideration is the energy associated with compressing hydrogen and delivering it to the vehicle. Like other gases, hydrogen releases heat when compressed. To avoid overheating the tank during refueling, compressed hydrogen is cooled to −20°C to −40°C beforehand. The energy required for precooling, as well as that required to compress hydrogen, reduce the WTW energy efficiency of FCEVs and, depending on the energy source, can result in upstream CO_2 emissions.

DOE has reported a gravimetric energy density of 1.48 kWh/kg (~4.5 wt%) and a volumetric energy density of 0.83 kWh/L for today's 700 bar compressed hydrogen storage systems. For comparison, Toyota has reported a gravimetric capacity of 5.7% for the Mirai hydrogen storage system; however, it is unclear if this includes the weight of the entire hydrogen storage system or just the tanks. The DFMA cost analysis conducted for DOE projects a system cost at high-volume production (500,000 systems per year) of $14.2/kWh, based on a single-tank configuration with a net usable hydrogen capacity of 5.6 kilograms in LDV applications; using the lower heating value of 33.3 kWh/kg of hydrogen, that translates to a tank cost of approximately $2,650. This analysis, in 2016$, is based on a tank design that uses aluminum BOP components (valves, regulators, piping, mounting brackets, insulation, etc.), a hoop-intensive winding pattern that reduces CF composite mass, and Toray T700S CF at a cost of $26/kg at volume. As indicated in the system cost breakdown shown in Figure 6.22, CF accounts for more than 50% of the system cost.

R&D efforts have led to a steady decrease in cost. Analysis of compressed hydrogen systems projected to annual manufacturing volumes of 100,000 and 500,000 (Ordaz et al., 2015; Adams et al., 2019), shown in

[13]One example is the webpage for the Toyota Mirai: https://www.toyota.com/mirai/Mirai_Fueling.pdf.

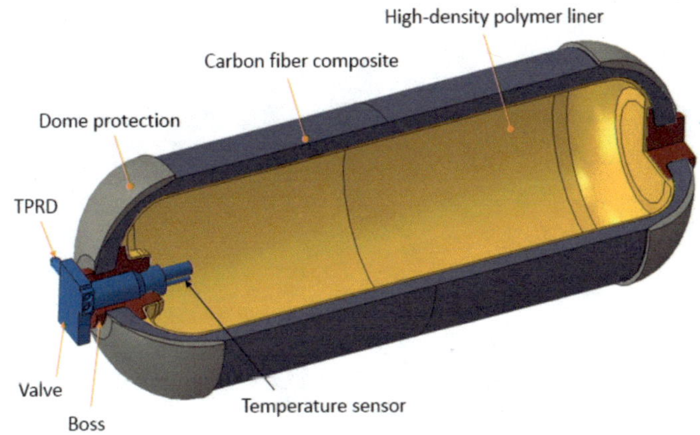

FIGURE 6.21 Schematic of a 700-bar Type-IV COPV for onboard FCEV hydrogen storage.
SOURCE: Process Modeling Group, Nuclear Engineering Division, ANL; reprinted from FCTO (2017).

Figure 6.23, indicates more than a 20% cost reduction since 2013. This is primarily owing to the development of lower-cost CF and resin, improved CF usage, and integrated balance of plant components (Houchins, 2019).

System design and architecture also have significant impact on cost. Automakers are storing hydrogen in two or three onboard tanks in today's commercial FCEVs. In both the Toyota Mirai and the Honda Clarity, the front hydrogen tank sits beneath the rear passenger seat, while the rear tank is behind the rear passenger seat. The Hyundai Nexo uses similar placement of its three-tank vehicle design. Table 6.3 shows the modeled cost of single- and two-tank configurations at production rates of 100,000 and 500,000 systems per year. The analysis

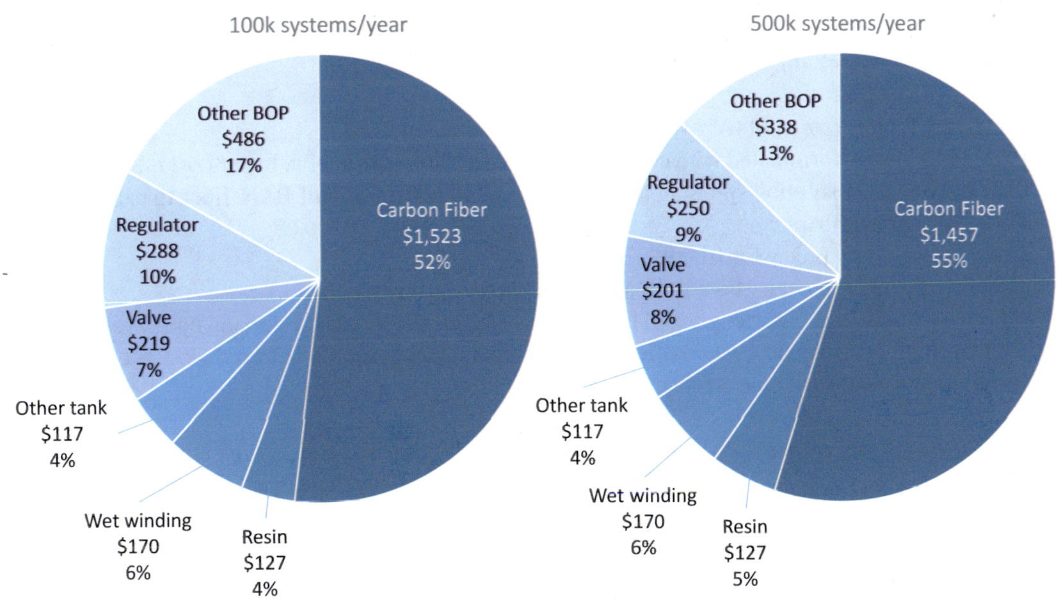

FIGURE 6.22 Storage system cost breakdown by percentage of the total cost (annual production of 100,000 units shown on left and 500,000 units shown on right).
SOURCE: Adams et al. (2019).

FIGURE 6.23 Comparison of storage system cost status in 2007$ and 2016$ as reported in 2013 (McWhorter and Ordaz, 2013), 2015 (Ordaz et al., 2015), and in 2019. Costs are for annual productions of 100,000 units (left) and 500,000 units (right). SOURCE: Adams et al. (2019).

indicates that two-tank storage systems are more expensive than single-tank systems primarily owing to a second set of in-tank valves required for the two-tank design. This accounts for the higher BOP costs in the two-tank design. The cost of the two-tank system also depends on the tank aspect ratios, which impact the hoop intensive winding pattern, and therefore, the mass of the tank. As shown in the table, the mixed-aspect design is projected to be lower cost.

DOE has ramped up R&D of compressed tanks, including initiating several new projects in July 2020, focused primarily on reducing the CF cost. High strength CF is almost exclusively produced from polyacrylonitrile (PAN) precursor fibers in a solution spinning process that requires significant capital expenditures (Das et al., 2016; Warren, 2016). The conversion of PAN precursor fiber to CF includes several moderate- to high-temperature processing steps that result in only 50% mass yield. Thus, R&D is focused on reducing cost of both the precursor and conversion processes, including the following (Miller and Stetson, 2019):

- Alternative CF precursors to lower fiber processing costs:
  - PAN-based fibers formulated with co-monomers and additives to enable lower cost melt spinning rather than conventional solution spinning and/or higher yield conversion of PAN-fiber to carbon;
  - Polyolefin-based fibers; and
  - Novel materials as precursor fibers.
- Fibers other than high-cost carbon, such as ultra-high strength fiber glass.
- Alternative resins with high strength and low weight—for example, vinyl ester and epoxy resin composites rather than high-cost epoxies. Alternative COPV manufacturing methods to reduce CF content, such as vacuum-assisted composite processing.

TABLE 6.3 Comparison of Tank Configurations Storing 5.6 kg Usable H_2 Showing Cost Impact of Single- versus Two-Tank Designs and Mixed versus Identical Aspect Ratio Two-Tank Configurations

Configuration	System Cost at 100,000 per Year	System Cost at 500,000 per Year
Single tank	$15.7/kWh	$14.2/kWh
Two-tank (identical aspect ratio: L/D = 2.8)	$20.0/kWh	$17.9/kWh
Two-tank (mixed aspect ratio: L/D = 2.8, 1.7)	$18.5/kWh	$16.4/kWh

SOURCE: Adams et al. (2019).

While the U.S. DRIVE 2025 target for gravimetric energy capacity for onboard hydrogen storage systems is within reach, achieving the volume and cost targets will be a challenge. The price for Toray T700S in 2019 ranged from \$26 to \$30/kg. Since the CF cost accounts for around 50% of total system cost, meeting the U.S. DRIVE system target of \$8/kWh requires CF at a cost of \$13–\$15/kg. Reducing cost while also reducing the weight and volume of the tanks presents a significant challenge. Nonetheless, other onboard storage options also face significant challenges and are earlier in their development. Compressed hydrogen tanks are likely to be the method of choice for FCEVs over the next 10–15 years.

Cold/cryo-compressed hydrogen. A potential alternative to 700 bar compressed hydrogen tanks is cold- or cryo-compressed hydrogen tanks, in which the hydrogen is stored at cryogenic temperatures—typically 70 K–200 K—and pressures of 100–700 bar. Interest in cryo-compressed hydrogen storage is driven by its potential for higher energy density, enabling a smaller tank size than 700 bar compressed hydrogen tanks; a lower cost than full liquefaction and a longer dormancy period than liquid hydrogen; and in some cases, a lower cost than 700 bar compressed hydrogen. For a 500 bar cold-compressed hydrogen system, one study estimated a 30% cost reduction and 38% mass reduction from a 700 bar system through material improvements, composite layup design and cold gas operation, even when the required onboard insulation for cold gas storage is included (Simmons, 2014). Onboard cold/cryo-compressed storage systems require vacuum insulation to reduce or eliminate hydrogen boil-off and achieve the dormancy target of 14 days.[14] R&D projects under way at a number of DOE national laboratories, some in collaboration with industry, are developing tank materials and BOP components, conducting burst tests, and modeling system designs to develop the technology further. Currently, there is more interest in these approaches for hydrogen storage at refueling stations and on board heavy-duty FCEVs than on board light-duty FCEVs.

Liquid hydrogen. Liquid hydrogen storage is a mature technology used for storing and delivering hydrogen in the industrial sector. The advantage of liquid hydrogen is its high energy density. While liquid hydrogen tanks do not require high pressure, at −253°C they do require double-walled tanks with multilayered vacuum insulation to minimize hydrogen boil-off, adding to the system weight, volume, and cost. Thus, there is more interest in liquid hydrogen for large bulk stationary storage rather than on board FCEVs.

Material-Based Hydrogen Storage

Longer-term hydrogen storage approaches are focused on developing chemical and solid-state materials with the potential to store hydrogen at near-ambient temperature, low-to-moderate pressures, and at energy densities greater than liquid or compressed hydrogen (Zacharia and Rather, 2015). Like physical-based hydrogen approaches, material-based methods—adsorbents, reversible metal hydrides, and chemical hydrogen storage materials—all have advantages and disadvantages. In adsorbents, the hydrogen molecule is weakly bound (physisorbed) to the surface of high-surface area, porous materials. The weak binding in adsorbents enables the hydrogen to release from and readsorb to the surface relatively easily compared to other materials-based storage systems; however, this weak binding interaction leads to lower storage capacities relative to metal hydrides and chemical hydrogen storage materials. Theoretically, the density of hydrogen physisorbed on the surface of materials can approach the density of liquid hydrogen at very low temperatures and relatively low pressures. Metal hydrides and chemical hydrogen storage materials, in which hydrogen is chemisorbed or chemically bound, have higher hydrogen binding energies, enabling hydrogen densities twice that of liquid hydrogen at ambient temperatures and low pressures. However, these stronger binding interactions lead to slower charge–discharge kinetics and poorer reversibility. In all cases, the hydrogen capacities of the materials must be sufficiently high to achieve the fully packaged and engineered system-level targets. There are currently no material-based hydrogen storage

[14] U.S. DRIVE dormancy time target is minimum 14 days for first release from initial 95% usable capacity. Boil-off target is 10% maximum reduction from initial 95% usable capacity after 30 days.

materials that meet all automotive requirements, and these approaches are not likely to be implemented in commercial FCEVs in 2025–2035.

Adsorbents. Adsorbent hydrogen storage materials include carbon-based materials (activated carbons, carbon nanotubes, nanofibers, and fullerenes), zeolites, MOFs, covalent organic frameworks, templated carbons, boron nitride materials, and porous polymers. Adsorbents have lower hydrogen capacities than other storage systems.

One of the approaches being explored to increase the hydrogen capacity of adsorbent materials is to incorporate a metal catalyst to lengthen the hydrogen bond and enable a stronger interaction between the hydrogen molecules and the metal catalyst. Referred to as the Kubas interaction (Boateng and Chen, 2020 and references therein), this approach has met with some success, in one case demonstrating reversible hydrogen adsorption of 10.5 wt% in a porous manganese hydride at 120 bar and ambient temperature (Morris et al., 2019).

Spillover techniques, in which a species adsorbed or formed on a surface migrates onto another surface, is another approach to increasing the hydrogen storage capacity of adsorbents. In this approach, hydrogen atoms migrate from a hydrogen-rich metal surface to a hydrogen-poor surface and, in some cases, into the bulk material. Hydrogen storage in carbon materials and MOFs has been significantly enhanced by spillover techniques. Functionalization of adsorbent materials, with heteroatom dopants such as boron, nitrogen, and phosphorus, is also being investigated to mitigate metal aggregation and enable more uniform metal dispersion.

These and other novel synthetic methods are improving the hydrogen storage properties of adsorption. For example, a vanadium MOF recently demonstrated a binding enthalpy in the range predicted to enable substantial ambient-temperature hydrogen adsorption (15–25 kilojoules per mole) (Stetson, 2020). Other methods are being developed to increase the surface area of adsorbents and control pore size. For example, zeolite-templated carbons are prepared by carbonizing an organic precursor in the nanospace of a zeolite (inorganic template), followed by dissolving the template to free the resulting carbon network (Masika and Mokaya, 2013). The goal is to make materials with narrow pore size distribution and high surface area and pore volume. MOFs are also being investigated as templates.

Reversible metal hydrides. In reversible metal hydrides, monatomic hydrogen is bound to other elements, usually metals or metalloids, within a solid. Metal hydrides are used in a variety of applications, including batteries and heat pumps, and although they have been well-studied,[15] none has the entire suite of properties needed to efficiently and affordably fuel an FCEV. Light metal hydrides, such as magnesium hydride (MgH_2) and aluminum hydride (also alane, or AlH_3), have high gravimetric capacities: 7.6 wt% and 10.1 wt%, respectively. However, these light metal hydrides require unsuitably high temperatures and/or high pressures to operate onboard a vehicle.

Two other types of metal hydrides have shown promise for hydrogen storage: (1) intermetallic metal hydrides, in which the hydrogen atoms occupy interstitial sites within an alloy, sometimes referred to as "solid solutions"; and (2) complex hydrides, in which the hydrogen is covalently bound to another atom to form a complex anion balanced by the presence of a cation. Like adsorbents, metal hydrides release hydrogen through reversible temperature-pressure equilibrium processes, enabling the dehydrogenated material to be rehydrogenated onboard the vehicle by applying pressurized hydrogen. Basic intermetallic metal hydride systems are categorized as AB5 (e.g., $LaNi_5$), AB (e.g., FeTi), A2B (e.g., Mg_2Ni), and AB2 (e.g., ZrV_2), all with varying degrees of chemical interaction with hydrogen. Substitutions in material composition can influence hydrogen absorption and desorption, and R&D efforts in this area are still under way.

Complex hydrides usually consist of alkali or alkaline earth elements ionically bonded to a complex anion. The anions typically contain hydrogen bound to a transition, main-group metal or metalloid (e.g., Fe, Ni, B, Al), or nitrogen. Examples are alanates—$NaAlH_4$, $LiAlH_4$, $MgAlH_4$; borohydrides—$NaBH_4$, $LiBH_4$; and alloyed combinations of them. Complex hydrides have very high hydrogen capacities; metal borohydrides, for example,

[15]A comprehensive database of published hydrogen alloys with properties relevant to hydrogen storage has been compiled by Sandia National Laboratories, and the resulting Hydride Information Center (Hydpark) has been incorporated into the U.S. DOE Hydrogen Storage Materials Database.

have gravimetric and volumetric energy densities that range from 14.9 wt% to 18.5 wt%, and from 2.72 kWh/L to 4.89 kWh/L, respectively (Rivard et al., 2019). Lithium borohydride ($LiBH_4$) has been widely studied; its current drawbacks for FCEV applications include high dehydrogenation temperatures, slow kinetics, and poor reversibility. One approach to addressing these shortcomings is lowering the high dehydrogenation enthalpy with additives that form new alloy or compound phases upon dehydrogenation, effectively destabilizing the component hydrides. To date, Li-N-H systems appear to have the most potential to meet DOE 2025 targets. Specifically, the reaction $Li_3N + 2\ H_2 \rightleftharpoons Li_2NH + LiH + H_2 \rightleftharpoons LiNH_2 + 2LiH$ has a total hydrogen capacity of about 10.5 wt%, and researchers have demonstrated 6.1 wt% reversibility at 250°C and 100 bar (Allendorf and Gennett, 2020). Other systems being investigated include TiN/MgH_2 and $TiO_x/Mg(BH_4)_2$, which have potential for light-activated hydrogen desorption at ambient temperature (Stetson, 2020).

To be suitable for FCEV applications, significant improvements are needed in metal hydride systems. R&D efforts are focused on increasing the reversible hydrogen storage capacity by modifying the composition of known materials or designing new alloys and improving the kinetics of hydrogen absorption and desorption.

Chemical hydrogen storage. Chemical hydrogen storage materials having potential for FCEV applications are typically solid or liquid molecules in which hydrogen is covalently bound to another element and released through a chemical reaction. They usually have the highest hydrogen storage capacities. However, because hydrogen is more strongly bound in chemical storage materials, it is released through non-equilibrium processes, which are more difficult to carry out onboard the vehicle. Examples of producing hydrogen via hydrolysis include lithium hydride (LiH), lithium aluminum hydride ($LiAlH_4$), and sodium borohydride ($NaBH_4$), which exothermically generate hydrogen gas when reacted with water. Chemical hydrogen storage systems, therefore, are generally not reversible onboard the vehicle and would be "recharged" offboard. Because of the exothermic nature of these reactions, systems must be designed to manage the heat that is generated. In addition, these chemical hydrides are costly and must be stored under an inert gas or liquid to protect them from water.

Endothermic chemical hydrogen materials release hydrogen when heated. Examples include the following:

- Decalin-to-naphthalene reaction: $C_{10}H_{18} \rightarrow C_{10}H_8 + 5H_2$, which can generate 7.3 wt% hydrogen at 210°C.
- Ammonia borane decomposition: $NH_3BH_3 \rightarrow NH_2BH_2 + H_2 \rightarrow NHBH + H_2$, which generates 6.1 wt% hydrogen at 120°C.

These hydrides eliminate the need for water and other equipment to manage heat. Research is currently directed at lowering dehydrogenation temperatures and improving reaction kinetics.

Another type of chemical hydrogen storage is liquid organic carriers—for example, N-ethylcarbazole and methyl-cyclopentane—which would enable a liquid refueling infrastructure. However, one of the drawbacks of current liquid hydrogen carriers is the tendency for the dehydrogenated product to solidify, which can make handling of the spent materials more difficult during removal from the vehicle and recharging. New liquid carrier materials currently being investigated include solutions of furans and pyrroles containing magnesium borane, and a system based on ammonium formate and captured CO_2. More details on novel liquid fuels can be found in Chapter 10.

In addition to the complex logistics of charging and recharging, chemical storage systems tend to be more costly than others, a challenge that will have to be addressed before developing a supporting infrastructure. The cost of building and operating regeneration plants to convert the spent material back to its fully loaded hydride form must also be considered.

Chemical and solid-state materials have the potential to meet vehicular hydrogen storage system requirements in the long term. The significant technical and economic challenges that must be overcome for their practical application in FCEVs make it unlikely they will be used in commercial LDVs before 2035.

In the United States, some hydrogen storage R&D is currently supported through DOE's HyMARC initiative—Hydrogen Materials Advanced Research Consortium, which shares its data through a public data hub (HyMARC Data Hub, 2020). DOE's 2020 Annual Merit Review site documents the latest developments through HyMARC. Progress in the development of hydrogen storage systems has been stymied by large fluctuations in applied research funding over the past 15 years, from a peak in 2007 to significant reductions in 2014 and beyond (Peterson and Farmer, 2017). These fluctuations have resulted in a loss of momentum and delays in advancement of vehicular hydrogen storage technologies.

6.5 HYDROGEN REFUELING INFRASTRUCTURE FOR FCEVS

A number of obstacles have limited the development of hydrogen fueling infrastructure for FCEVs. The cost of producing and delivering hydrogen to refueling stations is currently high, primarily owing to low volume demand (CaFCP, 2015; Connelly et al., 2019). Natural gas reforming is the source for most of the hydrogen produced today because it is less costly than producing hydrogen from renewable energy; this discourages some policy makers, who are increasingly focusing on decarbonization priorities, from supporting investments in hydrogen infrastructure. Although projections indicate that infrastructure costs per mile for FCEVs and BEVs could be comparable in the 2025 time frame (Melaina et al., 2014), the cost of building a hydrogen fueling station today is much higher than the cost of installing a BEV charging station, making investments in the latter seem more practical. However, *hydrogen refueling of light- and medium-duty FCEVs takes minutes, while even fast-charging of BEVs takes* hours. Therefore, an order of magnitude more FCEVs can be refueled at a hydrogen station. Given that the driving range of FCEVs is longer than that of BEVs, the total miles possible per day per fueling station for FCEVs allows the higher cost of the hydrogen station to be spread across a much larger number of vehicle miles driven.

Despite their complementary nature as ZEVs, FCEVs and BEVs are often cited as competing technologies and BEV/infrastructure investments and developments have outpaced those for FCEVs over the past 10 years. As a result, there is limited political awareness of and support for FCEVs and hydrogen infrastructure relative to BEVs and charging infrastructure, less public familiarity, a lower level of advocacy, and in some cases, disproportionate policy support (Trencher, 2020). California has overcome these obstacles to some extent and provides an important case study and resource for identifying and overcoming the challenges faced in building an early hydrogen infrastructure (Trencher, 2020).

6.5.1 Hydrogen Infrastructure Cost

A 2019 analysis by McKinsey and Company for Europe's Fuel Cells and Hydrogen Joint Undertaking puts the current infrastructure costs for FCEVs and BEVs at €4,000 ($4,524)[16] and €2,000 ($2,262) per vehicle, respectively (FCH JU, 2019). The hydrogen refueling infrastructure cost includes the cost of hydrogen production and distribution, and the hydrogen station. The analysis notes that, because FCEVs refuel in 5 minutes or less, one hydrogen station can serve 10 to 15 times more FCEVs than one BEV fast-charger, and projects that the hydrogen infrastructure will become less costly on a per vehicle basis compared to the charging infrastructure as the size of the FCEV fleet increases. Figure 6.24 describes projections for a phased deployment in which FCEV infrastructure cost decreases to an estimated €3,500 ($3,958) per vehicle after the initial hydrogen station network is built and station utilization increases, with additional economies of scale decreasing cost to €2,500 ($2,827) per vehicle or below. The analysis projects that grid upgrades for BEV charging, particularly an expanding fast-charging infrastructure, increase the BEV infrastructure cost to an estimated €2,500 per vehicle.

6.5.2 Hydrogen Infrastructure Development

In places where hydrogen refueling is available—primarily Japan, Germany, South Korea, and California—the markets for FCEVs are growing (Isenstadt and Lutsey, 2017). A significant driver for the hydrogen infrastructure in those regions is the availability of state or federal government subsidies for construction of hydrogen stations (Scheffler, 2019). Japan has 109 stations in operation and another 51 planned by the end of 2020; Japan's roadmap targets 320 stations in 2025 and 900 in 2030. Germany has 84 hydrogen refueling stations in operation (FuelCellsWorks, 2020). The goal of Germany's public-private partnership is 400 stations by the end of 2023 and

[16]Based on 1 Euro = 1.1310 U.S. dollar, Business Insider Currency Converter, July 6, 2020.

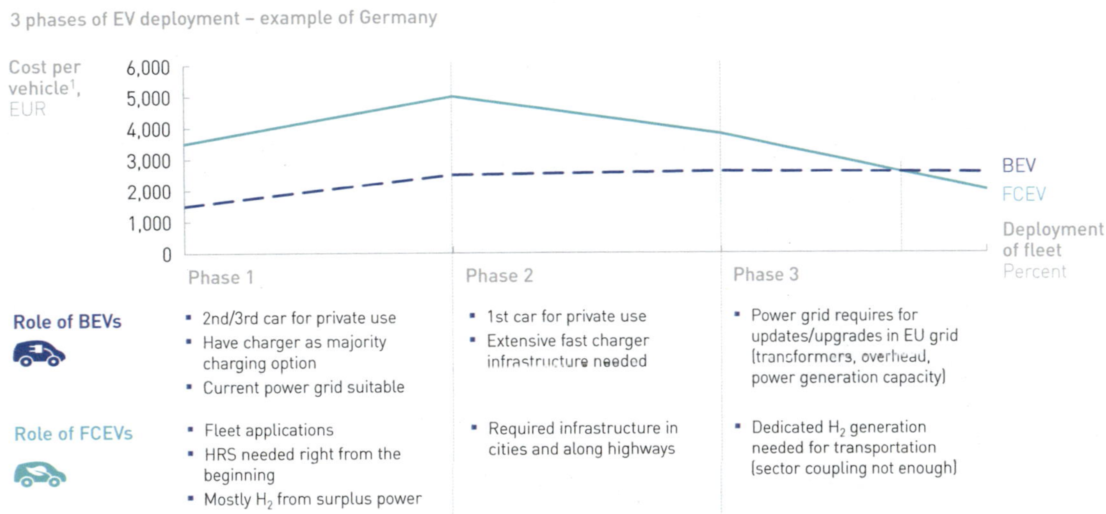

FIGURE 6.24 Estimated infrastructure costs per vehicle for FCEVs and BEVs in Germany. The hydrogen infrastructure phases are based on simulations done for Germany (Robinius et al., 2018). Phase 3 was estimated to start at ~13% electric vehicle penetration, and the break-even point (at ~$2,500 per vehicle) assumed to be at ~17 million ZEVs, or ~38% of the LDV fleet. NOTE: Cost per vehicle (1) includes refueling infrastructure and fuel generation and distribution infrastructure. SOURCE: FCH JU (2019).

900 by 2030. Throughout the rest of Europe, 31 stations are operational with an additional 21 scheduled to come online soon. South Korea had 14 stations in operation in 2018 with plans to open a total of 100 by 2022. China has 15 operating stations with another 33 in planning phase (Scheffler, 2019).

Using H2 Tools as a source (PNNL, n.d.), Greene et al. (2020) identified seven countries that account for more than 80% of hydrogen stations worldwide (Figure 6.25), including stations planned to be opened by the end of 2020. Of the stations indicated in Figure 6.25, 80% are open to the public while others are used by fuel cell bus companies or for other purposes (Greene et al., 2020).

In the United States, California is taking the lead. Through its Assembly Bill (AB) 8 program, the State committed $20 million per year over 10 years (2013–2023) to support the construction of 100 hydrogen stations, and to help support the stations' operations and maintenance (O&M) during the early stages of the infrastructure build-out (AB 8, 2013). The stations are projected to service up to 30,000 FCEVs sold to consumers by Toyota, Honda, and Hyundai.

California has more than 40 retail hydrogen stations in operation today, providing more than 11,800 kilograms per day (kg/day) and supporting more than 6,000 registered FCEVs (Baronas and Achtelik, 2019; Reed et al., 2020). With more than 20 additional stations under construction, this network is projected to provide 24,500 kg/day by the end of 2020. Average station capacity utilization (ratio of dispensed hydrogen to station capacity) during 2019 was around 34% (Baronas and Achtelik, 2019). According to industry experts, utilizations of 70%–80% are needed for profitability. The California Air Resources Board (CARB) and the California Energy Commission are developing a methodology to determine the cost and timeline to enable the California hydrogen fueling station network to be financially self-sufficient. The analysis will examine the cash flow and financial performance of the stations, including an assessment of the station installation and O&M costs, capacities and utilizations, and so on needed for profitability (CARB, 2019).

California's Low Carbon Fuel Standard Hydrogen Refueling Infrastructure credit program, launched in 2019, has encouraged hydrogen station operators to increase the renewable hydrogen content of their fuel and earn more credits. CARB reported that 39% of the hydrogen dispensed by the station network will come from renewable

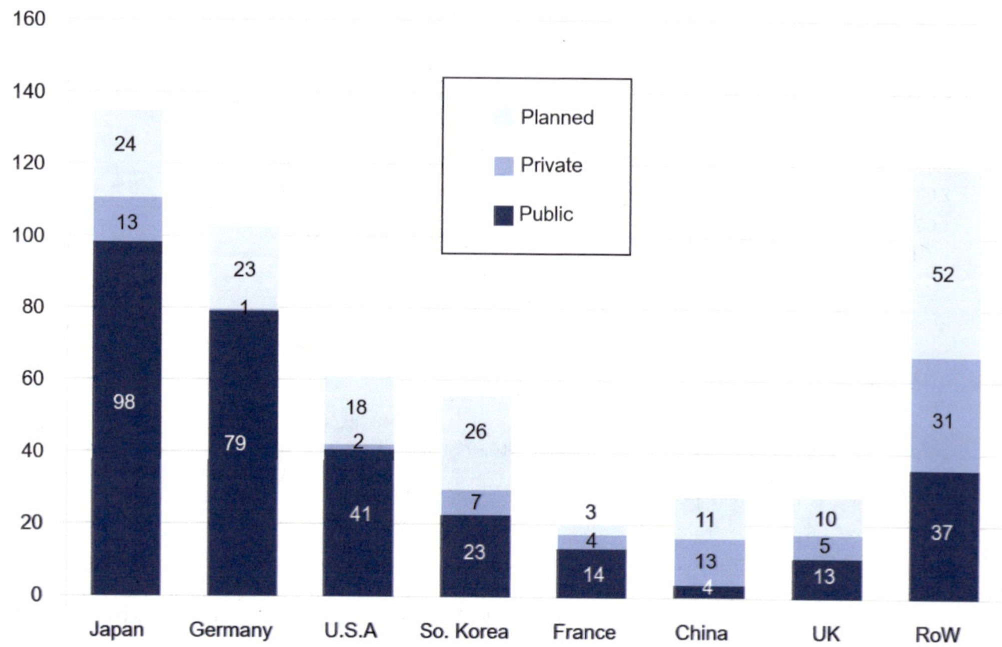

FIGURE 6.25 World hydrogen stations.
SOURCE: Originally published in *eTransportation*, 6, Greene et al., Challenges in the Designing, Planning and Deployment of Hydrogen Refueling Infrastructure for Fuel Cell Electric Vehicles, © Elsevier (2020).

sources and that some station operators have identified new hydrogen feedstock sources that will provide 100% renewable hydrogen (Baronas and Achtelik, 2019). California's Executive Order B-48-18 targets 200 hydrogen stations by 2025 (State of California, 2018). CARB coordinates with stakeholders through the public-private California Fuel Cell Partnership (CaFCP), which envisions 1 million FCEVs and 1,000 hydrogen fueling stations in California by 2030 (CaFCP, 2018).

Using its Hydrogen Financial Analysis Scenario Tool, H2FAST, NREL conducted financial assessments of California's hydrogen stations (Baronas and Achtelik, 2019, Appendix E; NREL, n.d.). It reported an average cost of $2.4 million for 180 kg/day stations supplied by delivered gaseous hydrogen, and $2.8 million for 350 kg/day stations supplied by delivered liquid hydrogen. These installed costs are all-inclusive—including equipment, design, permitting, engineering, construction, project management, and overhead. For comparison, a recent International Council on Clean Transportation (ICCT) study projects that single-dispenser H_2 stations having capacities of 1,100–1,700 kg/day for long-haul tractor trailers, drayage trucks, and delivery trucks will cost $2.0–$2.9 million in 2030, or less than $2,000 per kg of H_2 delivered (Hall and Lutsey, 2019).

For a $2 million, 180 kg/day California LDV station, costs are typically around $1.6 million for equipment and materials, and around $400,000 for permitting, site engineering, construction, commissioning, and general overhead (Baronas and Achtelik, 2019). To support O&M costs, CARB provided grants averaging $100,000 per year for up to 3 years for each station. A survey of the California hydrogen station owners indicated operating costs as high as $200,000 per year.

While installed costs are higher for larger, higher capacity stations than for smaller stations, the cost per kg of hydrogen delivered is typically lower for larger stations. Figure 6.26 shows installed station costs per kg of hydrogen delivered as a function of the station's daily capacity for California's stations (CARB, 2019). Costs ranged from about $5,000 per kg for a 500 kg/day station to roughly $25,000 per kg for a 180 kg/day station.

An analysis of station capital costs for various capacities over time is shown in Figure 6.27. The projected station costs are based on NREL's Hydrogen Station Capital Cost Calculator, which estimates cost reductions based on both economies of scale and the experience gained as more stations are built (Melaina and Penev, 2013).

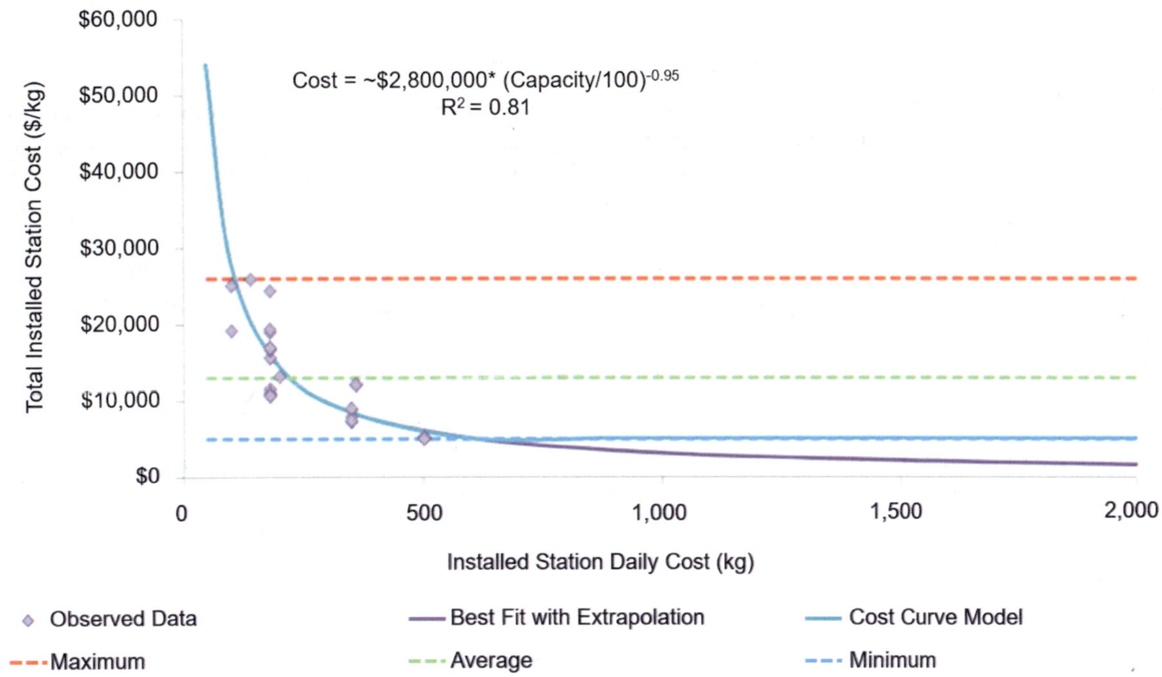

FIGURE 6.26 Actual and modeled station installation cost per kilogram of installed capacity for hydrogen stations in California's AB 8 program (based on 12-hour capacity).
SOURCE: CARB (2019).

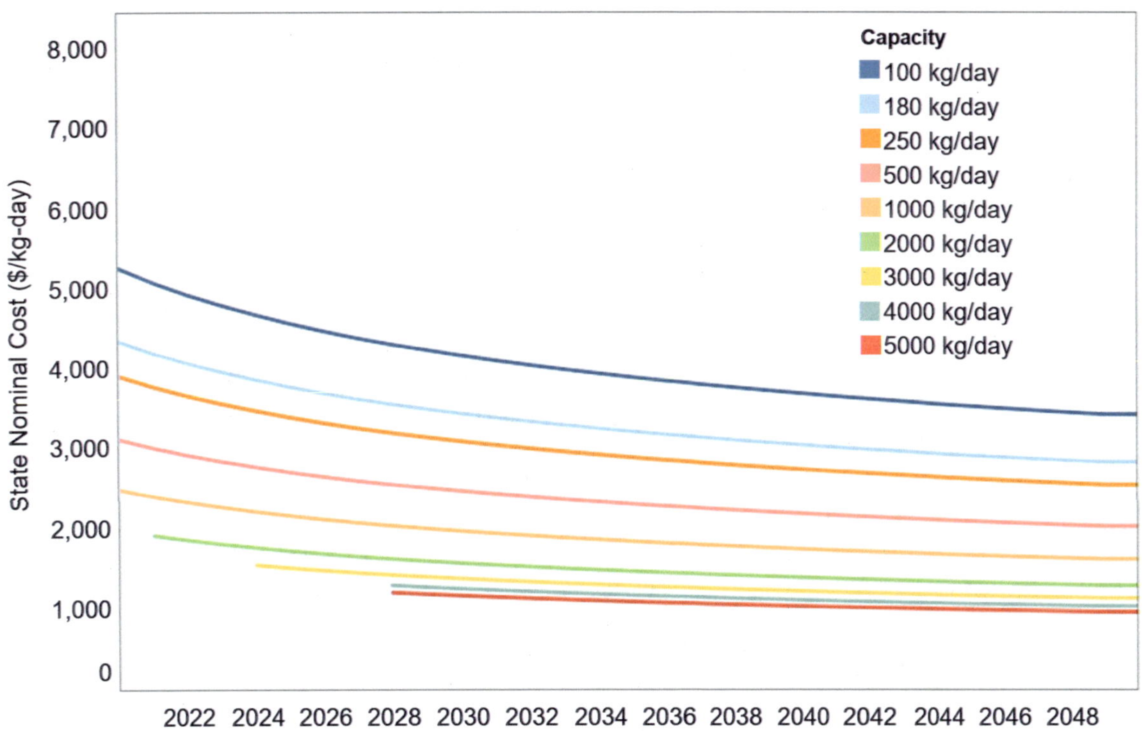

FIGURE 6.27 Hydrogen refueling station cost as a function of capacity and time.
SOURCE: Bush et al. (2019).

The number, size, and locations of the stations were estimated using NREL's Scenario Evaluation and Regionalization Analysis (SERA) Model and are based on meeting hydrogen demand from increasing deployment of FCEVs and providing the required hydrogen supply—production and delivery—to meet that demand (Bush et al., 2019).

Automakers and hydrogen suppliers have stated their intent to make the Northeast states the next market for FCEVs and refueling stations. Driven by the Multi-State Zero-Emission Vehicles (ZEVs) Programs Memorandum of Understanding, a collaboration of nine northeastern states and California, the Multi-State Zero-Emission Vehicle Action Plan was drafted to work toward the collective deployment of 3.3 million ZEVs, including FCEVs, by 2025, and the establishment of sufficient fueling infrastructure to enable this scale. Toyota and Air Liquide are collaborating to bring stations and vehicles to the Northeast; 12 stations are planned (Nied, 2015; Air Liquide, 2016a, 2016b). However, as mentioned previously, the prohibition of hydrogen-powered vehicles in tunnels and on the lower deck of two-tier bridges in the Northeast has delayed the introduction of FCEVs in that region (Port Authority of NY and NJ, 2016; State of New Jersey, 2019; Massachusetts Department of Transportation, 2019). (See above Section 6.3.1.2 for additional information on this topic.)

Table 6.4 summarizes three scenarios for FCEV adoption and infrastructure buildout in 2050 based on coordinated rollout of vehicles and stations that enables continued FCEV market growth in California and subsequent expansion into other regions (Bush et al., 2019; Melaina et al., 2017). Assumptions include various levels of consumer demand, policy drivers, and local and regional planning and coordination efforts. Recent review articles provide summaries of worldwide efforts to develop hydrogen refueling infrastructure for FCEVs, assessments of the challenges faced, and strategies to overcome the challenges, including the need for stronger and more consistent policy support (Greene et al., 2020; Trencher, 2020).

The buildout of hydrogen infrastructure is expected to benefit from development of fuel cells for MHDV applications and from fleet vehicles that need constant operation, quick refueling, and/or high daily VMT. The cost of hydrogen for all transport applications is expected to decrease in the next decade as existing hydrogen technologies are scaled up and hydrogen equipment and supply chain costs are reduced (Ogden, 2018). Current research is focused on reducing the costs of producing hydrogen from low-carbon sources and delivering hydrogen to the station and the vehicle.

TABLE 6.4 FCEV Adoption and Hydrogen Refueling Infrastructure Buildout Scenarios for 2050

Market Trend	Urban Markets	State Success	National Expansion
Dominant policy drivers	Support at local and municipal levels combines with strong early adopter demand.	ZEV mandate and other support policies.	Combination of strong local state and national policies.
Coordination and planning	Investments focused on most promising metropolitan markets.	Strong coordination across ZEV mandate states.	Strong coordination and planning across all regions.
Consumer adoption	High concentrations of early adopters guide market development.	FCEV adoption primarily driven by ZEV mandate.	Adoption moves quickly from concentrated early adopters and ZEV mandate states to broad megaregion markets.
HRS network expansion	Gradual expansion from promising urban markets to nearby cities.	Focus on ZEV mandate states, with gradual expansion into additional markets.	Strong policy drivers and coordination reduce investment risks, allowing rapid network expansion.
FCEV Sales per Year (millions) and Market Share (%) of Total Sales in Urban Areas in 2050[a]			
United States	**3.1 million (23%)**	**5.0 million (35%)**	**8.9 million (59%)**
California	1.0 million (49%)	1.3 million (64%)	1.7 million (84%)
Other ZEV states	0.9 million (26%)	1.9 million (56%)	1.9 million (57%)
Rest of country	1.2 million (10%)	1.9 million (14%)	5.3 million (41%)

[a]Total sales are based on EIA (2017). Future vehicle sales projections are taken from EIA Annual Energy Outlook. Within each census division, sales are allocated to different urban areas based on the proportion of current vehicle stock, from IHS automotive data. For example, if Arvin, California, has 1% of the current vehicles in the Pacific census division, it is assumed that 1% of the new vehicle sales in that division will occur in Arvin, California.

SOURCE: Bush et al. (2019).

6.5.3 Hydrogen Delivery

For FCEVs to be competitive with gasoline vehicles on a cost-per-mile basis in the LDV market, U.S. DRIVE has set a target of < $4/kg, untaxed and dispensed at the pump. CaFCP has reported that the average retail price of hydrogen at California stations from Q4 2018 through Q3 2019 was $16.51/kg (Baronas and Achtelik, 2019). Cost reductions are needed for low-carbon hydrogen production pathways and for delivering hydrogen, including the costs of compression, storage, and dispensing. For more information on hydrogen supply and future costs of H_2 production via PEM electrolysis see Chapter 10 (Section 10.2).

Hydrogen is commonly transported as a liquid by cryogenic tank truck or as a compressed gas by tube trailer (typically 180 bar) or by pipeline. Two additional approaches are being explored for the longer term: (1) transport in solid or liquid carrier form, using a material that chemically binds or physisorbs hydrogen; and (2) transport as a cryogenic gas at temperatures of around 80 K.

Roughly 75% of California's retail hydrogen stations are supplied by compressed H_2 delivery; the remainder use liquid H_2 or pipeline delivery, or generate hydrogen onsite using steam-methane reforming or electrolysis (Saur et al., 2019). Current delivery and dispensing costs (excluding production) for tube-trailer gaseous stations are projected to be $9.50/kg and $8/kg at 450 kg/day and 1,000 kg/day stations, respectively (2016$). For liquid tanker-based stations, delivery costs are estimated to be $11/kg at 450 kg/day and $8/kg at 1,000 kg/day stations.[17] R&D efforts currently under way could enable a reduction in hydrogen delivery and dispensing costs to $5/kg in 2025 at stations supplied by liquid hydrogen tanker trucks. To achieve a hydrogen cost of $4/kg at the pump, delivery costs of around $2/kg will be needed.

For hydrogen stations of 1,000 kg/day capacity, which are anticipated by 2025, liquid tanker delivery has been identified as the most viable approach (Martinez and Achtelik, 2017). For higher demand scenarios—hydrogen stations at 3,000 kg/day capacity—technoeconomic analyses suggest that pipelines are a relatively low-cost option for hydrogen delivery (Rustagi et al., 2018). More than 1,600 miles of hydrogen pipelines are in operation in the United States today, typically at 70 bar maximum pressure (U.S. DRIVE, 2017d). Higher pressure operation may be needed for economic distribution of hydrogen for refueling stations. Improvements in pipeline materials are expected to enable an operating pressure of 100 bar in the United States (Fekete et al., 2015). Higher pressure pipelines could also reduce the space and cost required by compression and storage equipment at the refueling station. A pipeline with a design pressure of 1,000 bar has been operating in a Germany industrial park since 2006; it currently delivers hydrogen directly to 350 and 700 bar dispensers at a hydrogen vehicle refueling station (Penev et al., 2019). Currently, only one hydrogen station in the United States is supplied via pipeline. In operation since 2011, the 55 bar pipeline delivers hydrogen to a station in Torrance, California (Air Products, 2016).

Improving the durability and reliability of station equipment is key to reducing hydrogen delivery costs (Rustagi, 2018). An assessment of equipment maintenance events from 2016 through 2018, 67% of which were unscheduled, indicated that hydrogen dispensers account for more than half (57%) of the required maintenance, followed by compressors (25%) and chillers (12%) (Saur et al., 2019). The assessment also showed that maintenance costs per kilogram decreased significantly during that time frame as more hydrogen was dispensed and as stations matured. Figure 6.28 shows that, in addition to increasing station capacity, improvements in compression, liquefaction, and dispenser technologies, and pipeline and storage materials, are needed to make hydrogen fuel cost-competitive in the marketplace (FCHEA, 2020).

Several OEMs and hydrogen providers have noted that deployment of fuel cells in heavy-duty transportation applications—drayage and long-haul trucks, buses, marine vessels, and large mining vehicles—will exponentially increase demand and drive down the cost of hydrogen production and distribution. More information on hydrogen production and its associated emissions can be found in Chapter 10. Figure 6.29 shows potential hydrogen cost reduction pathways for both LDV and heavy-duty vehicle (HDV) scenarios. DOE's Hydrogen at Scale program, summarized in Box 6.1, aims to reduce hydrogen cost by pursuing opportunities based on cross-sector synergies.

[17] Hydrogen Delivery Scenario Analysis Model (HDSAM) based on operational data from California stations; documented in Koleva and Rustagi (2020).

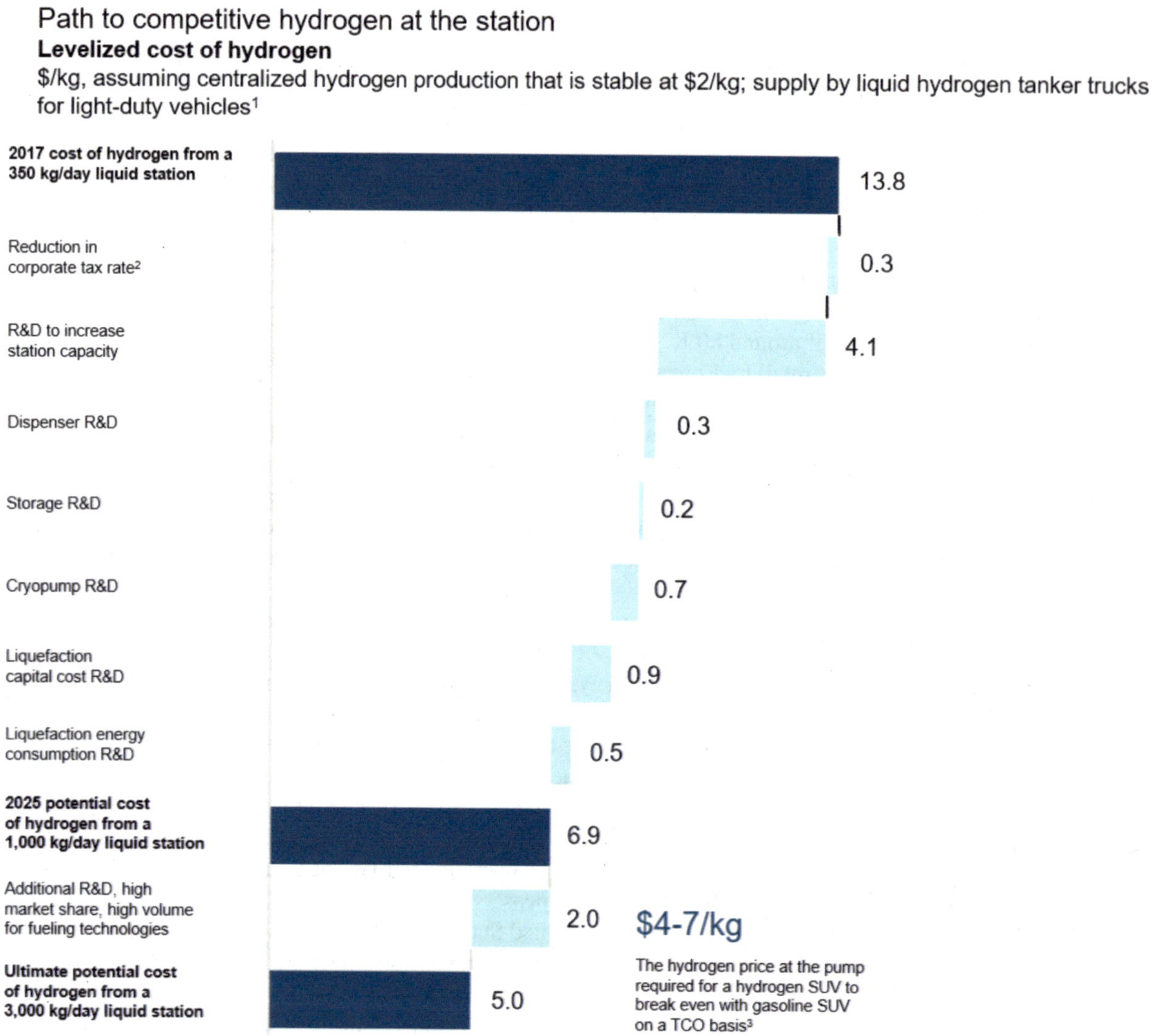

FIGURE 6.28 Pathway to lower cost hydrogen. NOTE: (1) Assumes a 7% discount rate representing the "marginal pretax rate or return on an average investment in the private sector in recent years." (2) Corporate rate assumed to decrease to 21% owing to the 2017 Tax Cuts and Jobs Act. (3) Assuming FCEV capex cost reduction owing to fuel cell production at scale, gasoline cost of $3.36/gallon from EIA 2030 outlook, a lifetime of 200,000 miles, ranges based on efficiency for SUV gasoline of 29 MPG (efficiency in 2019) and 39 MPG (efficiency in 2030) from EIA AEO2019 fuel efficiency outlook.
SOURCE: FCHEA (2020).

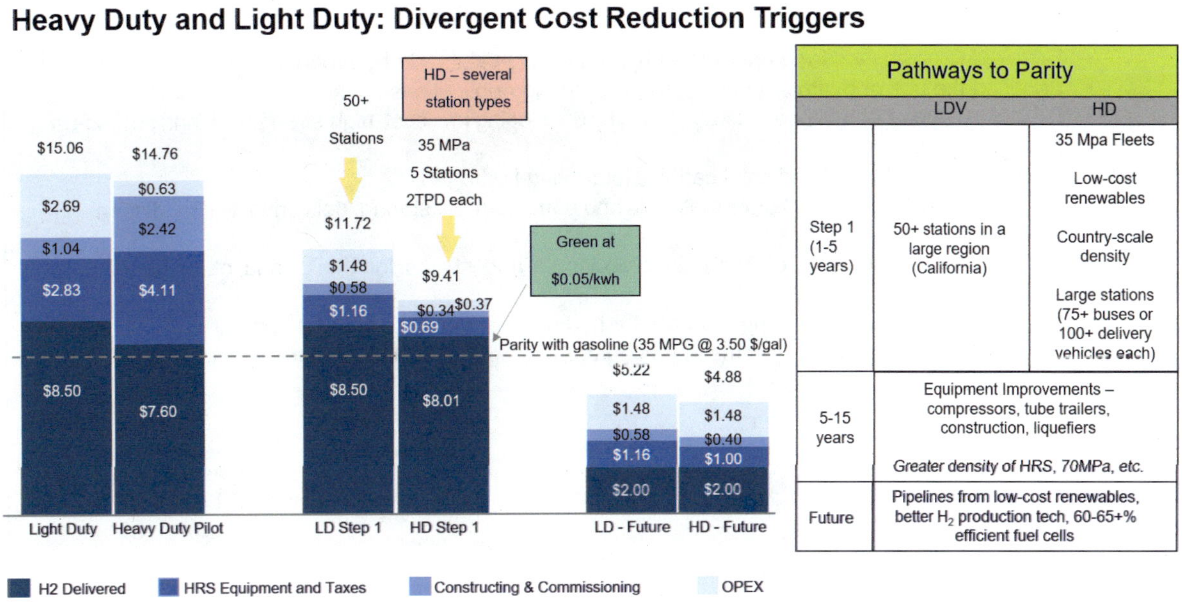

FIGURE 6.29 Potential hydrogen cost reduction pathways for transportation fuel cells in both LDV and HDV fuel cell vehicles. SOURCE: Munster (2019).

BOX 6.1
Hydrogen at Scale (H2@Scale)

Hydrogen demand for fuel cells will continue to increase as markets continue to grow in on-road vehicles, material handling equipment (forklifts) used in warehouses (Satyapal, 2019), stationary and backup systems used to provide clean and resilient power, and trains (Ruf et al., 2019) and marine vessels (ABB, 2020). At U.S. refineries, use of hydrogen has increased as demand for distillate fuel oil has increased and sulfur content regulations have become more stringent (EIA, 2019). As U.S. industries look to a lower carbon future, market opportunities will grow for renewable hydrogen in ammonia/fertilizer production, metal manufacturing and refining, and carbon-neutral synthetic liquid fuels. The hydrogen for those industrial processes can come from renewable electrolysis, while also serving as an energy storage medium for intermittent renewable energy sources such as solar or wind. By producing hydrogen when generation exceeds the load on the grid, curtailment of renewables can be reduced and the hydrogen can be stored, distributed, and/or used as a fuel.

DOE's Hydrogen at Scale concept seeks to align these multiple industries to enable affordable large-scale hydrogen generation, distribution, and utilization across all sectors. Facilitating the development of hydrogen uses in all sectors could reduce the cost of hydrogen production and accelerate build-out of a hydrogen distribution network, which would benefit the economics of hydrogen refueling stations and commercialization of FCEVs. The Hydrogen Council estimates that emerging applications for hydrogen can enable 10-fold growth in global demand (Hydrogen Council, 2017).

DOE's Hydrogen at Scale initiative is focused on the following R&D activities:

- Lowering the cost of hydrogen produced from electrolysis by developing improved electrolyzer stack materials/components and high-volume manufacturing processes.
- Developing affordable hydrogen production from biomass resources.

(continued)

- Developing processes for co-producing hydrogen and value-add by-products.
- Enabling the use of hydrogen in steel manufacturing applications.
- Reducing the cost of hydrogen storage tanks by developing low-cost, high-strength CF and scaling-up manufacturing.
- Developing novel liquid and solid carriers for storing hydrogen.
- Developing hydrogen-compatible materials and equipment for storing, delivering, and/or dispensing hydrogen.
- Advancing manufacturing of fuel cell components and stacks for trucks and other heavy-duty applications.
- Demonstrating new market opportunities for hydrogen in maritime and data center applications.
- Demonstrating pilot-scale systems that integrate hydrogen production, storage, and fueling.

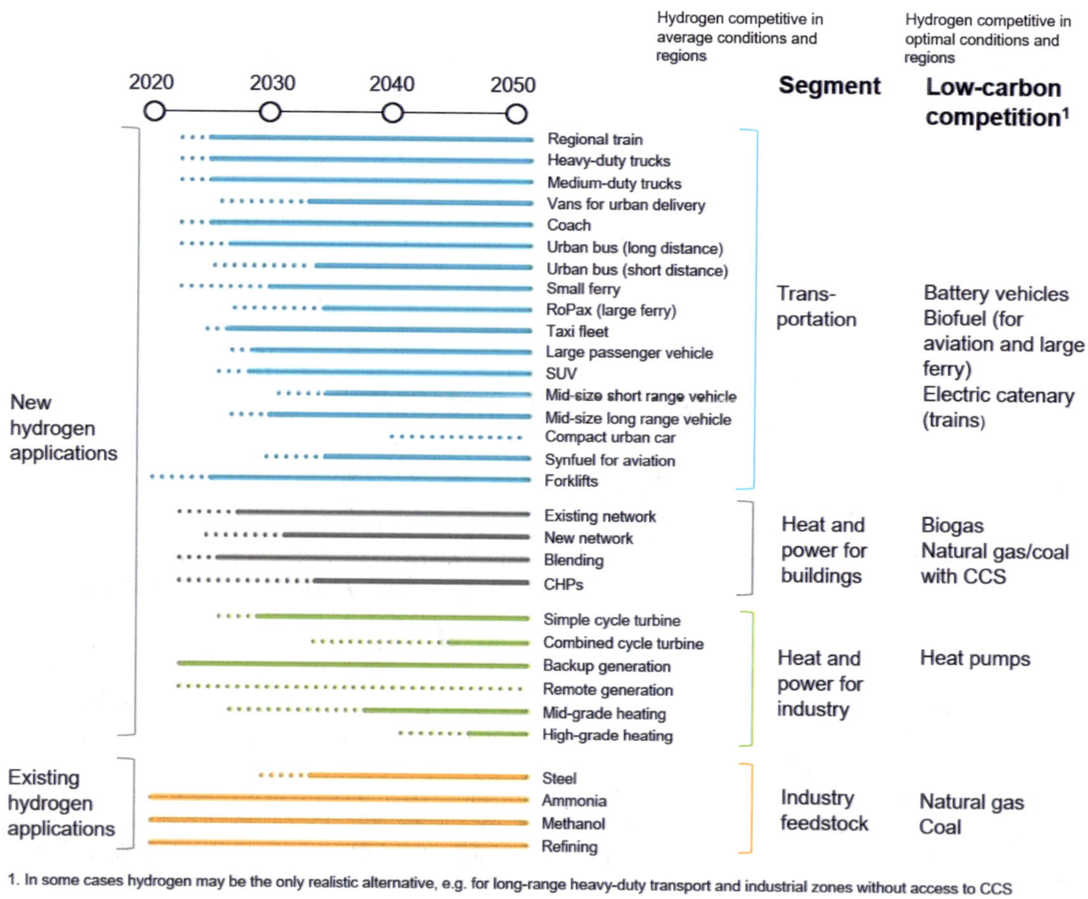

FIGURE 6.1.1 Applications in which hydrogen becomes a cost-competitive low-carbon solution and projected time frames. Dashed lines indicate cost competitiveness in some regions; solid lines in all regions analyzed. Timing depends on energy prices, infrastructure availability, and policies. Projected hydrogen demand is based on IEA Energy Technology Perspectives 2017, which projects sector energy demand under the two-degree scenario.
SOURCE: Hydrogen Council (2020); DOE (2020).

6.6 SUMMARY OF FUEL CELL VEHICLE COSTS

Continued technology advancements will lead to improved performance and reduced costs for FCEVs and hydrogen refueling stations in 2025–2035. Cost and effectiveness estimates for different components of the fuel cell system are outlined in Table 6.5.[18] Economies of scale will also bring down costs, and synergies with the renewable energy sector and industrial uses of hydrogen could also be an important factor. A recent analysis suggests that FCEVs could reach TCO parity with ICE SUVs in 2025–2030 at a hydrogen price of $4 to $7/kg at the pump (FCHEA, 2020). At current hydrogen prices, $4/kg by 2030 seems optimistic, especially for renewable hydrogen, which is likely to be required in a carbon-constrained world. A recent study from the ICCT projects a median price of $7.37/kg and a minimum price of $4.95/kg in 2030 for hydrogen produced by a grid-connected electrolyzer operating on renewable electricity only—and these prices exclude the cost of delivering the hydrogen to the station. However, as shown in Figure 6.28, pathways to $7/kg hydrogen in that time frame have been identified. On the vehicle side, a recent FCEV market study projects that more than 25 FCEV models could be available globally by 2030, accounting for 1%–1.5% of global passenger vehicle sales (or about 1.17–1.75 million cars; Wagner, 2020) (Research and Markets, 2020). The Fuel Cell Hydrogen and Energy Association projects sales of 1.2 million FCEVs in the United States by the end of 2030 and 4,300 hydrogen stations (FCHEA, 2020). However, for more widespread deployment in the United States, market expansion to urban areas outside California is needed, supported by policies that incentivize FCEV purchases and assist industry in building a hydrogen refueling network until stations become profitable. Three automakers have reiterated plans to continue development of fuel cell technology for the LDV market in the near term. While the shift by some automakers to a short-term focus on MHDV applications for fuel cells introduces uncertainties regarding widespread LDV deployment, the increased focus on those applications will enable continued fuel cell cost reductions, durability improvements, and hydrogen infrastructure build-out. While it is unlikely that FCEVs will have a significant impact on 2025–2030 CAFE standards, it is possible they will be a factor in 2035 and beyond.

Numbers for FCEV Cost and Effectiveness, shown in Table 6.5, were generated using ANL's Autonomie model,[19] a vehicle simulation tool used to evaluate a wide range of vehicle applications, powertrain configurations, and component technologies for different time frames. The tool estimates costs and projects potential future petroleum displacement. Table 6.5 reports simulation results for the "high technology progress" case, based on meeting DOE R&D targets established through U.S. DRIVE.

The simulations were performed for the following:

- Five powertrain configurations: ICEVs, hybrid electric vehicles, PHEVs, BEVs, and FCEVs;
- Five vehicle classes: compact car, midsize car, small SUV, midsize SUV, and pickup truck;
- Two performance categories: base (non-performance) and premium (performance);
- Different fuels: gasoline, diesel, hydrogen, and battery electricity; and
- Five different time frames: laboratory years 2015 (reference case), 2020, 2025, 2030, and 2045. The study assumes a 5-year delay between laboratory year and model year (the year the vehicle is first produced).

The results in Table 6.5 are from ANL's fifth revision of its Benefits and Scenario Analysis (Islam et al., 2020). The study used technical targets and other input from subject matter experts (technology development managers) in the DOE Vehicle Technologies Office and Hydrogen and Fuel Cell Technologies Office, and evaluated the impact of technology improvements on vehicle component sizes (i.e., power, energy, weight), fuel and electricity consumption, and manufacturing cost.

[18] These cost projections are based on DFMA analysis of modeled state-of-the-art fuel cell systems and assume that DOE cost and durability targets are met. They do not include the durability-adjusted cost estimates discussed in Section 6.4.

[19] Autonomie is a state-of-the-art vehicle system simulation tool used to assess the energy consumption, performance, and cost of multiple advanced vehicle technologies. Autonomie is packaged with a complete set of vehicle models for a wide range of vehicle classes, powertrain configurations, and component technologies, including vehicle level and component level controls. These controls were developed and calibrated using dynamometer test data.

TABLE 6.5 FCEV Cost and Effectiveness 2025 and 2035 Model Years (ANL's Autonomie)

FCEV Package Technologies	Technology Details by Vehicle Class (2025 MY)	Technology Cost by Class	Technology Effectiveness by Class (MPG combined cycle, effficiency, and kg usable H_2)	Technology Details by Vehicle Class (2035 MY)	Technology Cost by Class	Technology Effectiveness by Class (MPG combined cycle)
Total Package	Small, 436/425 mi range		74.7 MPG	Small, 441/433 mi range		92.9 MPG
Fuel Cell System	Midsize, 427/430		69.1 MPG	Midsize, 434/437		89.5 MPG
Battery	Small SUV, 434/418		59.4 MPG	Small SUV, 439/425		74.7 MPG
Motor	Midsize SUV, 439/411		55.2 MPG	Midsize SUV, 444/418		69.6 MPG
H_2 Storage System	Pickup, 441/418		44.2 MPG	Pickup, 447/422		56.3 MPG
Fuel Cell System	Small (67 kW max)	$3,383	63% efficiency	Small (60 kW max)	$2,246	68% efficiency
Fuel Cell Stack	Midsize (81 kW)	$4,094		Midsize (69 kW)	$2,565	
Balance of Plant	Small SUV (90 kW)	$4,502		Small SUV (79 kW)	$2,935	
	Midsize SUV (89 kW)	$4,455		Midsize SUV (77 kW)	$2,885	
	Pickup (143 kW)	$7,172		Pickup (122 kW)	$4,545	
Battery	Small (29 kW/1.2 kWh)	$491		Small (28 kW/1.2 kWh)	$430	
Li-ion	Midsize (36 kW/1.5 kWh)	$601		Midsize (28 kW/1.2 kWh)	$430	
	Small SUV (36 kW/1.5 kWh)	$601		Small SUV (31 kW/1.4 kWh)	$478	
	Midsize SUV (39 kW/1.6 kWh)	$655		Midsize SUV (31 kW/1.4 kWh)	$478	
	Pickup (46 kW/1.9 kWh)	$764		Pickup (37 kW/1.6 kWh)	$573	
Motor	Small (73 kW)	$736		Small (67 kW)	$335	
Induction Primary	Midsize (90 kW)	$903		Midsize (76 kW)	$383	
	Small SUV (100 kW)	$1,003		Small SUV (88 kW)	$444	
	Midsize SUV (98 kW)	$982		Midsize SUV (86 kW)	$430	
	Pickup (162 kW)	$1,626		Pickup (139 kW)	$695	
H_2 Storage Tank	Small (4.1 kg H_2)	$2,159	4.1 kg usable H_2	Small (3.3 kg H_2)	$1,259	3.3 kg usable H_2
	Midsize (4.4 kg)	$2,256	4.4 kg	Midsize (3.4 kg)	$1,275	3.4
	Small SUV (5.1 kg)	$2,475	5.1 kg	Small SUV (4.1 kg)	$1,369	4.1
	Midsize SUV (5.4 kg)	$2,596	5.4 kg	Midsize SUV (4.4 kg)	$1,410	4.4
	Pickup (6.9 kg)	$3,051	6.9 kg	Pickup (5.4 kg)	$1,564	5.4

SOURCE: Islam et al., 2020.

Uncertainties were addressed for both technology performance and cost by simulating two cases:

1. Low uncertainty case: assumed slow technology progress and was based on DOE technology manager estimates of OEM improvements driven by regulations and business as usual; and
2. High uncertainty case: assumed aggressive technology advancements based on achieving DOE R&D targets.

The costs in Table 6.5 are based on the best-case scenario—the high uncertainty case. Fuel cell power density was assumed to increase from 650 W/kg in the reference case (2015 laboratory year) to 900 W/kg in 2030 (MY 2035).[20] Fuel cell peak efficiency was assumed to increase from 61% in 2015 to 68% in 2030 (MY 2035). The simulation also projected decreases in fuel cell system peak power in 2025 and 2035 MY FCEVs owing to improved component efficiencies and vehicle lightweighting, which also led to reductions in FCEV fuel consumption. Manufacturing cost was assumed to decrease as a result of technology improvement as well.

Fuel cell estimates in the ANL study were based on the best available data in 2018 for fuel cells at high volume manufacturing (500,000 units per year). Since the conclusion of the study, more recent estimates of fuel cell technology performance and cost have been published by DOE, and some technical targets have been updated (Kongkanand, 2020). Cost reductions are dependent on several factors, including continued momentum in fuel cell R&D for LDVs, adequate R&D funding levels, and market opportunities for light-duty FCEVs to enable economies of scale.

6.7 FINDINGS AND RECOMMENDATIONS FOR FCEVs

FINDING 6.1: Limited volumes of fuel cell electric vehicles (FCEVs) have been introduced in California by Honda, Hyundai, and Toyota. Plans to introduce FCEVs in the Northeast United States have been delayed, largely owing to the prohibition of hydrogen-powered vehicles in tunnels and on the lower deck of two-tier bridges in that region. Recent studies of FCEV safety in tunnels have provided data responding to concerns of local officials. FCEVs will have minimal impact on the 2025–2030 Corporate Average Fuel Economy standards but are likely to become more important in a longer time frame. Focus on FCEV deployment is generally much stronger in Asia, particularly Japan, and in Germany, than in the United States.

FINDING 6.2: The lack of hydrogen fueling infrastructure is a significant obstacle to fuel cell electric vehicle (FCEV) deployment. The high cost of building hydrogen stations is often cited as a concern; however, there is a lack of up-to-date, detailed analyses on the cost of hydrogen infrastructure build-out, particularly in terms of cost per vehicle and cost per mile driven. Within the United States, hydrogen stations are mostly limited to California, driven by station subsidies provided by the State government and a coordinated rollout of FCEVs and stations. Plans to introduce hydrogen stations in the Northeast United States have been delayed owing to the delay in introduction of FCEVs there.

FINDING 6.3: While most automakers are continuing to develop fuel cell technology for light duty vehicles (LDVs), some automakers have shifted their short-term focus from LDV applications to medium- and heavy-duty vehicle (MHDV) applications. Fuel cells currently provide a lighter-weight electric propulsion system than batteries in MHDV applications as well as longer driving range between refueling and faster refueling times. MHDV fleets are expected to create early demand for hydrogen and facilitate development of the hydrogen refueling infrastructure.

FINDING 6.4: Materials and design and engineering improvements continue to lower the cost and improve the performance of fuel cell systems, hydrogen storage tanks, fuel cell electric vehicles, and hydrogen stations. The proton-exchange membrane fuel cell, containing platinum and platinum-alloy catalysts and perfluorosulfonic acid type membranes, is expected to be the automakers' technology of choice for 2025–2035 vehicles.

---

[20] For comparison, under the low uncertainty scenario, power density and peak efficiency increased to 675 W/kg and 65%, respectively, in 2030 (MY 2035), which the industry has already exceeded or is close to achieving.

FINDING 6.5: Current research and development efforts are focused on reducing cost and improving durability of fuel cell and hydrogen systems by (1) lowering fuel cell platinum content, (2) developing non-precious metal fuel cell catalysts, (3) developing higher temperature membranes to simplify fuel cell system design and engineering, (4) increasing the efficiency and reducing the capital cost of electrolyzers for producing renewable hydrogen, (5) developing lower cost carbon fiber manufacturing processes for compressed hydrogen storage tanks, and (6) developing lower cost hydrogen compression technologies.

FINDING 6.6: Economies of scale are critical to reducing cost of fuel cell electric vehicle (FCEV) and hydrogen technologies, yet manufacturing R&D efforts have been limited. Development of low-cost, high-throughput manufacturing processes for electrolyzers, fuel cells, and hydrogen storage tanks for all FCEV classes is needed to achieve economies of scale.

FINDING 6.7: Hydrogen R&D efforts are focused on producing renewable hydrogen and exploring synergies between hydrogen for transportation applications and hydrogen production and use in other industries and applications to lower cost and facilitate development of a hydrogen fuel infrastructure. These include efforts in developing renewable hydrogen as a feedstock for carbon-neutral synthetic liquid fuels such as methanol and dimethyl ether, and as a means of storing energy for the electric grid from intermittent wind and solar resources.

RECOMMENDATION 6.1: In addition to funding research and development (R&D) on hydrogen and fuel cell technologies for medium- and heavy-duty vehicle applications, the U.S. Department of Energy should continue funding R&D to reduce cost and improve performance of hydrogen and fuel cell technologies for light-duty vehicle applications. Funding for manufacturing R&D should be increased to enable reduced cost through economies of scale.

RECOMMENDATION 6.2: Coordinated state, regional, and national plans should be developed to enable successful, high-volume fuel cell electric vehicle and hydrogen station deployment through public-private partnerships. Data and analysis needs responding to concerns of state and local officials should be identified and addressed, particularly up-to-date information on the cost of hydrogen infrastructure build-out, cost of infrastructure per vehicle and per miles traveled, policy options to support initial infrastructure build-out, and safety concerns. The U.S. Department of Energy and U.S. Department of Transportation should coordinate with state governments to facilitate regional planning and provide independent, fact-based data to help guide local policy decisions.

6.8 REFERENCES

AB 8, Chapter 401, Statutes of 2013. Alternative fuel and vehicle technologies: Funding programs. (2013–2014). https://leginfo.legislature.ca.gov/faces/billNavClient.xhtml?bill_id=201320140AB8.

ABB. 2020. "ABB Brings Fuel Cell Technology a Step Closer to Powering Large Ships." ABB Group Release. April 8. https://new.abb.com/news/detail/60096/abb-brings-fuel-cell-technology-a-step-closer-to-powering-large-ships.

Adams, J. 2020. "DOE H_2 Heavy Duty Truck Targets." Presented at the Compressed Gas Storage for Medium and Heavy Duty Transportation Workshop. January 21.

Adams, J., C. Houchins, and R. Ahluwalia. 2019. "Onboard Type IV Compressed Hydrogen Storage System—Cost and Performance Status." 19008. DOE Hydrogen and Fuel Cells Program Record. Department of Energy. https://www.hydrogen.energy.gov/pdfs/19008_onboard_storage_cost_performance_status.pdf.

AFDC (Alternative Fuels Data Center). n.d.a. "How Do Fuel Cell Electric Vehicles Work Using Hydrogen?" Alternative Fuels Data Center. https://afdc.energy.gov/vehicles/how-do-fuel-cell-electric-cars-work.

AFDC. n.d.b. "Natural Gas Laws and Incentives in New York." Department of Energy, Office of Energy Efficiency and Renewable Energy. https://afdc.energy.gov/fuels/laws/NG?state=NY.

Ahluwalia, R., X. Wang, and J-K Peng. 2020a. "Performance and Durability of Advanced Automotive Fuel Cell Stacks and Systems with State-of-the-Art d-PtCo/C Cathode Catalyst in Membrane Electrode Assemblies." FY2019 Annual Progress Report. Argonne, IL: Argonne National Laboratory.

Ahluwalia, R., X. Wang, and J-K Peng. 2020b. "Fuel Cell System Modeling and Analysis." Presented at the U.S. DOE Hydrogen and Fuel Cells Program 2020 Annual Merit Review and Peer Evaluation Meeting, Washington, DC, May 19.

Air Liquide. 2016a. "Air Liquide Announces Locations of Several Hydrogen Stations in Northeast U.S.A." April 7. https://energies.airliquide.com/air-liquide-announces-locations-several-hydrogen-stations-northeast-usa.

Air Liquide. 2016b. "Air Liquide Celebrates Completion of New Hydrogen Fueling Stations in California and U.S. Northeast." October 8. https://energies.airliquide.com/new-hydrogen-fueling-stations.

Air Products. 2016. "Air Products Selected for Technology Upgrade at Shell Hydrogen Fueling Station in Torrance, California." Press Release. November 15. http://www.airproducts.com/Company/news-center/2016/11/1115-air-products-selected-for-technology-upgrade-at-shell-hydrogen-fueling-station.aspx.

Allendorf, M., and T. Gennett. 2020. "HyMARC: A Consortium for Advancing Hydrogen Storage Materials." Presented at the DOE Hydrogen Program: 2020 Annual Merit Review. https://www.hydrogen.energy.gov/pdfs/review20/st127_allendorf_gennett_2020_o.pdf.

Automotive News Europe. 2020. "Daimler Will End Development of Fuel Cell Cars." Automotive News Europe. April 22. https://europe.autonews.com/automakers/daimler-will-end-development-fuel-cell-cars.

Bach, P. 2019. "Bipolar Plates: Carbon or Metal for PEM Fuel Cells? [Infographic]." May 30. https://blog.ballard.com/bipolar-plates.

Banham, D., and S. Ye. 2017. Current status and future development of catalyst materials and catalyst layers for proton exchange membrane fuel cells: An industrial perspective. *ACS Energy Letters* 2(3):629–638. https://doi.org/10.1021/acsenergylett.6b00644.

Banham, D., J. Choi, T. Kishimoto, and S. Ye. 2019. Integrating PGM-free catalysts into catalyst layers and proton exchange membrane fuel cell devices. *Advanced Materials* 31(31):1804846. https://doi.org/10.1002/adma.201804846.

Baronas, J., and G. Achtelik. 2019. Joint Agency Staff Report on Assembly Bill 8: 2019 Annual Assessment of Time and Cost Needed to Attain 100 Hydrogen Refueling Stations in California. California Energy Commission and California Air Resources Board. Publication Number: CEC-600-2019-039.

Bizon, N. 2017. Energy optimization of fuel cell system by using global extremum seeking algorithm. *Applied Energy* 206(November):458–474. https://doi.org/10.1016/j.apenergy.2017.08.097.

BMW Group. 2019. "The BMW i Hydrogen NEXT. Our Fuel Cell Development Vehicle." September 10. https://www.bmwgroup.com/en/news/2019/bmw-i-hydrogen-next.html.

Boateng, E., and A. Chen. 2020 Recent advances in nanomaterial-based solid-state hydrogen storage. *Materials Today Advances* 6(June 1):100022. https://doi.org/10.1016/j.mtadv.2019.100022.

Borup, R., D. Leonard, K.C. Neyerlin, S. Kabir, and D. Cullen. 2019. "Low Cost Gas Diffusion Layer Materials and Treatments for Durable High Performance PEM Fuel Cells." Presented at the 2019 Annual Merit Review, Department of Energy. https://www.hydrogen.energy.gov/pdfs/review19/fc319_borup_2019_p.pdf.

Bush, B., M. Muratori, C. Hunter, J. Zuboy, and M. Melaina. 2019. Scenario Evaluation and Regionalization Analysis (SERA) Model: Demand Side and Refueling Infrastructure Buildout. Golden, CO: National Renewable Energy Laboratory. NREL/TP-5400-70090. https://www.nrel.gov/docs/fy19osti/70090.pdf.

CaFCP (California Fuel Cell Partnership). 2015. "Cost to Refill." https://cafcp.org/content/cost-refill.

CaFCP. 2018. *The California Fuel Cell Revolution: A Vision for Advancing Economic, Social, and Environmental Priorities*. July. https://cafcp.org/sites/default/files/CAFCR.pdf.

CaFCP. 2020. "H2 Station List." https://cafcp.org/sites/default/files/h2_station_list.pdf.

Car and Driver. 2019. "2020 Toyota Mirai Review, Pricing, and Specs." Car and Driver. December 3. https://www.caranddriver.com/toyota/mirai-2020.

CARB (California Air Resources Board). 2019. *2019 Annual Evaluation of Fuel Cell Electric Vehicle Deployment and Hydrogen Fuel Station Network Development*. July.

Castillo, G. 2017. "Hyundai Fuel Cell: NEXO Introduction." January. https://cafcp.org/sites/default/files/Hyundai%20%2-%20Gil%20Castillo%20-%20January%2023.pdf.

Connelly, E., A. Elgowainy, and M. Ruth. 2019. "Current Hydrogen Market Size: Domestic and Global." DOE Hydrogen and Fuel Cells Program Record, October 1. https://www.hydrogen.energy.gov/pdfs/19002-hydrogen-market-domestic-global.pdf.

Consumer Reports. n.d. "Toyota Mirai 2016–2020 Quick Drive." Consumer Reports Video. http://www.consumerreports.org/video/view/cars/auto-test-track/4179139813001/toyota-mirai-2016-2020-quick-drive/.

Crosse, J. 2020. "Fuel Cell Powertrains 'Could Be as Cheap as Petrol in Five Years,' Says BMW's Hydrogen VP." The Telegraph, January 25. https://www.telegraph.co.uk/cars/news/fuel-cell-powertrains-could-cheap-petrol-five-years-says-bmws/.

Daimler Truck AG. 2020. "The Volvo Group and Daimler Truck AG to Lead the Development of Sustainable Transportation by Forming Joint Venture for Large-Scale Production of Fuel Cells." Daimler Truck AG. April 21. https://media.daimler.com/marsMediaSite/en/instance/ko/The-Volvo-Group-and-Daimler-Truck-AG-to-lead-the-development-of-sustainable-transportation-by-forming-joint-venture-for-large-scale-production-of-fuel-cells.xhtml?oid=46192201.

Das, S., J. Warren, D. West, and S.M. Schexnayder. 2016. "Global Carbon Fiber Composites Supply Chain Competitiveness Analysis." NREL/TP-6A50-66071; ORNL/SR-2016/100. Golden, CO: National Renewable Energy Laboratory. https://doi.org/10.2172/1260138.

Dekel, D.R. 2018. Review of cell performance in anion exchange membrane fuel cells. *Journal of Power Sources* 375(January):158–169. https://doi.org/10.1016/j.jpowsour.2017.07.117.

Dekel, D.R., I.G. Rasin, and S. Brandon. 2019. Predicting performance stability of anion exchange membrane fuel cells. *Journal of Power Sources* 420(April):118–123. https://doi.org/10.1016/j.jpowsour.2019.02.069.

Deloitte China. 2020. "Fueling the Future of Mobility: Hydrogen and Fuel Cell Solutions for Transportation." Vol. 1. Deloitte China. https://www2.deloitte.com/content/dam/Deloitte/cn/Documents/finance/deloitte-cn-fueling-the-future-of-mobility-en-200101.pdf.

Dijoux, E., N.Y. Steiner, M. Benne, M. Péra, and B.G. Pérez. 2017. A review of fault tolerant control strategies applied to proton exchange membrane fuel cell systems. *Journal of Power Sources* 359(August):119–133. https://doi.org/10.1016/j.jpowsour.2017.05.058.

DOE (U.S. Department of Energy). 2017. "3.5 Manufacturing R&D." In *Fuel Cell Technologies Office Multi-Year Research, Development, and Demonstration Plan*. https://www.energy.gov/sites/prod/files/2015/06/f22/fcto_myrdd_manufacturing.pdf.

DOE. 2020. "H2@Scale." Department of Energy, H2@Scale. https://www.energy.gov/eere/fuelcells/h2scale.

DOE HTAC MSC. 2014. "Advanced Manufacturing Status and Opportunities to Accelerate Growth in Fuel Cell and Hydrogen Products." Manufacturing Subcommittee (MSC) Report to Hydrogen and Fuel Cell Technical Advisory Committee (HTAC). Department of Energy. https://www.hydrogen.energy.gov/pdfs/htac_msc_report_12-14.pdf.

Eberle, U., and R. von Helmolt. 2010. "CHAPTER NINE—Fuel Cell Electric Vehicles, Battery Electric Vehicles, and Their Impact on Energy Storage Technologies: An Overview." In *Electric and Hybrid Vehicles*, edited by Gianfranco Pistoia, 227–245. Amsterdam: Elsevier. https://doi.org/10.1016/B978-0-444-53565-8.00009-9.

Eberle, U., B. Müller, and R. von Helmolt. 2012. Fuel cell electric vehicles and hydrogen infrastructure: Status 2012. *Energy and Environmental Science* 5(10):8780. https://doi.org/10.1039/c2ee22596d.

Edelstein, S. 2014. "Audi A7 H-Tron Quattro 'Performance' Fuel-Cell Plug-In Car Unveiled At 2014 LA Auto Show." Green Car Reports. November 19. https://www.greencarreports.com/news/1095557_audi-a7-h-tron-quattro-performance-fuel-cell-plug-in-car-unveiled-at-2014-la-auto-show.

EIA (U.S. Energy Information Administration). 2017. *Annual Energy Outlook 2017 with Projections to 2050*. Washington, DC: U.S. Energy Information Administration.

EIA. 2019. "Today in Energy, U.S. Gulf Coast Refinery Demand for Hydrogen Increasingly Met by Merchant Suppliers." March 15. https://www.eia.gov/todayinenergy/detail.php?id=38712.

Eisenstein, P.A. 2020. "Q&A: Toyota's Hydrogen Chief Jackie Birdsall." Detroit Bureau. May 4. https://www.thedetroitbureau.com/2020/05/qa-toyotas-hydrogen-chief-jackie-birdsall/.

EPA. 2019. "2019 Fuel Economy Guide for DOE." fueleconomy.gov. https://www.fueleconomy.gov/feg/download.shtml.

Eudy, L. and M. Post. 2018. Fuel Cell Buses in U.S. Transit Fleets: Current Status 2018. NREL/TP-5400-72208 December. https://www.nrel.gov/docs/fy19osti/72208.pdf.

FCHEA (Fuel Cell and Hydrogen Energy Association). n.d. "Transportation." Fuel Cell and Hydrogen Energy Association. http://www.fchea.org/transportation.

FCHEA. 2020. Road Map to a US Hydrogen Economy. March. http://www.fchea.org/us-hydrogen-study.

FCH JU (Fuel Cells and Hydrogen Joint Undertaking). 2019. Hydrogen Roadmap Europe. Publications Office of the European Union, Luxembourg. https://www.fch.europa.eu/sites/default/files/Hydrogen%20Roadmap%20Europe_Report.pdf.

FCTO (Fuel Cells Technology Office). 2017. "Hydrogen Storage." Department of Energy. https://www.energy.gov/sites/prod/files/2017/03/f34/fcto-h2-storage-fact-sheet.pdf.

Fekete, J.R., J.W. Sowards, and R.L. Amaro. 2015. Economic impact of applying high strength steels in hydrogen gas pipelines. *International Journal of Hydrogen Energy* 40 (33). https://doi.org/10.1016/j.ijhydene.2015.06.090.

Figueroa-Santos, M.A., and A.G. Stefanopoulou. 2021. Fuel Cell Vehicle Optimization and Control. In: J. Baillieul and T. Samad (eds.), *Encyclopedia of Systems and Control*. London: Springer. https://doi.org/10.1007/978-1-4471-5102-9_100145-1.

Ford Edge HySeries—Ford. 2008. "World's First Drivable Fuel Cell Hybrid-electric Plug-in Vehicle." http://ophelia.sdsu.edu:8080/ford/10-21-2008/innovation/environmentally-friendly/hydrogen/ford-edge-hyseries/edge-fuel-cell-hybrid-346p.html.

Fox, M., B. Geller, T. Bradley, F. Kalhammer, B. Kopf, and F. Panik. 2012. Plug-in fuel cell vehicle technology and value analysis. *World Electric Vehicle Journal* 5(1):217–226. https://doi.org/10.3390/wevj5010217.

Freudenberg Sealing Technologies. 2018. "Seals for Compact Fuel Cells." https://www.fst.com/corporate/newsroom/press-releases/2018/freudenberg-gdl-press/.

Fuel Cells Bulletin. 2017. Ballard plans world's first PEMFC product with low-vost Nisshinbo non-precious metal catalyst. *Fuel Cells Bulletin* 2017(9):1. https://doi.org/10.1016/S1464-2859(17)30338-3.

FuelCellsWorks. 2020. "Germany—H2 Mobility Opens Hydrogen (H2) Station in Dortmund." May 7. https://fuelcellsworks.com/news/germany-h2-mobility-opens-hydrogen-h2-station-in-dortmund/.

Fumihiro, S., N. Yoshiyuki, S. Katsutoshi, S. Noriyuki, and S. Masato. 2015. "Development of Air Compressor for FCV." 豊田自動織機技報 66:8–13.

Galbach, P.C. 2020. "FedEx Express Hydrogen Fuel Cell Extended-Range Battery Electric Vehicles." DOE-FEDEX-06522, 1581967. https://doi.org/10.2172/1581967.

Gangloff, J.J. 2017. "Carbon Fiber Composite Material Cost Challenges for Compressed Hydrogen Storage Onboard Fuel Cell Electric Vehicles." Presented at the Fuel Cells Technology Office Webinar.Washington, DC. July 25. https://www.energy.gov/sites/prod/files/2017/07/f35/fcto_webinarslides_carbon_fiber_composite_challenges_072517.pdf.

GlobalAutoRegs. n.d. "Global Technical Regulation No. 13: Global Technical Regulation on Hydrogen and Fuel Cell Vehicles." https://globalautoregs.com/revisions?rule_id=141.

Glover, A.M., A.R. Baird, and C.B. LaFleur. 2020. Hydrogen Fuel Cell Vehicles in Tunnels. Sandia National Laboratories. SAND2020-4507 R.

Goodwin, A. 2019. "Audi Hasn't Forgotten About Fuel Cells." Roadshow. March 5. https://www.cnet.com/roadshow/news/audi-increasing-investment-in-hydrogen-fuel-cell/.

Green Car Congress. 2015. "Toyota Pops the Hood on the Technology of the Fuel Cell Mirai at SAE World Congress." Green Car Congress. April 29. https://www.greencarcongress.com/2015/04/20150429-mirai.html.

Green Car Congress. 2016. "Toyota Details Design of Fuel Cell System in Mirai: Work on Electrode Catalysts." Green Car Congress. April 19. https://www.greencarcongress.com/2016/04/20160419-toyota.html.

Green Car Congress. 2018a. "Producing the Mercedes-Benz GLC F-CELL Fuel-Cell SUV." Green Car Congress. March 21. https://www.greencarcongress.com/2018/03/20180321-glcfcell.html.

Green Car Congress. 2018b. "Mercedes-Benz Fuel Cell SUV PHEV in Production; First Units of GLC F-CELL to Be Delivered by End of October." Green Car Congress. October 10. https://www.greencarcongress.com/2018/10/20181010-glcfcell.html.

Greene, D.L., J.M. Ogden, and Z. Lin. 2020. Challenges in the designing, planning and deployment of hydrogen refueling infrastructure for fuel cell electric vehicles. *ETransportation* 6(November):100086. https://doi.org/10.1016/j.etran.2020.100086.

Gröger, O., H.A. Gasteiger, and J. Suchsland. 2015. Review—electromobility: Batteries or fuel cells? *Journal of the Electrochemical Society* 162(14):A2605. https://doi.org/10.1149/2.0211514jes.

Gurau, V., D. Fowler, D. Cox, and T. Taner. 2018. "Robotic Technologies for Proton Exchange Membrane Fuel Cell Assembly." In *Proton Exchange Membrane Fuel Cell*. InTech. https://doi.org/10.5772/intechopen.71470.

Hall, D., and N. Lutsey. 2019. Estimating the infrastructure needs and costs for the launch of zero-emission trucks. International Council on Clean Transportation. https://theicct.org/publications/zero-emission-truck-infrastructure.

Hames, Y., K. Kaya, E. Baltacioglu, and A. Turksoy. 2018. Analysis of the control strategies for fuel saving in the hydrogen fuel cell vehicles. *International Journal of Hydrogen Energy* 43(23):10810–10821. https://doi.org/10.1016/j.ijhydene.2017.12.150.

Hampel, C. 2020. "Hyundai Delivers the First H2 Trucks in Switzerland." electrive.com. July 6. https://www.electrive.com/2020/07/06/hyundai-delivers-the-first-h2-trucks-in-switzerland/.

Hanlin, J. 2019. "Fuel Cell Hybrid Electric Delivery Van Project." Presented at the 2019 DOE Annual Merit Review, Washington, DC, May 1. https://www.hydrogen.energy.gov/pdfs/review19/ta016_hanlin_2019_o.pdf.

Hasegawa, T., H. Imanishi, M. Nada, and Y. Ikogi. 2016. "Development of the Fuel Cell System in the Mirai FCV." SAE 2016 World Congress and Exhibition, 2016-01-1185. https://doi.org/10.4271/2016-01-1185.

Holton, O.T., and J.W. Stevenson. 2013. The role of platinum in proton exchange membrane fuel cells. *Platinum Metals Review* 57(4):259–271. https://doi.org/10.1595/147106713X671222.

Houchins, C. 2019. "Hydrogen Storage Cost Analysis (ST100)." Presented at the 2019 DOE Hydrogen and Fuel Cells Program Review, May 1. https://www.hydrogen.energy.gov/pdfs/review19/st100_james_2019_o.pdf.

Huya-Kouadio, J.M., B.D. James, and C. Houchins. 2018. Meeting cost and manufacturing expectations for automotive fuel cell bipolar plates. *ECS Transactions* 83(1):93–109. https://doi.org/10.1149/08301.0093ecst.

Hydrogen Council. 2017. "Hydrogen Scaling Up." November 13. https://hydrogencouncil.com/wp-content/uploads/2017/11/Hydrogen-Scaling-up_Hydrogen-Council_2017.compressed.pdf.

Hydrogen Council. 2020. "Path to Hydrogen Competitiveness: A Cost Perspective." Hydrogen Council. https://hydrogencouncil.com/wp-content/uploads/2020/01/Path-to-Hydrogen-Competitiveness_Full-Study-1.pdf.

HyMARC Data Hub. 2020. "Hydrogen Storage Materials Database." June 30. https://datahub.hymarc.org/dataset/hydrogen-storage-materials-db.

HyTunnel-CS. 2019. "About HyTunnel-CS." HyTunnel-CS. https://hytunnel.net/?page_id=31.

Hyundai Motor Group Newsroom. 2020. "Hyundai Motor Expands Partnership with U.S. Government to Support Further Exploration of Hydrogen Fuel Cell Technologies." Hyundai Motor Group Newsroom. February 10. https://www.hyundainews.com/en-us/releases/2980.

Hyundai USA. n.d. "The Hydrogen Economy Is Moving the World. Where Does Your Country Rank?" Bloomberg. https://sponsored.bloomberg.com/news/sponsors/features/hyundai/h2-economy-today/?adv=16713&prx_t=aXwFAAAAAAZ6MQA&prx_ro=s.

IIHS-HLDI (Insurance Institute for Highway Safety-Highway Loss Data Institute). 2019. "Hyundai's New Fuel Cell Vehicle Earns Safety Accolade." IIHS-HLDI Crash Testing and Highway Safety. August 8. https://www.iihs.org/news/detail/hyundais-new-fuel-cell-vehicle-earns-safety-accolade.

Isenstadt, A., and N. Lutsey. 2017. Developing hydrogen fueling infrastructure for fuel cell vehicles: A status update. International Council on Clean Transportation Briefing. October.

Islam, E., N. Kim, A. Moawad, and A. Rousseau. 2020. "Energy Consumption and Cost Reduction of Future Light-Duty Vehicles through Advanced Vehicle Technologies: A Modeling Simulation Study Through 2050." Report to the US Department of Energy, Contract ANL/ESD-19/10. June. https://www.autonomie.net/publications/fuel_economy_report.html.

James, B. 2017. "Bipolar Plate Cost and Issues at High Product on Rate." In ANL Report 17/12, 2017 Bipolar Plate Workshop Report. https://www.energy.gov/sites/prod/files/2017/11/f46/fcto_biploar_plates_wkshp_report.pdf.

James, B.D., J.M. Huya-Kouadio, C. Houchins, D.A. DeSantis, and Strategic Analysis. 2018. "Mass Production Cost Estimation of Direct H_2 PEM Fuel Cell Systems for Transportation Applications: 2018 Update." December. 355.

James, B.D. 2019. "Fuel Cell Systems Analysis." Project ID# FC163. Presented at the 2019 DOE Hydrogen and Fuel Cells Program Review Presentation. May 1. https://www.hydrogen.energy.gov/pdfs/review19/fc163_james_2019_o.pdf.

Jayakumar, A., S. Singamneni, M. Ramos, A. Al-Jumaily, and S. Pethaiah. 2017. Manufacturing the gas diffusion layer for PEM fuel cell using a novel 3D printing technique and critical assessment of the challenges encountered. *Materials* 10(7):796. https://doi.org/10.3390/ma10070796.

Jiang, H., L. Xu, J. Li, Z. Hu, and M. Ouyang. 2019. Energy management and component sizing for a fuel cell/battery/supercapacitor hybrid powertrain based on two-dimensional optimization algorithms. *Energy* 177(June):386–396. https://doi.org/10.1016/j.energy.2019.04.110.

Kaya, K., and Y. Hames. 2019. Two new control strategies: For hydrogen fuel saving and extend the life cycle in the hydrogen fuel cell vehicles. *International Journal of Hydrogen Energy* 44(34):18967–18980. https://doi.org/10.1016/j.ijhydene.2018.12.111.

Kendall, M. 2018. Fuel cell development for new energy vehicles (NEVs) and clean air in China. *Progress in Natural Science: Materials International* 28 (2): 113–120. https://doi.org/10.1016/j.pnsc.2018.03.001.

Kerviel, A., A. Pesyridis, A. Mohammed, and D. Chalet. 2018. An evaluation of turbocharging and supercharging options for high-efficiency fuel cell electric vehicles. *Applied Sciences* 8(12):2474. https://doi.org/10.3390/app8122474.

Kim, A., A. Moawad, R. Vijayagopal, and A. Rousseau. 2016. Impact of fuel cell and storage system improvement on fuel consumption and cost. *World Electric Vehicle Journal* 8(1):305–314. https://doi.org/10.3390/wevj8010305.

Kim, Y., M. Figueroa-Santos, N. Prakash, S. Baek, J.B. Siegel, and D.M. Rizzo. 2020. Co-optimization of speed trajectory and power management for a fuel-cell/battery electric vehicle. *Applied Energy* 260:114254. https://doi.org/10.1016/j.apenergy.2019.114254.

Kleen, G., and E. Padgett. 2021. Durability-Adjusted Fuel Cell System Cost, DOE HFTO Program Record. January. https://www.hydrogen.energy.gov/pdfs/21001-durability-adjusted-fcs-cost.pdf.

Kneer, A.P. 2019. "The Effect of Voltage Cycling on Catalyst Degradation in Automotive PEM Fuel Cells." Dissertation, Universität Ulm. https://doi.org/10.18725/OPARU-13598.

Koleva, M., and N. Rustagi. 2020. "Hydrogen Delivery and Dispensing Cost." 20007. DOE Hydrogen and Fuel Cells Program Record. https://www.hydrogen.energy.gov/pdfs/20007-hydrogen-delivery-dispensing-cost.pdf.

Kongkanand, A. 2017. Highly-Accessible Catalysts for Durable High-Power Performance. 2017 Annual Merit Review and Peer Evaluation Meeting. Washington, DC. https://www.hydrogen.energy.gov/pdfs/review17/fc144_kongkanand_2017_o.pdf.

Kongkanand, A. 2020. "Highly-Accessible Catalysts for Durable High-Power Performance." Presented at the 2020 DOE Hydrogen and Fuel Cells Program Annual Merit Review. May 30. https://www.hydrogen.energy.gov/pdfs/review20/fc144_kongkanand_2020_p.pdf.

Kongkanand, A., and M. Mathias. 2016. The priority and challenge of high-power performance of low-platinum proton-exchange membrane fuel cells. *The Journal of Physical Chemistry Letters*. 7:1127–1137. doi: 10.1021/acs.jpclett.6b00216.

Kurtz, J., S. Sprik, C. Ainscough, and G. Saur. 2017. "VII.A.1 Fuel Cell Electric Vehicle Evaluation." 2016 DOE Hydrogen and Fuel Cells Program Annual Progress Report. February.

Kurtz, J., S. Sprik, C. Ainscough, G. Saur, and M. Jeffers. 2016. "Spring 2016 FCEV Evaluation Results." Golden, CO: National Renewable Energy Laboratory. May.

La Fleur, C., G. Bran-Anleu, A.B. Muna, B.D. Ehrhart, M. Blaylock, and W.G. Houf. 2017. "Hydrogen Fuel Cell Electric Vehicle Tunnel Safety Study." SAND2017-11157. Sandia National Laboratories.

La Fleur, C.B., A.M. Glover, A.R. Baird, C.J. Jordan, and B.D. Ehrhart. 2020. Alternative Fuel Vehicles in Tunnels. Sandia National Laboratories, SAND2020-5466. May.

Lane, B. 2017. "Plug-in Fuel Cell Electric Vehicles: A Vehicle and Infrastructure Analysis and Comparison with Alternative Vehicle Types." Dissertation, University of California, Irvine. https://escholarship.org/uc/item/57h0j2sv#main.

Lane, B., B. Shaffer, and G.S. Samuelsen. 2017. Plug-in fuel cell electric vehicles: A California case study. *International Journal of Hydrogen Energy* 42(20):14294–14300. https://doi.org/10.1016/j.ijhydene.2017.03.035.

Lawrence, C. 2020. "Can Hydrogen Fuel Cells Solve the Pain Points of Electric Vehicles?" Intelligent Mobility Xperience. July 7. https://www.intelligent-mobility-xperience.com/can-hydrogen-fuel-cells-solve-the-pain-points-of-electric-vehicles-a-946358/?cmp=go-ta-art-trf-IMX_DSA-20200217&gclid=Cj0KCQjw59n8BRD2ARIsAAmgPmKMUJDFM1ZUpiweA9wJ7kToQpTrvl91o3irjGPAoae6S3yle3IamHMaAv6qEALw_wcB.

Lin, Z., J. Dong, and D.L. Greene. 2013. Hydrogen vehicles: Impacts of DOE technical targets on market acceptance and societal benefits. *International Journal of Hydrogen Energy* 38(19):7973–7985. https://doi.org/10.1016/j.ijhydene.2013.04.120.

Lohse-Busch, H., K. Stutenberg, M. Duoba, S. Iliev, M. Kern, B. Richards, M. Christenson, and A. Loiselle-Lapointe. 2018. "Technology Assessment of a Fuel Cell Vehicle: 2017 Toyota Mirai." Argonne, IL: Argonne National Laboratory. January 1. https://doi.org/10.2172/1463251.

Luth, M. 2019. "Fuel Cell Customers—Medium and Heavy-Duty Transportation." Fuel Cell and Hydrogen Energy Association. November 2. http://www.fchea.org/in-transition/2019/9/2/fuel-cell-customers-medium-and-heavy-duty-transportation.

Martinez, A., and G. Achtelik. 2017. "California's Hydrogen Fueling Network Progress and Growth Towards H2@SCALE." California Air Resources Board.

Masika, E., and R. Mokaya. 2013. Preparation of ultrahigh surface area porous carbons templated using Zeolite 13X for enhanced hydrogen storage. *Progress in Natural Science: Materials International* 23(3):308–316. https://doi.org/10.1016/j.pnsc.2013.04.007.

Massachusetts Department of Transportation. 2019. 700 CMR 7.00: Use of the Massachusetts Turnpike and the Metropolitan Highway System.

Mattuci. 2015. "Diagram of a PEM Fuel Cell." Diagram. Wikimedia Commons. https://en.wikipedia.org/wiki/Proton-exchange_membrane_fuel_cell#/media/File:Proton_Exchange_Fuel_Cell_Diagram.svg.

Mayyas, A., and M. Mann. 2019. Emerging manufacturing technologies for fuel cells and electrolyzers. *Procedia Manufacturing* 33:508–515. https://doi.org/10.1016/j.promfg.2019.04.063.

McWhorter, S., G. Ordaz. 2013. "Onboard Type IV Compressed Hydrogen Storage Systems—Current Performance and Cost." 13013. DOE Hydrogen and Fuel Cells Program Record. http://hydrogen.energy.gov/pdfs/13010_onboard_storage_performance_cost.pdf.

Melaina, M., and M. Penev. 2013. Hydrogen Station Cost Estimates: Comparing Hydrogen Station Cost Calculator Results with Other Recent Estimates. NREL/TP-5400-56412. Golden, CO: National Renewable Energy Laboratory. https://www.nrel.gov/docs/fy13osti/56412.pdf.

Melaina, M., M. Muratori, J. McLaren, and P. Schwabe. 2017. "Investing in Alternative Fuel Infrastructure: Insights for California from Stakeholder Interviews." TRB 96th Annual Meeting Compendium of Papers: 17-05279.

Melaina, M.W., Y. Sun, and B. Bush. 2014. "Retail Infrastructure Costs Comparison for Hydrogen and Electricity for Light-Duty Vehicles." In SAE 2014 World Congress and Exhibition, 2014-01-1969. https://doi.org/10.4271/2014-01-1969.

Miller, E., and N. Stetson. 2019. "Hydrogen Fuel R&D Subprogram Overview." DOE Hydrogen and Fuel Cells Program. FY 2018 Annual Progress Report. Department of Energy. https://www.hydrogen.energy.gov/pdfs/progress18/h2f_overview_2018.pdf.

Morris, L., J.J. Hales, M.L. Trudeau, P. Georgiev, J.P. Embs, J. Eckert, N. Kaltsoyannis, and D.M. Antonelli. 2019. A manganese hydride molecular sieve for practical hydrogen storage under ambient conditions. *Energy & Environmental Science* 12:1580e1591.

Morrison, G., J. Stevens, and F. Joseck. 2018. Relative economic competitiveness of light-duty battery electric and fuel cell electric vehicles. *Transportation Research Part C: Emerging Technologies* 87:183–196. https://doi.org/10.1016/j.trc.2018.01.005.

MotorTrend. 2017. "2016 Toyota Mirai Long-Term Verdict: Pulling Off the Hydrogen Highway." MotorTrend. September 9. https://www.motortrend.com/cars/toyota/mirai/2016/2016-toyota-mirai-review-long-term-verdict/.

Munster, J. 2019. "Hydrogen for Transport: Infrastructure Pathway to Parity and Below." Presented at the Hydrogen Fueling Infrastructure Webinar to Committee on Assessment of Technologies for Improving Fuel Economy of Light-Duty Vehicles—Phase 3. June 26. https://www.nationalacademies.org/event/06-26-2019/hydrogen-fueling-infrastructure-webinar.

Nakagaki, N. 2015. "The Newly Developed Components for the Fuel Cell Vehicle, Mirai." SAE Technical Paper 2015-01-1174. Warrendale, PA: SAE International. https://doi.org/10.4271/2015-01-1174.

Nied, D. 2015. "Hydrogen Fuel Cell Timeline." Toyota Today. January/February. https://www.toyotatoday.com/news/hydrogen-fuel-cell-timeline.htm.

NREL (National Renewable Energy Laboratory). n.d. Hydrogen Financial Analysis Scenario Tool (H2FAST). http://www.nrel.gov/hydrogen/h2fast/.

NREL. 2016. "Comparison of On-Road Fuel Economy | Fuel Cell Electric Vehicle Composite Data Products." National Renewable Energy Laboratory, Hydrogen and Fuel Cells. April 28. https://www.nrel.gov/hydrogen/assets/images/cdp_fcev_32.jpg.

O'Dell, J. 2018. "Anheuser-Busch Makes Record Order of 800 Nikola Fuel Cell Trucks." Trucks.Com. May 3. https://www.trucks.com/2018/05/03/anheuser-busch-nikola-truck-order/.

Ogden, J.M. 2018. "Prospects for Hydrogen in the Future Energy System." Davis, CA: Institute of Transportation Studies—University of California, Davis, March.

Ordaz, G., C. Houchins, and T. Hua. 2015. "Onboard Type IV Compressed Hydrogen Storage System—Cost and Performance Status 2015." 15013. DOE Hydrogen and Fuel Cells Program Record. Department of Energy. https://www.hydrogen.energy.gov/pdfs/15013_onboard_storage_performance_cost.pdf.

Padgett, E., and G. Kleen. 2020. "Automotive Fuel Cell Targets and Status." 20005. DOE Hydrogen and Fuel Cells Program Record.

Papageorgopoulos, D. 2019. "Fuel Cell R&D Overview." Presented at the 2019 Annual Merit Review and Peer Evaluation Meeting. April 29. https://www.hydrogen.energy.gov/pdfs/review19/plenary_fuel_cell_papageorgopoulos_2019.pdf.

Penev, M., J. Zuboy, and C. Hunter. 2019. Economic analysis of a high-pressure urban pipeline concept (HyLine) for delivering hydrogen to retail fueling stations. *Transportation Research Part D: Transport and Environment* 77: 92–105. https://doi.org/10.1016/j.trd.2019.10.005.

Peterson, D., and R. Farmer. 2017. "Historical Fuel Cell and Hydrogen Budgets." 17006. DOE Hydrogen and Fuel Cells Program Record. https://www.hydrogen.energy.gov/pdfs/17006_historical_fuel_cell_h2_budgets.pdf.

Plug Power. 2019. "Plug Power Signs Deal to Power StreetScooter Delivery Vans with ProGen Hydrogen Fuel Cell Engines." Plug Power, Inc. May 28. https://www.ir.plugpower.com/Press-Releases/Press-Release-Details/2019/Plug-Power-Signs-Deal-to-Power-StreetScooter-Delivery-Vans-with-ProGen-Hydrogen-Fuel-Cell-Engines/default.aspx.

PNNL (Pacific Northwest National Laboratory). n.d. "H2 Tools: International Hydrogen Fueling Stations." Available at h2tools.org, accessed March 30, 2021.

Port Authority of NY and NJ. 2016. Traffic Rules and Regulations for the Holland Tunnel, Lincoln Tunnel, George Washington Bridge, Bayonne Bridge, Goethals Bridge, Outerbridge Crossing. September.

Reed, J., E. Dailey, B. Shaffer, B. Lane, R. Flores, A. Fong, and G.S. Samuelsen. 2020. Roadmap for the Deployment and Buildout of Renewable Hydrogen Production Plants in California. CEC. Publication Number: CEC-600-2020-002.

Ren, X., X. Zhang, T. Teng, and C. Li. 2020. Research on estimation method of fuel cell health state based on lumped parameter model. *Energies* 13 (23): 6425. https://doi.org/10.3390/en13236425.

Research and Markets. 2020. "Global Fuel Cell Vehicles Market 2020–2030: Analysis of the Current and Future Scenarios of FCEV Sales, Stations, Hydrogen Fuel, and the Push from Governments." PRNewswire. October 19. https://www.prnewswire.com/news-releases/global-fuel-cell-vehicles-market-2020-2030-analysis-of-the-current-and-future-scenarios-of-fcev-sales-stations-hydrogen-fuel-and-the-push-from-governments-301154685.html.

Rivard, E., M. Trudeau, and K. Zaghib. 2019. Hydrogen storage for mobility: A review. *Materials* 12(12):1973. https://doi.org/10.3390/ma12121973.

Robinius, M., J. Linßen, T. Grube, M. Reuß, P. Stenzel, K. Syranidis, P. Kuckertz, and D. Stolten. 2018. Comparative analysis of infrastructures: Hydrogen fueling and electric charging of vehicles. *Energie and Umwelt / Energy and Environment* 408.

Ross, K. 2019. "Solid Phase Processing for Reduced Cost and Improved Efficiency of Bipolar Plates." Presented at the 2019 Annual Merit Review. Department of Energy. https://www.hydrogen.energy.gov/pdfs/review19/fc321_ross_2019_p.pdf.

Ruf, Y., T. Zorn, P. Akcayoz De Neve, P. Andrae, S. Erofeeva, F. Garrison, and A. Schwilling. 2019. *Study on the Use of Fuel Cells and Hydrogen in the Railway Environment*. Roland Berger for Shift2Rail Joint Undertaking and Fuel Cells and Hydrogen Joint Undertaking. https://shift2rail.org/wp-content/uploads/2019/05/Study-on-the-use-of-fuel-cells-and-hydrogen-in-the-railway-environment_final.pdf.

Rügheimer, U. 2014. "The Audi A7 Sportback H-Tron Quattro." Audi MediaCenter. November 19. https://www.audi-mediacenter.com:443/en/press-releases/the-audi-a7-sportback-h-tron-quattro-438.

Rustagi, N. 2018. "Overview of Status of Hydrogen Fueling Infrastructure in U.S." DOE FCTO International Hydrogen Infrastructure Workshop. September 11.

Rustagi, N., A. Elgowainy, and J. Vickers. 2018. Current Status of Hydrogen Delivery and Dispensing Costs and Pathways to Future Cost Reduction. DOE Hydrogen and Fuel Cells Program Record. December 17. https://www.hydrogen.energy.gov/pdfs/18003_current_status_hydrogen_delivery_dispensing_costs.pdf.

Satyapal, S. 2019. 2019 Annual Merit Review, Hydrogen and Fuel Cell Program Overview. April 28. https://www.hydrogen.energy.gov/pdfs/review19/plenary_overview_satyapal_2019.pdf.

Saur, G., S. Sprik, J. Kurtz, S. Onorato, S. Gilleon, and E. Winkler. 2019. NREL Hydrogen Station Data Collection and Analysis. DOE Hydrogen and Fuel Cells Program, 2019 Annual Merit Review and Peer Evaluation Meeting. https://www.hydrogen.energy.gov/pdfs/review19/ta014_sprik_2019_o.pdf.

Scheffler, R. 2019. "Costs Check Growth of Fuel-Cell Infrastructure." Ward's Auto. August 22. https://www.wardsauto.com/technology/costs-check-growth-fuel-cell-infrastructure.

Seredynski, P. 2018. "2019 Hyundai Nexo: Fuel-Cell Refinement, SUV Luxury." SAE International. November 13. https://www.sae.org/news/2018/11/2019-hyundai-nexo-fuel-cell-launch.

Sethuraman, V.A., J.W. Weidner, A.T. Haug, M. Pemberton, and L.V. Protsailo. 2009. Importance of catalyst stability vis-à-vis hydrogen peroxide formation rates in PEM fuel cell electrodes. *Electrochimica Acta* 54(23):5571–5582. https://doi.org/10.1016/j.electacta.2009.04.062.

Shinozaki, K., H. Yamada, and Y. Morimoto. 2011. Relative humidity dependence of Pt utilization in polymer electrolyte fuel cell electrodes: Effects of electrode thickness, ionomer-to-carbon ratio, ionomer equivalent weight, and carbon support. *Journal of the Electrochemical Society* 158(5):B467. https://doi.org/10.1149/1.3556906.

Sim, K., R. Vijayagopal, N. Kim, and A. Rousseau. 2019. Optimization of component sizing for a fuel cell-powered truck to minimize ownership cost. *Energies* 12(6):1125. https://doi.org/10.3390/en12061125.

Simmons, K.L. 2014. "IV.F.3 Synergistically Enhanced Materials and Design Parameters for Reducing the Cost of Hydrogen Storage Tanks." FY 2014 Annual Progress Report. https://www.hydrogen.energy.gov/pdfs/progress14/iv_f_3_simmons_2014.pdf.

Song, K., F. Li, X. Hu, L. He, W. Niu, S. Lu, and T. Zhang. 2018. Multi-mode energy management strategy for fuel cell electric vehicles based on driving pattern identification using learning vector quantization neural network algorithm. *Journal of Power Sources* 389(June):230–239. https://doi.org/10.1016/j.jpowsour.2018.04.024.

Spiegel, C. 2018. "Gas Diffusion Layer: Characteristics and Modeling." FuelCellStore. https://www.fuelcellstore.com/blog-section/gas-diffusion-layer-characteristics-and-modeling.

State of California. 2018. Executive Order B-48-18. https://www.ca.gov/archive/gov39/2018/01/26/governor-brown-takes-action-to-increase-zero-emission-vehicles-fund-new-climate-investments/index.html.

State of New Jersey. 2019. *2019 New Jersey Energy Master Plan—Pathway to 2050*. https://www.nj.gov/emp/docs/pdf/2020_NJBPU_EMP.pdf.

Stetson, N. 2015. Cold/Cryogenic Composites for Hydrogen Storage Applications in FCEVs, DOE Workshop. Dallas, TX. October 29. https://www.energy.gov/sites/prod/files/2015/11/f27/fcto_cold_cryo_h2_storage_wkshp_1_doe.pdf.

Stetson, N. 2020. "Hydrogen Fuel R&D Subprogram Overview." DOE Hydrogen and Fuel Cells Program. FY 2019 Annual Progress Report. U.S. Department of Energy. https://www.hydrogen.energy.gov/pdfs/progress19/h2f_overview_2019.pdf.

Sturgess, S. 2017. "Fuel Cells Gain Momentum as Range Extenders for Electric Trucks." Transport Topics. November 27. https://www.ttnews.com/articles/fuel-cells-gain-momentum-range-extenders-electric-trucks.

Sun, X., S.C. Simonsen, T. Norby, and A. Chatzitakis. 2019. Composite membranes for high temperature PEM fuel cells and electrolysers: A critical review. *Membranes* 9(7):83. https://doi.org/10.3390/membranes9070083.

Sundström, O., and A. Stefanopoulou. 2007. Optimum battery size for fuel cell hybrid electric vehicle with transient loading consideration—part II. *Journal of Fuel Cell Science and Technology* 4(2):176–184. https://doi.org/10.1115/1.2713779.

Taherian, R. 2014. A review of composite and metallic bipolar plates in proton exchange membrane fuel cell: Materials, fabrication, and material selection. *Journal of Power Sources* 265:370–390. https://doi.org/10.1016/j.jpowsour.2014.04.081.

Thomas, S., and M. Zalbowitz. 1999. "Fuel Cells—Green Power." Los Alamos National Laboratory. http://www.dartmouth.edu/~cushman/courses/engs171/FuelCells-article.pdf.

Thompson, S.T., and D. Papageorgopoulos. 2019. Platinum group metal-free catalysts boost cost competitiveness of fuel cell vehicles. *Nature Catalysis* 2:558–561. July. doi: 10.1038/s41929-019-0291-x.

Tomas, M., I.S. Biswas, P. Gazdzicki, L. Kullova, and M. Schulze. 2017. Modification of gas diffusion layers properties to improve water management. *Materials for Renewable and Sustainable Energy* 6(4):20. https://doi.org/10.1007/s40243-017-0104-6.

Toyota USA Newsroom. 2014. "The Future Has Arrived, and It's Called Mirai." Toyota USA Newsroom. November 17. https://pressroom.toyota.com/toyota-names-fuel-cell-vehicle-mirai/.

Toyota USA Newsroom. 2019a. "The Future of Zero-Emission Trucking Takes Another Leap Forward." Toyota USA Newsroom. April 22. https://pressroom.toyota.com/the-future-of-zero-emission-trucking-takes-another-leap-forward/.

Toyota USA Newsroom. 2019b. "Coupe-Inspired Design Modernizes All-New 2021 Toyota Mirai Sedan Concept." Toyota USA Newsroom. October 10. https://pressroom.toyota.com/coupe-inspired-design-modernizes-all-new-2021-toyota-mirai-sedan/.

Trencher, G. 2020. Strategies to accelerate the production and diffusion of fuel cell electric vehicles: Experiences from California. *Energy Reports* 6:2503–2519. https://doi.org/10.1016/j.egyr.2020.09.008.

Ul Hassan, N., M. Mandal, G. Huang, H.A. Firouzjaie, P.A. Kohl, and W.E. Mustain. 2020. Achieving high-performance and 2000 h stability in anion exchange membrane fuel cells by manipulating ionomer properties and electrode optimization. *Advanced Energy Materials* 10(40):2001986. https://doi.org/10.1002/aenm.202001986.

United Nations Economic and Social Council. 2017. "Draft Programme of Work (PoW) Under the 1998 Agreement." ECE/TRANS/WP.29/2017/144. United Nations. https://www.unece.org/fileadmin/DAM/trans/doc/2017/wp29/ECE-TRANS-WP29-2017-144e.pdf.

UPS. 2017. "UPS Unveils First Extended Range Fuel Cell Electric Delivery Vehicle." Automotive World. May 2. https://www.automotiveworld.com/news-releases/ups-unveils-first-extended-range-fuel-cell-electric-delivery-vehicle/.

U.S. DRIVE. 2017a. "Fuel Cell Technical Team Roadmap." United States Driving Research and Innovation for Vehicle Efficiency and Energy Sustainability (U.S. DRIVE) Partnership. https://www.energy.gov/sites/prod/files/2017/11/f46/FCTT_Roadmap_Nov_2017_FINAL.pdf.

U.S. DRIVE. 2017b. "Target Explanation Document: Onboard Hydrogen Storage for Light-Duty Fuel Cell Vehicles." United States Driving Research and Innovation for Vehicle Efficiency and Energy Sustainability (U.S. DRIVE) Partnership. https://www.energy.gov/sites/prod/files/2017/05/f34/fcto_targets_onboard_hydro_storage_explanation.pdf.

U.S. DRIVE. 2017c. "Hydrogen Storage Tech Team Roadmap." United States Driving Research and Innovation for Vehicle Efficiency and Energy Sustainability (U.S. DRIVE) Partnership. July. https://www.energy.gov/sites/prod/files/2017/08/f36/hstt_roadmap_July2017.pdf.

U.S. DRIVE. 2017d. "Hydrogen Production Tech Team Roadmap." November. https://www.energy.gov/sites/prod/files/2017/11/f46/HPTT%20Roadmap%20FY17%20Final_Nov%202017.pdf.

Vijayagopal, R., N. Kim, and A. Rousseau. 2017. "Fuel Cell Powered Vehicles: An Analysis of How Technology Progress Affects the Technical and Economic Feasibility." ANL-17/07. Argonne, IL: Argonne National Laboratory. https://www.autonomie.net/pdfs/ANL%2017_07%20Fuel%20Cell%20Powered%20Vehicles%20Analysis%20Report.pdf.

Wagner, I. 2020. "Global Vehicle Production Forecast Between 2025 and 2030." Statista. April 20. https://www.statista.com/statistics/266813/growth-of-the-global-vehicle-production-since-2009/.

Wang, X.X., M.T. Swihart, and G. Wu. 2019. Achievements, challenges and perspectives on cathode catalysts in proton exchange membrane fuel cells for transportation. *Nature Catalysis* 2(7):578–589. https://doi.org/10.1038/s41929-019-0304-9.

Warren, C.D. 2016. "Carbon Fiber Precursors and Conversion." Presented at the Department of Energy Physical-Based Storage Workshop: Identifying Potential Pathways for Lower Cost 700 Bar Storage Vessels. August 24. https://www.energy.gov/sites/prod/files/2016/09/f33/fcto_h2_storage_700bar_workshop_3_warren.pdf.

Whiston, M.M., I.L. Azevedo, S. Litster, K.S. Whitefoot, C. Samaras, and J.F. Whitacre. 2019. Expert assessments of the cost and expected future performance of proton exchange membrane fuel cells for vehicles. *Proceedings of the National Academy of Sciences* 116(11):4899–4904. https://doi.org/10.1073/pnas.1804221116.

Wilson, A.R., J. Marcinkoski, and D. Papageorgopoulos. 2016. "On-Road Fuel Cell Stack Durability—2016." 16019. DOE Hydrogen and Fuel Cell Technologies Program Record.

Wipke, K., S. Sprik, J. Kurtz, T. Ramsden, C. Ainscough, and G. Saur. 2012. "National Fuel Cell Electric Vehicle Learning Demonstration Final Report." Technical Report NREL/TP-5600-54860. National Renewable Energy Laboratory.

Wu, D., C. Peng, C. Yin, and H. Tang. 2020. Review of system integration and control of proton exchange membrane fuel cells. *Electrochemical Energy Reviews* 3(3):466–505. https://doi.org/10.1007/s41918-020-00068-1.

Yamashita, A., M. Kondo, S. Goto, and N. Ogami. 2015. "Development of High-Pressure Hydrogen Storage System for the Toyota 'Mirai.' " In SAE 2015 World Congress and Exhibition. 2015-01-1169. https://doi.org/10.4271/2015-01-1169.

You, W., E. Padgett, S.N. MacMillan, D.A. Muller, and G.W. Coates. 2019. Highly conductive and chemically stable alkaline anion exchange membranes via ROMP of trans-cyclooctene derivatives. *Proceedings of the National Academy of Sciences* 116(20):9729–9734. https://doi.org/10.1073/pnas.1900988116.

Yu, W., X. Sichuan, and H. Ni. 2015. Air compressors for fuel cell vehicles: A systematic review. *SAE International Journal of Alternative Powertrains* 4(1):115–122. https://doi.org/10.4271/2015-01-1172.

Yuan, W. 2008. "PEM Fuel Cell Fundamentals" in *PEM Fuel Cell Electrocatalysts and Catalyst Layers*, J. Zhang (ed.). Springer Nature.

Yue, M., S. Jemei, R. Gouriveau, and N. Zerhouni. 2019. Review on health-conscious energy management strategies for fuel cell hybrid electric vehicles: Degradation models and strategies. *International Journal of Hydrogen Energy* 44(13):6844–6861. https://doi.org/10.1016/j.ijhydene.2019.01.190.

Zacharia, R., and S. Rather. 2015. Review of solid state hydrogen storage methods adopting different kinds of novel materials. *Journal of Nanomaterials* 2015:1–18. https://doi.org/10.1155/2015/914845.

Zhou, D., A. Ravey, A. Al-Durra, and F. Gao. 2017. A comparative study of extremum seeking methods applied to online energy management strategy of fuel cell hybrid electric vehicles. *Energy Conversion and Management* 151:778–790. https://doi.org/10.1016/j.enconman.2017.08.079.

Züttel, A. 2003. Materials for hydrogen storage. *Materials Today* 6(9):24–33. https://doi.org/10.1016/S1369-7021(03)00922-2.

7

Non-Powertrain Technologies

Non-powertrain technologies, such as improved vehicle design, material substitution, and tire technologies, can reduce fuel consumption and greenhouse gas (GHG) emissions and improve energy efficiency through reduction in road loads of aerodynamic drag, vehicle mass, and tire rolling resistance. Such loads can represent 12%–19% of the vehicle energy consumption for a conventional vehicle, 16%–26% of total vehicle energy consumption for hybrid vehicles, and 64% for battery electric vehicles (BEVs) (see Figures 7.19–7.21 in Section 7.3.2, below). Additionally, more efficient accessories can reduce loads for climate control, power steering, and other power requirements. Accessory loads (not including air conditioning [A/C]) can represent up to 2% of total electricity loads for conventional vehicles, 3% for hybrids, and 4% for BEVs. A/C can represent up to 30 gallons/year with average vehicle use, about 5% of consumption (Kreutzer et al., 2017). Non-powertrain technologies that may be a part of the U.S. light-duty fleet in 2025–2035 are described below, and their costs and effectiveness estimated. Manufacturing issues and safety aspects of the technologies are also discussed.

7.1 AERO

The aerodynamic drag of a vehicle is most relevant for the portion of driving at high speed. Aerodynamic drag can be evaluated by multiplying vehicle frontal area (A) by a scaling factor, the drag coefficient (C_d). Table 7.1 reports the values for frontal area and C_d used in the Argonne National Laboratory Autonomie modeling for the 2020 Safer Affordable Fuel Efficient (SAFE) Vehicles Rule. Aerodynamic drag is primarily lowered by changes in vehicle body design but also through the addition of vehicle devices. In the 2020 SAFE Rule, aerodynamic drag reductions were separated into four categories for 5%, 10%, 15%, and 20% improvements (AERO5, AERO10, AERO15, and AERO20, respectively) with respect to a baseline, AERO0 (NHTSA/EPA, 2020). AERO levels were assigned to model year (MY) 2017 vehicles based on their percent reduction in C_d from the average value of the relevant body style in MY 2015 (Table 7.2). Aerodynamic improvements are typically achieved via incorporation of both passive and active aerodynamic technologies, examples of which are discussed further below.

A 2013 National Research Council committee estimated that under average driving conditions, a 10% reduction in drag resistance would reduce fuel consumption by about 2%. In that study's scenarios, reduction in new-vehicle-fleet aerodynamic drag resistance for the midrange case is estimated to average about 21% (4% reduction in fuel consumption) in 2030 and 35% (7% reduction in fuel consumption) in 2050. For the optimistic case, the aerodynamic drag reductions are estimated to average about 28% in 2030 and 41% in 2050 (NRC, 2013). The SAFE

TABLE 7.1 Values for C_d and Frontal Area Used in Autonomie Modeling

Vehicle Class	Drag Coefficient (C_d)	Frontal Area (A, m^2)	$C_d \times A$
Compact car	0.31	2.30	0.713
Midsize car	0.30	2.35	0.705
Small SUV	0.36	2.65	0.954
Midsize SUV	0.38	2.85	1.083
Pickup truck	0.42	3.25	1.365

NOTE: SUV = sport utility vehicle.
SOURCE: Islam et al. (2020).

Rule estimated that fleet-wide aerodynamic drag reductions of 10% to 20% could reduce fuel consumption by approximately 2% to 4% compared to the baseline fleet (NHTSA/EPA, 2020). Opportunities for aerodynamic drag reduction are shown in Figure 7.1.

There are two categories of aerodynamic drag reduction technologies, passive and active. Features that reduce aerodynamic drag via fixed changes to a vehicle's shape and size are termed passive aerodynamic technologies. Such features are implemented primarily during major model redesign cycles and include changing the frontal area or shape and lowering the vehicle height. However, the effects of these alterations on other vehicle attributes, such as utility, interior space, and engine cooling, must also be considered. Additional passive aerodynamic technologies can be employed during the midcycle refresh process. Vehicle components that can be added or modified to decrease aerodynamic drag include the exterior mirrors, underbody panels, front air dams, front and rear fascia, rear deck lips, and rear valances. Active aerodynamic technologies monitor the driving situation and deploy accordingly. Examples of active technologies include active grill shutters (AGS) and active ride height, which have sensors that cause the air dams or suspension systems to move in order to reduce aerodynamic drag.

7.1.1 Outside Mirrors (Replaced with Cameras)

Exterior side mirrors contribute to aerodynamic drag, thereby increasing vehicle fuel consumption. Replacing mirrors with cameras can reduce drag by an average of 2% to 7%, which results in a 0.3% to 1% improvement in fuel consumption (Yang, 2018). Additionally, the use of cameras removes the requirement to mount external mirrors and therefore decreases manufacturing complexity. Such cameras cost around $500, and mirror replacement can yield cost savings of $200–$400 depending on the complexity of the system. The wiring costs for cameras and external mirrors are approximately equal.

Substituting cameras for mirrors would require a change in vehicle safety regulation, as the National Highway Traffic Safety Administration (NHTSA) currently requires passenger vehicles to be equipped with at least two mirrors. In October 2019, NHTSA released a notice seeking public and industry input to inform a possible proposal to alter mirror requirements and allow camera monitoring systems to replace rear- and side-view mirrors. At the time of writing of this report, the proposed rule remains open for public comment.

TABLE 7.2 Percent of MY 2017 Fleet in Each AERO Level

AERO Level	% of MY 2017 Fleet
AERO0	41
AERO5	40
AERO10	13
AERO15	5
AERO20	1

SOURCE: Table VI-162 of NHTSA/EPA (2020).

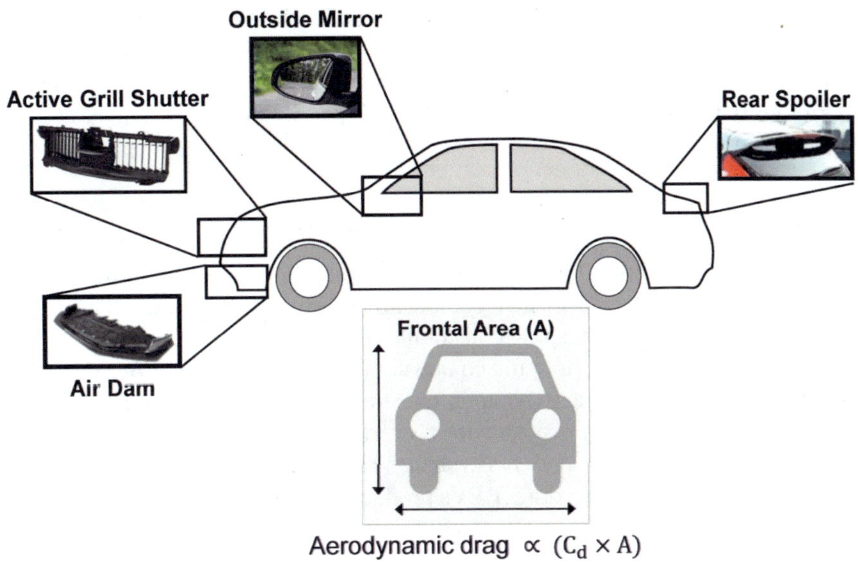

FIGURE 7.1 Opportunities for aerodynamic drag reduction include AGS, air dams, rear spoilers, and outside mirrors.
SOURCE: Committee generated using images from the public domain; Roechling Automotive (2020); and APR Performance.

7.1.2 Passive and Active Air Dams

At high speeds, the flow of air across a vehicle's underbody contributes to increased aerodynamic drag. Devices such as air dams, air scoops, and undercovers can be added onto the vehicle to control airflow around the underbody. At present, underbody panels are widely implemented with approximately 65% of 2015 vehicles equipped with this technology. Air dams, located beneath the front bumper, are used to redirect airflow to the sides of the vehicle, reducing turbulence underneath and thus reducing drag approximately 2% (NHTSA/EPA, 2020). In 2015, approximately one-quarter of new vehicles were equipped with front bumper air dams. Active air dams have even greater potential for reducing drag at high speeds (4–9%) and are ready for implementation but are not yet widely offered by manufacturers. Active air dams cost $30–$50, and passive air dams are less expensive, around $15–$25.

7.1.3 Active Grill Shutters

When a vehicle is in motion, air is drawn into a vehicle's engine through the grill located at the front of the vehicle. The majority of that air passes through the radiator located directly behind the grill, helping to keep the engine cool. However, there is often more air entering the engine bay than is needed to keep the engine cool. The unnecessary air entering the engine bay can add significant aerodynamic drag to the car. AGS selectively restrict airflow to the engine with an automatic opening and closing of shutters based on real-time needs. This reduces aerodynamic drag, thereby leading to improved fuel economy. AGS offers significant weight reduction up to 20%, owing to lower weight materials, and improvement in aerodynamic performance up to 3.0% compared to a non-AGS vehicle. Major automakers have been incorporating AGS into a wide variety of vehicle models. On average, AGS costs $300, with some variation owing to size and complexity.

7.1.4 Active Ride Height

Active ride height controls allow a vehicle to raise or lower its suspension depending on speed and road conditions, typically using hydraulic systems or air pressure pumps. Such controls are mainly used on premium vehicles to improve handling performance, maintaining comfort and stability in bumpy road conditions or off-road driving. Active ride height can also benefit fuel economy. The most aerodynamic drag reduction is achieved when a vehicle is lower to the ground; thus, active ride height systems automatically decrease vehicle height during smooth driving conditions (e.g., highway) to achieve the highest possible fuel efficiency (YourMechanic, 2015). These systems cost about $150 per vehicle.

7.1.5 Impact from Vehicle Electrification

Vehicles with electrified powertrains, especially those without an internal combustion engine (ICE), exhibit different aerodynamic behavior than vehicles with a combustion engine, owing to their thermal requirements, and prevalent vehicle designs based on different space constraints and opportunities. Vehicles with ICEs and fuel cell electric vehicles (FCEVs) require oxygen flow for combustion or hydrogen oxidation, which introduces intake and exhaust requirements and associated aerodynamic constraints. ICEs also produce a great deal of thermal waste energy that must be cooled, in part by exposing the powertrain components to ambient air under the vehicle. FCEVs produce heat at lower temperatures than ICE vehicles and therefore require larger radiators to reject that heat, which generally increases aerodynamic drag. BEVs produce much less heat from the propulsion system and require less cooling. BEVs therefore can have a smoother exterior and a flat underbody, which improves their aerodynamics. A challenge for aerodynamics of electric vehicles (EVs) is higher ride height caused by placement of their large battery packs at the base of the vehicle. For strong hybrids and BEVs, placing batteries at the base of the vehicle improves vehicle handling, with the heavy battery components lowering the vehicle center of gravity, and simplifies vehicle design, including for variable battery capacity. It also restores vehicle passenger and cargo room that would otherwise be taken up by battery packs. This, however, often leads to vehicle designs with increased frontal area, higher ride height, larger diameter wheels, and reduced essential overhang to accommodate passengers in the cabin. Tesla EVs are an exception to the higher ride height and show that EV designs do not necessarily have poor aerodynamics. As more BEV models are introduced in performance vehicles and the BEV skateboard concept becomes prevalent, the higher ride height issue will be negated.

7.1.6 Impact from CAVs

The potential impact of connected and automated vehicles (CAVs) on aerodynamic performance depends largely on the penetration and usage patterns of CAV technology, which are influenced by a variety of factors, as discussed in Chapters 8 and 9. CAVs implement sensors that may protrude from the vehicle, such as cameras and lidar. (See Chapters 8 and 9 for more detail about these technologies.) Early test implementations of lidar in particular have involved protruding sensors, but in commercial implementations, styling and aerodynamic design should limit the profile of sensors, although some aerodynamic effects may remain. Also, if fully automated vehicles are implemented, then vehicle shape options increase, as no driver needs to be facing forward and looking out a windshield. Changes to vehicle shape could influence the vehicle's frontal area and drag coefficient (C_d) as well as the design of the A, B, C, and D pillars. The usage pattern of CAVs will determine the impact of these changes on aerodynamic drag. Platooning is not expected to provide fuel savings for light-duty CAVs, in contrast to heavy-duty vehicles such as freight tractor-trailers. Additional discussion of the aerodynamics of CAVs can be found in Chapter 8.

7.1.7 Cost of Aerodynamic Technologies

The 2020 SAFE Rule provides direct manufacturing costs (DMC) for achieving aerodynamic improvements at each level, separated into categories for (1) passenger cars and sport utility vehicles (SUVs) and (2) pickup trucks (NHTSA/EPA, 2020). While the costs of specific technologies are not provided, sample lists of technologies to achieve each AERO level are reported, again separated by vehicle class. The rule attributes the higher costs for AERO15 and AERO20 to the required implementation of active as well as passive aerodynamic technologies. These sample technology lists and DMC are reported in Tables 7.3 and 7.4, respectively. Table 7.5 summarizes cost and effectiveness values for the specific aerodynamic technologies described in Sections 7.1.1–7.1.4.

TABLE 7.3 Example Technology Combinations to Achieve Each AERO Level

AERO Level	Improvements for Passenger Cars and SUVs	Improvements for Pickup Trucks
AERO5	Front styling, Roof line raised at forward of B pillar, Faster A pillar rake angle, Shorter C pillar, Low-drag wheels	Whole body styling (shape optimization), Faster A pillar rake angle, Rear spoiler, Wheel deflector/air outlet inside wheel housing, Bumper lip
AERO10	*AERO5 Technologies* + Rear spoiler, Wheel deflector/air outlet inside wheel housing, Bumper lip, Rear diffuser	*AERO5 Technologies* + Rear diffuser, Underbody cover (including rear axle cladding)
AERO15	*AERO10 Technologies* + Underbody cover (including rear axle cladding), Lowering ride height by 10 mm	*AERO10 Technologies* + Active grill shutters, Extended air dam
AERO20	*AERO15 Technologies* + Active grill shutters, Extended air dam	N/A

SOURCE: Adapted from Tables VI-160 and VI-161 of NHTSA/EPA (2020).

TABLE 7.4 DMC for Aerodynamic Drag Reductions

AERO Level	Passenger Cars and SUVs DMC (2018\$, MY 2017)	Pickup Trucks DMC (2018\$, MY 2017)
AERO0	\$0.00	\$0.00
AERO5	\$39.38	\$39.38
AERO10	\$80.51	\$80.51
AERO15	\$113.76	\$201.27
AERO20	\$201.27	\$525.06

SOURCE: Adapted from Tables VI-165 and VI-166 of NHTSA/EPA (2020).

TABLE 7.5 Cost and Effectiveness of Aerodynamic Technologies

Technology	Technology Cost	Reduction in Aerodynamic Drag	Reduction in Fuel Consumption
Camera (to replace outside mirrors)	\$500[a]	2%–7%	0.3%–1%
Passive air dam	\$15–\$25	2%	0.4%
Active air dam	\$30–\$50	4%–9%	0.8%–1.8%
Active grill shutter	\$300	3%	0.6%
Active ride height	\$150	3%	0.6%

[a]Cost per camera; note that replacing mirrors can also yield \$200–\$400 in cost savings.

7.1.8 Future of Aerodynamic Technologies

Manufacturers have widely deployed both active and passive aerodynamic drag reduction technologies. As of 2015, the most widely implemented aerodynamic drag reduction technologies were wheel dams, underbody panels, front bumper air dams, and AGS (NHTSA/EPA, 2020). The 2020 SAFE Rule notes that the prevalence of AGS, which can yield up to 3% aerodynamic improvement, has since increased further. Nonetheless, 80% of the MY 2017 fleet achieved less than 10% improvement in aerodynamic drag reduction relative to the baseline value (NHTSA/EPA, 2020). Some active aerodynamic drag reduction technologies, such as active ride height and active air dams, are available for implementation but have not been widely offered by manufacturers. However, the market for technologies is predicted to grow significantly in the next 5 to 10 years.

Even as technologies for aerodynamic drag reduction continue to be deployed, reduction in actual aerodynamic drag achieved in the fleet may not occur given the shift in consumer preference to crossover utility vehicles (CUVs) and SUVs over sedans. As noted above, aerodynamic drag is proportional to the product of frontal area and C_d, and pickup trucks, SUVs, and CUVs have a larger frontal area than sedans. For example, using the values in Table 7.1, even a 20% reduction in the drag coefficient of a small SUV, with no change to frontal area, would still give a larger ($C_d \times A$) than that of a midsize car with no aerodynamic drag reduction from the baseline. The general consumer shift to vehicles with larger frontal area, such as CUVs, SUVs, and pickup trucks, will dull the impact of reductions in C_d. Similar trends might be observed as frontal area increases with EV penetration and as electrification becomes more common on larger unibody vehicles and trucks, as described in Section 7.1.5 above. Ultimately, aerodynamic technologies will not be the only, or even perhaps the major, factor influencing the average aerodynamic drag properties of the light-duty fleet.

7.1.9 Findings and Recommendation for Aerodynamic Technologies

FINDING 7.1: Manufacturers have widely deployed both active and passive aerodynamic drag reduction technologies, including designing for low C_d and implementing grill and air dam shutters and low-drag underbodies. Further improvements in fuel economy from aerodynamic reductions through passive and active technologies will be somewhat limited, as the next options are relatively expensive.

FINDING 7.2: Aerodynamic drag is higher in sport utility vehicles (SUVs) and crossover utility vehicles (CUVs) than in sedans. Consumer preference for SUVs and CUVs will therefore limit the potential overall reduction of aerodynamic drag in the light-duty fleet. Electrification was initially incorporated on small vehicles but now is moving to CUVs and light-duty trucks. The EV emphasis on performance may profoundly shift the market (i.e., no grills and lower center of gravity).

FINDING 7.3: Regulation permitting, the elimination of outside mirrors has the potential for meaningful reductions in aerodynamic drag.

RECOMMENDATION 7.1: Pending the results of a full safety review, the National Highway Traffic Safety Administration should alter mirror requirements to allow camera monitoring systems to replace rear- and side-view mirrors.

7.2 MASS REDUCTION

From 2025–2035, mass reduction will be implemented in the context of increasingly electrified powertrains; developments in alternative manufacturing technologies and processes; varying raw material availability and cost; and company goals for technology leadership, meeting customer needs, corporate sustainability, and regulatory compliance. Lightweighting is one of the significant levers that can be utilized to increase fuel economy and is also used to improve aspects of performance like acceleration time and handling. It is most effective when holistically incorporated into the vehicle design through design optimization. Assuming no mass add-back, a 10% mass reduction generally produces an increase in fuel economy of 6%–7% for passenger cars and 4%–5% for light trucks (NRC, 2015).

Automakers take a holistic approach when determining the best methods to achieve fuel economy targets for a particular vehicle, meaning that all opportunities are considered and then valued in terms of overall effectiveness. Lightweighting is evaluated similarly, and there are key elements to be considered: (1) primary mass reduction (actual component lightweighting); (2) secondary mass reduction (owing to the lighter vehicle load, downsizing the powertrain or suspension components to deliver carryover performance); (3) mass add-back (to meet market requirements); and (4) market shifts in vehicle type and class, which greatly impact the corporate average fuel economy (CAFE).[1] (See Chapter 11 for further discussion of consumer choice.) The committee considers primary and secondary mass reductions in estimating the cost and fuel economy effectiveness of material substitution and design optimization for manufacturers' compliance options. Mass add-back and shifts in vehicle type and class impact the total effectiveness of the standards in meeting the goals of fuel consumption and GHG reduction. However, these effects are under the management of test weight class and overall CAFE by each manufacturer, as they involve an understanding of the model's baseline assumptions and its footprint-based standard.

In recent years, lightweighting via materials substitution has occurred primarily in a shift away from mild steel and toward medium- and high-strength steel and aluminum. Trends in material use in the North American light-duty vehicle fleet from 2008 to 2018 are shown in Table 7.6, both as average pounds per vehicle and as percentage of total vehicle weight.

Advances in materials, design, and manufacturing will continue to lead to new options for lighter materials in light-duty vehicles. Table 7.7 reports material use by vehicle component for a MY 2020 baseline fleet, which was determined by analyzing the 33 highest-selling vehicles in the 2019 U.S. fleet (Bailo et al., 2020). The material penetration in this MY 2020 fleet is consistent with a 5% mass reduction from the MY 2016 baseline, for primary mass reduction alone and not including mass add-back for customer comfort, safety features, and so on. The MY 2020 fleet includes a mix of mild and high-strength steel, aluminum, magnesium, and composites. Structural components such as frames tend toward steel, while closures are more likely to be aluminum. Figure 7.2 shows an example of projected material progression in the U.S. light-duty fleet from 2020 to 2040, in which material composition of the body-in-white and closures shifts from steel to lighter weight materials including generation-3 steel, aluminum, magnesium, and polymeric materials.

In addition to materials substitution, automakers will use design optimization to reduce mass in vehicles. About 40% of the vehicles included in the MY 2020 baseline fleet described above are expected to be redesigned before MY 2025. In 2025–2035, vehicle design for lightweighting will occur in the context of increased mass that comes with electric powertrains and comfort and safety features associated with driver assist and CAVs. Mass increase owing to electrification is expected to be much larger than that for CAV technologies. Figure 7.3 shows an estimate of mass added for advanced driver assistance system (ADAS) and autonomous driving features, which amounts to 22 kg (50 lb). In contrast, in comparing EVs to ICEs today, the mass added for the battery in the Chevy Bolt amounts to 436 kg, leading to a Bolt that is about 300 kg (660 lb) or 22% heavier than a VW Golf (UBS, 2017).

In 2025–2035, the mass increase relative to conventional ICE vehicles is expected to be approximately 300 kg (660 lb) for EV propulsion and 22.4 kg (50 lb) for driver assist and connected automated vehicle technology in a small or medium-size car (UBS, 2017; Figure 7.3; Gawron et al., 2018). Larger vehicles, or those with longer ranges, will have even greater increases owing to the battery. Because the majority of the energy of a BEV's propulsion system goes to moving the vehicle, a reduction in vehicle weight, including reducing the weight/energy of the battery, is key to increasing the range of the vehicle, rather than improvements in the efficiency of the electric drive system.

[1]Consider a base vehicle that an automaker wants to optimize for mass reduction during a redesign. Primary mass reduction is the mass reduction taken first, through design optimization and materials substitution, primarily of the body and closures. Secondary mass reduction is taken in the powertrain, suspension, and other vehicle components, made possible by the primary mass reduction. In practice, both primary and secondary mass reduction are considered, and ideally optimized, together. Mass add-back is mass that is added to the lightweighted vehicle, which would have also been added to the base, unlightweighted vehicle. For example, this includes the mass of technologies that the automaker must or wants to add during the redesign for competitiveness. These include mandatory items related to regulatory changes as well as improved features related to human-machine interface, climate control features, and so on. Market shifts address the impact of consumer and automaker choices on the mix of heavy and light vehicles in the market. Market shifts consider the shift in sales between models, as well as between vehicle classes.

TABLE 7.6 Average Material Use in the North American Light-Duty Fleet 2008–2018, Reported in Pounds per Vehicle and as Percent of Total Vehicle Weight

	2008	2009	2010	2011	2012	2013	2014	2015	2016	2017	2018
Average weight (pounds/vehicle)	3965	3860	3865	3914	3806	3812	3834	3889	3929	3960	3979
Regular steel	1596	1462	1422	1405	1335	1322	1308	1293	1295	1222	1215
High- and medium-strength steel	513	510	541	594	604	612	632	681	720	765	772
Stainless steel	74	67	70	71	66	72	71	73	72	72	71
Other steels	32	30	31	31	29	31	31	31	31	31	30
Iron castings	248	201	236	255	263	264	271	260	242	244	249
Aluminum	310	319	332	337	342	348	361	387	404	415	427
Magnesium	11	11	11	11	10	10	9	9	9	10	10
Copper and brass	69	70	72	71	70	69	67	65	67	69	69
Lead	43	41	40	38	35	34	35	35	35	37	34
Zinc castings	9	9	9	9	8	8	8	8	8	9	9
Powder metal	42	40	40	41	43	44	45	44	43	44	44
Other metals	5	5	5	5	5	5	4	5	5	5	5
Plastics/polymer composites	334	368	343	336	319	317	317	324	325	348	351
Rubber	202	246	228	223	205	197	194	196	196	204	205
Coatings	31	35	35	32	27	27	28	28	28	30	28
Textiles	47	57	54	49	48	49	48	44	44	46	46
Fluids and lubricants	211	214	215	217	215	218	220	221	222	222	223
Glass	97	87	90	96	93	94	94	93	92	95	97
Other	89	88	90	91	89	90	91	93	91	92	95
	2008	2009	2010	2011	2012	2013	2014	2015	2016	2017	2018
As a percent of total weight	100.0	100.0	100.0	100.0	100.0	100.0	100.0	100.0	100.0	100.0	100.0
Regular steel	40.2	37.9	36.8	35.9	35.1	34.7	34.1	33.2	33.0	30.9	30.5
High- and medium-strength steel	12.9	13.2	14.0	15.2	15.9	16.1	16.5	17.5	18.3	19.3	19.4
Stainless steel	1.9	1.7	1.8	1.8	1.7	1.9	1.9	1.9	1.8	1.8	1.8
Other steels	0.8	0.8	0.8	0.8	0.8	0.8	0.8	0.8	0.8	0.8	0.8
Iron castings	6.3	5.2	6.1	6.5	6.9	6.9	7.1	6.7	6.1	6.2	6.2
Aluminum	7.8	8.3	8.6	8.6	9.0	9.1	9.4	10.0	10.3	10.5	10.7
Magnesium	0.3	0.3	0.3	0.3	0.3	0.3	0.2	0.2	0.2	0.2	0.3
Copper and brass	1.8	1.8	1.9	1.8	1.8	1.8	1.7	1.7	1.7	1.7	1.7
Lead	1.1	1.1	1.0	1.0	0.9	0.9	0.9	0.9	0.9	0.9	0.9
Zinc castings	0.2	0.2	0.2	0.2	0.2	0.2	0.2	0.2	0.2	0.2	0.2
Powder metal	1.1	1.0	1.0	1.0	1.1	1.2	1.2	1.1	1.1	1.1	1.1
Other metals	0.1	0.1	0.1	0.1	0.1	0.1	0.1	0.1	0.1	0.1	0.1
Plastics/polymer composites	8.4	9.5	8.9	8.6	8.4	8.3	8.3	8.3	8.3	8.8	8.8

NOTE: Polypropylene is also used in the thermoplastics polyolefin elastomers (TPO) as well and its use in that area is reported separately under rubber. Average TPO use is nearly 35 pounds per vehicle.
SOURCE: ACC (2019).

TABLE 7.7 Materials Use by Vehicle Component for MY 2016 and MY 2020 Baseline

Component	MY 2016 Baseline	MY 2020 Baseline
Fender	Mild/BH steel	BH steel and aluminum (50:50)
A-pillar	AHSS	UHSS 1500 hot formed
Floor	Mild steel	HSS 440–590 with UHSS reinforce
Front bumper structure	AHSS	Mostly aluminum with some steel
Roof panel	Mild steel	Mild/BH steel
Door outer	Mild/BH steel	LSS and aluminum
Hood	Aluminum	95% aluminum
Decklid	Mild steel	LSS, Al, Mg, composite
Engine cradle/front frame	HSLA	HSS 400–600
Steering knuckle	Iron and aluminum	HSS 400–500 and aluminum
IP beam	Mild steel/HSLA	HSS and two magnesium

NOTE: AHSS = advanced high-strength steels; BH = bake hardenable; HSLA = high-strength low alloy; HSS = high-strength steels; LSS = low strength steel; UHSS = ultra-high-strength steels.
SOURCE: Bailo et al. (2020).

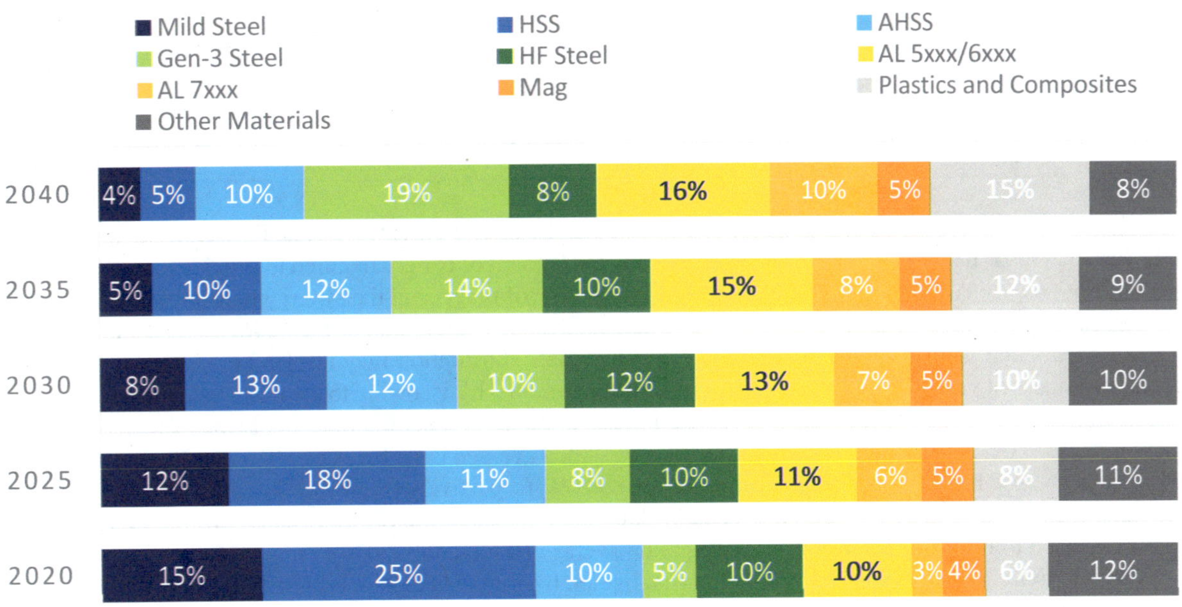

FIGURE 7.2 Sales-weighted percent of different materials estimated to be implemented over time in the body-in-white and closures in light-duty vehicles. Between 2020 and 2035, growth is seen in the use of generation-3 steel, magnesium, plastics, and composites. Reductions are seen in the use of mild steel, high-strength steel, hydrogen fluoride (HF) steel, and other materials. NOTE: Generation-3 steel is an advanced high-strength steel with relatively high formability as well as strength (Billur and Altan, 2013). NOTE: Other materials include dampeners, static sealers, adhesives, and glass. 100% includes body-in-white and closures only. Not included are powertrain/chassis, interiors, windshield, and dynamic sealers.
SOURCE: Modi and Vadhavkar (2019).

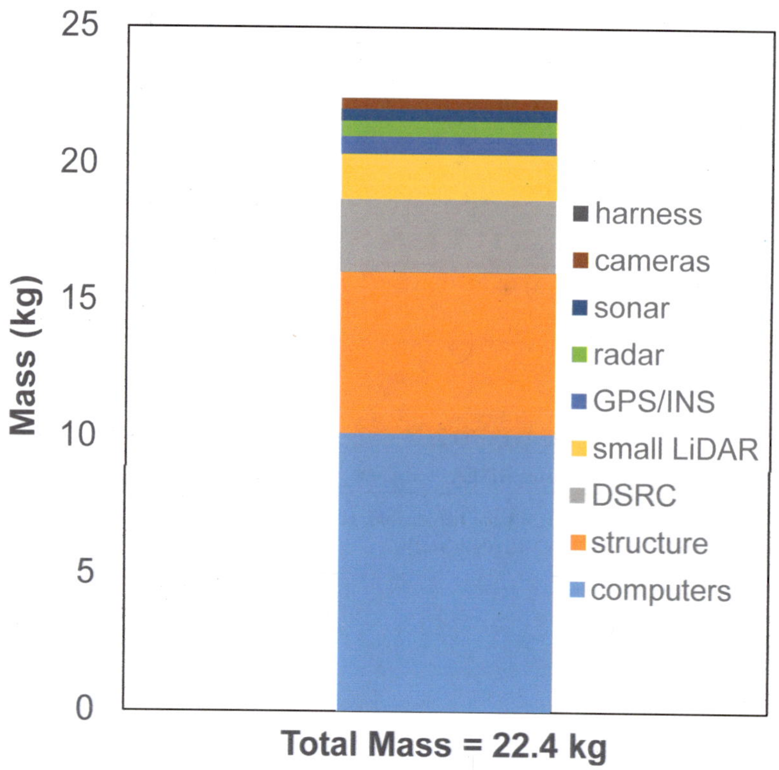

FIGURE 7.3 Step chart showing mass in ADAS technologies.
NOTE: DSRC = dedicated short-range communication; INS = inertial.
SOURCE: Committee generated based on data for a medium CAV subsystem from Gawron et al. (2018).

Reducing the weight of an EV by 20% will increase the range by up to 14% (Bull, 2011). Reductions in the weight and other road loads have a synergistic effect, decreasing the amount of battery required to provide the same range, which further reduces the weight of the vehicle. The importance of range, and the synergy between vehicle weight reduction and battery reduction, indicates that BEVs may implement mass reduction technologies earlier than ICEs.

Mass reduction implemented through design and materials substitution may not lead to significant (if any) reduction in mass of vehicles on the road on a per-class or per-model basis, owing to the mass-increasing pressures that are occurring in parallel. For instance, vehicle masses of the largest selling models of small cars and pickup trucks decreased 0% and 4%, respectively, on a sales-weighted average basis between 2016 and 2020, while their footprints increased by a sales weighted average of 2% and 6%, respectively (Figures 7.4 and 7.5). During vehicle redesign for both cars and light trucks, mass add-back owing to the addition of safety and performance technologies is expected to be nearly 5% of the curb weight on average (Bailo et al., 2020).

While individual models have become somewhat lighter, the mass of the new vehicle fleet has overall become heavier, as the market has shifted away from sedans to crossovers, SUVs, and trucks. Figure 7.6 shows that between the years of 2010 and 2016, the mass increase for cars has occurred owing to an increase in vehicle footprint, rather than an increase in weight within a given footprint. Figure 7.7 shows a similar pattern for trucks of low footprint (which are often crossovers built on car platforms); however, any pattern of changes in footprint and weight is harder to discern in the larger footprints of the truck fleet, where more diverse vehicle types are present. The time period 2010–2016 also saw a trend to more purchases of CUVs and SUVs (Figure 7.8), which tend to have less aerodynamic shapes and more mass for the same footprint.

The lack of absolute mass reduction translates to a lack of mass-related absolute fuel economy improvement. There would still be mass and fuel economy improvement from implementation of mass reduction technologies

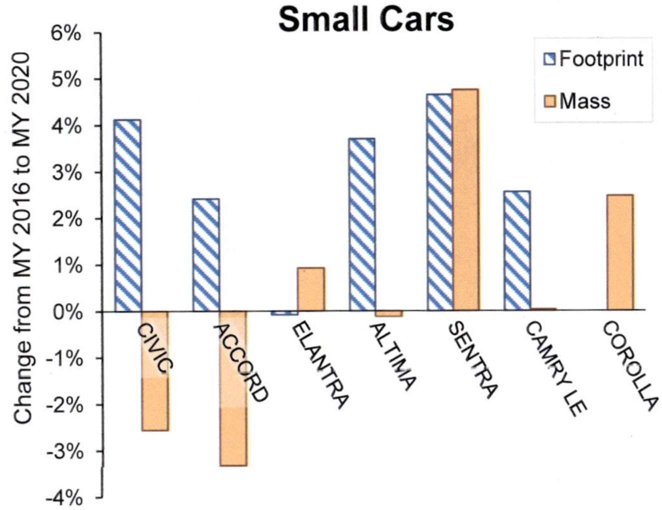

FIGURE 7.4 Mass reduction and change in footprint for the top-selling small car models in their most recent redesign. When sales-weighted, the top-selling small cars average 0% mass reduction and a 2% increase in footprint.
SOURCE: Committee generated using data from Bailo et al. (2020).

and optimization, relative to a counterfactual where those technologies were not used, and the mass increases for ADAS and other advanced technologies were still occurring. The following sections will describe the current technology development status and expected future breakthroughs for materials and design processes in 2025–2035.

7.2.1 Material Opportunities for Mass Reduction

7.2.1.1 Steel

Steel is used in the automotive body and for a variety of vehicle components, typically structural elements, reinforcements, and high-formability parts. Different grades of steel span a wide range of ultimate tensile strengths,

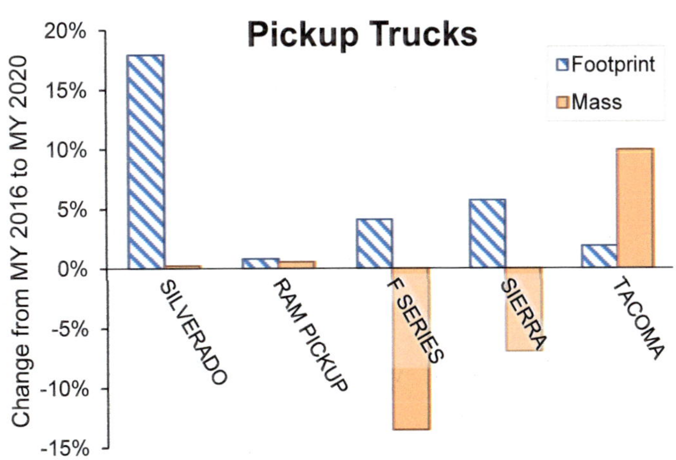

FIGURE 7.5 Mass reduction and change in footprint for the top-selling pickup truck models in their most recent redesign. When sales-weighted, the top-selling pickup trucks average 4% mass reduction and a 6% increase in footprint.
SOURCE: Committee generated using data from Bailo et al. (2020).

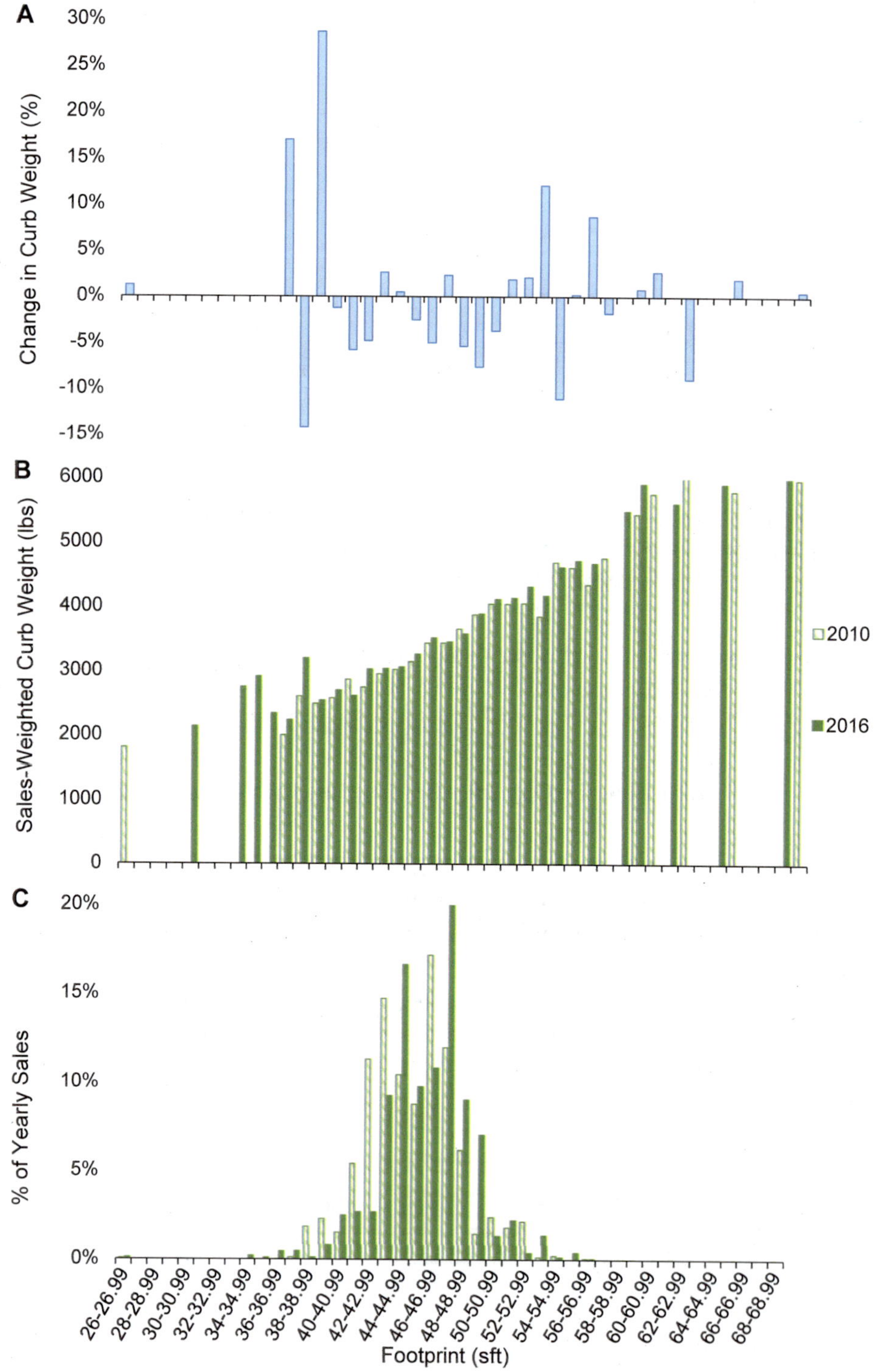

FIGURE 7.6 Comparison of weight and footprint of vehicles classified as passenger cars in the MY 2010 and 2016 fleets. All values are sales-weighted. The figure illustrates that vehicles generally get heavier with footprint (B); that there is no trend in individual footprints getting heavier over time (A); and that there is a shift in sales to cars with larger footprints between 2010 and 2016 (C). SOURCE: Committee generated using model-by-model 2010 and 2016 MY data released as part of NHTSA and U.S. Environmental Protection Agency (EPA) rulemakings.

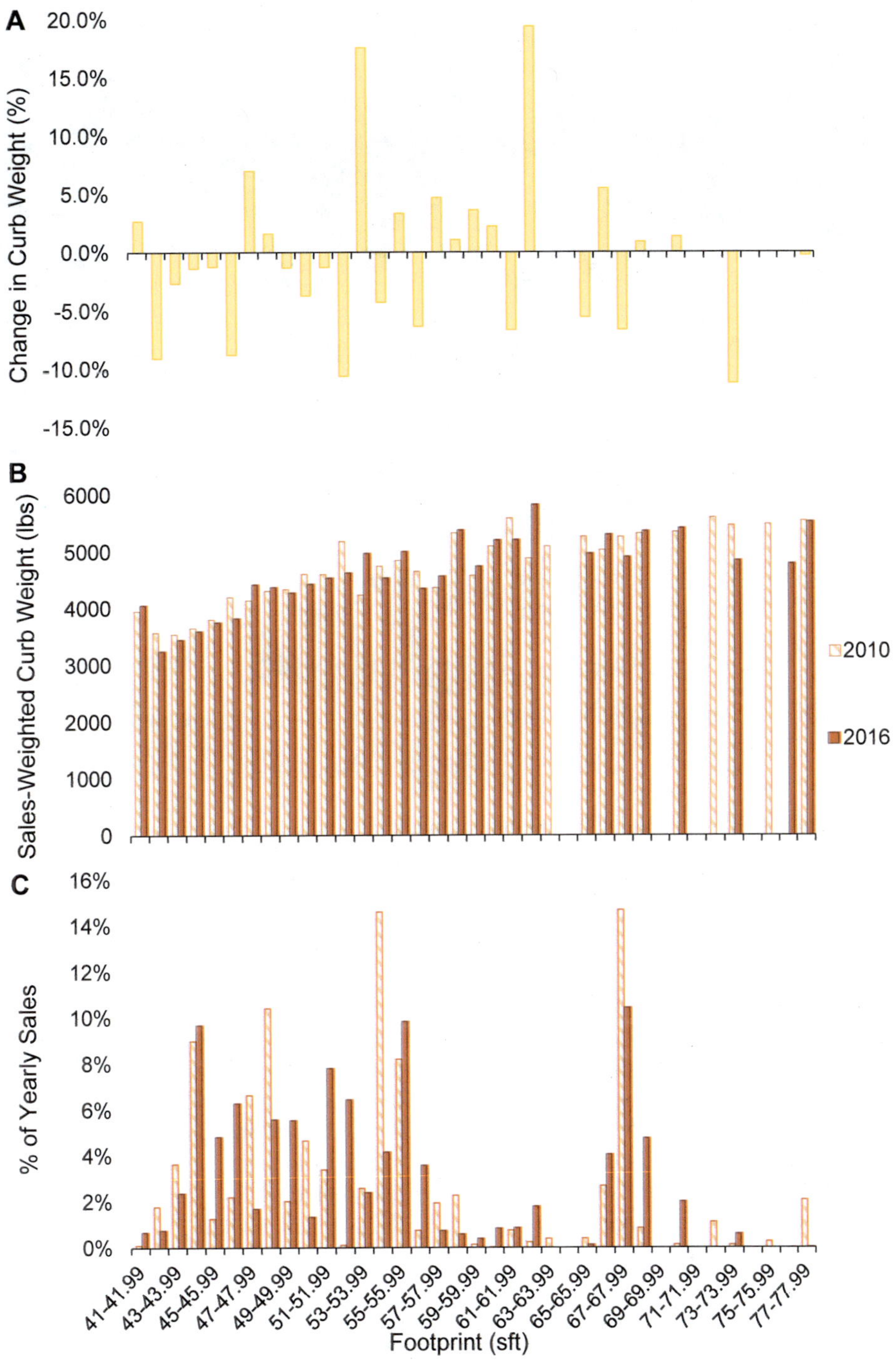

FIGURE 7.7 Comparison of weight and footprint of vehicles classified as light trucks in the MY 2010 and 2016 fleets. All values are sales-weighted. The figure illustrates that light trucks generally get heavier with footprint at low footprints (which includes many crossovers), leveling off at higher footprints (which includes most pickup trucks) (B); that there is no trend in individual footprints getting heavier over time (A); and that there is a shift in sales to trucks with larger footprints between 2010 and 2016 (C), although the details of the trend are not as clear as with passenger cars.

SOURCE: Committee generated using model-by-model 2010 and 2016 MY data released as part of NHTSA and EPA rulemakings.

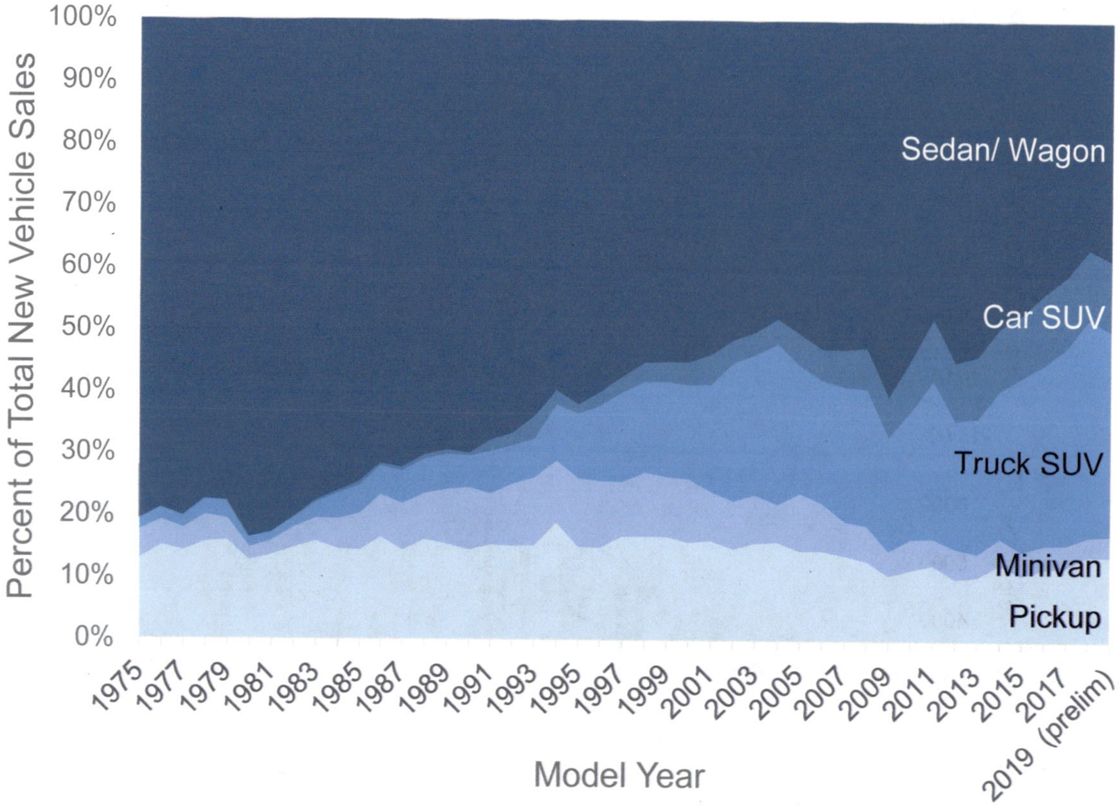

FIGURE 7.8 Vehicle classes over time, showing the reduction in market share of sedans/wagons and minivans, and the increase in car SUVs and truck SUVs. The total share of vehicles classified as trucks (truck SUV, minivan, and pickup) was approximately 50% of vehicles in 2019, up from about 20% in 1975.
SOURCE: Committee generated using data from EPA (2020).

from ~200 to 2000 Mpa (Figure 7.9), and all grades have lower manufacturing costs than other advanced materials (Bailo et al., 2020). Use of advanced high-strength steel (AHSS) and ultra-high-strength steel (UHSS) provide 10% and 25% weight savings, respectively, compared to mild steel (Bailo et al., 2020). The auto industry's history and familiarity with steel manufacturing is another incentive for continued widespread use of the material.

Looking ahead, the steel industry aims to improve both technologies and sustainability in their manufacturing processes. In particular, the industry will focus on improving material formability simultaneously with increased material strength. Most steelmakers are also working to decrease carbon emissions through infrastructure updates such as incorporation of electric arc furnaces. Considering the planned implementation of more UHSS and improved forming technologies, incorporation of 50%–55% of steel in vehicles can be expected in 2025–2035.

However, the penetration of steel in the future fleet will also depend heavily on the status of advancements in other materials and technologies. For example, significant decreases in the cost of aluminum and composite polymers may entice automakers to prioritize those materials, because they can provide 30%–60% more mass reduction than steel. Mass add-back from advanced technologies could also force automakers to move away from steel for lightweighting purposes (Bailo et al., 2020).

7.2.1.2 Aluminum and Aluminum Alloys

Aluminum (Al) provides 35%–40% reduction in mass compared to mild steel (Bailo et al., 2020), and aluminum technology is likely to support 1.5–2.0 miles per gallon fuel savings (Summe, 2019). Aluminum's high

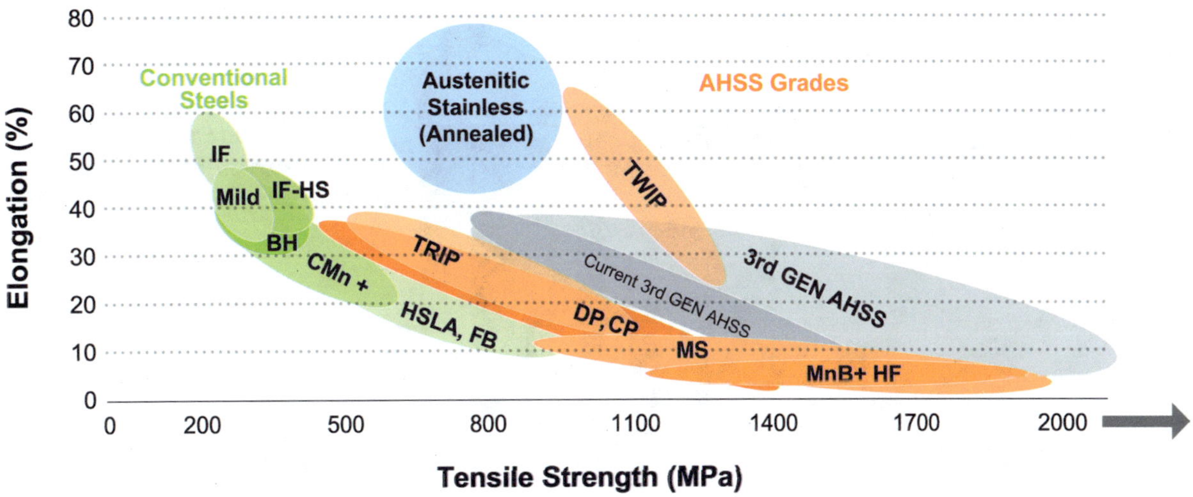

FIGURE 7.9 Percent elongation versus tensile strength for various generations of high-strength steel. First-generation provided strength; second-generation provided strength and ductility; and third-generation provided strength, ductility, and usability. SOURCE: WorldAutoSteel (n.d.).

recyclability also makes it attractive to manufacturers, who are increasingly committed to meeting sustainability targets. There has been a 63% increase in aluminum use in vehicles from 2012 to 2020 (Ducker Worldwide and Aluminum Association, 2017), and this amount is expected to double by 2040 (Modi and Vadhavkar, 2019). The main opportunities for increased Al use are in doors and other bolt-on components, and Al is projected to comprise 20%–25% of body-in-white and closures by 2035 (Bailo et al., 2020). The Ford F-150 is the best example of the role Al can support in the lightweighting, incorporating 90% Al or Al alloys in an aluminum-intensive body with a combined steel/Al frame and corresponding weight reductions in the engine (often by enabling a smaller engine displacement) and chassis. These modifications result in a 700 lb curb weight reduction, which improves fuel economy by 19%. As depicted in Figure 7.3 above, EVs will require similar amounts of lightweighting to compensate for their increased propulsion system weight. Al is being implemented in those cases—for instance, in EV battery boxes and vehicle bodies.

Advances in manufacturing during 2020–2035 could impact the likelihood of using Al for lightweighting. For instance, ongoing efforts in industry are developing 6xxx and 7xxx grades of aluminum (Bailo et al., 2020). By 2035, ultra-high-strength aluminum is projected to be commercially available (Figure 7.10), which will increase the applicability of Al throughout the vehicle and potentially produce about 45% weight reduction in certain components (Figure 7.11). Improvements to continuous casting processes could reduce the manufacturing costs of Al sheets, providing further incentive to incorporate Al materials. However, significant technological advances and investments in other areas, such as batteries and ADAS systems, could make automakers less willing to pay for lightweighting with aluminum (Bailo et al., 2020).

7.2.1.3 Polymer Composites and Carbon Fiber

Polymer composites and carbon fiber provide significant mass reduction opportunities, weighing about 50% less than steel and 30% less than aluminum. Composites are corrosion-free and can be designed to be long-lasting, and have no fatigue, high stiffness, high tensile strength, excellent thermal properties, and low thermal expansion. Furthermore, the use of polymer composites and carbon fiber allows for part consolidation and decreases tooling costs (Bailo et al., 2020).

Polymer composites have been used in light-duty vehicles since the 1960s (Figure 7.12; ACC, 2019), and carbon fiber saw initial implementation in the 1990s. In the North American light-duty fleet, usage of polymer

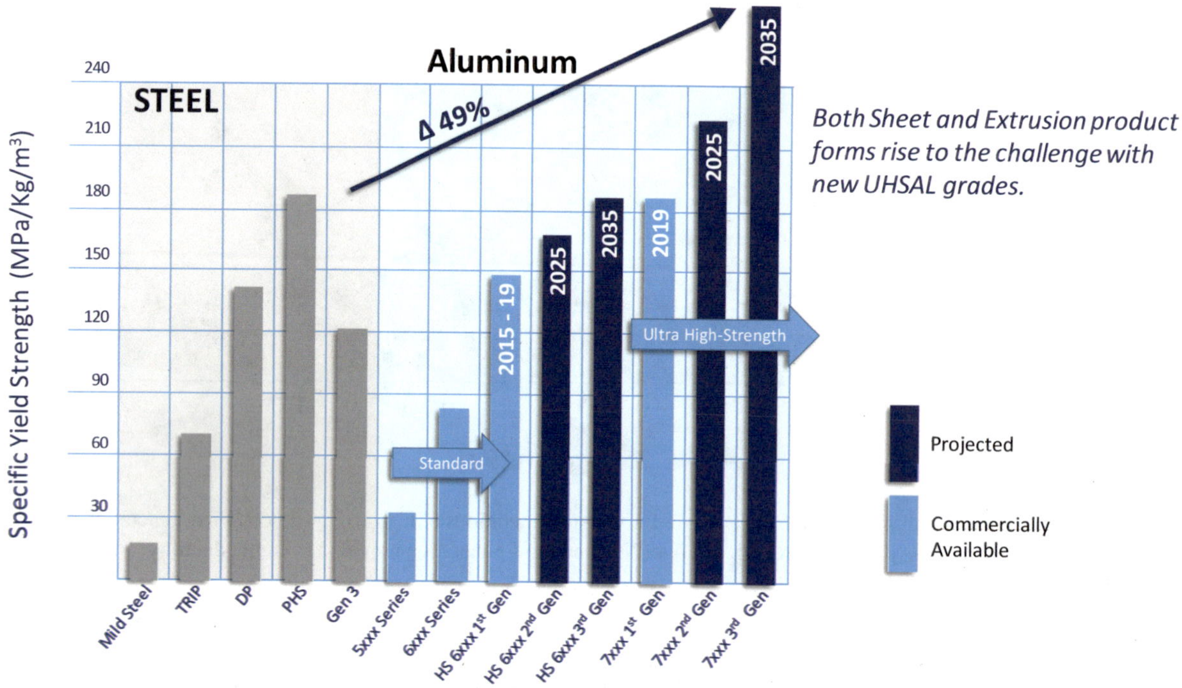

FIGURE 7.10 By 2035, high-strength Al will enable lightweighting of more structural and safety-critical components, leading to up to 45% lightweighting of those components, relative to generation-3 HSS. NOTE: HSS = high-strength steel; UHSAL = ultra-high-strength aluminum.
SOURCE: Summe (2019).

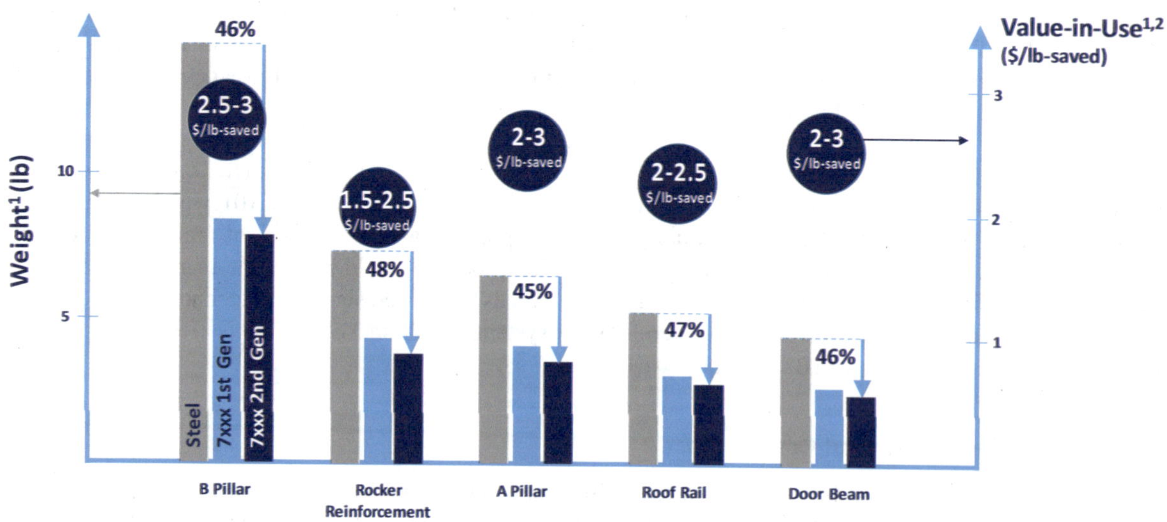

FIGURE 7.11 Components, weight-saved, and value in use for the primary mass reduction of several vehicle components in transitioning from steel to first- and second-generation 7xxx aluminum alloys. Aluminum use leads to approximately 45% weight reduction at a cost of approximately $2.5/lb. NOTE: (1) High Strength 7xxx Aluminum Alloys: Design and Business Case for Automotive Applications, Bad Nauheim, April 2019 (2) Does not include possible cumulative secondary weight savings such as engine reduction.
SOURCE: Summe (2019).

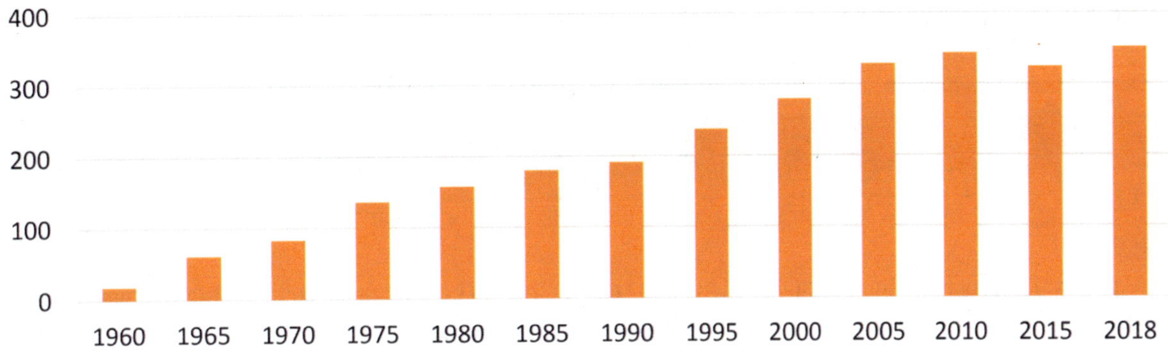

FIGURE 7.12 Usage of polymer composites and plastics in the North American light-vehicle (pounds/vehicle) fleet since 1960.
SOURCE: ACC (2019).

composites and plastics has remained fairly constant at 300–350 lb/vehicle, or 8%–10% of total vehicle weight, since 2008 (Table 7.6, above). Figure 7.13 depicts the variety of polymer and plastic materials used in vehicles in 2018 and their average usage by weight (ACC, 2019). Carbon fiber was first used primarily in sports cars and at low production volume; however, the introduction of BMW's i3 in 2013 moved these composite materials into mass production, and opportunities for carbon fiber and other polymer composites continue to grow. Composite materials have the potential to be incorporated into many vehicle components, including liftgate, door inner, fender, roof panel, front bulkhead, floor reinforcement, A/B pillar reinforcement, truck bed, and seats, and could account for 8%–12% of vehicle composition in 2025–2035 (Bailo et al., 2020).

The production processes for polymer and carbon fiber composites have improved over the past 30 years. The first processes were batch with "Autoclave" production process at low volume with high cost. Resin transfer molding is used today for mass production, but it comes with a large amount of unused waste material. Further recyclability and reuse will be vital to the growth in incorporation of polymer materials in automotive applications.

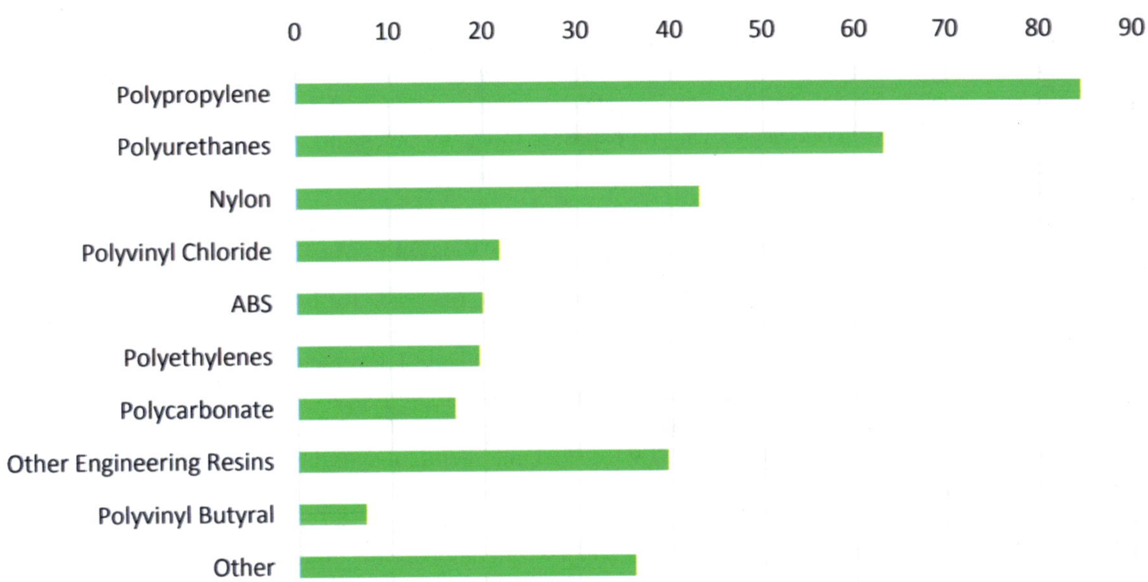

FIGURE 7.13 Average polymer and plastic use by material and weight (pounds/vehicle) in the 2018 North American light-duty vehicle fleet.
SOURCE: ACC (2019).

TABLE 7.8 Material Cost of Automotive Grade Carbon Fiber

Year	$/lb
2005	18
2015	11
2030	5.5

Future production processes include "pultrusion" and "tape laying." These methods combine high volume and no waste of material, which will reduce the component cost and allow the auto industry to introduce more lightweight material in the future.

Nonetheless, there remain several major barriers to incorporating polymer composite and carbon fiber materials. Although the material cost of automotive grade carbon fiber has decreased significantly in recent years (Table 7.8), both the raw material and manufacturing costs of composites, in $/lb of material, are expected to be significantly higher than those for metals in 2025–2035. The current precursor material to carbon fiber, polyacrylonitrile (PAN), is hydrocarbon-based, so the cost of carbon fiber materials depends largely on oil and gas prices. The price of oil is expected to stay in the $30 per barrel of oil range through 2030 (PrimeXBT, 2020). In the coming decade, PAN could be replaced step-by-step with an alternative material like lignin, a naturally occurring compound found in trees. This replacement should result in lower material costs, because lignin is a by-product of the papermaking process and therefore more widely available and less expensive than PAN. In addition to overcoming cost barriers, implementation of carbon fiber and polymer composites will require advancements in tooling, joining, and design (Bailo et al., 2020).

7.2.1.4 Magnesium

Magnesium is typically a die-cast part with relatively good strength and ductility and offers 60%–70% mass reduction compared to mild steel (Bailo et al., 2020). Magnesium is also plentiful and fully recyclable, which is attractive in terms of sustainability. However, current usage is low, at around 1% of the vehicle's total material. To date, magnesium has primarily been utilized on higher end products, and components include engine parts, steering components, instrument panel, and seats. Future opportunities for magnesium include incorporation into vehicle front-end components and powertrain castings, but magnesium is not expected to exceed 4%–6% of the body-in-white and closures. This lack of opportunity relates in part to concerns about corrosion resistance, which limits its application to internal components with no exposure to weather elements. Other barriers are low formability, high cost, limited supply chain, and challenges with joining.

7.2.2 Manufacturing Issues and Opportunities Related to Mass Reduction

Manufacturers consider many factors in their decisions about materials for lightweighting. The choice of material depends not only on that material's physical properties but also on its sustainability and availability, particularly given the increasing globalization of the automaker industry. For each material, manufacturers must develop new design models, forming and joining technologies, and tooling processes. These technologies and processes then need to be scaled up to achieve high production volumes. Current and projected manufacturing costs per lb mass reduction are reported in Table 7.9 for steel, aluminum, magnesium, and polymer composites. Incorporating new materials also requires increased capital expenditure and additional safety considerations to minimize the risk of part failure. The lightweighted cars must continue to meet consumer demand for performance and noise, vibration, and harshness. In addition to material considerations, a manufacturer's willingness to invest in mass reduction depends on regulatory issues and technological developments in other areas. Interviews with a variety of automakers identified the primary drivers for lightweighting decisions as CAFE/GHG regulations, the amount of electrification in the fleet, battery cost and density, and mass add-back from advanced technologies (Bailo et al., 2020).

TABLE 7.9 Manufacturing Costs for Different Materials, 2020–2035

	Manufacturing Cost ($/lb)			
Material	2020	2025	2030	2035
Mild steel	0.34	0.34	0.34	0.34
HSS	0.30	0.30	0.30	0.30
AHSS	0.39	0.37	0.36	0.35
UHSS (HF)	0.40	0.39	0.38	0.37
Aluminum	0.73	0.68	0.64	0.60
Magnesium	0.65	0.62	0.60	0.57
Composites/carbon fiber	14.76	13.10	11.63	10.32

SOURCE: Bailo et al. (2020), costs adjusted from 2019$ to 2018$ and from $/kg to $/lb (1 kg = 2.2 lb).

7.2.3 Overview of Materials for Mass Reduction

The projected changes in implementation of materials across all U.S. light-duty vehicles in 2020–2040 are shown in Figure 7.2 above. The costs per lb of mass reduction for unibody cars/SUVs and pickup trucks with various possible material substitution types are depicted in Figures 7.14 and 7.15, respectively.[2] These plots of cost per percent mass reduction consider lightweighting from materials substitution only, not from modification or removal of vehicle components as would be done in a full design optimization. The effectiveness of material substitution depends on the deployment of different materials and the resulting mass reduction (taking into account secondary mass reduction opportunities).

Mapped onto Figures 7.15 and 7.16, and summarized in Table 7.10, are estimates for material penetration (and the corresponding mass reduction and cost) based on scenarios[3] that describe potential future fleets. The future fleets differ in electrification volume, battery pack cost, and battery energy cell density, three variables identified by automakers as key driving factors in their willingness to pay for lightweighting (Bailo et al., 2020). Definitions of the variables and scenarios are given in Table 7.11. For both vehicle classes, scenario one is predicted to represent the mass market in 2025–2030, and scenario three is predicted to represent the mass market in 2030–2035. However, it should be noted that the scenarios do not report on what automakers could do to achieve the most lightweighting but rather what they might do in the context of other available technology options and regulatory standards.

7.2.4 Design Optimization

The automotive industry has emphasized that design optimization is key to selecting the "right" material for the "right" application. The following elements must be considered: (1) fuel economy importance in the vehicle class, (2) price sensitivity for vehicle class, (3) volume and profit margin of the vehicle class, (4) safety and regulatory considerations, (5) customer expectation, (6) manufacturing methodology and cost, and (7) sustainability cost (scrap/recycle/reuse). All of these factors are taken into consideration when designing the initial vehicle. Weight reduction after initial design is very difficult, given the need to meet other durability and safety requirements, and is not usually a cost improvement.

[2]The data for these plots of cost per percent of mass reduction were provided by interviews with automakers and independent organizations such as the American Iron and Steel Institute and the American Composites Corporation, and the numbers were validated with other documents. Both these data and engineering judgment by the Center for Automotive Research were used in developing the figures. The plots are meant to provide guidelines, not guarantees or standards. Several factors could influence prices and allow different materials to be utilized.

[3]These scenarios provide estimates for when certain material trends might be observed in the fleet, either in premium or mass-market vehicles; however, the mass reduction levels and corresponding costs are not limited to the year(s) indicated by the scenarios.

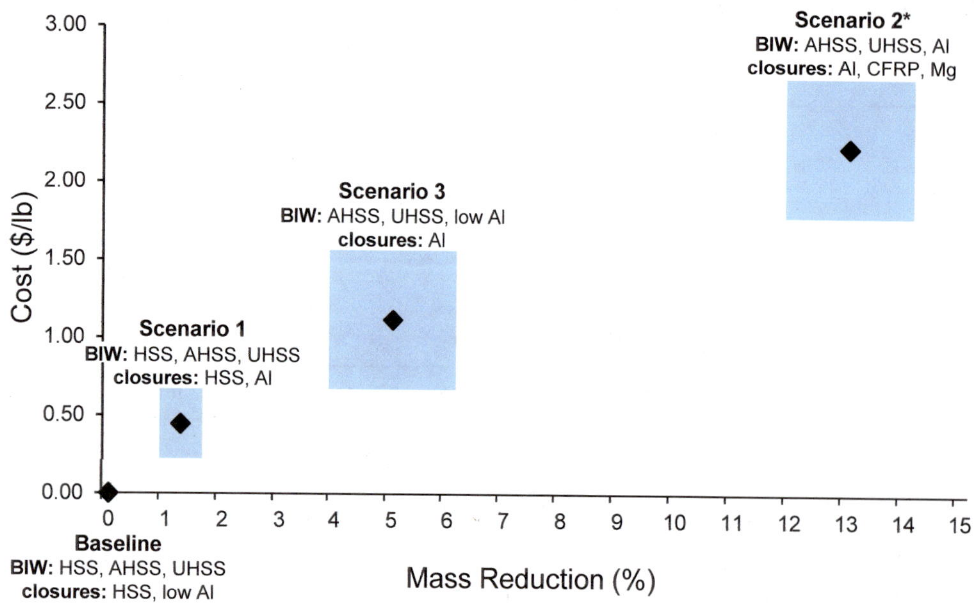

FIGURE 7.14 Cost ($/lb of material) for different levels of mass reduction from materials substitution in unibody cars and SUVs. NOTE: Black diamonds indicate a representative vehicle in each scenario, and blue boxes denote uncertainty in percent mass reduction and cost within that scenario. * indicates inclusion of secondary mass reduction, calculated as described in NRC (2015). SOURCE: Committee generated, adapted from Bailo et al. (2020).

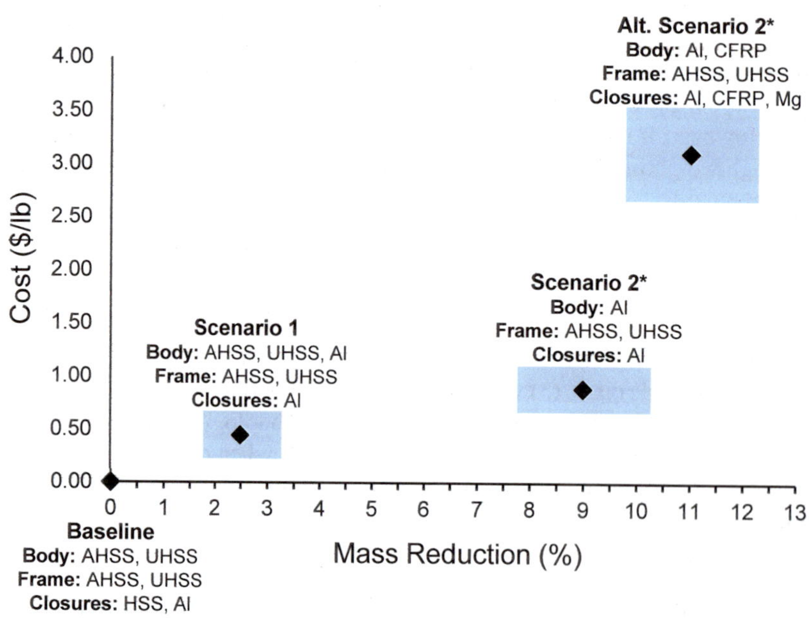

FIGURE 7.15 Cost ($/lb of material) for different levels of mass reduction from material substitution in pickup trucks. NOTE: Black diamonds indicate a representative vehicle in each scenario, and blue boxes denote uncertainty in percent mass reduction and cost within that scenario. * indicates inclusion of secondary mass reduction, calculated as described in NRC (2015). SOURCE: Committee generated, adapted from Bailo et al. (2020).

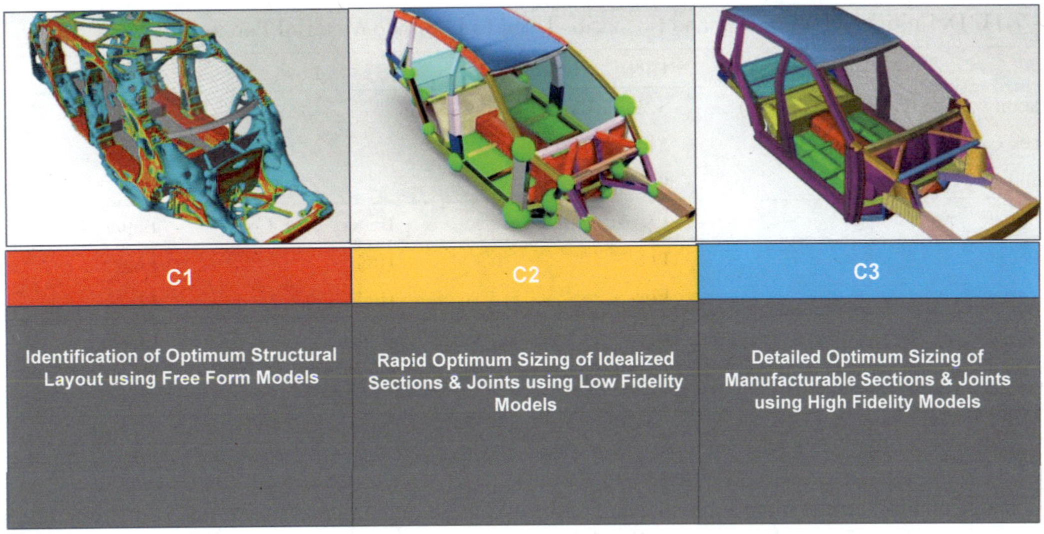

FIGURE 7.16 Design process showing the use of computer modeling to identify the structural layout, and low-detail and high-detail optimum sizing of joints and sections for a vehicle body-in-white.
SOURCE: Yen (2020).

TABLE 7.10 Projected Costs, Mass Reduction, and Material Trends for Potential Scenarios in 2025–2035

Vehicle Class	Scenario	Cost[a]	Mass Reduction[b]	Expected Material Trend
Cars and Unibody SUVs	Baseline	N/A	N/A	Body: HSS, AHSS, UHSS Closures: HSS, low Al
	One	\$0.22–\$0.67	1.0%–1.5%	Body: HSS, AHSS, UHSS Closures: HSS, Al
	Two	\$1.78–\$2.67	12%–14%[c]	Body: Al, AHSS, UHSS Closures: Al, comp, Mg
	Three	\$0.67–\$1.56	4%–6%	Body: AHSS, UHSS, low Al Closures: Al
Pickup Trucks	Baseline	N/A	N/A	Body: AHSS, UHSS Frame: AHSS, UHSS Closures: HSS, Al
	One	\$0.22–\$0.67	2%–3%	Body: AHSS, UHSS, Al Frame: AHSS, UHSS Closures: Al
	Two	\$0.6 –\$1.11	8%–10%[c]	Body: Al Frame: AHSS, UHSS Closures: Al
	Two (alternative)	\$2.67–\$3.56	10%–12%[c]	Body: Al, comp Frame: AHSS, UHSS Closures: Al, comp, Mg
	Three	\$0.05–\$0.45	2%–3%	Body: AHSS, UHSS, Al Frame: AHSS, UHSS Closures: Al

[a]Costs reported per lb mass reduction, converted from reported per kg values (1 kg = 2.2 lb) and from 2019\$ to 2018\$.
[b]Mass reduction reported as percent reduction in curb weight from MY 2020 baseline.
[c]Includes secondary mass reduction, calculated as described in NRC (2015).
SOURCE: Bailo et al. (2020).

TABLE 7.11 Definitions of Variables and Scenarios Used to Estimate Material Penetration

Variable	High	Low	
Electrification volume (CAFE/GHG proxy)	>25% BEV, 30%–50% hybrids	<15% BEV, 20%–25% hybrids	
Battery pack cost	$145–$170/kWh	<$100/kWh	
Battery cell energy density	900 Wh/liter	700 Wh/liter	
Scenario	**Electrification Volume**	**Battery Pack Cost**	**Battery Density**
Baseline	Low	High	Low
One	Low	High	Low
Two	High	High	Low
Three	High	Low	Low

SOURCE: Bailo et al. (2020).

Vehicle design has become more sophisticated, with computer modeling and simulation being used throughout the design process and with both design and validation moving to virtual vehicles. Use of modeling and simulation in vehicle design allows performance targets to be met for the whole vehicle design while optimizing cost and weight reductions. Figure 7.16 shows the design of vehicle components based on a structural layout identified using modeling. Figure 7.17 shows details of the process for vehicle and component designs that meet specifications while minimizing cost, weight, or other parameters.

The automotive industry must continue to improve fuel economy and/or EV range for a customer who expects these improvements but is unwilling to pay for them. To address this challenge, automakers will look for weight reduction materials/methods with a corresponding cost reduction or at least level cost. For example, using carbon fiber in lieu of metals may reduce both weight and cost owing to its ability to greatly reduce the complexity of the component, and hence the assembly costs.

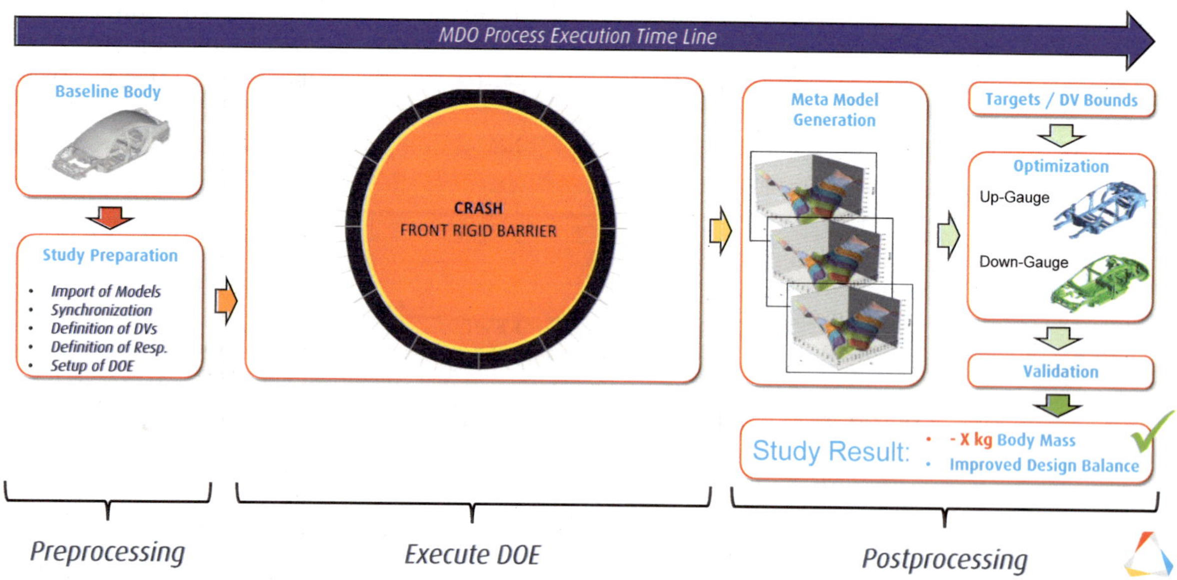

FIGURE 7.17 A process to simulate vehicle requirements, generate meta models, and optimize vehicle and component design to result in improved performance with mass reduction.
NOTE: DOE = design of experiment; DV = design variable; MDO = multi-disciplinary optimization.
SOURCE: Yen (2020).

7.2.5 Summary of Mass Reduction Opportunities in 2025–2035

In 2025–2035, as the fleet trends toward greater electrification and a higher penetration of advanced technologies, mass reduction through lightweighting and design optimization will be critical for meeting fuel economy and performance targets. Material use is expected to shift away from mild and high-strength steel and toward primarily advanced and UHSS and aluminum. Magnesium, polymer composites, and carbon fiber are projected to contribute 12%–18% of vehicle weight. An automaker's material choice and willingness to pay for lightweighting depend on a variety of factors, including material availability, manufacturing cost, technological developments in other areas, and regulatory requirements. Design optimization, particularly in the initial vehicle design, is also critical for maximizing the benefits of using advanced materials for mass reduction. These potential changes in mass from electrification, ADAS, and advanced materials are further discussed in the context of vehicle safety in Section 7.5.

7.2.6 Findings and Recommendation for Mass Reduction

FINDING 7.4: Lightweighting represents the greatest opportunity for fuel economy improvement in road load and accessory reduction. There have been many breakthroughs in high-strength steel, aluminum alloys, and composites, as well as manufacturing methodologies, to allow further implementation.

FINDING 7.5: Conventional internal combustion engine and hybrid models gain the most improvement from primary and secondary mass reduction. BEV platforms, also known as skateboards, have the opportunity to be utilized across many body styles that can communize lightweighting strategies and optimize aerodynamic parameters.

FINDING 7.6: The key to achieving the benefits of substitution with lighter weight materials is early consideration in the design process, relevant for high-strength steel, aluminum, magnesium, and composites. Design optimization in the planning stage of a new model far outweighs subsequent model year opportunities, because changes post-design finalization are often costly and not nearly as substantial.

FINDING 7.7: Lightweighting can offset increased mass resulting from the addition of electrification and advanced driving technologies (mass increase for battery electric vehicles can be approximately 500 lb). Electric vehicles are incentivized to reduce mass because doing so also improves their range, a key parameter for consumers. In order to offset increased mass, electrified, high profit/volume vehicle models will likely integrate alternative materials first. After technology improvement allows the cost to be reduced, the new materials will be installed on lower volume/profit models. To date, it has proven difficult to fully offset the weight of the battery, but implementation on higher performance, higher-margin products will allow for more discretionary budget allocated for lightweighting. Furthermore, significant research efforts are ongoing to reduce the mass of the battery pack itself. (See Chapter 5.)

RECOMMENDATION 7.2: The U.S. Department of Energy should support the development of alternative materials and manufacturing methodologies to allow mass reduction in electrified and safety system applications.

7.3 TIRES

Rolling resistance accounts for 4%–7% of the energy use of a typical gasoline ICE vehicle and greater shares for hybrids and BEVs (DOE/EPA, 2019b, c, d). Rolling resistance in automotive tires is primarily caused by the energy dissipated when tires are deformed by contact with the road. The force, F, required to overcome rolling resistance is equal to the downward force perpendicular to the road surface owing to the weight of the vehicle, N, multiplied by the coefficient of rolling resistance, C_{rr} (NRC, 2015):

$$F = C_{rr}N$$

For a given C_{rr}, rolling resistance is a function of vehicle weight and changes little with vehicle speed (EPA/DOT/CARB, 2016). A tire's C_{rr} is determined by the materials of which it is made, as well as its structural design, aspect ratio, tread pattern, and depth (NRC, 2015).

7.3.1 Trade-Offs

When designing tires, manufacturers must consider many factors beyond rolling resistance that are important to safe handling and braking or to consumer satisfaction. In addition to braking and traction on wet and dry pavement, tires affect steering and must stand up to potholes and other deviations from ideal road conditions. In addition to safety, consumer concerns include durability, handling, ride comfort, noise, and cost (EPA/DOT/CARB, 2016). Although there can be trade-offs among these attributes, tire manufacturers point out that trade-offs can be overcome at a cost either in terms of higher prices for new tires or reduced durability, requiring more frequent replacement (NRC, 2015). The SAFE Rule for 2021–2026 asserts that there are currently no data connecting low rolling resistance tires to accident or fatality rates and that tire makers are able to balance stopping distance and rolling resistance (NHTSA/EPA, 2020).

7.3.2 Impact on Fuel Economy

A widely used rule of thumb is that a 10% reduction in rolling resistance (C_{rr}) can improve the fuel economy of an ICE vehicle by 1%–2% (TRB, 2006; EPA/DOT/CARB, 2016), and if the powertrain can be downsized to maintain equal performance, the benefit could be as large as 3% (NRC, 2015). The fuel economy benefit of reducing rolling resistance depends to a great extent on the fraction of energy input to the vehicle that is used to provide power to its wheels to overcome inertia, aerodynamic drag, and rolling resistance. The energy requirements and losses vary with vehicle class, as illustrated in Figures 7.18–7.20 for conventional gasoline, hybrid, and electric-only vehicles, respectively. For a typical gasoline vehicle, only 16%–25% of the energy in gasoline goes to powering the wheels, with approximately 25% of that energy (or 4%–7% of the total energy input) going to overcoming rolling resistance (DOE/EPA, 2019b; Pannone, 2015). The energy losses of a hybrid vehicle are smaller: 24%–38% of the energy in gasoline powers the wheels and so a larger fraction, 6%–11%, of the total energy input goes to overcoming rolling resistance (DOE/EPA, 2019c). Electric-only vehicle powertrains are much more efficient. Because some of the energy losses can be recovered by regenerative breaking, 77%–82% of the energy to the vehicle ends up powering the wheels.[4] As a result, a much higher fraction, 22%–23%, of the energy input into an EV is used to overcome rolling resistance (DOE/EPA, 2019d), making the energy efficiency of EVs 3 to 5 times more sensitive to the C_{rr} of their tires. Therefore, the importance of reducing rolling resistance in hybrid vehicles is about 50% greater than for an ICE vehicle, and the importance in BEVs is about 3 to 4 times as great.

7.3.3 Coefficients of Rolling Resistance

The SAFE Rule (NHTSA/EPA, 2020) used 0.009 as the baseline value for average rolling resistance[5] but assigned individual rolling resistance coefficients to specific vehicles based on confidential business information provided by manufacturers. With this baseline C_{rr} value, a 10% reduction (ROLL10) corresponds to a C_{rr} of 0.0081, and a 20% reduction (ROLL20) corresponds to a C_{rr} of 0.0072. The choice of 0.009 was based on information from several sources, including the Rubber Manufacturers Association and Pannone (2015) (Figure 7.21). The C_{rr} estimates developed by Pannone reflect MY 2014 vehicles. As shown in Table 7.12, the C_{rr} values vary by the design intent of vehicles but also show wide variability across makes and models within a design category.

[4]The estimated electricity losses during charging assume a 40% state of charge, 42A EV service equipment, breakers and transformer and 50A AC current (Apostolaki-Iosifidou et al., 2017, Tables 2–4).

[5]Rolling resistance can be expressed by the rolling resistance coefficient (RRC or C_{rr}), which is the value of the rolling resistance force divided by the wheel load.

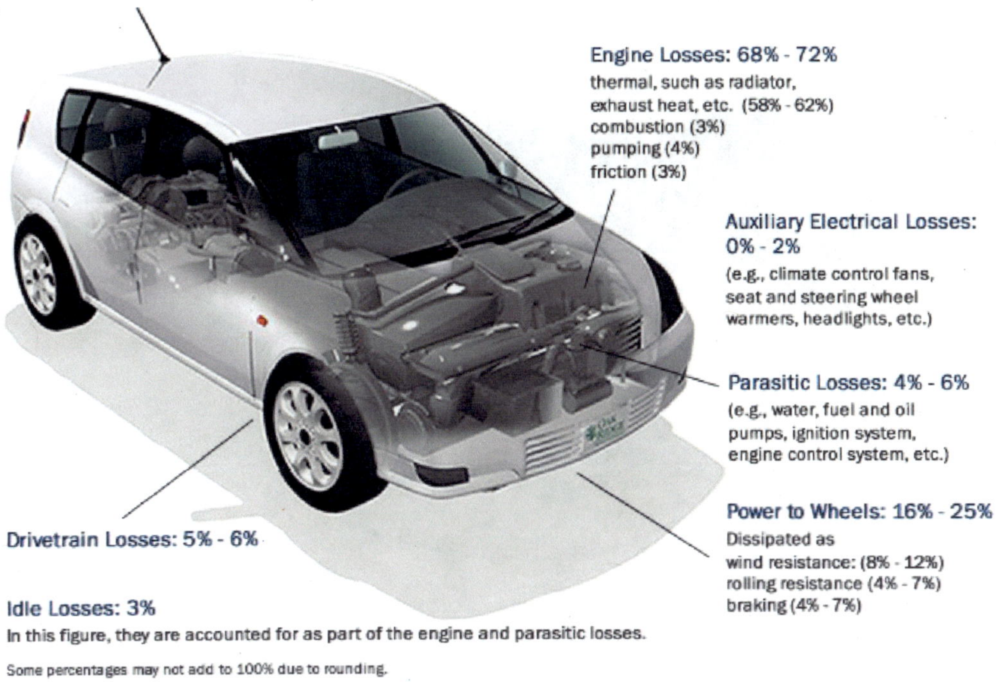

FIGURE 7.18 Diagram of energy requirements for combined city/highway driving for gasoline vehicles, showing the power to the wheels after engine losses, parasitic losses, drivetrain losses, auxiliary electric losses, and idle losses.
SOURCE: DOE/EPA (2019b).

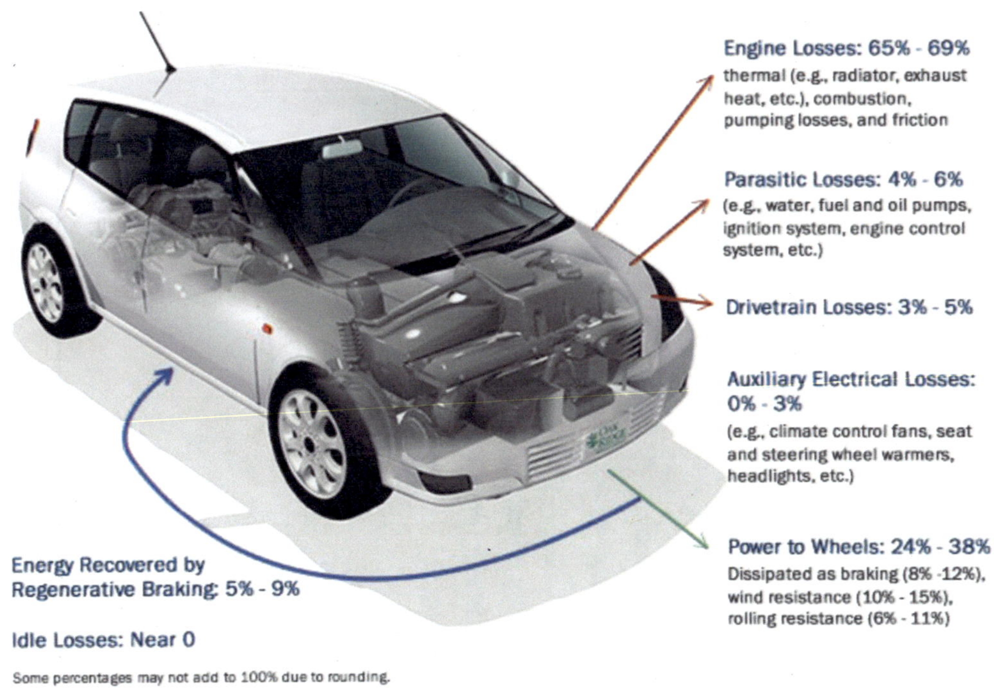

FIGURE 7.19 Diagram of energy requirements for combined city/highway driving for hybrid vehicles, showing the power to the wheels including regenerative braking energy after engine losses, parasitic losses, drivetrain losses, auxiliary electric losses, and idle losses.
SOURCE: DOE/EPA (2019c).

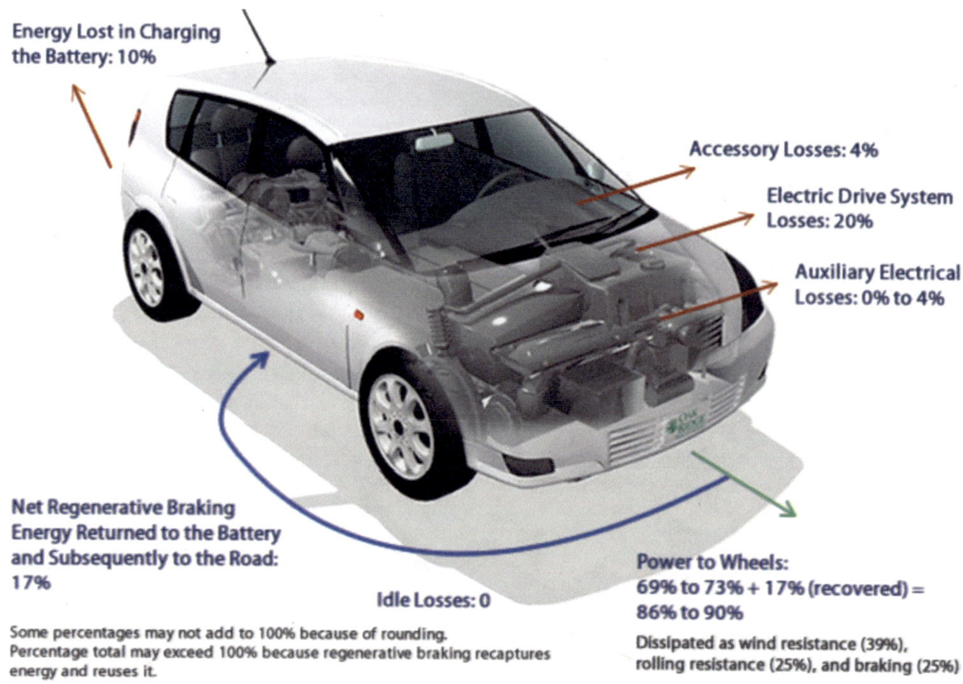

FIGURE 7.20 Diagram of energy requirements for combined city/highway driving for EVs, showing the power to the wheels including regenerative braking energy after losses from battery charging inefficiency, parasitic losses, electric drive losses, auxiliary electric losses, and idle losses.
SOURCE: DOE/EPA (2019d).

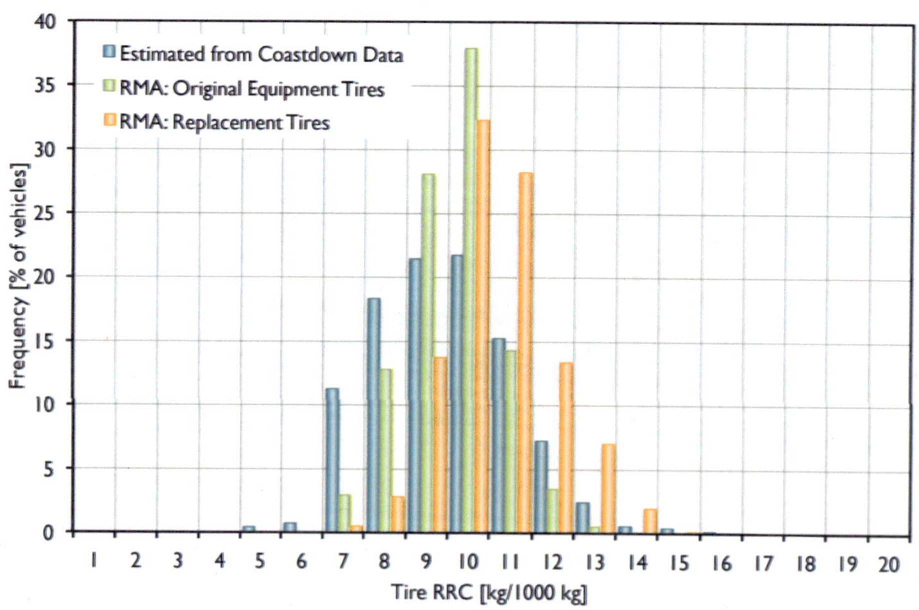

FIGURE 7.21 Tire rolling resistance values (RRC, in kg/1,000 kg) for both original equipment and replacement tires.
SOURCE: Pannone (2015).

TABLE 7.12 Estimated Distributions of Tire Rolling Resistance Coefficients (in kg/1,000 kg) by Design Category

Vehicle/Tire Category	Sample Size	Average RRC	Minimum RRC	90th Percentile RRC	75th Percentile RRC	Maximum RRC
All vehicles	1,358	9.0	4.4	6.9	7.8	15.1
Fuel economy oriented	74	8.1	4.4	6.2	7.4	10.8
Balanced	1,083	8.9	4.8	6.9	7.7	15.1
Off-road oriented	17	9.4	7.1	7.7	8.3	12.0
Performance oriented	184	10.1	6.1	7.8	8.9	14.5

SOURCE: Pannone (2015).

The base year C_{rr} value chosen by the agencies (0.009) corresponds exactly to the average for all vehicles in Table 7.12. Pannone (2015) suggests that the most extreme values in Table 7.12 likely reflect estimation errors rather than actually achieved C_{rr} values.

Although low rolling resistance tires are installed on new vehicles by automakers, the choice of replacement tires is at the discretion of the vehicle owner. According to Pannone (2015) and information provided by the Rubber Manufacturers Association, replacement tires have higher levels of rolling resistance than original equipment tires (see Figure 7.21, above). To encourage vehicle owners to choose low rolling resistance tires, the Tire Efficiency, Safety, and Registration Act of 2015 required the U.S. Department of Transportation to develop minimum fuel-efficiency standards for tires and to create a consumer information program for tire fuel efficiency, traction, and durability (PL 114-94, Part III, 2015). However, current tire labeling by the U.S. government does not yet include information on rolling resistance.

7.3.4 Near-Term Technologies

The rolling resistance of automobile tires can be improved in a variety of ways, including increasing the inflation pressure, changing materials, optimizing tire construction for low hysteresis, changing tire geometry (e.g., lower aspect ratio), and reducing sidewall and tread deformation. NRC (2015) reported that some tire manufacturers had lowered their tires' C_{rr} by 2% per year for at least 30 years, which would make the C_{rr} of 2015 tires about 45% lower than that of tires available in 1985. The 2015 NRC report also reported that C_{rr} values measured in 2005 ranged from 0.00615 to 0.01328 with a mean of 0.0102 (NRC, 2015). Research supporting the rulemaking for MYs 2017–2025 considered two levels of reductions in rolling resistance, assuming a combination of design and material changes: (1) a 10% reduction in C_{rr} (ROLL10), giving a 1.9% reduction in fuel consumption over the base 2017 tire; and (2) a 20% C_{rr} reduction (ROLL20), giving a 3.9% reduction in fuel consumption over the base tire (EPA, 2016). The 10% reduction was considered achievable through a combination of increased tire diameter and sidewall stiffness and reduced aspect ratio (also reducing rotational inertia). ROLL20 was assumed to require more advanced materials and complete tire redesign. Silica tread technology in combination with a new silica, polymer, and coupling agent were considered key technologies for ROLL20.

According to EPA/NHTSA (2012), ROLL10 first entered the market in 1993 and had achieved a degree of widespread adoption by 2008. At the time that the 2017–2025 rule was finalized (EPA/NHTSA, 2012), ROLL20 tires were not available in the marketplace. ROLL20 was assumed to be available in MY 2017 and to begin replacing ROLL10 afterward. ROLL20 tires were assumed to become widely available in the marketplace in 2022–2023 and to penetrate the new vehicle market quickly, reaching a share of 73% by 2021 and 97% by 2025. The DMC of ROLL10 tires was estimated to be $6 per vehicle ($1.20 per tire for five tires). ROLL20 was estimated to add $66 per vehicle, assuming that only four tires would be provided by automakers per vehicle (both estimates are 2013$) (NHTSA/EPA/CARB, 2016). In the SAFE Rule (NHTSA/EPA, 2020), the agencies redefined the base year tire (ROLLO) and changed the base year market penetrations of ROLL10 and ROLL20

TABLE 7.13 Estimated Market Penetrations of ROLL10 and ROLL20 Tire Rolling Resistance Technologies

ROLL	Draft TAR (MY 2015 Baseline)	NPRM (MY 2016 Baseline)	Final SAFE Rule (MY 2017 Baseline)
ROLL0	99.80%	64%	59%
ROLL10	0.1%	10%	21%
ROLL20	0.1%	26%	20%

NOTE: NPRM = Notice of Proposed Rulemaking; SAFE = Safer Affordable Fuel-Efficient [Vehicles]; TAR = Technical Assessment Report.
SOURCE: Table VI-167 of NHTSA/EPA (2020).

tires, substantially increasing the MY 2016 and MY 2017 year estimates versus the previous MY 2015 estimates (Table 7.13).

The SAFE Rule (NHTSA/EPA, 2020) adopted the DMC estimates shown in Table 7.14. Both ROLL10 and ROLL20 tires were assumed to require replacement after 40,000 miles.

7.3.5 Advanced Tire Technologies

The 2012 rule (EPA/NHTSA, 2012) also considered the possibility of a 30% reduction in rolling resistance by 2025. Although it reports that tire suppliers believed that there were innovations that could enable such a reduction, the rule concluded that there was "little evidence supporting improvements beyond LRRT2 by 2025."[6] The U.S. Department of Energy has sponsored tire research with the objective of improving vehicle fuel economy by 3% and reducing tire weight by 20% through a combination of six technological advances (Donley, 2014):

1. Partial replacement of carbon black and silica with nano-fiber materials;
2. Ultra-lightweight tire bead bundle;
3. Ultra-lightweight tire belt package;
4. Ultra-lightweight inner liner (barrier film liner);
5. Formulation options for ultra-long-wearing and low-hysteresis tread compound; and
6. New design of low-hysteresis, energy-efficient tire profile.

The research project concluded that combinations of these technologies had the potential to reduce rolling resistance by 27% to 31% with relatively low commercial and performance risk (Donley, 2014). NHTSA/EPA (2020) judged that a 30% reduction in rolling resistance would require changes to tire profiles, strengthening of tire walls, changes in tread design, integration of tire designs with active chassis control, and development of new materials to replace silica. Active chassis control systems are required to offset the slippage and handling concerns that result from the lower tractive forces of ROLL30. According to NHTSA, no ROLL30 tires are currently commercially available, nor are they expected to become available by 2025 (NHTSA/EPA, 2020). On the other hand, HD Systems asserted that ROLL30 tires could be available by 2025 (NHTSA/EPA, 2020).

Tires with two or more separate air chambers inflated to different pressures have the potential to reduce material hysteresis damping, which accounts for 80%–95% of a tire's rolling resistance during steady-state driving on a level road (Aldhufairi and Olatunbosun, 2017). The effects of alternative multichamber designs were investigated by Aldhufairi et al. (2019) by means of finite element analysis. The most fuel efficient multichamber design reduced rolling resistance by 40% in the simulations with minor trade-offs in grip and ride comfort. Although multichamber designs appear to be able to substantially reduce tire rolling resistance, they are in an experimental stage of development and face challenges with respect to manufacturability, cost, and maintenance.

[6]The 2012 rule referred to a 20% reduction relative to the base year tire as LRRT2.

TABLE 7.14 Estimated Costs (2018$) of Near-Term Tire Technologies in the SAFE Rule

Technology	DMC	Total Cost (includes Retail Price Equivalent Correction and Learning)
ROLL0	$0.00	$0.00
ROLL10	$5.186	$7.78
ROLL20	$40.54	$60.81

NOTE: Costs for each technology are incremental to a baseline vehicle (Base V). Costs for MY 2017, incremental to Base V.
SOURCE: Table VI-168 of NHTSA/EPA (2020).

Tire manufacturers have introduced several advanced tire concepts, ranging from "airless" non-pneumatic tires to magnetically levitated, connected, intelligent tires. While these concepts offer potential improvements such as no blow-outs or flats, greater recyclability, improved handling, and even the ability to be regenerated via 3D printing, they do not claim reduced rolling resistance versus advanced pneumatic tires.

7.3.6 Findings for Tires

FINDING 7.8: Low rolling resistance tires with 0.009 coefficient of rolling resistance (ROLL0) have been implemented to the fullest possible extent in MY 2017. Reductions of a further 10% and 20% are also partially implemented, although potential remains for ROLL10, ROLL20, and even ROLL30 implementation in 2025–2035.

FINDING 7.9: Noninflatable tires are being developed, specifically for urban, shared vehicles, but the impact to fuel economy is not likely to improve upon adoption of pneumatic tires.

7.4 ACCESSORIES AND OTHER OFF-CYCLE TECHNOLOGIES

Additional improvements to accessories and related technologies are off-cycle technologies, meaning that their fuel economy benefits are not captured on the Federal Test Procedure (FTP) or that they impact emissions of non-carbon dioxide (CO_2) GHGs. These technologies mainly reduce the accessory load on the engine or affect the thermal management of the cabin, and they can earn credits that are applied to fuel economy or GHG compliance. Such crediting schemes are primarily discussed in Chapter 12.

7.4.1 Accessories Electrification

Electrifying accessories such as motors and fans improves fuel consumption primarily by reducing the mechanical load on the engine. The most advantageous opportunities for converting from mechanical to electrical devices are with devices that operate only intermittently, such as power steering and the A/C compressor (NRC, 2015). Similar opportunities exist in other areas—for instance, by improving alternators, converting hydraulic water pumps to electricity, and improving the efficiencies of electric cooling fans. While these technologies may not be reflected in the FTP, some benefits from accessory electrification are reflected in the tests used to develop fuel economy labels.

Electric power steering (EPS) provides reductions in fuel consumption by eliminating the need for belt-driven power steering pumps that draw load from the engine even when the wheels are not being turned. EPS is also needed for hybrid and plug-in electric vehicles. The current penetration is around 80% (NHTSA/EPA, 2020; Els, 2017), and EPS will likely be implemented in most of the fleet before the years that are the focus of this study.

The most recent National Research Council fuel economy technology report (NRC, 2015) and recent regulatory documents (NHTSA/EPA, 2020) have lumped several other accessories together, including improved alternators, electrified hydraulic water pumps, and improved electric cooling fans. These documents report two different levels of efficiencies for these accessories. Level one (IACC1) includes a high-efficiency alternator, an electric water pump, and electric cooling fans, whereas level two (IACC2) includes a higher efficiency alternator and improved cooling fans (NRC, 2015). Cost and effectiveness estimates have not changed since the earlier documents. The penetrations of these technologies and their contribution to meeting current fuel economy regulations is being debated in the context of the current fuel economy regulatory activities (NHTSA/EPA, 2018; ICCT, 2018). However, the consensus is that most of these accessory improvements will be implemented in the majority of the fleet before 2025. In fact, the baseline fleet of the 2020 SAFE Rule assumes full incorporation of IACC1 technologies; thus, the reported cost and effectiveness values for IACC in the SAFE Rule are equivalent to IACC2 in previous rules, which represents "high-level" improvements to electric water pumps and alternators (NHTSA/EPA, 2020).

7.4.2 Air Conditioning

Outside of mass reductions, improvements in A/C provide the next largest source of non-powertrain improvements in fuel economy efficiency and GHG emissions. Energy impacts of A/C technologies are particularly important given the high (>95%) penetration of A/C systems in U.S. cars and light trucks (NHTSA/EPA, 2020). A/C improvements stem from reducing engine loads, which improves fuel efficiency, and reducing leakage of coolants and using coolants with lower global warming potentials (GWPs), which reduces non-CO_2 GHG impacts. As noted in the previous National Research Council report, A/C contributes significantly to the on-road efficiency gap between CAFE certification values and real-world fuel consumption because the air conditioner is turned off during the FTP but used by drivers during vehicle operations (NRC, 2015). The technologies used to reduce A/C engine loads focus on the compressor, which circulates the refrigerant within the system; electric motor controls; and system controls. Although reductions in A/C leakage and alternative low-GWP refrigerants do not affect fuel economy, reducing coolant leaks through improved hoses, connectors, and seals and replacing current coolants with lower GWP refrigerants do reduce overall GHG impacts from light-duty vehicle operations.

7.4.3 Tire Off-Cycle Technologies

Tire rolling resistance is sensitive to inflation pressure. A rule of thumb is that each 1 psi reduction in inflation pressure of all tires reduces fuel economy by 0.2% (DOE/EPA, 2019a). According to NHTSA, only 19% of motorists correctly inflate their tires (NHTSA, 2019). A greater concern with underinflated tires, however, is the impact on vehicle safety of loss of handling and traction. NHTSA reports that underinflated tires and other tire maintenance issues contributed to 738 tire-related fatalities in 2017 (NHTSA, 2019). Since 2008, all new passenger cars and light trucks have come equipped with tire pressure monitoring systems that warn motorists when tires are dangerously underinflated (NHTSA, 2019). Self-inflating tire systems have been developed to keep tires inflated to the proper pressure automatically. Such systems are used in some heavy-duty vehicles but are not yet optional or standard equipment for light-duty vehicles (NASEM, 2019).

7.4.4 Other Off-Cycle Technologies

A host of other off-cycle technologies are discussed in prior National Academies and regulatory reports. These include low-drag-resistant brakes, which reduce friction of brake pads on rotors when brakes are not engaged, and secondary axle disconnect, which disconnects an axle from all-wheel drive vehicles in some driving conditions when the torque of a second axle is not needed. Both technologies provide about 1% reduction in fuel consumption at less than about $100. Additional off-cycle technologies, including high-efficiency exterior lighting, solar roof panels, passive and active cabin ventilation, and solar reflective paint, can further influence the thermal control of the cabin or have other potential off-cycle impacts on fuel economy. A more extensive list of these technologies can be found in Tables VI-173 and VI-174 of NHTSA/EPA (2020), and their definitions are provided in 40 CFR 86.1869-12(b)(4). The associated credits will be discussed in Chapter 12.

7.4.5 Issues for Off-Cycle Technologies

As discussed further in Chapter 12, the objective for developing and crediting potential reduction for technologies that do not provide fuel efficiency benefits on the FTP is to recognize that there may be cost-effective approaches to reducing the fuel consumption and GHG impacts of vehicles that are not represented on the FTP. However, Lutsey and Isenstadt (2018) point out that a high off-cycle credit use scenario, where credits are provided at levels over and above the ones on the predetermined list and there is no 10 g/mi limit, off-cycle technologies could provide a significant fraction of compliance with the current standards. This situation could be problematic owing to the high uncertainty in the extent of actual emissions reductions from off-cycle technologies and the potential use of off-cycle technologies in place of the cost-effective technologies assumed in setting the standards. Consequently, the Lutsey and Isenstadt analysis questions how the off-cycle technologies program might be constructed in the future. There will most likely still be additional technologies that provide additional fuel economy and GHG benefits outside the FTP. However, the testing to ensure that such technologies deliver the anticipated benefits in the real world is needed, both for the current setting and undoubtedly in the future.

7.4.6 Findings and Recommendation for Accessories and Off-Cycle Technologies

FINDING 7.10: Heating and cooling efficiencies are an area of active research and hold the greatest promise in the electrification of vehicles, where the power draw for heating and cooling is a motivation to improve these efficiencies for consumer acceptance.

FINDING 7.11: Several automakers are using cooling systems that incorporate the powertrain, battery, and cabin heating and cooling to optimize use of and exhaust of heat from all components. The recovery and conversion of thermal energy from batteries and exhaust into electric energy to charge the battery is an area of active research and development.

FINDING 7.12: While significant thermal loss remains in ICE propulsion systems, recovering this energy and converting it into electrical energy is not a cost-effective focus for automakers.

RECOMMENDATION 7.3: The U.S. Department of Energy should provide research funding for the assessment of thermal improvements in electrified systems.

7.5 CONSIDERATIONS FOR MASS AND SAFETY IN LIGHT OF INCREASED PENETRATION OF ADAS AND ELECTRIFICATION

Motor vehicle safety is important. Preliminary estimates indicate that more than 36,000 people died in motor vehicle crashes in 2019, and every year vehicle crashes lead to millions of significant injuries and billions of dollars in medical care and lost wages (NHTSA, 2020; CDC, 2020). Many factors affect motor vehicle safety, including those associated with the driver—unsafe driving behaviors and driver error—as well as system-wide deficiencies in road design, traffic environment, and vehicle maintenance. Other factors include vehicle crashworthiness and vehicle-to-vehicle mass disparity and structural and geometric compatibility. While any implications of fuel economy regulations for vehicle safety have been small compared to the primary determinants of vehicle safety, understanding and addressing the potential unintended consequences is important. This section describes some factors related to both fuel economy and safety that NHTSA should pay particular attention to in 2025–2035.

This report considers the technologies that can be implemented in 2025–2035 to improve fuel economy. From this evaluation, the committee has identified two key areas for NHTSA to consider in order to better understand the relationship between vehicle safety and fuel economy technologies: changes in crash prevalence owing to ADAS and changes in mass disparity that could occur in that time period. ADAS implementation and mass disparity may or may not change in response to fuel economy standards themselves. Regardless of the driving force for ADAS implementation and mass disparity, however, NHTSA should examine their impact on the broader relationship between fuel economy technologies and safety.

To improve fuel economy, automakers are expected to redesign about 40% of vehicles in the MY 2020 baseline fleet by MY 2025 using both design optimization and materials substitution to reduce vehicle mass, as noted earlier in this chapter. During 2025–2035, these lightweighting efforts will likely occur concurrently with increased adoption of electric powertrains and comfort and safety features associated with driver assist and CAVs, both of which increase vehicle mass. This will result in new vehicle designs in 2025–2035 that will differ from current vehicles in the fleet. Based on electrification volume, battery pack cost, and battery energy cell density as variables, several potential lightweighting scenarios using advanced materials (high-strength steel, aluminum, and magnesium) in the future fleets were shown Table 7.10. In the scenarios considered, all vehicle classes can achieve curb weight reductions, although the reductions have a wide range, 1%–14% for cars and unibody SUVs and 2%–10% for pickup trucks.

In light of the significant number of crashes and potential for injury or death, the federal government, consumers, and automakers all recognize the importance of vehicle crashworthiness. All new vehicles must meet the relevant Federal Motor Vehicle Safety Standards (FMVSS) for crashworthiness, regardless of what technologies they implement (e.g., powertrain or lightweighting). Consumers expect automakers to continue delivering safer vehicles. In response, automakers attempt to go above and beyond the required FMVSS and also make improvements for individual models year by year. Solutions to improve crashworthiness include better design and better materials, which are implemented to increase safety regardless of lightweighting or powertrain technology. Current crashworthiness standard tests, however, do not consider crash compatibility between vehicles of different sizes and weights. Because the tests require a vehicle to protect itself, large vehicles are more protected in crashes with smaller vehicles, exacerbating the potential problems with mass disparity in the fleet.

NHTSA should study the relationship between lightweighting and safety in the case of reduced or changed accident type as a result of ADAS or, similarly, in the case of different levels of mass disparity in the fleet. The increasing implementation of ADAS systems is intended to reduce crashes and intends to reduce injuries, property damage, and deaths from those crashes. ADAS implementation will not eliminate all crashes, however, including for vehicles not implementing ADAS, and will also change the prevalence of some crash types. A study that examined the effects of transportation trends, safety initiatives, and new technology on crashes in 2020–2030 forecasted significant decreases in injured occupants from road departures and control loss crashes, mainly owing to the penetration of electronic stability control into the fleet, but smaller decreases in injured occupants in lane change, opposite direction, and other crashes mitigated by a lesser penetration and effectiveness of ADAS systems in these crash scenarios (Mallory et al., 2019). This study accounted for numerous trends including population growth by age group; proportion of occupants in future crashes by vehicle class; seatbelt use; and penetration of ADAS systems such as automatic braking with forward crash warning and crash imminent braking, blind spot detection, lane and road departure warning, lane-keeping support, and Level 3–5 automated driving systems. NHTSA should continue to study how crashes change in an ADAS-enabled fleet, and if changes in crash propensity or severity affect the total societal safety risk when more vehicles also incorporate improved fuel economy technologies, such as new materials or advanced powertrains.

Mass disparity exists in the current light-duty fleet owing to the wide range of vehicle sizes as well as the different technologies implemented in vehicles. There is even greater disparity when pedestrian, bicycle, motorcycle, and medium- and heavy-duty vehicles and modes are considered. There will continue to be mass and size disparity in the fleet in 2025–2035, but it is not clear if the disparity will increase or decrease relative to that in the current fleet. Furthermore, it is not clear how much of the changes in mass disparity will be a direct result of the fuel economy standards rather than occurring independently.

Statistical analyses of historical crash data have been performed to understand the relationship between average mass across the fleet, mass disparity, and traffic fatalities. There is a consensus that increasing the mass disparity of the fleet increases societal fatality risks (Farmer, 2019; Wenzel, 2019). However, the specific relationships identified in previous studies may not be relevant for understanding the future fleet because of new vehicle designs and the change to footprint-based standards. Potential changes in mass disparity in 2025–2035 should be studied, particularly those that may arise from a shift from sedans to CUV/SUV/pickup trucks; mass increases in one vehicle class but not another; lightweighting to improve fuel efficiency and performance; and increases in mass from electrified powertrains, ADAS, and other safety and comfort features. Particular attention should be paid to how these changes in fleet mass would affect societal safety risk. Examples of current efforts toward this goal include NHTSA's computer-aided engineering modeling simulations, which compute societal occupant injury risk in the vehicle fleet from crashes with lightweight vehicle concept designs (NHTSA, 2011; Samaha et al., 2014; Radwan, 2015), and Transport Canada's recent crash test series, which compares occupant protection in EVs to the protection offered by equivalent ICE vehicle models (Tylko et al., 2019).

Last, the committee recommends that FMVSS improve testing protocols for crash compatibility, as better crash compatibility will reduce the adverse effect of mass disparity on crash safety for passengers of all vehicles. The committee also recommends that NHTSA develop standard naming and test protocols for safety and fuel economy benefits from ADAS. Furthermore, education of consumers on the benefits of ADAS would likely increase consumer acceptance and adoption of systems that will increase safety, and potentially reduce fuel consumption, through reduced crashes, reduced congestion, and improved vehicle operation.

7.5.1 Findings and Recommendations for Mass and Safety

FINDING 7.13: There exists mass disparity in the current fleet, and that mass disparity may increase or decrease in the future, depending on changes in vehicle technologies and other attributes implemented for fuel economy.

FINDING 7.14: Current crashworthiness standard tests do not consider crash compatibility between vehicles of different sizes and weights, leading to heavier weight vehicles being more protected in crashes with lighter weight vehicles.

RECOMMENDATION 7.4: The National Highway Traffic Safety Administration should continue to study how crashes change in an advanced driver assist systems-enabled fleet, and if changes in crash propensity or severity affect the total societal safety risk when more vehicles also incorporate improved fuel economy technologies, such as new materials or advanced powertrains.

RECOMMENDATION 7.5: The National Highway Traffic Safety Administration should study potential changes in mass disparity in 2025–2035, particularly disparities that may arise from a shift from sedans to crossover vehicle/sport utility vehicle/pickup trucks, mass increases in one vehicle class but not another, lightweighting to improve fuel efficiency and performance, and increases in mass from electrified powertrains and other safety and comfort features. Particular attention should be paid to how these changes in fleet mass would affect societal safety risk. This could be achieved by conducting relevant crash tests and/or further development of computer-aided engineering fleet modeling to simulate real-world crash interactions of new vehicle designs and vulnerable users at different impact speeds and impact configurations.

RECOMMENDATION 7.6: The National Highway Traffic Safety Administration should develop testing protocols and corresponding Federal Motor Vehicle Safety Standards with frontal crash compatibility requirements to address disparities in mass, stiffness, and geometries in vehicles designs.

7.6 TOTAL OPPORTUNITIES FOR ROAD LOAD AND ACCESSORY POWER DRAW REDUCTION

Road load reduction improves the efficiency of a vehicle by reducing the energy required to move the vehicle. Accessory loads are similar, although their reduction improves the efficiency of providing the accessories, such as climate control, headlights, and infotainment systems. Reducing both road and accessory loads not only directly reduces vehicle energy use but also allows other aspects of the vehicle propulsion system to be resized. For example, less required energy means the engine or battery and motor can be reduced in size, which then allows further reduction in mass (Lovins, 2020). Table 7.15 gives the committee's estimates for costs and effectiveness of implementing various road load reduction technologies in 2025–2035.

In 2025–2035, automakers will pursue road load reduction strategies to meet customer expectations and fuel economy and emissions standards. Automakers will continue to develop and incorporate new technologies and designs to reduce road load on individual vehicle models, at the very least to offset mass add-back from electrification and safety and performance technologies. At the same time, however, certain trends in the automotive market may reduce the impact of road load reduction technologies on total fuel consumption and GHG emissions reductions. For instance, as discussed in Section 7.2 above, both the mass and footprint of vehicles are generally increasing. The path and degree to which automakers choose to implement these technologies will be informed by their overall brand portfolio strategy and larger technology and market trends, such as the extent and cost of fleet electrification.

7.6.1 Summary Cost and Effectiveness Table

TABLE 7.15 Cost and Effectiveness of New Technologies for Mass Reduction, Aerodynamic Drag Reduction, and Tire Rolling Resistance Reduction

	Technology Cost (2018$)			Technology Effectiveness
Technology	2025	2030	2035	(% Fuel Consumption Reduction)[a]
Mass reduction—unibody[b]				
1%–1.5%	0.45–0.67	0.22–0.67	0.05–0.22	0.6–1.05
4%–6%	0.67–1.78	0.67–1.56	0.58–0.89	2.4–4.2
12%– 14%[c]	1.78–2.67	1.78–2.67	1.34–1.78	7.2–9.8
Mass reduction—truck[b]				
2%–3%	0.45–0.67	0.22–0.67	0.05–0.45	0.8–1.5
8%–10%[c]	0.67–1.27	0.89–1.34	0.45–0.89	3.2–5.0
10%–12%[c]	4.01–4.90	3.34–4.45	2.67–3.56	4.0–6.0
Aerodynamic drag reduction[d]				
5%	35.50	30.28	27.37	1.3
10%	68.49	61.91	55.97	2.3
15%	96.78	87.48	79.08	3.5
20%	171.23	154.78	139.91	4.8
Tire rolling resistance reduction[e]				
10%	4.24	4.00	3.89	2
20%	27.19	24.80	24.32	4

[a]Fuel consumption reduction is relative to a baseline vehicle, and not incremental to previous technology.
[b]MR percent is curb weight reduction from a MY 2020 baseline; costs ($/lb) for materials substitution only, as reported in Bailo et al. (2020), converted from 2019$ to 2018$ and from $/kg to $/lb (1 kg = 2.2 lb).
[c]Includes secondary mass reduction, calculated as described in NRC (2015).
[d]Cost values are 2% per year reductions from MY 2017 values reported in NHTSA/EPA, 2020. Effectiveness values are committee estimates based on NRC (2015) and NHTSA/EPA (2020).
[e]Percent reduction from baseline C_{rr} of 0.009; cost and effectiveness values are committee estimates based on NHTSA/EPA (2020).

7.7 REFERENCES

Aldhufairi, H.S., K. Essa, and O. Olatunbosun. 2019. Multi-chamber tire concept for low rolling resistance. *SAE International Journal: Passenger Cars—Mechanical Systems* 12(2):111–126. Article ID: 06-12-02-0009.

Aldhufairi, H.S., and O.A. Olatunbosun. 2017. Developments in tire design for lower rolling resistance: A state of the art review. *Proceedings of the Institution of Mechanical Engineers, Part D: Journal of Automobile Engineering* 232(14): 1865–1882.

American Chemistry Council. 2019. "Plastics and Polymer Composites in Light Vehicles." August. https://www.automotiveplastics.com/wp-content/uploads/Plastics-and-Polymer-Composites-in-Light-Vehicles-2019-REV-Sm.pdf.

Apostolaki-Iosifidou, E., P. Codani, and W. Kempton. 2017. Measurement of power loss during electric vehicle charging and discharging. *Energy* 127:730–742.

Bailo, C., S. Modi, M. Schultz, T. Fiorelli, B. Smith, and N. Snell. 2020. "Vehicle Mass Reduction Roadmap Study 2025–2035." Ann Arbor, MI: Center for Automotive Research.

Billur, E., and T. Altan. 2013. Three generations of advanced high-strength steels for automotive applications. *Part I, Stamping Journal*: 16–17.

Bull, M. 2011. "Mass Reduction Performance of PEV and PHEV Vehicles." In *22nd International Technical Conference on the Enhanced Safety of Vehicles (ESV)*, 8.

CDC (Centers for Disease Control and Prevention, National Center for Injury Prevention and Control). 2020. "Cost Data and Prevention Policies | Motor Vehicle Safety." Centers for Disease Control and Prevention. November 6. https://www.cdc.gov/transportationsafety/costs/index.html.

DOE/EPA (U.S. Department of Energy and U.S. Environmental Protection Agency). 2019a. "Keeping Your Vehicle in Shape." https://www.fueleconomy.gov/feg/maintain.jsp.

DOE/EPA. 2019b. "Where the Energy Goes: Gasoline Vehicles." https://www.fueleconomy.gov/feg/atv.shtml.

DOE/EPA. 2019c. "Where the Energy Goes: Hybrids." https://www.fueleconomy.gov/feg/atv-hev.shtml.

DOE/EPA. 2019d. "Where the Energy Goes: Electric Cars." https://www.fueleconomy.gov/feg/atv-ev.shtml.

Donley, T. 2014. "Improving Vehicle Fuel Efficiency Through Tire Design, Materials, and Reduced Weight." Department of Energy Annual Merit Review 2014, Vehicle Technologies, Project ID: VSS083. June.

Ducker Worldwide and Aluminum Association. 2017. "Unprecedented Growth Expected for Automotive Aluminum as Multi-Material Vehicles Ascend, New Survey of Automakers Says." Ducker Worldwide. August 1. https://www.ducker.com/news-insights/unprecedented-growth-expected-automotive-aluminum-multi-material-vehicles-ascend-new.

Els, P. 2017. "Electric Power Steering Systems (EPS) Have Never Been Safer." Automotive IQ. March 14. https://www.automotive-iq.com/chassis-systems/articles/electric-power-steering-systems-eps-have-never-been-safer.

EPA (U.S. Environmental Protection Agency). 2020. *The 2019 EPA Automotive Trends Report: Greenhouse Gas Emissions, Fuel Economy, and Technology since 1975*. Washington, DC: U.S. Environmental Protection Agency. March. https://nepis.epa.gov/Exe/ZyPDF.cgi?Dockey=P100YVFS.pdf.

EPA/NHTSA (U.S. Environmental Protection Agency and National Highway Traffic Safety Administration). 2012. *Joint Technical Support Document: Final Rulemaking for the 2017–2025 Light-Duty Vehicle Greenhouse Gas Emission Standards and Corporate Average Fuel Economy Standards*. EPA-420-R-12-901. http://www.epa.gov/otaq/climate/documents/420r12901.pdf.

EPA/NHTSA/CARB (U.S. Environmental Protection Agency, National Highway Traffic Safety Administration, and California Air Resources Board). 2016. "Draft Technical Assessment Report: Midterm Evaluation of Light-Duty Vehicle Greenhouse Gas Emission Standards and Corporate Average Fuel Economy Standards for Model Years 2022–2025." EPA-420-D-16-901. https://www.nhtsa.gov/staticfiles/rulemaking/pdf/cafe/Draft-TAR-Final-Executive-Summary.pdf.

Farmer, C.M. 2019. "Fuel Economy and Highway Safety." Presentation to the Committee on Assessment of Technologies for Improving Fuel Economy of Light-Duty Vehicles—Phase 3." Insurance Institute for Highway Safety. September 19. https://www.nationalacademies.org/event/09-19-2019/safety-webinar.

Gawron, J., G.A. Keoleian, R.D. De Kleine, T.J. Wallington, and H.C. Kim. 2018. Life cycle assessment of connected and automated vehicles: Sensing and computing subsystem and vehicle level effects. *Environmental Science and Technology* 52(5):3249–3256.

ICCT (International Council on Clean Transportation). 2018. International Council on Clean Transportation Comments on the Safer Affordable Fuel-Efficient (SAFE) Vehicles Rule for Model Years 2021–2026 Passenger Cars and Light Trucks.

Islam, E.S., A. Moawad, N. Kim, and A. Rousseau. 2020. *A Detailed Vehicle Simulation Process to Support CAFE and CO_2 Standards for the MY 2021–2026 Final Rule Analysis*. ANL/ESD-19/9. March 31. https://doi.org/10.2172/1608044.

Kreutzer, C., B. Kekelia, J.P. Rugh, and G. Titov. 2017. "U.S. Light-Duty Vehicle Air Conditioning Fuel Use and the Impact of Four Solar/Thermal Control Technologies." Presented at the SAE 2017 Thermal Management Systems Symposium, Plymouth, MI. October. https://www.nrel.gov/docs/fy18osti/69047.pdf.

Lovins, A.B. 2020. "Integrative Design of Automobiles." Presented at the Design Optimization and Lightweighting Webinar to Committee on Assessment of Technologies for Improving Fuel Economy of Light-Duty Vehicles—Phase 3. January 6. https://www.nationalacademies.org/event/01-06-2020/design-optimization-webinar.

Lutsey, N., and A. Isenstadt. 2018. How will off-cycle credits impact U.S. 2025 efficiency standards? ICCT White Paper. https://theicct.org/sites/default/files/publications/Off-Cycle-Credits_ICCT-White-Paper_vF_20180327.pdf

Mallory, A., A. Kender, E. Hutter, and K. Moorhouse. 2019. "Crashes and Injuries in 2020–2030: Development of a Crash Data Projection Model," Injury Biomechanics Research, Proceedings of the Forty-Seventh International Workshop, sponsored by NHTSA, November 2019, San Antonio, Texas.

Modi, S., and A. Vadhavkar. 2019. "Technology Roadmap: Materials and Manufacturing." Ann Arbor, MI: Center for Automotive Research. https://www.cargroup.org/wp-content/uploads/2019/10/Technology-Roadmap_Materials-and-Manufacturing.pdf.

NASEM (National Academies of Sciences, Engineering, and Medicine). 2019. *Reducing the Fuel Consumption and Greenhouse Gas Emissions of Medium- and Heavy-Duty Vehicles, Phase Two: Final Report*. Washington, DC: The National Academies Press. https://doi.org/10.17226/25542.

NHTSA (National Highway Traffic Safety Administration). 2011. "Finite Element Modeling in Fleet Safety Studies," presentation to the NHTSA Mass-Size-Safety Symposium, February 25. Washington, DC. https://one.nhtsa.gov/Laws-&-Regulations/CAFE-—Fuel-Economy/NHTSA-Workshop-on-Vehicle-Mass–Size–Safety.

NHTSA. 2019. "Tires." https://www.nhtsa.gov/equipment/tires.

NHTSA. 2020. "Early Estimates of 2019 Motor Vehicle Traffic Data Show Reduced Fatalities for Third Consecutive Year." May 5. https://www.nhtsa.gov/press-releases/early-estimates-traffic-fatalities-2019.

NHTSA/EPA (National Highway Traffic Safety Administration and U.S. Environmental Protection Agency). 2018. *The Safer and Affordable Fuel-Efficient (SAFE) Vehicles Rule for Model Year 2021–2026 Passenger Cars and Light Trucks: Preliminary Regulatory Impact Analysis*. October 16. https://www.nhtsa.gov/sites/nhtsa.dot.gov/files/documents/ld_cafe_co2_nhtsa_2127-al76_epa_pria_181016.pdf.

NHTSA/EPA. 2020. *Final Regulatory Impact Analysis: The Safer Affordable Fuel-Efficient (SAFE) Vehicles Rule for 2021–2026 Passenger Cars and Light Trucks*. March. https://www.nhtsa.gov/sites/nhtsa.dot.gov/files/documents/final_safe_fria_web_version_200330.pdf.

NRC (National Research Council). 2013. *Transitions to Alternative Vehicles and Fuels*. Washington, DC: The National Academies Press

NRC. 2015. *Cost, Effectiveness, and Deployment of Fuel Economy Technologies for Light-Duty Vehicles*. Washington, DC: The National Academies Press.

Pannone, G. 2015. *Technical Analysis of Vehicle Load Reduction Potential for Advanced Clean Cars*. Report by CONTROLTECH, LLC., for the California Air Resources Board. https://ww2.arb.ca.gov/sites/default/files/classic//research/apr/past/13-313.pdf.

PrimeXBT. 2020. "Crude Oil Prices Forecast and Predictions for 2020, 2025, and 2030." Price Prediction. October 21. https://primexbt.com/blog/oil-price-prediction-forecast/.

Radwan, R., 2015. "Real World Derived Simulation Methodology for the Evaluation of Fleet Crash Protection of New Vehicle Designs." George Washington University, ProQuest Dissertations Publishing, Pub. No. 3686081.

Röchling. n.d. "Active Grille Shutter." Röchling. https://www.roechling.com/automotive/products-solutions-1/aerodynamics/active-grille-shutter.

Samaha, R., R. Radwan, P. Prasad, D. Marzougui, C. Dui, K. Digges, S. Summers, L. Zhao, and A. Farsan-Anelli, 2014. "Methodology for Evaluating Fleet Protection of New Vehicle Designs: Application to Lightweight Vehicle Designs." DOT HS 812 051A. National Highway Traffic Safety Administration. https://www.nhtsa.gov/crashworthiness/vehicle-aggressivity-and-fleet-compatibility-research.

Summe, T. 2019. "Fuel Economy with Aluminum." Presented at Materials Webinar to Committee on Assessment of Technologies for Improving Fuel Economy of Light-Duty Vehicles—Phase 3. May 17. https://www.nationalacademies.org/event/05-17-2019/materials-webinar.

TRB (Transportation Research Board). 2006. *Tires and Passenger Vehicle Fuel Economy*. TRB Special Report 286. Washington, DC: Transportation Research Board.

Tylko, S.J., K. Tang, and A. Bussières. 2019. Comparison of Occupant Protection in Electric vs. Internal Combustion Vehicles. IRC-19-10. IRCOBI Conference 2019.

UBS. 2017. "UBS Evidence Lab Electric Car Teardown—Distruption Ahead?" Q-Series. May 18. https://neo.ubs.com/shared/d1wkuDlEbYPjF/.

Wenzel, T. 2019. "Relationships Between Mass, Footprint, and Societal Fatality Risk in Recent Light-Duty Vehicles." Presentation to the National Academies Committee on Assessment of Technologies for Improving Fuel Economy of Light-Duty Vehicles—Phase 3. September 19. https://www.nationalacademies.org/event/09-19-2019/safety-webinar.

WorldAutoSteel. n.d. "Advanced High-Strength Steel (AHSS) Definitions." World Steel Association. https://www.worldautosteel.org/steel-basics/automotive-advanced-high-strength-steel-ahss-definitions/.

Yang, Z. 2018. Fuel-efficiency technology trend assessment for LDVs in China: Vehicle technology. San Francisco, CA: The International Council for Clean Transportation. September 17. https://theicct.org/sites/default/files/publications/PV_Tech_Trend_Vehicle_20180917.pdf.

Yen, R. 2020. "Simulation-Driven Lightweight Design for Automotive Structures." Presented at the Design Optimization and Lightweighting Webinar to the Committee on Assessment of Technologies for Improving Fuel Economy of Light-Duty Vehicles—Phase 3. January 6. https://www.nationalacademies.org/event/01-06-2020/design-optimization-webinar.

YourMechanic. 2015. "How Does Adjustable Suspension Work?" Autoblog. https://www.autoblog.com/2015/11/17/how-does-adjustable-suspension-work/.

8

Connected and Automated Vehicles

8.1 INTRODUCTION

The rapid growth in the application of electronic controls and information and communications technologies in automobiles goes beyond improving internal vehicle operations. Advances in sensing, control, and communication technologies will allow vehicles to respond to external information and to take increasing control of aspects of driving previously handled by the driver.

The primary motivation for the development of vehicle automation to date has been improved safety. In combination with connectivity, it can bring other benefits as well, such as convenience, equity and accessibility, productivity, and commerce/entertainment. If designed with efficiency in mind, automated vehicles could substantially improve fuel efficiency, lowering drivers' fuel costs and increasing driving range for electric vehicles (EVs) while delivering societal benefits through reduced fuel use and emissions. Table 8.1 lists the benefits that may be realized from increasing automation and connectivity, the vehicle features that enable that benefit, and the primary beneficiaries. The term "autonomous" refers to those highly automated systems that can operate without

TABLE 8.1 Potential Benefits of Connected and Automated Vehicle Technologies

Effect	Automation and Connectivity Features That Enable the Benefit	Primary Beneficiary
Enhanced safety	Crash avoidance	Drivers, passengers, other road users
Greater mobility	Autonomous driving	People unable to drive
Convenience	Autonomous parking, automation of driving	Drivers, passengers
Less stressful driving experience	Automation of driving tasks, optimized driving behavior	Drivers, passengers
Higher fuel efficiency	Optimized driving behavior, powertrain operation, and routing	Drivers, passengers
Reduced emissions per mile	Optimized driving behavior, powertrain operation, and routing	Society

NOTE: The vehicle efficiency and emissions benefits that are the focus of this study are highlighted in grey.

a human driver (Society of Automotive Engineers [SAE] Levels 4 and 5; see Figure 8.2, below). Autonomous vehicles are the subject of Chapter 9.

This chapter is concerned with how connected and automated vehicle (CAV) technologies could affect the fuel efficiency of individual vehicles or groups of vehicles in close proximity. The impacts on energy use from fully autonomous vehicles, such as increased travel owing to accessibility for the non-driving population and greater productivity for vehicle occupants, are discussed in Chapter 9.

8.2 CAV TECHNOLOGIES

8.2.1 Automation Technologies and Operating Modes

The push for safer vehicles to reduce the number of road fatalities and injuries has been behind the introduction of electronically controlled systems to enhance human driving capabilities and supplement or replace mechanical- and hydraulic-based vehicular systems. These systems, known as advanced driver assistance systems, or ADAS, use sensors like radio detection and ranging (radar), light detection and ranging (lidar), cameras, sonar, global positioning system (GPS), digital maps, and actuators, along with complex software control systems to warn the driver of potentially dangerous behavior or conditions and assist the driver in case of imminent danger. Figure 8.1 shows some of these sensors and the safety features they are used for. These sensors, described in detail in Section 8.2.1.1, have overlapping capabilities and have their own strength and limitations.

ADAS systems provide the basis for future driving modes involving increased levels of automation up to and including the fully autonomous vehicle. SAE has defined various levels of automation in SAE international standard J3016 (SAE International, 2019). It identifies six distinct levels of driving automation from Level 0 to

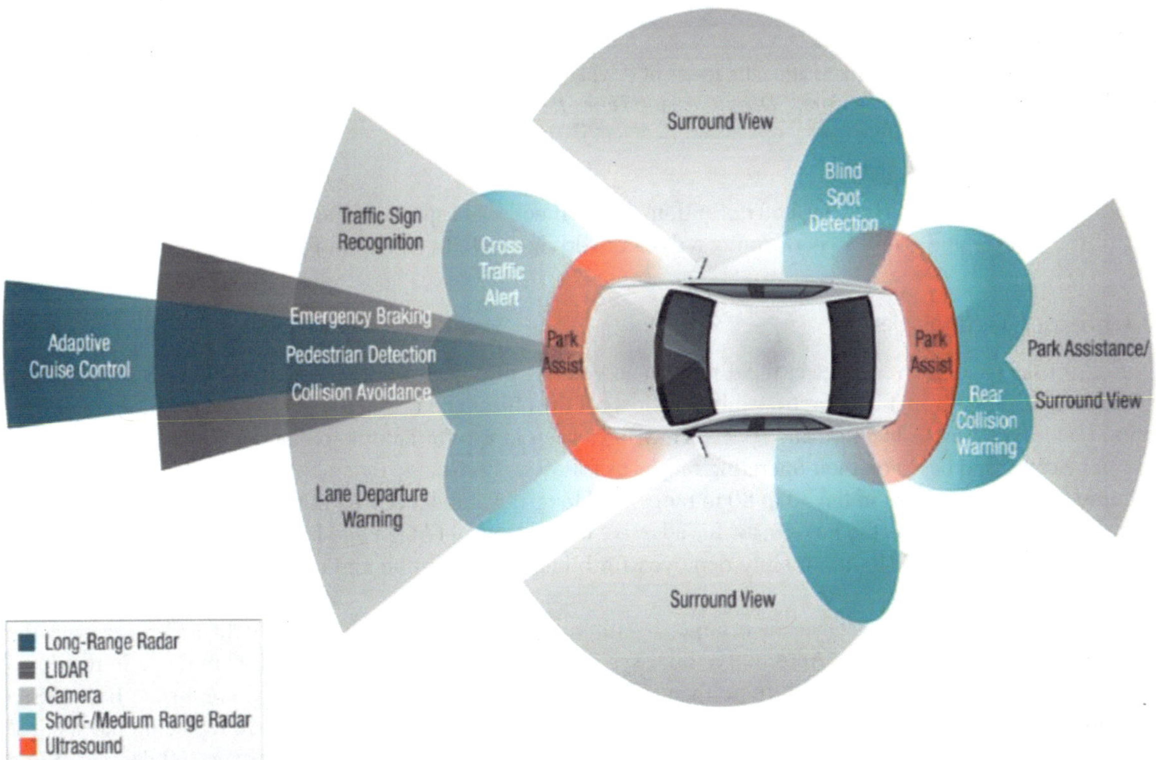

FIGURE 8.1 ADAS 360-degree vision sensors and applications. SOURCE: Courtesy Texas Instruments.

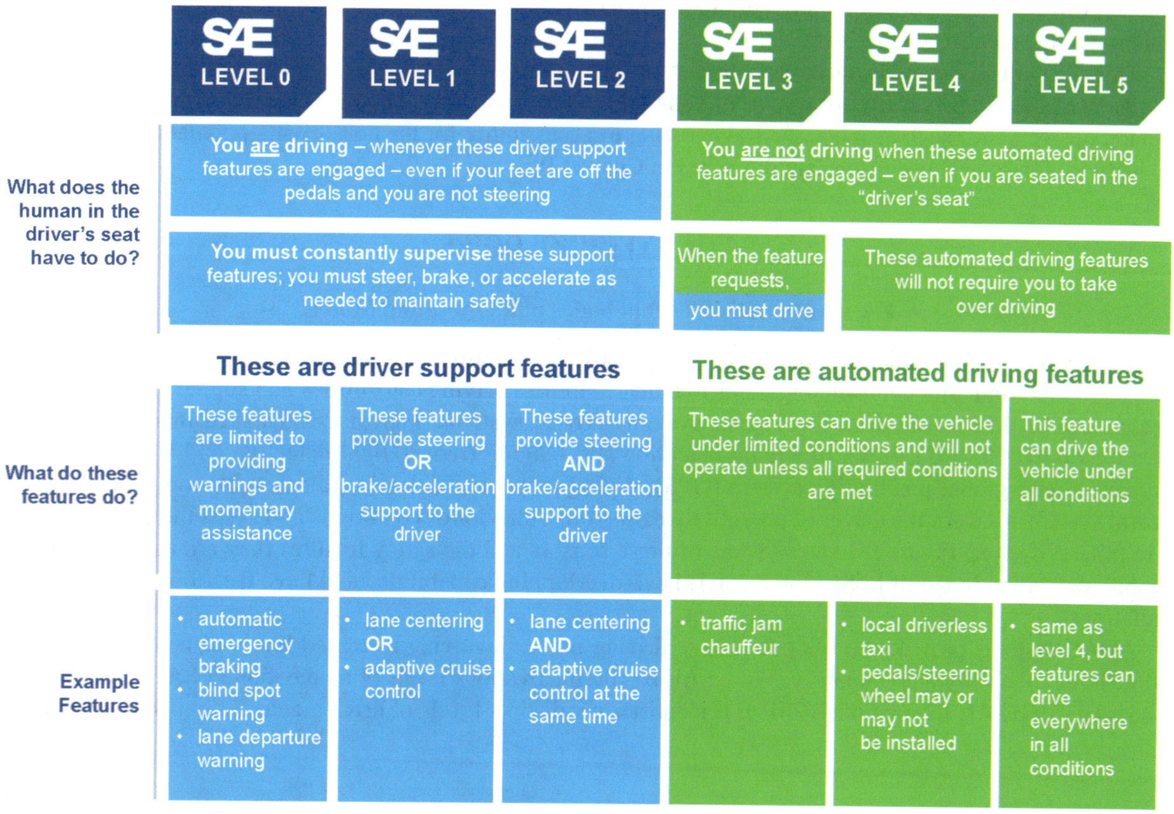

FIGURE 8.2 Summary of SAE International's levels of driving automation for on-road vehicles. SOURCE: © SAE International from SAE J3016™ *Taxonomy and Definitions for Terms Related to Driving Automation Systems for On-Road Motor Vehicles* (2018-06-05), https://www.sae.org/standards/content/j3016_201806/.

Level 5 and provides classifications and definitions of the functional aspects of the technology and the respective roles of the vehicle and driver in performing a dynamic driving task, as shown in Figure 8.2.

8.2.1.1 Sensor Technologies

Radar uses radio waves to detect the distance and velocity of objects in almost all environmental conditions. Radio waves emitted from the radar transmitter reflect off the object and return to the radar receiver, giving information about the object's location and speed. They have been in use for automotive applications for many years and can be classified based on their operating ranges as follows: short-range radar (SRR) in the 0.2 to 30 m range; medium-range radar (MRR) in the 30 to 80 m range; and long-range radar (LRR) in the 80 m to more than 200 m range. For example, LRR is the key sensor for adaptive cruise control (ACC) and highway automatic emergency braking systems, while MRRs are mainly deployed for blind zone detection and rear cross traffic and SRRs for rear emergency breaking.

Lidar sensors measure distance to a target by illuminating the target with laser light and measuring the return time of the reflected light. Automotive lidar systems typically can provide up to 200 m range in restricted field-of-view (FOV). They cost much more than radar or camera sensors. More recently, solid-state lidar (SSL) technologies that have no moving parts and are therefore more reliable and lower cost are replacing mechanical scanning lidar, which physically rotates the laser and receiver assembly to collect data over 360 degrees. SSLs currently have lower FOV, but their low cost allows for using multiple sensors to expand coverage zone. They are also more compact and can be integrated with the vehicle body, reducing drag.

TABLE 8.2 Comparison of Sensors Used for Vehicle Automation, Strengths and Limitations

	Ultrasonic	Camera	Radar	Lidar
Cost	Low	Low	Medium	High
Size	Small	Medium	Small–medium	Medium–large
Speed detection	Low	Low	High	Medium
Sensitivity to color	No	High	No	No
Robust to weather	High	Low	High	Medium
Robust to day and night	High	Low	High	High
Dimensional resolution	Low (0.01 m)	High (1 Mp)	Medium (0.4, 0.4, 1.8 m)	High (horizontal 0.125 degrees, vertical 0.6 degrees)
Range (m)	Short (8)	Medium–long (60–250)	Short, medium, long (20, 70, 250)	Long (150–200)

SOURCE: Michigan Tech Research Institute (2017).

Cameras, in contrast, are passive systems, hence much simpler and lower cost. They are the only sensor technology capable of capturing high levels of detail such as texture, color, and contrast information, and therefore they represent the technology of choice for object classification. They play a key role in many of the ADAS systems in the market today, such as lane departure warning (LDW) and lane keeping systems. On the other hand, cameras are susceptible to adverse weather conditions and variations in lighting.

Ultrasound technology is used to detect objects in a range of a few centimeters to a few meters. It generates sound waves that bounce off objects and return to the car's sensor, with a delay that is related to the objects' distance from the vehicle. Ultrasound works in low-light conditions and poor weather, but it cannot classify objects; it only identifies their presence and distance. It also has relatively poor time resolution and is primarily used for slow speed conditions such as slow traffic or parking.

A side-by-side comparison of the four sensor technologies showing their strengths and limitations is given in Table 8.2.

To ensure safety in view of sensor limitations, sensor redundancies are necessary. The use of cameras in combination with radar and/or lidar in what is known as sensor fusion provides for a more reliable situational analysis of the environment around the vehicle. Since camera systems provide the most application coverage they are projected to see the largest volume growth close to 400 million sensor units by 2030. With radar and lidar costs coming down, they too are expected to see large percentage growth and volumes reaching 40–50 million sensor units by 2030 (Ors, 2017).

Last, keeping externally mounted sensors (cameras and lidar) clean is essential for reliable interpretation of the scene. To this end, several technologies for automating the cleaning process, ranging from liquid-jet wiper-based cleaning systems to ultrasonic-based, are at different stages of development (Brooke, 2020b; MS Foster, n.d.).

8.2.1.2 Data and Mapping Technology

In addition to onboard sensors, automakers are acquiring and using digital map, traffic, and other data to help their vehicles operate safely and efficiently. Such information can be used for routing and navigation, for predictive optimization of the powertrain operation, and for safe operation of the vehicle. High-definition (HD) digital mapping is an important addition to vehicle automation technology at all levels. While map accuracy for navigation purposes is 10 m and 1–2 m for ADAS purposes, HD maps have an accuracy of 10 cm. The market potential for using such maps particularly for Level 2 and 3 automation is substantial and provides the driving force behind this costly development (Markets and Markets, 2019). HD maps complement information from other localization sensors like GPS, lidar, and vehicle-to-infrastructure (V2I) communication and allow a vehicle to adjust to upcoming conditions in advance. It is important to note that HD maps do not require high accuracy GPS to provide the cm-level localization. High accuracy GPS systems (Differential GPS-DGPS or Real-Time Kinematic-RTK) using a fixed ground station to provide error correction present other alternatives for providing cm-level localization but are considered more expensive and less accurate options.

Data for digital maps come from third-party sources, such as HERE, Google Maps, Waze, or TomTom, as well as from the fleet of vehicles produced by the automaker itself, if equipped with the ability to share data collected by the vehicles with the automaker. Data about routes, infrastructure, and traffic may be stored on board the vehicle or transmitted to the vehicle as needed. Vehicles' use of HD maps is not expected to be power-intensive. While maps contain a huge amount of data, what is delivered to the client for use is a small subset of the total and requires limited onboard storage. Computation needs are also modest; predictive analytics requires less computing power than real-time processing of sensor data.

Developing these maps is both labor- and equipment-intensive, and hence costly. Mapping may cost hundreds of dollars per lane mile for data acquisition, and the data must be updated frequently. Moreover, this cost does not include map production, which requires the extraction of features with high accuracy. Costs are expected to decline over the next decade, however, as cars themselves increasingly contribute to the collection of data and artificial intelligence (AI) allows machine reading of lidar data, replacing manual work. Despite the high costs of creating HD maps, automakers thus far have worked independently to create or acquire these capabilities. There may be a role for a federal agency such as the Federal Highway Administration to oversee and/or host the data collection effort and perhaps develop a national specification. This could greatly improve the efficiency of map creation and ensure consistency across mapping products.

FINDING 8.1: High-definition maps are increasingly important elements of automation packages that complement information from sensors. By allowing the vehicle to anticipate and adjust to roadway features well in advance, they can improve safety and efficiency. However, they are expensive to produce and maintain and there has been little collaboration among automakers to establish common map specifications or to reduce their cost.

8.2.2 Connectivity Technologies

Vehicle connectivity is emerging as an important feature, alone or as a complement to automation, to promote greater safety through advanced knowledge of other vehicles' sudden stopping and other potential hazards. It can bring multiple other benefits as well, including better fuel efficiency. Connected vehicles can use a combination of different technologies to communicate with their surroundings including other road users, roadside infrastructure, and other devices, which may be near or far.

8.2.2.1 Vehicle-to-Everything Connectivity

The capability for direct data exchange between a vehicle and other objects, called vehicle-to-everything or V2X connectivity, includes vehicle-to-vehicle (V2V) and V2I communications. V2V enables the exchange of information from any onboard device such as the brake system, steering, and so on that is operated by a controller area network (CAN bus) or other electronic system. Each vehicle broadcasts its location, heading, and speed in a secure and anonymous manner. For dedicated short-range communications (DSRC) this information must be transmitted 10 times per second in uncongested conditions (NHTSA, 2016). All surrounding vehicles, provided they are suitably equipped, receive the message, and each estimates the risk imposed by the transmitting vehicle. V2I allows equipped vehicles to share information with similarly equipped components that support a road system and operate on the same frequency, which may include traffic lights, lane markers, streetlights, signage, or parking meters, for example.

V2X communication can be viewed as a sensor capable of 360-degree coverage and of providing high-level data that require only simple decoding to process rather than complex signal processing techniques required of other sensors (V2X Core Technical Committee, 2020). V2X enables bidirectional information flow and operates well in poor weather and lighting. It also has a range of 300 m or more, which exceeds that of sensors in ADAS systems (Table 8.2).

TABLE 8.3 Per Vehicle Cost of Onboard DSRC Equipment Cost—Delphi Survey Average

	2017	2022
Direct manufacturer cost of embedded DSRC	$175	$75
Total cost (direct + indirect) of embedded DSRC	$350	$300
Consumer cost of aftermarket DSRC	$200	$75

SOURCE: CAR (2012).

V2X is especially useful for safety-related communications because it can support very low-latency, secure transmissions (DOT, 2020a). It also provides the connectivity for efficiency opportunities such as vehicle platooning. Signal phase and timing (SPaT) communications from traffic signals can be leveraged to generate safety (red light violation warning), mobility (traffic optimization, bus priority), and vehicle efficiency (acceleration optimization) benefits. V2X is not intended to provide the long range of cellular networks that is needed for other types of communication. V2X requires installation of roadside equipment, which could hinder implementation across the vast areas needed to allow vehicle communication in all parts of the United States.

Two wireless communication standards for V2X connectivity are DSRC and cellular-based communications (C-V2X), both for radio communications.[1] The two systems are not interoperable, although both can be used on the same vehicle. In the United States, the Federal Communications Commission (FCC) allocated a dedicated spectrum of 75 MHz in the 5.9 GHz band for Intelligent Transportation Systems (ITS) specifically for DSRC technology (FCC, 2019a). While few vehicle models have been sold with V2X capability to date, at least 25 state transportation departments are already using this dedicated band for a variety of safety projects involving vehicles equipped with aftermarket connectivity devices. Projects include V2V and pedestrian collision avoidance, transit priority, and emergency vehicle traffic signal preemption (DOT, 2020b). In December 2019, the FCC issued a Notice of Proposed Rulemaking to propose that the lower 45 MHz of the band be repurposed for unlicensed uses including Wi-Fi and the upper 20 MHz of the band be dedicated to C-V2X, which currently has no spectrum allocation. The FCC also sought comment on whether the remaining 10 MHz should also be dedicated to C-V2X or be retained for DSRC (FCC, 2019b). Many automotive industry groups commented on the proposed rulemaking, noting that if frequencies were removed from the original 75 MHz band, ITS communications would face greater likelihood of congestion and disruption. The U.S. Department of Transportation expressed concern that this action would choose a winning technology (C-V2X) prematurely and would have adverse safety implications, and requested that the FCC pause its consideration of the proposal (DOT, 2020c). Nonetheless, on November 18, 2020, the FCC voted to adopt its proposal to release the lower 45 MHz of the band, retaining the upper 30 MHz for ITS uses utilizing the C-V2X protocol (FCC, 2020a).

DSRC equipment comprises onboard and roadside transceivers. A 2012 report of the Center for Automotive Research reported DSRC per vehicle cost projections from a connected vehicle technology industry Delphi study, as shown in Table 8.3 (CAR, 2012).

C-V2X, now under development, offers an alternative to DSRC for direct communication between vehicles and other road users, vehicles, and infrastructure. While C-V2X equipment is currently somewhat more expensive (ABI Research, 2018), proponents claim a cost advantage for C-V2X over DSRC based on private sector expertise and integration of road-side units with the cellular network (Qualcom, 2017).

General Motors (GM) is the only automaker to have equipped any vehicles in the United States with DSRC, and deployment was limited to CTS sedans, which were discontinued following model year (MY) 2019 (Abuelsamid, 2019; Lopez, 2019). Similarly, Europe initially planned to adopt DSRC but is now moving toward C-V2X. China adopted C-V2X at the outset and is now the world leader in testing and implementation of the technology. Multiple pilots are now under way. Ford has announced plans to launch C-V2X in China by 2021 and in all vehicles sold in the United States starting in 2022 (News Wheel, 2019).

[1]Here "cellular" refers not to the use of cellular networks but to the electronics used in cellular radios.

8.2.2.2 Cellular Network Connectivity

While direct, short-range connectivity is essential for safety-critical communications, connectivity through cellular networks is used for a rapidly expanding universe of other vehicle capabilities. Most new vehicles today offer internet connectivity for infotainment and in-vehicle apps. They also include telematics, which provides detailed in-use vehicle operating data to automakers.

Connectivity through cellular networks can also add to the safety and efficiency benefits of short-range V2X connectivity by providing wide-area road condition information, real-time map updates, and routing alternatives. When combined with C-V2X short-range connectivity, it can produce cost savings as well as new business opportunities in the realm of ecommerce. Such activity could help manufacturers fund the development of advanced sharing programs, which is key for automakers and others entering the shared mobility space.

Until very recently, cellular devices in the United States used 4G-LTE (Long-Term Evolution), the fourth-generation standard for wireless cellular communications. 4G-LTE was designed to provide up to 10 times the speeds of 3G networks for mobile devices using the existing cellular infrastructure. LTE is the most modern standard that is readily available in most areas of the United States. The latency of LTE is variable depending on many factors that can decrease speed of communication on the network. Signal interference from obstructions can make usage in cities unpredictable.

5G is the fifth generation of cellular mobile communications. It succeeds the 4G, 3G, and 2G systems. 5G performance targets high data rate, reduced latency, energy saving, cost reduction, higher system capacity, and massive device connectivity. Major carriers have now reached nationwide 5G coverage in the United States. 5G makes possible peer-to-peer communications through the network using mobile "edge computing," providing the necessary improvements in bandwidth and latency as well as the ability to separate safety-critical and non-safety critical messages (ABI Research, 2018; Qualcomm, 2019). Additional capabilities for connected vehicles will include the ability to share sensor data and planned trajectory and to build local HD maps in real time (Green Car Congress, 2020). However, to achieve such benefits 5G requires many more towers than its predecessors. A recent study forecasts 41 million 5G C-V2X vehicles globally by 2030 and 83 million by 2035 (Green Car Congress, 2020).

8.2.3 Specifications of CAV System Components

Table 8.4 shows representative specifications for current (Level 2) automated vehicle system components.

TABLE 8.4 Onboard Technologies That Enable Automation and Connectivity and Their Capabilities as Implemented in 2020 Vehicles

	Specs in 2020			
Technology	Number per Vehicle	Weight per Unit (kg)	Power Draw per Unit (W)	Cost per Unit (US$)
Stereo camera	7–11	0.06	1–2	45
Lidar[a]	0–2	0.83–12.25 (small–large)	8–60	2,000
Radar	1–5	0.25	4–18	55
Ultrasonic sensor	0–12	0.04	0.1	10
DSRC onboard unit	1	2.7	6	75
C-V2X radio/chipset	1			90
Inertial measurement unit w/GNSS[b]	1		1–7	
Computer	1–2	3.13–5.075	80–96	800

[a]Lidar not typically included in Level 2 systems but shown here for completeness.
[b]Global Navigation Satellite System.
SOURCE: Gawron et al. (2018); Baxter et al. (2018); CAR (2012); ABI Research (2018); committee estimates.

Cost and power draw of these individual components are generally decreasing. Automative lidar costs in particular are projected to fall rapidly, from about $2,000 in 2020 to hundreds of dollars within a period of 3 to 5 years. Some lidar suppliers have already commercialized units at well under $1,000, although with limited range and resolution. For example, Valeo's SCALA is available at high volumes at a cost of $600 (Rangwala, 2020). Velodyne, which demonstrated a compact, solid-state $100 lidar for applications such as drones and robots at CES 2020, has announced that its solid-state, automotive grade Velarray H800 has a target price of less than $500 at high-volume production (Krok, 2020; Velodyne Lidar, 2020). At the same time, as automation advances, the number and sophistication of devices will increase. Tesla's Autopilot V3 computer uses chips capable of 36 trillion operations per second (Abuelsamid, 2019); by comparison, chipmaker Nvidia reports that its next-generation chip will be able to process 200 trillion operations per second. The trajectories of total cost and power consumption for automation systems are not yet clear, and are discussed further in Section 8.3.8.1.

In addition to these onboard technologies, CAVs rely to varying degrees upon cloud computing, traffic and map data, infrastructure sensors, and security systems. These offboard systems will be costly and will consume considerable amounts of energy, although on a per-vehicle basis requirements may be low in the future if most vehicles on the road are using their services. For both on- and offboard systems, sophisticated algorithms, models, and data acquisition are required, at high cost to automakers and suppliers.

8.2.4 Onboard Computing

Vehicle automation systems require substantial onboard computing. The continued expansion of their functionality to systems with greater vehicle driving authority requires dramatic increases in electronic content and interconnected, embedded systems. These are complex systems, in contrast to the relatively simple, stand-alone computing systems that once controlled basic engine and vehicle functions. These systems are also evolving to become highly cyber-physical in nature, and perhaps represent some of the most sophisticated, widely distributed cyber-physical systems that exist today. These are characterized by the following:

- Deep physical interactions (sensors, actuators, controllers);
- Deeply embedded electronics (nonlinear multilayered interactions);
- High degrees of computation (hundreds of trillions of operations per second);
- Rich needs to communicate (high speed, rich data communication is a must for time critical decisions);
- Pervasive integration (cyber and physical); and
- High degree of coupling with driver behavior.

Because of these systems' complexity and their need for tight control of multiple interconnected, embedded systems with high levels of security, accuracy, and reliability, there are many challenges facing the designer and developer of such systems. For example, components and subsystems can no longer be designed and developed in isolation and then integrated into the vehicle. Now, complete systems have to be integrated at the outset of the design process and in setting system requirements, in order to allow mutual interactions at deeper and deeper levels. Also, efforts are ongoing to centralize and integrate some control functions in order to optimize the number of the electronic control units wiring and connectors involved, reducing cost. In addition to the high cost and large power demand—more than 2 kilowatts (kW) reported for prototypes—for intensive computing, this complex cyber-physical system requires new computing hardware and software architectures and new system engineering and design tools to be integrated into the vehicle. Examples include the following:

- The integration of sophisticated control algorithms involving a large number of software code lines makes it increasingly difficult to verify and validate these algorithms and their interaction under a plurality of inputs and conditions (states) using conventional trial and error approaches. The use of emerging, systematic, and automated "formal methods" techniques for requirements checking of model and software generations are becoming essential for the design of reliable software. Furthermore, the integration of algorithms developed by multiple sources requires strict adherence to software architecture standards such as AUTOSAR.

- Advanced control strategies and architectures are used to ensure "fail-soft" and "fail-operational" requirements, needed for Level 2–Level 5 automated driving.
- Information security is critical with the increasing usage of wireless communications involving data transfer, which also interacts with vehicle control devices (throttle, steering, and braking). Therefore, a strong emphasis on automotive cybersecurity is key to the success of advanced safety critical systems.
- To achieve the control accuracy and reliability required for advanced active safety and autonomous driving systems there is an increasing need for fast, reliable, onboard communications, and computational approaches at affordable cost. These typically involve modern field programmable gate array capable of parallel processing high volumes of data on a single chip utilizing sophisticated error correction algorithms.
- Diagnostics and prognostics of cyber-physical systems present a challenge owing to their complexity, but at the same time they are considered key enablers for systems service and repair and customer peace of mind.

The sophisticated technologies addressing the above challenges have already found their way into the design and development process of automakers and suppliers as a means of ensuring the safety and integrity of these new complex products. Furthermore, the large electric power demanded by onboard computing experienced by prototype autonomous vehicles, as well as its cost, is expected to greatly diminish by the time autonomous vehicles are more widely available (2030) owing to intensive efforts by chip manufacturers (Stewart, 2018).

8.2.5 CAV Systems in Production

Several vehicle models with Level 2 CAV systems are in production and many more, as well as some Level 3 systems, are in the preproduction phase or in planning in the United States and other locations. Table 8.5 details several of the available and upcoming models.

8.3 IMPACTS OF CAV TECHNOLOGIES ON VEHICLE EFFICIENCY

Vehicle automation can increase or decrease fuel consumption owing to a variety of factors. Even at low levels (Level 1 to Level 2), automated driving can adjust vehicle operation (speed, acceleration, deceleration) to achieve an eco-driving mode. Existing electronic vehicle controls will then minimize unnecessary or unexpected braking and high acceleration (unless there is a safety need), reducing high-torque and fuel-inefficient maneuvers.

Even during human driving (Level 0), the integration of short-term (5–20 second) load preview (prediction of upcoming need for acceleration and deceleration) using onboard sensing and long-term (1–20 minute) forecast of driving patterns (route estimated time of arrival, traffic ahead, signal lights, signs, intersection congestion, accidents, etc.), through historic data or rough real-time traffic information available from popular navigation tools, will provide opportunities for further optimizing powertrain efficiency. As will be discussed below, anticipating the load demand offers unique opportunities to improve powertrain efficiency while satisfying these load demands by calibrations that judiciously manage torque (power) and energy reserves.

Connectivity can complement automation to help a vehicle run more efficiently. Automation does not require connectivity, as the car can replicate the driver via AI and algorithms. However, if an algorithm is short sighted in optimizing (short horizon), has poor perception (noisy or delayed signals), or causes aggressivity (short time headway), it may increase fuel consumption (Zhu et al., 2019; Prakash et al., 2016). The full benefits of vehicle automation can be achieved only with connectivity and there are many benefits that connectivity can provide sooner, and perhaps at lower cost, than can be provided through automation. Particularly in the early years, using both automation algorithms and connectivity will allow for the best decisions by the computer control module. Optimization of powertrain operation for efficiency will be enhanced by information on upcoming activity from other vehicles, traffic lights, and other infrastructure. The ability to engage in more co-operative driving scenarios or to form tightly controlled clusters of vehicles as in platooning requires connectivity and automation features.

TABLE 8.5 Example Automated Vehicles Deployed, Including Their Level of Automation, Technology, Automation Features, and Details on the Year Introduced and the Technology Penetration

Make and Model	Level of Automation	Technology	Automation Features	Year Introduced (MSRP + Additional Cost of Automated Tech)	Technology Penetration
Mercedes-Benz[a]	2 (Driver Assistance Package)	• Radar • Lidar • Cameras • Ultrasonic sensors • Digital HD map technology	• Adaptive cruise control • Active steering assist • Active blind spot assist • Active lane keeping assist • Active lane change assist • Active brake assist with cross-traffic function • Active emergency stop assist • Active speed limit assist • Evasive steering assist • Congestion emergency braking • Route-based speed adaptation	Introduced 2013 (updated 2016, 2017) (+$2,250)	Available on most Mercedes Benz models
Nissan ProPilot Assist Altima, Rogue, Leaf	2 (Pro Pilot Assist with Navi-link)	• Centrally mounted radar module • Single camera behind the windshield • 3D HD map data • Front, side, and rear sonar • Driver monitoring sensors	• Lane centering and blind spot intervention • Accelerating and braking in response to vehicle ahead • Advanced steering assist • Automatic emergency braking with pedestrian and cyclist detection	MY 2016 (Serena, Japan) ($22,000–$32,000) MY 2018 (Rogue, United States) ($25,000–$33,000 + $0–$2,500)	High for both luxury and economy models (Japan) Luxury models in the United States
Cadillac CT6	2 (Super Cruise)	• Lidar maps • In-vehicle infrared cameras • Driver monitoring infrared camera • Radar sensors • Driver attention system • DSRC transceivers (CT5)	• Lane centering • Adaptive cruise control • Forward collision system	MY 2018 ($75,000 + $2,500)[b]	CT6 discontinued in 2020
Audi e-tron	2	• Central driver assistance control unit (zFAS) • Side, front, and rear sensors • Top-view camera with 360-degree 3D	• Active lane assist with emergency assist • Adaptive cruise assist with turn assist, maneuver assist, and efficiency assistant	MY 2019 ($74,000 + $3,500)	

continued

TABLE 8.5 Continued

Make and Model	Level of Automation	Technology	Automation Features	Year Introduced (MSRP + Additional Cost of Automated Tech)	Technology Penetration
BMW X5 with Extended Traffic Jam Assistant[b]	2 (Extended Traffic Jam Assistant)	• Driver-monitoring optical camera • Front-racing radar • Shorter-range radar sensors at the vehicle's corners • Three-camera bundle with microchip controls mounted ahead of the rear-view mirror	• Hands-free and pedal-free driving on limited-access highways and on surface streets at low speeds	MY 2019 ($59,000–$82,000 + $0–$1,700)	Debuted on 2019 X5. Available as standard or optional on these 2020 models: • 3 Series • 7 Series • 8 Series • X5 • X6 • X7
Hyundai Ioniq, Sonata, Nexo, Palisade	2 (SmartSense with Highway Driving Assist)	• Radar sensors • Cameras	• Highway driving assist with speed regulation and lane centering • Collision avoidance (front and rear) • Driver attention warning • Lane-keeping assist • Adaptive cruise control	MY 2019 ($23,000–$60,000 + $0–$3,100)	Standard in limited trims, available in add-on packages for lower-tier trims
Tesla (Model S, 3, X)	2	• Forward facing radar • Rear, side, and forward cameras • Ultrasonic sensors	• Automated steering, acceleration, and braking within its lane • Auto park • Auto lane change • Matches speed to driving conditions	MY 2019 Model S ($89,200) Model 3 ($35,000) Model X ($81,000)	Available worldwide
Lincoln Corsair, Aviator, Nautilus	2 (Lincoln CoPilot360 Plus)	• Cameras • Sensors • Driver monitoring system • Radar	• Lane keeping • Adaptive cruise control with stop-and-go capability and speed sign recognition • Traffic jam assistance • Precollision assist and pedestrian detection • Evasive steering assist	MY 2020 ($42,000–$55,000 + $4,150–$4,950)	Available as an add-on in luxury trims
Toyota Mobility Teammate	2 (Mobility Teammate)	• Radar • Lidar • Cameras • Ultrasonic sensors	• Automatically change lanes, follow lanes, and pass vehicles on limited-access roads • Automatic parking	MY 2020 (not for sale)	LS1 Concept in Japan

Cadillac CT5, CT4, Escalade	2+ (Enhanced Super Cruise)	• Lidar maps • In-vehicle infrared cameras • Driver monitoring infrared camera • Radar sensors • Driver attention system • DSRC transceivers (CT5)	• Lane centering • Adaptive cruise control • Forward collision system • Automated lane change on demand	MY 2021 ($40,000–$75,000 + $2,500–$6,150)[c]	MY 2021 triples number of models offering Super Cruise, along with additional capabilities
Nissan ProPilot 2.0 Ariya	3 (Pro Pilot 2.0)	• 360-degree sensing—front camera, front and side sonar, front and side radar, AVM cameras • Driver monitoring camera • Intelligent dynamic suspension • Direct adaptive steering	• Lane centering • Accelerating and braking in response to vehicle ahead • Advanced steering assist • Hands-free single lane, highway driving • Hands-on guided lane changing abilities • Assist in passing, lane diversions and lane exiting	MY 2020 (Skyline, Japan, $40,000)	Standard with hybrid Skyline in Japan Will be available in United States with MY 2021
BMW 2021 iNext	3	• 5G	• In "ease mode" the steering wheel retracts away from the person and the display shifts and some level of self-driving activates	2021 (highway pilot)	
Mercedes-Benz[d]	3 (Drive Pilot) 4 (Intelligent Parking Pilot)	• Lidar • Cameras • Digital HD map technology • Positioning system (more precise than typical GPS) • external microphones • Infrared driver sensors • Redundant steering, braking and electrical systems	• Drive pilot: Level 3 conditionally automated • Intelligent parking pilot: self-driving using V2X in capable parking garages	Late 2021 (Germany only)	
Cruise Origin[e] (GM-Honda)	4	• Sensors • Cameras • Radar • Acoustics • Lidar	• No steering wheel or gas pedal	2022 (not for sale)	Purpose-built for "automated taxi" and mobility as a service pilot deployments, beginning in San Francisco
Argo AI (Ford-Volkswagen)[f]	4	• Sensors • Cameras • Lidar (short and long range)		2022 (not for sale)	Commercial service expected in 2022, test Fusion Hybrids and Escape Hybrids in six U.S. cities

NOTE: AVM = around-view monitor; MSRP = manufacturer's suggested retail price. MSRP and technology package costs sourced from automaker websites and online catalog.
[a]Mercedes-Benz USA (2017); Autotrader (2020).
[b]Paukert (2018).
[c]But must already have Driver Assist and Technology Package, $3,650. Patel (2020); Cadillac Pressroom (2020).
[d]Wardlaw (2020).
[e]Brooke (2020a).
[f]Korosec (2020); Wayland (2020); Ford Motor Company (2020).

CAV technologies may also bring indirect energy savings through system effects. For example, safety improvements from automated emergency braking or lane-keeping assist can save fuel by reducing the congestion associated with vehicle crashes. On the other hand, automation technologies can increase vehicle miles traveled: in a 2018 survey conducted by University of California, Davis researchers, Tesla buyers reported large increases in weekend travel and travel under congested conditions owing to availability of the Autopilot system (Hardman et al., 2019). Fully autonomous vehicles could further increase or decrease fuel consumption overall by changing vehicle miles traveled and vehicle size and weight, as discussed in Chapter 9. This section focuses on automation technologies' direct impacts on individual vehicles' fuel efficiency. They may vary substantially across implementations of a given technology and can depend strongly on driving conditions and adoption levels. Understanding of these factors is improving as a result of modeling and real-world experience, however.

8.3.1 Velocity Optimization

Using automation to optimize vehicle speed can produce substantial fuel savings. For most powertrains, minimization of tractive energy to transport a vehicle from a point A to point B corresponds to a "pulse and glide" operation which involves three phases—namely, short but high accelerations to a cruising speed, followed by a fixed cruising speed, and finalized by short, sharp braking. The resulting change in acceleration, also known as jerk, associated with this energy-optimum pattern increases passenger discomfort and is limited by the automakers. Instead, they aim to deliver smooth velocity transitions and minimization of accelerations that exemplifies eco-driving style.

ACC provides an opportunity for velocity optimization. ACC was primarily designed for convenience (cruising) and safety (maintaining a safe headway from proceeding vehicles) and not for fuel efficiency. The ACC functionality involves low-level automation and could be adjusted with a short preview for fuel efficiency creating the commonly known eco-ACC functionality. The optimal velocity trajectory maintains a distance from the vehicle ahead that minimizes acceleration events and aerodynamic drag and is short enough to prevent cut-ins from other lanes while ensuring a safe distance from the lead vehicle. Prediction of the velocity of the vehicle in front of the subject vehicle is necessary since optimal decisions depend on future acceleration demands. When the subject vehicle is on an open road, the eco-driving approach may use the speed limit for that road segment based on existing navigation tools.

To assess the benefit of these optimal following velocity profiles in a laboratory setting, a lead vehicle is typically represented by a federal testing drive cycle. Implementing the optimization in the laboratory on an Escape 1.6 L Ecoboost engine with full drive cycle preview showed up to a 17% increase in fuel efficiency (Prakash et al., 2016). Real-time implementation with model predictive control achieves improvements in fuel efficiency proportional to the prediction horizon. When the prediction horizon reaches 20 seconds, most of the fuel efficiency benefits of full preview are achieved. A preview of less than 5 seconds, however, may eliminate most of the benefits and can even make the automated driving efficiency worse than that of a human-driven vehicle unless special robustification measures are applied. With those more robust algorithms, some increase in fuel efficiency was obtained with as little as 1.5 seconds' preview of the lead vehicle velocity (Hellstrom et al., 2018).

8.3.2 Added Benefit of Connectivity

In a vehicle without connectivity, eco-ACC strategies assume a constant acceleration or constant velocity up to a predefined prediction horizon (Guanetti et al., 2018). These methods result in large errors in expected velocity at the prediction horizon that can degrade the eco-ACC performance (He and Orosz, 2017; Vajedi and Azad, 2016). Connectivity enables a preview of the trajectory of a vehicle in front of the subject vehicle or the distance to a stop sign or an upcoming traffic light. Traffic signal information would be available through transmission of SPaT messages.

V2V communication can dramatically improve the accuracy at the prediction horizon used for fuel efficiency optimization, as shown in Figure 8.3 (Hyeon et al., 2019).

If an automated vehicle using eco-driving algorithms is tested for fuel economy by "following" a federal testing drive cycle, it will exceed the allowable deviation from the standard drive cycles for fuel consumption testing.

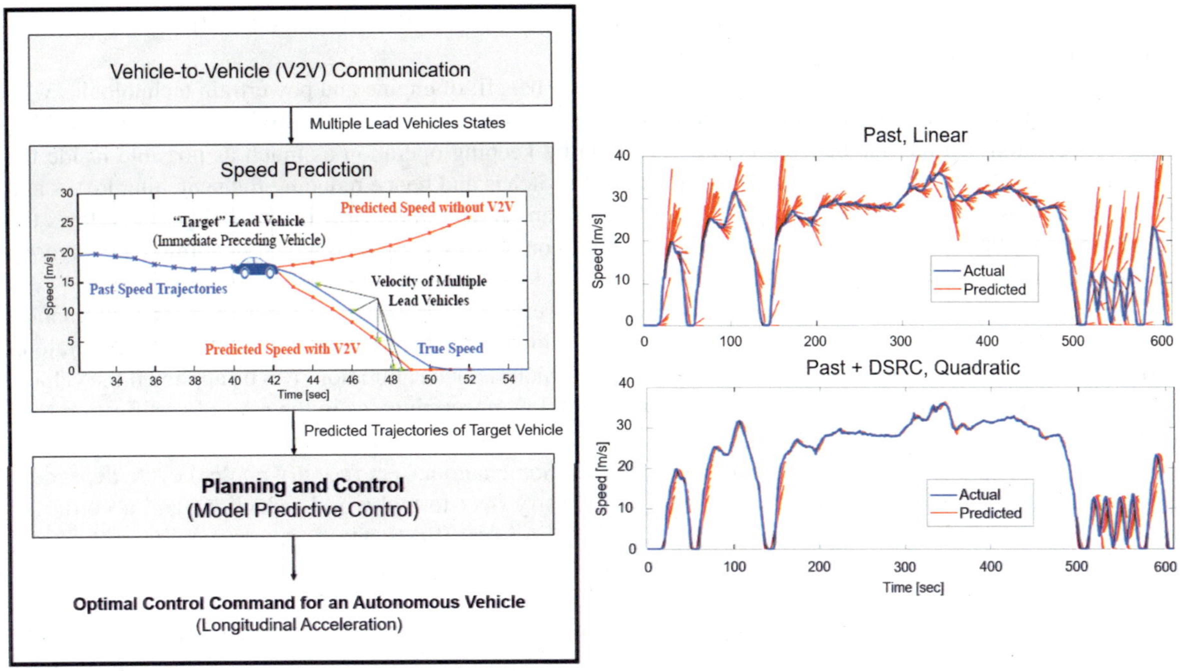

FIGURE 8.3 The prediction of the lead vehicle trajectory improves dramatically with information from preceding vehicles. Error in the prediction horizon decreases the fuel efficiency potential of eco-driving. SOURCE: © 2019 IEEE. Reprinted, with permission, from E. Hyeon, Y. Kim, N. Prakash, and A.G. Stefanopoulou, 2019, "Short-Term Speed Forecasting Using Vehicle Wireless Communications." In *2019 American Control Conference (ACC)*, 736–741. https://doi.org/10.23919/ACC.2019.8814400.

Hence the automated vehicle cannot use eco-driving during testing, so the fuel savings go undetected, removing an incentive for automakers to adopt the technology.

> **FINDING 8.2:** Connectivity technologies can add substantially to the fuel savings benefits of automation technologies.

8.3.3 Efficiency in Internal Combustion Engine-Based Powertrains

The preceding sections discuss how automation can be used to optimize a vehicle's velocity profile to save fuel. Connectivity that enables the prediction of upcoming acceleration or braking (load preview) will also enable the full realization of the benefits of many powertrain technologies whose calibration involves significant trade-offs between efficiency and drivability by reserving torque for unexpected accelerations. These two efficiency opportunities in co-optimizing velocity profiles and powertrain control enabled by connectivity and automation can achieve benefits that are nearly additive.

Forseeing the need for acceleration or braking will enable currently automated powertrain technologies such as high dilution, advanced ignition, and turbocharging to be calibrated in their most efficient value without drivability compromises. Nazari et al. (2017) showed a 4% reduction in fuel consumption from high exhaust gas recirculation on the US06 drive cycle with full load preview compared to a strategy that reserves boost pressure for fast transients. This significant fuel efficiency gain can be attained with low-level (Level 1 to Level 2) automated driving that plans load transitions; much of the benefit can be achieved with even 1-second load preview. The benefit is much smaller (0.6% reduction in fuel consumption) if there is load anticipation of only 300 milliseconds, however

(Hellstrom et al., 2018). The loss in fuel savings is indicative of the sensitivity of these benefits to the extent of unanticipated load transients.

Connected and automated driving can also attain the full benefit of engine and powertrain technologies with discrete operating modes such as cylinder deactivation and gear switching (Olin et al., 2019; Sciarretta and Vahidi, 2020). Automation can provide benefits in two ways: (1) by keeping operation as much as possible inside the most efficient mode and (2) by minimizing the number of switches and hence reducing the switching losses and wear. The mode-switching process is in general inefficient, and having a preview of upcoming load allows the powertrain controller to decide if it is worth switching based on how long the load will stay within a given mode and how long it takes to recover the switching losses.

While anticipating the load can benefit high efficiency powertrains, automakers do not currently take advantage of this opportunity for added efficiency because the acceleration/deceleration anticipation will not be accurate in traffic with low penetration of automated and/or connected vehicles. Such calibrations can be applied if the vehicle recognizes that it is driving in a lane or an environment with low uncertainty, as in the case of certification testing. However, such adjustments in calibration based on drive cycle would need to be well documented to declare and explain their legitimate need and benefits to regulators. Some automakers recently applied cycle-dependent adjustments to improve emissions test results that did not carry over to real-world reductions and accordingly incurred "cycle-beating" fines. Being unable to claim the benefits in certification testing, automakers may find that adjusting calibration using load anticipation to improve fuel efficiency is not worthwhile, at least until automated vehicles are more prevalent in the fleet.

Furthermore, while connectivity and automation permit optimal operation of many powertrain efficiency technologies, high penetration of CAVs would lessen the need for complex technologies such as engines with turbocharging, lean combustion, or dynamic cylinder deactivation, or high-gear number transmissions. That is because when traffic is largely comprised of automated vehicles, eco-driving (minimization of acceleration) will be prevalent, greatly reducing the need for high torque. Right-sized, therefore small, naturally aspirated engines will be adequate. While downsized engines and overall powertrains will have poor drivability and fuel efficiency when they are used in abrupt or uncertain traffic, the deterministic traffic associated with a high penetration of automated driving would enable aggressive powertrain downsizing for vehicles not requiring towing or load-carrying capability.

At least in the near term, automakers will be reluctant to design general purpose light-duty vehicles that can function well only under automated driving conditions. This creates a near-term barrier to their taking advantage of this opportunity to employ low-cost, fuel-efficient, downsized powertrains in CAVs. Dedicated vehicles for use in enclosed and low speed areas, such as the shuttles being used in various closed-campus services, are good candidates for such aggressive downsizing, however.

FINDING 8.3: Connected and automated driving can allow some engine and powertrain efficiency technologies to achieve their full savings potential. However, high penetration of highly automated vehicles that create a deterministic traffic environment would limit the need for high torque capability and thus make small, naturally aspirated engines with stop-start and limited torque-assist the most cost-effective and efficient option for engine-powered vehicle powertrains where towing and load-carrying capabilities are not required. At least in the near term, however, automakers will be reluctant to design general purpose, light-duty vehicles that can function well only under automated driving conditions.

8.3.4 Power Management of Hybrid Powertrains

CAV technologies offer additional efficiency opportunities in hybrid electric vehicles (HEVs) by optimizing their power management through engine load leveling and regenerative braking when the battery can accept the energy. Optimal decisions on how to divide the power delivered between an engine and a battery depend on continuous forecasting of future efficiency and opportunities to replenish the energy used from the battery. Automation can improve HEV fuel efficiency by providing a planned load profile (removing uncertainty from the optimization horizon) or by co-optimizing power management and the vehicle velocity profile in an eco-driving scenario.

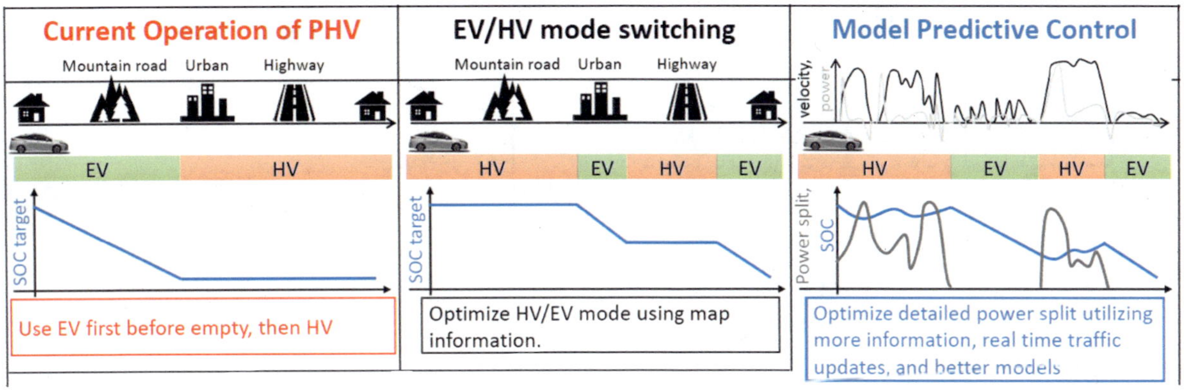

FIGURE 8.4 Optimal power management takes into account trip information to schedule charge depletion instead of the default practice of charge depleting then charge sustaining (CDCS) operation.
NOTE: HV = hybrid vehicle; PHV = plug-in hybrid vehicle; SOC = state of charge.
SOURCE: Butts for Huang et al. (2019).

For example, an algorithm can determine that the battery will be able to provide the demanded power with the engine off for a longer period if the system anticipates or can induce a deceleration command later to recharge the battery.

These efficiency gains (2%–3%) from power management in a strong HEV by automated driving or load preview can be smaller than the efficiency gains obtained in mild or micro-HEVs (4%–6%) for two reasons (Karbowski et al., 2020; Nazari et al., 2019). First, efficiency gains from automation can be larger in powertrains having lower baseline efficiency. Second, automation can effectively manage over-constrained systems such as micro-HEVs, where torque assist and regenerative braking are limited by the small battery size.

In plug-in HEVs (PHEVs), optimizing the charge depletion and power management depends on the vehicle speed and accelerations associated with urban or highway traffic, as shown in Figure 8.4. Generally, a strategy of charge depleting mode followed by charge sustaining mode (CDCS) is not optimal when the required trip energy is greater than the energy stored in a battery. Huang et al. (2019) find 3%–7% fuel efficiency benefits from real-time optimization of power management, relative to the CDCS strategy. The optimal battery discharge trajectory depends on the traffic or speed profile of the entire trip. Some traffic information can be obtained from historical travel data through onboard GPS and mobile navigation applications (Wang et al., 2019). Real-time information through V2X can also be integrated, but Huang et al. (2019) showed that using historical data to optimize power management can also be effective.

Through decades of development, the energy management and optimization algorithms for HEVs and PHEVs have become a mature field. Nonetheless, the computational burden of this optimization problem of planning a long trip without intermediate charging availability (for PHEVs) is still evolving from the first few sources (Paganelli et al., 2002; Guzzella and Sciarretta, 2013) with different algorithms thoroughly reviewed for PHEVs (Martinez et al., 2018) and HEVs (Sciarretta and Vahidi, 2020). Handling the engine on-off switching systematically by penalizing the cranking fuel loss provides 1%–2% fuel efficiency benefits. Additional benefits of up to 6% can be derived by co-optimizing the power management and the velocity profile leading to so-called pulse-and-glide behavior (Chen et al., 2018). This introduces passenger discomfort, however, and hence will be more appropriate for unoccupied vehicles used for delivery purposes.

Realizing the full potential of real-time engine, gear, and power management relies on accurate load prediction. Prediction to a short (1 second) horizon is extremely important for choosing correct gear values, boosting level, and engine on-off decisions, whereas a medium horizon (5 seconds) load preview is needed to avoid chattering (switching among optimum values). Longer time horizons and communication from signalized intersections improve the fuel efficiency of conventional or hybrid powertrains (Figure 8.5; Pozzi et al., 2020).

However, uncertainties at some time ahead (similar to adding an unplanned trip to a grocery store on your way home from work) will need to be accommodated in a PHEV (or battery electric vehicle [BEV]) where there

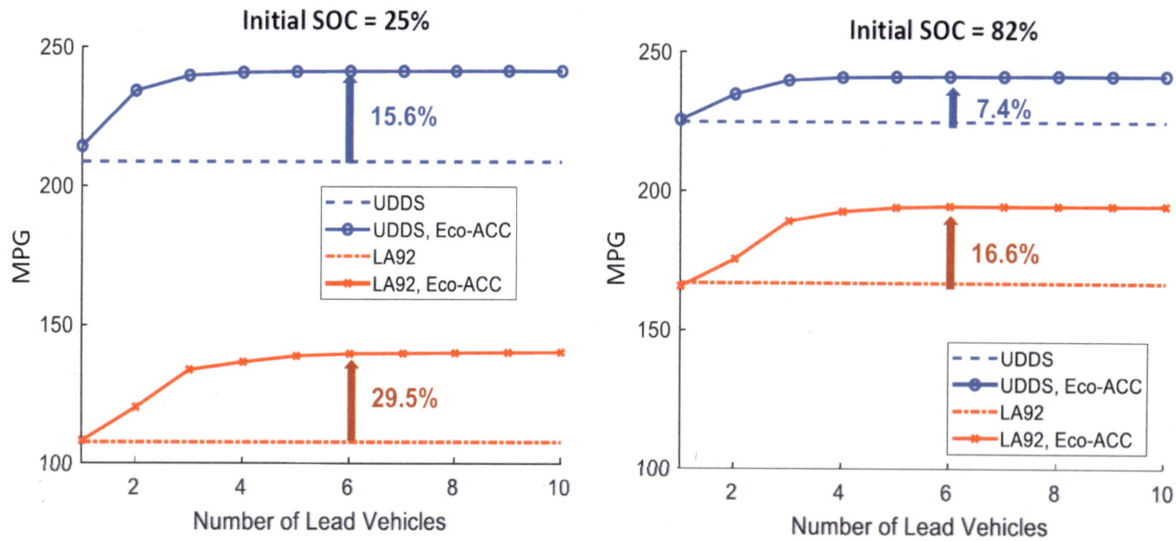

FIGURE 8.5 Communication from lead vehicles provides information for the traffic (velocity and position) ahead of the automated vehicle, which enables accurate prediction and optimization of its acceleration. The simulation results, performed with the Autonomie model using a Prius Prime, show that automation with good communication can improve a PHEV's fuel efficiency as much as 15.6% on the UDDS cycle with low initial state of charge (SOC, left figure) and 7.4% (right figure) with high initial SOC. The results show that the 10 second horizon achieved with V2V from five vehicles ahead (assuming approximately 2 second headway for safety) achieves the full potential efficiency gain from prediction. Here the urban cycle UDDS and the more aggressive cycle LA92 are two dynamometer driving schedules used for regulatory emissions testing. SOURCE: Team Presentation by SWRI at NEXTCAR ARPA-E 2019 Annual Meeting.

is a desired final state like a depleted battery. Automated driving of a PHEV or BEV is most efficient with a long prediction horizon, which connectivity can provide, to accommodate slow dynamics (battery state and thermal) (Lee et al., 2020).Optimization accounting for a long horizon creates a large computational load, however, which in turn draws power and reduces the benefits an EV may experience from automated driving.

In conclusion, PHEVs have the potential to achieve the highest percentage increase in energy efficiency from automation. Automated driving of PHEVs that combines optimal eco-driving and power management incorporates and even enhances all the above benefits and can achieve as much as 14% higher fuel efficiency on trips that significantly exceed the battery range (Chen et al., 2018). The efficiency gain is drive-cycle-dependent; over a combination of repeated highway cycles (similar to US06 and Highway Fuel Economy Test [HWFET]) followed by cycles that represent congested urban driving (similar to the EPA Urban Dynamometer Driving Schedule [UDDS]), fuel efficiency gains of more than 14% over a simple CDCS operation strategy are achievable (Chen et al., 2020). These benefits are not currently detected in the certification testing, however, so customer satisfaction is the sole incentive to the manufacturer to implement these optimization strategies.

> **FINDING 8.4:** The look-ahead information provided by connected and automated vehicle technologies can be used to improve power management in hybrids, increasing efficiency by 2%–3% in strong hybrids and 4%–6% in mild hybrids. PHEVs have the potential for the highest percentage gains, with as much as 14% higher fuel efficiency on trips that significantly exceed the battery range.

8.3.5 Interactions with Electrification

The interactions between automation and connectivity technologies and electrification are diverse. Energy efficiency benefits of CAV technologies are smaller for BEVs than for other vehicles because BEV powertrains are highly efficient and not subject to many of the energy losses that CAVs can mitigate in other vehicle types through strategies such as eco-driving. In addition, BEVs stand to lose range owing to CAV system power draw, at least until those loads have been greatly reduced. Indeed, given that BEVs will ultimately be equipped with CAV technologies to improve safety, functionality, and convenience, one might view the primary energy efficiency objective for a BEV's CAV technologies as offsetting any range loss arising from the system's power requirements.

On the other hand, BEVs' electrical and auxiliary systems may otherwise be a good match for CAV packaging requirements. Eco-routing, which relies heavily on real-time traffic information, could be highly beneficial to BEV efficiency in hilly terrains, although it should be noted that travel time may increase as a result. Weather forecasting integrated with traffic could lower cabin and battery thermal loads by 2%–4% (Sun et al., 2019) and thus reduce the cold-weather impact on BEV range. Connectivity could allow BEVs to seamlessly choose routes and operating modes tailored to individual vehicles to maximize range and extend battery life (Taiebat et al., 2018). Fully networked charging infrastructure will optimize routes and advise drivers or the autonomous vehicle of available charging stations. Connectivity could also enable planning the demand response and various functionalities such as V2G to support the grid or V2B to support buildings (peak management, or renewable energy storage), which reduces offboard fuel consumption and emissions. It could also help to address range anxiety by providing accurate, route-dependent information on battery state of charge and charging options.

In developing more advanced CAV systems, and in particular for fully autonomous vehicles, automakers vary in their powertrain approaches. For example, Ford is using a hybrid platform for its first autonomous vehicle prototypes, citing range loss and battery degradation of autonomous BEVs (Iaconangelo, 2020). However, other automakers including GM, as well as autonomous vehicle startups and ride hailing companies, are moving directly to a BEV platform for autonomous vehicles (Mohan et al., 2020). Indeed, full autonomy can complement and facilitate vehicle electrification in additional ways, as discussed in Chapter 9.

FINDING 8.5: Compared to other powertrains, all-electric powertrains will see the lowest efficiency gain from connected and automated vehicle (CAV) technologies because they are already the most efficient. However, electric vehicles will benefit from other synergies with CAV technologies.

8.3.6 Effects by Powertrain and Drive Cycle

Karbowski et al. (2020) used vehicle simulation to quantify the effects of many of the CAV-enabled optimization strategies discussed above for conventional, hybrid, and battery electric vehicles. Figure 8.6 shows the percent energy savings they found from eco-driving (velocity control) with and without powertrain controls, and with and without connectivity (V2I), on multiple drive cycles. On the highway cycle, all vehicle types and control strategies achieved modest benefits of 4%–5% except for a small added gain for hybrids from powertrain control. As expected, urban cycle benefits were much larger, with biggest gains for conventional and hybrid vehicles employing speed and powertrain optimization as well as V2I communication (traffic SPaT). Velocity optimization alone showed savings of more than 10% on a suburban cycle for some powertrain types.

The results in Figure 8.6 may differ from real-world and test results for various reasons: traffic is not considered; city and highway cycles do not coincide with standard test cycles and are not statistically representative of U.S. driving; imperfections in calibration and future horizon knowledge; and unaddressed trade-offs among energy consumption, drivability, and travel time (Karbowski et al., 2020). Nonetheless, the results are instructive in that they provide a consistent basis for comparing savings across vehicle types and CAV capabilities.

8.3.7 Congestion Avoidance and Reduction

A connected vehicle has the potential to avoid or mitigate congestion and thereby reduce energy consumption. It can modify routes to avoid congested areas based on traffic flow data from the cloud. Connected vehicles may also reduce congestion by using data to drive more smoothly, stop less often, and optimize speed. This can improve the fuel efficiency not only of the connected vehicle but of following vehicles as well (Karbowski et al., 2020).

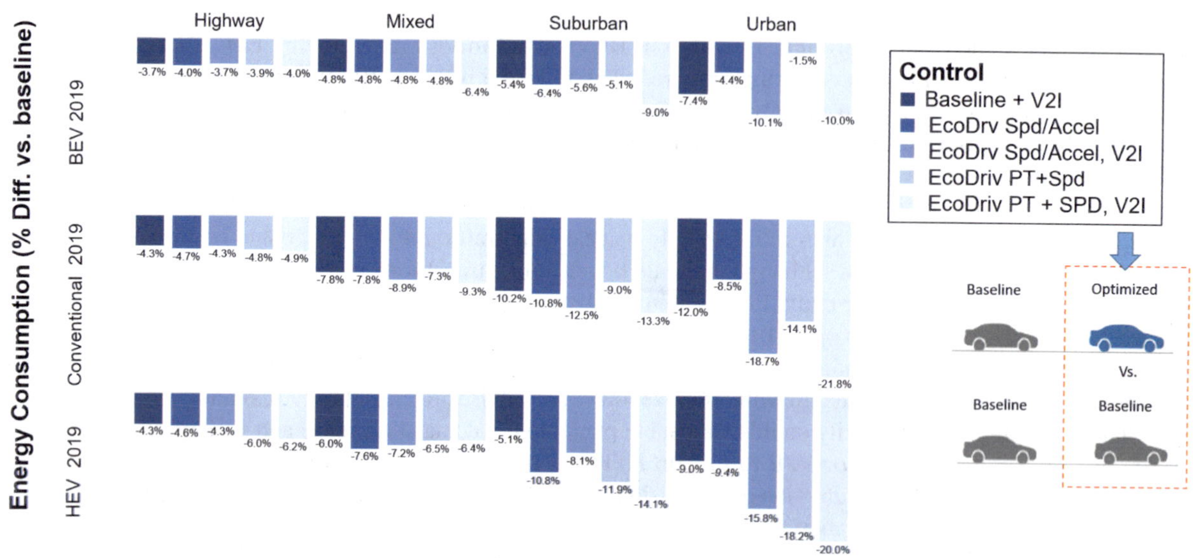

FIGURE 8.6 Percent energy savings from simulation of various automation and connectivity capabilities.
SOURCE: Karbowski et al. (2020).

Congestion reduction strategies such as traffic flow metering and traffic light management are more effective if the car can communicate with the traffic light and respond appropriately. Last, connectivity and automation can reduce the number of crashes, eliminating the congestion owing to those incidents.

8.3.8 Vehicle Load Issues for CAVs

8.3.8.1 Electric Equipment Power Demand

The electric power demanded by the computers, sensors, and actuators used in vehicle automation can be considerable. This power must be generated by a larger engine-driven alternator, resulting in an increase in fuel consumption. Table 8.4 shows approximate per-unit power draw of CAV components; the numbers and combinations of the sensor types will vary by systems at each level of automation. Total power needs for sensors are estimated to be 200–250 watts (W) for a system of Level 3–4 (Baxter et al., 2018; Gawron et al., 2018).

However, moving beyond Level 3 requires substantially more computing power owing to the variety of circumstances the systems must respond to and the ability to respond reliably without human back-up. This is likely to increase power demand by an order of magnitude, resulting in a load of up to a few kilowatts (Baxter et al., 2018). This load would far exceed that of the car's heating, ventilation, and air conditioning system, for example, and could substantially reduce fuel efficiency and, for EVs, range as well.

Hamza et al. (2019) simulated Levels 4 and 5 automated driving systems for real-world vehicle trips from the California Household Travel Survey assuming constant system power consumption of 2.5 kW above that of a non-automated vehicle, consistent with "the likely present-day value" (Hamza et al., 2019). They found that such a load would reduce efficiency on these real-world trips by up to 35% and lower range for EVs by 23%–27%. However, targets for system power requirements in the future are several times lower (0.5 kW) (Hamza et al., 2019).

Most compact electronic equipment in the range of 500 W and above needs cooling which increases the weight, volume, and power draw. Novel cooling systems with optimized secondary-loop for heat pipes, heat pumps, and other thermal management systems are being explored intensively, primarily driven by the reduced range that such accessory systems can cause to BEVs.

The equipment and data processing required to support CAV systems will also consume energy in servers processing the data streams to and from vehicles, roadway infrastructure, and the cloud, creating large new electricity loads in the aggregate. Such electricity loads are not part of the onboard vehicle energy use but will impact total transportation system energy use. However, DOE (2020) estimate that these off-vehicle energy requirements are likely to be very small on a per-vehicle basis.

8.3.8.2 Vehicle Mass and Aerodynamic Drag

Automation equipment (sensors, actuators, power needs, computing resources, and associated thermal management equipment) will increase weight. However, the increase will likely be quite small. Examples of systems in production (Level 2) or in prototype (Level 4) showed total equipment weight of 6–16.5 kg, which would increase fuel consumption by less than 1% (Gawron et al., 2018).

Automation technologies could dramatically improve vehicle safety. Level 1 to 3 vehicles are likely to crash less frequently than conventional vehicles, and Level 4 and 5 vehicles presumably will crash very rarely. If these highly automated vehicles were to achieve full penetration, crashworthiness requirements could be substantially reduced. In this scenario, reductions in vehicle mass could be achieved by eliminating some safety equipment such as air bags (passive systems) as well as metal reinforcements required in today's vehicles. Weight de-compounding effects, including the ability to use smaller engine and auxiliary systems while maintaining performance, would further reduce vehicle mass and fuel consumption. However, full penetration of these higher levels of automation throughout the vehicle stock will not occur by 2035.

Sensors for some early CAV systems increased drag substantially (Gawron et al., 2018), but these systems are increasingly streamlined to improve aerodynamics and appearance. If a much-reduced risk of crashes were to cause some jurisdictions to increase highway speed limits, however, fuel use would increase owing to higher drag.

Automated and connected vehicles can operate cooperatively in a convoy-like "platoon." Because of the fast response afforded by the electronically controlled braking, acceleration, and steering systems of autonomous vehicles, especially if the vehicles are connected to one another, cars can drive closer together with much smaller interspacing. This shields cars from aerodynamic drag, saving fuel. Platooning is being tested for tractor trailers, for which each percentage point improvement in fuel efficiency brings large fuel savings. While platooning has been projected to reduce fuel consumption by several percent (NACFE, 2016), the industry is still debating the cost-effectiveness of the technology in light of the likelihood of forming platoons (Menzies, 2019). For light-duty vehicles, where per-vehicle saving would be much smaller and a mechanism for platoon formation not yet clear, effectiveness is unproven.

FINDING 8.6: Power needs of sensors and onboard computers can be considerable, reaching multiple kilowatts in prototype vehicles with high levels of automation. Power draw for a given function will decline rapidly over time as electronic systems evolve and refine, but total electrical load of these systems may remain significant as their functionality increases, due especially to growing computing requirements. Other negative efficiency impacts of automation technology owing to increased load (e.g., increased weight or drag) are likely to be minimal by the time it is commercialized.

8.4 ESTIMATES OF FUEL EFFICIENCY EFFECTS

8.4.1 Combined Potential for Fuel Economy Improvement

Table 8.6 shows estimates of combined savings potential for three generic CAV technology packages. While the package descriptions reference the SAE levels of automation, they are meant to group CAVs according to the nature of the fuel savings opportunity rather than by the human/system division of driving responsibility. The first package has Level 2 automation, basically ACC. The second package adds powertrain controls, which optimize internal combustion engine (ICE) technology calibrations, and vehicle connectivity, which expands the vehicle's look-ahead capability and can adjust operation accordingly. Here, Level 3 is not called out, because the fuel efficiency opportunities of capabilities like automated lane change are unclear. The third package includes

TABLE 8.6 Fuel Consumption Potential and Direct Manufacturing Cost for Three CAV Packages in 2020

Technology Package Fuel Savings Principles	Technologies in Package and Technology Assumptions	2020 Package Cost	Package Effectiveness (Fuel Consumption Reduction)					
			ICE	HEV	PHEV	BEV	Power Draw
Level 2 automation: Optimized velocity, minimized acceleration events	• Sensors ∘ Radar (1–5) ∘ Cameras (7–11) ∘ Ultrasound (0–12) • Data and mapping technology • Onboard computing and vehicle controls • Wiring	Total package cost: $1,520 Itemized costs: Camera: $45 Radar sensor: $55 Ultrasonic sensor: $10 ADAS Control Module: $800 Wiring: $90	5% urban/ 5% hwy	4% urban/ 3% hwy	8% in combined hwy and urban driving (longer than battery range)[a]	4% urban/ 4% hwy	~200 W; minor increase in fuel consumption
Level 2 automation w/ PT controls + connectivity: Above plus: • V2X to extend prediction horizon • Optimization of engine and transmission controls + power management (HEV, PHEV)	Previous package plus: • Communications technology (DSRC or C-V2X transceiver) • Additional computer	Total package cost: $2,410 Above itemized costs plus: Transceiver: $75–$90 CPU: $800	9% urban/ 5% hwy	6% urban/ 3% hwy	20% in some driving conditions (longer than battery range)	5% urban/ 4% hwy Additional 5% with optimum thermal and state-of-charge management	~200 W; minor increase in fuel consumption
Level 4/5 + connectivity: Above plus: Fully autonomous driving, permitting low-cost ride hailing, lower car ownership, and more high-efficiency vehicles	Same as above, plus: • Sensor ∘ Lidar (2–7) • Additional computer Level 5 can be used anywhere, while Level 4 may be used only in geofenced or other limited areas.	Total package cost: $7,210–$17,210[b] Itemized costs: Lidar: $2,000 CPU: $800	Wide range of energy outcomes including savings from increased BEV adoption, vehicle "rightsizing," and ridesharing, as well as higher consumption from autonomous driving system power draw and more vehicle miles traveled				Currently ~2.5 kW; 47% increase in fuel consumption for ICE/HEV, 30% for BEV300 (small car). Industry aim is to reduce power needs to < 1 kW.

NOTE: Technology requirements and costs were developed by the committee from its information gathering. CAV technologies are added primarily for safety and other non-energy related purposes, so costs should not be entirely attributed to fuel economy. Package effectiveness estimates do not reflect operation over the standardized test cycles, but rather are mostly based on testing or simulation reflecting driving patterns closer to actual conditions and optimization for individual vehicles. The technology effectiveness represents an upper bound with respect to a baseline without the CAV technology packages. Level 3 automation does not appear in the table because no additional fuel efficiency benefits have been established for Level 3 capabilities. Hwy = highway.

[a] High-efficiency benefits from PHEV CAVs are observed when the entire trip before an opportunity to charge is approximately known and this trip involves highway driving when the battery state of charge is high, followed by urban driving.

[b] Vehicle not yet commercially available; range shown reflects hardware costs only. High costs of development, data acquisition, software, etc. are not included but should be much reduced once vehicles reach substantial sales volumes.

connected Level 4 and Level 5 vehicles capable of autonomous driving (though not yet commercially available). These vehicles may produce a wide range of impacts on vehicle fuel consumption, which are largely behavior-driven and are not quantified in the table but are further discussed in Chapter 9. Table 8.7 shows cost estimates for the three CAV packages in 2025, 2030, and 2035.

FINDING 8.7: Automation technologies can produce or enable diverse fuel efficiency impacts for the vehicle in which they are installed. Individual automation technologies may result in fuel savings of more than 10% in some driving conditions, although they can also increase fuel use if not implemented for efficiency.

FINDING 8.8: With reliable vehicle-to-infrastructure information, connected and automated vehicle technologies in combination could increase fuel efficiency by as much as 20% in some driving conditions. Unreliable communication could be detrimental and lead to increased fuel consumption.

8.4.2 Technology Adoption

The aggregate energy impacts of CAV technologies will depend in large part on their pace of adoption by the consumer and on public sector investment and policy to guide and support their use. Level 1 and Level 2 automated vehicle systems are available today, at a cost of $1,500 to $3,000 (EIA, 2018). These systems are beginning to achieve substantial take rates even on high volume vehicles. In the second quarter of 2019, 10% of new cars sold in the United States had Level 2 driving capability, up from 2% a year earlier (Canalys, 2019). Toyota and Nissan have overtaken premium car brands in volume of Level 2 vehicles sold. Canalys (2019) notes that in the second quarter of 2019, more than 60% of cars sold with Level 2 automation cost less than $40,000.

Figure 8.7 shows projected penetration of safety-oriented ADAS features such as automatic emergency braking and LDW increasing dramatically over the next 20 years, starting at 10%–50% in 2020 and increasing to at least 80% in 2035 for almost all technologies shown. While these features are generally distinct from those that deliver energy efficiency benefits, they share equipment with some of the eco-driving capabilities discussed above and hence suggest a very high penetration of lower levels of automation by 2035.

It should be noted that none of the systems deployed to date include powertrain optimization, and connectivity, whether V2V or V2I, has very limited implementation at this point. Many automakers are hesitant to deploy these technologies owing to weak return on investment, especially while penetration, and hence driver benefits, of many CAV capabilities are small. Moreover, the systems may not be implemented to achieve maximum fuel savings. Hence the energy efficiency benefits of today's Level 2 vehicles are likely at the low end of the potential gains shown above.

The effects of connectivity on fuel economy depend on its prevalence in the vehicle population. Although studies have shown that connected vehicles have an influence on non-connected vehicles, the overall improvement would be larger if all vehicles were communicating. There are several companies working on aftermarket solutions for "legacy" vehicles and this work is expected to be emphasized as the benefits for connectivity related to safety predominately are clarified. The ability for all vehicles to communicate has the potential to greatly enhance safety and fuel economy.

Level 3 systems are emerging on luxury vehicles. Mercedes has announced a Level 3 version of its Drive Pilot system, and BMW plans to introduce the all-electric iNext crossover with Level 3 automation (Davies, 2020). BMW's mid-2021 release will be for a limited highway pilot only, however. The system may not be capable of adjusting to pedestrians, for example, and will not operate during lightning events (Fuerst, 2020). More generally, the auto industry has raised safety and liability concerns in connection with Level 3 deployment, and there is currently no consensus on when, or even whether, Level 3 will become widespread in the vehicle market (Szymkowski, 2020).

While many automakers previously announced that they would deploy fully autonomous vehicles before 2020, most early deadlines have fallen by the wayside (Figure 8.8). Now automakers are planning for Level 4 releases in the mid-2020s. Level 4 systems are far more expensive, and automakers have expressed uncertainty about consumers' willingness to pay for them. However, those costs are declining rapidly as lidar and computing system design advance.

TABLE 8.7 Technology Costs for the Three CAV Cost Packages as of 2020, and Cost Projections in 2025, 2030, and 2035

Technology Package and Fuel Savings Principles	Technologies in Package and Technology Assumptions	Estimated and Projected Technology Cost (DMC, 2018$)			
		2020	2025	2030	2035
Level 2 automation Optimized velocity, minimized acceleration events	• Sensors ∘ Radar (1–5) ∘ Cameras (7–11) ∘ Ultrasound (0–12) • Data and mapping technology • Onboard computing and vehicle controls • Wiring	$1,520	$1,446	$1,375	$1,307
Level 2 automation w/ PT controls + connectivity Above plus: • V2X to extend prediction horizon • Optimization of engine and transmission controls 1 power management (HEV, PHEV)	Previous package plus: • Communications technology (DSRC or C-V2X transceiver) • Additional computer	$2,410	$2,292	$2,180	$2,073
Level 4/5 + connectivity Above plus: Fully autonomous driving, permitting low-cost ride hailing, lower car ownership, and more high-efficiency vehicles	Same as above, plus: • Sensor ∘ Lidar (2–7) • Additional computer L5 can be used anywhere, while L4 may be used only in geofenced or other limited areas.	$7,210–$17,210	$3,725–$5,406	$3,511–$5,030	$2,545–$4,683

NOTE: The cost projections assume a steep cost reduction of 30% annually between 2020 and 2025, and 2% annually from 2025 to 2035 for lidar technology, and 1% annually for all years for other technologies.

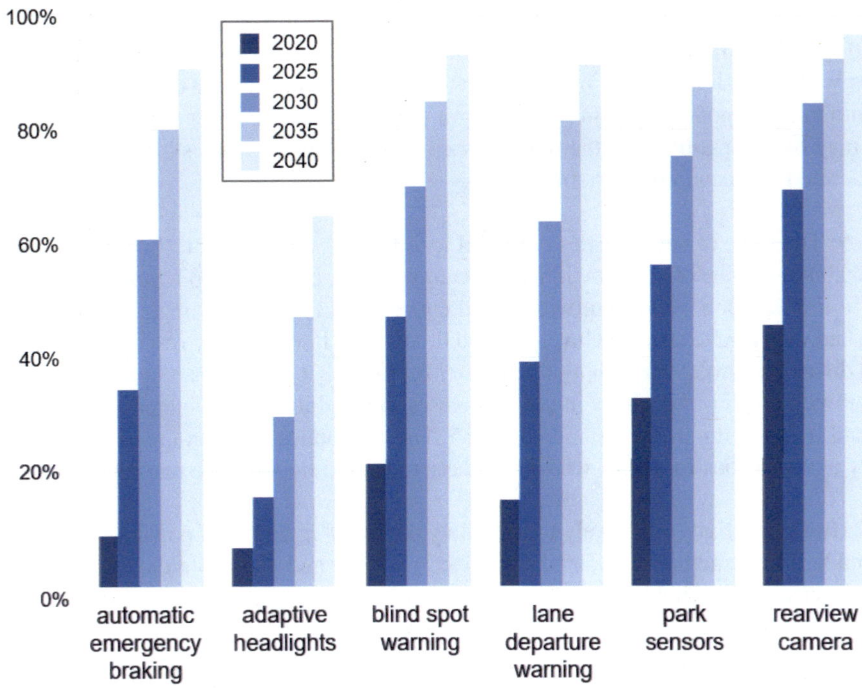

FIGURE 8.7 Predicted penetration of ADAS technologies in all on-road vehicles, 2020–2040.
SOURCE: Farmer (2019).

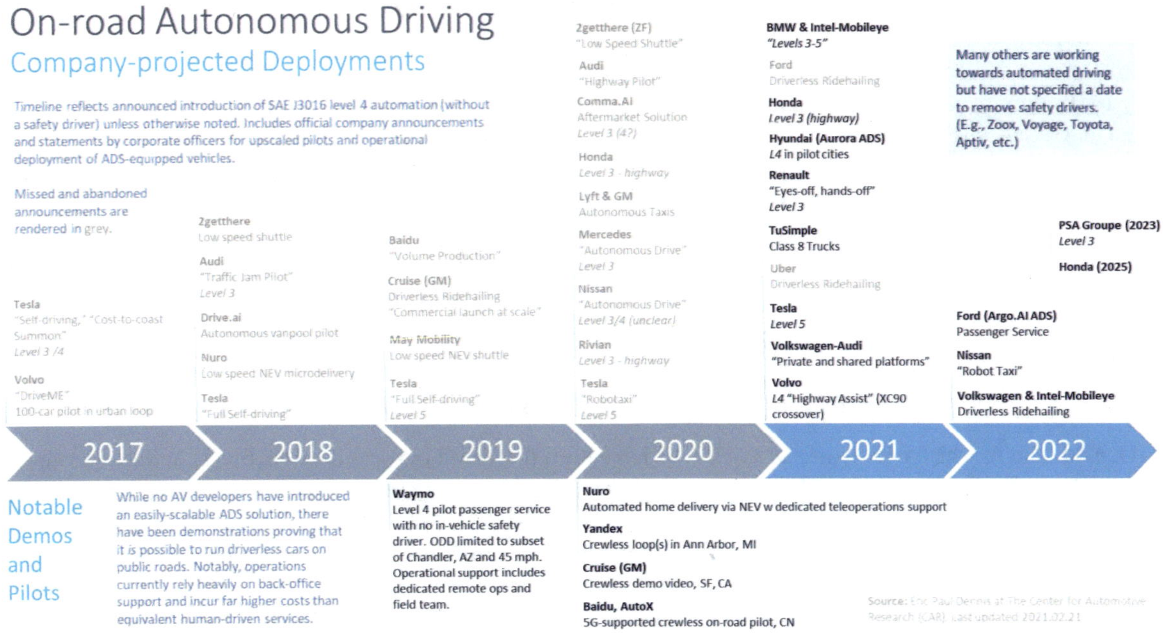

FIGURE 8.8 Company announcements of deployment of highly automated vehicles.
SOURCE: Dennis (2020).

Ongoing trends in cost reduction for CAV hardware and software suggest that cost of even high levels of automation could be within reach of many new car buyers before 2035. However, technology challenges and liability concerns will likely linger. This means limitations on the locations or conditions in which the vehicle can operate in autonomous mode. This in turn will limit these vehicles' market appeal to consumers. However, such limitations may be compatible with the requirements of some applications such as ride hailing, where these vehicles may generate large economic benefit.

FINDING 8.9: The share of new vehicles equipped with Level 2 automation technology first exceeded 10% in early 2019, with a rapid rate of adoption in mainstream vehicles. Level 3 is going to appear in the market in the early 2020s in limited applications, but widespread concerns in the industry regarding the safety and liability issues associated with a system in which vehicle and driver each have primary responsibility under some conditions make it difficult to project the adoption trajectory for Level 3. Consumer acceptance of Level 4 is uncertain at present, but with continued declines in cost, increases in capabilities, and increases in consumer familiarity with automated features, it may be common by 2035. The economics of autonomous (Levels 4 and 5) vehicles will be highly advantageous for fleets, which are likely to lead in building the market for these vehicles.

FINDING 8.10: Connectivity is unlikely to be widely deployed in 2025 but could reach high adoption levels by 2035 if public infrastructure is updated to collect, process, and distribute data and if useful, affordable, connectivity services are available.

8.5 POLICY ISSUES RELATED TO CAV ENERGY IMPACTS

New technologies often necessitate changes to policy. Vehicle safety standards, for example, can be expected to change in response to CAV technologies and could ultimately eliminate equipment requirements that these new, safer vehicles will render obsolete. At the same time, proactive changes to policy can guide deployment of new technology to ensure that it delivers on potential societal benefits.

Policies for CAVs are in the early stages of development. The National Highway Traffic Safety Administration proposed Federal Motor Vehicle Safety Standard No. 150 mandating V2V connectivity on all new light-duty vehicles in 2016, but this rule was never adopted (NHTSA, 2016). The FCC's recent decision to reduce the width of the spectrum dedicated to ITS communications and to specify that the remaining 30 MHz is to be used exclusively for C-V2X may end the option to move ahead with DSRC V2X and could limit the use of V2X altogether in the United States (FCC, 2020b). The few policies that have been adopted or explored at the federal level have focused on safety, cybersecurity, and privacy issues and standardization of equipment and protocols, not energy use implications (US Senate, 2017; DOT, 2017). At least seven states have adopted policies to promote testing and deployment of CAVs, and most others have taken steps in that direction (EIA, 2017). Some local governments have developed policies or programs to guide how CAVs will be used in urban and suburban areas. However, these policies are generally concerned with fully autonomous vehicles and their potential implications for travel behavior rather than vehicle efficiency. These include changes in vehicle miles traveled, which will be discussed in Chapter 9.

8.5.1 CAV Technologies in Fuel Economy Standards

One domain in which energy impacts of CAVs have been discussed in some detail is fuel economy and emissions standards. While some CAV technologies can substantially improve energy efficiency, these savings would not generally raise a vehicle's test fuel economy, because the test protocols are designed to ensure consistency of driving across all vehicles. Hence, precisely the features that would distinguish CAVs—changes to driving behavior—go undetected during the test. Consequently, various parties have advocated credits for CAV technologies under the standards, either specifically to recognize fuel savings not evident in testing or more generally as a way to incentive their adoption (Bin-Nun, 2018).

However, CAV technology fuel efficiency impacts can be large, as previously discussed, so accommodating these technologies outside the normal test protocols could have major effects on the Corporate Average Fuel

Economy program. Projected savings for Level 2 automation plus connectivity can exceed 10%. This is comparable to credits available under the standards for air conditioning improvements, which are a major element of manufacturer compliance. Furthermore, CAV technologies are typically driven by non-fuel economy benefits, so providing fuel economy credits for them would not necessarily result in further efficiency improvements but rather would displace other, on-cycle cost-effective efficiency technologies. One way to avoid this problem would be for the agencies to consider CAV technologies that can save energy in setting the level of the standards.

The velocity trajectories through which CAVs can achieve fuel efficiency improvements may differ significantly from the velocity profile of the standard fuel economy test cycles. Relaxing the constant velocity error margins around the standard velocity trajectories dictated by the federal test procedures would allow CAV fuel efficiency impacts over the certification test cycle to be quantified (Prakash et al., 2016). This could encourage manufacturer adoption and optimization for fuel efficiency. However, if the error margin in the velocity trajectory were to be relaxed in the testing of all vehicles, fuel economy would likely increase for all vehicles, reducing any detected efficiency benefit for CAVs.

As discussed in Chapter 12 (Regulatory Structure and Flexibilities), automakers can earn off-cycle credits under the standards for deploying technologies whose fuel savings benefits are not fully reflected on the official vehicle test cycles. The purpose of off-cycle credits is to help bring into the market technologies that improve real-world fuel economy but do not affect tested fuel economy. In principle, CAV technologies could be eligible for off-cycle credits when they do save fuel. While these technologies are generally offered to provide benefits other than fuel savings, automakers have indicated that in some instances the ability to obtain off-cycle credits increases the likelihood that they will deploy a CAV technology.

Fuel savings from CAV technologies may be difficult to document, especially at the confidence level required for off-cycle credits, or may differ greatly across vehicles. For example, ACC can provide fuel savings by smoothing acceleration. However, benefits may vary sharply from implementation to implementation (Zhu et al., 2019). An early analysis found that ACC could increase fuel use by up to 3% or decrease it by up to 10%, depending on the system's time headway and following algorithms (Mersky and Samaras, 2016). As an added complication, corridor-level simulations have found that CAV eco-driving strategies can dramatically increase fuel consumption owing to traffic instabilities when penetration of the technology is high (DOE, 2020). Savings are restored and enhanced when there is connectivity between vehicles (Zhang and Cassandras, 2019). Delays or interruptions in information associated with low penetration of connected vehicles will also reduce the fuel efficiency benefits of CAVs (Hyeon et al., 2019). Other efficiency benefits from simultaneously managing velocity profiles in approaching and departing stop lights or signs rely on connectivity and the reliability of information broadcasting from signalized intersections (Bae et al., 2019). In such cases, credits would need to be justified separately for each implementation of technology in a vehicle model in order to properly represent the system's fuel efficiency impact. In addition, the energy demands of CAV systems would need to be fully accounted for in determining the amount of any credits.

To date, the agencies have stated that, in order to be eligible for credits, an off-cycle technology must have benefits that are rigorously and fully documented and must reduce emissions from the vehicle receiving the credit. Technologies that reduce emissions primarily by changing the operation of vehicles other than the one in which the technology is installed (through collision avoidance or other means of improving traffic flow) are not eligible for off-cycle credit.[2] This is reasonable, because the benefits of such technologies are subject to great uncertainty, and their effects on fuel consumption and emissions are qualitatively different from those the standards programs were designed to measure and promote.

Another novel issue raised by CAVs is that their fuel efficiency impacts may depend upon the availability of related technology in nearby vehicles and, in the case of connectivity, infrastructure. If the agencies were to consider credits for these technologies, the amount of credit should reflect realistic assumptions regarding technology

[2] "Thus, for a technology to be 'counted' under the credit provisions, it must make direct improvements to the performance of the specific vehicle to which it is applied." *Federal Register* 77(199):62733. "Off-cycle credits may not be approved for crash-avoidance technologies, safety critical systems or systems affecting safety-critical functions, or technologies designed for the purpose of reducing the frequency of vehicle crashes." 40 CFR §86.1869-12(a).

adoption levels over the life of the vehicle. However, determining the appropriate level of credit in these cases would likely add considerable complexity to the program.

FINDING 8.11: Automation and connectivity enable, but do not ensure, fuel efficiency improvement over present vehicle technologies. Current fuel economy test cycles and procedures generally will not detect these improvements. Allowing limited departures from standard test cycles would permit some of these technologies' fuel efficiency gains—and losses—to be measured in testing.

FINDING 8.12: While connected and automated vehicle (CAV) technologies today are generally offered to provide safety, comfort, or convenience, the ability to obtain off-cycle credits for a CAV technology would in some cases increase the likelihood that automakers would deploy it.

RECOMMENDATION 8.1: The U.S. Environmental Protection Agency, U.S. Department of Transportation, and U.S. Department of Energy should conduct research on current driving patterns in the United States *sufficient to support sound estimates of the energy impacts of off-cycle fuel efficiency technologies including* connected and automated vehicle technologies. This research should be updated at regular intervals.

RECOMMENDATION 8.2: In setting the stringencies of the fuel efficiency and greenhouse gas emissions standards, the U.S. Department of Transportation and U.S. Environmental Protection Agency should consider the contribution that connected and automated vehicle technologies could make to saving energy. Off-cycle credits should be available for connected and automated vehicle technologies only to the extent they improve the fuel efficiency of the vehicle on which they are installed. Credits should be based on realistic assumptions, where needed, regarding technology adoption on other vehicles or infrastructure.

8.5.2 Other Policies to Promote CAV Technology Adoption

Thus far, automakers have been developing technologies that use only information acquired on the vehicle, information shared from other vehicles within the company, or through information sharing platforms such as HERE, or existing commercial and public information, including various maps and data on traffic. Advanced Transportation and Congestion Management Technologies Deployment Program Grants have been funded from an experimental perspective to be able to ascertain safety and fuel economy improvements. Also, in 2020, the Federal Highway Administration released a notice of funding opportunity for "up to $60 million in Federal Funding to provide grants to eligible entities to develop model deployment sites for large scale installation and operation of advanced transportation technologies to improve safety, efficiency, system performance, and infrastructure return on investment" (FHWA, 2020). Individual automakers or partnerships between automakers and others can provide a basis for automated and connected vehicle technologies, but efforts at the national, state, and local level will be required to instrument infrastructure for connectivity and provide consistent regulatory environment for CAVs nationwide.

RECOMMENDATION 8.3: The U.S. Department of Transportation should consider how it could promote consistency and efficiency in creating and maintaining high-definition maps of U.S. roads and develop its role accordingly.

8.6 REFERENCES

ABI Research. 2018. "V2X System Cost Analysis: DSRC+LTE and C-V2X+LTE." https://unex.com.tw/public/uploads/shortcuts/ABI-DSRC-price-comparison.pdf.

Abuelsamid, S. 2019. "Nvidia Unveils Next-Generation Orin Chip to Power Self-Driving Vehicles." Forbes. December 17. https://www.forbes.com/sites/samabuelsamid/2019/12/17/nvidia-announces-next-generation-orin-chip-for-automated-vehicles/#7acc35f4f373.

Autotrader. 2020. "What Is the Mercedes-Benz Driver Assistance Package?" Autotrader. August 10. https://www.autotrader.com/car-tech/what-is-the-mercedes-benz-driver-assistance-package.

Bae, S., Y. Choi, Y. Kim, J. Guanetti, F. Borrelli, and S. Moura. 2019. "Real-Time Ecological Velocity Planning for Plug-in Hybrid Vehicles with Partial Communication to Traffic Lights." *ArXiv:1903.08784 [Cs]*, September. http://arxiv.org/abs/1903.08784.

Baxter, J. A., D.A. Merced Cirino, D.J. Costinett, L.M. Tolbert, and B. Ozpineci. 2018. Review of Electrical Architectures and Power Requirements for Automated Vehicles. Oak Ridge National Laboratory. doi:10.1109/ITEC.2018.8449961.

Bin-Nun, A. 2018. "Using Fuel Efficiency Regulations to Conserve Fuel and Save Lives by Accelerating Industry Investment in Autonomous and Connected Vehicles." Securing America's Energy Future. https://secureenergy.org/AVsandFuelEconomy/.

Brooke, L. 2020a. "Origin of the Species." SAE International. May 11. https://www.sae.org/news/2020/05/cruise-origin-chief-engineer-qa.

Brooke, L. 2020b. "Ultrasonics to Keep Lidar Clean." SAE International. Magazine Article 20AVEP11_08, 2020-11-06. https://saemobilus.sae.org/content/20AVEP11_08/.

Butts, K., for M. Huang, S. Zhang, and Y. Shibaike. 2019. "Real-time Long Horizon Model Predictive Control of a Plug-in Hybrid Vehicle Power-Split Utilizing Trip Preview." Paper 20199015. PFL 2019 Proceedings. Tokyo: Society of Automotive Engineers of Japan.

Cadillac Pressroom. 2020. "Cadillac to Roll Out Enhanced Super Cruise." media.gm.com. January 28. https://media.gm.com/media/us/en/cadillac/home.detail.html/content/Pages/news/us/en/2020/jan/0128-cadillac.html.

Canalys. 2019. "10% of New Cars in the US Sold with Level 2 Autonomy Driving Features." September 9. https://www.canalys.com/newsroom/canalys-level-2-autonomy-vehicles-US-Q2-2019.

CAR (Center for Automotive Research). 2012. "Connected Vehicle Technology Industry Delphi Study." https://www.michigan.gov/documents/mdot/09-27-2012_Connected_Vehicle_Technology_-_Industry_Delphi_Study_401329_7.pdf.

Chen, D., N. Prakash, A. Stefanopoulou, M. Huang, Y. Kim, and S. Hotz. 2018. "Sequential Optimization of Velocity and Charge Depletion in a Plug-in Hybrid Electric Vehicle." Conference: 14th International Symposium on Advanced Vehicle Control, Bejing, China.

Chen, D., N. Prakash, and A. Stefanopoulou. 2020. "An Iterative and Hierarchical Approach to Co-optimizing the Velocity Profile and Power-split of Plug-in Hybrid Electric Vehicles." Conference: 2020 American Control Conference, Denver, CO, USA.

Davies, A. 2020. "BMW Takes Self-Driving to the Next Level." Autoweek. June 23. https://www.autoweek.com/news/technology/a32852529/bmw-takes-self-driving-to-the-next-level/.

Dennis, E.P. 2020. Announced Deployment Timeline. Center for Automotive Research.

DOE (U.S. Department of Energy). 2020. "SMART Mobility Connected and Automated Vehicles Capstone Report." Office of Energy Efficiency and Renewable Energy.

DOT. 2017. Automated Driving Systems 2.0: A Vision for Safety. National Highway Traffic Safety Administration. https://www.nhtsa.gov/sites/nhtsa.dot.gov/files/documents/13069a-ads2.0_090617_v9a_tag.pdf.

DOT. 2020a. "Connected Vehicles and Cybersecurity." Department of Transportation. https://www.its.dot.gov/factsheets/pdf/cv_%20cybersecurity.pdf.

DOT. 2020b. "The 5.9 GHZ Safety Band." https://www.transportation.gov/sites/dot.gov/files/2020-02/Safety_Band_Where_Is_the_Safety_Band_in_Use.pdf.

DOT. 2020c. "First Report and Order, Further Notice of Proposed Rulemaking, and Order of Proposed Modification from the Federal Communications Commission in the Matter of Use of the 5.850–5.925 GHz Band (ET Docket No. 19-138)." https://ecfsapi.fcc.gov/file/1109637413744/2020.11.06%20DOT%20Letter%20to%20FCC%20Chairman%20re%20Comments%20on%20Safety%20Band%20Decision%20(Signed).pdf.

EIA (Energy Information Administration). 2017. "Study of the Potential Energy Consumption Impacts of Connected and Automated Vehicles." https://www.eia.gov/analysis/studies/transportation/automated/pdf/automated_vehicles.pdf.

EIA. 2018. "Autonomous Vehicles: Uncertainties and Energy Implications." Annual Energy Outlook 2018: Issues in focus. https://www.eia.gov/outlooks/aeo/section_issues.php#av.

Farmer, C.M. 2019. "Fuel Economy and Highway Safety." Presentation to the Committee on Assessment of Technologies for Improving Fuel Economy of Light-Duty Vehicles—Phase 3." Insurance Institute for Highway Safety. September 19.

FCC (Federal Communications Commission). 2019a. "Dedicated Short Range Communications (DSRC) Service." FCC web page. https://www.fcc.gov/wireless/bureau-divisions/mobility-division/dedicated-short-range-communications-dsrc-service.

FCC. 2019b. "FCC Seeks to Promote Innovation in the 5.9 GHz Band." Press release. December 12. https://docs.fcc.gov/public/attachments/DOC-361339A1.pdf.

FCC. 2020a. "FCC Modernizes the 5.9 GHz Band for Wi-Fi and Auto Safety." https://docs.fcc.gov/public/attachments/DOC-368228A1.pdf.

FCC. 2020b. "Use of the 5.850–5.925 Band." Proposed rule. *Federal Register*. Vol. 85, No. 25.

FHWA (Federal Highway Administration). 2020. "Advanced Transportation and Congestion Management Technologies Deployment." https://www.fhwa.dot.gov/fastact/factsheets/advtranscongmgmtfs.cfm.

Ford Motor Company. 2020. "How Ford's Next-Gen Test Vehicle Lays the Foundation for Our Self-Driving Business." Medium. October 20. https://medium.com/self-driven/how-fords-next-gen-test-vehicle-lays-the-foundation-for-our-self-driving-business-aadbf247b6ce.

Fuerst, S. 2020. "Automated Vehicle Technology, Public Policy and BMW's Level 3 AV System." Webinar. Eno Center for Transportation. July 14. https://www.enotrans.org/event/webinar-automated-vehicle-technology-public-policy-and-bmws-level-3-av-system/.

Gawron, J., G.A. Keoleian, R.D. De Kleine, T.J. Wallington, and H.C. Kim. 2018. Life cycle assessment of connected and automated vehicles: Sensing and computing subsystem and vehicle level effects. *Environmental Science and Technology* 52(5):3249–3256.

Green Car Congress. 2020. "ABI Research Forecasts 83M 5G C-V2X Connected Cars by 2035." March 16. https://www.greencarcongress.com/2020/03/20200316-abi.html.

Guanetti, J., Y. Kim, and F. Borrelli. 2018. Control of connected and automated vehicles: State of the art and future challenges. *Annual Reviews in Control* 45:18–40. https://doi.org/10.1016/j.arcontrol.2018.04.011.

Guzzella, L., and A. Sciarretta. 2013. *Vehicle Propulsion Systems: Introduction to Modeling and Optimization*. 3rd ed. Berlin Heidelberg: Springer-Verlag. https://doi.org/10.1007/978-3-642-35913-2.

Hamza, K., J. Willard, K. Chu, and K.P. Laberteaux. 2019. Modeling the effect of power consumption in automated driving systems on vehicle energy efficiency for real-world driving in California. *Transportation Research Record* 2673(4):339–347. https://doi.org/10.1177/0361198119835508.

Hardman, G., J.H. Lee, and G. Tal. 2019. "Partially Automated Vehicles and Travel Behavior." University of California, Davis. Presentation to NASEM Fuel Economy Committee. January 24.

He, C.R., and G. Orosz. 2017. "Saving Fuel Using Wireless Vehicle-to-Vehicle Communication." In *2017 American Control Conference (ACC)*, 4946–4951. https://doi.org/10.23919/ACC.2017.7963721.

Hellstrom, E., M. Jankovic, M.H. Shelby, P. Pietrzyk, A. Richards, and J. Rollinger. 2018. "Improving Transient Torque Response for Boosted Engines with VCT and EGR." SAE 2018-01-0861. https://doi.org/10.4271/2018-01-0861.

Huang, M., Zhang, S., and Shibaike, Y. 2019. "Real-time Long Horizon Model Predictive Control of a Plug-in Hybrid Vehicle Power-Split Utilizing Trip Preview." SAE Technical Paper 2019-01-2341. https://doi.org/10.4271/2019-01-2341.

Hyeon, E., Y. Kim, N. Prakash, and A.G. Stefanopoulou. 2019. "Short-Term Speed Forecasting Using Vehicle Wireless Communications." In *2019 American Control Conference (ACC)*, 736–741. https://doi.org/10.23919/ACC.2019.8814400.

Iaconangelo, D. 2020. "Will Self-Driving Cars Slow the EV Boom? It Depends." Energywire. July 1. https://www.eenews.net/energywire/stories/1063483527/print.

Karbowski, D., J. Jeong, D. Shen, J. Han, N. Kim, and Y. Zhang. 2020. "Energy-Efficient Connected and Automated Vehicles—EEMS016." Presentation at the Department of Energy Vehicle Technologies Office Annual Merit Review.

Korosec, K. 2020. "Ford Postpones Autonomous Vehicle Service Until 2022." TechCrunch. April 28. https://social.techcrunch.com/2020/04/28/ford-postpones-autonomous-vehicle-service-until-2022/.

Krok, A. 2020. "Velodyne's Tiny Velabit Packs a Big Lidar Punch for Just $100." CNET Road Show. January 7. https://www.cnet.com/roadshow/news/velodyne-velabit-small-inexpensive-lidar-ces/.

Lee, S., Y. Kim, H. Kahng, S. Lee, S. Chung, T. Cheong, K. Shin, et al. 2020. Intelligent traffic control for autonomous vehicle systems based on machine learning. *Expert Systems with Applications* 144. https://doi.org/10.1016/j.eswa.2019.113074.

Lopez, J. 2019. "The Cadillac CTS Sedan Officially Discontinued." Cadillac Society (blog). July 1. https://cadillacsociety.com/2019/07/01/the-cadillac-cts-sedan-is-now-officially-discontinued/.

Markets and Markets. 2019. "HD Map for Autonomous Vehicles Market by Solution (Cloud-Based and Embedded), Level of Automation, Usage (Passenger and Commercial), Vehicle Type, Services (Advertisement, Mapping, Localization, Update and Maintenance), and Region—Global Forecast to 2030." October. https://www.marketsandmarkets.com/Market-Reports/hd-map-autonomous-vehicle-market-141078517.html.

Martinez, C.M., M. Heucke, F. Wang, B. Gao, and D. Cao. 2018. Driving style recognition for intelligent vehicle control and advanced driver assistance: A survey. *IEEE Transactions on Intelligent Transportation Systems* 19(3):666–676. March. doi: 10.1109/TITS.2017.2706978.

Menzies, J. 2019. "Daimler Abandons Platooning to Focus on Automation." Truck News.com. https://www.trucknews.com/technology/daimler-abandons-platooning-to-focus-on-automation/1003089387/.

Mercedes-Benz USA. 2017. "Driver Assistance Tech FAQ." Mercedes-Benz USA Online Newsroom. October 9. https://media.mbusa.com/releases/driver-assistance-tech-faq?firstResultIndex=0&sortOrder=PublishedDescending.

Mersky, A., and C. Samaras. 2016. Fuel economy testing of autonomous vehicles. *Transportation Research Part C: Emerging Technologies* 65:31–48. doi:10.1016/j.trc.2016.01.001.

Michigan Tech Research Institute. 2017. "Benchmarking Sensors for Vehicle Computer Vision Systems." Michigan Technological University. https://mtri.org/automotivebenchmark.html.

Mohan, A., S. Sripad, P. Vaishnav, and V. Viswanathan. 2020. Trade-offs between automation and light vehicle electrification. *Nature Energy* 5(7):543–549. https://doi.org/10.1038/s41560-020-0644-3.

MS Foster. n.d. "Camwash Vehicle Camera Washing System." https://msfoster.com/products/cameras-cleaning/camwash/camwash/#:~:text=The%20vehicle%20camera%20washing%20system%2C%20CamWash%2C%20cleans%20away,It%20eliminates%20dirt%20without%20scratching%20the%20camera%20lens.

NACFE (North American Council for Freight Efficiency). 2016. Confidence report: Two-truck platooning. https://nacfe.org/wp-content/uploads/2018/02/TE-Platooning-CR-FINAL-_0.pdf.

Nazari, S., R.J. Middleton, J. Martz, and A.G. Stefanopoulou. 2017. "The Elusive Consequences of Slow Engine Response on Drive Cycle Fuel Efficiency." In *2017 American Control Conference (ACC)*, 5379–5385. https://doi.org/10.23919/ACC.2017.7963791.

Nazari, S., N. Prakash, J. Siegel, and A. Stefanopoulou. 2019. "On the Effectiveness of Hybridization Paired with Eco-Driving." In *2019 American Control Conference (ACC)*, 4635–4640. Philadelphia: IEEE. https://doi.org/10.23919/ACC.2019.8814975.

News Wheel. 2019. "Ford Plans First C-V2X Vehicles for China by 2021." https://thenewswheel.com/first-ford-c-v2x-vehicles-china-2021/.

NHTSA (National Highway Traffic Safety Administration). 2016. "Federal Motor Vehicle Safety Standards; V2V Communications." Notice of Proposed Rulemaking. https://www.govinfo.gov/content/pkg/FR-2017-01-12/pdf/2016-31059.pdf.

Olin, P., K. Aggoune, L. Tang, K. Confer, J. Kirwan, S.R. Deshpande, S. Gupta, et al. 2019. "Reducing Fuel Consumption by Using Information from Connected and Automated Vehicle Modules to Optimize Propulsion System Control." SAE 2019-01-1213. https://doi.org/10.4271/2019-01-1213.

Ors, A. 2017. "RADAR, Camera, Lidar, and V2X for Autonomous Cars." NXP website blog. May 24. https://blog.nxp.com/automotive/radar-camera-and-lidar-for-autonomous-cars.

Paganelli, G., S. Delprat, T.M. Guerra, J. Rimaux, and J.J. Santin. 2002. "Equivalent Consumption Minimization Strategy for Parallel Hybrid Powertrains." In *Vehicular Technology Conference. IEEE 55th Vehicular Technology Conference. VTC Spring 2002 (Cat. No.02CH37367)* 4(4):2076–2081. https://doi.org/10.1109/VTC.2002.1002989.

Patel, J. 2020. "Cadillac Super Cruise Pricing Starts At $2,500 On 2021 Escalade." CarsDirect. May 15. https://www.carsdirect.com/automotive-news/cadillac-super-cruise-pricing-starts-at-2-500-on-2021-escalade.

Paukert, C. 2018. "Why the 2019 Audi A8 Won't Get Level 3 Traffic Jam Pilot in the US." CNET Road Show. May 14. https://www.cnet.com/roadshow/news/2019-audi-a8-level-3-traffic-jam-pilot-self-driving-automation-not-for-us/.

Pozzi, A., S. Bae, Y. Choi, F. Borrelli, D.M. Raimondo, and S. Moura. 2020. "Ecological Velocity Planning through Signalized Intersections: A Deep Reinforcement Learning Approach." In *2020 59th IEEE Conference on Decision and Control (CDC)*, 245–52. Jeju, South Korea: IEEE. https://doi.org/10.1109/CDC42340.2020.9304005.

Prakash, N., G. Cimini, A.G. Stefanopoulou, and M.J. Brusstar. 2016. "Assessing Fuel Economy from Automated Driving: Influence of Preview and Velocity Constraints." In *Volume 2: Mechatronics; Mechatronics and Controls in Advanced Manufacturing; Modeling and Control of Automotive Systems and Combustion Engines; Modeling and Validation; Motion and Vibration Control Applications; Multi-Agent and Networked Systems; Path Planning and Motion Control; Robot Manipulators; Sensors and Actuators; Tracking Control Systems; Uncertain Systems and Robustness; Unmanned, Ground and Surface Robotics; Vehicle Dynamic Controls; Vehicle Dynamics and Traffic Control*, V002T19A001. Minneapolis: American Society of Mechanical Engineers. https://doi.org/10.1115/DSCC2016-9780.

Qualcomm. 2017. "Top 10 Things You May Not Know about C-V2X." October 31. https://www.qualcomm.com/news/onq/2017/10/31/top-10-things-you-may-not-know-about-c-v2x.

Qualcomm. 2019. "How 5G Low Latency Improves Your Mobile Experiences [video]." OnQBlog. May 13. https://www.qualcomm.com/news/onq/2019/05/13/how-5g-low-latency-improves-your-mobile-experiences.

Rangwala, S. 2020. "Money for Everythin'" Forbes. November 23. https://www.forbes.com/sites/sabbirrangwala/2020/11/23/money-for-everythin/?sh=2a8dbbf44e1c.

SAE International. 2019. "SAE J3016 Levels of Driving." https://www.sae.org/news/2019/01/sae-updates-j3016-automated-driving-graphic.

Sciarretta, A., and A. Vahidi. 2020. *Energy-Efficient Driving of Road Vehicles: Toward Cooperative, Connected, and Automated Mobility*. Lecture Notes in Intelligent Transportation and Infrastructure. Cham: Springer International. https://doi.org/10.1007/978-3-030-24127-8.

Stewart, J. 2018. "Self-Driving Cars' Massive Power Consumption Is Becoming a Problem." Wired. February 6. https://www.wired.com/story/self-driving-cars-power-consumption-nvidia-chip/.

SWRI (Southwest Research Institute) Team Presentation. 2019. NEXTCAR ARPA-E 2019 Annual Meeting.

Sun, J., Y. Feng, I. Kolmanovsky, D. Zhao, and C. Mi. 2019. "Integrated Power and Thermal Management for Connected and Automated Vehicles (IPTM-CAV) Through Real-Time Adaptation and Optimization." ARPA-E review meeting. https://arpa-e.energy.gov/sites/default/files/1_2019_ARPA-E_AnnualMeeting_Slides_iPTM_CAVs_PublicRelease.pdf.

Szymkowski, S. 2020. "Audi Hangs up Hopes for Level 3 Partial Automation System." CNET Road Show. https://www.cnet.com/roadshow/news/audi-a8-level-3-automation-traffic-jam-pilot-system/.

Taiebat, M., A.L. Brown, H. Safford, S. Qu, and M. Xu. 2018. A review on energy, environmental, and sustainability implications of connected and automated vehicles. *Environmental Science & Technologyz.* 52: 11449–11465.

U.S. Senate. 2017. S.1885. To support the development of highly automated vehicle safety technologies, and for other purposes. https://www.congress.gov/bill/115th-congress/senate-bill/1885/text.

V2X Core Technical Committee. 2020. "V2X Communications Message Set Dictionary." 6th edition. SAE International. https://doi.org/10.4271/J2735_202007.

Vajedi, M., and N.L. Azad. 2016. Ecological adaptive cruise controller for plug-in hybrid electric vehicles using nonlinear model predictive control. *IEEE Transactions on Intelligent Transportation Systems* 17(1):113–122. https://doi.org/10.1109/TITS.2015.2462843.

Velodyne Lidar. 2020. "Velodyne Lidar Unveils Solid State Lidar Sensor." Press Release. November 13. https://velodynelidar.com/press-release/velodyne-lidar-unveils-breakthrough-solid-state-lidar-sensor/.

Wang, P., Y. Li, S. Shekhar, and W.F. Northrop. 2019. "Actor-Critic Based Deep Reinforcement Learning Framework for Energy Management of Extended Range Electric Delivery Vehicles." In *2019 IEEE/ASME International Conference on Advanced Intelligent Mechatronics (AIM),* 1379–1384. Hong Kong, China: IEEE. https://doi.org/10.1109/AIM.2019.8868667.

Wardlaw, C. 2020. "Mercedes Takes '21 S-Class to Next Level of Autonomous Driving." Wards Auto. September 23. https://www.wardsauto.com/vehicles/mercedes-takes-21-s-class-next-level-autonomous-driving.

Wayland, M. 2020. "Volkswagen Closes $2.6 Billion Investment in Self-Driving Start-up Argo AI." CNBC. June 2. https://www.cnbc.com/2020/06/02/vw-closes-2point6-billion-investment-in-self-driving-startup-argo-ai.html.

Zhang, Y., and C.G. Cassandras. 2019. An impact study of integrating connected automated vehicles with conventional traffic. *Annual Reviews in Control* 48:347–356. https://doi.org/10.1016/j.arcontrol.2019.04.009.

Zhu, L., J. Gonder, E. Bjarkvik, M. Pourabdollah, and B. Lindenberg. 2019. An automated vehicle fuel economy benefits evaluation framework using real-world travel and traffic data. *IEEE Intelligent Transportation Systems Magazine* 11(3):29–41.

9

Autonomous Vehicles

9.1 INTRODUCTION

Self-driving vehicles have been a frequent topic in automotive news articles and auto executive talks around the world for more than a decade. They have captured the interest of investors and suppliers small and large and have become a strong motive force behind many start-ups. The high level of automaker and supplier investments, mergers and acquisitions, and active programs in this area, including several automaker announcements of production of fully autonomous vehicles in the near future, led to a widespread belief that cars that drive themselves would soon be commonplace. Level 4 automated vehicles, which operate without human engagement in specified areas or modes, are already in revenue service by fleets in Arizona, Texas, and Florida (Bloomberg, 2020). More recently, however, many have questioned the readiness of the technology for volume commercialization.

Meanwhile parties focused on climate stability, urban livability, and transportation equity have begun to weigh in on the role of autonomous vehicles in a changing mobility landscape. They have raised concerns, independent of the technical challenges to deployment, that autonomous driving could aggravate many of the problems arising from a transportation system highly dependent upon inexpensive travel in personal vehicles. Hence, while the arrival of autonomous vehicles seems inevitable, there is considerable uncertainty regarding the timing and consequences of their arrival.

Chapter 8 discusses ways in which connected and automated vehicle (CAV) technologies can affect fuel efficiency. This chapter is about fully autonomous vehicles—that is, Level 4 and Level 5 CAVs. It is concerned with the energy implications not only of the properties of the vehicles themselves but also of changes in vehicle ownership, travel choices, and driving modes that would result from the use of vehicles that drive themselves. Fully autonomous vehicles would introduce qualitative changes in vehicle use and could allow wholesale transformation of the transportation system and travel behavior. People previously unable to drive will have the ability to travel by car unassisted. Some people currently owning and driving their own cars may choose to share rides in autonomous fleet vehicles instead, if mobility services make it easy and cheap to do so. Others will choose to own an autonomous vehicle and may drive more miles as a result, because they can reclaim travel time for other purposes. In that case commuting distances could be expected to increase as some people choose to live farther from their places of work and other common trip destinations. As these technologies are deployed, the system may operate with autonomous vehicles typically carrying multiple passengers in urban areas or, alternatively, with autonomous vehicles driving long distances with no passengers at all, depending on cost and convenience.

Possible impacts on transit, walking, and biking vary widely as well: autonomous vehicles could divert trips from other modes or complement them, for example by filling transit service gaps with ride hailing options made more affordable by not requiring a driver.

Autonomous vehicles could also be used as public transit vehicles and shuttles, with savings from reduced labor costs facilitating expansions in service and perhaps more comprehensive coverage with on-demand, flexible route services. Another likely early application of this technology is urban delivery vehicles. The rise of e-commerce and growth in consumption of prepared food have increased the demand for home delivery and the cost of providing it would be greatly diminished without the cost of drivers. In all these applications, autonomous vehicle deployment raises concerns about loss in driver jobs, which could serve as a barrier to acceptance. Such services will create new jobs in fulfillment and logistics, however, so net job impacts are unclear. The COVID-19 pandemic highlights other dynamics in the prospects for autonomous vehicles: while the pandemic has greatly increased demand for home delivery, it has also reduced demand for ride hailing and public transit service. The long-term implications of these developments remain to be seen.

Interest in autonomous vehicles for personal use is based on prospects for expanded mobility and greater convenience. These vehicles could fundamentally alter how people choose to travel and how transportation systems are designed, however, as well as affect levels of congestion, the number of miles driven, and levels of vehicle emissions. Autonomous vehicles' role in urban areas raises its own set of challenges and opportunities, and if these vehicles are to help achieve transportation and climate objectives, cities will need to lay the groundwork for their arrival.

Much of the discussion in this chapter is necessarily speculative, given the enormous uncertainties in the evolution of the autonomous vehicles market and how these vehicles will be used. Autonomous vehicles' impacts on transportation systems and the corresponding energy use and carbon emissions implications are far less certain and potentially much larger than vehicle-level fuel efficiency impacts of automation and connectivity technologies.

9.2 VEHICLE MILES TRAVELED

The availability of personal autonomous vehicles could result in increased vehicle miles traveled (VMT) in several ways: by making time spent in a car more productive or relaxing by allowing non-driving activities, by enabling people who cannot drive to travel in vehicles unaccompanied, by shifting trips from non-automobile modes to private automobile, and by allowing cars to drive without occupants. Taiebat et al. (2019) estimated an increase of 2% to 47% in average household VMT through rebound and induced demand associated with a complete shift to personal CAVs.

Autonomous vehicles could allow reduced VMT in certain activities, as in the case of searching for parking. Parking in urban areas is often a challenge to find, in addition to being costly if available. In many centers of activity, people typically circle the roads near their destination for some time in hope of finding a parking spot. Cookson and Pishue (2017) found that the average American driver spends 17 hours looking for parking every year, resulting in 1.7 billion gallons of additional fuel spent per year. A personal autonomous vehicle can drop its owners off at their destination, drive itself to any place where reasonably priced parking is available, and pick them up on demand. This practice would eliminate the miles driven and congestion caused in looking for a convenient parking spot, although the net impact on miles traveled would depend on the location of the parking identified by the autonomous vehicle and the potential change in the traveler's destination based on a perceived level of difficulty in the trip.

On the whole, however, personal autonomous vehicles are expected to increase VMT. A scenario in which autonomous vehicles result in reduced VMT is one in which autonomous vehicles are largely fleet vehicles that carry more than one passenger. Modeling of the potential to reduce VMT through shared rides includes Magill (2018), which found an opportunity for 30% reductions in VMT, emissions, and transportation costs through a transition to ridesharing from the use of single-occupant vehicles. However, some industry analysts assert that electric autonomous vehicles could dramatically reduce the cost of ride hailing trips relative to today's levels (UBS, 2017), which would lessen the incentive for ride hailing customers to share rides. Shared ride services are now jeopardized by the COVID-19 pandemic as well. Uber and Lyft suspended this user option in March 2020.

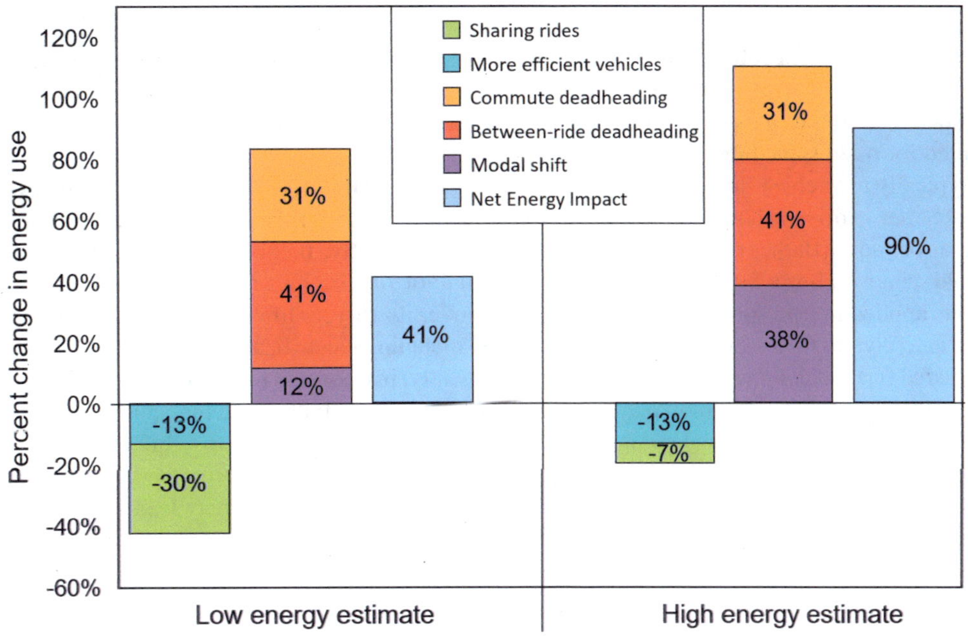

FIGURE 9.1 Range of estimated effects on energy use owing to changes in vehicle usage when ride hailing is implemented. SOURCE: Reprinted from *Transportation Research Part D: Transport and Environment*, 70, Wenzel et al., Travel and Energy Implications of Ridesourcing Service in Austin, Texas, 18–34, Copyright (2019), with permission from Elsevier.

Several other effects may tend to increase autonomous ride hailing vehicles' VMT. Based on experience to date with ride hailing companies using drivers, VMT may tend to increase when these services enter a new market owing to diversion of trips from transit and other modes as well as owing to miles driven without passengers at the beginning and end of the day and between fares. For example, Wenzel et al. (2019) estimate using data from RideAustin in Austin, Texas, that the net effect of ride hailing on energy use is a 41%–90% increase (Figure 9.1).

It should be noted that the Austin study's conclusions that ride hailing services are likely to produce an increase in energy use relies upon (1) data from a service (RideAustin) in its early years of operation, and (2) assumptions regarding the level of modal shift and ride-sharing that draw from a nascent literature. The ride-sharing assumptions are crucial, in that a high level of ride-sharing is one of the key mechanisms that has been identified to allow autonomous vehicles to contribute to energy use reductions.

Anair et al. (2020) estimated that ride hailing results in 69% higher emissions than the rides it replaces. As in the RideAustin study, this result is driven by the prevalence of deadheading and the displacement of more energy-efficient transportation modes. On the other hand, Anair et al. find that in an alternative scenario of 50% pooled rides and electric ride hailing vehicles, ride hailing trips would reduce carbon emissions of the trips they replace by more than 50%.

9.3 VEHICLE OWNERSHIP MODELS

In ride hailing, delivery, and transit fleets, autonomous vehicles' ability to operate without a driver could substantially reduce the cost of the transportation services they provide. Especially in high-density areas, the convenience and low cost of such services could not only expand their use but also induce many people to give up personal vehicles and use shared or other energy-efficient modes in place of driving.

Even automakers, whose growth has relied for decades on increasing levels of personal vehicle ownership, have reason to promote and respond to these fleet applications. Autonomous vehicles are likely to be expensive for some years after their introduction owing to their sensors, computers, and embodied intellectual property. They will also need frequent software updates and maintenance, at least at the outset. They also may be subject to limitations in where they can operate and raise privacy and security issues. They will certainly change the "driving" experience. Hence, despite the presumed safety and convenience of these vehicles, personal ownership may be limited for quite some time.

Fleets typically use vehicles much more intensively than do most owners. While an average light-duty vehicle in the United States is driven about 12,000 miles per year, an autonomous ride hailing vehicle might drive 60,000 miles per year or more (Barber et al., 2019; EIA, 2017). Consequently, fuel efficiency improvements to a fleet vehicle would achieve after a 3-year lifetime a present value of fuel savings almost 50% higher than the same improvements applied to a personal vehicle would achieve over its 15-year life (Barber et al., 2019). Hence fleets have greater incentive to invest in efficiency technologies, including electrification, for their vehicles.

These shared fleet vehicles will be refreshed more frequently, in either the top hat (upper vehicle body components that can be placed atop a common platform) or the powertrain, or both. With higher cost vehicles requiring regular updates of certain systems and components, fleets may choose to hold on to vehicles longer while swapping out powertrain and electronic components. Vehicles may become more like commercial aircrafts, in which "interiors/infotainment" are updated while the "shell" is reused. This refresh rate may lead to vehicles that begin to utilize more recycled material, reducing their overall carbon footprint.

9.4 VEHICLE CHARACTERISTICS

9.4.1 Vehicle Size and Weight

Autonomous vehicles present an opportunity to offer a new mobility model that can potentially be much cheaper (cost per mile traveled) than other means of transportation, particularly for trips with limited size and performance requirements, such as commuting. Such a vehicle could, for example, be configured for car sharing, autonomously transporting a single passenger with a small propulsion system. It might have a lower insurance premium (being safer than a human driven vehicle) and could offer a more economical alternative to using a full size, multiple-passenger car for commuting to work. This could result in significant fuel savings at this individual vehicle level, although the system-level impacts of such a mobility model would depend upon its effects on transit use, congestion levels, land use patterns, and how households met their needs for non-commute trips.

While such a mobility model may be speculative in U.S. passenger transport, there are numerous products being developed now in Asia and Europe with these features. Such vehicles are termed "quadricycles" in the European Union and, similar to U.S. low-speed vehicles, have limited vehicle weight, propulsion system power, and maximum speed, in general not exceeding 30 miles per hour (Regulation (EU) No 168, 2013). Task-specific choice of vehicle size and powertrain will most likely occur first in delivery fleets, which will deploy vehicles so as to minimize cost by optimizing size and performance for the given load.

Ride hailing fleets will also have an incentive to use vehicles with size and performance characteristics matching demand. An analysis of the fuel economy implications of this "rightsizing" effect found that these fleet vehicles would be smaller on average than today's vehicles and, if compliant with the current size-based fuel economy standards, would have 20% higher average fuel economy as a result (Barber et al., 2019). Furthermore, as the passenger experience becomes more important than the driving or ownership values of consumers, other vehicle features will shift as well.

9.4.2 Fuel Efficiency

As discussed in Chapter 8, fuel efficiency improvements for individual vehicles from CAV technologies are largely realized at low levels of automation in tandem with connectivity. The vehicle fleet could achieve greater fuel efficiency if all vehicles were equipped and controlled so as to optimize the operation of the entire network rather than individual vehicle operation. In this case, travel times for individual vehicles might increase even while the efficiency of the system grows. Further energy use and cost reductions could follow from much reduced need for complexity and power in vehicles used in this way. While such systems may be technologically achievable with CAVs at low levels of automation, it is not clear that drivers would be prepared to accept reduced capabilities in their personal vehicles or externally imposed travel time increases before their vehicle is fully autonomous or substituted by a fleet vehicle. Moreover, such fundamental changes in people's relationship with vehicles and driving might be tolerated only in congested areas.

While such a scenario is highly speculative at this point, it does illustrate how the adoption of autonomous vehicles could result in much higher average fuel efficiency. Modeling exercises are providing insight into the magnitude of the resulting energy savings.

9.5 RELATIONSHIPS AMONG AUTONOMY, CONNECTIVITY, SHARING, AND ELECTRIFICATION OF VEHICLES

Rapid electrification of vehicles along with the required charging infrastructure and a zero-emissions source of electricity are increasingly acknowledged to be essential to timely decarbonization of the transport sector. In a world of shared and autonomous vehicles, self-charging induction systems or self-docking systems will be needed. These vehicles will also need to be fast charged since downtime will be costly to their owners.

Autonomous vehicles do not require electrification, especially in rural areas. In urban areas, however, electric autonomous vehicles are likely to be the best option for ridesharing services. Urban planning and building codes will need to be modified to ensure charging infrastructure is available throughout the city. Furthermore, roadways need to be redesigned properly for multimodality. Autonomous electric vehicles can be also used as mobile energy storage and can be deployed to support and strengthen the grid in a utility-managed scenario, or enable micro-gridding through vehicle-to-building connectivity in emergency scenarios.

Electrification of CAVs may be motivated by certain synergistic effects present when a vehicle is connected, autonomous, and electric. Mass reduction enabled by CAV safety improvements presents an opportunity to reduce energy storage requirements without sacrificing range. Fully autonomous electric vehicles gain additional benefits such as the ability to refuel without a driver present where contactless charging is available and, in fleet applications, the ability to optimally assign vehicles according to the length of requested trips. Wireless charging is expected to improve substantially in the important attributes of power transfer, charging efficiency, and position accuracy. A notable example of the improvements is the 120 kilowatt wireless charging with 97% efficiency achieved in late 2018 (ORNL, 2018). Safety issues and other practical considerations will need attention. Robotic assistance and self-docking will be implemented to assist autonomous vehicle fleets, leveraging learnings from such implementations for hydrogen refueling stations.

9.6 COMBINED ENERGY IMPACTS OF AUTONOMOUS VEHICLES

9.6.1 Factor Analysis

A substantial body of literature on possible energy use impacts of autonomous vehicles has accumulated since 2014. A recent analysis by Argonne National Laboratory, Oak Ridge National Laboratory, and National Renewable Energy Laboratory synthesized that literature to identify a range of plausible energy impacts of autonomous vehicles owing to an array of factors (DOE, 2020). Results from that meta-analysis are the basis for the discussion in this section. It should be noted that this study assumes 100% penetration of autonomous vehicles in the fleet and hence does not represent realistic scenarios for 2025–2035. Projecting the effects of automation on the various factors at lower levels of penetration is difficult because these effects can be highly nonlinear in the technology penetration (Rios-Torres, 2020). The findings of the analysis are valuable nonetheless, because they demonstrate how the various factors are likely to interact in the long run. These insights can be used to help guide deployment of autonomous vehicles in the meantime so as to achieve energy savings ultimately, along with other beneficial outcomes.

The national laboratories' approach was to use the results of many earlier studies to create probability distributions for the impacts of 24 individual factors on energy use. After adjusting for interactions between factors, they

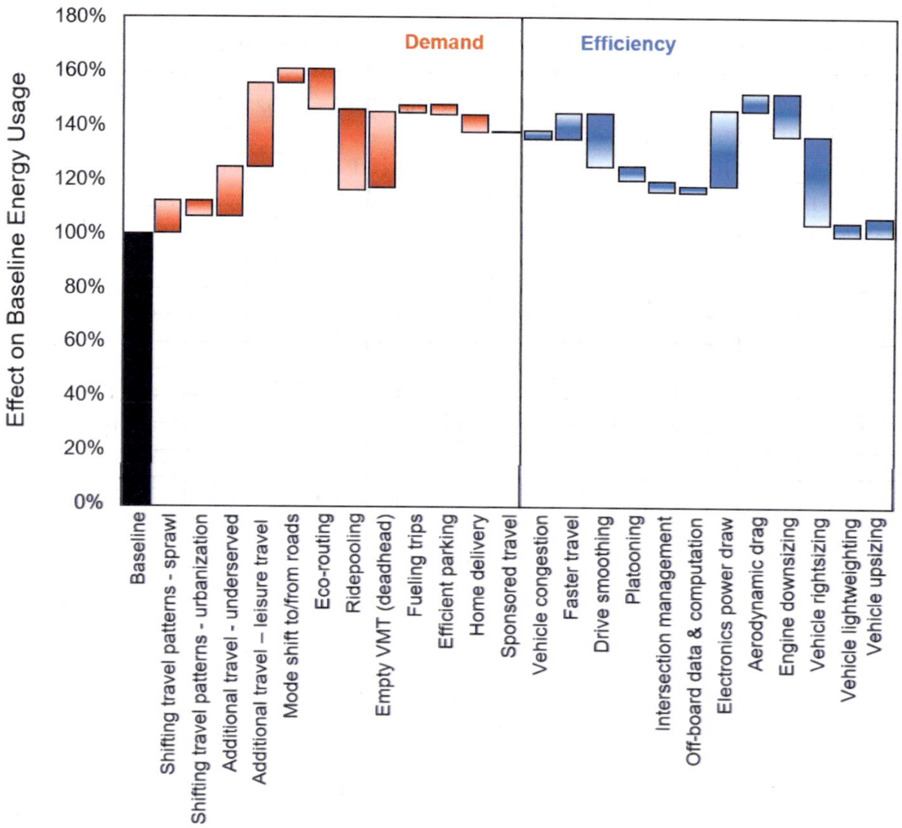

FIGURE 9.2 Energy changes from each factor.
SOURCE: Gohlke (2020).

carried out a Monte Carlo analysis to evaluate the combined impact of all factors. Ninety percent of scenarios generated in this analysis produced a change in total energy use between 40% reduction and 70% increase, with an average increase in energy use of 10% (DOE, 2020). While substantially smaller than the range of possibilities found in an earlier national laboratory analysis (Stephens et al., 2016), this range of outcomes underscores the continuing uncertainty about the size and direction of the likely energy impacts of CAV adoption.

Of the 24 factors considered in the analysis, half were related to CAVs' effects on travel demand and the other half to their effects on vehicle efficiency. Both travel demand and energy efficiency impacts included factors that increase energy use and factors that reduce it. Average effects of each of these factors are shown in Figure 9.2.

The national laboratory study also explored the sensitivity of the results with respect to assumptions about autonomous vehicles' properties and/or travel behaviors, including the following: whether autonomous vehicles are battery electric vehicles, whether vehicles continue to be privately owned or are replaced by fleet vehicles, and whether rides are shared. Of particular relevance to the discussion in Chapter 8, the authors also investigated the effect of limiting vehicles to Level 2 automation and, separately, the effects of eliminating connectivity. Findings from these scenarios include the following:

- Autonomous vehicles will increase energy use far more (24%) if the underlying fleet is electrified than if it is not (3%). This is primarily because energy efficiency benefits of CAV technologies for electric vehicles are lower than for internal combustion engine vehicles, as discussed in Chapter 8. Hence, VMT increase is the dominant effect in this case. Absolute energy use would still be relatively low in the electrified autonomous scenario given the efficiency of electric vehicles relative to internal combustion engine vehicles.

- If autonomous vehicles are fleet vehicles, the total VMT increase is smaller than if they are privately owned; and the fleet vehicles benefit more than private vehicles from the direct energy efficiency improvements of the CAV technologies. The net result is that total energy increases by 5% in an all-fleet scenario, compared with a 29% energy use increase in an all-private-autonomous vehicle scenario.
- If CAVs were limited to Level 2 automation, total energy use would decline by 34% relative to a non-CAV status quo. That is because these vehicles would achieve the full fuel efficiency benefit of connectivity and automation while VMT would actually be 9% lower than in the non-CAV scenario.
- Alternatively, with fully autonomous vehicles but no connectivity, VMT would increase somewhat less than in the baseline but vehicle efficiency would improve by only 6% (versus 21% in the baseline). Energy use would increase by 25%.

9.6.2 Technology Adoption

The energy impacts of autonomous vehicles will depend on their pace of adoption as well as the many factors described in Section 9.6.1. This applies not only to the increases or decreases that follow directly from the autonomous capabilities themselves but to an even greater extent to the system effects of autonomous vehicles. Speed harmonization and other congestion-reducing effects, for example, will follow only when a substantial fraction of the fleet is autonomous.

9.6.2.1 Determinants of Adoption

Autonomous vehicles offer potential buyers an array of benefits that could help drive their adoption. The continued innovation and fast pace of technology advances in autonomous vehicle development encourages automakers and suppliers to continue investing in the technology and to investigate revenue generation opportunities as offshoots of the technology, such as new modes of urban tourism (Cohen and Hopkins, 2019).

However, the enthusiasm on the part of consumers and automakers is not without reservation, as there are still several issues to resolve that, along with commercialization at a reasonable price point, are impeding a quick introduction of the technology. These include the following:

- *Technical issues affecting safety:* For example, achieve more robust object identification and precise positioning.
- *Cybersecurity issues:* Ensure complete hardening of defense against any possible malicious attack.
- *Regulatory issues:* Approve policies and regulations governing the operation of autonomous vehicles and sharing roads with conventional vehicles and other users.
- *Infrastructure readiness:* Achieve adequate coverage by digital maps and/or roadway connectivity devices, including in rural areas, and dedicate lanes as needed for autonomous operation.
- *Privacy issues:* Establish protocols for automakers', governments', and third parties' access to and use of the highly detailed data generated by connected vehicles.
- *Legal and liability issues:* Establish a consistent legal framework for assignment of liability in case of crash or malfunction that is acceptable to industry and consumer interests.
- *Customer acceptance:* Expand fraction of public that trusts the operation and security of the technology and values its benefits over the "driving pleasure" of non-autonomous vehicles.

In the area of consumer acceptance alone, researchers have identified multiple factors relevant to autonomous vehicle adoption rates, including safety, performance-to-price value, mobility benefits, value of travel time, symbolic value, and environmental friendliness (Jing et al., 2020). Behavioral approaches introduce additional factors such as perceived ease of use, perceived usefulness, and social norms. Determinants of autonomous vehicles' adoption for personal use, ride hailing service, or transit services include attitudes toward the environment, collaborative consumption, and car ownership (Acheampong and Cugurullo, 2019).

Litman (2020) cites consumer travel and housing preferences as well as development practices and other public policies as further determinants of autonomous vehicle adoption rates. Thus, the range of relevant factors is wide, and many are difficult to predict given the dramatic departure autonomous vehicles represent from current driving and travel options. However, understanding the roles and relationships of these many factors is important to anticipating and guiding the trajectory of autonomous vehicles so as to realize their benefits and avoid adverse consequences.

9.6.2.2 Market Penetration

Several automakers have postponed their projected release dates for fully autonomous vehicles in recent years, and the COVID-19 pandemic is causing further delays and attrition among the suppliers and engineers working on these technologies (Bloomberg, 2020). The industry's timeline has lengthened accordingly. Substantial sales are still anticipated over the next decade, however, with fleet sales starting to ramp up by 2025 and personal vehicles following around 2030.

Figure 9.3 shows several scenarios of automated vehicle market penetration from McKinsey (Gao et al., 2016), including a "low-disruption" scenario in which fully autonomous vehicles reach only a few% of the market by 2035 and a "high disruption" scenario in which they reach two-thirds market penetration by 2035. More recent commercial projections from IHS Markit, Deloitte, and others continue to include a wide range of sales trajectories (IHS Markit, 2018; Schiller et al., 2020; Murray, 2014; Alexander and Gartner, 2014; Lanctot, 2017; Gibson, 2018; Forsgren et al., 2018), with some even anticipating an autonomous-only vehicle market by the early 2030s (Mayor et al., 2018). These projections can be difficult to interpret absent stated assumptions regarding the full range of adoption factors, including changes in vehicle ownership patterns, use of shared ride services, and practices in home delivery.

Owing to the multiple dimensions of uncertainty, much of the academic research on autonomous vehicle adoption stops short of projecting the trajectories of sales or fleet penetration (Talebian and Mishra, 2018; Shabanpour et al., 2018). Some such projections do exist, however. For example, Bansal and Kockelman (2016) simulated CAV adoption scenarios defined by consumers' willingness to pay, technology price reductions of 5% or 10% per year,

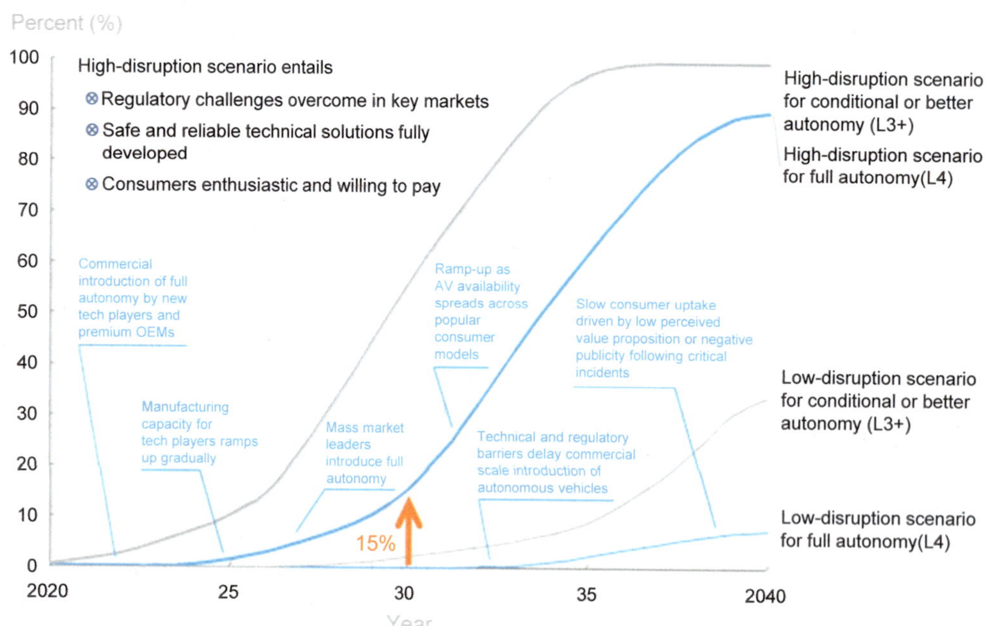

FIGURE 9.3 Market share of fully autonomous vehicles. SOURCE: Exhibit from Gao et al., "Disruptive trends that will transform the auto industry," January 2016, McKinsey & Company, www.mckinsey.com. Copyright (c) 2021 McKinsey & Company. All rights reserved. Reprinted by permission.

and technology adoption regulations. The simulation was calibrated with results from a consumer survey. Across eight scenarios, they found that sales share of Level 4 automation would reach 10%–34% by 2030, 15%–44% by 2035, and 19%–75% by 2040. A subsequent analysis drawing on the theory of diffusion of innovations and using results from a survey of university employees found lower sales shares of 1%–5% in 2030, 5%–25% in 2035, and 8%–60% in 2040, based on annual reductions in price of 5%–20% (Talebian and Mishra, 2018). It is worth noting that both cited academic analyses gave very wide ranges in the projected sales of autonomous vehicles in 2030, 2035, and 2040.

9.7 AUTONOMOUS VEHICLES AND ENERGY USE: POLICY ISSUES

After several years of study by researchers and assessments by practitioners in various fields, the range of plausible energy impacts of the adoption of autonomous vehicles includes large positive and large negative values. While some part of this uncertainty can be attributed to the fact that autonomous vehicles are not in general use today and hence their impacts are speculative, the large, indeterminate energy and emissions impact of their deployment is also indicative of the need for public policies to promote favorable outcomes. Factors affecting VMT and vehicle efficiency will contribute significantly to the net energy impacts, so policies regarding both usage and efficiency merit consideration. This section begins with a discussion of issues relating specifically to fuel economy regulation and concludes with an overview of other areas where policies might be considered.

9.7.1 Autonomous Vehicles and Fuel Economy Standards

Commercialization of autonomous vehicles will raise a variety of issues relevant to fuel economy standards. These relate not only to the fuel economy of the vehicles themselves but also to possible changes in vehicle ownership models and usage.

If autonomous vehicles experience very low crash incidence, there could be an opportunity to dramatically lightweight vehicles upon full transition to an autonomous fleet. That will not occur within the time horizon of this study (2035), however. A study from the Insurance Institute for Highway Safety finds also that two-thirds of crashes could still occur in an all-autonomous environment unless autonomous vehicles are programmed to give priority given to safety protocols over occupant preferences when the two conflict (Mueller et al., 2020). Automakers consider safety heavily in their autonomous vehicle programs, however, so they are highly likely to program their vehicles accordingly.

9.7.1.1 Ownership Models

As noted above, to the extent that autonomous vehicles contribute to a shift away from personal ownership of vehicles and toward fleet ownership, they could alter the profile of the future fleet, moving it toward smaller, less powerful vehicles on average, with vehicles having special capabilities or high carrying capacity largely dedicated to applications requiring those capabilities. The current structure of fuel economy standards can accommodate shifts in the sales distribution of vehicle classes, in that the standard for each automaker self-adjusts to the size and type of vehicles sold each model year.

There is no similar accommodation for a shift in performance needs, however, so the agencies will need to factor any such shift into their calculation of achievable levels of fuel economy. Recent fuel economy and greenhouse gas emissions rulemaking analyses have segmented the market into "performance" and "non-performance" vehicles for purposes of assessing technology effectiveness and penetration. A similar approach could be applied to account for increasing fleet ownership of vehicles, assuming the agencies can make reasonable projections of such trends. Alternatively, fleet vehicles might be regulated under separate standards, given that both vehicle characteristics and usage patterns will differ substantially from those of personal vehicles. The high annual mileage and resultant accelerated payback of incremental costs associated with fleet usage should generally improve the cost-effectiveness of fuel economy technologies, raising achievable fuel economy levels.

A shift from personal to fleet ownership would also mean a smaller vehicle stock, since each vehicle would meet the needs of multiple users. This would not necessarily mean reduced vehicle sales, since fleet vehicles would be replaced more frequently. However, if these fleets achieved high average occupancy, the vehicle stock would presumably be further reduced and sales would be lower. These factors will warrant consideration in future standards-setting if they substantially affect the dynamics of vehicle sales.

If high annual miles and other characteristics of fleet service were to alter the relative lifetimes of vehicle body, powertrain, and electronic systems and lead to large-scale reuse of major vehicle parts and systems affecting efficiency, implications for fuel economy regulation could be substantial. The definition of "new vehicle" and of regulated parties would need to be reconsidered to prevent deterioration of the standards' relevance for real-world fuel economy.

9.7.1.2 Usage Patterns

Autonomous vehicles' possible effects on VMT raises several questions of potential relevance to fuel economy standards. In personal use, autonomous vehicles could induce additional travel by reducing the cost of driving, especially in the form of time freed up for other activities. This phenomenon is similar to the rebound effect associated with improved fuel economy. However, Taiebat et al. (2019), using a microeconomic model to estimate elasticities of VMT demand with respect to fuel and time costs, found that households had much greater sensitivity to time costs than to fuel costs. If autonomous vehicles in fact are found to have substantially higher VMT than the vehicles they replace, this should be reflected in the analysis of achievable fuel economy, since present value of fuel savings from an increase in efficiency will be higher for a vehicle that accumulates miles more quickly. Furthermore, if fuel economy standards were found to affect autonomous vehicles' sales share, rebound associated with autonomous vehicle time savings should be considered in the analysis of the standards' effects.

In fleet use, autonomous vehicles' effects on VMT are indeterminate, but some have advocated that the high potential for shared rides and or high mileage accumulation in ride hailing fleets should be rewarded in fuel efficiency standards. In the Safer Affordable Fuel Efficient (SAFE) Vehicles Notice of Proposed Rulemaking, the National Highway Traffic Safety Administration and the U.S. Environmental Protection Agency requested comment on the idea that autonomous vehicles "placed in ridesharing or other high mileage applications" might be eligible for credits because the "per-mile emission reduction benefits would accrue across a larger number of miles for shared-use vehicles" (NHTSA/EPA, 2018). It is not clear that lifetime mileage for these vehicles would be higher than for personal vehicles, however; they might instead move to the resale market in a few years and be scrapped at an earlier age than privately owned cars are, as is the case with rental cars today. With regard to credits for shared-ride vehicles, predicting the rate of sharing could be quite difficult and the extent to which these vehicles divert riders from transit and non-motorized modes remains to be seen. An additional consideration related to fleet ownership of autonomous vehicles is that these vehicles may be sold as personal vehicles in the secondary market. Hence, much more data on autonomous vehicle usage patterns would be needed to support any assumptions regarding their VMT-based effects on energy use.

9.7.2 Other Energy-Related Policy Options for Autonomous Vehicles

Policies already being pursued or considered at various levels of government to slow or reverse VMT growth will be relevant to autonomous vehicles. These policies include modernization and expansion of transit services and land use planning to ensure accessibility to most destinations by non-auto modes and minimize the need to drive. They also include mileage-based user fees, which could be easily implemented for autonomous vehicles to address a variety of special considerations and circumstances using their data and communications capabilities. Mileage fees could be used, for instance, to promote efficient use of autonomous vehicles by increasing rates for zero-occupant vehicles or reducing them for high-occupant vehicles.

Such policies are already in use for ride hailing vehicles. For example, as of January 2020, the city of Chicago collects surcharges on ride hailing trips in the central business district of $3.00 for solo rides and $1.75 for shared rides (Uber, 2019). The charges are intended to address the congestion caused by ride hailing vehicles and generate revenue for mass transit upgrades (Spielman, 2019). Such considerations will become more pressing with the advent of autonomous vehicles in these fleets. A group of international nongovernmental organizations working to promote livable cities developed the Shared Mobility Principles for Livable Cities (2017), among them the principle that autonomous vehicles must be shared in urban areas. Cities could also help to ensure that autonomous ride hailing supports transit services by reducing charges for trips accessing transit.

The state of California has adopted targets to reduce greenhouse gas emissions per-passenger-mile for ride hailing companies to push these companies to prioritize shared rather than single-passenger ride hailing trips and to promote the use of low emissions vehicles in their fleets. Additional goals of California's program include supporting usage of transit and micro-mobility, and maximizing equity of access to transportation services (CARB, 2019).

Other strategies to ensure that autonomous vehicle adoption reduces energy consumption include policies to discourage ownership of autonomous vehicles for personal use; giving priority access to curb space, parking facilities, and designated highway lanes to multiple-occupant vehicles; reducing travelers' reluctance to share rides by providing advanced information about fellow riders and installing personal safety measures; creating integrated systems of "Mobility as a Service" as the local level; maximizing the convenience of travel without personal vehicles; and prioritizing the deployment of autonomous vehicles for transit and micro-transit services (Greenwald and Kornhauser, 2019).

9.8 FINDINGS AND RECOMMENDATIONS

FINDING 9.1: The energy implications of autonomous vehicles will be determined to a large degree by their effects on peoples' mode choices, vehicle miles traveled, and other travel behaviors. Research to date indicates that at full penetration autonomous vehicles could plausibly produce impacts ranging from a 40% reduction to a 70% increase in energy consumption. Absent new policies, autonomous vehicles will tend to reduce the cost of driving and therefore increase miles driven, perhaps very substantially. To the extent that they are used for shared rides and/or they are more likely than other vehicles to be electric, they will reduce transportation energy use.

FINDING 9.2: A second major determinant of the energy impacts of autonomous vehicles will be expectations of vehicle performance and features. Purchasers of autonomous vehicles are likely to prioritize comfort, convenience, and affordability rather than engine horsepower or acceleration. Fleet-owned autonomous vehicles will be right-sized, based on their intended purpose. Autonomous vehicles that are operated cooperatively with the surrounding traffic in urban or congested areas can achieve very high fuel economy, although perhaps with a cost in travel time for some individuals.

FINDING 9.3: Autonomous driving capability is likely to add at least $5,000–$7,500 to the cost of any vehicle sold with such capability in the next decade. Ensuring safety under all conditions, resolving cybersecurity issues, developing appropriate regulations, and gaining consumer acceptance of a radically different driving experience is likely to take even longer. Consequently, fleets and other users with special needs are likely to drive the market for autonomous vehicles through 2030; earlier industry projections of substantial sales before 2025 were overly optimistic. Autonomous vehicles' share of the market in 2035 is highly uncertain but likely to fall in the 0%–40% range, with ride hailing and delivery fleets accounting for 40%–60% of those sales.

FINDING 9.4: Fleet autonomous vehicles will be purpose-built and will reflect the needs of ride hailing and delivery companies. They will differ from typical vehicles for personal use in terms of size, body type, power, and luxury. They may be more likely to be electric as well, given high power needs, high urban usage, and ability to guide themselves to a charging station. Usage patterns (annual mileage, scrappage rates, etc.) also will differ from those for personal vehicles. In dense urban areas, micro-mobility products not currently subject to Corporate Average Fuel Economy standards may replace many personal automobiles.

RECOMMENDATION 9.1: Prior to the advent of autonomous vehicles, the National Highway Traffic Safety Administration (NHTSA) should consider in detail the ways in which autonomous vehicle properties and usage will differ from non-autonomous vehicles and how these differences should be reflected in the stringency and structure of fuel economy standards. NHTSA should consider regulating fuel efficiency of autonomous vehicles for fleet use differently from personally owned vehicles. Maximum feasible standards for these vehicles could be substantially more stringent than standards for personally owned vehicles; a requirement that autonomous vehicles be zero-emission vehicles should be considered, especially in urban areas.

RECOMMENDATION 9.2: To achieve the fuel-savings potential of autonomous driving and avoid its unintended consequences, the U.S. Department of Transportation (DOT) should consider actions to guide the effects of autonomous driving on the U.S. transportation system. This includes pricing strategies that promote sharing of autonomous vehicles and their complementarity to less energy-intensive modes. DOT should begin now to develop and provide information to other agencies and to Congress to highlight the need for policies to guide autonomous vehicle deployment.

RECOMMENDATION 9.3: While developing requirements and protocols to address cybersecurity and privacy concerns associated with autonomous vehicles, the U.S. Department of Transportation should also ensure that data generated by these vehicles is used to understand driving behavior, usage patterns including occupancy and relationship to other modes, and real-world fuel efficiency.

RECOMMENDATION 9.4: Given potential implications of autonomous vehicle adoption for energy use, emissions, and land use development patterns, the U.S. Department of Transportation should work with the U.S. Department of Energy, U.S. Environmental Protection Agency, and U.S. Department of Housing and Urban Development to support research and policies that advance the simultaneous achievement of the safety, economic, environmental, and equity benefits that autonomous vehicles can provide.

9.9 REFERENCES

Acheampong, R.A., and F. Cugurullo. 2019. Capturing the behavioural determinants behind the adoption of autonomous vehicles: Conceptual frameworks and measurement models to predict public transport, sharing and ownership trends of self-driving cars. *Transportation Research Part F: Traffic Psychology and Behaviour* 62:349–375. https://doi.org/10.1016/j.trf.2019.01.009.

Alexander, D., and J. Gartner. 2014. "Self-driving vehicles, advanced driver assistance systems, and autonomous driving features: Global market analysis and forecasts." Navigant Consulting, Inc. https://www.navigantresearch.com/research/autonomous-vehicles.

Anair, D., J. Martin, M.C.P. de Moura, and J. Goldman. 2020. *Ride Hailing's Climate Risks: Steering a Growing Industry Toward a Clean Transportation Future.* Cambridge, MA: Union of Concerned Scientists.

Bansal, P., and K. Kockelman. 2016. Forecasting Americans' long-term adoption of connected and automated vehicle technologies. https://www.caee.utexas.edu/prof/kockelman/public_html/trb16cavtechadoption.pdf.

Barber, E., W. Chernicoff, and D. MacKenzie. 2019. "Fleet Right-Sizing: The Corporate Average Fuel Economy Effect of a Transition to a Shared Autonomous Fleet," 98th Annual Transportation Research Board Meeting, paper [extended abstract] 19-03931.

Bloomberg. 2020. "The State of the Self-Driving Car Race 2020." May 15. https://www.bloomberg.com/features/2020-self-driving-car-race/.

CARB (California Air Resources Board). 2019. "Clean Miles Standard: About." https://ww2.arb.ca.gov/our-work/programs/clean-miles-standard/about.

Cohen, S., and D. Hopkins. 2019. Autonomous vehicles and the future of urban tourism. *Annals of Tourism Research* 74:33–42.

Cookson, G., and B. Pishue. 2017. "The Impact of Parking Pain in the US, UK, and Germany." Kirkland, WA: INRIX Research. http://inrix.com/press-releases/parking-pain-us/.

DOE (U.S. Department of Energy). 2020. "SMART Mobility Connected and Automated Vehicles Capstone Report." Systems and Modeling for Accelerated Research in Transportation (SMART) Mobility Laboratory Consortium. Office of Energy Efficiency and Renewable Energy. https://www.energy.gov/sites/prod/files/2020/08/f77/SMART-CAVS_Capstone_07.22.20.pdf.

EIA (Energy Information Administration). 2017. "Study of the Potential Energy Consumption Impacts of Connected and Automated Vehicles." https://www.eia.gov/analysis/studies/transportation/automated/pdf/automated_vehicles.pdf.

Forsgren, K.E., D.R. Shah, D.L. Lum, N. Madlani, and L. Orlowski. 2018. "The Road Ahead for Autonomous Vehicles." RatingsDirect. S&P Global Ratings. https://www.ibtta.org/sites/default/files/documents/SP%20Global%20Ratings%20-%20Road%20Ahead%20For%20Autonomous%20Vehicles-Enhanced%20May-14-2018.pdf.

Gao, P., H.-W. Kaas, D. Mohr, and D. Wee. 2016. "Disruptive Trends That Will Transform the Auto Industry." McKinsey & Company. January 1. https://www.mckinsey.com/industries/automotive-and-assembly/our-insights/disruptive-trends-that-will-transform-the-auto-industry.

Gibson, K. 2018. "Forecast: Autonomous-Vehicle Sales to Top 33 Million in 2040." Drive. January 2. https://www.thedrive.com/article/17298/forecast-autonomous-vehicle-sales-to-top-33-million-in-2040.

Gohlke, D. 2020. "Nationwide Energy and Mobility Impacts of CAV Technologies." Presented at the 2020 DOE Vehicle Technologies Office Annual Merit Review About Energy Efficiency Mobility Systems. June 3. https://www.energy.gov/sites/prod/files/2020/06/f75/eems081_gohlke_2020_o_5.13.20_1012AM_LR.pdf.

Greenwald, J., and A. Kornhauser. 2019. It's up to us: Policies to improve climate outcomes from automated vehicles. *Energy Policy* 127(2019):445–451.

IHS Markit. 2018. "Autonomous Vehicle Sales to Surpass 33 Million Annually in 2040, Enabling New Autonomous Mobility in More Than 26 Percent of New Car Sales, IHS Markit Says." IHS Markit. January 2. https://news.ihsmarkit.com/prviewer/release_only/slug/automotive-autonomous-vehicle-sales-surpass-33-million-annually-2040-enabling-new-auto.

Jing, P., G. Xu, Y. Chen, Y. Shi, and F. Zhan. 2020. The determinants behind the acceptance of autonomous vehicles: A systematic review. *Sustainability* 12:1719. https://www.mdpi.com/2071-1050/12/5/1719.

Lanctot, R. 2017. "Accelerating The Future: The Economic Impact of the Emerging Passenger Economy." https://newsroom.intel.com/newsroom/wp-content/uploads/sites/11/2017/05/passenger-economy.pdf.

Litman, T. 2020. Autonomous vehicle implementation predictions: Implications for transport planning. Victoria Transport Policy Institute. https://www.vtpi.org/avip.pdf.

Magill, J. 2018. A Nationwide Ridesharing Analysis. Junior Independent Paper. Princeton University. May. Presented at Smart Driving Car Summit.

Mayor, T., T. Dubner, B. Lakshman, and Y. Agarwal. 2018. "Will This Be the End of Car Dealerships as We Know Them?" KPMG LLP. https://advisory.kpmg.us/content/dam/advisory/en/pdfs/the-end-of-car-dealerships.pdf.

Mueller, A., J. Cicchino, and D. Zuby. 2020. What humanlike errors do autonomous vehicles need to avoid to maximize safety? *Journal of Safety Research* 75:310–318. S0022437520301262. https://doi.org/10.1016/j.jsr.2020.10.005.

Murray, C. 2014. "Study: Autonomous Cars Headed for Massive Sales by 2035." DesignNews.com. July 30. https://www.designnews.com/study-autonomous-cars-headed-massive-sales-2035.

NHTSA/EPA (National Highway Traffic Safety Administration and U.S. Environmental Protection Agency). 2018. "The Safer Affordable Fuel-Efficient (SAFE) Vehicles Rule for Model Years 2021–2026 Passenger Cars and Light Trucks." *Federal Register*. August 24. P. 43464. https://www.federalregister.gov/documents/2018/08/24/2018-16820/the-safer-affordable-fuel-efficient-safe-vehicles-rule-for-model-years-2021-2026-passenger-cars-and.

ORNL (Oak Ridge National Laboratory). 2018. "ORNL Demonstrates 120-Kilowatt Wireless Charging for Vehicles." Oak Ridge National Laboratory. October 19. https://www.ornl.gov/news/ornl-demonstrates-120-kilowatt-wireless-charging-vehicles.

Regulation (EU) No 168. 2013. Regulation (EU) No 168 of the European Parliament and of the Council of 15 January 2013 on the approval and market surveillance of two- or three-wheel vehicles and quadricycles. *Official Journal* 60:52.

Rios-Torres, J. 2020. "Multi-scenario Assessment of Optimization Opportunities due to Connectivity and Automation." Project ID# EEMS020. Presentation for U.S. DOE Annual Merit Review.

Schiller, T., P. Kummer, A. Berdichevskiy, M. Weidenbach, and J. Sadoun. 2020. "The Future of Car Sales in 2035." Deloitte. https://www2.deloitte.com/global/en/pages/consumer-business/articles/future-of-car-sales-in-2035.html.

Shabanpour, R., A. Shamshiripour, and A. Mohammadian. 2018. Modeling adoption timing of autonomous vehicles: Innovation diffusion approach. *Transportation* 45(6):1607–1621. https://doi.org/10.1007/s11116-018-9947-7.

Shared Mobility Principles for Livable Cities. 2017. https://www.sharedmobilityprinciples.org/.

Spielman, F. 2019. "Lightfoot Unveils $40 Million Congestion Fee on Solo Ride-hail Trips; Downtown Rides take Biggest Hit." *Chicago Sun Times*. October 18.

Stephens, T., J. Gonder, Y. Chen, Z. Lin, C. Liu, and D. Gohlke. 2016. "Estimated Bounds and Important Factors for Fuel Use and Consumer Costs of Connected and Automated Vehicles." National Renewable Energy Laboratory Technical Report NREL/TP-5400-67216.

Taiebat, M., S. Stolper, and M. Xu. 2019. Forecasting the impact of connected and automated vehicles on energy use: A microeconomic study of induced travel and energy rebound. *Applied Energy* 247:297–308. https://doi.0rg/10.1016/j.apenergy.2019.03.174.

Talebian, A., and S. Mishra. 2018. Predicting the adoption of connected autonomous vehicles: A new approach based on the theory of diffusion of innovations. *Transportation Research Part C: Emerging Technologies* 95:363–380.

Uber. 2019. "New Surcharges Mandated by the City of Chicago May Make Your Trip More Expensive." Uber blog. December 18. https://www.uber.com/blog/chicago/new-surcharges-mandated-by-the-city-of-chicago/.

UBS 2017. "How Disruptive Will a Mass Adoption of Robotaxis Be?" September 28. https://neo.ubs.com/shared/d1RIO9M-kGM/ues83702.pdf.

Wenzel, T., C. Rames, E. Kontou, and A. Henao. 2019. Travel and energy implications of ridesourcing service in Austin, Texas. *Transportation Research Part D*:70. https://doi.org/10.1016/j.trd.2019.03.005.

10

Energy and Emissions Impacts of Non-Petroleum Fuels in Light-Duty Vehicle Propulsion

Automakers are planning for a passenger vehicle fleet that will be predominantly powered with non-petroleum fuels, including electricity, hydrogen, and low-carbon synthetic fuels. This chapter describes the potential for non-petroleum fuels to provide larger amounts of energy for light-duty vehicle propulsion and the resulting impacts of using such fuels. Specifically, the chapter addresses the opportunities and challenges of using these fuels to provide power for light-duty transportation needs, the possible developments for these fuels in 2025–2035, and the impacts of the fuels on energy use and emissions. The chapter includes findings and recommendations about the use of alternative fuels in 2025–2035, as well as a discussion of their treatment in energy efficiency regulations.

10.1 INTRODUCTION

The source of energy to power light-duty vehicles has been predominantly petroleum fuel since the first long-distance internal combustion engine (ICE) vehicles were produced in the late 19th century. In the most recent 2019 data, approximately 90% of the energy to power light-duty transportation is in the form of gasoline or diesel, with the remainder being primarily ethanol and other biofuels blended with petroleum fuels (Davis and Boundy, 2020). Petroleum is an easily transported, easily transferred liquid with high energy density, providing some of the most efficient energy per volume and energy per mass of any energy carrier. Historically, gasoline and diesel have been inexpensive and readily available, making them ideal for energy consumption onboard a vehicle. Despite these advantages, petroleum fuels continue to have disadvantages in combustion-related greenhouse gas (GHG) and criteria emissions and also have limits to efficiency given their onboard combustion in engines. Light-duty vehicles contribute about 16% of U.S. GHG emissions (EPA, 2020a), nearly all from fuel combustion on board the vehicle.

Although petroleum fuels have dominated vehicle propulsion, additional fuel options have provided varying shares of energy capacity for on-road passenger transportation over time. Ethanol (notably at 10%, 15%, and 85% blends), biodiesel, propane, and natural gas have all been used in ICEs. Current research efforts are developing low-carbon synthetic "drop-in" fuels that would have the same or improved combustion properties as gasoline or diesel fuel but lower life cycle GHG emissions. Electricity and hydrogen are also increasingly being used to propel vehicles using motors, batteries, power electronics, and in the case of hydrogen, fuel cells. Some vehicles have been designed to use only one type of fuel, such as dedicated natural gas or electric fuel, and some have been designed for a fuel mix, such as E85, mixed gasoline and ethanol, or for fuel switching, such as plug-in hybrid electric vehicles (PHEVs). Today, biofuels represent the most significant alternative fuels used in light-duty vehicles, with approximately 10% of the energy used by vehicles being ethanol blended into gasoline. Natural gas and electricity for vehicle propulsion each represent less than 1% of the energy used by the U.S. light-duty fleet (Davis and Boundy, 2020; DOE-EIA, 2020).

The motivations for using alternative fuels have varied with fuel and technology capability, the price of fuel, as well as incentives for reduced petroleum use, improved energy efficiency, reduced emissions, and greater use of domestically produced fuel. For example, with developments like longer-range battery electric vehicles (BEVs) and technologies that enable biofuel use, alternative fuels are better able to meet light-duty transportation needs. The development of low-carbon synthetic fuels, although primarily aimed at decarbonizing hard-to-electrify transport such as aviation and long-haul freight, could also benefit light-duty vehicles by decarbonizing vehicle use, especially for the legacy combustion fleet. However, widespread penetration of alternative fuels will depend highly on their cost in comparison to that of petroleum. Continued research and development (R&D) on enabling technologies and/or economic incentives for low-carbon processes could allow alternative fuels to become cost-competitive with gasoline and diesel (Davis et al., 2018; De Luna et al., 2019).

The motivation for fueling vehicles with energy stores other than petroleum can be attributed in part to their potential to increase the vehicle's energy efficiency, dramatically reduce transportation criteria and GHG emissions to the atmosphere, and decrease total system energy use. These reductions in GHG emissions arise from reducing the combustion of fossil carbon, both off and on board the vehicle. Onboard, fuel-based GHG emissions are primarily from combustion of carbon-based fuels in engines. Offboard, fuel-based GHG emissions are called upstream emissions, and include the GHG emissions associated with all of the processes to make, transport, and store the fuel before it is used on board the vehicle. Biofuels and low-carbon synthetic fuels contribute to the goal of reducing total fuel-based GHG emissions to varying degrees, depending on their upstream emissions, which include emissions from feedstock recovery, fuel production, and the transportation, storage, and distribution of both feedstock and fuel. Switching away from onboard combustion entirely, for instance by using battery electric power to drive motors or hydrogen to power fuel cells, not only reduces onboard emissions (often effectively to zero) but generally also reduces net emissions on a well-to-wheels basis, given that electric propulsion also provides efficiency gains that reduce overall transportation energy consumption, and the electric grid is often cleaner than onboard combustion. Nonetheless, these advantages for electric propulsion in higher energy efficiency and decreased emissions are often accompanied by challenges of lower energy density and a nascent fueling infrastructure, as discussed further below and in Chapters 5 and 6.

This chapter focuses on *emerging* alternative fuels, specifically electricity,[1] hydrogen, and low-carbon synthetic fuels, rather than more widely implemented alternative fuels like ethanol. All of these fuels have the potential for reduced GHG emissions relative to gasoline and diesel, and alternative-fuel vehicles can have decreased energy use compared to conventional vehicles. However, the relationships between these metrics of energy efficiency, GHG emissions, and petroleum consumption are complex, vary by fuel and powertrain, and require life cycle analyses of the fuel-vehicle system, as detailed in Section 10.3. In evaluating the current and future incorporation of these fuels in the light-duty fleet, this chapter considers both technology developments and regulatory issues. Findings and recommendations about the use of these emerging alternative fuels in 2025–2035 are provided.

10.2 ELECTRICITY, HYDROGEN, AND LOW-CARBON SYNTHETIC FUELS

The sections below describe the current technology for generating and using electricity, hydrogen, and low-carbon synthetic fuels, as well as the implications of their use onboard vehicles for fuel consumption, energy consumption, GHG emissions, fueling cost, vehicle cost, fuel infrastructure, and fuel production. Table 10.1 summarizes some key metrics for each alternative fuel in comparison to conventional gasoline and diesel fuels. Note that some upstream fuel production processes emissions are reduced using carbon capture and storage (CCS).

[1]While electricity is, strictly speaking, an energy carrier and not a fuel, it is considered as an alternative fuel throughout this report, in line with the definition of alternative fuels provided in the Energy Policy Act of 1992 (see Section 10.3.1).

TABLE 10.1 Comparison of Vehicle Energy Sources, Including Gasoline, Electricity, Hydrogen, and Low-Carbon Liquid Fuels

Fuel	Gasoline with 10% Ethanol	Conventional Diesel	Electricity	Hydrogen		Low-Carbon Liquid Fuels	
Example vehicle	2019 Toyota Camry	2019 Chevrolet Cruze	2019 Chevrolet Bolt	2020 Hyundai Nexo	2020 Hyundai Nexo	2019 Chevrolet Cruze Diesel	2019 Toyota Camry
Example fuel	U.S. average gasoline	U.S. average diesel	U.S. average grid	Fossil-fuel derived H_2[a]	Low-carbon H_2[b]	Synthetic low-carbon Fischer-Tropsch diesel[c]	Co-Optima isobutanol (20% blend in gasoline)
On-board fuel consumption (unit relevant to fuel type)	0.029 gal/mi[d]	0.027 gal/mi[d]	n/a	0.018 kg/mi[d]	0.018 kg/mi[d]	0.027 gal/mi[d]	0.028 gal/mi[e]
Well-to-wheels energy use (kWh/mi)[f]	1.23	1.22	0.55	0.91	"blue" H_2: 0.94 "green" H_2: 0.96	GTL w/CCS: 1.74 "e-fuel": 3.65	1.22[g]
Vehicle operation energy use (kWh/mi)[h]	0.97	1.03	0.28	0.58	0.58	1.03	0.94[g]
Well-to-wheels GHG emissions (g/mi)[i]	315	338	124[j]	197	"blue" H_2: 57 "green" H_2: 36	GTL w/CCS: 323 "e-fuel": 4	272[g]
Tailpipe GHG emissions (g/mi)[k]	256	281	0	0	0	GTL w/CCS: 271 "e-fuel": 263	247[g]
Feedstock GHG emissions (g/mi)[l]	5	28	12	13	"blue" H_2: 13 "green" H_2: 0	GTL w/CCS: 27 "e-fuel": –260	–18
Fuel GHG emissions (g/mi)[m]	54	29	112	184	"blue" H_2: 44 "green" H_2: 36	GTL w/CCS: 24 "e-fuel": 1	43
Fueling cost per 12,000 miles	$901[n]	$988[o]	$437[p]	$3,538[q]	>$3,538[q]	GTL w/CCS: $1470[r] "e-fuel": >$1470[r]	$2,024[s]
Vehicle component costs in 2025–2035 relative to ICE vehicle	Same as ICE	Same as diesel	$6,116[t]	$8,581[u]	$8,581[u]	Same as diesel	Same as ICE
Fuel production infrastructure	Existing international petroleum drilling, refining transportation infrastructures	Existing international petroleum drilling, refining transportation infrastructures	Existing electrical generation, transmission, and distribution facilities with possible capacity expansion	Existing facilities for steam methane reforming; new hydrogen transportation infrastructures	Expansion of electrolysis and/or CCS capabilities; new hydrogen transportation infrastructures	New synthetic fuel inputs, fuel synthesis facilities, and transportation infrastructures	New biological inputs, fuel synthesis facilities, and transportation infrastructures

continued

TABLE 10.1 Continued

Fuel	Gasoline with 10% Ethanol	Conventional Diesel	Electricity	Hydrogen		Low-Carbon Liquid Fuels	
Fuel use and distribution infrastructure	Existing gas station model	Existing gas station model	New fueling infrastructure for private and public fueling	New fueling infrastructure	New fueling infrastructure	Existing gas station model	Existing gas station model with possible updates based on fuel properties

[a]Defined as gaseous hydrogen produced in central plants by steam reforming of natural gas.

[b]Example fuels are "blue" and "green" hydrogen. "Blue" hydrogen defined as gaseous hydrogen produced in central plants by steam reforming of natural gas with carbon capture and storage (90% CCS rate). "Green" hydrogen defined as gaseous hydrogen produced in central plants by electrolysis with electricity from solar photovoltaics.

[c]Example fuels are gas-to-liquid Fischer-Tropsch diesel with carbon capture and storage (GTL w/CCS) and "e-fuel," produced using H_2 from solar electrolysis and corn ethanol by-product CO_2. GTL w/CCS uses energy input from U.S. average grid electricity, and "e-fuel" uses energy input from renewable electricity.

[d]Value from www.fueleconomy.gov (DOE/EPA, n.d.), includes assumptions from Alternative Fuels Data Center.

[e]Estimated as 3.3% more efficient than engine with conventional gasoline, per default parameters in GREET1_2020rev1 for a Co-Optima vehicle with 20% isobutanol blend in gasoline (ANL, 2021).

[f]Total energy use from GREET1_2020rev1 Model (ANL, 2021) converted from Btu/mile to kWh/mile, unless otherwise noted. These values include energy use for feedstock recovery, transportation, and storage; fuel production, transportation, storage, and distribution; and vehicle operation. Fuel economy inputs are the combined adjusted fuel economy values for the relevant example vehicle (2019 Toyota Camry, 2019 Chevrolet Cruze, 2020 Hyundai Nexo) from the EPA/DOE Fuel Economy Guide dataset, unless otherwise noted.

[g]Per default parameters in the GREET1_2020rev1 Model (ANL, 2021) for a Co-Optima vehicle, the fuel economy input is 103.3% of the baseline gasoline vehicle value (in this case, the combined adjusted fuel economy value for the 2019 Toyota Camry from the EPA/DOE Fuel Economy Guide dataset).

[h]Vehicle operation energy use values from GREET1_2020rev1 Model (ANL, 2021) converted from Btu/mile to kWh/mile, unless otherwise noted. Fuel economy inputs are the combined adjusted fuel economy values for the relevant example vehicle (2019 Toyota Camry, 2019 Chevrolet Cruze, 2020 Hyundai Nexo) from the EPA/DOE Fuel Economy Guide dataset, unless otherwise noted.

[i]Includes tailpipe emissions and emissions from feedstock (recovery, transportation, and storage) and fuel (production, transportation, storage, and distribution); from GREET1_2020rev1 Model (ANL, 2021) using combined adjusted fuel economy values for example vehicles (2019 Toyota Camry, 2019 Chevrolet Cruze, 2020 Hyundai Nexo) from the EPA/DOE Fuel Economy Guide dataset as input, unless otherwise noted.

[j]If input is 100% renewable electricity rather than U.S. average grid electricity, well-to-wheels emissions are approximately 0 g/mi.

[k]Emissions from fuel combustion onboard the vehicle calculated with GREET1_2020rev1 Model (ANL, 2021) using combined adjusted fuel economy values for example vehicles (2019 Toyota Camry, 2019 Chevrolet Cruze, 2020 Hyundai Nexo) from EPA/DOE Fuel Economy Guide dataset as input, unless otherwise noted.

[l]Emissions associated with feedstock recovery, storage, and transportation; from GREET1_2020rev1 Model (ANL, 2021), with default inputs for each example fuel type unless otherwise noted (see footnotes [a–c]).

[m]Emissions associated with fuel production, storage, transportation, and distribution; from GREET1_2020rev1 Model (ANL, 2021), with default inputs for each example fuel type unless otherwise noted (see footnotes [a–c]).

[n]Calculated using U.S. national average gasoline price in January 2020 of $2.59/gallon (DOE, 2020a).

[o]Calculated using U.S. national average diesel price in January 2020 of $3.05/gallon (DOE, 2020a).

[p]Calculated using U.S. national average price of electricity from (AFDC, 2020).

[q]Calculated using average retail price of hydrogen from Q4 2018 – Q3 2019, reported by the California Energy Commission (CEC/CARB, 2019). Price likely higher for "green" or "blue" hydrogen given their higher production costs.

[r]Projected dispensed cost of gas-to-liquid Fischer-Tropsch diesel with CCS in 2025 reported in Elgowainy et al. (2016), converted from 2013$ to 2019$. Cost for "e-fuel" likely higher given the higher production costs of renewable H_2 and corn ethanol by-product CO_2.

[s]State-of-technology minimum fuel selling price from Cai et al. (2018), updated from 2014$ to 2019$.

[t]Estimated combined cost of battery, motor, and inverter for 200-mile BEV in small car class in 2025; see Chapter 5.

[u]Estimated combined cost of fuel cell system, battery, motor, and H_2 storage tank for fuel cell vehicles in the small SUV class in 2025; see Chapter 6.

10.2.1 Electricity

Electrical energy can be very efficiently transformed into vehicle movement using motors, with approximately 60%–73% of the electrical energy stored on board a vehicle transferred to the wheels (DOE, 2020b). In comparison, about 12%–30% of gasoline's energy is provided to the wheels in ICE vehicles, and 21%–40% of gasoline's energy is provided to the wheels in hybridized powertrains (DOE, 2020b). (On a well-to-wheels basis, electric vehicles [EVs] show less efficiency benefit, with an energy conversion efficiency of approximately 30%–36%, compared to 9%–24% for ICE vehicles and 16%–31% for hybrid vehicles.[2]) Some vehicles use electricity exclusively (BEVs), and some vehicles can switch between using electricity and petroleum fuels (PHEVs). The necessary technologies for using electricity on board vehicles, including electric motors, batteries, and power electronics, and the infrastructure requirements for EV deployment, are discussed in Chapter 5. This chapter focuses on the production and use of low-carbon electricity to power light-duty vehicles.

10.2.1.1 Electricity Production

Electric propulsion systems use less energy than combustion engine powertrains and produce zero emissions onboard the vehicle. When considering overall energy use of the transportation system, the energy and emissions used in generating grid electricity must be taken into account, especially if EVs become a significant share of the fleet, since that would lead to substantial increases in electricity use in transportation. Emissions from the U.S. grid have been decreasing in recent years, and the most current data available indicate a nationwide average of 433 grams of carbon dioxide equivalent per kilowatt-hour (gCO_2e/kWh) in 2018, down from 517 gCO_2e/kWh in 2012, or an average decline of 12 gCO_2e/kWh per year (EPA, 2020b). This decrease can be largely attributed to a shift from coal to natural gas fueled generation and an increasing share of renewable electricity generation. In 2019, the majority of U.S. electricity generation came from fossil fuels, 23% from coal and 38% from natural gas. Nuclear power accounted for 20% of electricity generation, renewables contributed 17%, and other sources produced the remaining 1% (EIA, 2020c). At the current average emissions rate, EV propulsion typically emits less than a conventional ICE vehicle on a well-to-wheels basis (see Table 10.1). A future U.S. grid in 2025–2035 is likely to be lower emitting than the 2019 grid, primarily owing to projected reductions in electricity generation from coal and increases in electricity generation from wind and solar (see Section 10.2.1.3).

10.2.1.2 Electricity Enabling Technology and Commercialization Needs

Increased deployment of EVs will be enabled primarily by advancements in battery technologies, improvements in charging infrastructure, changes to consumer behavior, and new policies and regulations, as discussed in Chapters 5, 11, and 12. Unlike other alternative fuel options, electricity can already be produced on a commercial scale; however, further efforts are required to decrease grid emissions and make electricity a truly low-carbon fuel option on a well-to-wheels basis in all regions of the United States. Technologies that enable low-carbon electricity generation are, for the most part, already mature and include wind turbines, solar PV, hydroelectric power plants, and nuclear power plants. Decarbonizing electricity generation will rely on increasing the capacity of those technologies. Additionally, owing to the intermittency of renewable energy sources like wind and solar, continued investment in research, development, and deployment for energy storage could help to improve the reliability and economics of a low-carbon grid. Current and emerging energy storage technologies include pumped hydro, compressed air, lithium-ion batteries, flow batteries, solid-state batteries, thermal storage, chemical storage (e.g., hydrogen), and flywheels (Zablocki, 2019). Fuel synthesis powered by renewable electricity, as described in later sections of this chapter, could also act as a sort of storage for intermittent renewable electricity if the fuel production methods do not require constant operation and if the capital costs for intermittently used equipment remain economic.

---

[2]Calculated using well-to-pump efficiency values for E10 gasoline (78.8%) and U.S. average electricity mix (49.9%) from Argonne National Laboratory's GREET1_2020rev1 Model.

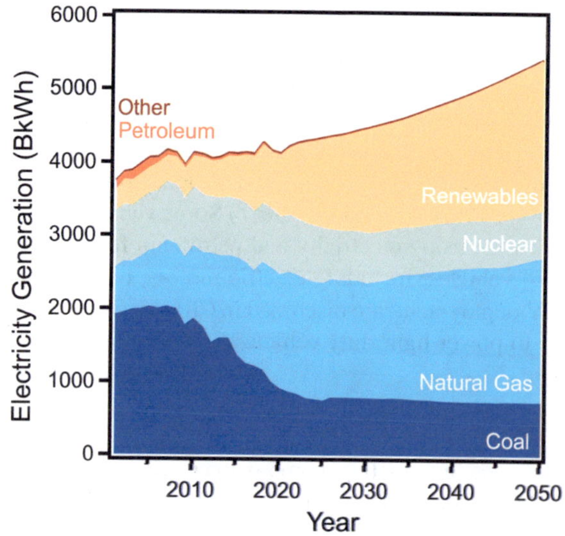

FIGURE 10.1 U.S. electricity generation by source, 2000–2050. Projections are from AEO2020 Reference Case.
SOURCE: Committee generated using data from EIA (2020a).

10.2.1.3 Electricity Potential in Medium to Long Term

The U.S. Energy Information Administration estimates that the grid in 2025 and 2035 will produce approximately 77% and 74%, respectively, of the GHG emissions per kWh as the 2019 grid (EIA, 2020a) (Figure 10.1). Incorporating policies aimed at achieving net-zero carbon emissions by mid-century would result in further reductions in average U.S. grid emissions during 2025–2035, since these scenarios often prioritize decarbonization of the electric grid in the near-term (DOE, 2017; Lawson, 2018; Mahajan, 2019; Larson et al., 2020; NASEM, 2021). Such reductions in grid emissions would further incentivize the use of electricity as a vehicle fuel by decreasing the well-to-wheels emissions of EVs, which already benefit from higher drivetrain efficiency and lower tailpipe emissions relative to petroleum-fueled vehicles. Ultimately, electrification of the light-duty fleet, and hence electricity use in transportation, will depend on many factors, including build-out of the charging infrastructure, changes to consumer behavior around vehicle fueling, and adoption of policies to incentivize the manufacturing and purchase of EVs. However, given the considerable efforts to increase efficiency and reduce well-to-wheel GHG emissions, the use of electricity as a vehicle fuel is expected to significantly increase in 2025–2035.

> **FINDING 10.1:** Technologies for the generation of low-carbon electricity are already mature. They are increasingly becoming cost-effective and more widely deployed, which will help to decarbonize the electric grid and reduce well-to-wheel greenhouse gas emissions from electric vehicles. Large-scale storage of renewable electricity remains a challenge.

10.2.2 Hydrogen

Total global demand for hydrogen (H_2) is around 115 million metric tons per year. Around 70 million metric tons of pure hydrogen are used per year in oil refining and ammonia production for fertilizers, and another 45 million metric tons per year are used in industrial processes that do not require hydrogen to be separated from other gases (IEA, 2019). The United States alone produces around 10 million tons of hydrogen per year (Connelly et al., 2019). Governments and industries worldwide project that demand for hydrogen will grow substantially in the coming years owing to its potential to decarbonize the transportation and industrial sectors (Hydrogen Council, 2017). Chapter 6 provides more detail about the technologies for the use of hydrogen in transportation.

10.2.2.1 Hydrogen Production

Worldwide, hydrogen generation comes almost entirely from fossil fuel sources (roughly 75% from natural gas and 23% from coal), with the remaining 2% from water electrolysis (IEA, 2019). The least expensive way to produce hydrogen today is through steam methane reforming (SMR), which accounts for nearly all of the commercially produced hydrogen in the United States (Connelly et al., 2019). Platts Analytics estimates the cost of hydrogen production from SMR in the United States at under $1 per kilogram (kg) without carbon capture and at $1.40 per kg with carbon capture, using a natural gas price of $3.50 per million British thermal unit (MMBtu) (Robinson, 2020). The actual production costs may be lower, as natural gas prices in the United States ranged from $2.22 to $3.11 per MMBtu in 2019 (EIA, 2020b). Renewable sources of hydrogen today include reformation of biomethane produced through anaerobic digestion, thermochemical conversion of biomass through processes such as gasification, and water electrolysis using electricity generated from renewable energy (Reed et al., 2020). Recent analyses have suggested that the extraction of naturally occurring hydrogen from geologic formations could also be a viable source of zero-carbon, low-cost hydrogen (Zgonnik, 2020; NH2E, 2019). Globally, the cost of hydrogen production from renewable electrolysis ranges from $2.50 to $6.00 per kg H_2 (in 2018$) and depends largely on the price of electricity (IEA, 2019). Current costs of hydrogen production in the United States are summarized in Table 10.2.

TABLE 10.2 Current Costs of Hydrogen Production in the United States, Excluding Delivery, Reported in 2018$

	Hydrogen Production Methods[a]		
Source	Capacity (kg H_2/day)	Capital Cost ($/kg H_2)	Hydrogen Cost ($/kg H_2)
Steam Methane Reforming[b]			
Central production	380,000	$0.16	$1.21
Central production w/CCS	380,000	$0.43	$1.64
Distributed production	1,500	$0.38	$1.58
Coal Gasification[c]			
Central production	620,000	$0.67	$1.83
Central production w/CCS	620,000	$0.90	$2.21
Biomass Gasification[d]	155,000	$0.33	$2.53
Bio-Derived Liquid (e.g., ethanol) Reforming[e]	1,500	$0.56	$7.09
Biomass Fermentation[f]	50,000	$44.00	$82.61
Water Electrolysis			
Proton Exchange Membrane			
Central production[g]	56,000	$0.42	$5.07
Distributed production[h]	1,695	$0.58	$5.23
Alkaline[i]	52,300	$0.73	$5.00
Solid Oxide[j]	50,000	$1.22	$4.89

[a]From Hydrogen Analysis Production Case Studies (NREL, n.d.), unless otherwise noted. All costs updated to 2018$.
[b]Assumes $3.92/mmBtu natural gas, $0.073/kWh electricity (central), $0.113/kWh (distributed), 90% CCS efficiency.
[c]Assumes $0.049/kg coal, $0.073/kWh electricity, 87% CCS efficiency.
[d]Assumes $0.105/kg woody biomass feedstock, $8.56/mmBtu natural gas, $0.113/kWh electricity.
[e]Assumes $2.27/gal ethanol and $0.113/kWh electricity.
[f]Baseline 2015 status from Randolph and Studer (2017), assuming no by-product credit, fermentation broth concentration of 12.8 g/L, and feedstock (corn stover) cost of $0.096/kg; all costs updated to 2018$.
[g]Assumes $0.073/kWh electricity.
[h]Assumes $0.077/kWh electricity.
[i]Assumes $0.070/kWh electricity.
[j]Assumes $0.073/kWh electricity.

10.2.2.2 Hydrogen Enabling Technology and Commercialization Needs

For fuel cell electric vehicles (FCEVs) to be competitive with gasoline vehicles on a cost-per-mile basis in the light-duty vehicle market, the retail price of hydrogen, which incorporates costs of production, delivery, and taxes, must decrease from its current value. In 2018–2019, the California Energy Commission and California Air Resources Board reported the average retail price of hydrogen as $16.51 per kg H_2 (CEC/CARB, 2019). U.S. DRIVE has set an ultimate target of <$4 per kg H_2,[3] untaxed and dispensed at the pump, and a target of $7 per kg for 2025 (targets expressed in 2016$) (Ramsden and Joseck, 2018). For R&D planning purposes, apportioned cost targets are <$2.00 per kg H_2 for producing hydrogen and <$2.00 per kg H_2 for delivering hydrogen, including the costs of compression, storage, and dispensing. At the target cost for H_2 production and delivery ($4.36 per kg H_2 in 2020$) and using the fuel consumption value of the 2020 Hyundai Nexo (see Table 10.1), the fueling cost is $0.078 per mile. In comparison, fueling the 2019 Toyota Camry with conventional gasoline at the U.S. average price in January 2020 costs $0.075 per mile.

The growing focus on carbon-free energy strategies has increased interest in and development of technologies that produce hydrogen from renewable energy sources, particularly water electrolysis using low-temperature polymer electrolyte membrane (PEM) electrolyzer systems. As mentioned above, hydrogen generation from renewable electricity is currently up to 6 times more expensive than hydrogen production by SMR. At the current low demand for hydrogen in transportation, distributed hydrogen production is likely the most viable and economic approach for this application (DOE-EERE, n.d.). The most recent analysis from the U.S. Department of Energy's (DOE's) Hydrogen and Fuel Cells Program estimates the cost for distributed production of hydrogen from PEM electrolysis at $4–$6 per kg (2016$, with an electrolyzer capital cost of $1,000 per kW and renewable electricity costs of $0.03–$0.04 per kWh) (Vickers et al., 2020). Increases in hydrogen demand across transportation and other energy sectors could make centralized production the more economically favorable option, although this will require higher upfront capital costs and significant build-out of a hydrogen infrastructure (DOE-EERE, n.d.). For centralized hydrogen production, DOE's Hydrogen Production Analysis tool[4] recently estimated a baseline cost of $4.83 per kg using data provided by several manufacturers and projecting to centralized hydrogen production at 50,000 kg per day, an electrolyzer manufacturing volume of 700 megawatts per year (about 7× the current production capacity), and an electricity cost of $0.074 per kWh (Peterson et al., 2020). The same analysis sought possible pathways for meeting the target production cost of $2.00 per kg H_2, examining the projected impacts of reduced electricity price, reduced electrolyzer capital cost, improved performance, and enhanced durability. The results, shown in Figure 10.2, project a hydrogen cost of $1.87 per kg at $0.03 per kWh and $1.77 per kg at $0.02 per kWh using curtailed renewable electricity. Both of these future cases also require increases in stack efficiency, decreases in capital cost and degradation rate, and increases in stack lifetime relative to the current baseline scenario. Other studies project significantly higher renewable hydrogen costs—for example, a median price of $5.92 per kg for H_2 from curtailed electricity in 2050 (Christensen, 2020)—which reflects the variability in assumptions and data inputs used in such analyses.

The high capital costs of PEM electrolyzers largely result from their use of noble metal catalysts: platinum-based catalysts for the cathode and iridium-based catalysts for the anode. However, these electrolyzers are expected to benefit significantly from materials and manufacturing R&D aimed at improving the durability and reducing the cost of automotive PEM fuel cells. Other electrolyzers under development use alkaline or solid oxide electrolytes (DOE, 2019a; Schalenbach et al., 2018). Alkaline electrolyzers can use lower cost nickel- and cobalt-based catalysts and have similar performance to PEM electrolyzers. High-temperature electrolyzers, such as solid oxide, operate at higher efficiency than PEM or alkaline systems and therefore use less electricity, but are less durable (Reisert et al., 2018). Additional R&D efforts by DOE's HydroGEN program focus on advanced water splitting technologies, including materials for photoelectrochemical, solar thermochemical, and low- and high-temperature electrolysis. Nonetheless, even with significant reduction in electrolyzer capital costs, the

[3] 1 kg H_2, on a lower heating value basis, is approximately equal to 1 gallon of gasoline equivalent.

[4] Hydrogen Production Analysis is a discounted cash-flow model providing transparent reporting of process design assumptions and a consistent cost analysis methodology for H_2 production at central and distributed facilities. Hydrogen Production Analysis utilizes data from the Energy Information Administration Annual Energy Outlook 2017 Report, where 2016$ is the standard cost basis. See http://www.hydrogen.energy.gov/h2a_prod_studies.html.

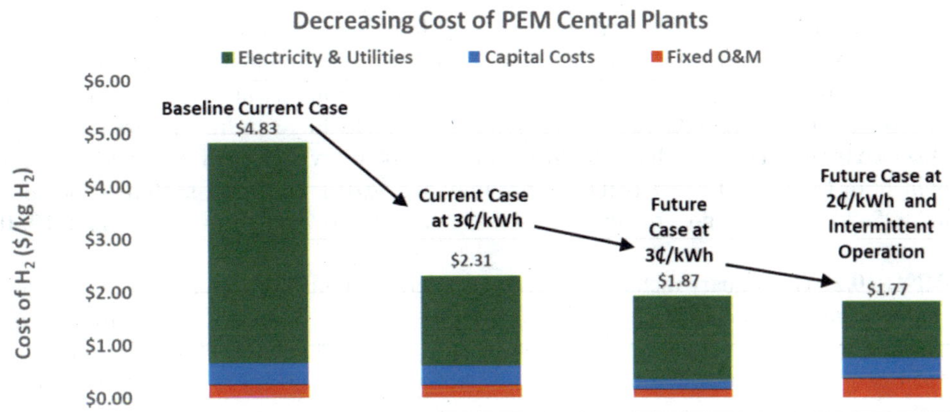

Capacity Factor	97%	97%	97%	40%
Stack Electrical Efficiency	66.1%	66.1%	69.8%	69.8%
System Cap Cost	$460/kW	$460/kW	$233/kW	$233/kW
Effective Electricity Cost	7.4¢/kWh	3.0¢/kWh	3.0¢/kWh	2.0¢/kWh
Degradation Rate	1.5 mV/khrs	1.5 mV/khrs	1 mV/khrs	1 mV/khrs
Stack Lifetime	7 yrs	7 yrs	10 yrs	10 yrs

FIGURE 10.2 Pathway toward low-cost H_2 production via PEM electrolysis, showing the impacts of reduced electricity cost, including an intermittent scenario with low-cost variable electricity pricing, and improvements in electrolyzer capital cost, lifetime, and efficiency. All costs reported in 2016$.
SOURCE: Peterson et al. (2020).

primary factors dictating the cost-effectiveness of H_2 production from electrolysis will be electrical efficiency and electricity cost (James et al., 2019).

10.2.2.3 Hydrogen Potential in Medium to Long Term

SMR with carbon capture, utilization, and storage (CCUS) and water electrolysis are currently the primary routes for producing low-carbon hydrogen, and their production volume has remained fairly constant at around 0.36 megatonnes per year (Mt/yr) since 2015 (IEA, 2020a). New projects for both technologies are reported to begin operation in the 2020s and would increase global low-carbon hydrogen production to 1.45 Mt/yr by 2023 (IEA, 2020a). Through its H2@Scale initiative, DOE recently announced funding for three multimillion-dollar R&D projects on electrolyzer manufacturing (DOE-EERE, 2020). However, the decrease in global hydrogen demand as a result of the COVID-19 pandemic may delay some of the progress in low-carbon hydrogen production (IEA, 2020b).

Other processes and technologies to generate hydrogen without carbon emissions have the potential to become more widespread in the coming decades. For example, in some advanced nuclear reactor designs, the process heat could be used to produce H_2 via steam electrolysis or thermochemical water splitting (IAEA, 2013). These Generation IV reactors are largely still in the development phase but could see initial deployment in the 2030s. Methods to convert biomass to hydrogen—anaerobic digestion, fermentation, and thermochemical gasification—are currently more expensive and less technologically mature than other low-carbon options but have the potential to be negative-emitting processes if combined with CCUS (IEA, 2019). Another proposed route to low-carbon H_2 is methane splitting, in which methane is converted to H_2 and elemental carbon under anaerobic conditions and high temperatures. Two commercial plants for methane splitting exist in the United States, and the technology may gain more traction if the market for solid carbon materials increases as expected (IEA, 2019).

The current low production volumes and high cost of low-carbon hydrogen are not the only factors limiting the use of H_2 as a low-carbon fuel for light-duty vehicles, however. Substantial penetration of FCEVs in the light-duty fleet will require significant build-out of the hydrogen fueling infrastructure and reductions in fuel cell and storage tank costs, as discussed in Chapter 6. The expansion of a hydrogen infrastructure for heavy-duty vehicles and long-distance transportation—which are difficult to decarbonize via electrification alone—could prompt further deployment of light-duty FCEVs (MIT Energy Initiative, 2019). However, such developments are unlikely to occur in 2025–2035 without government incentives or subsidies (MIT Energy Initiative, 2019; IEA, 2019).

FINDING 10.2: The primary methods for generating low-carbon hydrogen are water electrolysis and steam methane reforming with carbon capture, utilization, and storage; however, both currently suffer from high costs and low production volumes. Federal research programs and industry efforts, including the U.S. Department of Energy's H2@Scale initiative, fund research and development to decrease the cost of low-carbon hydrogen generation.

10.2.3 Low-Carbon Synthetic Fuels

Historically, synthetic fuels are energy carriers, typically liquid, manufactured from a source such as coal or natural gas, and used as a substitute for conventional petroleum fuels. Synthetic fuels are chemically similar to gasoline and diesel fuels but are produced from carbon sources other than petroleum. The chemical properties and high energy density of such fuels make them similar enough to existing petroleum fuels that they can "drop-in" to existing and future ICE engines designed for petroleum. Synthetic fuels have been manufactured for many decades through gas-to-liquid (GTL) processes like the Fisher-Tropsch synthesis, methanol synthesis, and methanol-to-gasoline (MTG) process. In established implementations, these processes convert fossil sources of carbon, such as coal, oil, and natural gas, into a mixture of carbon monoxide (CO) and H_2 called synthesis gas (syngas). In the presence of metal catalysts under high temperatures, syngas can then be converted into hydrocarbon chains (Fischer-Tropsch process) or methanol (methanol synthesis). Methanol can further be dehydrated to dimethylether and then converted into gasoline over a zeolite catalyst. GTL processes are widely used to produce synthetic liquid fuels at commercial scale. However, the resulting fuels are not low-carbon and in fact have higher well-to-wheels emissions than petroleum fuels owing to the emissions generated in their production.

In order to produce low-carbon synthetic fuels, the feedstocks in the Fischer-Tropsch reaction, methanol synthesis, or MTG process must be replaced with non-fossil sources of carbon, such as biomass or captured atmospheric carbon, and low-carbon hydrogen. This use of "de-fossilized" carbon fuels results in a closed carbon cycle, where combustion of the fuel leads to low or zero increases in the concentration of CO_2 in the atmosphere; therefore, low-carbon synthetic fuels could help to decarbonize the transportation sector. For the fuel to be *net-zero* carbon, all aspects of the feedstock recovery and fuel production processes, including their transportation, storage, and distribution, must be decarbonized, an effort that spans a variety of energy sectors.

Unlike some other low-carbon fuels like electricity or hydrogen, use of synthetic fuels often does not have a vehicle energy efficiency or fuel economy benefit. If the synthetic fuels are chemically identical to gasoline and used in an ICE, they offer no inherent fuel economy improvement relative to petroleum-derived gasoline used in the same engine. Synthetic fuels, however, can be designed with advantageous properties that do allow for improved fuel economy or reduced criteria emissions. Sometimes realizing improved fuel economy or reduced criteria emissions benefits from specially designed synthetic fuels requires engine modifications. Despite a lack of efficiency benefit, low-carbon synthetic fuels provide an opportunity to decarbonize legacy fleet vehicles and may be particularly relevant in light-duty vehicle applications ill-suited to electrification, such as long-distance and constant-operation road transport.

10.2.3.1 Low-Carbon Synthetic Fuels Production

Today, no large-scale, low-carbon synthetic fuels are available for light-duty vehicle transportation. To produce a low-carbon synthetic fuel, fundamental aspects of GTL processes must be modified to utilize non-fossil carbon

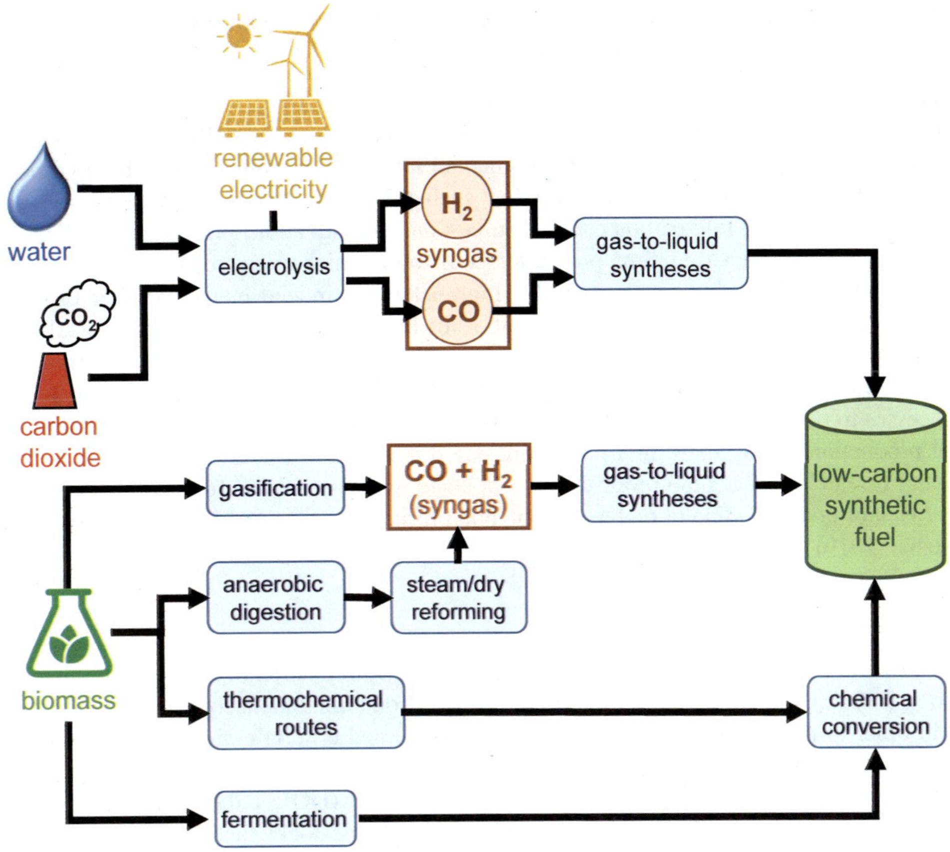

FIGURE 10.3 Pathways for the production of low-carbon synthetic fuel.
SOURCE: Committee generated, adapted from Royal Society (2019).

and low-carbon hydrogen, or new processes must be developed and commercialized at large scale. Figure 10.3 depicts potential pathways to the production of low-carbon synthetic fuels using various feedstocks and (electro) chemical processes.

Low-Carbon Fischer-Tropsch Fuel Synthesis Pathways

Two primary approaches have been developed to produce low-carbon fuels via Fischer-Tropsch pathways. One approach involves initial electrochemical reduction of captured CO_2 to CO and then, in a second thermal step, reaction of the generated CO with "blue" or "green" hydrogen[5] via conventional Fischer-Tropsch methods to produce hydrocarbon fuels. Alternatively, current research aims to develop systems that can perform Fischer-Tropsch chemistry starting from CO_2 in a single reactor using a single catalyst. One of the challenges associated with this pathway is the low steady-state concentration of CO present during the reaction, which limits chain growth and yields a product distribution rich in light hydrocarbons that are not suitable as liquid fuels. Therefore, further research into catalysts that give improved product distributions is necessary (NASEM, 2019; Royal Society, 2019).

[5]"Blue" hydrogen refers to hydrogen production in which the resultant carbon emissions are captured and stored or reused. "Green" hydrogen refers to hydrogen produced from renewable energy sources, with no corresponding carbon emissions.

Other Chemical Catalytic, Electrocatalytic, and Biological Fuel Synthesis Pathways

In addition to Fischer-Tropsch synthesis pathways, other thermochemical, electrochemical, and biological pathways are being pursued to produce hydrocarbon fuels or precursors to chemical fuels, such as CO, H_2, and oxygenated hydrocarbons (Zeman and Keith, 2008; NETL, 2011; NASEM, 2019; Royal Society, 2019). For any chemical or biological pathway to lead to a commercial low-carbon synthetic fuel, the hydrocarbon target must be synthesized from CO_2 or a CO_2-derived product such as biological molecule, with low or zero upstream emissions in its synthesis, using low-carbon energy sources, and with all additional inputs similarly low-carbon. One primary example is DOE's Co-Optima program (see below), which evaluates biomass-derived feedstocks for incorporation into vehicle fuels, particularly for high-octane-optimized engines. Other examples include biomass-to-gasoline processes, involving the gasification of biomass and subsequent chemical conversion to fuel, and thermochemical conversion of biomass via pyrolysis or hydrothermal liquefaction followed by chemical refining steps (Phillips et al., 2011; Royal Society, 2019). Modifying the commercial methanol synthesis and MTG processes to use low-carbon feedstocks and renewable electricity would also be a path to low-carbon synthetic fuel production. As of 2019, fuel production via direct chemical conversion of CO_2 was considered at a fundamental research or benchtop-proof-of-concept stage, with barriers to fuels production including low selectivity and lack of understanding of carbon-carbon bond formation steps (NASEM, 2019; Basic Energy Sciences Roundtable, 2019).

Co-Optima Program

In 2016, DOE initiated a collaborative effort toward the "Co-Optimization of Fuels and Engines," or the Co-Optima program (DOE, n.d.). Research in the program focuses on simultaneously optimizing fuels and engine technologies in order to increase engine efficiency, reduce GHG and criteria emissions, and decrease spending on fuel. The Co-Optima targets in the light-duty sector are a 35% increase in fuel economy by 2025 relative to a 2015 baseline and \$20–\$30 billion savings on fuel expense per year (DOE, n.d.). To meet its goal of reducing GHG emissions, the Co-Optima initiative focuses on identifying blendstocks that can be derived from biomass rather than from fossil sources. Blendstocks that meet health and fuel-quality standards are further evaluated using a "merit function" developed by Co-Optima researchers, which quantifies the efficiency of the blendstock relative to conventional fuel based on its octane number, sensitivity, heat of vaporization, flame speed, particulate matter index, and catalyst light-off temperature (Farrell et al., 2018). This evaluation has so far identified 10 promising blendstocks (Figure 10.4) that have higher merit function values than E10 premium gasoline, meet fuel-quality requirements, and could be commercially available by 2025–2030. However, 4 of these blendstocks—methanol, prenol, the furan mixture, and cyclopentanone—have higher barriers to commercialization owing to issues with chemical stability and compatibility, volatility, and/or toxicity (Gaspar et al., 2019). More detailed analyses of the blendstocks are ongoing and include investigating the relationship between molecular composition and fuel properties; evaluating blendstock performance in different engine combustion modes; and performing techno-economic, life cycle, and refinery integration analyses of the blendstocks and engine technologies (Farrell et al., 2018; DOE, 2019b).

10.2.3.2 Low-Carbon Synthetic Fuels Enabling Technology and Commercialization Needs

Several developments will be required for wide-scale commercialization of low-carbon synthetic fuels. These may include improved technology for low-carbon hydrogen generation, CCS, and electrochemical conversion of CO_2.

Hydrogen Generation

Hydrogen is a fundamental input in many potential low-carbon synthetic fuels pathways. As described in Section 10.2.2, at present, hydrogen is primarily generated from natural gas through SMR and coal gasification, resulting in significant carbon emissions. To produce low-carbon synthetic fuels, the hydrogen source must be

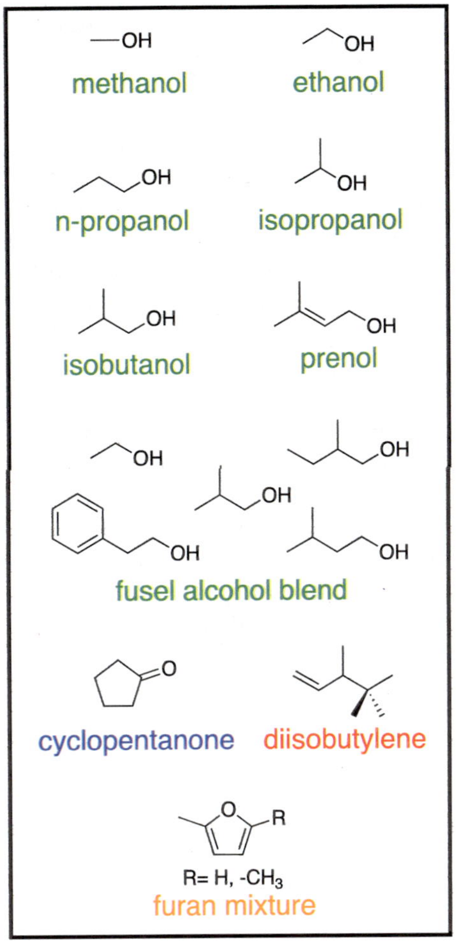

FIGURE 10.4 Ten promising blendstocks identified by the Co-Optima program. These blendstocks have higher merit function values than E10 premium gasoline, meet fuel-quality requirements, and could be commercially available by 2025–2030. The six with the fewest barriers to adoption are ethanol, n-propanol, isopropanol, isobutanol, the fusel alcohol blend, and diisobutylene. SOURCE: Committee generated, adapted from Gaspar et al. (2019).

carbon-free or low-carbon. Section 10.2.3.2 above describes several methods for generating low-carbon hydrogen, including SMR with CCUS, water electrolysis (low- and high-temperature pathways), biomass conversion, and methane splitting.

Carbon Capture

Captured carbon necessary for the production of low-carbon synthetic fuels can come from many sources, including industrial waste streams, combustion in power generation, and ambient air. The concentration and pressure of the gas stream, as well as its onsite availability or transport requirements, govern the choice of removal technology and the resulting cost. High concentration CO_2 sources such as those from industrial processes or power generation provide a cheaper source of carbon. Carbon obtained through direct air capture is more dilute and therefore more difficult to separate and requires 2–4 times more energy than more concentrated mixtures like flue gases. Direct air capture approaches are technologically feasible, but because of additional separation processes, they will be more expensive than capture from fossil fuel power plants or other concentrated sources. While several companies are currently working to commercialize direct air capture, the capital costs of these facilities are highly uncertain, as only small-scale units have been built so far (Carbon Engineering, n.d.; Climeworks, n.d.; Soletair Power, 2020). More research and pilot plants are needed to optimize the technology.

CO_2 Electrolyzers

CO_2 electrolyzers could provide an abundant source of non-fossil CO as an input to Fischer-Tropsch pathways for the production of low-carbon fuels. Few CO_2 electrolyzer technologies have been commercialized to-date, although several companies have reported plans for scale-up (NASEM, 2019; Sanchez et al., 2019). Multiple technologies and cell designs have been proposed, including low-temperature PEM electrolysis and high-temperature molten carbonate or solid oxide electrolysis (NASEM, 2019; Sanchez et al., 2019; Küngas, 2020). In laboratory-scale demonstrations, solid-oxide electrolyzers have shown the best overall performance in terms of efficiency, durability, and selectivity for CO (Küngas, 2020). However, the ability of PEM electrolyzers to operate at ambient temperatures and pressures is attractive for many industrial applications, and these electrolyzers have been the subject of most technoeconomic analyses on CO_2 electroreduction (Sanchez et al., 2019). For example, De Luna et al. (2019) determined that electrochemically generated CO would be cost-competitive with fossil-derived CO in a PEM electrolyzer with 90% product selectivity and 70% energetic efficiency at an electricity cost of $0.04 per kWh. Achieving this goal will require simultaneous optimization of catalyst and cell design to improve rates and selectivities and to minimize challenges associated with product separation, low CO_2 solubility, mass transport, and system durability (Weekes et al., 2018; NASEM, 2019; De Luna et al., 2019; Sanchez et al., 2019). In the long term, engineering principles developed for commercial water electrolysis systems could be adapted to facilitate the industrial deployment of CO_2 electrolyzers.

10.2.3.3 Low-Carbon Synthetic Fuels Potential in Medium to Long Term

Production of low-carbon synthetic fuels is currently limited by high costs and inefficiencies owing to energy losses from the many processes involved in manufacturing (Li et al., 2016; Cai et al., 2018; Royal Society, 2019). High costs and low efficiencies may be acceptable for low volume, high value commodities, but they are untenable for very high volume, low margin products like fuels, especially in comparison to inexpensive and readily available gasoline, diesel, and electricity. Heavier, diesel-like fuels used in compression ignition engines have more near-term options for commercial drop-in fuels, as compared to lighter, gasoline-like spark-ignition engine fuels (AFDC, n.d.). In the medium term, low-carbon synthetic fuels will likely be first introduced as blends with existing fossil fuels (Farrell et al., 2018). Examples of this are already available for diesel blends (Neste, 2016; Renewable Energy Group, 2020), and recent studies indicate high potential benefits of incorporating low-carbon synthetic fuels into conventional gasoline. For instance, Dunn et al. (2020) determined that using a biomass-derived isopropanol blendstock (31% by volume in conventional gasoline) could lead, on a life cycle basis, to 4%–7% reduction in GHG emissions, 3%–4% reduction in water consumption, and 3% reduction in particulate emissions from 2025 to 2050 compared to a business as usual case. The Co-Optima program targets the 2025–2030 time frame for commercialization of a bio-blendstock fuel (Farrell et al., 2018), but significant barriers still exist to the widespread adoption of these fuels, such as increasing feedstock supply and constructing new biorefineries (Dunn et al., 2020). Longer-term cost reductions and improvements to process efficiencies will improve cost competitiveness for all low-carbon synthetic fuels; however, this is unlikely to be sufficient without policy or regulatory interventions to give value to low-carbon fuels.

FINDING 10.3: Other than biomass-derived ethanol, there are currently no large-scale commercial low-carbon synthetic fuels available in the light-duty vehicle sector. The U.S. Department of Energy's Co-Optima program aims to develop low-carbon fuels comprised of biomass-derived blendstocks mixed with conventional gasoline for commercialization by 2025–2030. Another promising route to low-carbon liquid fuels is Fischer-Tropsch synthesis from non-fossil CO_2 and renewable H_2; however, the technologies underlying this pathway still require development and scale-up.

FINDING 10.4: Low-carbon liquid fuels, which can serve as drop-ins for conventional gasoline and diesel, present an opportunity to decarbonize both the existing and future light-duty vehicle fleets. Producing a net-zero carbon liquid fuel will require all aspects of feedstock recovery and fuel production to be decarbonized. Incentivizing the production and use of low- and net-zero carbon liquid fuels may require changes to the current regulations for vehicle and fuel systems, since these synthetic fuels have no advantage over gasoline when considering only tailpipe greenhouse gas emissions.

10.3 LOW-CARBON FUELS IN THE 2025–2035 FLEET

Petroleum fuels dominate passenger vehicle propulsion because they are inexpensive, easily distributed, and have high energy density. Improvements in engines, powertrain technologies, and other vehicle technologies may not be able to achieve sufficient improvements in energy efficiency, reductions in petroleum use, and reductions in emissions to meet fuel economy and GHG emissions standards. Therefore, automakers are increasingly developing technologies to use non-petroleum fuels. This section discusses the potential role of low-carbon fuels in the future fleet, considering regulatory issues, metrics for energy efficiency and emissions, and techno-economic and market factors.

10.3.1 Regulatory Issues

The corporate average fuel economy (CAFE) standards as originally envisioned sought reduction in petroleum fuel use. Concern about the growing reliance on petroleum, especially *imported* petroleum fuel, for transportation energy led to the first energy efficiency regulations on passenger vehicles, enacted in the Energy Policy and Conservation Act of 1975. U.S. legislators identified fuel economy, in miles per gallon of fuel, as the appropriate energy efficiency metric, where fuel was defined as gasoline and diesel oil. The legislation also gave the Secretary of Transportation the flexibility to include by rule any other liquid or gaseous fuel "if he determines that such inclusion is consistent with the need of the Nation to conserve energy" (EPCA, 1975). The Energy Policy Act of 1992 defined alternative fuels as "pure methanol, ethanol, and other alcohols; blends of 85% or more of alcohol with gasoline; natural gas and liquid fuels domestically produced from natural gas; propane; coal-derived liquid fuels; hydrogen; electricity; pure biodiesel (B100); fuels, other than alcohol, derived from biological materials; and P-Series fuels" (Energy Policy Act, 1992). The same act authorized DOE to "designate other fuels as alternative fuels, provided that the fuel is substantially non-petroleum, yields substantial energy security benefits, and offers substantial environmental benefits" (Energy Policy Act, 1992). These expanding definitions of alternative fuels, coupled with the projected increase of alternative-fuel vehicles in the light-duty fleet, highlight a need to reevaluate the methods by which such fuels are regulated.

As introduced in Chapter 2 and further discussed in Chapter 12, the current CAFE and GHG standards have credits and incentives for alternative-fueled vehicles. Per a provision in Public Law 103-272, the National Highway Traffic Safety Administration (NHTSA) cannot estimate the use or availability of such vehicles in evaluating and setting the stringency of fuel economy standards (103rd Congress, 1994). In recognition of their value in displacing petroleum, NHTSA instead incentivizes deployment of technologies that use alternative fuels by artificially enhancing their fuel economy, considering one gallon equivalent of the alternative fuel to equal 0.15 gallons of gasoline. For dual-fueled vehicles, the percent operation on gasoline versus alternative fuel will be estimated using the Society for Automotive Engineers' utility factors beginning in model year (MY) 2020. In regulating GHG emissions, as a temporary measure, the U.S. Environmental Protection Agency (EPA) does not consider upstream emissions from the electric grid, and thus BEVs (and the electric operating portions of PHEV use) are assigned emissions of 0 g/mile. As further incentive, the GHG standards use multipliers for BEVs, FCEVs, and PHEVs that count them as more than one vehicle, thereby helping manufacturers meet their fleet-wide emissions targets. However, these multipliers will phase out after MY 2021. Neither the CAFE nor GHG emissions standards include provisions for low-carbon synthetic fuels, as they are still emerging technologies. As mandated in the Energy Independence and Security Act, the U.S. Department of Transportation must reassign fuel economy standards at least every 5 years (EISA, 2007). NHTSA further interprets the statute to suggest that standards cannot be set for more than 5 years in the future. Thus, reevaluation and possibly changes to the incentives and crediting schemes for alternative fuels are likely to occur during this study's time period of 2025–2035.

If low-carbon synthetic liquid fuels become a significant part of the light-duty vehicle and fuels system, the CAFE and GHG regulations will need to account for their use when considering vehicle efficiency and emissions standards. Such fuels have tailpipe emissions and onboard energy and fuel use similar to those of conventionally fueled vehicles. However, by definition low-carbon synthetic fuels have lower full-fuel-cycle emissions than conventional gasoline or diesel. Depending on the production method, their full-fuel-cycle energy use can be higher or lower than that of conventional fuels, as described in Section 10.3.2 below. Capturing the emissions benefits of low-carbon synthetic fuels within the CAFE and GHG standards would require a vehicle-fuel system approach, including a full-fuel-cycle assessment. Such an approach would need to be applied to all vehicle-fuel systems for equivalent comparison across the fleet.

One example of a policy that regulates and incentivizes low-carbon fuel production is California's Low Carbon Fuel Standard (LCFS). The California state legislature mandated a 20% reduction of carbon intensity, measured in grams of CO_2 equivalent per megajoule (gCO_2e/MJ), in the light-duty vehicle fuels sold in the state over 15 years (2015–2030). A life cycle approach is used to evaluate fuel pathways, and a credit generating and trading scheme allows industry-wide compliance with the yearly decreasing standard (CARB, 2020b). The LCFS is a low-carbon fuel standard, not a zero-carbon fuel standard, and it includes drop-in liquid fuels as well as fuel blends and alternative powertrain fuels like hydrogen and electricity. Ethanol, renewable diesel, and biodiesel alone or in fuel blends are the highest volume contributors to the standard implementation (CARB, 2020a). Inspired by California's LCFS, similar programs are in use or under development in Oregon, Washington, Canada, and Brazil (CARB, 2020b).

10.3.2 Energy Efficiency and Emissions Metrics

Inherently tied to the regulatory issues discussed above is the choice of metrics used to evaluate the energy efficiency and emissions of vehicle fuels. The current regulatory standards consider petroleum fuel consumption and tailpipe GHG emissions, both on a per-vehicle mile basis. In a future fleet with significant penetration of alternative-fuel vehicles, standards based on fuel consumption in gallons of liquid fuel per mile may become less relevant. The use of only tailpipe rather than full-fuel-cycle GHG emissions incentivizes the deployment of zero-emission vehicles, but it misrepresents the actual carbon emissions associated with energy use in a light-duty fleet with high penetration of zero-emission vehicles and/or low-carbon liquid fuels. In such a fleet, a more complete picture of a vehicle's carbon footprint would be captured in its well-to-wheels GHG emissions, which consider the entire life cycle of the fuel, including both production and use onboard the vehicle. Further discussion of vehicle life cycle emissions can be found in Chapter 12.

In the long term, NHTSA and EPA might reconsider the appropriate metrics for setting regulatory standards to align with the goals of energy efficiency, energy security, cost savings, and emissions reductions, or any other goals that may arise in 2025–2035. For example, the relationship between energy use and GHG emissions could become more complex in a future fleet with high prevalence of low-carbon alternative fuels. With most of the proposed low-carbon fueling options, such as electricity and hydrogen, vehicles achieve both reduced GHG emissions and decreased energy use relative to conventional gasoline vehicles on a well-to-wheels basis (Gao, 2011; Ramachandran and Stimming, 2015; CaFCP, 2016; Liu et al., 2020). However, in some cases, the use of a low-carbon synthetic fuel can reduce GHG emissions with no benefit to overall energy use, for instance, if the fuel is synthesized from CO originally derived from captured CO_2 and transformed into liquid fuel via traditional Fischer-Tropsch methods. In the near term, some high-energy options might be more cost effective than developing and deploying lower energy technologies for the production of low-carbon feedstocks. Given consumer preference for least-cost pathways, regulatory standards for low-carbon synthetic fuels should consider accounting for these potential trade-offs between GHG emissions and total system energy use.

10.3.3 Techno-Economic and Market Factors

As of 2020, there are ever-growing options for plug-in EVs available to consumers, and infrastructure for charging at home, and in some locations at businesses and public chargers. FCEVs are available, although only in very limited markets where refilling options exist. Currently, net-zero carbon synthetic drop-in fuels are not available for customers, but low-carbon blended fuels are available in certain markets. The deployment of vehicles (plug-in EVs and FCEVs) and of fuels (low-carbon drop-in liquid fuels) will determine the impact of these non-petroleum options on vehicle energy use, petroleum fuel use, and GHG emissions. For electric fuels to have a greater impact on reducing energy consumption and GHG emissions, improvements are required in the energy density and cost of the vehicle battery, the capability of the charging infrastructure to meet consumers' needs and comfort level, and the well-to-wheels emissions of the vehicle-fuel system (achieved by decarbonizing electricity generation). An increase in hydrogen fuel use could occur with reduced costs of FCEVs, build-out of the hydrogen fueling infrastructure to meet consumers' needs, and decreased emissions for hydrogen generation. Low-carbon synthetic fuels can be incorporated into the existing fueling infrastructure and reduce GHG emissions from the current fleet as well as future vehicles; however, their implementation will depend on improving capabilities and scaling up fuel synthesis, decreasing well-to-wheels emissions, and reducing vehicle and fuel costs relative to other low-carbon options.

10.3.4 Outlook for Non-Petroleum Fuels

The various regulatory, techno-economic, and market factors described above will dictate the extent of non-petroleum fuel availability in the light-duty vehicle fleet during 2025–2035. The use of such fuels, and the amount of resulting reductions in GHG emissions in the light-duty sector, will depend on those factors as well as the vehicle technologies deployed. The current regulatory standards incentivize the deployment of zero-emission vehicles by artificially enhancing their fuel economy value and considering only tailpipe GHG emissions. Additional regulations such as a nationwide LCFS or other decarbonization policies could further increase the development and use of low-carbon fuels. In the long term, it will be important to consider how to incorporate low-carbon fuels into the existing CAFE program, particularly with regard to metrics for fuel consumption, GHG emissions, and energy use, as well as their incorporation into the stringency of the standards. Non-petroleum fuels present opportunities to decrease GHG emissions, both on- and offboard the vehicle, but in some cases result in increased energy use depending on the choice of low-carbon pathway.

FINDING 10.5: Greenhouse gas (GHG) emissions are a relevant metric for all light-duty vehicle fuels, including diesel, gasoline, biofuels, low-carbon synthetic fuels, electricity, and hydrogen. Considering the full-fuel-cycle emissions of such fuels will become increasingly important in understanding a vehicle's impact on GHG emissions given the expected growth in non-petroleum fuel use in the light-duty fleet during 2025–2035.

10.4 RECOMMENDATIONS FOR NON-PETROLEUM FUELS

RECOMMENDATION 10.1: The U.S. Department of Energy, in partnership with the U.S. Department of Defense and the private sector, should facilitate the deployment and commercialization of low-carbon fuels by increasing the capacity of renewable electricity generation and providing more research and development funding for technologies that enable low-carbon hydrogen generation and low-carbon synthetic fuel production.

RECOMMENDATION 10.2: As low-carbon fuels become more prevalent in the light-duty fleet, the National Highway Traffic Safety Administration and U.S. Environmental Protection Agency should consider a full-fuel-cycle approach to setting regulatory standards, which would take into account both upstream and onboard energy use and emissions. One approach to incorporating fuels into a regulatory regime might be to explore a nationwide low-carbon fuel standard, such as those currently used in California and Oregon.

10.5 REFERENCES

103rd Congress. 1994. To revise, codify, and enact without substantive change certain general and permanent laws, related to transportation, as subtitles II, III, and V-X of title 49, United States Code, "Transportation," and to make other technical improvements in the Code, Pub. L. No. 103–272, H.R. 1758 (1994). https://www.congress.gov/103/statute/STATUTE-108/STATUTE-108-Pg745.pdf.

AFDC (Alternative Fuels Data Center). n.d. "Renewable Hydrocarbon Biofuels." https://afdc.energy.gov/fuels/emerging_hydrocarbon.html.

AFDC. 2020. Alternative Fuel Price Report. April. https://afdc.energy.gov/fuels/prices.html.

ANL (Argonne National Laboratory). 2021. *The Greenhouse Gases, Regulated Emissions, and Energy Use in Technologies (GREET) Model* (version 2020 rev1). GREET1 Model. Lemont, Illinois, 2021. https://greet.es.anl.gov/greet.models.

Basic Energy Sciences Roundtable. 2019. "Report of the Basic Energy Sciences Roundtable on Liquid Solar Fuels." 1615599. https://doi.org/10.2172/1615599.

CaFCP (California Fuel Cell Partnership). 2016. "Air, Climate, Energy, Water Security: A Guide to Understanding the Well-to Wheels Impact of Fuel Cell Electric Vehicles." https://cafcp.org/sites/default/files/W2W-2016.pdf.

Cai, H., J. Markham, S. Jones, P.T. Benavides, J.B. Dunn, M. Biddy, and L. Tao. 2018. Techno-economic analysis and life-cycle analysis of two light-duty bioblendstocks: Isobutanol and aromatic-rich hydrocarbons. *ACS Sustainable Chemistry and Engineering* 6(7):8790–8800. https://doi.org/10.1021/acssuschemeng.8b01152.

CARB (California Air Resources Board). 2020a. "Data Dashboard." Last updated May 29, 2020. https://ww3.arb.ca.gov/fuels/lcfs/dashboard/dashboard.htm.

CARB. 2020b. "Low Carbon Fuels Standard." Last updated June 24, 2020. https://ww2.arb.ca.gov/sites/default/files/2020-03/basics-notes.pdf.

Carbon Engineering. n.d. "Our Story." https://carbonengineering.com/our-story/.

CEC/CARB (California Energy Commission and California Air Resources Board). 2019. Joint Agency Staff Report on Assembly Bill 8: 2019 Annual Assessment of Time and Cost Needed to Attain 100 Hydrogen Refueling Stations in California. December. CEC-600-2019-039. https://ww2.energy.ca.gov/2019publications/CEC-600-2019-039/CEC-600-2019-039.pdf.

Christensen, A. 2020. "Assessment of Hydrogen Production Costs from Electrolysis: United States and Europe." International Council on Clean Transportation. June 18. https://theicct.org/publications/assessment-hydrogen-production-costs-electrolysis-united-states-and-europe.

Climeworks. n.d. "The Climeworks Story: From Lab Scale to Climate Relevant." https://www.climeworks.com/page/story-to-reverse-climate-change.

Connelly, E., A. Elgowainy, and M. Ruth. 2019. "Current Hydrogen Market Size: Domestic and Global." DOE Hydrogen and Fuel Cells Program Record, October 1. https://www.hydrogen.energy.gov/pdfs/19002-hydrogen-market-domestic-global.pdf.

Davis, S.C., and R.G. Boundy. 2020. *Transportation Energy Data Book: Edition 39*. Oak Ridge, TN: Oak Ridge National Laboratory. https://doi.org/10.2172/1767864.

Davis, S.J., N.S. Lewis, M. Shaner, S. Aggarwal, D. Arent, I.L. Azevedo, S.M. Benson, et al. 2018. Net-zero emissions energy systems. *Science* 360(6396):eaas9793. https://doi.org/10.1126/science.aas9793.

De Luna, P., C. Hahn, D. Higgins, S.A. Jaffer, T.F. Jaramillo, and E.H. Sargent. 2019. What would it take for renewably powered electrosynthesis to displace petrochemical processes? *Science* 364(6438):eaav3506. https://doi.org/10.1126/science.aav3506.

DOE (U.S. Department of Energy). n.d. "Co-Optima Research." Energy.Gov. https://www.energy.gov/eere/bioenergy/co-optima-research.

DOE. 2017. "Chapter 3: Building a Clean Electricity Future." In *Transforming the Nation's Electricity System: The Second Installment of the QER*, 100. Washington, DC. https://www.energy.gov/sites/prod/files/2017/01/f34/Chapter%203%20Building%20a%20Clean%20Electricity%20Future_0.pdf.

DOE. 2019a. Hydrogen and Fuel Cells Program, 2019 Annual Progress Report. https://www.hydrogen.energy.gov/annual_progress19_h2fuel.html.

DOE. 2019b. "Co-Optimization of Fuels and Engines: FY19 Year in Review." Washington, DC: U.S. Department of Energy. https://www.energy.gov/sites/prod/files/2020/06/f75/beto-co-optima-fy19-yir-report-june-2020.pdf.

DOE. 2020a. "Clean Cities Alternative Fuel Price Report, January 2020." January. U.S. Department of Energy.

DOE. 2020b. "Where the Energy Goes." https://www.fueleconomy.gov/feg/atv.shtml.

DOE-EERE. (U.S. Department of Energy Office of Energy Efficiency and Renewable Energy) n.d. "Central Versus Distributed Hydrogen Production." Hydrogen and Fuel Cell Technologies Office. https://www.energy.gov/eere/fuelcells/central-versus-distributed-hydrogen-production.

DOE-EERE. 2020. "H2@Scale New Markets Funding Opportunity Announcement (FOA)." Funding Opportunity Announcement DE-FOA-0002229. Washington, DC: Department of Energy.

DOE-EIA (U.S. Department of Energy and Energy Information Administration). 2020. Monthly Energy Review, March 2020.

DOE/EPA (U.S. Department of Energy, U.S. Environmental Protection Agency). n.d. "Find and Compare Cars." www.fueleconomy.gov. https://www.fueleconomy.gov/feg/findacar.shtml.

Dunn, J.B., E. Newes, H. Cai, Y. Zhang, A. Brooker, L. Ou, N. Mundt, et al. 2020. Energy, economic, and environmental benefits assessment of co-optimized engines and bio-blendstocks. *Energy and Environmental Science* 13(8):2262–2274. https://doi.org/10.1039/D0EE00716A.

EIA (Energy Information Administration). 2020a. "Annual Energy Outlook 2020." Annual Energy Outlook. Washington, DC: U.S. Energy Information Administration. https://www.eia.gov/outlooks/aeo/.

EIA. 2020b. "Henry Hub Natural Gas Spot Price." https://www.eia.gov/dnav/ng/hist/rngwhhdM.htm.

EIA. 2020c. "Electric Power Monthly—Table 1.1. Net Generation by Energy Source: Total (All Sectors), 2010–June 2020." U.S. Energy Information Administration. https://www.eia.gov/electricity/monthly/epm_table_grapher.php.

EISA (Energy Independence and Security Act). 2007. Pub. L. 110-140, 121 Stat. 1492, enacted December 19, 2007.

Elgowainy, A., J. Han, J. Ward, F. Joseck, D. Gohlke, A. Lindauer, T. Ramsden, et al. 2016. "Cradle-to-Grave Lifecycle Analysis of U.S. Light Duty Vehicle-Fuel Pathways: A Greenhouse Gas Emissions and Economic Assessment of Current (2015) and Future (2025–2030) Technologies." https://doi.org/10.2172/1254857.

Energy Policy Act. 1992. Pub. L. 102-486, 106 Stat. 2776, enacted October 24, 1992.

EPA (U.S. Environmental Protection Agency). 2020a. "Fast Facts: U.S. Transportation Sector Greenhouse Gas Emissions 1990–2018." Washington, DC. June 2020. https://nepis.epa.gov/Exe/ZyPDF.cgi?Dockey=P100ZK4P.pdf.

EPA. 2020b. "Emissions and Generation Resource Integrated Database (EGRID)." Collections and Lists. US EPA. July 27. https://www.epa.gov/egrid.

EPCA (Energy Policy and Conservation Act). 1975. Pub.L. 94-163, 89 Stat. 871, enacted December 22, 1975.

Farrell, J.T., J. Holladay, and R. Wagner. 2018. "Co-Optimization of Fuels and Engines: Fuel Blendstocks with the Potential to Optimize Future Gasoline Engine Performance; Identification of Five Chemical Families for Detailed Evaluation." NREL/TP—5400-69009, DOE/GO—102018-4970, 1434413. https://doi.org/10.2172/1434413.

Gao, L. 2011. Well-to-wheels analysis of energy use and greenhouse gas emissions for alternative fuels. *International Journal of Applied Science and Technology* 1(6):1–8. http://www.ijastnet.com/journals/Vol_1_No_6_November_2011/1.pdf.

Gaspar, D.J., B.H. West, D. Ruddy, T.J. Wilke, E. Polikarpov, T.L. Alleman, A. George, et al. 2019. "Top Ten Blendstocks Derived from Biomass for Turbocharged Spark Ignition Engines: Bio-Blendstocks with Potential for Highest Engine Efficiency." https://doi.org/10.2172/1567705.

Hydrogen Council. 2017. "Hydrogen Scaling Up." November 13. https://hydrogencouncil.com/wp-content/uploads/2017/11/Hydrogen-Scaling-up_Hydrogen-Council_2017.compressed.pdf.

IAEA (International Atomic Energy Agency). 2013. "Hydrogen Production Using Nuclear Energy." Technical Report NP-T-4.2. Nuclear Energy. Vienna: International Atomic Energy Agency (IAEA). http://www-pub.iaea.org/MTCD/Publications/PDF/Pub1577_web.pdf.

IEA (International Energy Agency). 2019. *The Future of Hydrogen*, 37. June. https://webstore.iea.org/the-future-of-hydrogen.

IEA. 2020a. *Hydrogen*. IEA, Paris. https://www.iea.org/reports/hydrogen.

IEA. 2020b. *The Covid-19 Crisis and Clean Energy Progress*. IEA, Paris. https://www.iea.org/reports/the-covid-19-crisis-and-clean-energy-progress.

James, B.D., D. DeSantis, and G. Saur. 2019. "2019 Analysis of Advanced Hydrogen Production Pathways." DOE Hydrogen and Fuel Cells Program, 2019 Annual Progress Report. https://www.hydrogen.energy.gov/pdfs/progress19/h2f_p102_james_2019.pdf.

Küngas, R. 2020. Review—electrochemical CO_2 reduction for CO production: Comparison of low- and high-temperature electrolysis technologies. *Journal of the Electrochemical Society* 167(4):044508. https://doi.org/10.1149/1945-7111/ab7099.

Larson, E., C. Greig, J. Jenkins, E. Mayfield, A. Pascale, C. Zhang, J. Drossman, et al. 2020. "Net-Zero America: Potential Pathways, Infrastructure, and Impacts." Princeton, NJ: Princeton University. December 15. https://environmenthalfcentury.princeton.edu/sites/g/files/toruqf331/files/2020-12/Princeton_NZA_Interim_Report_15_Dec_2020_FINAL.pdf.

Lawson, A. 2018. "Decarbonizing U.S. Power." C2ES.

Li, X., P. Anderson, H.M. Jhong, M. Paster, J.F. Stubbins, and P.J.A. Kenis. 2016. Greenhouse gas emissions, energy efficiency, and cost of synthetic fuel production using electrochemical CO_2 conversion and the Fischer–Tropsch process. *Energy and Fuels* 30(7):5980–5989. https://doi.org/10.1021/acs.energyfuels.6b00665.

Liu, X., K. Reddi, A. Elgowainy, H. Lohse-Busch, M. Wang, and N. Rustagi. 2020. Comparison of well-to-wheels energy use and emissions of a hydrogen fuel cell electric vehicle relative to a conventional gasoline-powered internal combustion engine vehicle. *International Journal of Hydrogen Energy* 45(1):972–983. https://doi.org/10.1016/j.ijhydene.2019.10.192.

Mahajan, M. 2019. "How to Reach U.S. Net Zero Emissions by 2050: Decarbonizing Electricity." https://www.forbes.com/sites/energyinnovation/2019/11/12/how-to-reach-us-net-zero-emissions-by-2050-decarbonizing-electricity/#445d68ac49e7.

MIT Energy Initiative. 2019. Insights into Future Mobility. Cambridge, MA: MIT Energy Initiative. http://energy.mit.edu/insightsintofuturemobility.

NASEM (National Academies of Sciences, Engineering, and Medicine). 2019. *Gaseous Carbon Waste Streams Utilization: Status and Research Needs*. Washington, DC: The National Academies Press. https://doi.org/10.17226/25232.

NASEM. 2021. *Accelerating Decarbonization of the U.S. Energy System*. Washington, DC: The National Academies Press. https://doi.org/10.17226/25932.

Neste. 2016. "Renewable Diesel Handbook." Neste. https://www.neste.com/sites/default/files/attachments/neste_renewable_diesel_handbook.pdf.

NETL (National Energy Technology Laboratory). 2011. "Affordable, Low-Carbon Diesel Fuel from Domestic Coal and Biomass."

NH2E (Natural Hydrogen Energy LLC). 2019. "Energy of Natural Hydrogen." http://nh2e.com/.

NREL (National Renewable Energy Laboratory). n.d. "H2A: Hydrogen Analysis Production Case Studies." Hydrogen and Fuel Cells. https://www.nrel.gov/hydrogen/h2a-production-case-studies.html.

Peterson, D., J. Vickers, and D. DeSantis. 2020. "Hydrogen Production Cost from PEM Electrolysis-2019." DOE Hydrogen and Fuel Cells Program Record, February 3. https://www.hydrogen.energy.gov/pdfs/19009_h2_production_cost_pem_electrolysis_2019.pdf.

Phillips, S.D., J.K. Tarud, M.J. Biddy, and A. Dutta. 2011. "Gasoline from Wood via Integrated Gasification, Synthesis, and Methanol-to-Gasoline Technologies." January 1. https://doi.org/10.2172/1004790.

Ramachandran, S., and U. Stimming. 2015. Well to wheel analysis of low carbon alternatives for road traffic. *Energy and Environmental Science* 8(11):3313–3324. https://doi.org/10.1039/C5EE01512J.

Ramsden, T., and F. Joseck. 2018. "Hydrogen R&D Cost Target Calculation—2018 Update." DOE Program Record (Fuel Cell Technologies Office). September. https://www.hydrogen.energy.gov/pdfs/18004_h2_cost_target_calculation_2018.pdf.

Randolph, K., and S. Studer. 2017. "Hydrogen Production Cost from Fermentation." DOE Hydrogen and Fuel Cells Program Record. U.S. Department of Energy. February 27. https://www.hydrogen.energy.gov/pdfs/16016_h2_production_cost_fermentation.pdf.

Reed, J., E. Dailey, B. Shaffer, B. Lane, R. Flores, A. Fong, and G.S. Samuelsen. 2020. Roadmap for the Deployment and Buildout of Renewable Hydrogen Production Plants in California. CEC. Publication Number: CEC-600-2020-002.

Reisert, M., A. Aphale, and P. Singh. 2018. Solid oxide electrochemical systems: Material degradation processes and novel mitigation approaches. *Materials* 11(11):2169.

Renewable Energy Group. 2020. "REG Ultra Clean." https://www.regi.com/products/transportation-fuels/reg-ultra-clean-diesel.

Robinson, J. 2020. "Cost, Logistics Offer 'Blue Hydrogen' Market Advantages over 'Green' Alternative." https://www.spglobal.com/platts/en/market-insights/latest-news/electric-power/031920-cost-logistics-offer-blue-hydrogen-market-advantages-over-green-alternative.

Royal Society. 2019. *Sustainable Synthetic Carbon Based Fuels for Transport: Policy Briefing*. September. https://royalsociety.org/-/media/policy/projects/synthetic-fuels/synthetic-fuels-briefing.pdf.

Sánchez, O.G., Y.Y. Birdja, M. Bulut, J. Vaes, T. Breugelmans, and D. Pant. 2019. Recent advances in industrial CO_2 electroreduction. *Current Opinion in Green and Sustainable Chemistry* 16:47–56. April. https://doi.org/10.1016/j.cogsc.2019.01.005.

Schalenbach, M., A.R. Zeradjanin, O. Kasian, S. Cherevko, and K.J.J. Mayrhofer. 2018. A perspective on low-temperature water electrolysis—challenges in alkaline and acidic technology. *International Journal of Electrochemical Science*. 13:1173–1226. doi: 10.20964/2018.02.26.

Soletair Power. 2020. "Sales of First Commercial Power-to-X Unit." Soletair Power (blog). June 1. https://www.soletairpower.fi/2020/06/sales-of-first-commercial-power-to-x-unit/.

Vickers, J., D. Peterson, and K. Randolph. 2020. "Cost of Electrolytic Hydrogen Production with Existing Technology." DOE Hydrogen and Fuel Cells Program Record. U.S. Department of Energy, September 22. https://www.hydrogen.energy.gov/pdfs/20004-cost-electrolytic-hydrogen-production.pdf.

Weekes, D.M., D.A. Salvatore, A. Reyes, A. Huang, and C.P. Berlinguette. 2018. Electrolytic CO_2 reduction in a flow cell. *Accounts of Chemical Research* 51(4):910–918. https://doi.org/10.1021/acs.accounts.8b00010.

Zablocki, A. 2019. "Fact Sheet: Energy Storage (2019)." Environmental and Energy Study Institute (EESI). https://www.eesi.org/papers/view/energy-storage-2019.

Zeman, F.S., and D.W. Keith. 2008. Carbon neutral hydrocarbons. *Philosophical Transactions of the Royal Society A: Mathematical, Physical and Engineering Sciences* 366(1882):3901–3918. https://doi.org/10.1098/rsta.2008.0143.

Zgonnik, V. 2020. The occurrence and geoscience of natural hydrogen: A comprehensive review. *Earth-Science Reviews* 203(April 1):103140.

11

Consumer Acceptance and Market Response to Standards

When and how different fuel-saving technologies are incorporated into vehicles depends on multiple market and vehicle system factors in addition to the requirement of fuel economy and greenhouse gas (GHG) standards—including consumer demand and willingness to pay, how manufacturers respond to the standards, how the technologies affect other aspects of the vehicle consumers' value, and the state of infrastructure supporting the technologies. These factors are also important determinants of the full costs of the standards.

Manufacturers and consumers can respond to increased fuel economy and GHG standards in various ways. As discussed extensively in the previous chapters, manufacturers can incorporate various powertrain and non-powertrain technologies into their vehicles to increase their fuel efficiency while maintaining levels of other attributes (e.g., acceleration performance, capacity, and amenities). Manufacturers can also apply technologies partially to improving fuel economy and partially to improving other attributes, creating a trade-off between improvements in fuel economy and other performance attributes. They can earn, buy, and sell credits for things such as alternative-fuel vehicle sales, or for overcomplying with the standards. Because compliance depends on the sales-weighted average of a manufacturer's vehicles' fuel economies, a manufacturer can increase its average fuel economy by adjusting prices and shifting sales to vehicles that exceed their footprint-based fuel economy standards and away from those that do not. Last, if a manufacturer were to increase the footprint of its vehicles or redesigned cars so that they are classified as light trucks, it would reduce the overall standard the manufacturer needs to attain.

Based on recent automaker decisions and their announced plans, each automaker can be expected to pursue a combination of these options, taking into consideration consumer preferences, technology implementation costs, competition with other automakers, and other market factors. A principal consideration of automakers is consumer preferences for fuel economy and other vehicle attributes, which may be affected by changes in the standards.

The effectiveness of the standards at reducing fuel consumption and emissions depends on consumer behavior, including vehicle choices and driving decisions. Consumer preferences for vehicle fuel economy and other attributes affect which vehicles are purchased and the extent to which technologies are adopted by the market. Similarly, consumers' decisions about driving behavior affects the extent to which fuel-efficiency technologies lead to reductions in overall fuel use.

Consumers considering a new vehicle have a wide range from which to choose. When manufacturers offer a vehicle with particular fuel-saving technologies at a certain price, a potential consumer will determine whether the vehicle and its technologies are worth the price. The term *willingness to pay for fuel economy* denotes the amount the consumer is willing to pay for a vehicle with higher fuel economy that is otherwise identical to another vehicle. Because consumers have many options available to meet their transportation needs, if a manufacturer increases a vehicle's price and fuel economy at the same time, a consumer may purchase that vehicle if willingness to pay for increased fuel economy exceeds the price increase. If not, the consumer may purchase another new vehicle, may purchase a used vehicle, or may not purchase any vehicle. Because vehicles are bundles of attributes, comparison is more difficult in practice, as often consumers will not see vehicles that are entirely equivalent except for in fuel economy.

From the consumer's perspective, it is useful to classify technologies according to their visibility to drivers and passengers. Technologies that can be considered invisible, such as multivalve engines and additional speed transmissions, increase fuel economy at an extra cost to the manufacturer but do not affect the vehicle's operation in a way that is noticeable to the driver. Other technologies, such as start/stop and alternative fueling, are visible to drivers. These technologies raise fuel economy but affect the vehicle's performance or desirability in other ways. For example, stop-start ignition may bother drivers who are not accustomed to the technology turning off and on while stopped at a red light, and plugging in a vehicle is a procedurally different way to fuel.

Invisible technologies affect consumer purchase decisions by affecting the vehicle's price and fuel economy. Adding fuel-saving technology raises the cost of producing the vehicle, which may increase the price of the vehicle as the manufacturer tries to recover at least some of the costs. Additionally, invisible technologies may present trade-offs with vehicle attributes (e.g., increased fuel economy but decreased acceleration performance). For these types of technologies, a key consideration is how much consumers are willing to pay for the increase in fuel economy resulting from the technologies.

For visible technologies, perceptions of the technologies or trade-offs with other vehicle attributes also affect consumer purchasing decisions. For example, if certain powertrain technologies increase noise or vibrations when driving the vehicle, or negatively affect the vehicle's handling or "drivability," this could reduce a consumer's desire to purchase vehicles with the technologies (Helfand et al., 2016). These negative aspects (whether actual or perceived) are generally referred to as "hidden costs" of the technologies (Gillingham and Palmer, 2014). Visible technologies may also have positive attributes. For example, some consumers may be attracted to the novelty, quietness, or smooth acceleration of an electric or fuel cell vehicle. For visible technologies, the consumers' valuation (positive or negative) of the technology is a key consideration in addition to willingness to pay for the increase in fuel economy.

To set the stage for a discussion of consumer acceptance, this chapter first reviews historical trends in light-duty vehicle fuel economy and related vehicle attributes. These trends reflect responses to previous standards together with other market developments and technological change. Then, key economic concepts for understanding consumer acceptance are presented: rebound effects in response to fuel economy changes, how consumers value fuel economy and vehicle attributes, and the economic approaches for understanding this consumer valuation, from a traditional and behavioral perspective. These are important for informing how the agencies must consider a wide range of potential benefits and costs when evaluating future standards—estimations that depend critically on how consumers respond to fuel economy improvements and related changes in vehicle cost, technology, and design. This chapter then discusses the role of consumer acceptance in the transition toward new technology and closes with a consideration of electric vehicle (EV) incentives and adoption.

11.1 HISTORICAL MARKET TRENDS

The experience of the past 45 years illustrates how innovation, changing consumer preferences, and regulatory standards have influenced the light-duty vehicle market, providing a useful reference point for considering how the market may change in the future. From 1975 to 2019, the sales-weighted average per-mile fuel consumption and GHG emissions of new light-duty vehicles fell by 49%; the ratio of horsepower to vehicle weight increased 74%; and the combined market share of light trucks increased from 19% to 50% (EPA, 2020, tabulation by vehicle type). Examining these vehicle trends, and technology changes, can provide insights into how the market has achieved fuel economy increases in the past. Fuel economy and GHG standards play a role in influencing these vehicle trends, but they are also driven by changes in consumer demand, fuel prices, market conditions, and other factors. Where studies have endeavored to separate these effects and causally attribute changes in technology or vehicle attributes to the fuel economy or GHG standards, these findings are discussed.

11.1.1 Changes in Vehicle Attributes and Technologies

Trends in sales-weighted average fuel economy and GHG emissions reveal how the average efficiency of new vehicles produced in a year has changed over time. In 1975, when the first fuel economy standards were passed by Congress, the sales-weighted average estimated real-world fuel economy was 13.1 miles per gallon (MPG) across all light-duty vehicles. It increased quickly to 22 MPG by 1987. Then, although the sales-weighted average fuel economy of both passenger car and light truck segments remained flat, the sales-weighted average across the full fleet decreased as the market share of light trucks grew. By 2004, the average across all light duty vehicles had fallen to 19.3 MPG. As light truck standards began to tighten, and Congress passed higher efficiency standards in 2006, the sales-weighted average again began to rise, reaching 25.5 MPG in 2019 (EPA, 2020). It has been noted in the literature that tightened standards have resulted in vehicles with higher fuel economy and lower performance than would have been realized otherwise (Klier and Linn, 2016). The key features of the fleet related to fuel efficiency, GHG emissions, and performance measures are summarized in Table 11.1.

TABLE 11.1 Vehicle Attributes for the U.S. Light-Duty Fleet in 1975 and 2019, Including Fuel Economy, GHG Emissions, Footprint, Curb Weight, Power, Power-to-Weight Ratio, and Production Share

Vehicle Attribute (Sales Weighted, Fleet Average)	1975	2019	% Change
Estimated real-world fuel economy (MPG)	13.1	25.5	95%
Passenger cars	13.5	29.9	120%
Light trucks	11.6	22.3	92%
GHG emissions (g/mi)	681	346	–49%
Passenger cars	661	293	–6%
Light trucks	764	399	–48%
Footprint (ft^2)	N/A	50.2	N/A
Passenger cars	N/A	46.7	N/A
Light trucks	N/A	53.6	N/A
Curb weight (lb)	4060	4110	1%
Passenger cars	4057	3624	–11%
Light trucks	4073	4592	13%
Power (hp)	137	244	78%
Passenger cars	136	213	56%
Light trucks	142	276	94%
Power-to-weight ratio (hp/1,000 lb)	0.034	0.059	76%
Passenger cars	0.034	0.059	75%
Light trucks	0.035	0.060	72%
Production share			
Passenger cars	80.7%	49.8%	–38%
Light trucks	19.3%	50.2%	160%

NOTE: Fuel economy has risen, while GHG emissions have fallen, in concert with increases in weight, power, power-to-weight ratio, and production share of light trucks over passenger cars.
SOURCE: EPA (2020).

GHG emissions largely parallel the trends in fuel consumption over this time period. Sales-weighted average estimated tailpipe GHG emissions fell from 681 grams per mile (g/mi) in 1975 to 405 g/mi in 1987, then began to increase until 2004 as sales shifted toward more light trucks relative to passenger cars. In 2004, the sales-weighted average GHG emissions were 461 g/mi. Since then, they have decreased to 346 g/mi in 2019.

Improvements in fuel efficiency, and corresponding reductions in GHG emissions, have varied across different market segments. The largest improvements were in the passenger vehicle segment, which increased fuel economy by approximately 95%, from a sales-weighted average of 13.5 MPG in 1975 to 25.5 MPG in 2019. The increase in pickup trucks was the smallest among the segments, from 11.9 MPG to 19.4 MPG, or 63%. Trends in estimated real-world MPG and carbon dioxide (CO_2) for different vehicle classes are displayed in Figure 11.1.

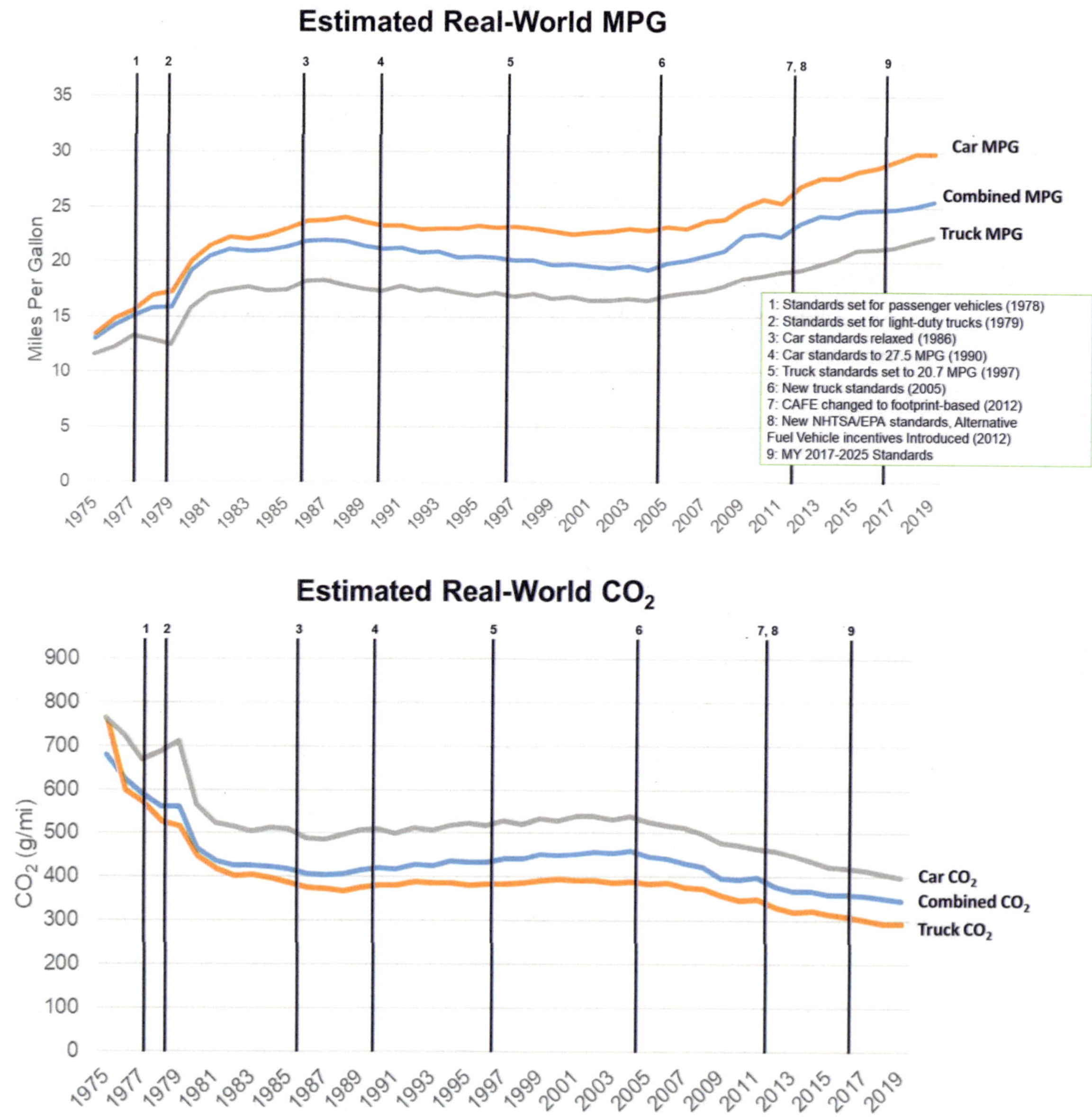

FIGURE 11.1 Changes in estimated real-world fuel economy and per-mile CO_2 emissions for new light-duty cars and trucks as they correspond with changes in regulations.
SOURCE: Adapted from Jenn et al. (2016); data from EPA (2019).

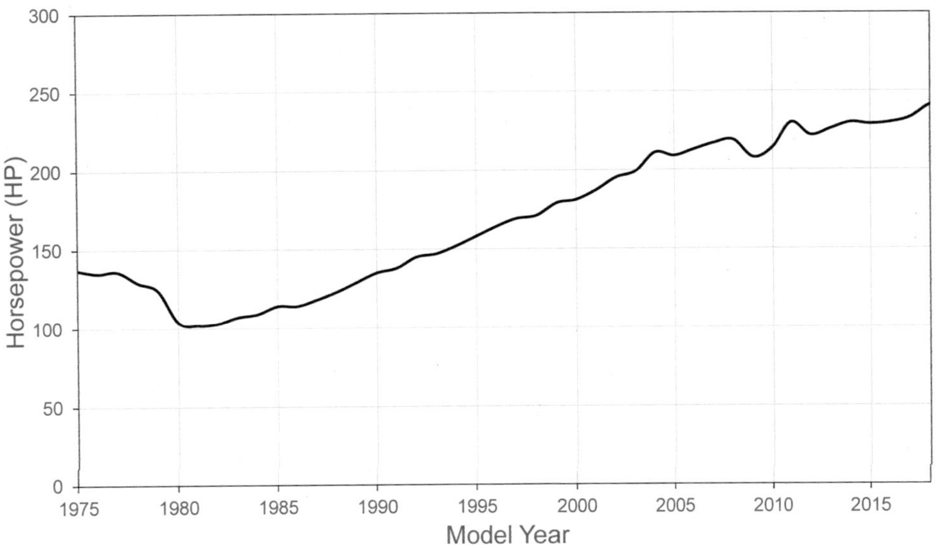

FIGURE 11.2 Change in sales-weighted average horsepower over model years 1975–2019.
SOURCE: EPA (2020).

Sales-weighted average horsepower (hp) has also increased significantly over time, from 137 hp in 1975 to 244 hp in 2019, depicted in Figure 11.2. All segments of vehicles have seen increases in horsepower over this time, with pickup trucks experiencing the greatest increase (141 hp to 337 hp, 139%), and passenger cars experiencing the smallest increase (approximately 136 hp to 213 hp, 57%).

Overall, sales-weighted average vehicle weight is approximately the same as it was in 1975, although passenger car weight has decreased, while light truck weight has increased. The average weight of light-duty vehicles decreased by 22% from 1976 to 1981, but by 2007 all the lost weight had been regained. Since then, average weight has fluctuated somewhat from year to year without a clear upward or downward trend. However, the overall trend masks significant differences at the segment level. The increase in light truck weight is largely driven by pickup trucks, which have increased in weight 26% since 1975. Over this same time, the average weight of sedans and wagons decreased by 12%.

The ratio of horsepower to weight (hp/lb, a measure of acceleration potential) increased 76% during 1975 to 2019 for the average light-duty vehicle. From 1978 to 1985, the average hp/lb ratio changed little. With the exception of small, temporary reductions, the fuel economy standards remained constant from 1985 until 2005, with gradual increases in light truck standards. Passenger car standards were constant through this period, until increasing in 2011. From 1985 to 1995, light-duty hp/lb increased at an average annual rate of 2.35%, but at a declining rate of increase from 1995 onward. For the 9 years prior to the increase in fuel economy standards in 2011, passenger car hp/lb increased at the rate of 0.82% per year. When fuel economy standards were increasing in 2010 to 2019, passenger car hp/lb increased 0.93% per year (EPA, 2020). For the 15 years prior to the increase in light truck standards (1989–2004), light truck hp/lb increased at an average rate of 2.0%/yr. Trends in average new vehicle weight by type are shown in Figure 11.3.

Trends in vehicle attributes show that for most vehicle segments, improvements in both fuel efficiency and power over the past 30 years have been accomplished, while sales-weighted average vehicle weight increased. This indicates that adoption of fuel economy technologies has improved the operating efficiency of these vehicles. Indeed, there has been significant growth in the adoption of higher efficiency powertrain technologies, including multivalve and variable valve engines, high-speed transmissions, direct injection, and others. Multivalve and variable valve engines started entering the market in the mid 1980s and have reached more than 90% share of produced vehicles (EPA, 2020). Until the early 2000s, almost all vehicles had only four or five gears, and by 2011, more than 50% of vehicles had six gears or more. In 2019, 48% of vehicles had seven or more gears and 24% had continuously variable transmissions (CVTs). Some technologies, such as CVTs, turbocharging, and cylinder

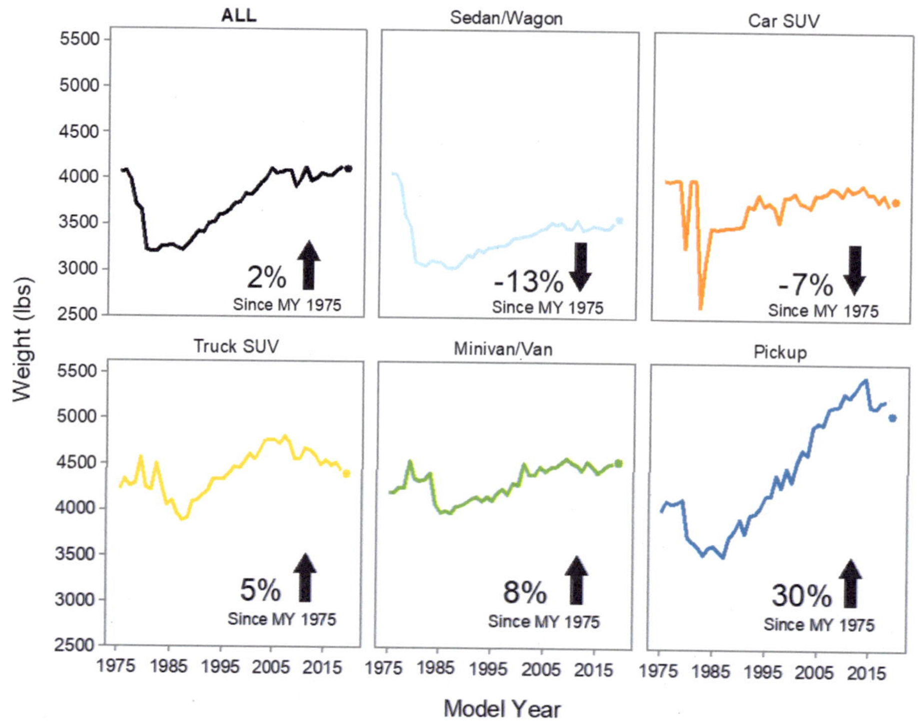

FIGURE 11.3 Change in average new vehicle weight by vehicle type since model year 1975.
SOURCE: EPA (2020).

deactivation, started entering the market in the early 2000s but at a slower rate of growth. Start/stop capabilities were introduced in 2012 and grew to approximately 42% by 2019. EVs remain a small percentage of vehicles produced, at 2.6% in 2019. After two decades in the market, gasoline hybrid electric vehicles accounted for only 5% of light-duty vehicle sales in 2019. Trends in powertrain technologies over model years are displayed in Figure 11.4.

11.1.2 Changes in Market Share

Examining the market share of vehicle segments, it is clear that the share of light trucks relative to passenger cars has been increasing. Since 1975, the market share of sedans and wagons has declined, and the share of sport utility vehicles (SUVs) has increased significantly. The largest increase comes from "truck SUVs," which are SUVs that are classified as light trucks under the corporate average fuel economy (CAFE) Congressional classifications because they have 4-wheel drive and off-road capabilities as defined by vehicle dimensions such as ground clearance and approach and departure angles. These truck SUVs have increased from less than 2% of sales in 1975 to 33% of sales in 2019. Pickup trucks, in contrast, have held a roughly constant market share, hovering between 10% and 18%. Production share by vehicle type is displayed in Figure 11.5. Owing to these shifts in market share and because trucks have lower fuel economy than cars, the sales-weighted average fuel economy of all new light-duty vehicles sold in the United States increased by only 1.4 MPG from 2014 through 2019. These changes in market share have motivated concerns regarding mass disparity in the light-duty fleet.

FINDING 11.1: Vehicle horsepower and acceleration performance have continued to improve over the past two decades, while standards were tightening. However, the tightening of the standards has caused manufacturers to offer internal combustion engine (ICE)-based vehicles that have higher fuel economy and less performance on average than otherwise would have been the case. The standards have affected vehicle attributes other than fuel economy, and the overall welfare effects of the program on consumers are still subject to study. Whether forgone performance

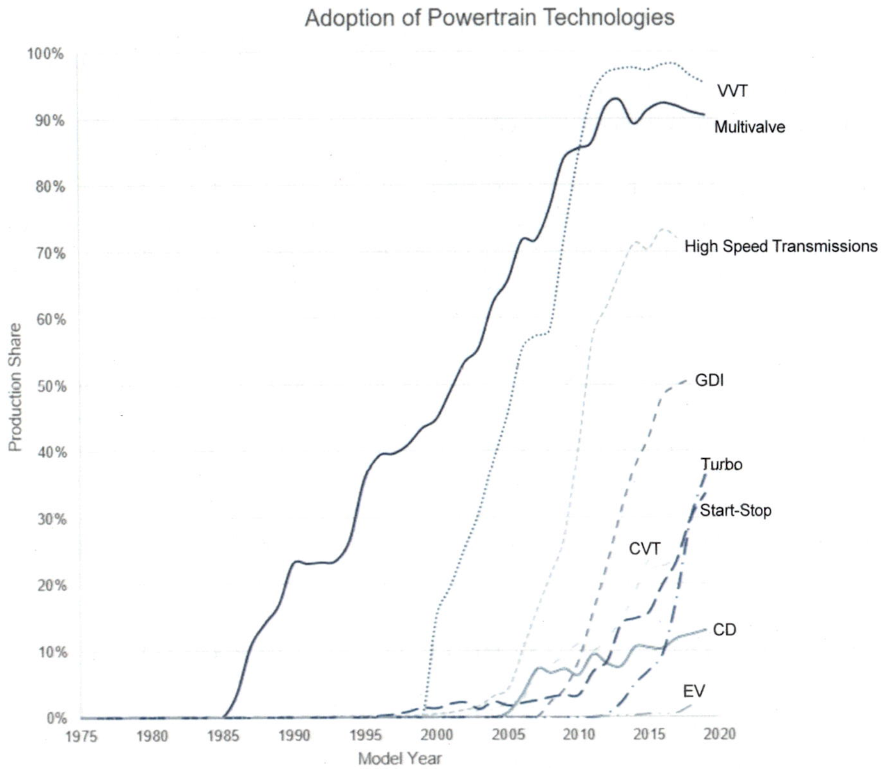

FIGURE 11.4 Production share of powertrain technologies from model years 1975–2019.
SOURCE: Committee generated using data from EPA (2020).

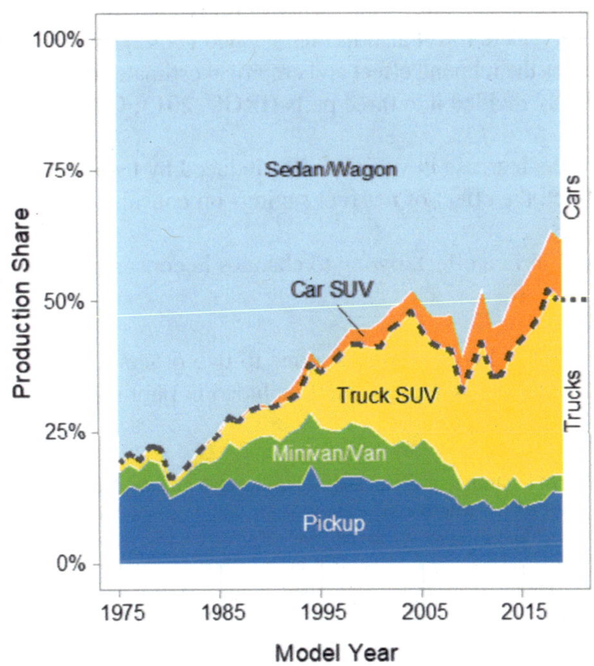

FIGURE 11.5 Production share of car and truck categories over model years 1975–2018.
SOURCE: EPA (2020).

or other attribute improvements will occur in the future in response to tightened standards depends on technology adoption; the extent to which fuel economy trades off with performance in hybrid, plug-in hybrid, battery electric, and fuel cell powertrains; the impact on production costs; and consumer valuation of fuel economy and performance attributes. Currently policy makers do not have enough information on which actions are most effective.

RECOMMENDATION 11.1: The agencies should collect further evidence on the influence of vehicle performance trade-offs on automaker compliance strategies and consumers, and reassess whether forgone performance improvements should be included in benefit-cost analysis of the standards. The agencies should assess how new technologies penetrating the market will affect the trade-offs among greenhouse gas emissions rates, performance, and other attributes.

FINDING 11.2: The sales-weighted average vehicle footprint has not changed significantly since the footprint-based standards went into effect. However, the sales-weighted average footprint and weight of pickup trucks has increased. Additionally, with growth in consumer demand for SUVs and crossovers, and dual standards for passenger cars and light trucks, there has been an incentive to shift model offerings to light trucks, per regulatory definition.

RECOMMENDATION 11.2: The National Highway Traffic Safety Administration should ensure that its standards do not incentivize increases in the mass of heavier vehicles, exacerbating mass disparity.

11.2 FUEL ECONOMY AND VEHICLE TRAVEL: REBOUND EFFECTS

Fuel economy affects how much driving consumers choose to do, which in turn affects the extent to which increased fuel economy reduces overall fuel use and emissions. Increasing the energy efficiency of energy-using equipment reduces the cost of energy required for its use. Lower energy cost induces increased use, all else equal. This phenomenon is known as the "rebound effect." It is not in dispute that there is a fuel economy rebound effect for motor vehicle travel, and there is a consensus in the literature that the rebound effect should be included in cost-benefit analyses of fuel economy and GHG standards. However, there is still consequential disagreement about its magnitude. The size of the rebound effect matters because it affects the ability of fuel economy improvements to reduce externalities like GHG emissions and because vehicle travel also has unintended consequences like traffic congestion and local air pollution. This section discusses the rebound effect and empirical estimates of it in behavioral science research.

The rebound effect is typically divided into three parts (IRGC, 2013; Gillingham et al., 2016):

1. A *direct* rebound effect, the increase in vehicle miles induced by the reduced cost of travel.[1]
2. An *indirect* rebound effect, the effect of net fuel savings on consumers' income, increasing purchases of other goods and services.[2]
3. An *economy-wide* or equilibrium effect, owing to changes in consumption patterns and prices throughout the economy.

For vehicle travel, the second and third components are one to two orders of magnitude smaller than the direct rebound effect (Greene et al., 2019) and are typically not included in published studies.[3]

---

[1]Like all price reductions, this consists of a substitution effect and an income effect on the demand for travel.

[2]Increasing fuel economy by adding fuel economy technologies to vehicles will, in general, increase vehicle prices, which would tend to diminish the rebound effect to some degree. Many estimates of the rebound effect do not take associated increases in vehicle price into account.

[3]On average, 3.3% of U.S. households' expenditures are for "gasoline, other fuels and motor oil" (BLS, 2019). If fuel economy were doubled and half of the fuel savings were offset by the cost of the fuel economy improvement, average household income would increase by 0.825%. If consumers allocated 3.3% of that to more vehicle travel, vehicle miles would increase by only 0.03%. Economy-wide effects can be approximated by estimating the effect on world oil prices. Greene et al. (2019) estimated that without the 70% increase in U.S. light-duty fuel economy from 1975 to 2018, increased petroleum demand would have raised world oil prices by about $2 to $3 per barrel, increasing the price of gasoline by 2–5%. If the elasticity of vehicle travel with respect to the price of gasoline is about –0.2, then the reduction in the price of gasoline owing to increased fuel economy would increase travel by 0.4–1%, implying an increase of 0.06–0.14% for a 10% increase in fuel economy.

Comprehensive international reviews of estimates of the rebound effect can be found in Sorrell and Dimitropoulos (2007), Gillingham et al. (2016), and Dimitropoulos et al. (2018). Dimitropoulos et al. (2018) report the results of a meta-analysis of 74 primary studies of the direct rebound effect, 41 of which were based on U.S. data, containing a total of 1,120 international estimates, based on data from various periods covering more than half a century, and including studies that estimate the elasticity of miles traveled to fuel economy, fuel prices, or fuel costs.[4] Among the findings were that the international average short-run (approximately 1 year) rebound effect was about 10%–12%, with a larger long-run effect of 26%–29%. In Table V of their paper, they present estimates that vary by the price of gasoline, population density, and gross domestic product (GDP) per capita. Consistent with theory, their meta-analysis concluded that the rebound effect increased with the price of gasoline and population density and decreased with per capita GDP, making rebound estimates from different countries not directly comparable.[5] Linearly interpolating the Table V estimates, using 2018 U.S. values for gasoline price ($0.63/liter), population density (33.75/km^2), and GDP per capita ($51,552), Dimitropoulos et al.'s (2018) international meta-analysis model predicts a long-run rebound elasticity of 0.20 (20%) for the United States for the year 2017.

Other findings by Dimitropoulos et al. (2018) include the fact that studies based on microdata (e.g., household survey data) typically produce long-run rebound estimates twice or more as large as studies based on aggregate data (p. 169). However, the magnitude of the differences depended on the number of years covered by the microdata: studies based on 15 to 25 years of microdata produced results similar to studies based on aggregate data. In addition, studies using fuel price rather than fuel economy or fuel cost per mile as the explanatory variable also produced higher rebound elasticity estimates.

The meta-analysis also indicated that the rebound effect had been declining, worldwide, at the rate of about 0.7 percentage points per year. A declining rebound effect in the United States was first observed by Small and Van Dender (2007), who concluded that rising income and falling gasoline prices caused the rebound effect to fall to 2.2% in the short run and 10.7% in the long run for the 1997–2001 period. Hymel and Small (2015) updated and extended that analysis, using state-level data for the years 1966–2009. They again found a declining rebound effect over time but a somewhat higher rebound effect in the 2002–2009 period, possibly caused by higher gasoline prices. They also found decisively asymmetric responses to rising versus falling fuel cost per mile and reasoned that the effect when fuel costs per mile decreased would be a better measure of the rebound effect of increased fuel economy. The asymmetric model produced a long-run rebound effect of 4.2% with respect to fuel efficiency for the 2000–2009 period and a long-run effect of 18.4% for the full sample period of 1966–2009 (Hymel and Small, 2015, Table 8).

Using microdata from the 2009 U.S. National Household Travel Survey, Linn (2016) demonstrated that assumptions about the correlations between fuel economy and other vehicle and household characteristics affected rebound estimates. Considering all factors, Linn (2016) concluded that the rebound effect was between 20% and 40%. Assessing 14 recent U.S. estimates of the rebound effect for vehicle use, Gillingham (2020) concluded that they indicated a central estimate of 0.1. Although recent studies and reviews have narrowed the range of plausible values for future U.S. rebound effects, whether the rebound is, say, 10% rather than 20% will have an important impact on the net social benefits (i.e., benefits net of costs) of fuel economy and GHG standards.

11.3 HOW MUCH DO CONSUMERS VALUE FUEL COST SAVINGS AND WHAT ARE THE IMPLICATIONS FOR BENEFIT-COST ANALYSIS?

A large body of research over the past 40 years has helped to better understand consumer choices and market equilibrium in response to policies such as the fuel economy regulations. This body of research has empirically measured consumer and market behavior, and developed multiple theories that seek to explain and predict

[4]Consistent with Office of Management and Budget Circular A-4, which provides guidance on benefit-cost analysis performed by U.S. regulatory agencies, estimates of rebound using U.S. data are most relevant to analyzing U.S. fuel economy and GHG standards.

[5]"However, variation of estimates is large and can mainly be explained by differences in the time horizon considered, the elasticity measure used, and the type of data and econometric approach employed in primary studies. We also find that the rebound effect is declining over time and that lower per capita incomes, higher gasoline prices and higher population density are associated with larger rebound effects." (Dimitropoulos et al., 2018, p. 163).

this behavior. Currently, there is not a consensus across the research literature on a single theoretical framework that best represents consumer choices, although there is a general consensus on some aspects of representing consumer behavior. Differing theories of consumer choice have important implications for the rationale for regulating fuel economy and for estimating the costs and benefits of fuel economy and GHG standards. Although there is a lack of consensus about how consumers value fuel economy when making car-buying decisions, there is general agreement that the actual fuel savings realized over time should be fully valued in cost-benefit analyses. The greatest area of disagreement is how standards will affect consumers' satisfaction with new vehicles and, as a consequence, the sales of new vehicles and the retirement of used vehicles. This section first describes the U.S. Environmental Protection Agency's (EPA's) and the National Highway Traffic Safety Administration's (NHTSA's) assumptions about consumer and market behavior, and whether these assumptions are consistent with evidence and theoretical frameworks in behavioral research. It then presents two alternative theoretical frameworks that describe consumer choices: the traditional economic approach and the behavioral economics approach.

11.3.1 How Much Do Consumers Value Fuel Cost Savings?

A key question for fuel economy and GHG regulation is whether the unregulated market provides the socially optimal level of fuel economy and GHG emissions. Prior to the Safer Affordable Fuel Efficient (SAFE) Vehicles Rule, EPA and NHTSA had argued that standards are justified not just by externalities—in particular, energy security and climate externalities—but also the energy paradox, the apparent tendency of markets to undersupply energy-efficient technology. The agencies argued that the unregulated market does not always adopt technologies for which the value of fuel cost savings exceeds the cost of technology adoption. In fact, past benefit-cost analysis of standards could have justified the standards chosen based entirely on the value to consumers of the fuel cost savings—without even counting the societal benefits from improved energy security and lower GHG emissions.

One possible explanation for the energy paradox is that consumers do not count the full value of the fuel cost savings when choosing a vehicle. Because of this undervaluation, manufacturers have insufficient incentive to adopt technology and reduce fuel costs, giving rise to the energy paradox.

In this context, a sizable amount of literature has attempted to estimate how much consumers are willing to pay for fuel cost savings. The 2015 National Research Council light-duty fuel economy report summarizes this literature (NRC, 2015), concluding that the literature has failed to converge on a clear answer. Because the literature has not evolved much since that report, the literature is briefly summarized here, and readers are referred to the previous report for more detail. While there are several metrics one can use to describe consumer preferences for fuel cost savings, the notion of the valuation ratio is intuitive and has been used widely. The valuation equals the amount that a consumer is willing to pay for a small reduction in fuel at the time of vehicle purchase, divided by the present discounted value of those fuel cost savings over the lifetime of the vehicle. For example, a valuation ratio of 0.5 means that a consumer is willing to pay $0.50 for $1 present value of future fuel cost savings. A valuation ratio less than 1 is referred to as undervaluation.

Some papers in this literature, such as McFadden and Train (2000), estimate discrete choice models that attempt to predict the specific vehicle a consumer chooses based on the vehicle's price, fuel economy, and other attributes, and perhaps on the consumer's demographics. Other papers, such as Busse et al. (2013), estimate the effects of a vehicle's fuel costs on vehicle prices and market shares.

Many papers found undervaluation, and many have found full or even overvaluation. Both earlier studies and more recent ones have found undervaluation. Studies using either methodology (discrete choice or otherwise) have found undervaluation.

11.3.2 Explanations of Potential Consumer Undervaluation

Next, several alternative explanations are discussed for why consumers may undervalue fuel cost savings when choosing a new vehicle. In short, the literature has not yet resolved which explanation best explains any apparent consumer undervaluation.

One set of explanations is consistent with a utility maximization framework, in which it appears to researchers that consumers undervalue fuel cost savings. For example, if fuel economy improvements caused by standards are not visible to consumers, they do not realize that they are paying more for higher fuel economy. Alternatively, consumers may misunderstand that fuel costs are inversely related to fuel economy (sometimes referred to as MPG illusion), or there may be negative attributes of fuel-saving technologies that consumers consider and that the analyst ignores (sometimes referred to as hidden costs). When assessing whether consumers undervalue fuel cost savings, analysts must make assumptions on the discount rate that consumers use to assess future fuel cost savings. Consequently, explanations that fall under behavioral economics, such as present bias, may cause consumers to use different discount rates from those assumed by the analyst. Likewise, manufacturers may misoptimize if, for example, their forecasts of consumer demand are systematically biased.

The behavioral understanding of consumers' decision making differs from the traditional economic understanding.[6] "To a psychologist, it is self-evident that people are neither fully rational nor completely selfish, and that their tastes are anything but stable. Our two disciplines seemed to be studying different species, which the behavioral economist Richard Thaler later dubbed Econs and Humans" (Kahneman, 2011, p. 269).[7]

In many behavioral models, consumers' decisions are not based on well-behaved utility functions. Instead, consumers' preferences can be strongly influenced by the context, or framing of choices. Context dependence is considered a fundamental aspect of human decision making and one that contradicts choice models that rely on simple scalability, such as consistent choice of the option that provides maximum utility (Trueblood, 2012; Trueblood et al., 2014). Moreover, human decision-making processes often systematically differ from the traditional economic model in ways that are not appropriately characterized as mistakes. The behavioral model maintains that human beings have two fundamentally different modes of thinking, an intuitive (fast, system 1) and a deliberative (slow, system 2), and their decisions may differ depending on which mode of thinking dominates:

- System 1 operates automatically and quickly, with little or no effort and no sense of voluntary control.
- System 2 allocates attention to effortful mental activities that demand it, including complex computations. (Kahneman, 2011, p. 21)

The traditional economic model is similar to the behavioral model's system 2, which is capable of overruling system 1 but typically does not. Which system is used to make a decision is affected by the context, or framing, of the decision.

Behavioral researchers have identified many important ways in which the behavior of human beings systematically differs from the model of rational economic behavior (e.g., Thaler, 2015; Kahneman, 2011; Dellavigna, 2009; Starmer, 2000). Among these are bounded rationality; satisficing[8] rather than optimizing when faced with complex choices (like the choice among makes and models of automobiles); mental accounting (the consideration of value in relative rather than absolute terms); the endowment effect; loss aversion (causing potential losses to weight approximately twice as much as potential gains when consumers are faced with choices); present bias when weighing current versus future consumption; the compromise effect (leading consumers to avoid extreme options [Kivetz et al., 2004]); inadequate information or decision-making skills; inattentiveness; salience (information framed to attract attention is given greater weight); and the MPG illusion (Larrick and Soll, 2008). These differences may be in play when consumers make decisions about fuel economy and novel technologies, with important implications for standards and other policies (Allcott and Greenstone, 2012; Allcott et al., 2014, p. 73).

One behavioral difference, loss aversion, may play an especially important role in consumers' choices concerning fuel economy technology because of the direction and potential magnitude of its effects. System 1 is loss averse. Faced with a risky choice, humans typically weigh potential losses about twice as much as potential gains

[6] In what appears to be the only peer-reviewed study investigating the actual decision making of U.S. households about fuel economy, Turrentine and Kurani (2007) found no evidence to support the economic theory of rational expectations.

[7] Daniel Kahneman was awarded the Nobel Prize in Economics in 2002 for his work in behavioral economics including Cumulative Prospect Theory and loss aversion. His book cited here, *Thinking Fast and Slow*, won the National Academies' Best Book Award for 2012. Richard Thaler won the 2017 Nobel Prize in Economics for his work in behavioral economics.

[8] Satisficing is the assessment that available options are adequate.

and overestimate the probability of loss (Thaler, 2015; Kahneman, 2011; Camerer, 2005; Kahneman and Tversky, 1979).[9] However, the degree of loss aversion varies considerably from one individual to another (e.g., Nuemann and Böckenholt, 2014; Gächter et al., 2007). Critical factors that determine the value of fuel economy to a car buyer are substantially uncertain, including actual fuel economy versus label value (Greene et al., 2017; Wali et al., 2018) and the future price of gasoline (Hamilton, 2009). Simulations applying loss aversion to consumers' fuel economy choices have shown that it can account for undervaluing of fuel economy improvements by half or more (Greene et al., 2013; Greene, 2011a).

Loss aversion is a general characteristic of human behavior, confirmed in numerous experimental as well as psychological and neurological studies (Sokol-Hessner and Rutledge, 2019) and observed by means of magnetic resonance imaging of the human brain (Tom et al., 2007). However, loss aversion has only recently been documented in the context of consumers choosing among new vehicles (Mrkva et al., 2019).

A crucial feature of loss aversion is its context dependency. Most of consumers' choices do not induce a loss averse response. However, when a choice is framed with the following attributes, loss aversion is expected (Novemsky and Kahneman, 2005):

1. A simple choice between a risky alternative and doing nothing—for example, buy versus do not buy a fuel economy technology (Ert and Erev, 2013).
2. A choice that is infrequently encountered—for example, purchase a vehicle every few years (Erev et al., 2017).
3. A choice involving substantial gains and losses—for example, in the hundreds or thousands of dollars (Kahneman and Tversky, 1979).

The choice to buy or not buy an energy-efficient or novel technology (e.g., the choice between a hybrid or standard version of the same car) is a simple risky bet framed to induce loss aversion. The choice between a large car and a small car or light truck, or the choice to buy a new car or hold on to one's used vehicle is not framed to induce loss aversion.

The direction of loss aversion's effect is clear. The upfront cost of an option to buy a fuel economy technology is known (e.g., the hybrid versus nonhybrid), but the value of future fuel savings is uncertain, chiefly owing to uncertainty about the actual fuel economy that will be achieved but also owing to uncertainty in future fuel prices, the amount of driving that will be done, discount rates, and so on. The payoff to the risky bet is the difference between the perceived uncertain value of future fuel savings and its cost. Loss aversion says that the perceived losses (cost > value) will be weighed twice the perceived gains. This unequivocally results in a bias toward undervaluing the fuel economy technology relative to its expected value. Only if the probability of losses is zero, in which case the choice is not risky, will there be no downward bias. Using typical uncertainties in the relevant variables, Greene (2011) showed that loss aversion would result in undervaluing fuel economy technologies by half or more, resulting in implicit payback periods of 3 to 4 years. If manufacturers understand that consumers will undervalue fuel economy technologies, they would be reluctant to add them to new vehicles.

According to the National Research Council (NRC, 2015) and the SAFE rulemaking (NHTSA/EPA, 2020), manufacturers believe that consumers significantly undervalue fuel economy and are typically willing to pay for only around 2.5 years of expected future fuel savings, consistent with the predictions of loss aversion (Greene, 2019).[10] Greene et al. (2013) present evidence from four national random sample surveys from 2008 to 2013 that consistently supports the manufacturers' view of consumers' willingness to pay for technology-based fuel economy improvements. Loss aversion implies that manufacturers would be reluctant to offer their customers the option to buy or not buy technology that improves fuel economy unless the fuel savings quickly repaid the upfront cost. However, although loss aversion is consistent with an average payback period of 2.5 years, research has not yet demonstrated that loss aversion of new vehicle buyers explains such a payback period.

[9]Humans overweight losses relative to gains in riskless choices, which is termed the "endowment effect." Loss aversion has been found not to apply to the exchange of money for goods in normal, riskless market transactions (Novemsky and Kahneman, 2005).

[10]Apparent undervaluing of energy efficient technology has been observed across a wide variety of energy using durable goods and has been termed the "energy efficiency gap" or "energy efficiency paradox" (Gerarden et al., 2015; Gillingham and Palmer, 2014, p. 1486).

Consumers' fuel economy choices in the context of fuel economy standards are very different. Because all makes and models gradually become more fuel efficient, the decision to buy a more efficient vehicle is no longer a simple risky bet, a choice to buy or not buy added fuel economy technology. Because all vehicles with similar attributes are more fuel efficient, MPG differences among similar vehicles tend to be small and may not be salient to consumers' vehicle choices (e.g., Leard, 2018; Sallee, 2014). With fuel economy improving gradually over a period of several years, consumers have the opportunity to learn whether the fuel savings promised by window stickers and advertisements are being realized, reducing uncertainty. Under these circumstances, loss aversion is not expected for gradual and across-the-board fuel economy improvements. Consumers finding that they get more miles per gallon are likely to fully value the fuel savings. If the value of those savings exceeds their upfront cost, consumers are likely to prefer the newer, more efficient vehicles to comparable but less efficient used models.

The behavioral model provides a possible explanation for several facts that otherwise might appear anomalous. From the perspective of the behavioral model, the fact that dozens of studies conducted over the past 30 years are almost evenly divided with respect to whether consumers substantially undervalue fuel economy or value at approximately its expected, discounted present value (Greene, 2010; Helfand and Wolverton, 2011; Greene et al., 2018) is a reflection of inferences based on models that typically do not allow consumers' preferences to vary according to the framing of choices. For example, fuel prices have varied dramatically over this period, which could affect preferences. On the other hand, estimating consumers' willingness to pay for fuel economy is a particularly difficult problem for statistical inference and the variation in estimates may simply reflect that. From the perspective of the behavioral, it is also reasonable that previous assessments of the technological potential to increase light-duty vehicle fuel economy by National Research Council committees and the regulatory agencies have found substantial potential to increase fuel economy at costs considerably less than the discounted present value of expected future fuel savings (e.g., NRC, 1992, 2011, and 2015; TRB and NRC, 2002). On the other hand, the studies may have underestimated costs by not quantifying the foregone value of other attributes, particularly acceleration. From the perspective of the behavior model, the fact that fuel economy standards have consistently enjoyed overwhelming popular support (NRC, 2015; CRSG, 2018, 2017) is also not an anomaly but an expected result. If loss aversion does not apply to fuel economy improvements required by standards, as discussed above, there could be other reasons why consumers undervalue technology-based fuel economy improvements.

11.3.3 Implications for New Vehicle Sales and Used Vehicle Retirements

If fuel economy and GHG standards require manufacturers to produce and sell vehicles that consumers consider less desirable than the vehicles that would have been produced in the absence of standards, new vehicles sales would be lower than they would have been without the standards. If new vehicles are less desirable, demand would shift in favor of used vehicles, increasing their prices above what would have been the case without standards. As a result of the higher used vehicle prices, consumers will hold onto their used vehicles longer, slowing the rate of turnover of the vehicle stock. This phenomenon, known as the "Gruenspecht effect," was first identified as a possible outcome of vehicle emissions regulations (Gruenspecht, 1982). The traditional economic model implies that the Gruenspecht effect is a highly likely, if not certain, unintended consequence of fuel economy and GHG regulations. With respect to cost-benefit analysis, the Gruenspecht effect has two negative consequences. First, there will be a loss of consumers' surplus because new vehicles are more expensive but not as desirable as they would have been without the standards, and used vehicles are more expensive. Manufacturers will suffer a loss of profits owing to lower new vehicle sales. Second, slowing the rate of turnover of the used vehicle stock will slow the penetration of new safety technologies into the on-road vehicle fleet, resulting in higher levels of traffic fatalities and injuries than would have been the case without the regulations.

On the other hand, the Gruenspecht effect is not predicted by the behavioral model, under which it is not only possible but likely that if the fuel savings from increased fuel economy exceed its cost, consumers will find the more fuel-efficient vehicles required by regulation to be preferable to those that would otherwise have been produced. Under the behavioral model, consumers undervalue fuel economy technology when it is offered as a simple risky choice to purchase or not to purchase. The framing of the choice induces a predominantly system 1 response rather than a more rational system 2 response. This discourages manufacturers from using technologies that have payback periods longer than about 2 or 3 years. By changing the framing of fuel economy technology choices (in general, all new vehicles are more fuel efficient and only a little more fuel efficient than the previous model year), standards can deliver vehicles whose fuel savings substantially exceed their increased prices and that will be accepted by consumers. Absent loss aversion, consumers are likely to appreciate the value of future fuel savings and therefore be pleased with the new, more fuel-efficient vehicles. It is possible that sales would increase rather than decrease and likewise manufacturers' profits. In that case, increased new vehicle sales would reduce used vehicle prices, benefiting buyers of used vehicles and accelerating the turnover of the vehicle stock.

11.3.4 Implications of Potential Undervaluation for Benefit-Cost Analysis

Earlier, it was noted that the literature has not settled on a single explanation for potential consumer undervaluation of fuel cost savings. Here, it is pointed out that the interpretation of undervaluation has implications for benefit-cost analysis of fuel economy and GHG standards. Traditional economics provides a framework for evaluating the welfare effects of fuel economy and GHG standards. These welfare effects *include consumers and manufacturers in the vehicle market and other markets affected by the regulation* (e.g., used vehicles), and social welfare effects that are external to those markets, such as GHG emissions reductions. The welfare effects are estimated by comparing scenarios that differ only in the level of the standards. For example, one might compare the fuel economy and GHG standards for model years 2022–2025 that the Obama administration adopted with the standards the Trump administration proposed for the same model years. Implementing this framework requires assumptions about how manufacturers and consumers make decisions. Recent papers usually assume that manufacturers comply with standards using multiple options. All include sales mix, many include technology adoption, and some include trade-offs among fuel economy, horsepower, torque, and footprint.

The vehicle market is concentrated, with the top-5 firms accounting for 66% of the market and the top-10 firms accounting for 90% of the market in 2018 (EPA, 2020, Table 5.11). It is readily apparent that manufacturers pursue different pricing and technology strategies and that their strategies are commonly known.

There are relatively few automakers, in part owing to the large capital investments necessary to enter the automotive market. These few automakers sell differentiated products, making this an impure oligopolistic market. This market structure presents imperfect competition, meaning that manufacturers account for the effects of their own decisions on the decisions of other manufacturers. For example, suppose that a manufacturer ignores competitive responses and decides to raise the horsepower of its vehicles. The manufacturer would expect to attract consumers from other manufacturers. The situation is different if the manufacturer anticipates competitive responses by other manufacturers. That is, if the first manufacturer raises horsepower of its vehicles, the manufacturer expects its competitors to respond by raising horsepower of their vehicles. Consequently, all vehicles would have higher horsepower and the first manufacturer would not have a competitive advantage over the other manufacturers. Therefore, the manufacturer is less likely to raise horsepower if it anticipates competitive responses. The assumption of imperfect competition accounts for these feedbacks and competitive interactions among manufacturers.

On the consumer side, economic models include the assumption that each consumer chooses the vehicle that maximizes subjective utility. This is usually formulated as a discrete choice problem, where a household chooses whether to purchase a vehicle, and if so, which vehicle. Modelers may embed this choice in decisions about selling or scrapping other vehicles and the amount those vehicles are driven. Models usually allow for the possibility that preferences for vehicle attributes vary across individuals. For example, some may have higher willingness to pay for fuel economy than others. As a result, the market will be segmented, with high willingness to pay consumers typically purchasing vehicles with high fuel economy. Accounting for consumer preference variation is important for characterizing how manufacturers' price and attribute choices affect sales. For example, one would expect that if a manufacturer raises the price of a luxury vehicle, most consumers will substitute to other luxury vehicles.

Other markets may be affected, particularly the used vehicle market. If consumers have low willingness to pay for fuel economy, tighter standards could increase demand for used vehicles because they are not subject to the standards. Higher demand would raise prices of used vehicles. In turn, the decision to scrap a vehicle depends on its price in the used vehicle market. For example, if a repair is needed, a consumer may decide to make the repair and sell the vehicle rather than scrap it if the price of the vehicle exceeds the repair cost. Under this interpretation, the choices about vehicle purchase, scrappage, and miles traveled should be modeled jointly so that the model is based on a coherent set of assumptions that is consistent with the literature.

It is important to recognize that even if consumers choose a vehicle to maximize, consumers may make systematic mistakes when choosing a vehicle. For example, the possibility that fuel economy improvements caused by the standards are not visible to consumers was noted earlier. In that case, the consumer's willingness to pay would be less than the present discounted value of the fuel cost savings. The analysis would use the consumer's preferences to characterize consumer choices in each scenario, and then would conduct the welfare analysis accounting for these mistakes. To take an extreme example, suppose consumers have zero demand for fuel economy, meaning that they completely ignore fuel economy when purchasing a vehicle. In that case, standards would reduce manufacturer profits, because if manufacturers attempted to raise prices, consumers would substitute to used vehicles. Welfare analysis would account for the reduction in new vehicle sales and effects on manufacturer profits. It would also include actual fuel cost savings to consumers who purchase new vehicles.

Welfare effects are sensitive to the assumptions made. Accounting for various compliance options generally reduces costs to manufacturers. Accounting for consumer preferences and variation in those preferences across consumers has implications for total costs and for distribution costs across manufacturers and consumers.

In short, the utility maximization approach assumes that consumers maximize utility and that manufacturers are attempting to maximize profits. It is important to recognize that this approach can allow for the possibility that either manufacturers or consumers make mistakes (as distinct from a situation in which the researcher fails to correctly specify the model consumers use to make decisions). If consumers or manufacturers are misoptimizing, they make decisions according to their (incorrect) beliefs, and benefits and costs are evaluated using the realized outcomes. For example, if consumers undervalue fuel cost savings, the undervaluation would be accounted for when they are choosing which vehicle to buy, but their mistake would be corrected for when evaluating consumer benefits of lower fuel costs.

If the agencies adopt the utility maximization approach, they would model consumers as maximizing their own subjective well-being, while subject to a budget constraint and given the prices and attributes of vehicles in the new and used markets. If the benefit-cost analysis of the standards includes the effects of the standards on miles driven, accidents, scrappage, and other outcomes, it would be best that the agencies model consumer choices about vehicle purchase, scrappage, and miles traveled in an internally consistent manner. As of the SAFE Rule, although the agencies analyze the effects of standards on scrappage and miles traveled, they do not use an internally consistent framework. For example, they assume that fuel economy standards affect total new vehicle purchases but not the purchases of individual vehicle types, which is inconsistent with utility maximization. These inconsistencies yield some perverse results and likely cause the agencies to misestimate the effectiveness of the standards at reducing fuel consumption and GHG emissions, as well as the aggregate welfare effects, which include changes in manufacturer profits, consumer well-being, and other factors (Bento et al., 2018).

As noted above, some of the behavioral economics explanations for undervaluation can be incorporated in a standard utility maximization framework. If research reveals that other behavioral explanations are at work, it may be necessary to reformulate the welfare framework.

11.4 TRANSITIONS TO NEW TECHNOLOGY

As new technologies such as EVs and autonomous vehicles (AVs) become more relevant for improving the fuel use and emissions of vehicles, it will become more important to incorporate consumer and market barriers to new technologies into frameworks that aim to understand and predict vehicle market and driving behavior. This section describes relevant market failures and consumer barriers to transitioning to new technologies.

The transition to new vehicle technology platforms like EVs and AVs will affect and disrupt every segment of the motor vehicle ecosystem, including automakers, suppliers, dealers, fleet owners and operators, repair and maintenance facilities, ride share services, insurers, and consumers. Section 11.4.2 addresses some of the impediments to adoption of new vehicle technologies such as EVs and AVs by existing market participant companies. Section 11.5.2 discusses impediments and challenges to consumer adoption and acceptance of these novel vehicle technologies.

FINDING 11.3: A lack of consumer understanding and familiarity as well as risk aversion to novel technology is an important barrier to novel technologies. Innovative approaches including education, consumer incentives, and dealership training may be required to overcome these barriers. Additional studies and information are needed to inform the most effective implementation of these actions.

RECOMMENDATION 11.3: More data and research on the relative effectiveness on vehicle adoption of different actions (e.g., education, consumer incentives, regulations, building supporting infrastructure) is needed to inform policy decisions, as well as whether effectiveness depends on who is undertaking the actions.

11.4.1 Market Failures and Company Responses to Disruptive Change

All companies in the motor vehicle industry face challenges in the shift to new vehicle technologies such as battery electric vehicles (BEVs), fuel cell vehicles, and connected and automated vehicles. These new technologies will require fundamental changes in the way vehicles are designed, manufactured, serviced, fueled, and operated. These changes will create opportunities for some companies but will represent a major change in the way existing companies across the vehicle ecosystem currently do business. Suppliers will need to produce and market new vehicle technologies. Manufacturers and their engineers will need to design, test, and build vehicles with very different core technologies. Dealers will need to sell vehicles that sales staff may be unfamiliar with, and that may provide a different driving experience for consumers and much lower maintenance costs (a major revenue stream for dealers). Energy suppliers will need to shift their supply and infrastructure to new fuel modalities. Fleet owners will need to adjust to new vehicle types with novel operation and fueling requirements. Fuel infrastructure will need to shift from gasoline refueling stations to electricity and hydrogen refueling stations. Not unexpectedly, companies in all these sectors are likely to exhibit some conscious or subconscious resistance to these technology changes.

The history of technology change teaches us that resistance to new technologies is a standard response of the incumbents (Juma, 2016). Joseph Schumpeter described how "creative destruction" is the engine of economic development and our market economy, in which "new consumers, goods, the new methods of production or transportation [and] the new markets" improve our economy and society at the expense of, and over the objections by, existing industries and businesses that are displaced by the new technologies (Schumpeter, 1942, p. 83). This resistance to new technologies is further extended by Clayton Christensen's concept of "disruptive innovation," in which entrenched market leaders stay focused on serving their customers with existing technologies while underestimating the disruptive impacts of new technologies that are initially less in demand but that may eventually displace their existing products, and if they are not careful, existing companies (Christensen, 1997).

Another relevant factor is that technological development is characterized by long periods of gradual, incremental change, punctuated by rare periods of rapid technology change that can disrupt existing markets and industries (Tushman and Anderson, 1986). The shift from ICEs to battery electric and fuel cell vehicles may represent one of these rare technological discontinuities. In such periods of rapid technology change, "[s]kills that brought product-class leaders to preeminence are rendered largely obsolete; new firms founded to exploit the new technology will gain market share at the expense of organizations that, bound by traditions, sunk costs, and internal political constraints, remain committed to outmoded technology" (Tushman and Anderson, 1986, p. 444).

In addition to these well-established general dynamics of technology change, several other aspects tend to "lock in" the incumbent petroleum/ICE paradigm and impede the transition to battery electric, fuel cell, and connected and automated vehicles. Like many other cleaner energy technologies, the private companies who need to invest in the development and deployment of the new cleaner technologies do not capture all the societal benefits of the cleaner energy, thus creating a classic externality problem that will result in companies under-investing from a societal good perspective in these cleaner technologies (Jaffe et al., 2005). A special case of this problem is that a company that invests in such technologies will not be able to capture all the learning-by-doing benefits of its investments, but rather some of the resulting knowledge and experience spillover to competing companies. This will create private incentives for firms to delay and free ride on the investments of others (Nemet, 2012). Additionally, the manufacturers of electric and fuel cell vehicles, the primary movers in the transition away from ICEs, are dependent on complementary technologies such as battery or fuel cell manufacturers and infrastructure developers, which are mostly outside of their direct control. This further enhances the risk and deters investment in cleaner technology vehicles (Marchant, 2014). The transition to new vehicle powertrains that use new types of energy will require enormous capital investments throughout the vehicle life cycle, which may delay or deter vehicle manufacturers from going "all in," especially in the post-COVID-19 era. An example of the historic lack of automaker commitment is limited advertising of EVs: in 2019, only 0.3% of automaker advertising dollars were spent on EVs (Plumer and Popovich, 2020), as well as anecdotal lack of availability of BEV models at dealerships, especially outside of the states that have a zero-emission vehicle (ZEV) mandate. Last, consumer unfamiliarity and hesitance about alternative fueled vehicles, discussed elsewhere in this chapter, presents a further barrier for vehicle manufacturers to invest in these technologies.

Notwithstanding all these barriers and disincentives, vehicle manufacturers are not opposed to battery electric, fuel cell, and connected and automated vehicles in principle. All major vehicle manufacturers now recognize the need to address climate change and reduce fossil fuel consumption, and have publicly stated that they want to be part of the solution rather than the problem (Toyota Motor Company, 2020; Volkswagen Group, 2020; Ford Motor Company, 2020; Honda, 2020; General Motors, 2016; Nissan Motor Coporation, n.d.; among others). Many manufacturers have adopted a long-term goal of zero-emission fleets. Vehicle manufacturers have invested billions into developing hybrid, battery electric, fuel cell, and connected and automated vehicles. Moreover, to the extent existing vehicle manufacturers do not develop these vehicles fast enough, start-ups such as Tesla and Waymo will quickly occupy that space, and they have already become leaders in electric and automated vehicles, respectively. Yet, despite these important industry efforts and support for transitioning away from ICEs, vehicle manufacturers are unlikely to move as fast and as far as society requires on their own volition, given the factors identified above. Moreover, other private participants in the vehicle ecosystem, such as energy suppliers and dealers, may be even more resistant to technology change (NRC, 2015).

The net effect of these factors is that from a societal perspective rational vehicle manufacturers and other private actors in the vehicle ecosystem will under-invest and delay in the transition to alternative fueled vehicles, even if they understand and in principle support that transition. Two general types of policy interventions can help to address these barriers, loosely grouped into "carrots" and "sticks." Carrots can consist of government research funding and initiatives, tax credits or other subsidies for the sale or purchase of ZEVs, and infrastructure investments such as refueling stations (Nemet, 2012). Sticks usually come in the form of technology-forcing regulations. Of course, sticks and carrots are not exclusive, and the best policy strategy would be an integrated policy mix of regulatory requirements and incentives (Kivimaaa and Kerna, 2016; NRC, 2013).

11.4.2 Consumer Acceptance Barriers and Challenges

In addition to the complex consumer response to energy-saving vehicle technologies discussed in the previous sections of this chapter, another important element of consumer response is how the public responds to new technologies, going beyond vehicle fuel economy attributes. Consumers have important and often complex responses to new technologies that disrupt or change important aspects of their daily routines and may present new physical, economic, or other risks. EVs and AVs are likely to raise such reactions given their key differences from traditional vehicles that consumers are familiar with. Section 11.4.2.1 summarizes what is known about how consumers respond to novel technologies generally. Section 11.4.2.2 discuses consumer responses to EVs. Section 11.4.2.3 describes how consumers are likely to respond to AVs.

11.4.2.1 Consumer Responses to Novel Technologies

Consumers have diverse and complex reactions to novel technologies that disrupt or change their daily lives. Some consumers are attracted to novel products, and these individuals tend to be "early adopters" of new technologies (Rogers, 2003; Hirunyawipada and Paswan, 2006). The experience of these early adopters can provide important positive or negative feedback to the product manufacturers and their fellow, more reticent but mainstream consumers (Rogers, 2003). Negative experiences reported by early adopters can depress the willingness of other consumers to try the new technology. For example, initial purchasers of diesel cars in the United States experienced high diesel costs and new maintenance problems, and word of these problems spread throughout the rest of the population and deterred other consumers from purchasing diesel vehicles (Ram and Sheth, 1989).

While some consumers seek out and desire new innovations, many others affirmatively resist new technologies that disrupt their daily routine or present new risks. This phenomenon of "innovation resistance" has been defined as "the resistance offered by consumers to an innovation, either because it poses potential changes from a satisfactory status quo or because it conflicts with their belief structure." (Ram and Sheth, 1989, p. 6). This innovation resistance is in many cases not simply the lack of willingness to adopt but rather is driven by one or more practical or psychological barriers that actively dissuade a consumer from adopting a new technology (Kleijnen et al., 2009). A highly discontinuous innovation based on a new technology, such as the first computer, creates major changes in a consumer's way of life and is most likely to trigger innovation resistance (Ram and Sheth, 1989). "The higher the discontinuity of an innovation, the higher the resistance is likely to be" (Ram and Sheth, 1989, p. 6).

Consumer innovation resistance has a number of potential contributing factors, including functional barriers and risk barriers (Ram and Sheth, 1989). The most common functional barrier is when a new product alters product usage patterns by changing existing routines, practices, or habits of a consumer (Ram and Sheth, 1989). "[C]onsumers have an intrinsic desire for psychological equilibrium" and "[a]ny change imposed on their behavior has the potential to disturb this equilibrium"; "the consumer thus more often opts for resisting the change than going through a disturbing process of readjustment" (Ram, 1987, p. 208). An example is consumer resistance to carpooling, which required significant changes in consumer schedules, convenience, and comfort levels (Ram and Sheth, 1989).

A second type of functional barrier is when the innovation does not provide a superior performance-to-price value, so consumers see no benefit in transitioning to the new technology (Ram and Sheth, 1989). An example of such a barrier was the introduction of video discs, which consumers did not perceive as better value than existing video cassette recordings (VCRs) because, unlike the VCR, the videodiscs could not be recorded or reused.

Consumers may also resist new technologies out of concern for the physical, economic, performance, or social risks of an innovative technology (Conchar et al., 2004). A novel technology may present uncertain or unique physical risks, which makes consumers reluctant to try the new product. Resistance may be based on perceived or uncertain risks of the new product. For example, many consumers initially resisted microwave ovens because they feared the (non-ionizing) radiation might physically harm them (Ram and Sheth, 1989).

Consumers may also hold off buying a new product, owing to economic risks—in particular, the concern that a novel technology is still improving—and thus a consumer may be better off waiting until the technology gets better and cheaper (Ziamou and Veryzer, 2005). Consumers are concerned about being stranded with obsolete or inferior products, such as when early purchasers of Sony Betamax felt aggrieved when their video players and videos were displaced by the superior Video Home System (VHS) technology (Kleijnen et al., 2009). "Products based on new technologies are especially susceptible to this risk" (Ram and Sheth, 1989, p. 8).

A third and related type of risk barrier is concerns about the performance of a new technology, based on fears that the effectiveness or safety of the new product may not have been fully validated. Consumers will postpone their purchase of the new technology until its safety and effectiveness have been proven by widespread consumer use (Kleijnen et al., 2009). Such concerns about performance apply to any new vehicle model (Ram and Sheth, 1989).

A variety of strategies are available to mitigate or overcome the functional and risk barriers that contribute to innovation resistance. One strategy is to design the product so that it better fits within consumers' existing lifestyle and infrastructure. For example, when Chrysler first started marketing minivans, the commercial vans that previously existed would not fit into many consumers' garages, so the vehicles were redesigned to fit into garages and this broke down the usage barrier (Ram and Sheth, 1989). Another strategy for breaking down usage barriers is mandating the technology through legislation, which has been effective for a variety of technologies such as unleaded gasoline, seat belts, and smoke detectors (Ram and Sheth, 1989).

Probably the most effective strategy for overcoming adoption barriers is to provide superior performance value over existing products (Ram and Sheth, 1989). An example is the introduction of electronic calculators, which could perform functions that their predecessor electromechanical calculators could not, and hence consumers quickly were willing to buy and even pay more for this superior performance (Ram and Sheth, 1989). Another effective strategy is to offer the product on a trial basis, so consumers can experience the new technology and get comfortable with the performance and perceived risks of the new technology (Rogers, 2003). An example is when the herbicide 2,4-D was first introduced, the manufacturers offered farmers a free trial for use on a 5- or 10-acre field, which successfully assured the farmers of the utility and safety of the new product (Ram and Sheth, 1989). Consumers are also likely to overcome their resistance to a technology if their costs are sufficiently inexpensive (Kleijnen et al., 2009).

Last, public education is an important strategy for overcoming consumer barriers to new technologies (Kleijnen et al., 2009). Much consumer resistance to new technologies is based on "suspicion" of unproven technologies, even when that suspicion is not supported by evidence (Kleijnen et al., 2009). While education of the public can be a slow and laborious process, it can significantly improve public acceptance over time, especially if supported by government or credible opinion-leaders (Ram and Sheth, 1989). Companies have used advertising showing products being used by unexpected types of people to help expand acceptance of a product—for example, Honda ran ads of respectable people such as priests and an elderly lady in tennis shoes riding their motorcycles with the caption "The nicest people ride on a Honda" (Ram and Sheth, 1989).

11.4.2.2 Consumer Resistance to BEVs

BEVs present two of the major factors that lead to innovation resistance. First, "innovations which *conflict with the usage patterns* of competing and well-established . . . or that contradict well-established workflows, practices, or habits, will face resistance" (Kleijnen et al., 2009, p. 346). "[R]esistance would seem to be a normal response of consumers when confronted with innovations" (Ram, 1987, p. 208). BEVs represent a significant change in how consumers fuel and use their vehicles. One particular issue is the longer fueling time (recharging compared with filling a gas tank), especially where fast-charging infrastructure is not available. Marketing or educational communications that show how EVs fit easily into consumers' daily routines are likely to be effective (Kleijnen et al., 2009). For example, Toyota successfully marketed its Prius hybrid by focusing much of its "marketing communications on showing how the Prius is still able to deliver a driving experience consistent with current usage habits (e.g., ability to cover long distances without refueling)" (Kleijnen et al., 2009, p. 354). Lessons from this effort could be drawn upon to inform BEV marketing efforts.

Informational and word-of-mouth campaigns may also be effective in educating consumers about the ways that BEVs may improve the consumer's quality of life, such as better acceleration, quieter ride, lower operational costs, reduced maintenance, and the convenience of being able to refuel at home or work without having to make regular trips to gas stations. Word-of-mouth and media propagation of good experiences with a new technology play an important role in reducing consumer resistance (Ram, 1987). Additionally, evidence suggests that consumers may view BEVs more positively after having experiences with the vehicle type (Liu et al., 2020; Schmalfuß et al., 2017). Perhaps the nation that has first achieved mainstream adoption of BEVs is Norway, where BEVs represented more than 46% of new vehicle sales in 2019 (EAFO, 2020). Analysis of public experiences in that country indicate that the two biggest factors for the mainstream uptake of BEVs was government subsidies and tax breaks that made BEVs cost-competitive with ICEs,[11] and improvements in performance, addressing consumers'

[11]The Norwegian government provides purchase incentives for BEVs by exempting purchases of such vehicles from the value added tax (VAT) and registration fees, which are usually quite substantial in Norway. In addition, local incentives include exemptions from road tolls and parking charges (Figenbaum and Nordbakke, 2019).

concerns over range (Figenbaum and Nordbakke, 2019). Greater awareness of these benefits should help overcome the resistance of many consumers owing to the changes that BEVs will create to their daily routines. Education on the differences among BEVs, plug-in hybrid electric vehicles (PHEVs), and other powertrain types may also present consumers with the opportunity to correct misperceptions and evaluate the BEV technology with a more-informed understanding of the technology. In addition to educational and marketing programs, making the recharging capacity more easily available and user friendly will also help to overcome resistance based on unfamiliarity and disruption of daily routines.

Second, BEVs create resistance based on risk—in this case, primarily fear of inadequate range of BEVs and the risk of being stranded by dead batteries. This "range anxiety" has been a long-standing concern about BEVs but should be diminished by consumers' gradual understanding of how much range they need, EV models' ability to meet those needs, and the increased range on the order of 300 miles on a single charge. In addition, educational and firsthand experiences showing how the ranges of modern BEVs easily fit into most consumers' daily routines and vehicle use patterns can also help relieve some consumer anxiety. As the range of BEVs has increased, the range anxiety that some consumers continue to express is becoming more psychological than real (Franke and Krems, 2013). Moreover, resistance to technologies is often exacerbated when consumers base their perceptions on "stereotypes, rumor, or other indirect, non-experiential, sources" rather their own firsthand experience (Kleijnen et al., 2009, p. 346).

Another consumer barrier is not whether to buy a BEV but when to buy such a product (Ziamou and Veryzer, 2005). As discussed above, products based on new and emerging technologies are particularly subject to this consumer hesitancy on when to purchase (Ram and Sheth, 1989). A rapid pace of technological change causes consumers of the technology to weigh in favor of delaying purchase decisions as the technology rapidly improves (Balcer and Lippman, 1984), so the perception that BEV technology is rapidly improving may cause consumers to wait for better versions of the technology to appear. Two strategies have been identified to address this delayed purchase barrier. The first is to provide the product on a trial basis (Rogers, 2003; Ram, 1987), which in the case of BEVs may be achieved through vehicle leasing. Price reductions through mechanisms such as tax credits may also help overcome consumers' reluctance to purchase a product at the present time (Kleijnen et al., 2009).

While early adopters represent the initial market for BEVs in the United States today, the success of these vehicles will require such vehicles to become more mainstream.

11.4.2.3 Consumer Responses to Autonomous Vehicles

Connected and automated vehicles (CAVs) also involve a significant technology change that consumers are not familiar with and may resist. The bulk of this discussion applies to fully autonomous vehicles, although consumer barriers to connected and partially autonomous vehicles include issues of trust, particularly pertaining to automatic lane keeping or adaptive cruise control.

The major source of resistance to CAVs is likely to be concern about the safety and physical risks of such technologies. Although CAVs are likely to be deployed only if they are significantly safer than human-driven vehicles, public opinion polls show that many, even the majority of, consumers have concerns about the safety of CAVs that may deter them from adopting such vehicles (AAA, 2019; Hewitt et al., 2019). Consumers often overestimate the risk and danger from unfamiliar, exotic technologies such as artificial intelligence-driven AVs (Kaplan, 2018; Slovic and Peters, 2006). In addition, disproportionate and sensational media coverage of a new technology, including any accident involving a CAV, can induce negative consumer perceptions of new technologies (Kleijnen et al., 2009).

Fears of physical risks can be addressed by providing information that is responsive to consumer concerns—but rather than pushing the information to the consumers, it is better for the information to be made available so that consumers can easily find useful information when they seek it (Kleijnen et al., 2009). CAVs may provide a benefit for fuel economy and GHG emissions, although the extent to which consumers may over- or undervalue those fuel savings is unknown. Public education programs and resources on CAV safety may be needed to overcome consumer fears and resistance to CAVs. One additional factor influencing adoption will be the cost of a CAV, with details regarding price and ownership models still to be seen.

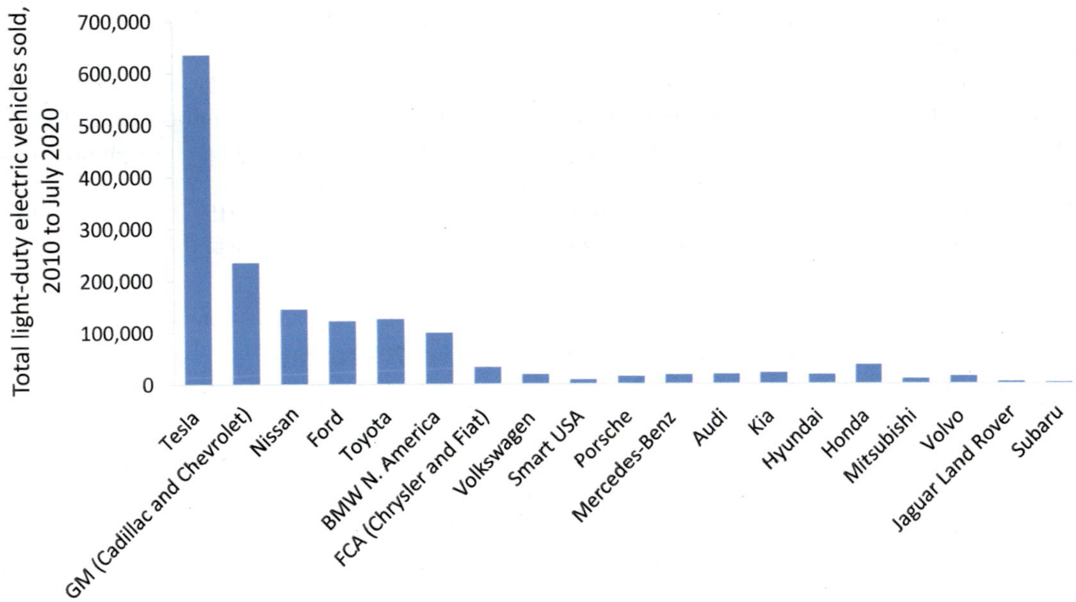

FIGURE 11.6 Cumulative light-duty electric drive vehicle sales from 2010 to July 2020.
SOURCE: Committee generated, using data from Argonne National Laboratory (2020).

11.5 ROLE OF EV INCENTIVES, IMPACT OF INCENTIVE EXPIRATION, AND WHETHER TO CONTINUE EV INCENTIVES

11.5.1 Rationale of EV Incentives

New technology subsidies like EV incentives[12] complement environmental regulations such as CAFE and ZEV and play two major roles—assisting in compliance with these regulations and stimulating technology innovation and transition (Jaffe et al., 2005). EV incentives have contributed to early plug-in electric vehicle (PEV) market development. A National Academies report on barriers to PEV deployment identified vehicle cost as a meaningful barrier to deployment, recommending that the federal government continue vehicle purchase incentives and funding for battery research, as well as invest in research on the relationship between charging infrastructure availability and adoption (TRB and NRC, 2015). Cumulative EV sales by automaker are displayed in Figure 11.6. Depending on how fast the technological and market factors improve, EV incentives, especially purchase subsidies, may continue to be important in assisting with CAFE compliance, if the CAFE regulations for 2025–2035 are designed to depend on large sales shares of PEVs.

11.5.2 Major Types of EV Incentives, Their Roles, and Transition to Focusing on Purchase Incentives

There are two main types of EV incentives: purchase subsidies and operational incentives. Purchase subsidies, such as tax credits, are one-time monetary benefits that are directly perceived by consumers to offset the price of the vehicle and typically paid by the government. Operational incentives are the tangible or intangible benefits repeatedly received by PEV owners during the ownership of the vehicle, typically in forms of access to privileged lanes (e.g., high-occupancy vehicle [HOV] or bus lanes), parking benefits (free, discounted, or preferred parking),

[12]Purchase subsidies in this section refer to policies that directly stimulate monetary benefits to buyers of PEVs. Examples are instant price discount, tax credit, and reduced fees that would have been posed for gasoline vehicle purchase.

charging benefits (free or discounted workplace or public charging), and/or road fee waivers (e.g., waiver of road toll). Hardman (2019) reviewed 41 relevant studies of operational incentives for PEVs. Of the 41 studies reviewed, 23 out of 30 studying HOV access found that HOV lane access (reducing travel time during congestion) has a positive impact on PEV adoption. All 18 studies that investigate the role of charging infrastructure found that availability and improvement of charging infrastructure is correlated with PEV adoption, although the causality direction is less clear. Seven out of 10 studies that analyze road fees found that toll fee waivers correlate with PEV sales, and the remaining 3 could not confirm or exclude the effect (Hardman, 2019). It is difficult to rank operational incentives with respect to their impact on PEV sales, but it seems quite evident that operational incentives are correlated with PEV sales.

11.5.3 Overview of EV Incentives from 2011 to Now

PEV purchase subsidies in the United States mainly include federal tax credit and state incentives. The federal tax credit provides up to $7,500, depending on battery capacity (a proxy for vehicle capability) and number of sales of the manufacturer. Most PHEVs on the market in July 2020 were eligible for $4,000–$4,500 tax credits, and almost all BEVs are eligible for the maximum $7,500. If an automaker's cumulative PEV sales have reached 200,000 units, its products are ineligible for the tax credit. As of July 2020 (Figure 11.7), Tesla and General Motors (GM) have exceeded this limit, which means that their PEV products will not be eligible for the federal tax credits.[13] Nissan, Ford, Toyota, and BMW have 100,000 to 200,000 cumulative PEV sales,[14] with 13 other automakers' cumulative PEV sales still far below 100,000 (AFDC, n.d.a).

State PEV purchase subsidies vary widely. Since the introduction of PEVs to the U.S. market in the late 2010s, some states have provided purchase subsidies for PEVs, ranging from $750 to $6,000 per vehicle. Not all states have provided purchase subsidies, and some have ended their subsidies. The eligible amount can depend on electric range and income, with the specific rules set by each state. The maximum available purchase subsidies by states from 2010 to 2015 are summarized in Figure 11.7. As of July 2020, New Jersey had the largest maximum purchase subsidy of $5,000. The actual amount is $25 per mile of electric range. Colorado, Delaware, and Maryland provided up to $4,000, $3,500, and $3,000, respectively. Massachusetts, Oregon, California, Connecticut, and New York provided up to $2,000–$2,500 per BEV. Purchase subsidies have been generally smaller for PHEVs and greater for fuel cell electric vehicles, if available. Details of these state purchase subsidies can be found on the National Council of State Legislatures website (Hartman and Dowd, 2017) and the DOE Alternative Fuels Data Center (AFDC, n.d.b).

11.5.4 U.S. Purchase Subsidies Compared to Other Global Top Markets

The U.S. PEV purchase subsidies, including both federal and state incentives, are comparable to the top PEV markets around the world. According to Hardman et al. (2017), PEV purchase subsidies around the world exist in four forms—point of sale grant, sales tax and value added tax (VAT) exemptions, post-purchase rebates, and income tax credits. The U.S. federal PEV incentive as previously discussed is the only known example of income tax credits. The U.S. state purchase subsidies are usually in the form of post-purchase rebates.

As shown in Table 11.2, the U.S. maximum eligible amount for the PEV purchase subsidy varies by state, from $7,500 to $10,000. This appears to be more generous than some countries such as Canada or Germany, and less than the amount in the Netherlands and Norway. Although less studied, there is another type of EV purchase incentive in the form of a relative penalty for purchasing a gasoline vehicle. In principle, this kind of incentive can be combined with the "sales tax and VAT exemptions" in Table 11.2 and renamed as "avoided cost," as these monetary values are not received as revenue by the consumer but as the payment avoided by not buying a conventional ICE vehicle. For example, in some major cities in China, regulations limiting new vehicle registration give

[13] The tax credit will not be totally eliminated; instead, it will be phased out in the quarters following the one during which the 200,000 cumulative PEV sales is reached.

[14] BMW's PEV cumulative sales are 99,017 as of July 2020.

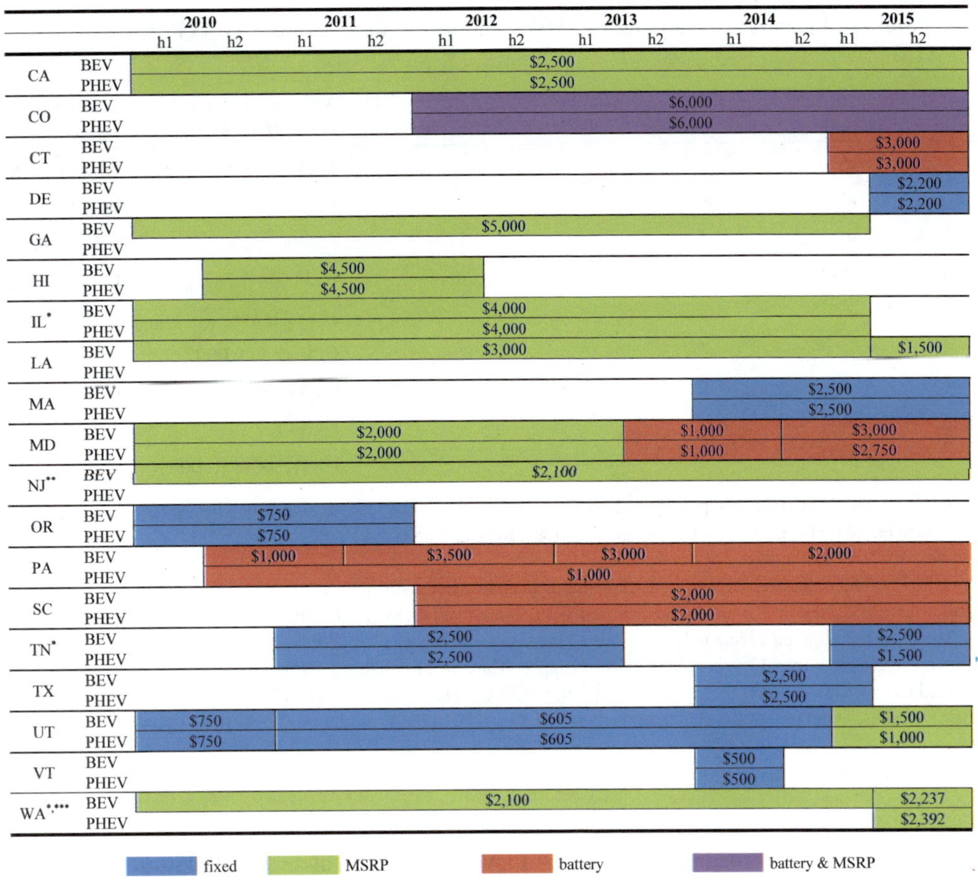

FIGURE 11.7 Summary of maximum state EV purchase incentives.
NOTE: *Subsidy for PHEVs applies only to extended range electric vehicles (EREVs). In Tennessee, this was the case for the first round of subsidies (2011h1–2012h2). **Based on an average manufacturer's suggested retail price (MSRP) of $30,000 for the Nissan Leaf and 7% sales tax. ***Assumes a minimum 7% sales tax (varies by county). For illustration, the BEV amount between 2010h1 and 2015h1 is 7% of $30,000. In 2015h2, the sales tax exemption expanded to include EREVs but restricted eligible vehicles (both BEVs and EREVs) to those valued at less than $35,000. As a result, the only eligible EREV is the Chevrolet Volt. Based on the vehicles that were actually sold in 2015h2, the maximum BEV amount corresponds to a vehicle in the price range of a Kia Soul EV. SOURCE: Wee et al. (2019).

exception to PEVs. Such regulations have created an additional cost of about $13,000 based on the license plate bidding price in Shanghai in 2017. This cost penalty for a gasoline vehicle purchase acts virtually as a purchase incentive for PEVs and is of the same order of magnitude as, and provided in addition to, the combined central and local government purchase subsidies (Ou et al., 2017, 2019). As a result, the value of incentives in Table 11.1 for China should be approximately doubled to account for the PEV purchase subsidies in major cities of China that restricts purchase of conventional ICE vehicles.

11.5.5 Effectiveness of EV Incentives

Studies of EV policies in different countries and regions largely agreed that purchase subsidies are necessary for nurturing the early market for EVs (Hardman et al., 2017; Hao et al., 2014; Lutsey et al., 2018; Zhou et al., 2015). A study in 2016 based on an ex-post stated preference survey of 2,882 PEV owners in 11 states

TABLE 11.2 Purchase Incentives Available for BEVs and Their Values in the Top Global Markets

	Point of Sale Grant	Sales Tax and VAT Exemptions	Post-Purchase Rebates	Income Tax Credits	Value of Incentives (Local Currency)	Value of Incentives (US$)
Canada	✓				CA$5,000–$8,500[a]	US$3,850–$6,850
China	✓	✓			CN¥65,000	US$9,800
France	✓	✓			€6,300	US$1,000–$7,000
Germany	✓				€5,000	US$5,500
Japan	✓	✓			JP¥800,000	US$7,800
Netherlands		✓			€1,000–€20,000[b]	US$1,110–$22,000
Norway		✓	✓		90,000kr	US$11,000–$20,000[c]
United Kingdom	✓				£4,500	US$5,800
United States			✓	✓	US$7,500–$10,000[d]	US$7,500–$10,000[d]

[a]Rebates in Canada are administered at the provincial level, and different incentives are available between provinces.
[b]These estimates are based on the difference in sales tax paid for a BEV and an ICE vehicle.
[c]Saving based on 25% VAT Exemption and Purchase Tax.
[d]Based on the US$7,500 federal tax credit and US$2,500 that is available in California.
NOTE: The value of incentives does not consider other incentives that are available when owning BEVs—for example, free parking or yearly tax exemptions. The table therefore considers only the value of incentives related to the purchase of a BEV. Incentives differ between vehicle sizes, and whether a vehicle older than 13 years old is being scrapped. They also include a 2.4% VAT reduction.
SOURCE: Reprinted from Renewable and Sustainable Energy Reviews, 80, Hardman et al., The effectiveness of financial purchase incentives for battery electric vehicles—A review of the evidence, 1100–1111, Copyright (2017), with permission from Elsevier.

attributes more than 30% of PEV sales to the federal PEV taxi credit (Tal and Nicholas, 2016). The federal tax credit, the California state rebate, and HOV lane access are found to be the most important incentives for PEV purchase (Jenn et al., 2020). A systemic review (published in 2017) of 35 studies on whether purchase subsidies had effectively promoted PEV sales found that availability of purchase incentives and EV sales are correlated, availability of purchase incentives increases EV purchase probability using choice experiments, and incentives are an important factor in the EV purchase decision process based on post purchase surveys (Hardman et al., 2017). Quantitative studies have reached a general agreement, according to Jenn et al. (2020), on the magnitude of EV purchase subsidy effectiveness—$1,000 purchase subsidy resulting in about 2.6% to 4.3% increase in PEV sales (DeShazo et al., 2017; Jenn et al., 2018; Tal and Nicholas, 2016). Another general agreement is that point-of-sale purchase subsidies are more efficient than post-sale rebates (Gallagher and Muehlegger, 2011; De Shazo, 2016).

Some studies recommended phase-down subsidies along with the reduction of battery cost, a logical recommendation based on the theoretical intent of the technology subsidy: to overcome high initial battery costs and increase scale to help lower battery costs. However, this recommendation implicitly relies on two assumptions—that the original level of subsidies is ideal or adequate and that consumers have homogenous willingness to pay for PEVs. If the incentive level is inadequate, it can become more adequate and thus be more effective with reduction of battery cost. In such cases, incentive phase-down should start only after the battery cost decreases to the extent that makes the incentive level adequate. By comparison, U.S. PEV purchase subsidies are not as generous as those in countries with leading PEV sales shares such as China and Norway, while U.S. consumers have been found to have a lower willingness to pay for BEVs than the consumers in China (Helveston et al., 2015). These together suggest that the current PEV purchase subsidies may need to be continued and improved.

U.S. PEV sales have been low except in a few states with complementary state-level policies such as California. The comparison of PEV sales between California and the other parts of the country underscores the importance of PEV purchase subsidies and other incentives for promoting PEV adoption (Li et al., 2017).

Comparing incentives and market aspects among international markets, California, and the rest of the United States shows differences in gasoline prices, charging infrastructure, and ZEV regulations. Gasoline prices are relatively low in the United States as a whole, as compared to California and international markets, possibly an obstacle for PEV sales. The deployment of charging infrastructure in the United States is less aggressive than the PEV sales in leading countries. The United States does not have a national ZEV program that can provide nationwide policy support for PEV sales, unlike the dual-credit policy in China that combines a ZEV mandate and a CAFE program at the national level. All of these factors point to the need for continuing and even strengthening the purchase subsidies, unless the aforementioned factors can be quickly improved for facilitating PEV adoption.

Consumer heterogeneity is another important factor and perspective for redesigning the PEV purchase incentives. A survey study of 14,000 PEV buyers using multinomial logistic regression suggests that incentives are becoming more important when the PEV market progresses beyond early adopters (Jenn et al., 2020). This may be counterintuitive but also important. Willingness to pay for PEVs could be lower with mainstream consumers than with early adopters, which calls for greater incentives according to the optimal incentive design recommendation by DeShazo et al. (2017), unless battery cost reduction accelerates. The actual outcome of incentive phase-out will depend on how fast the PEV technologies will improve and their costs will decrease; how fast charging infrastructure will be developed; and how other policies contribute to forming implicit, indirect, or internal subsidies for PEVs, such as the dual-credit policy in China, the ZEV mandate in California, the CAFE flexibilities for PEVs, and road priority (access to bus lanes and HOV lanes). Nevertheless, the finding that the mainstream consumers may have a lower willingness-to-pay for PEVs and thus more likely need incentives as PEV purchase motivation is important for envisioning the role of purchase incentives for large-volume sales of PEVs in 2025–2035.

11.5.6 Criticism, Skepticism, and Concerns Over EV Incentives Provision or Their Design

Criticisms over PEV incentives can stem from an overall objection to PEVs based on life cycle GHG emissions or loss of gas tax revenues, but a more common concern is related to the policy design of the PEV incentive, purchase subsidies in particular, rather than the provision of incentives. Incentives are viewed as a good motivator to encourage consumer consideration of PEVs and to increase their cost-competitiveness. The objective of a PEV subsidy design is to maximize PEV adoption subject to the consumer's economic budget constraint (DeShazo et al., 2017), although it is debatable whether PEV sales or electrified vehicle miles traveled (eVMT) should be targeted. One consensus is that the same amount of subsidy should be provided at point of sale to consumers, rather than as post-sale rebates, in order to increase incentive effectiveness or reduce fiscal burdens. Not all PEV shoppers are aware of the purchase subsidies, which potentially inhibits the incentive impact and calls for consumer education. It is also widely accepted that the purchase subsidy should be given to low- or medium-income consumers for consideration of social equity and the better policy effectiveness, as high-income consumers are less price-elastic. In November 2016, California started excluding high-income residents from the $2,500 PEV rebate and adding an additional $2,000 for low-income consumers. The federal PEV tax credit and the other state incentive policies are yet to consider income eligibility. Some studies suggest that the purchase subsidy should give priority to low-end BEVs rather than high-end BEVs (Hardman et al., 2017), but this may be achieved already with consideration of income eligibility. Some suggest prioritizing purchase subsidies for long-range PHEVs, as short-range PHEV users have been found to skip charging and mainly use the vehicle as a hybrid. Electric range of PEVs is correlated with vehicle price and thus the income and price elasticity of the candidate buyers, but also affects the amount of distance traveled on electricity. How to design the subsidy scheme to maximize effectiveness on both PEV sales and eVMT requires sophisticated consideration of consumer heterogeneity and is not yet well studied. Another knowledge gap is the conditions to end the incentives. In principle, the incentives can be ended when the PEV sales have entered the late majority consumers and can be entirely market driven (Hardman et al., 2017). However, there is no clear way to define such conditions, also called a "tipping point," which could be a concern for lack of a clear exit strategy for government expenditure commitment. Last, purchase subsidies and incentives may not be sufficient to overcome barriers such as a lack of supporting infrastructure (e.g., charging for EVs), vehicle attributes (e.g., range for EVs), or overall price.

FINDING 11.4: Numerous studies have found that purchase subsidies have effectively increased plug-in hybrid and battery electric vehicle sales. Studies from California have indicated similar benefits from subsidizing fuel cell vehicles. Subsidies available at time of sale are more effective than tax credits or tax deductions owing to their immediacy and because not all car buyers can receive the value from tax credits and deductions. Studies have argued that purchase subsidies should be given to low- or medium-income EV buyers for consideration of both equity and policy effectiveness, as high-income consumers are less sensitive to the subsidy. California has implemented both point-of-sale rebates and income eligibility. The purchase subsidy amount overall in the United States is above the average but not the highest, relative to top zero-emission vehicle markets in the world.

RECOMMENDATION 11.4: The U.S. federal plug-in hybrid, battery electric vehicle, and fuel cell electric vehicle purchase subsidies should be continued beyond the current cumulative sales cutoffs to further increase the size of the market including lower-income, more price-responsive, and probably late-majority consumers.

RECOMMENDATION 11.5: The extended purchase subsidies should be changed into point-of-sale rebates to increase effectiveness. Income eligibility should be considered for both policy equity and effectiveness. The combination of point-of-sale rebates and income eligibility does not necessarily require dealerships to verify consumers' income. As implemented in California, consumers can apply for the rebate right after buying a plug-in electric vehicle or even before (in San Diego County). The application process includes income verification through the government website. There may be better ways of implementing both point-of-sale rebates and income eligibility. Studies are recommended to optimize which type of plug-in hybrid, battery electric vehicle, and fuel cell electric vehicle and how much electric range should receive more or less subsidy, with considerations of equity and policy effectiveness in promoting sales or electrified vehicle miles traveled.

FINDING 11.5: The federal plug-in electric vehicle (PEV) incentives have expired for only 2 (Tesla and GM) out of 19 manufacturers that are selling PEVs in the United States. This suggests still a large number of PEVs to be sold and be eligible for the subsidy, but it is unclear how soon those 17 players can improve their product competitiveness and catch up on sales, given current overall PEV demand.

FINDING 11.6: Consumer unfamiliarity and misperceptions about the disruptive impacts and performance of battery electric vehicles (BEVs) are a significant impediment to consumer demand for such vehicles. Improved performance of BEVs—in particular, longer range through better batteries—will help to overcome such impediments. Empirical evidence on consumer responses to other technologies and BEV sales in other jurisdictions suggest that better infrastructure, tax credits, and public education campaigns can overcome consumer demand barriers.

RECOMMENDATION 11.6: Battery electric vehicle and fuel cell vehicle sales can contribute to meeting more stringent corporate average fuel economy standards, so governments should seek to stimulate consumer demand for such vehicles through infrastructure investments, purchase subsidies, public education programs (for both consumers and dealerships), and other incentives. Research should be conducted to determine which of these strategies are more cost-effective than the others (and in what contexts).

RECOMMENDATION 11.7: The 2025–2035 corporate average fuel economy standard should be set and designed to depend on and incentivize a significant market share of zero-emission vehicles (plug-in hybrid electric vehicle, battery electric vehicle, and fuel cell electric vehicle).

11.6 REFERENCES

AAA (American Automobile Association). 2019. "Three in Four Americans Remain Afraid of Fully Self-Driving Vehicles." March 14. https://newsroom.aaa.com/2019/03/americans-fear-self-driving-cars-survey.

AFDC (Alternative Fuels Data Center). n.d.a. "Qualified Plug-In Electric Vehicle (PEV) Tax Credit." Alternative Fuels Data Center, U.S. Department of Energy Office of Energy Efficiency and Renewable Energy. https://afdc.energy.gov/laws/409.

AFDC. n.d.b. "State Laws and Incentives." Alternative Fuels Data Center, U.S. Department of Energy Office of Energy Efficiency and Renewable Energy. https://afdc.energy.gov/laws/state.

Allcott, H., and M. Greenstone. 2012. Is there an energy efficiency gap? *Journal of Economic Perspectives* 26(1): 3–28. https://doi.org/10.1257/jep.26.1.3.

Allcott, H., S. Mullainathan, and D. Taubinsky. 2014. Energy policy with externalities and internalities. *Journal of Public Economics* 112: 72–88.

Argonne National Laboratory. 2020. "Light Duty Electric Drive Vehicles Monthly Sales Updates." https://www.anl.gov/es/light-duty-electric-drive-vehicles-monthly-sales-updates.

Balcer, Y., and S.A. Lippman. 1984. Technological expectations and adoption of improved technology. *Journal of Economic Theory* 34(4):292–318. December.

Bento, A.M., K. Gillingham, M.R. Jacobsen, C.R. Knittel, B. Leard, J. Linn, V. McConnell, et al. 2018. Flawed analyses of U.S. auto fuel economy standards. *Science* 362(6419):1119–1121. https://doi.org/10.1126/science.aav1458.

BLS (U.S. Bureau of Labor Statistics). 2019. "Consumer Expenditure Surveys: CE Tables." Table 1110. https://www.bls.gov/cex/2017/combined/decile.pdf.

Busse, M.R., C.R. Knittel, and F. Zettelmeyer. 2013. Are consumers myopic? Evidence from new and used car purchases. *American Economic Review* 103(1):220–56. DOI: 10.1257/aer.103.1.220.

Camerer, C., G. Loewenstein, and D. Prelec. 2005. Neuroeconomics: How neuroscience can inform economics. *Journal of Economic Literature* 43(1):9–64. https://doi.org/10.1257/0022051053737843.

Christensen, C.M. 1997. *The Innovator's Dilemma: When New Technologies Cause Great Firms to Fail.* Boston, MA: Harvard Business School Press.

Conchar, M.P., G.M. Zinkhan, C. Peters, and S. Olavarrieta. 2004. An integrated framework for the conceptualization of consumers' perceived-risk processing. *Journal of the Academy of Marketing Science* 32(4):418–436.

CRSG (Consumer Reports Survey Group). 2017. "2017 Automotive Fuel Economy Survey Report." provided by Shannon Baker-Brandstetter, Consumers Union, 9/13/2018.

CRSG. 2018. "2018 Automotive Fuel Economy Survey Report." provided by Shannon Baker-Brandstetter, Consumers Union, 9/13/2018.

Dellavigna, S. 2009. Psychology and economics: Evidence from the field. *Journal of Economic Literature* 47(2):315–372.

DeShazo, J.R. 2016. Improving incentives for clean vehicle purchases in the United States: Challenges and opportunities. *Review of Environmental Economics and Policy* 10(1):149–165. December 1.

DeShazo, J.R., T.L. Sheldon, and R.T. Carson. 2017. Designing policy incentives for cleaner technologies: Lessons from California's plug-in electric vehicle rebate program. *Journal of Environmental Economics and Management* 84:18–43. July 1.

Dimitropoulos, A., W. Oueslati, and C. Sintek, 2018. The rebound effect in road transport: A meta-analysis of empirical studies. *Energy Economics* 75:163–179.

EAFO (European Alternative Fuels Observatory). 2020. "AF Market Share New Registrations M1 (2019)." BEV, Norway. https://www.eafo.eu/vehicles-and-fleet/m1#.

EPA (U.S. Environmental Protection Agency). 2019. *Greenhouse Gas (GHG) Emission Standards for Light-Duty Vehicles: Manufacturer Performance Report*, and excel tables, accessed on 11/22/19 at https://www.epa.gov/automotive-trends/download-data-automotive-trends-report.

EPA. 2020. "EPA Automotive Trends Report: Greenhouse Gas Emissions, Fuel Economy, and Technology Since 1975." https://nepis.epa.gov/Exe/ZyPDF.cgi?Dockey=P100YVK3.pdf.

Erev, I., E. Ert, O. Plonsky, D. Cohen, and O. Cohen. 2017. From anomalies to forecasts: Toward a descriptive model of decisions under risk, under ambiguity, and from experience. *Psychology Review* 124(4):369–409.

Ert, E., and I. Erev, 2013. On the descriptive value of loss aversion in decisions under risk: Six clarifications. *Judgment and Decision Making* 8(3):214–235.

Figenbaum, E., and S. Nordbakke. 2019. Battery Electric Vehicle User Experiences in Norway's Maturing Market. Institute of Transport Economics, Norwegian Centre for Transport Research Report 11719/2019.

Ford Motor Company. 2020. "Ford Expands Climate Change Goals, Sets Target to Become Carbon Neutral By 2050: Annual Sustainability Report." Ford Media Center. June 24. https://media.ford.com/content/fordmedia/fna/us/en/news/2020/06/24/ford-expands-climate-change-goals.html.

Franke, T., and J.F. Krems. 2013. Interacting with limited mobility resources: Psychological range levels in electric vehicle use. *Transportation Research Part A: Policy and Practice* 48:109–122.

Gächter, S., A. Herrmann, and E.J. Johnson. 2007. "Individual-Level Loss Aversion in Riskless and Risky Choices." Discussion Paper Series. Institute for the Study of Labor, Bonn.

Gallagher, K.S., and E. Muehlegger. 2011. Giving green to get green? Incentives and consumer adoption of hybrid vehicle technology. *Journal of Environmental Economics and Management* 61(1):1–15. January 1.

General Motors. 2016. "GM Recognized as World Leader for Climate Change Action." General Motors Green. October 31. https://www.generalmotors.green/product/public/us/en/GMGreen/energy_efficiency.detail.html/content/Pages/news/us/en/gm_green/2016/1031-climate-change-action.html.

Gerarden, T.D., R.G. Newell, R.N. Stavins, and R.C. Stowe. 2015. "An Assessment of the Energy-Efficiency Gap and Its Implications for Climate-Change Policy." Working Paper 20905. Cambridge, MA: National Bureau of Economic Research.

Gillingham, K., and K. Palmer. 2014. Bridging the energy efficiency gap: Policy insights from economic theory and empirical evidence. *Review of Environmental Economics and Policy* 8(1):18–38.

Gillingham, K., D. Rapson, and G. Wagner. 2016. The rebound effect and energy efficiency policy. *Review of Environmental Economics and Policy* 10(1):68–88.

Gillingham, K. 2020. The rebound effect and the proposed rollback of U.S. fuel economy standards. *Review of Environmental Economics and Policy* 14(1):136–42. https://doi.org/10.1093/reep/rez015.

Greene, D.L. 2010. "How Consumers Value Fuel Economy: A Literature Review." EPA-420-R-10-008. Washington, DC: Environmental Protection Agency.

Greene, D.L. 2011a. Uncertainty, loss aversion and markets for energy efficiency. *Energy Economics* 33:608–616.

Greene, D.L. 2011b. What's greener than a VMT tax? The case for an indexed energy user fee to finance U.S. surface transportation. *Transportation Research D: Environment* 16:451–458.

Greene, D.L. 2019. Implications of Behavioral Economics for the Costs and Benefits of Fuel Economy Standards. Current Sust./Renew. Energy Reports. https://doi.org/10.1007/s40518-019-00134-4.

Greene, D.L., D.H. Evans, and J. Hiestand. 2013. Survey evidence on the willingness of U.S. consumers to pay for automotive fuel economy. *Energy Policy* 61:1539–1550.

Greene, D.L., A.J. Khattak, J. Liu, X. Wang, J.L. Hopson, and R. Goeltz. 2017. What is the evidence concerning the gap between on-road and Environmental Protection Agency fuel economy ratings? *Transport Policy* 53:146–160.

Greene, D., A. Hossain, J. Hofmann, G. Helfand, and R. Beach. 2018. Consumer Willingness to Pay for Vehicle Attributes: What Do We Know? *Transportation Research Part A: Policy and Practice* 118 (December): 258–79. https://doi.org/10.1016/j.tra.2018.09.013. eene, D.L., C. Sims, and M. Muratori. 2019. "Two Trillion Gallons." Howard H. Baker, Jr. Center for Public Policy, University of Tennessee, Knoxville.

Gruenspecht, H.K. 1982. Differentiated regulation: The case of auto emissions standards. *American Economic Review* 72:328–331.

Hamilton, J.D. 2009. Understanding crude oil prices. *Energy Journal* 30(2):179–206.

Hao, H., X. Ou, J. Du, H. Wang, and M. Ouyang. 2014. China's electric vehicle subsidy scheme: Rationale and impacts. *Energy Policy* 73:722–732. October 1.

Hardman, S. 2019. Understanding the impact of reoccurring and non-financial incentives on plug-in electric vehicle adoption—A review. *Transportation Research Part A: Policy and Practice* 119:1–14.

Hardman, S., A. Chandan, G. Tal, and T. Turrentine. 2017. The effectiveness of financial purchase incentives for battery electric vehicles—A review of the evidence. *Renewable and Sustainable Energy Reviews* 80:1100–1111.

Hartman, K., and E. Dowd. 2017. "State Efforts to Promote Hybrid and Electric Vehicles." National Conference of State Legislatures. September 26. https://www.ncsl.org/research/energy/state-electric-vehicle-incentives-state-chart.aspx#hybrid,%20accessed%20on%20Sep%201,%202020.

Helfand, G. and A. Wolverton. 2011. Evaluating the consumer response to fuel economy: A review of the literature. *International Review of Environmental and Resource Economics* 5(2):103–46. https://doi.org/10.1561/101.00000040.

Helfand, G., M. McWilliams, K. Bolon, L. Reichle, M. Sha, A. Smith, and R. Beach. 2016. Searching for hidden costs: A technology-based approach to the energy efficiency gap in light-duty vehicles. *Energy Policy* 98(November):590–606. https://doi.org/10.1016/j.enpol.2016.09.014.

Helveston, J.P., Y. Liu, E.M.D. Feit, E. Fuchs, E. Klampfl, and J.J. Michalek. 2015. Will subsidies drive electric vehicle adoption? Measuring consumer preferences in the U.S. and China. *Transportation Research Part A: Policy and Practice*. 73:96–112. March 1.

Hewitt, C., I. Politis, T. Amanatidis, and A. Sarkar. 2019. "Assessing Public Perception of Self-Driving Cars: The Autonomous Vehicle Acceptance Model." Microsoft. March. https://www.microsoft.com/en-us/research/uploads/prod/2019/03/assessing-public-perception.pdf.

Hirunyawipada, T., and A.K. Paswan. 2006. Consumer innovativeness and perceived risk: Implications for high technology product adoption. *Journal of Consumer Marketing* 23(4):182–198.

Honda. 2020. "Honda's Approach to Climate Change." Honda Corporate Social Responsibility (blog). December 23. https://csr.honda.com/sub-feature/designing-for-the-environment/.

Hymel, K.M., and K.A. Small. 2015. The rebound effect for automobile travel: Asymmetric response to price changes and novel features of the 2000s. *Energy Economics* 49:93–103.

IRGC (International Risk Governance Council). 2013. *The Rebound Effect: Implications of Consumer Behavior for Robust Energy Policies*. Geneva, Switzerland.

Jaffe, A.B., R.G. Newell, and R.N. Stavins. 2005. A tale of two market failures: Technology and environmental policy. *Ecological Economics* 54:164–174.

Jenn, A., I.M.L. Azevedo, and J.J. Michalek. 2016. Alternative fuel vehicle adoption increases fleet gasoline consumption and greenhouse gas emissions under United States corporate average fuel economy policy and greenhouse gas emissions standards. *Environmental Science and Technology* 50(5):2165–2174. https://doi.org/10.1021/acs.est.5b02842.

Jenn, A., K. Springel, and A.R. Gopal. 2018. Effectiveness of electric vehicle incentives in the United States. *Energy Policy* 119(August):349–56. https://doi.org/10.1016/j.enpol.2018.04.065.

Juma, C. 2016. *Innovation and Its Enemies: Why People Resist New Technologies*. Oxford, UK: Oxford University Press.

Kahneman, D. 2011. *Thinking Fast and Slow*. New York: Farrar, Straus, and Giroux.

Kahneman, D., and A. Tversky. 1979. Prospect theory: An analysis of decision making under risk. *Econometrica* 47: 263–291.

Kaplan, J. 2018. "Why We Find Self-Driving Cars So Scary." Wall Street Journal. May 31. https://www.wsj.com/articles/why-we-find-self-driving-cars-so-scary-1527784724.

Kivimaaa, P., and F. Kerna. 2016. Creative destruction or mere niche support? Innovation policy mixes for sustainability transitions. *Research Policy* 45(1):205–217.

Kleijnen, M., N. Lee, and M. Wetzels. 2009. An exploration of consumer resistance to innovation and its antecedents. *Journal of Economic Psychology* 30:344–357.

Klier, T., and J. Linn. 2016. Technological change, vehicle characteristics, and the opportunity costs of fuel economy standards. *Journal of Public Economics* 133:41–63.

Larrick, R.P., and J.B. Soll. 2008. The MPG illusion. *Science* 320(5883):1593–94. https://doi.org/10.1126/science.1154983.

Leard, B. 2018. Consumer inattention and the demand for vehicle fuel cost savings. *Journal of Choice Modeling* 29:1–16.

Li, X., P. Chen, and X. Wang. 2017. Impacts of renewables and socioeconomic factors on electric vehicle demands—panel data studies across 14 countries. *Energy Policy* 109:473–478. October. https://doi.org/10.1016/j.enpol.2017.07.021.

Linn, J. 2016. The rebound effect for passenger vehicles. *Energy Journal* 37(2):257–288.

Liu, R., Z. Ding, X. Jiang, J. Sun, Y. Jiang, and W. Qiang. 2020. How does experience impact the adoption willingness of battery electric vehicles? The role of psychological factors. *Environmental Science and Pollution Research* 27(20):25230–25247. https://doi.org/10.1007/s11356-020-08834-w.

Lutsey, N., M. Grant, S. Wappelhorst, and H. Zhou. 2018. *Power Play: How Governments Are Spurring the Electric Vehicle Industry*. San Francisco, CA: International Council on Clean Transportation.

Marchant, G.E. 2014. Complexity and anticipatory socio-behavioral assessment of government attempts to induce clean technologies. *UCLA Law Review* 61:1858–1894.

McFadden, D., and K. Train. 2000. Mixed MNL models for discrete response. *Journal of Applied Econometrics* 15(5):447–470.

Mrkva, K., E.J. Johnson, S. Gächter, and A. Herrmann. 2020. Moderating loss aversion: Loss aversion has moderators, but reports of its death are greatly exaggerated. *Journal of Consumer Psychology* 30(3):407–28. https://doi.org/10.1002/jcpy.1156.

Nemet, G.F. 2012. Subsidies for new technologies and knowledge spillovers from learning by doing. *Journal of Policy Analysis and Management* 31(3):601–622.

NHTSA/EPA (National Highway Traffic Safety Administration and U.S. Environmental Protection Agency). 2020. *The Safer Affordable Fuel-Efficient (SAFE) Vehicles Rule for 2021-2026 Passenger Cars and Light Trucks*. April 30, 2020. https://www.govinfo.gov/content/pkg/FR-2020-04-30/pdf/2020-06967.pdf.

Nissan Motor Corporation. n.d. "Climate Change." Environmental Activities. https://www.nissan-global.com/EN/ENVIRONMENT/GREENPROGRAM/CLIMATE/.

Novemsky, N., and D. Kahneman. 2005. The boundaries of loss aversion. *Journal of Market Research* XLII:119–128.

NRC (National Research Council). 1992. *Automotive Fuel Economy: How Far Can We Go?* Washington, DC: The National Academies Press. https://doi.org/10.17226/1806.

NRC. 2011. *Assessment of Fuel Economy Technologies for Light-Duty Vehicles*. Washington, DC: The National Academies Press. https://doi.org/10.17226/12924.

NRC. 2013. *Transitions to Alternative Vehicles and Fuels*. Washington, DC: The National Academies Press. https://doi.org/10.17226/18264.

NRC. 2015. *Cost, Effectiveness, and Deployment of Fuel Economy Technologies for Light-Duty Vehicles*. Washington, DC: The National Academies Press. https://doi.org/10.17226/21744.

Nuemann, N., and U. Bockenholt. 2014. A meta-analysis of loss aversion in product choice. *Journal of Retailing* 90(June). https://doi.org/10.1016/j.jretai.2014.02.002.

Ou, S., Z. Lin, Z. Wu, J. Zheng, R. Lyu, S.V. Przesmitzki, and X. Hu. 2017. *A Study of China's Explosive Growth in the Plug-in Electric Vehicle Market*. Oak Ridge, TN: Oak Ridge National Laboratory. January.

Ou, S., X. Hao, Z. Lin, H. Wang, J. Bouchard, X. He, S. Przesmitzki, et al. 2019. Light-duty plug-in electric vehicles in China: An overview on the market and its comparisons to the United States. *Renewable and Sustainable Energy Reviews* 112:747–761. September. https://doi.org/10.1016/j.rser.2019.06.021.

Plumer, B., and N. Popovich. 2020. "Super Bowl Ads Hyped Electric Cars. But Will Anyone Buy Them?" New York Times. February 2, sec. Climate. https://www.nytimes.com/interactive/2020/02/02/climate/super-bowl-ads-electric-car-hummer .html.

Ram, S. 1987. A model of innovation resistance. *Advances in Consumer Research* 14:208–212.

Ram, S., and J.N. Sheth. 1989. Consumer resistance to innovation: The marketing problem and its solutions. *Journal of Consumer Marketing* 6(2):5–14.

Rogers, E.M. 2003. *Diffusion of Innovations*. 5th ed. New York: Free Press.

Sallee, J.M. 2014. Rational inattention and energy efficiency. *Journal of Law and Economics* 57:781–820.

Schmalfuß, F., K. Mühl, and J.F. Krems. 2017. Direct experience with battery electric vehicles (BEVs) matters when evaluating vehicle attributes, attitude and purchase intention. *Transportation Research Part F: Traffic Psychology and Behaviour* 46:47–69. April. https://doi.org/10.1016/j.trf.2017.01.004.

Schumpeter, J.A. 1942. *Capitalism, Socialism and Democracy*. New York: Harper and Brothers.

Slovic, P., and E. Peters. 2006. Risk perception and affect. *Current Directions in Psychological Science* 15:322.

Small, K.A., and K. Van Dender. 2007. Fuel efficiency and motor vehicle travel: The declining rebound effect. *Energy Journal* 28(1):25–51.

Sokol-Hessner, P., and R.B. Rutledge. 2019. The psychological and neural basis of loss aversion. *Current Directions in Psychological Science* 28(1):20–27. https://doi.org/10.1177/0963721418806510.

Sorrell, S., and J. Dimitropoulos. 2007. "UKERC Review of Evidence for the Rebound Effect. Technical Report 2: Econometric Studies." UKERC/WP/TPA/2007/010. UK Energy Research Centre, Sussex Energy Group, University of Sussex, UK.

Starmer, C. 2000. Developments in non-expected utility theory: The hunt for a descriptive theory of choice under risk. *Journal of Economic Literature* 38:332–382.

Tal, G., and M. Nicholas. 2016. Exploring the impact of the federal tax credit on the plug-in vehicle market. *Transportation Research Record Journal of the Transportation Research Board* 2572(1):95–102. January 1.

Thaler, R.H. 2015. *Misbehaving: The Making of Behavioral Economics*. New York: W.W. Norton and Company.

Tom, S.M., C.R. Fox, C. Trepel, and R.A. Poldrack. 2007. The neural basis of lass aversion in decision-making under risk. *Science* 315:515–518.

Toyota Motor Company. 2020. "Environmental Initiatives." Toyota Motor Corporation Official Global Website. https://global.toyota/en/sustainability/esg/environmental/index.html.

TRB and NRC (Transportation Research Board and National Research Council). 2002. *Effectiveness and Impact of Corporate Average Fuel Economy (CAFE) Standards*. Washington, DC: The National Academies Press. https://doi.org/10.17226/10172.

TRB and NRC. 2015. *Overcoming Barriers to Deployment of Plug-in Electric Vehicles*. Washington, DC: The National Academies Press. https://doi.org/10.17226/21725.

Trueblood, J.S. 2012. Multialternative context effects obtained using an inference task. *Psychonomic Bulletin & Review* 19(5):962–68. https://doi.org/10.3758/s13423-012-0288-9.

Trueblood, J.S., S.D. Brown, and A. Heathcote. 2014. The multiattribute linear ballistic accumulator model of context effects in multialternative choice. *Psychological Review* 121(2):179–205. https://doi.org/10.1037/a0036137.

Turrentine, T.S., and K.S. Kurani. 2007. Car buyers and fuel economy? *Energy Policy* 35:1213–1223.

Tushman, M.L., and P. Anderson. 1986. Technological discontinuities and organizational environments. *Administrative Science Quarterly* 31(3):439–465.

Volkswagen Group. 2020. "Climate Change: Nine Facts." Volkswagen. https://www.volkswagenag.com/en/news/stories/2019/09/nine-facts-on-climate-change.html.

Wali, B., D.L. Greene, A.J. Khattak, and J. Liu. 2018. Analyzing within garage fuel economy gaps to support vehicle purchasing decisions–A copula-based modeling and forecasting approach. *Transportation Research Part D* 63:186–208.

Wee, S., M. Coffman, and S. LaCroix. 2019. Data on U.S. state-level electric vehicle policies, 2010–2015. *Data in Brief* 23:103658. April 1.

Zhou, Y., M. Wang, H. Hao, L. Johnson, H. Wang, and H. Hao. 2015. Plug-in electric vehicle market penetration and incentives: A global review. *Mitigation and Adaptation Strategies for Global Change* 20(5):777–795. June 1.

Ziamou, P., and R.W. Veryzer. 2005. The influence of temporal distance on consumer preferences for technology-based innovations. *Journal of Product Innovation and Management* 22:336–346.

12

Regulatory Structure and Flexibilities

As the U.S. fuel economy program approaches a half-century of accomplishment, it is useful to examine how the premises and context of the program have changed, and how it may need to adapt going forward, especially with its statutory reauthorization authority scheduled to expire in 2030. This chapter considers key elements of the Corporate Average Fuel Economy (CAFE) program, whose effects will be especially meaningful as fuel economy regulation evolves in the context of climate change. Given the scope of this study, the committee focuses on policies affecting passenger vehicles rather than broader policies such as carbon or fuel taxes. This chapter begins by describing the history of U.S. vehicle regulation, followed by a discussion of the test cycles, off-cycle corrections, and discrepancies with real-world fuel economy. As alternative powertrain technologies continue to be adopted, considerations of life cycle environmental impacts and how vehicle policies may affect multiple sectors become especially important, and are described here. Regulatory flexibilities and the credit trading system are then presented, along with the rationale for these flexibilities, a review of credit trading, and how they affect compliance costs and are affected by changes in utilization. The final section of the chapter discusses global issues related to the U.S. program, including several case studies of programs administered by other countries.

12.1 HISTORY OF VEHICLE FUEL ECONOMY REGULATION

As introduced in Chapter 2, vehicle fuel economy regulations began with the passage of the Energy Policy and Conservation Act of 1975 (EPCA), providing the National Highway Traffic Safety Administration (NHTSA) with the ability to regulate manufacturer fleet-averaged fuel economy of light-duty vehicles beginning with MY 1978. The legislation set a goal of 27.5 miles per gallon (MPG) for passenger cars for each vehicle manufacturer, to be met by MY 1985. The key U.S. light-duty vehicle fuel economy and greenhouse gas (GHG) emissions statutes, regulations, and features are summarized in Table 12.1.

Under the EPCA mandate, NHTSA set a standard for light trucks which peaked at 20.5 MPG in 1987. In the late 1980s, Congress relaxed both the passenger car and light truck standards by a few MPG but returned to 27.5 MPG for passenger cars by 1990 and to 20.7 MPG for light trucks by 1996. From 1996 until 2004, standards stayed flat. Light truck efficiency standards were improved to 22.2 MPG between 2005 and 2007, and to 23.5 MPG by 2010. Fuel economy standards as set by NHTSA can be complied with by achieving the appropriate weighted average of tested fuel economy, by paying a fine, and through various credit mechanisms, described further below (EPA/NHTSA, 2012).

TABLE 12.1 Vehicle Fuel Economy and GHG Emissions Statutes, Regulations, and Key Features

Statutory Mandates	Regulated Periods and Key Regulatory Features
Energy Policy and Conservation Act of 1975 (EPCA) EPCA established a federal program to set energy targets for consumer products, including the U.S. Department of Transportation (DOT) regulation of vehicles. The goal was to reduce energy and petroleum consumption. Statutorily required fuel economy of 27.5 MPG by MY 1985.	Passenger cars 1978–1985: no standard, 27.5 MPG 1986–1989: relaxed standards, 26–26.5 MPG 1990–2010: 27.5 MPG Light trucks 1978–1987: no standard, 20.5 MPG 1988–1993: relaxed standards, 20–20.4 MPG 1994–2010: 20.5–23.5 MPG Major regulatory features • Fleet average fuel economy • Passenger car and light truck fleets • Domestically manufactured fleet standard • Compliance with standards by average achieved fuel economy, credit earning, banking and trading, or fines

Energy Independence and Security Act of 2007 (EISA)
 	
Updated EPCA with the first statutory increase in fuel economy standards since EPCA in 1975. Mandated average of 35 MPG for each manufacturer for the total of the passenger car and light truck fleet by 2020. Then maximum feasible average fuel economy from 2021 to 2030. DOT must issue regulations for at least 1 but not more than 5 MYs.
 	
Clean Air Act of (CAA) 1963, Endangerment Finding of 2007
 	
CAA gives the U.S. Environmental Protection Agency (EPA) authority to set standards for any air pollutant which "may reasonably be anticipated to endanger public health." Supreme Court judgment in *Massachusetts v. EPA* (2007) led to EPA regulation of GHGs from light-duty vehicles, under the CAA mandate.
 	
Clean Air Act of 1963, Section 202, 177
 	
Waiver for California Air Resources Board (CARB) to set higher emissions standards for vehicles. Option for other states to join in the more stringent California standards, which 13 states and District of Columbia did.
	Passenger cars
	2012–2016: 33.3–37.8 MPG
	2017–2020 (originally 2017–2025 for GHG, 2017–2022 for fuel economy): 3.5%/year 2017–2021, 5%/year 2022–2025
	
	Light trucks
	2008–2010, unreformed optional footprint-based standard: 22.4–23.4 MPG
	2011: 24.3 MPG
	2012–2016: 25.3–28.8 MPG
	2017–2020 (originally 2017–2025 for GHG, 2017–2022 for fuel economy): 3.5%/year 2017–2021, 5%/year 2022–2025
	
	Major regulatory features
	• Agreement for one national program including DOT, EPA, and CARB regulations
	• Footprint-based standards
	• Fuel economy:
	  ◦ Continuation of fleet average fuel economy, passenger car and light truck fleets, domestic fleet, and compliance by fleet average fuel economy, credits, or fines
	  ◦ EISA: Maximum feasible fuel economy for each MY, for at least 1 but not more than 5 MYs at a time
	• GHG:
	  ◦ Credit earning, banking, and trading based on GHG performance, on advanced technologies, on off-cycle technologies, and on low global warming potential (GWP) pollutants

EPCA, EISA, and CAA, withdrawal of California waiver	Passenger cars
	2021–2026: 1.5%/year increase
	Light trucks
	2021–2026: 1.5%/year increase
	
	Major regulatory features
	• Agreement for one national program including DOT, EPA, but eliminating CARB GHG regulations
	• Footprint-based standards
	• Fuel economy:
	  ◦ Continuation of fleet average fuel economy, passenger car and light truck fleets, domestic fleet, and compliance by fleet average fuel economy, credits, or fines
	  ◦ EISA: Maximum feasible fuel economy for each MY, for at least 1 but not more than 5 MYs at a time
	• GHG:
	  ◦ Credit earning, banking, and trading based on GHG performance, on advanced technologies, on off-cycle technologies, and on low GWP pollutants

NOTE: Years reported in model year (MY).

With the passage of the Energy Independence and Security Act of 2007 (EISA), NHTSA established regulations that maintained vehicle performance while raising fuel economy of light-duty vehicles to a minimum of 35 MPG by 2020 and to "maximum feasible average fuel economy" from 2021 to 2030 (110th Congress, 2007). Beginning in model year (MY) 2008, the structure of the fuel economy standards changed. The first change was to use vehicle footprint as an attribute to adjust the standard requirements, implemented for the light truck standards optionally from 2008 to 2010 and required from 2011. Previously, manufacturers had to comply with one sales-weighted average fuel economy for passenger cars and one sales-weighted average fuel economy for light trucks. Under the footprint standards beginning in 2008 for light trucks and 2012 for passenger cars, the required fuel economy target for a manufacturer was weighted not only on its sales of passenger cars and light trucks but also on the sales-weighted average footprint of the vehicles it sold in a given model year.

The *Massachusetts v. EPA* Supreme Court ruling determined that carbon dioxide (CO_2) and other GHGs from motor vehicles must be regulated under Section 202 of the Clean Air Act (CAA; United States Code, 1990). Following from this ruling, California was enabled to request a waiver to regulate such emissions under CAA Section 209. With EISA and motor vehicle GHG emissions falling under the CAA, the MY 2012–2016 and MY 2017–2025 single standards were produced by NHTSA, the U.S. Environmental Protection Agency (EPA), and the California Air Resources Board (CARB). The new standards were structured around a footprint-based requirement and attempted to harmonize standards for fuel economy under the EISA mandate with GHG emissions standards under the CAA. These differed from the original CAFE structure, which had three groups of domestic passenger cars, imported passenger cars, and light trucks (Lattanzio et al., 2020). The national program standards included increased flexibilities in trading and banking of credits earned for overcompliance with the CAFE standards. There were credits for dual-fueled vehicles such as electricity and 85% ethanol blends, and in the GHG program, for improved refrigerants and for use of advanced technologies, such as electrification. It should be noted that NHTSA cannot consider alternative-fueled vehicles. The national program standards through MY 2018 reached an average of 25.1 MPG, with 29.9 MPG for passenger cars and 21.9 MPG for light trucks (EPA, 2019). These values were projected to reach 56 MPG for passenger cars and 40.3 MPG for light trucks by MY 2025, presenting an average fuel economy value of 49.6 MPG (EPA/NHTSA, 2012). EPA undertook a mid-term evaluation of the light-duty GHG standards for 2022–2025, with the proposed determination finding that automakers had outperformed the standards for the first 4 years of the program (MY 2012–2015) and that automakers were poised to meet the standards at lower costs than were previously estimated, in addition to the consumer and environmental benefits provided. However, in 2018, the EPA Mid-Term Evaluation Final Determination concluded that the MY 2022–2025 emissions standards were not appropriate and were in need of revision (EPA Press Office, 2020), prompting the development of the Safer Affordable Fuel Efficient (SAFE) Vehicles Final Rule. As noted in Chapter 2, the SAFE Rule results in lower fuel economy and emissions standards for MY 2021–2026 than previous regulations, with a projected 40.4 MPG required fuel economy in 2026, compared with 46.7 MPG previously (NHTSA, 2020).

U.S. automakers have typically expressed a preference for consistent and predictable regulations, given long product planning cycles (MacDuffie and Orts, 2019). Additionally, automakers do not want to create different vehicle models for different markets (Temple, 2019), making consistent regulations across markets desirable.

12.2 MEASURING FUEL ECONOMY AND GHG EMISSIONS

12.2.1 Discrepancies with Real-World Fuel Consumption

It has long been recognized that the combined city and highway test cycle estimates on which compliance with fuel economy and GHG standards are based are neither accurate nor unbiased estimates of what is achieved in actual vehicle use (McNutt et al., 1978). After extensive analysis of real-world data, the EPA proposed adjustment factors in 1984 that lowered the combined estimates by about 15%, on average (Hellman and Murrell, 1984). The adjusted MPG estimates were used to provide approximately unbiased fuel economy information to the public on window stickers, gas mileage guides, distribution to the media, and websites.

The problem of accuracy for individual drivers remained; as the vehicle labels say, "Actual results will vary for many reasons. . . ." For consumers, the chief limitation of existing fuel economy information is inaccuracy not bias. Analysis of almost 75,000 MPG estimates by individual motorists found that the variability around the label MPG estimates was substantial: root-mean-square errors of 21% for gasoline vehicles, 16% for hybrids, and 14% for diesels (Greene et al., 2017). In addition, the same data indicate that deviations from the label estimates are only weakly correlated among vehicles owned by the same household, suggesting that the variability decreases the usefulness of label estimates for comparing makes and models (Wali et al., 2018).

Given that changes in driving behavior, traffic conditions, and vehicle technologies could potentially increase the divergence between the combined cycle ratings and average real-world fuel economy, the EPA again revised adjustment factors in 2008 (EPA, 2006).The EPA revised its methodology again in 2017 to adapt to changes in vehicle technologies (EPA, 2017a). The relationship between combined cycle and label fuel economy is illustrated in Figure 12.1. A recent analysis of a large but self-selected database of on-road fuel economy collected by the website fueleconomy.gov suggests that the original 1984 correction factors may still be valid, but this cannot be known for certain without a statistically rigorous sample design and validity checks (Greene et al., 2017).

Technological advances have the potential to greatly reduce the cost and improve the accuracy of survey-based fuel economy information. Most vehicles now monitor their own fuel consumption and can report fuel economy estimates to the driver, log estimates in the onboard vehicle diagnostic, or even send them back to the manufacturer via an internal internet connection. There are issues with the accuracy of onboard estimates, but they are relatively small and solvable (e.g., Posada and German, 2013). GHG standards adopted in 2019 require the European Commission to monitor the real-world fuel and electricity use of light-duty vehicles and prevent the real-world emissions gap from growing, with some studies of options being carried out (e.g., Dornoff, 2019). Similar studies for the United States could help formulate a cost-effective plan for meeting a variety of needs for real-world fuel economy and GHG emissions data. One example is CARB's pilot study using onboard diagnostic and global positioning system data to assess vehicle emissions (CARB, 2020).

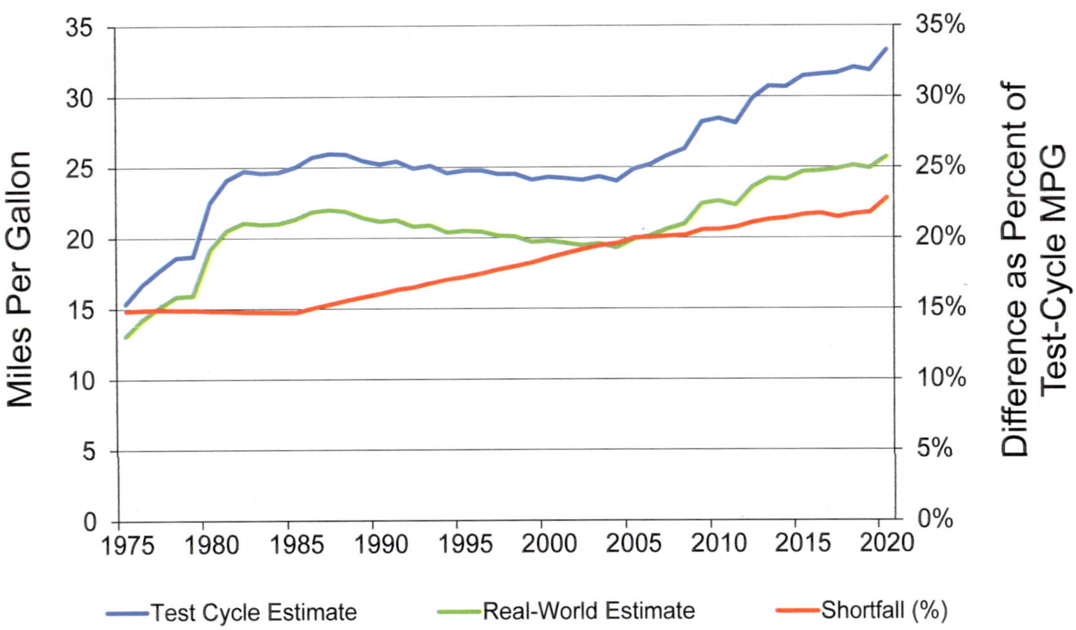

FIGURE 12.1 EPA test cycle versus real-world MPG estimates and shortfall: 1975–2019.
SOURCE: Committee generated using data from EPA (2020).

Given the magnitude of costs and benefits, a modest expenditure on real-world outcomes to validate fuel savings is called for. Given the importance of consumer choices for the direct and indirect consequences of the programs, research to develop more accurate fuel economy information for individual consumers seems appropriate. Traditionally, crediting has been based on test cycle, per the law governing the U.S. Department of Transportation (DOT) fuel economy standards. However, vehicles currently being produced/sold in the U.S. market can record fuel consumption over specific periods of time, which provides the capabilities for verifying performance and could enable a shift from the test cycle-based approach of estimating emissions to an approach of directly measuring emissions.

FINDING 12.1: There have been substantial improvements in onboard technology, allowing vehicle performance to be monitored and providing data that can be used to validate performance assumptions.

RECOMMENDATION 12.1: The agencies should undertake a large-scale data collection process, possibly in collaboration with automakers, which measures, records, and reports the real-world fuel economy of vehicles to evaluate fuel consumption and emissions reductions caused by certain off-cycle technologies. The agencies should consider a transition to using on-road fuel consumption and emissions data to adjust future crediting with the standards.

12.2.2 Test Cycles and Off-Cycle Corrections

The procedure of the two-cycle test for measuring vehicle fuel economy is described in Chapter 2 (Section 2.4). The test cycles, codified for fuel economy in EPCA 1975, were designed to measure fuel consumption while the vehicle operated a given speed versus time on a dynamometer. They were based on test cycles originally developed to measure vehicle tailpipe emissions. A comprehensive description of the Federal Test Procedure for city and highway driving is available in Kühlwein et al. (2014).

The regulations include an off-cycle credit program to account for technologies that improve the fuel efficiency and GHG emissions of a vehicle in real-world driving conditions but are not able to be measured in the test cycle conditions. For example, if vehicles have higher efficiency exterior lights, these can result in some fuel efficiency and GHG benefits on the road, but because lights are not turned on during the test-cycle, they would not be accounted for in the tests. To recognize the benefits of these types of technologies and incentivize automakers to adopt them, the agencies provide the ability for manufacturers to earn off-cycle credits for relevant technologies. Additionally, automakers can apply for additional off-cycle credits for other technologies if they present sufficient supporting data, which are evaluated by the agencies. EPA allows manufacturers to provide data justifying off-cycle credits either using the 5-cycle tests, or by using their own test procedure that is approved by EPA and subject to public comment. In 2012, EPA streamlined the off-cycle crediting process by preapproving credits in 13 technology areas. EPA converts approved credits to corresponding fuel consumption credits for NHTSA.

Provided that off-cycle credits are assigned accurate values representing real-world benefits of the technologies in terms of reducing fuel consumption and GHG emissions, they offer a mechanism of correcting for the inability of the test conditions to simulate real-world operation. However, in practice it is difficult to accurately estimate the real-world benefit, or verify whether the fuel consumption and GHG reductions assigned by the credits are realized on the road. The benefits of several off-cycle technologies such as solar/thermal control and active climate-controlled seats depend on human behavior to realize fuel consumption and GHG benefits. For example, actively cooled seats reduce fuel consumption by more effectively cooling the driver or passengers such that they would operate the air conditioning less or on a lower setting. These effects of human behavior are not easy to examine in test settings, and it is unclear if the manufacturers' estimated effects accurately represent real-world behavior. Currently, the evidence manufacturers present to the agencies is often not described in enough detail so that it is clear how the test procedures were conducted. Furthermore, the submission of off-cycle applications, and the agencies' approval process, is done on an ad-hoc basis. It is not clear what the agencies' criteria are for approval, creating an environment where automakers are developing detailed requests for how to change off-cycle crediting.

FINDING 12.2: Provided that off-cycle credits are assigned accurate values that represent the actual real-world benefits of the technologies in terms of reducing fuel consumption and greenhouse gas (GHG) emissions, they offer a mechanism of correcting for the inability of the test conditions to simulate real-world operation. However, in practice it is difficult to accurately estimate the real-world benefit, or verify whether the fuel consumption and GHG reductions assigned by the credits are realized on the road. Furthermore, the submission of off-cycle applications, and the agencies' approval process, is often done on an ad-hoc basis.

FINDING 12.3: Telemetric technologies exist that would allow agencies to collect fuel consumption and emissions data from vehicles on the road.

12.2.3 Emerging Technologies and Test Cycles

12.2.3.1 Crediting Promoting New Technologies

In the GHG regulations, flex-fuel and alternative fuel vehicles receive credits based on their potential to lower emissions in the long run. However, crediting likely raises emissions in the short term by effectively weakening the standard that manufacturers must attain. Over crediting may foster other social goals like promoting new technologies.

Electric and fuel cell vehicles are overcredited in two ways. First is that upstream emissions from electricity or hydrogen generation are not counted. Second is the multiplier, which counts each vehicle as more than one. Since these vehicles have emission rates below standard (partly because of the first type of over crediting), the multiplier effectively weakens the GHG standards. Jenn et al. (2016) show that these over crediting provisions raise emissions substantially. These provisions also mean that any other policy promoting these vehicles will raise emissions. For example, a federal tax credit for EVs increases sales of those vehicles, effectively loosening the standards and raising emissions.

The provisions could reduce emissions in the long term if they successfully boost near-term market shares, and if these higher market shares reduce market failures associated with new technology. For example, there may be spillovers in learning across firms in electric vehicle (EV) production. These spillovers would cause firms to underinvest in producing EVs. Boosting EV sales via over crediting could generate spillovers, reducing costs in the long term. In turn, lower costs could enable agencies to set tighter GHG standards in the future, reducing emissions in the long term.

This argument for over crediting may be valid in theory, but there are a few open questions. First, whether the trade-off between short-term and long-term emissions benefits society depends on the magnitude of market failures, and the committee is not aware of research assessing whether the market failures are sufficiently important (Linn and McConnell, 2019).

Second, there is not a strong argument for using the GHG standards to boost EV market shares rather than being technology neutral. Standards may not be the most economically efficient means through which to resolve market failures. That being said, to the extent that market failures would cause market shares under neutral standards to be lower than optimal, federal and state governments can subsidize them and use other policies to boost market shares. Some of the market failures, such as network externalities for charging infrastructure, likely vary at local levels, making local and state governments an appropriate level for policy. Eliminating over crediting would prevent perverse outcomes where states attempting to reduce emissions in the short term may actually increase U.S. emissions.

12.2.3.2 Battery Electric Vehicles

Calculations of EV range differ depending on the test cycle procedure used (Pike, 2012). The extent of the difference between test miles per gallon of gasoline-equivalent (MPGe) and real MPGe (and the resulting electric range calculated) influences whether a different test cycle procedure should be considered for battery electric vehicles (BEVs). Currently, most EVs use an EPA test cycle adjustment where values are adjusted by 0.7 to derive fuel economy label estimates reflecting real-world performance (Good, 2017). The effects of real-world use on BEV performance and exposure to weather are not captured as part of the test cycle. As described in Chapter 5, BEV efficiency and aging patterns depend on their use and on climate. In colder climates, extra energy is used for battery warm-up (Hu et al., 2020) and range will be negatively affected. The EPA test procedures drive each cycle several times until the battery is depleted. By this time (particularly for >250 mile range packs), the pack will be thermally equilibrated, so the average electricity consumption over time will disguise the MPGe one gets for short trips. In particular, short commuting trips and limited access to overnight parking with plug-in capability (allowing for slow charging and thermal maintenance) will be an important factor in colder climates. Similar efficiency and aging issues in addition to safety-related issues will need to be addressed for hot-weather use of BEVs, although EPA testing can modify the MPGe testing to do a hot/cold soak in between every cycle to check the discrepancy. Additionally, electric and fuel cell vehicles are also differently impacted by congestion than internal combustion engine (ICE) vehicles, as they are found to perform proportionally better in congestion, compared to free-flow conditions (Bigazzi and Clifton, 2015).

12.2.3.3 Fuel Cell Vehicles

As with BEVs, fuel cell electric vehicles (FCEVs) face performance challenges in cold climates, which would affect vehicle performance. As noted in Chapter 6, proton-exchange membrane fuel cells produce water as a by-product, and the fuel cell membrane must be properly hydrated for efficient ionic conduction and operation. At very high temperatures, water may evaporate, and the membrane may dry out, leading to degradation in performance. However, at sub-freezing temperatures, water in the pores of the membrane, catalyst, and gas diffusion layers may freeze. This can obstruct the flow of air (oxygen) and hydrogen, causing a decrease in power generation. Ice expansion can also lead to membrane electrode assembly (MEA) degradation, particularly over many freeze-thaw cycles. During normal operation, the heat generated by the fuel cell prevents the MEA components from freezing; however, control strategies are required during shutdown and start-up of FCEVs to prevent damage and enable quick start-up in sub-freezing climates. The details of control strategies implemented by automakers are proprietary; however, a number of reports have indicated their success. Fuel cell and vehicle manufacturers have developed successful strategies to mitigate the effects of sub-freezing temperatures on FCEV performance and prevent damage to the fuel cell over the short term. What is not clear in the public literature is the impact on fuel cell durability and performance of many freeze-thaw cycles that occur over a long period of time and the impacts of climate control and cold start strategies on FCEV fuel economy and range. However, researchers continue to study these issues—cold-start impacts (Gaoi et al., 2019; Wan et al., 2014), effects of freezing on fuel cell durability (Macauley et al., 2016), and climate effects on drive cycle and fuel economy (Henning et al., 2019; Lee et al., 2019).

12.2.3.4 Connected and Automated Vehicles

Connected and automated vehicles (CAVs) are not subject to substantial regulation at the federal level. As discussed in Chapter 7, in principle CAV technologies could be eligible for earning off-cycle credits; however, doing so would require effective documentation of the fuel savings that result from CAV technologies. Additionally, while many of the potential energy-saving benefits of CAVs would be undocumented on the test cycle, some CAV technologies could negatively influence fuel economy as measured on the test through drag or energy demands for CAV devices. These considerations present the need to think about how CAVs would integrate into the test cycle procedure and fuel economy standards. In the most recent guidance issued by the National Science and Technology Council and U.S. DOT, tests to study the performance of advanced driver assistance systems and automated driving systems performed by EPA are discussed in the context of informing fuel economy regulations with high-quality information (NSTC/DOT, 2020). In 2017, Senate legislation was introduced that would have established (among other committees) an expert panel on transportation and the environment that included fuel consumption in its scope (Canis, 2020). However, outside of these mentions, government discussion of how CAVs may be treated in fuel economy regulations has been limited.

12.2.4 Policies Affecting Passenger Vehicles

Other policy types can affect crediting, compliance, benefits, and costs of standards in complex ways. To explain how these policies interact with one another, it is helpful to distinguish between quantity- and price-based policies.

A quantity-based policy targets a quantity such as the level or rate of emissions. For example, a cap-and-trade program fixes the quantity of emissions that are permitted over a certain period of time. The fuel economy and GHG standards are quantity-based policies that determine the rate of fuel consumption or emissions per mile traveled.

In contrast, a price-based policy affects the cost of emissions or the cost of using a technology that depends on its emissions. For example, a carbon tax imposes a cost of emitting carbon. A subsidy for constructing new wind or solar generators is also a price-based policy because it reduces the cost of those investments. Note that market-based quantity policies introduce a price on emissions, such as the emissions credit price in a cap-and-trade program, but the distinction between quantity- and price-based policies depends on whether the policy targets the quantity or price of emissions.

A price-based policy that affects consumer demand for vehicles will also affect the compliance costs of the standards. For example, given the fuel economy and GHG standards, introducing a carbon tax on transportation fuels raises the cost of driving vehicles in proportion to their emissions. The tax reduces demand for vehicles with high emissions and raises demand for vehicles with low emissions, reducing the societal costs of achieving a particular emissions objective.

Likewise, price-based policies for EVs reduce the costs of the standards, including policies that reduce the cost or raise the benefit of having an EV, such as permitting access to HOV lanes, to the extent that they affect consumer demand for EVs. Information campaigns that increase demand for EVs have similar effects on the costs of the emissions standards.

Scrappage programs, such as Cash-for-Clunkers, that subsidize scrapping older vehicles and incentivize purchases of low-emitting new vehicles would reduce the costs associated with meeting fuel economy standards. Such programs would not affect the average emissions rate of new vehicles, but they would likely reduce economy-wide emissions by hastening the turnover of the vehicle. In short, any price-based policy that raises demand for low-emitting vehicles reduces the cost of achieving the standards.

The situation is more complex with quantity-based policies that affect new vehicles. A policy that affects only vehicles that are covered by the standards does not affect total emissions of new vehicles but does reduce the incremental costs of the standards. This is the situation with the zero-emission vehicle (ZEV) policy, which could mandate a combined market share of BEVs, plug-in hybrid electric vehicles (PHEVs), or FCEVs. (Strictly speaking, because of a ZEV crediting system, the program does not fix a market share of ZEVs, but this was ignored for expositional reasons. Accounting for the crediting provisions does not affect the main conclusions). Compared to a hypothetical case in which there are federal standards and no ZEV program, adding ZEVs does not affect the overall national emissions rate of new vehicles (again, putting aside the over crediting in the national programs), as long as the national program remains binding. Mandating a particular technology, such as ZEV, increases costs relative to a purely (non-zero-emissions) performance-based standard. If the federal programs contain a uniform set of standards across the car and truck classes and do not restrict trading across firms or classes, the total cost of ZEV and the federal programs exceeds the cost of the federal programs without this program. This is because the federal program without ZEV would be cost effective, minimizing the cost of achieving the standard (putting aside long-run technology considerations discussed above). Consequently, since market shares and technology choices would differ when ZEV is layered on top of the standards, costs per ton of emissions reduction must be higher than with federal standards only. Note that the existence of ZEV appears to reduce the estimated costs of fuel economy and GHG standards because the agencies estimate the incremental costs of their policies, assuming other policies are included in the baseline scenario considered. It should also be noted that ZEV and price-based policies, while they do not influence emissions in the contemporaneous CAFE regime, could lead to tighter standards in the next regime because of their ability to reduce compliance costs.

Policy making to promote ZEV sales and speed the retirement of older, higher-emitting vehicles may raise the possibility of establishing accelerated scrappage programs. These programs could increase the removal of older vehicles from the fleet, providing progress toward emissions reductions goals. However, the impact of accelerated scrappage programs on emissions reduction, used car markets, shift in travel to other vehicles and other modes, and equity between new and used car buyers is not well understood. Before an accelerated scrappage policy were to be implemented, further study would be required to analyze the ideal policy design of such a program.

RECOMMENDATION 12.2: Because the effectiveness and impacts of an accelerated scrappage program are not well understood, the U.S. Environmental Protection Agency, in conjunction with the U.S. Department of Energy and U.S. Department of Transportation, should study the effectiveness of accelerated scrappage programs at emissions reduction, increasing zero-emission vehicle sales, and addressing equity considerations. Such a scrappage study could assess factors including which vehicles should qualify for the program, the impact of altering vehicle age mix on vehicle miles traveled, the shift of travel to other vehicles and modes from the scrapped vehicles, and the impact on the used vehicle market and on travel in used vehicles, especially for lower-income vehicle buyers and users. The cost and effectiveness of accelerated scrappage programs should be compared to other means of reducing greenhouse gas emissions from the light-duty vehicle sector.

12.2.5 Policies Affecting Other Sectors

Federal standards and policies may affect multiple sectors. Regulations pertaining to EV emissions may relate to the electricity sector, for example. Emissions reduction is one of the three primary goals of the CAFE/GHG regulations. When considering all three goals of fuel consumption, energy, and emissions reduction, there are other policies that relate to and influence the light-duty sector's impact on these goals. Notably, if deep GHG emissions reduction is a goal, then there will need to be consideration of not only onboard vehicle emissions but also the emissions from related sectors, like electricity (for vehicle charging) and manufacturing (of vehicles and their materials and components). This motivates the need for life cycle thinking.

12.2.5.1 Background on Vehicle Life Cycle

The life cycle of a vehicle encompasses the creation of its materials, the assembly process, use of the vehicle, and disposal and recycling at its end of life. The life cycle energy or environmental impacts are the total of these impacts from each of these stages. Typical systems boundaries for vehicle life cycle assessments and what processes are included in their system boundaries are depicted in Figure 12.2. The well-to-wheels life cycle for a vehicle encompasses stages from resource extraction through onboard vehicle energy use. The well-to-tank life cycle encompasses resource extraction through the distribution of energy carriers but does not include vehicle use.

Life cycle thinking becomes particularly important when considering the total fuel consumption and environmental impacts from different vehicle types, especially when some do not directly combust or emit themselves. In the 1990s, California among other states set regulations to attain 2% ZEV sales by 1998. At the time, EVs were constructed with lead acid batteries, and when comparing the life cycle environmental impacts of an electric car using a lead acid battery with those from a conventional automobile, the ZEV resulted in 60 times more lead releases per km of vehicle use than a conventional car using leaded gasoline (Lave et al., 1995). CARB did not follow through with this early ZEV mandate, influenced in part by life cycle studies (Matthews et al., 2014).

For ICE vehicles, the majority of a vehicle's environmental impacts are created during the use phase of its life cycle. For example, a life cycle assessment of a Volkswagen Golf VII (assumed to be driven 200,000 km) found the use phase comprising 79% of total GHG emissions, compared with 20% of the total coming from vehicle production, and 1% from end-of-life impacts (Broch et al., 2015). This finding indicates that the GHG emissions from fuel production, transportation, and combustion from driving 200,000 km outnumber those from manufacturing, assembling, and end of life for vehicle components by nearly a factor of four. However, the breakdown of life cycle emissions is very different for an EV, whose manufacturing phase emissions exceed those of conventional

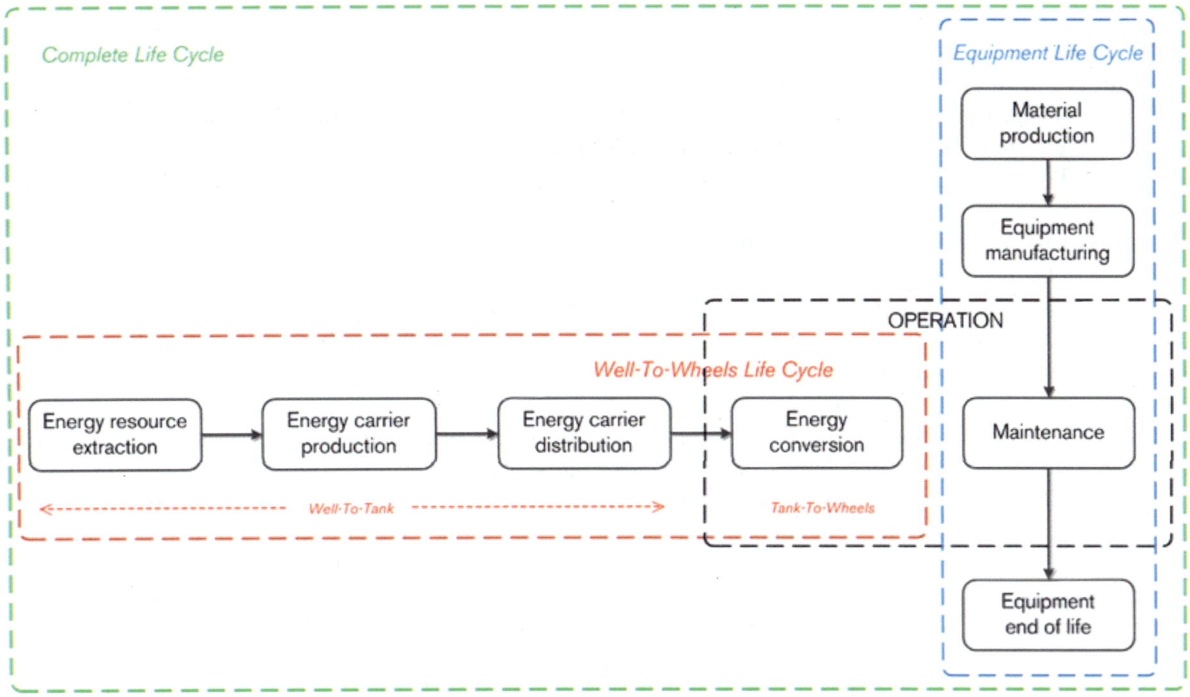

FIGURE 12.2 Simplified depiction of life cycle stages and boundaries for a vehicle.
SOURCE: Nordelöf et al. (2014).

vehicles owing to battery production (although often with lower total life cycle emissions, owing to the use phase) (Hall and Lutsey, 2018).

With emissions from fuel combustion typically creating the large majority of a vehicle's life cycle burdens, the composition of a vehicle's fuel takes on an important role. Policies like the Renewable Fuels Standard draw on life cycle comparisons of fuel GHG emissions, with EPA's approved fuel pathways requiring life cycle GHG reductions in substitute fuels ranging from 20% for ethanol derived from corn starch to a 60% reduction for cellulosic biofuel, when compared with a 2005 petroleum baseline, excluding older facilities grandfathered into the program (EPA, 2017b). Regulating fuels can provide meaningful GHG emissions reductions, with an analysis of California's Low Carbon Fuel Standards finding them resulting in a 10% reduction in CO_2 in California's transportation sector (Huseynov and Palma, 2018). This standard also treats electricity as fuel, with the discussion of different fuels' emissions presented in Chapter 10.

The regulatory accounting for use-phase energy and environmental impacts from vehicles without direct combustion during use is more complex. A review of life cycle assessments of BEVs finds them to be often more energy efficient and less polluting than conventional vehicles, and their GHG emissions are highly sensitive to the carbon-intensity of the electricity mix (Helmers and Weiss, 2017). In the SAFE Rule, there is an extension of EPA assigning EVs 0 grams of upstream emissions per mile through MY 2026 (NHTSA/EPA, 2020), excluding the burdens from electricity used to power the vehicle. This assumption will be revisited beginning for MY 2027. For PHEVs and BEVs, NHTSA applies a petroleum equivalency factor to the measured electrical consumption to determine the gasoline equivalent fuel economy for operation on electricity, although with an incentive only counting 15% of the energy consumed as electricity.[1] The extent

[1]The petroleum equivalency factor is derived from physics, and this 15% is based on a carryover of identical incentives for all alternative fuels from E-85, without any physical meaning except for E-85 vehicles.

to which automakers should or should not be held accountable for emissions attributable to the electricity grid mix remains a contentious topic.

With vehicle use driving the bulk of a vehicle's environmental impacts, lightweighting presents a potential means for reducing the amount of fuel consumed per mile driven, with an extensive discussion present in Chapter 7. However, the manufacturing of materials to attain vehicle lightweighting does influence the life cycle impacts of the vehicle. For example, an analysis comparing lightweighting using high strength steel or aluminum found that the impacts varied notably depending on the location of aluminum production and whether secondary (recycled) aluminum could be used (Kim et al., 2010). While this analysis found the vehicle's use-phase to still contribute 87%–95% of life cycle GHG emissions, the choice of lightweighting material did still matter, with it taking longer for use-phase emissions reductions to offset the added manufacturing emissions for aluminum than for high strength steel. The SAFE Rule does not account for upstream materials production emissions, citing the complexity of processes involved and in providing comparisons across materials (NHTSA/EPA, 2020). Automakers have sustainability goals, which often include reducing manufacturing emissions. Twelve green design principles have been proposed to guide applications of lightweighting and attain environmental improvements, including resolving technical, economic, and environmental performance trade-offs while maintaining vehicle safety; sourcing abundant and low-impact materials; and designing for material recovery, among others (Lewis et al., 2019). In general, effective recycling programs for vehicle components could decrease life cycle energy and environmental burdens.

The SAFE Rule's Final Environmental Impact Statement reviews the relevant life cycle assessment literature, concluding that most efficiency-increasing technologies would decrease GHG emissions, energy use, and other environmental impacts on a life cycle basis (NHTSA/EPA, 2020). Uncertainty related to the upstream production is noted, with its potential to reduce environmental improvements attained. Regulations pertaining to upstream and connected sectors will have implications for vehicle life cycle emissions, with the electric sector and EVs being a prominent and timely example.

12.2.5.2 Scenarios of GHG Regulation for the Electricity Sector

As EVs become a more prominent element of the fleet, the emissions-intensity of the electricity sector will have a larger influence on light-duty vehicle emissions. Given recent history of GHG regulation from the electricity sector, several cases are considered. The simplest would be a binding cap on national GHG emissions from the electricity sector. In this case, charging an EV has no effect on electricity sector emissions because any increase in fossil-fuel fired generation caused by the charging would be offset by an equivalent decrease in emissions somewhere else in the electricity system, leaving total electricity sector emissions unchanged. In other words, the cap determines total emissions, and an increase in consumption caused by charging an EV cannot affect total emissions. Consequently, if there is a national, binding cap on electricity sector emissions, EPA should consider EVs to have zero emissions for crediting purposes (and likewise for the electric portion of a PHEV). A similar argument would pertain to a binding national cap on multiple sectors including electricity.

The second case is a regional cap on emissions, such as the Regional Greenhouse Gas Initiative or AB 32. In the context of an electricity sector cap, emissions leakage refers to a situation in which pricing emissions from one region cause electricity generation to increase in other unregulated regions because the cap raises the cost of generating electricity inside the region, relative to the cost outside the region. As long as there is transmission connecting the two regions, electricity generation could increase outside the region, causing an increase in emissions. For example, AB 32 covers California but not Arizona, and California often imports electricity generated in Arizona. Capping emissions in California raises the cost of fossil fuel-fired generation in California relative to Arizona, which could increase fossil fuel-fired generation in Arizona and imports from Arizona to California. This would cause total emissions across California and Arizona to increase, compared to a hypothetical in which there is no transmission between the two states. In other words, the cap is attained by shifting emissions from California to Arizona, which does not reduce aggregate emissions.

With a regional cap, if one ignores emissions leakage outside of the program, charging EVs in the region does not affect total emissions for the same reason as with a national cap. In that case, EVs charged in that region should be considered zero-emission vehicles for crediting purposes. In principle, EPA could allow for emissions leakage in its crediting calculations based on scenario analysis of electricity sector models or a review of the literature on emissions leakage.

The final case is a regional or national electricity emissions rate standard, rather than a cap. For example, the Clean Power Plan (CPP) and Affordable Clean Energy (ACE) rule set GHG emissions rates for certain fossil fuel-fired generators. If the standard includes all generators (which it could have with the CPP but not ACE), then the emissions from an EV equal the amount of electricity consumed in charging multiplied by the emissions rate set by the standard.

Well-to-wheels and the equipment lifetime emissions can also be addressed through regulations that do not pertain to light-duty vehicle fuel economy. For example, certain fuels may be used in a variety of sectors, so attaining emissions reductions via a fuels-based light-duty vehicle standard may not be the most straightforward or effective method.

12.3 REGULATORY FLEXIBILITIES

12.3.1 Credit Trading System

Both the CAFE and GHG regulations allow manufacturers to earn credits for exceeding the standards, transfer credits within their firm, carry credits forward and backward within specific time windows, trade credits with other firms, and use credits to comply with the standards. Rules and penalties differ between the two regulations, and credits earned under one regulatory regime may not be used in another. Further background on the historical regulations and current CAFE and GHG targets is provided in Chapter 2.

The objective of the credit systems is to improve the regulations' economic efficiency and to reduce the chances of severe harm to any manufacturer. Banking and borrowing credits makes it easier for manufacturers to deal with shocks such as changes in fuel prices, and to achieve a more-uniform marginal cost of compliance over time. In the global automotive market, many fuel economy technologies are available to all manufacturers. However, some technologies are proprietary to a particular manufacturer, and others are patented, which can increase their cost to competing manufacturers. Firms may also differ with respect to their internal expertise or assessment of the technological prospects or market acceptance of different strategies. Firms serving different market segments may also find that their customers' preferences for fuel economy, technological solutions, or other vehicle attributes differ. Because vehicle design decisions must be made years in advance and manufacturers' expectations may differ, some strategies for meeting standards may prove to be more successful than others. For all these reasons, the costs of meeting CAFE and GHG standards may vary from one manufacturer to another.

When manufacturers face different costs of compliance, credit trading can enable firms with higher costs to purchase credits from firms with lower costs at favorable prices. Credit trading thereby allows firms to meet the standards at lower costs than would be possible if every manufacturer were individually required to meet the standards. NHTSA CAFE credits (or deficits) are calculated as 10 times the difference between the manufacturer's sales-weighted harmonic mean fuel economy (MPG) and its regulatory requirement, multiplied by the number of vehicles sold in the model year in question. Because there are separate car and truck standards, car and truck credits are earned separately.[2] Because the credits are defined in terms of fuel economy, the car and truck credits are not equivalent in terms of expected fuel savings or GHG emissions. Credits traded between car and truck fleets must be adjusted to approximate equivalence in terms of expected gallons of fuel saved. The amount of credits that can be transferred from one category to another is capped at 2 MPG from 2018 on (He, 2014). The EPA GHG standards, on the other hand, are calculated as the expected difference in CO_2 emissions over a vehicle's lifetime, meaning these credits can be accurately directly traded between car and truck fleets.

According to (49 CFR 536.6), CAFE credits earned before 2008 could be carried forward for only 3 years. EPA credits GHG emissions and NHTSA credits fuel economy, but fuel economy credits can be traded across vehicles after converting to fuel consumption. In the CAFE program, firms can pay a penalty for non-compliance

[2]A manufacturer's sales-weighted harmonic mean fuel economy is equal to the inverse of the sales-weighted arithmetic average of rates of fuel consumption.

TABLE 12.2 Comparison of Credit Provisions under NHTSA and EPA Programs

Regulation	NHTSA CAFE Program	EPA GHG Program
Definition of a credit	1/10 MPG above manufacturer's required MPG standard for fleet.	1 Mg of CO_2 below the manufacturer's required standard.[a]
Credit banking (carry forward)	5-year banking period.	From 2009 to 2011, companies banked credits through the Early Crediting Program; 5-year banking period, with the exception that credits earned between 2010 and 2016 can be carried forward through 2021.
Credit borrowing (carry back)	3-year carry back period.	3-year carry back period.
Limits on manufacturers' credit transfers between car and truck fleets	Limits on credits[b] that can be transferred between cars and trucks: MY 2011–2013, 1.0 MPG MY 2014–2017, 1.5 MPG MY 2018 on, 2.0 MPG	No limits on transfers between cars and trucks in each manufacturer's fleet.
Monetary cost of noncompliance	Fee of $5.50/tenth mile over standard, per vehicle; starting 2019, $14/tenth mile over standard.	Unknown penalty, but could be as high as $37,500 per car for violation of the CAA.
Provisions for alternative fuel vehicles	Credits for ethanol and methanol in fuels are being reduced. For electric vehicles, electricity use is converted to equivalent gallons of gasoline and only 15% of that is counted for compliance.	Allows manufacturers to count each alternative fuel vehicle as more than a single vehicle. Multipliers range from 2.0 to 1.3, depending on the extent of alternative fuel used and the MY. Emissions from battery electric vehicles assumed to be zero.
Exemptions	No exemptions for manufacturers with limited product lines; fines can be paid.	Temporary Lead-Time Allowance Alternative Standards for manufacturers with limited product lines through 2015.

[a]Vehicle and fleet average compliance for EPA's GHG program is based on a combination of CO_2, hydrocarbons, and carbon monoxide emissions, which are the carbon containing exhaust constituents. These GHG emissions are referred to here as CO_2 emissions for shorthand.

[b]There are also some restrictions by NHTSA on transfers and trades between imported and domestically produced car fleets and truck fleets.

SOURCE: Leard and McConnell (2017).

of $55 per vehicle per MPG. Non-compliance with the GHG standards puts the firm in violation of the CAA, and applicable fines are assessed in court on a case-by-case basis. For example, in 2014 Hyundai-Kia paid a penalty of $100 million for violating the CAA and emitting an estimated excess of 4.75 million metric tons CO_2-equivalents from their MY 2012–2013 vehicle fleets (EPA, 2014). Under the CAA, EPA also has the authority to revoke a non-complying manufacturer's ability to continue to sell vehicles in the United States.

The value of fuel economy and GHG credits may differ because of the different penalties for non-compliance with the two standards. The fine for failure to meet the CAFE standards (49 CFR § 578.6(h)) was initially set at $5 per 0.1 MPG per vehicle and was raised in 1997 to $5.50 per 0.1 MPG per vehicle (62 Fed. Reg. 5167). Authority to do so was given by the Debt Collection Improvement Act of 1996 (P.L. 104-134 Sec. 31001). This fine was increased to $14 per 0.1 MPG as directed by the Federal Civil Penalties Inflation Adjustment Act Improvements Act of 2015 (P.L. 114-74 Sec. 701), which requires fines to be adjusted for inflation (Laing, 2018), but then later reduced to $5.50, although this remains in litigation at the time of writing (81 Fed. Reg. 43524; modified by 81 Fed. Reg. 95489).

The key differences between the CAFE and GHG credit provisions are summarized in Table 12.2.

12.3.2 Rationales for Flexibilities and Trading

The crediting system allows manufacturers to average the emissions rates and fuel economy across their vehicles. Consequently, each vehicle sold by a manufacturer does not have to achieve its particular requirement, and the manufacturer can sell vehicles that fall short of their requirement as long as the manufacturer sells a sufficient number of vehicles that exceed their respective requirement or else the manufacturer obtains sufficient credits by other means (e.g., purchasing from another firm). Firms can transfer credits across car and truck fleets, and under the CAFE program, firms can transfer credits to increase fuel economy up to 2 MPG. There is no limit on credit transfers in the EPA program.

This section considers how the agencies could implement a uniform standard that applies across cars and light trucks, where the term "uniform" is used to mean that a vehicle's fuel economy requirement depends only on its footprint and not on its class. NHTSA has argued that the law prevents it from considering compliance credits when setting standards, and therefore it cannot set uniform standards across cars and light trucks. NHTSA uses a computational model to estimate costs and benefits of particular levels of standards, and the agency uses the modeling results to help choose levels of standards. If NHTSA were to allow cross-class credit trading when it models the standards, the agency would estimate lower costs for any level of standards, compared to modeling costs without cross-class trading. Consequently, the agency could justify more stringent standards by including cross-class credit trading when determining the level of the standards. In contrast with NHTSA, there does not appear to be anything in the statutes preventing EPA from setting a uniform standard and allowing unlimited credit trading across classes, which would be cost effective.

Compliance flexibility reduces the overall cost of achieving a particular fuel economy or GHG emissions rate target, compared to a hypothetical policy in which each vehicle has to achieve its particular standard. Marginal abatement costs—the change to the manufacturer in cost per unit of emissions or fuel consumption reduction—may vary across the vehicles a manufacturer sells. For example, suppose a manufacturer specializes in large cars and sport utility vehicles (SUVs), and that the cost of adding fuel-saving technology is higher for the SUVs than for the large cars—either because the cost of installing the technologies is higher or because consumers of the SUVs are less willing to pay for fuel economy improvements than consumers of its mid size cars. In this hypothetical, the manufacturer can offer a fuel economy in the large cars that exceeds the fuel economy requirement for those cars and can offer a fuel economy for the SUVs that falls short of the requirement for the SUVs. This compliance strategy would be less costly than choosing fuel economy of both vehicle types that exactly equal the respective requirements, because in that case the manufacturer would have to add extra technology to the SUV, which is relatively expensive.

Aside from offering flexibility to average across a manufacturer's vehicles to determine overall compliance with the standards, the standards have required lower emissions and fuel consumption rates for cars than for light trucks. The rationale appears to be that technology costs vary across classes and that trucks often have different uses. For example, pickup trucks and SUVs may be used for towing heavy loads or off-roading capabilities.

However, this argument is faulty, particularly for EPA (which does not impose restrictions on credit trading across classes), and it places too much burden on regulatory agencies to estimate the costs of reducing emissions and fuel consumption across the two classes. As explained here, setting separate standards raises the costs of meeting standards compared to setting uniform standards across the two classes.

Consider a hypothetical regulator that knows the marginal abatement costs of each vehicle. The regulator would minimize the cost of achieving a given emissions reduction by setting standards to equate marginal abatement costs across vehicles. This is true because if the regulator did not equalize marginal costs, it could always adjust standards to reduce total costs. For example, suppose the regulator sets the standard such that the marginal cost of raising fuel economy is higher for light trucks than for cars. The regulator could reduce the total cost of compliance by weakening light truck standards and tightening car standards. Therefore, if the regulator is setting standards separately for cars and light trucks, and if marginal costs are expected to be higher for light trucks than for cars, the regulator would impose weaker standards on light trucks than cars (in this case, weaker means a lower fuel economy and higher GHG emissions rate conditional on footprint). Of course, it should be noted that a single set of standards would harm some manufacturers and help others.

Although EPA and NHTSA have a lot of information about technology costs and performance, they do not have perfect foresight, and their predictions of costs may turn out to be inaccurate. In that case, setting standards based on expectations will cause realized marginal costs to be unequal across vehicles. Hence, with incomplete information, setting different standards across vehicles is not cost effective.

Cost effectiveness is only achieved if there are uniform standards with unlimited trading; allowing one but not the other would raise compliance costs. With unlimited trading across classes, a firm could minimize compliance costs by equalizing marginal costs across vehicles. For example, if EPA and NHTSA set standards for future model years and costs turn out to be higher for light trucks than for cars, the firm could over-comply for cars and under-comply for trucks, transferring the excess credits for cars to its trucks. Such overcompliance would not be allowed without credit trading across classes. Likewise, restrictions on the amount of credit trading a firm can use, as currently exist in the NHTSA program, raise compliance costs.

The recent fluctuations in gasoline prices—such as the 25% decrease that occurred in March and April 2020 owing to COVID-19—illustrate the importance of setting uniform standards and allowing unlimited trading. Low gas prices generally reduce demand for fuel economy and raise compliance costs, and probably by different amounts across vehicles (Leard et al., 2019). Standards finalized in 2020 were set based on gas price projections from 2019. If actual gas prices turn out to differ from those expected prices, actual compliance costs will differ from expectations. Consequently, if agencies had tried to equate marginal costs across vehicles when they finalized the standards, those standards would likely not be cost effective. If low gas prices persist past 2020, unlimited trading across firms and classes would allow manufacturers to minimize compliance costs given the realized lower gasoline prices. That is, unlimited trading gives manufacturers the flexibility to respond to unexpected changes in market conditions.

To be cost effective, marginal costs would have to be equated not just across vehicles within a firm but also across firms. Trading credits across firms could equalize marginal costs, as low-cost firms over-comply and sell to high-cost firms. But as shown below, credit trading has been fairly limited, and it is unlikely that marginal costs are equal across firms.

In principle, the agencies could use the fact that there has been limited trading to justify standards that vary across classes. The argument would be that because cross-firm trading does not equalize marginal costs across firms, the agencies could at least reduce variation in marginal costs across firms by equating marginal costs across classes. This would be a crude approach to trying to equalize costs across firms, because although it might reduce variation across firms, it would increase variation across classes. Another way to reduce the variation in marginal costs across firms would be to include a cost containment mechanism—sometimes called a safety valve—which would be similar conceptually to features of many cap-and-trade programs and renewable portfolio standards. Under this approach, a firm with high compliance costs could purchase credits from EPA at a specified price.

FINDING 12.4: Compared to setting separate fuel economy and greenhouse gas standards for cars and light trucks, a single standard across cars and light trucks would reduce the total societal costs of achieving a particular level of fuel consumption or emissions, if cross-class trading is included. Equivalently, a single standard would cause greater fuel consumption and emissions reductions than separate standards that have the same cost as the single standard.

12.3.3 Review of Credit Trading to Date

12.3.3.1 Data on Fuel Economy Trading

NHTSA (2019) publishes an annual Credit Status Report listing the number of positive or negative credits for each manufacturer by three categories: (1) domestic passenger cars, (2) imported passenger cars, and (3) light trucks. At the time of writing, the most recent year data were available is 2017. An estimate of credits earned by MPG performance only from 2008 to 2012 can be calculated from NHTSA's Manufacturer Performance Report. NHTSA does not report credit trades between manufacturers, nor does it report the prices of credits traded. NHTSA CAFE reports also provide manufacturer CAFE numbers with and without the credits earned by selling dual and flex fueled vehicles. These allow some inferences to be made about credit trades and transfers by a manufacturer but, in general, do not allow complete estimation of trades and transfers.

From 2008 to 2011, the industry accumulated credits at an annual rate of 35–40 million per year. Manufacturers with deficits reduced the net credits earned by 1.6%, 0.9%, 1.2%, and 7.4% in 2008, 2009, 2010, and 2011, respectively.

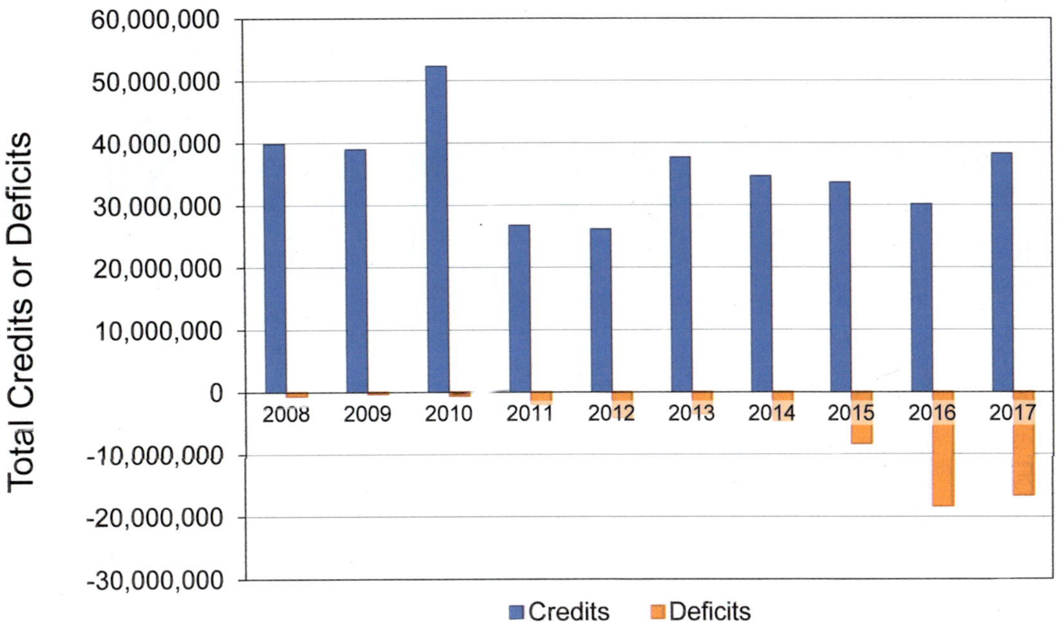

FIGURE 12.3 Annual MPG performance credits and deficits held by automakers.
SOURCE: Committee generated using data from NHTSA (2019).

The number of manufacturers with annual deficits increased after 2008, and total deficits grew to 61% of earned credits in 2016 followed by 44% in 2017 (Figure 12.3). Possible explanations for the increase in deficits include the sudden drop in gasoline prices of approximately $1 per gallon in 2015, the increasing stringency of the standards, and the expiration of 75 million credits earned in 2009 at the end of 2014.

12.3.3.2 Data on GHG Trading

EPA publishes data on GHG credits and credit sales and purchases in annual performance reports for manufacturers (EPA, 2020), as well as a substantial amount of related data. Compliance numbers (achieved MPG) are reported by manufacturer, vehicle class, and model year. Figure 12.4 shows the industry average GHG emissions (blue bars) and standard (green bars) for 2009–2018. EPA reports that in 2014 74,843,471 (76%) of the 98,520,511 early credits earned in 2009 expired unused. All credits earned between 2010 and 2016 expire in 2021; all credits earned in subsequent years expire after 5 years.

Although the industry's net deficit decreased from 2016 to 2018, all but five manufacturers experienced deficits in 2018, and the great majority of credits (91%) were earned by just two manufacturers: Tesla and Honda. The manufacturers shown in Figure 12.5 are arranged from left to right in order of sales volume and account for 99% of 2018 sales.

EPA reported credit sales and purchases by manufacturers from 2010 to 2015, but has reported only cumulative purchases and sales by manufacturer at year end since 2015. The data reveal a very thin market for trades between 2010 and 2015 (Figure 12.6). In 2015, only one trade for MY 2017 was reported, for Fiat Chrysler Automobiles (FCA). From 2012 to 2015, less than 1% of the credit balance existing in any given year was traded. In 2018, 7.1% of the existing credit balance was traded, but FCA alone accounted for 71% of credits purchased. They were the dominant buyer and Honda the dominant seller. Although the published EPA data provide sales and purchases,

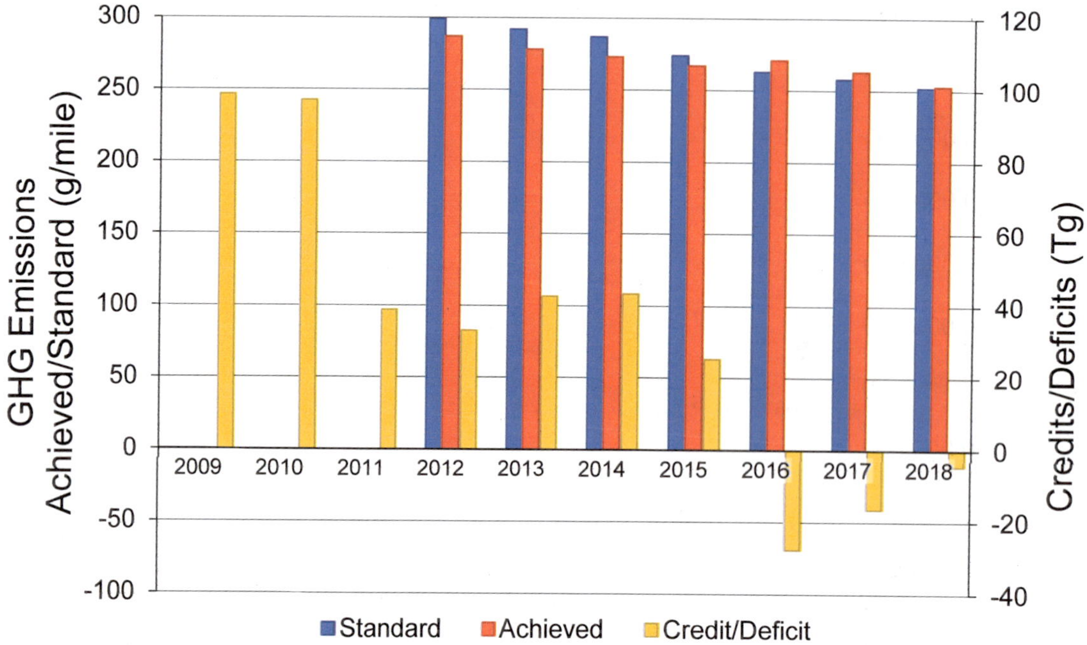

FIGURE 12.4 Industry GHG performance and standards: 2009–2018.
SOURCE: Committee generated using data from EPA (2020).

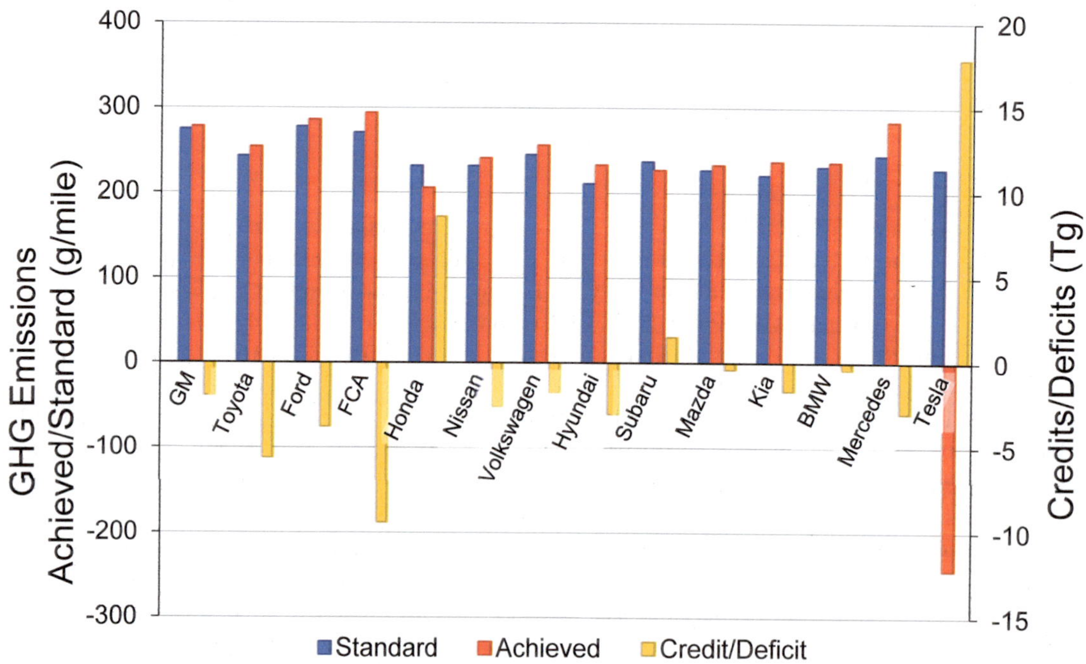

FIGURE 12.5 GHG emissions achieved, standards, and credits/deficits by manufacturer: 2018.
SOURCE: Committee generated using data from EPA (2020).

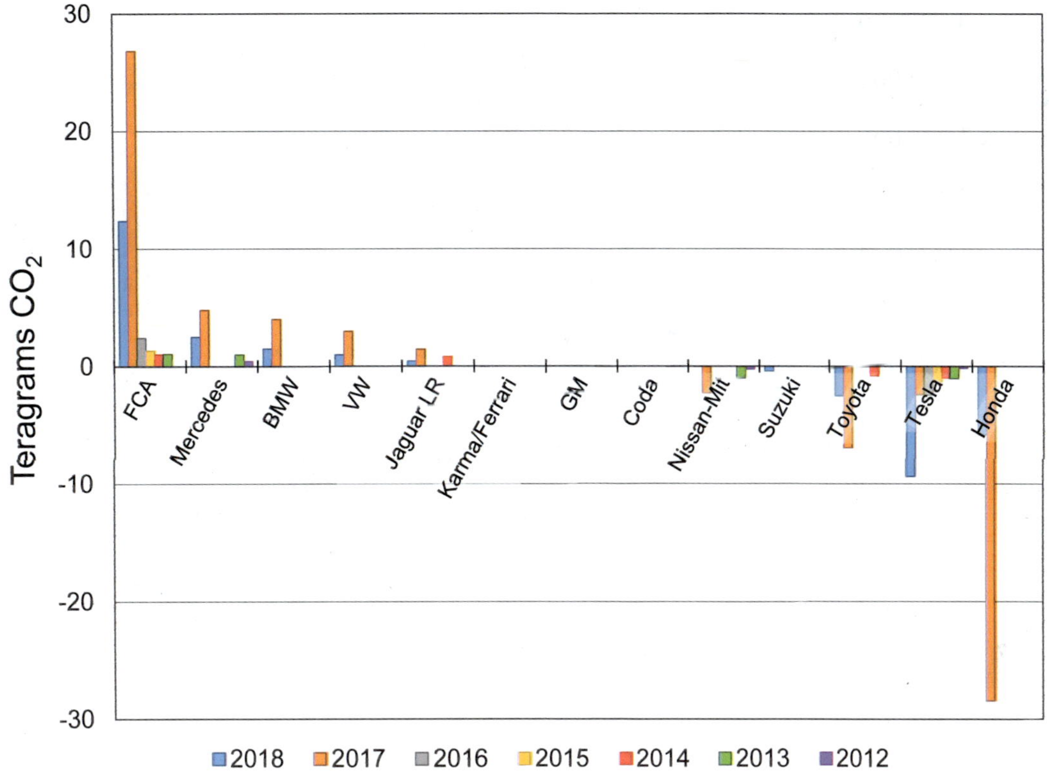

FIGURE 12.6 GHG credit purchases (+) and sales (–) by model year (2012–2018).
SOURCE: Committee generated using data from EPA Automotive Trends Reports and Manufacturer Performance Reports.

they do not include bilateral trades or prices paid. Over 2012–2018, nine manufacturers did not buy or sell GHG credits, including Ford, Hyundai/Kia, Mazda, Mitsubishi, Subaru, and Volvo. Figure 12.6 shows only the manufacturers that actually bought or sold credits.

Forty-eight Teragrams (Tg) of CO_2 credits were traded in 2017, a substantial increase over the 12.7 Tg traded in 2016. However, 76 Tg of credits were allowed to expire. Toyota allowed 29.7 Tg of credits to expire, while 14.1 Tg of Honda's credits expired—58% of total expired credits in MY 2017. Overall, the industry had a credit deficit of 18 Tg CO_2 in MY 2017.

The largest credit balance is held by Toyota, with credits of more than 60 million tons of CO_2 banked at the end of MY 2017 (Figure 12.7). Although Tesla has earned a large number of credits through its sales of EVs, it had sold most of the credits it had earned by the end of 2018.

12.3.3.3 Credit Prices

Leard and McConnell (2017) used SEC filings by Tesla to estimate the price of GHG credits in 2012 and 2013. They divided the reported proceeds from the sales by the quantity of credits Tesla traded. They estimated prices of $36 per credit in 2012 and $63 per credit in 2013 (2014$). They also used the results of a 2014 settlement between the U.S. government and Hyundai/Kia to derive a credit price estimate. The settlement required the firm to forego 4.75 million GHG credits that the EPA estimated to be worth in excess of $200 million, resulting in an estimated credit price of $42 (2014$). The NHTSA fine is $5.50 per extra

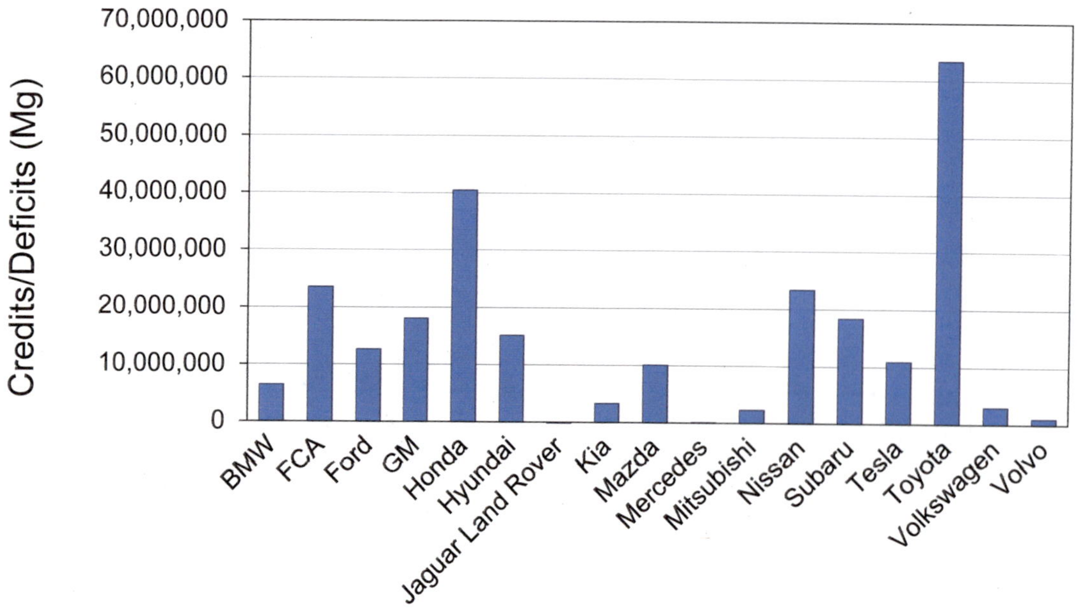

FIGURE 12.7 GHG credit balances after MY 2018 by manufacturer.
SOURCE: Committee generated using data from EPA (2020).

0.1 MPG exceeded, recently reduced from $14. Given the very small number of trades, these estimates should be interpreted with caution.

FINDING 12.5: Credit trades are not transparent (i.e., prices not observed), which makes evaluation of manufacturer compliance costs more difficult.

RECOMMENDATION 12.3: The National Highway Traffic Safety Administration should publicly report credit trade quantities and prices between manufacturers. This reporting would increase transparency and provide useful information for economic analysis of the regulations.

12.3.4 Effect of Credit Trading on Compliance Costs

Several studies have estimated that the flexible use of credits provides substantial benefits to manufacturers. Using simulation modeling and the CAFE standards for 2001 as the base case, Kiso (2017) estimated that CAFE credit trading lowered the total cost of compliance by $110–$140 million. Liu et al. (2014) used optimization modeling to estimate the effects of air conditioning and flex fuel vehicle credits and banking of credits on the shadow price of compliance with GHG standards from 2011 to 2020. They found that removing the ability to use air conditioning and flex fuel vehicle credits induced non-zero shadow prices in years in which the standards were otherwise not binding and increased shadow prices by 75% to 133% in years in which the standards were binding in their base case (e.g., from $12.60 to $29.40 in year 2019; 2007$). Rubin et al. (2009) estimated the benefits to manufacturers of credit transfers across vehicle classes within a firm and credit trading among firms. Recognizing the relatively small number of firms that would participate in a credit trading market, they assumed an imperfect market characterized by oligopoly and oligopsony. Assuming perfectly competitive credit markets, they found that allowing within firm transfer of credits among vehicle classes reduced the average cost of compliance by 7%–10%, depending on assumptions about the cost of fuel economy technology, gasoline prices, and consumer payback periods. Adding inter-firm trading reduced costs by another 3%–8%, depending on assumptions. Assuming

Cournot-Nash oligopolistic competition had almost no effect on the average benefits of credit trading. The effects of an oligopolistic versus competitive credit market on individual small producers could be large, however. Rubin and Leiby (2000) estimated the value per vehicle of fuel economy credits for alternative fuel vehicles during the 1991 to 1998 period to be $1,100 for a dedicated alternative fuel vehicle and $550 for a dual or flexible fuel vehicle.

FINDING 12.6: Credit trading across a manufacturer's car and truck fleets appears to have reduced overall compliance costs. Trading across firms is limited, however, so it is not clear to what extent it is lowering the costs of compliance.

12.3.5 How Changing Utilization Affects Crediting

EPA and NHTSA estimate the changes in fuel consumption and emissions caused by standards using assumptions on vehicle use over the life of the vehicle. The agencies make assumptions about the relationship between vehicle age and probability of being scrapped, as well as miles traveled by age, conditional on surviving to that age. These estimates are the basis for crediting lifetime GHG emissions in the EPA program and for trading fuel economy credits in the NHTSA program.

New technologies like CAVs could affect vehicle use, as well as the timing and magnitude of fuel consumption/emissions changes. Magnitude matters because crediting is based on lifetime emissions and fuel consumption. Timing matters because of discounting. If two vehicles have the same lifetime emissions and one vehicle is used more heavily at beginning of life, discounting at a strictly positive rate would mean that this vehicle has a higher present-day value of lifetime fuel costs and emissions.

The possibility that new technologies affect vehicle utilization really is not any different conceptually from changes in driving and scrappage. Agencies have been updating scrappage and vehicle miles traveled (VMT) assumptions using newer and better data, such as accounting for longer vehicle lifetimes and higher VMT. In principle, the agencies could adjust assumptions as technologies affect utilization and scrappage. A better alternative would be to track on-road emissions as the basis for evaluating compliance. Considerations and the basis for tracking on-road emissions have been presented further in Section 12.2.1 Discrepancies with Real-World Fuel Consumption.

12.4 INTERNATIONAL CONTEXT OF REGULATORY ENVIRONMENT

The U.S. programs to regulate fuel economy and GHG emissions from light-duty vehicles are not taking place in a vacuum. Rather, they are part of an increasingly globalized motor vehicle market where regulatory requirements in one jurisdiction can affect and even drive the availability and cost of vehicle technologies in other jurisdictions. This section examines international efforts to promote (1) more efficient and alternative fueled vehicles and (2) autonomous and connected vehicles, and the impacts and consequences these international efforts may have on the U.S. fuel economy and GHG emissions programs for light-duty vehicles.

International standards and automakers' announcements on global technology deployment provide important broader context on expected vehicle technology deployment in the time frame for this analysis. Many governments around the world (including Brazil, Canada, China, the European Union, India, Japan, Mexico, Saudi Arabia, South Korea, and the United States) have established fuel economy or GHG emission standards for passenger vehicles typically going through 2025 or later. The standards cover more than 80% of global passenger vehicle sales (ICCT, 2020). These vehicle regulations influence the technology investment and deployment decisions of all major global vehicle manufacturers. As a result, even with the uncertainty in the U.S. regulatory situation, global vehicle manufacturing companies plan for continuous year-on-year technology improvements that lower the per-mile CO_2 emissions, and increase the efficiency, of new vehicles. The 2025 automaker targets for EVs based on government regulations are shown in Figure 12.8.

Although regulatory agencies around the world have taken different approaches with the design, test cycles, certification, and crediting procedures in their regulations, there are similarities in their approaches. The standards generally require annual reductions of 2% to 5% in new vehicles' per-mile CO_2 emissions. Increasingly, the

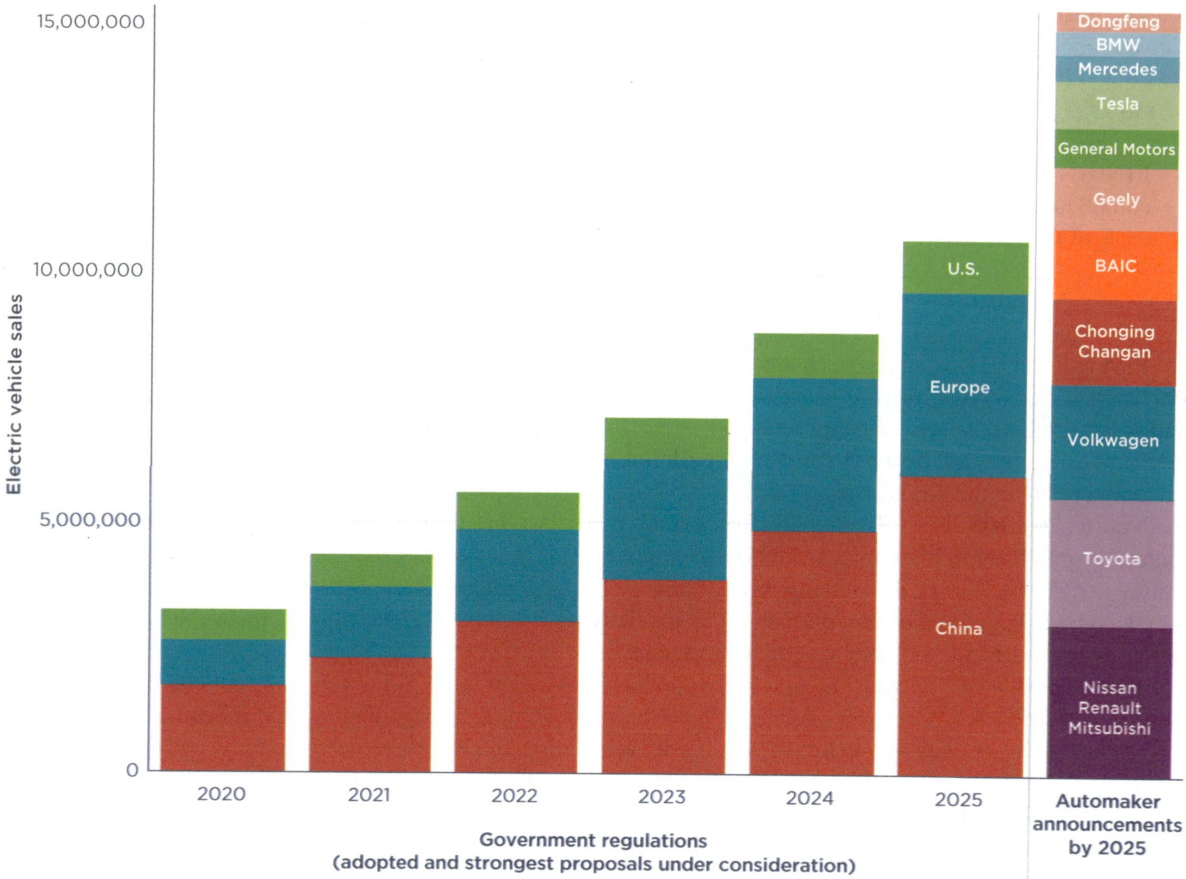

FIGURE 12.8 Estimated electric vehicle government regulations and 2025 automaker targets for electric vehicles.
SOURCE: Lutsey (2018).

regulatory agencies around the world are taking advantage of declining battery costs and integrating EV within their regulatory frameworks and their setting of regulatory stringency. China also has direct regulations requiring EVs be deployed in greater percentages of new vehicles sales over time (Cui, 2018; Ministry of Industry and Information Technology, 2019). In the European Union, there are more stringent CO_2 standards through 2030 and built-in incentives for EVs (Mock, 2019). In addition to the 11 U.S. states with ZEV regulations, Washington state has recently passed legislation to adopt a ZEV regulation (Washington State Legislature, 2020). Minnesota and New Mexico are considering similar regulations (State of Minnesota, 2019; State of New Mexico, 2019), but ZEV state authority is being litigated. Quebec has adopted a similar ZEV regulation (Gouvernement du Québec, 2019). British Columbia's in-development ZEV Act as proposed will also include an enforceable requirement for 100% ZEV sales by 2040 (British Columbia, 2019).

12.4.1 Fuel Efficiency and Alternative-Fueled Vehicles

Several other jurisdictions have taken, or announced plans to take, aggressive regulatory actions to improve light-duty vehicle fuel economy and to advance ZEV technologies such as BEVs. To the extent that these international programs are implemented successfully, they could have significant impacts on the costs and feasibility of more fuel efficient or alternative vehicle technologies in the United States, thus affecting both regulatory feasibility and manufacturers' plans and capabilities in the United States.

For example, strong regulatory targets for EVs or ZEVs in some countries can serve as demonstration programs that can inform other jurisdictions what strategies are successful or unsuccessful in encouraging manufacturers to offer such vehicles and consumers to purchase or lease these vehicles. Mandates to sell EVs or ZEVs in some countries may force manufacturers to invest in the platforms and technologies to make such vehicles succeed in those jurisdictions, and that learning and technology can then be transferred by those same manufacturers to the United States and other markets. Mandates or incentives in other countries may help overcome economies of scale in the production of more efficient or alternative-fueled vehicles.

On the negative side, strong requirements for ZEVs in other countries may increase demand and prices for some inputs to EVs and other efficient vehicle technologies, and may divert and occupy the engineering resources and budgets of some global vehicle manufacturers. Additionally, for U.S. automakers, a potential negative outcome of other countries enacting these ZEV regulations prior to the United States is that these countries will have the opportunity to establish and shape relevant supply chains, which may put them in a stronger position to supply the U.S. market if ZEV policy is enacted. For instance, China is the biggest market for vehicle sales in the world and a jurisdiction that has staked out an ambitious program to convert its fleets to EVs. The policies of China and other jurisdictions with significant programs are summarized in Sections 12.4.1.1 through 12.4.1.7.

12.4.1.1 China

China is the largest passenger vehicle market. In 2018, new vehicle sales reached 28.1 million (including 23.7 million passenger vehicles), adding to a vehicle population of 240 million (China Daily, 2019; Li et al., 2019). Although vehicle sales have decreased since 2018 (Kubota, 2020), the vehicle population in China is expected to grow substantially in the next decade, as the per-capita vehicle ownership is still less than approximately 20% of the U.S. level and Chinese income growth is expected to continue (Li et al., 2019). The rapid growth of vehicle population has posed serious challenges to energy security, with around 70% of China's oil consumption imported (Azihu and Xu, 2020). Promoting vehicle efficiency and new energy vehicles (NEVs)[3] has become a national strategy (CATARC, 2019).

China's passenger vehicle fuel efficiency is regulated at both the vehicle level and the corporate level. Each new vehicle must not exceed the maximum fuel consumption limit based on the vehicle curb weight, as regulated by the policy GB19578-2014 (IEA, 2019). Starting from 2012, vehicle manufacturers need to meet the production-weighted corporate average fuel consumption (CAFC) target, also based on curb weight.[4] China's fuel economy standards could exceed those in the United States and even California by 2025. However, it is unclear which country's fuel economy standards have more compliance flexibilities and how the two countries may compare with respect to the actual ICE vehicle fuel economy. China's CAFC target compared to other countries' fuel economy targets is displayed in Figure 12.9.

CAFC regulation is integrated in the Dual-Credit policy[5] for 2018–2020 with NEV quota policies through very generous flexibilities. The CAFC compliance flexibility with NEVs is realized in three ways: each NEV is treated as 0 liters per 100 kilometers (L/100 km), unit sales of NEVs are treated as multiple units, and NEV surplus credits can be used to offset CAFC deficits. The achieved fuel consumptions in 2018 were 6.60 L/100 km for ICE passenger vehicles only and 5.80 L/100 km with NEV flexibility, a 12.12% difference. These flexibilities strongly indicate the intention of promoting NEVs and accepting the potential temporary leakage effect on ICE vehicle average efficiency. In theory, generous NEV flexibility may cause higher sales of large, heavy, and less efficient vehicles and therefore higher average fuel consumption of new ICE vehicles than without such flexibilities. Although such leakage effect has been demonstrated by counterfactual modeling results (Lin and Ou, 2019), no definitive evidence would indicate that such a leakage effect has actually occurred in the real world. In fact,

[3]In China, NEV is defined to include plug-in hybrid electric, battery electric, and fuel cell vehicles.

[4]In contrast to footprint, as used in the United States.

[5]The official name of the policy is in Chinese and has been translated into Parallel Administration of Passenger Car Enterprise Average Fuel Consumption and New Energy Vehicle Credits.

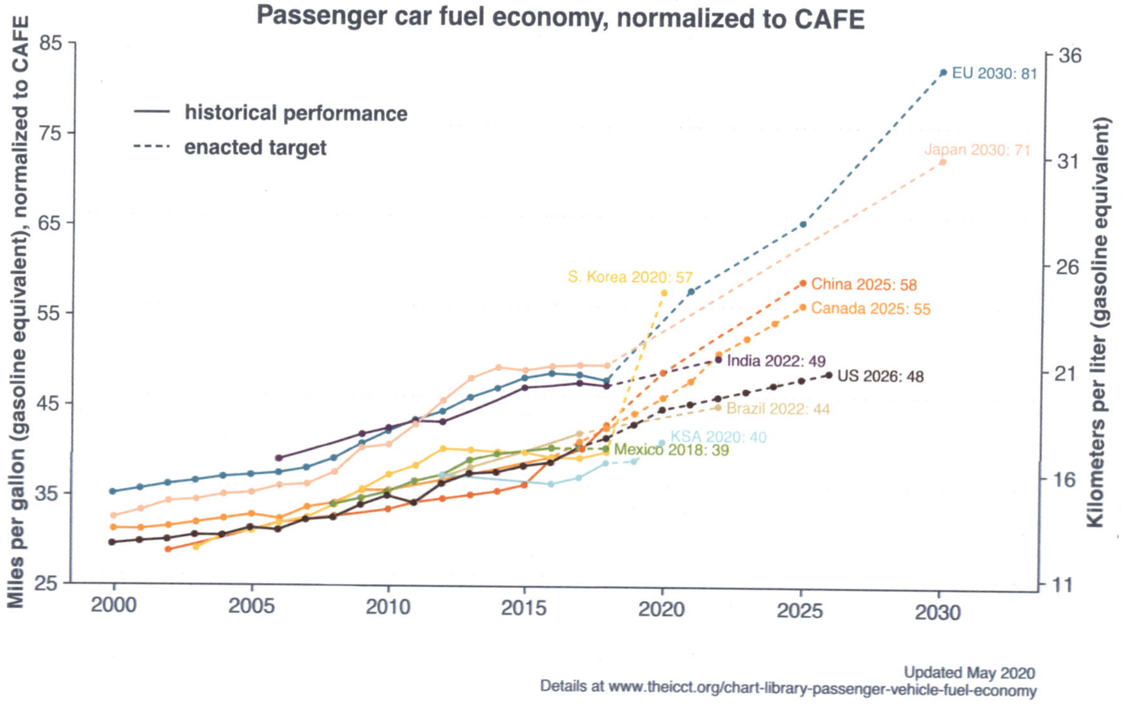

FIGURE 12.9 Comparison of China's passenger vehicle fuel economy standards compared with other nations.
SOURCE: ICCT (2020).

average fuel consumption of Chinese ICE passenger vehicles continues to decrease, despite increases in vehicle footprint, weight, and power and the CAFC compliance flexibility with NEV sales, as shown in Figure 12.10.

The 2018–2020 dual-credit policy has been significantly over-complied. According to compliance data released by the Chinese government (Ministry of Industry and Information Technology of the People's Republic of China, 2020a), in 2019, among the 144 vehicle manufacturers or importers responsible for a total of 20.93 million passenger vehicles (exports excluded), 6.4 million positive CAFC credits and 5.1 million negative CAFC credits were achieved (overcompliance by 25%). Meanwhile, 4.2 million positive NEV credits and 0.86 million negative NEV credits were generated (overcompliance by 388%). This overall overcompliance, however, is accompanied by a significant number of undercomplying manufacturers or importers. The achieved production-weighted average fuel consumption in 2019 was 5.56 L/100 km, down by 4% from 5.8 L/100 km in 2018.[6] Achieving the 5 L/100 km goal (non-binding) by 2020, a 10% reduction from 2019, has become extremely difficult, if not impossible.

China is still pursuing the non-binding goal of 4 L/100 km by 2025 on the average fuel consumption of new passenger vehicles, which requires an average 5.3% annual reduction from the 2019 level. From 2013 to 2018, the average fuel consumption of new ICE passenger vehicles decreased from 7.33 L/100 km to 6.60 L/100 km, an annual reduction of 2.08% (CATARC, 2019). Including NEVs, the average fuel consumption of new passenger vehicles for 2016–2019 are 6.43, 6.05, 5.80, and 5.56 L/100 km, corresponding to an annual reduction of 8.7%, 5.9%, 4.1%, and 4.1%, respectively. It appears that achieving the 2025 goal of 4 L/100 km will require an accelerated market penetration of NEVs in addition to continued improvements in the efficiency of new ICE passenger vehicles.

Given the over-compliance results and the challenging 2025 goal, the Chinese government in June 2020 released an extended dual-credit policy for 2021–2023 with some important modifications that tighten the rules (Ministry of Industry and Information Technology of the People's Republic of China, 2020b). The NEV quota is

[6] 2018, ICE vehicle 6.6 L/100 km; all including plug-in electric vehicles is 5.8 L/100 km.

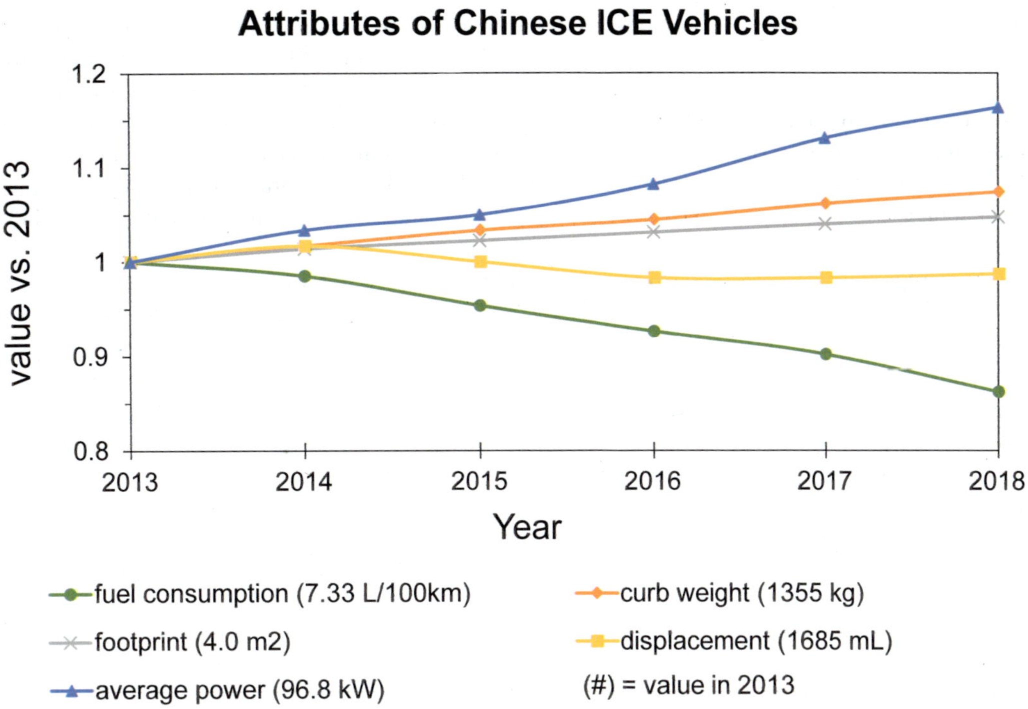

FIGURE 12.10 China's ICE vehicle fuel consumption, curb weight, footprint, average power, and displacement over 2013–2018 (normalized to 2013 level).
SOURCE: Committee generated using data from CATARC (2019).

increased from 12% for 2020 to 14%, 16%, and 18% for 2021, 2022, and 2023, respectively, while the credits for a given NEV type have overall decreased. For example, credits per PHEV decrease from 2 in the 2018–2020 rules to 1.6 in the 2021–2023 rules. Credits per BEV are tightened from 2 to 5 in the 2018–2020 rules to 1 to 3.4 in the 2021–2023 rules, depending on the electric range. In the old rules, a BEV could be treated as up to 1.2 units if its electric consumption per mile was low enough. Such a multiplication treatment is also less generous in the new rules. Another important change is the introduction of low-fuel-consumption (LFC) vehicles. Each conventional ICE vehicle has a curb-weight-based fuel consumption target. An ICE vehicle with a fuel consumption rate lower than a given ratio of this target is qualified as an LFC vehicle. This ratio is stipulated as 123%, 120%, and 115% for 2021, 2022, and 2023, respectively. Each LFC vehicle will be treated as a partial vehicle in the calculation of total production and multiplied by the NEV quota to calculate the required number of NEV credits. The partial vehicle ratios are stimulated to be 0.5, 0.3, and 0.2 for 2021, 2022, and 2023, respectively. The LFC concept clearly intends to encourage adoption of efficiency technologies for ICE vehicles for the return of a relaxed NEV credit requirement. The 2021–2023 rules also modified the rules for credit carryover and trade and explicitly include alcohol ether fuels in the category of conventional vehicle fuels.

12.4.1.2 European Union

Since 2009, the EU has set standards for the average CO_2 emissions of each vehicle manufacturer, with the initial standards set at 130 grams of CO_2 per kilometer (g CO_2/km) by 2015 and subsequent standards that take effect in 2021 set at 95 g CO_2 /km (EU, 2009). In April 2019, the European Union adopted a new regulation that set out new CO_2 emission performance standards for new passenger cars and for new vans in the EU (EU, 2019). This regulation began to apply in 2020, maintaining the original regulatory targets set forth in 2009 for 2020, but setting new targets that apply from 2025 and 2030. These targets are defined as a percentage reduction from the

2021 starting points, with passenger cars required to achieve a 15% reduction in CO_2 emissions by 2025 and a 37.5% reduction by 2030. Vans are required to achieve a 15% reduction by 2025 and a 31% reduction by 2030 from their 2021 targets. The specific emission targets for each manufacturer to comply with are based on the EU fleet-wide targets, taking into account the average test mass of a manufacturer's new vehicle fleet. The new regulations also include provisions to verify CO_2 emissions of vehicles in-service and to ensure that the emission test procedure yields results that are representative of real-world emissions. According to an analysis by the International Council on Clean Transportation (ICCT), the EU's CO_2 standards are the most stringent in the industrial world (Yang and Rutherford, 2019).

The new EU regulatory program also incentivizes electric and other low carbon alternative fuel technologies, defined in the regulations as zero- and low-emission vehicles (ZLEVs) (EU, 2019). A ZLEV is a passenger car or a van with CO_2 emissions between 0 and 50 grams per kilometer (g/km). A crediting system for ZLEVs commences in 2025, in which the CO_2 emission target of a manufacturer is relaxed if its share of ZLEV registered in a given year exceeds the following benchmarks: (1) cars—15% ZLEV from 2025 on and 35% ZLEV from 2030 on; and (2) vans—15% ZLEV from 2025 on and 30% ZLEV from 2030 on. A 1 percentage point exceedance of the ZLEV benchmark will increase the manufacturer's CO_2 target by 1%, up to a maximum of 5%.

Individual EU nations have reinforced the EU regulations, and in particular the promotion of ZEVs, with robust financial incentives. France has implemented a "bonus-malus" that combines a tax charge on high-emitting CO_2 models with a bonus payment of €3,000–6,000 (depending on vehicle value and year) for low-carbon vehicles, which as of 2020 will be restricted to BEVs and fuel cell vehicles (Randall, 2019). Similarly, Germany announced an increased financial incentive program in February 2020 that will extend at least through 2025, that will pay €5,000–6,000 for EVs and €3,750–5,000 for plug-in hybrids (Randall, 2020).

12.4.1.3 Non-EU European Nations

The most proactive nation in Europe, and in fact globally, with respect to BEV implementation is Norway. BEVs represented 31% of new vehicle sales in Norway in 2018, which raised BEVs to 7% of the total national on-road fleet (Figenbaum and Nordbakke, 2019). Norway has set a goal in its national vehicle policy to require all new vehicles sold from 2025 on to be ZEVs (Figenbaum and Nordbakke, 2019). The United Kingdom, which recently left the EU through Brexit, has set forth a goal of 100% ZEVs by 2040 (U.K. Department for Transport, 2018). It has set an interim goal for 2030 of at least 50%, and as many as 70%, of new car sales and up to 40% of new van sales being ultra-low emission (U.K. Department for Transport, 2018).

12.4.1.4 Japan

Japan's current fuel efficiency standards for passenger cars were adopted in 2009 and took effect in 2020. In June 2019, Japan promulgated new fuel efficiency standards to take effect in 2030. The new standards require a fleet-wide average fuel efficiency of 25.4 kilometers per liter (km/L), which represents a 32.4% improvement compared to the actual performance in 2016 (Japanese Ministry of Economy, Trade, and Industry, 2020). The 2020 standards were based on different vehicle weight bins, with the heavier vehicle bins having the least stringent fuel efficiency targets, whereas the 2030 standards are based on a continuous curve that plots fuel efficiency targets against vehicle weight (Yang and Rutherford, 2019). In addition, the 2020 standards only applied to gasoline and diesel vehicles, while the 2030 standards will also apply to BEVs and plug-in hybrid vehicles. In order to calculate the energy consumption efficiency of electric and plug-in hybrid vehicles relative to gasoline vehicles, a well-to-wheels model is used that considers the energy consumption efficiency on the upstream side of the supply of gasoline, electricity, or other fuel source (Japanese Ministry of Economy, Trade, and Industry, 2020). The standards estimate that 20% of new passenger car sales in 2030 will be BEVs and PHEVs (Yang and Rutherford, 2019). A comparative analysis by the ICCT concluded that the new Japanese fuel efficiency standards are less stringent than the European standards but more stringent than those of the United States, Canada, and China (Yang and Rutherford, 2019).

12.4.1.5 India

India has adopted fuel economy norms in two phases. The CAFC standard that took effect in 2017–2018 is 130 g CO_2/km (or 19.2 km/L), decreasing to 113 g/km (or 20.96 km/L) in 2022–2023 (Vishvak and V V, 2019). India has indicated consistent and strong support for EVs, although the specific goals and requirements have fluctuated in recent years. The government announced in 2017 that the goal will be 100% EVs by 2030 but subsequently backed off to a target of 30%. The government is currently offering financial incentives for EV purchases, recently cutting the Goods and Services tax on EVs from 12% to 5% and cutting the tax on electric chargers and charging stations from 18% to 5% (Carpenter, 2019). The government has also recently committed $1.4 billion (10,000 crore rupees) in EV subsidies in its Faster Adoption and Manufacturing of Hybrid and Electric Vehicles, or FAME, program (Carpenter, 2019).

12.4.1.6 Canada

Canada has sought to align its fuel economy and GHG emissions standards for LDVs with those of the United States. The most recent Canadian standards were adopted in 2014, adopting standards similar to those previously adopted by the United States for MY 2017 and beyond (Government of Canada, 2014). Canada has not yet adopted standards equivalent to the U.S. standards for 2020–2025 but is expected to do so soon. The federal government has established a goal of 30% ZEVs in new passenger car sales by 2030, and 100% by 2040. In May 2019, the Canadian province of British Columbia legislated the world's first 100% ZEV mandate for passenger vehicles. The Zero-Emission Vehicles Act of 2019 enacted 10%, 30%, and 100% ZEV sales targets by 2025, 2030, and 2040, respectively (Government of British Columbia, 2019).

12.4.1.7 Overview

Many industrial countries have now adopted more aggressive fuel economy standards than the United States, including more ambitious targets for ZEV sales (Cui et al., 2020). This is a goal shared by many automotive companies as well, with General Motors announcing plans to manufacture one million EVs by 2025, as an example (Pyper, 2020). Of course, some of these targets may be more aspirational than realistic; for example, Canada targeted selling 500,000 EVs by 2018, but less than 100,000 were sold (Rabson, 2019). Nonetheless, these international regulatory pressures for more and sooner EV sales will influence the timing and volume of commercial EV sales in the United States, with one of the heaviest and largest mass fleets. This influence comes both from providing learning and experience on consumer, infrastructure, and incentive strategies to enhance EV sales, and from providing economy of scale and technology leverage for vehicle manufacturers.

FINDING 12.7: The global regulatory push for more efficient vehicles—in particular, electric vehicles—and lower greenhouse gas emissions by many industrialized nations will provide an important impetus for both technology availability and adoption by manufacturers for the U.S. market and valuable learning experience for incentivizing consumer adoption of electric vehicles.

RECOMMENDATION 12.4: The 2025–2035 Corporate Average Fuel Economy standard should be set and designed to depend on and incentivize a significant market share of zero-emission vehicles (plug-in hybrid electric vehicles, battery electric vehicles, and fuel cell electric vehicles).

12.4.2 "Internationalization" of Vehicle Platforms/Technologies

Automakers are global companies that sell their vehicles in multiple countries around the world. Vehicle models will often need to be customized for individual countries according to their specific regulatory specifications and unique demand characteristics. However, automakers do use many shared components across their vehicles sold around the world in global platforms and powertrain families.

The use of the same platforms and powertrain families can reduce production costs through economies of scale over multiple international markets. The global integration of automotive development means that regulations in international markets can influence an automaker's vehicle production in the United States and their capabilities to implement various vehicle technologies. For instance, advances in powertrain and non-powertrain technologies that an automaker achieves in order to meet international emission regulations could then be available to transfer to the U.S. market. Examples of these advances include mass reduction through design and material substitution; battery cell design and cathode materials; fuel cell materials and manufacturing; and power and control electronics. Furthermore, cost reductions made possible by learning-by-doing and manufacturing improvements from implementing these technologies in other countries may also be able to transfer to the U.S. context.

12.5 FUEL ECONOMY REGULATION IN A WARMING WORLD

Over the course of the CAFE program, its objectives and scope have evolved. The initial drivers of the CAFE program, including energy security and energy conservation, remain relevant public policy objectives, although their relative importance has diminished as the United States has become self-sufficient in energy supply. At the same time, new priorities underlying the program have emerged, especially minimizing GHG emissions through increased energy efficiency and substitution with clean energy sources and technologies. The alignment of the EPA motor vehicle GHG emission control program with the NHTSA CAFE program reflects this evolving policy focus.

These shifting objectives of the CAFE program have been accompanied by an expansion in the scope and diversity of vehicles regulated by the program. When the program commenced in the mid-1970s, virtually all regulated vehicles had gasoline-fueled, ICEs. Over the years, the program has expanded to encompass engines fueled by other liquid fuels including diesel, natural gas, and ethanol-blended fuels. More importantly, alternative powertrains are now becoming an increasing percentage of the regulated fleet, including hybrid engines, BEVs, and fuel cell vehicles. CAVs will further add to the diversity of light-duty vehicles subject to the CAFE standards. As the diversity of fuels, powertrains, and technologies continues to expand, developing metrics to compare fuel economy across these various vehicle types becomes a challenge. The most common and relevant metric is GHG emissions, which have become the most useful and widespread metric for measuring overall fuel efficiency.

Another implication of the focus on GHG emissions is that light-duty vehicles are only a part of the GHG problem. It is quite possible that during the time frame of this study, 2025–2035, the United States will adopt a more comprehensive program to regulate GHG emissions from all sources, such as a carbon tax or a national CO_2 cap-and-trade program. Such a national program may encompass GHG emissions from light-duty vehicles, which will raise the issue of whether such a national program will or should complement or substitute for the CAFE program.

FINDING 12.8: Vehicle efficiency has been regulated since 1978 by the U.S. Department of Transportation in liquid fuel used per mile, which met two goals of reduced energy use and reduced petroleum use onboard the vehicle. The U.S. Environmental Protection Agency began to regulate greenhouse gas (GHG) emissions from the light-duty fleet in 2009, and the regulations have been paired ever since, as they have significant shared features. In the period of 2025–2035, where climate goals are continually increasing in importance and the diversity of light-duty vehicle fueling sources is growing, GHG emissions reduction will become an important metric for meeting the goals of both vehicle efficiency and GHG emissions reduction.

RECOMMENDATION 12.5: The National Highway Traffic Safety Administration should continue to align the Corporate Average Fuel Economy program with the U.S. Environmental Protection Agency greenhouse gas emissions program for light-duty vehicles, seeking to reduce or eliminate any inconsistencies between the programs.

12.6 REFERENCES

110th Congress. 2007. "H.R.6—110th Congress (2007–2008): Energy Independence and Security Act of 2007." Legislation. 2007/2008. December 19, 2007. https://www.congress.gov/bill/110th-congress/house-bill/6.

Azihu, C., and M. Xu. 2020. "China Opens Up Oil and Gas Exploration, Production for Foreign, Domestic Firms." Reuters. January 9. https://www.reuters.com/article/us-china-oil-mining-idUSKBN1Z806Q.

Bigazzi, A.Y., and K.J. Clifton. 2015. Modeling the effects of congestion on fuel economy for advanced power train vehicles. *Transportation Planning and Technology* 38(2):149–161. https://doi.org/10.1080/03081060.2014.997449.

British Columbia. 2019. *Zero-Emission Vehicles Act*. https://www2.gov.bc.ca/gov/content/industry/electricity-alternative-energy/transportation-energies/clean-transportation-policies-programs/zero-emission-vehicles-act.

Broch, F., J. Warsen, and S. Krinke. 2015. "Implementing Life Cycle Engineering in Automotive Development as a Helpful Management Tool to Support Design for Environment." In *Life Cycle Management*, edited by G. Sonnemann and M. Margni, 319–329. Dordrecht: Springer Netherlands. https://doi.org/10.1007/978-94-017-7221-1_23.

Canis, B. 2020. "Issues in Autonomous Vehicle Testing and Deployment." Congressional Research Service. https://fas.org/sgp/crs/misc/R45985.pdf.

CARB (California Air Resources Board). 2020. "Light-Duty Vehicle Research." https://ww2.arb.ca.gov/resources/documents/light-duty-vehicle-research.

Carpenter, S. 2019. "India's Plan to Turn 200 Million Vehicles Electric in Six Years." Forbes. December 5. https://www.forbes.com/sites/scottcarpenter/2019/12/05/can-india-turn-nearly-200-million-vehicles-electric-in-six-years/?sh=73f4ad7915db.

CATARC (China Automotive Technology and Research Center). 2019. *Annual Report on Energy-Saving and New Energy Vehicle in China 2019*. https://www.sanmin.com.tw/Product/index/007569348.

China Daily. 2019. "China Ahead in Car Sales Race for 10 Consecutive Years." January 16. http://www.china.org.cn/business/2019-01/16/content_74377606.htm.

Cui, H. 2018. "China's New Energy Vehicle Mandate Policy." International Council on Clean Transportation. https://theicct.org/publications/china-nev-mandate-final-policy-update-20180111.

Cui, H., D. Hall, and N. Lutsey. 2020. "Update on the Global Transition to Electric Vehicles through 2019." International Council on Clean Transportation. https://theicct.org/sites/default/files/publications/update-global-EV-stats-20200713-EN.pdf.

Dornoff, J. 2019. One goal, multiple pathways: A review of approaches for transferring on-board fuel consumption meter data to the European Commission. White Paper. International Council on Clean Transportation. https://theicct.org/sites/default/files/publications/OBFCM_data_transfer_20191022.pdf.

EPA (U.S. Environmental Protection Agency). 2006. Fuel Economy Labeling of Motor Vehicles: Revisions to Improve Calculation of Fuel Economy Estimates: Final Technical Support Document. EPA420-R-06-017. Office of Transportation and Air Quality.

EPA. 2014. "Hyundai and Kia Clean Air Act Settlement." Other Policies and Guidance. October 31. https://www.epa.gov/enforcement/hyundai-and-kia-clean-air-act-settlement.

EPA. 2017a. "Derived 5-Cycle Coefficients for 2017 and Later Model Years." Guidance memorandum. National Vehicle and Fuel Emissions Laboratory, Plymouth, MI. https://iaspub.epa.gov/otaqpub/display_file.jsp?docid=35113&flag=1.

EPA. 2017b. "Overview for Renewable Fuel Standard." Overviews and Factsheets. US EPA. June 7. https://www.epa.gov/renewable-fuel-standard-program/overview-renewable-fuel-standard.

EPA. 2019. *Greenhouse Gas (GHG) Emission Standards for Light-Duty Vehicles: Manufacturer Performance Report*, and Excel tables. https://www.epa.gov/automotive-trends/download-data-automotive-trends-report.

EPA. 2020. *The 2019 Automotive Trends Report*. EPA-420-R-20-006, Detailed Data. https://www.epa.gov/automotive-trends/explore-automotive-trends-data#DetailedData.

EPA/NHTSA (U.S. Environmental Protection Agency and National Highway Traffic Safety Administration). 2012. "2017 and Later Model Year Light-Duty Vehicle Greenhouse Gas Emissions and Corporate Average Fuel Economy Standards." *Federal Register* 77(99). https://www.govinfo.gov/content/pkg/FR-2012-10-15/pdf/2012-21972.pdf.

EPA Press Office. 2020. "U.S. DOT and EPA Put Safety and American Families First with Final Rule on Fuel Economy Standards." Speeches, Testimony, and Transcripts. US EPA. March 31. https://www.epa.gov/newsreleases/us-dot-and-epa-put-safety-and-american-families-first-final-rule-fuel-economy-standards.

EU (European Parliament and Council of the European Union). 2009. Regulation (EC) No 443/2009. https://eur-lex.europa.eu/legal-content/EN/TXT/PDF/?uri=CELEX:32009R0443&from=EN.

EU. 2019. Regulation (EU) 2019/631. https://eur-lex.europa.eu/legal-content/EN/TXT/PDF/?uri=CELEX:32019R0631&from=EN.

Figenbaum, E., and S. Nordbakke. 2019. *Battery Electric Vehicle User Experiences in Norway's Maturing Market*. Oslo, Norway: Norwegian Centre for Transport Research. August. https://www.toi.no/getfile.php?mmfileid=50956.

Gaoi, J., M. Li, Y. Hu, H. Chen, and M. Yan. 2019. Challenges and developments of automotive fuel cell hybrid power system and control. *Science China Information Sciences* 62:51201. http://scis.scichina.com/en/2019/051201.pdf.

Good, D. 2017. "EPA Test Procedures for Electric Vehicles and Plug-in Hybrids." Draft Summary—Regulations Take Precedence. US EPA. https://www.fueleconomy.gov/feg/pdfs/EPA%20test%20procedure%20for%20EVs-PHEVs-11-14-2017.pdf.

Gouvernement du Québec. 2019. The Zero-Emission Vehicle (ZEV) Standard. http://www.environnement.gouv.qc.ca/changementsclimatiques/vze/index-en.htm.

Government of British Columbia. 2019. "Zero-Emission Vehicles Act." Government of British Columbia. May 30. https://www2.gov.bc.ca/gov/content/industry/electricity-alternative-energy/transportation-energies/clean-transportation-policies-programs/zero-emission-vehicles-act.

Government of Canada. 2014. Regulations Amending the Passenger Automobile and Light Truck Greenhouse Gas Emissions Regulations. Pub. L. No. SOR/2014-207 (2014). https://gazette.gc.ca/rp-pr/p2/2014/2014-10-08/html/sor-dors207-eng.html.

Greene, D.L., A.J. Khattak, J. Liu, X. Wang, J.L. Hopson, and R. Goeltz, 2017. What is the evidence concerning the gap between on-road and Environmental Protection Agency fuel economy ratings? *Transport Policy* 53:146–160.

Hall, D., and N. Lutsey. 2018. "Effects of Battery Manufacturing on Electric Vehicle Life-Cycle Greenhouse Gas Emissions." Briefing. International Council on Clean Transportation. https://theicct.org/sites/default/files/publications/EV-life-cycle-GHG_ICCT-Briefing_09022018_vF.pdf.

He, H. 2014. "Credit Trading in the US Corporate Average Fuel Economy (CAFE) Standard." Briefing. International Council on Clean Transportation. March. https://theicct.org/sites/default/files/publications/ICCTbriefing_CAFE-credits_20140307.pdf.

Hellman, K.S., and J.D. Murrell. 1984. "Development of Adjustment Factors for EPA City and Highway MPG Values." SAE Technical Paper, 840496. Warrendale, PA: Society of Automotive Engineers.

Helmers, E., and M. Weiss. 2017. Advances and critical aspects in the life-cycle assessment of battery electric cars. *Energy and Emission Control Technologies* 2017(5):1–18. https://doi.org/10.2147/EECT.S60408.

Henning, M., A.R. Thomas, and A. Smyth. 2019. An analysis of the association between changes in ambient temperature, fuel economy, and vehicle range for battery electric and fuel cell electric buses. *Urban Publications*. https://engagedscholarship.csuohio.edu/urban_facpub/1630.

Hu, X., Y. Zheng, D.A. Howey, H. Perez, A. Foley, and M. Pecht. 2020. Battery warm-up methodologies at subzero temperatures for automotive applications: Recent advances and perspectives. *Progress in Energy and Combustion Science* 77: 100806. March. https://doi.org/10.1016/j.pecs.2019.100806.

Huseynov, S., and M.A. Palma. 2018. Does California's low carbon fuel standard reduce carbon dioxide emissions? *PLOS ONE* 13(9):e0203167. https://doi.org/10.1371/journal.pone.0203167.

ICCT (The International Council on Clean Transportation). 2020. "Chart Library: Passenger Vehicle Fuel Economy." May. https://theicct.org/chart-library-passenger-vehicle-fuel-economy.

IEA (International Energy Agency). 2019. *GB 19578-2014 Fuel consumption limits for passenger cars (stage VI)*. Policy of the People's Republic of China. Updated October 22, 2019. https://www.iea.org/policies/2999-gb-19578-2014-fuel-consumption-limits-for-passenger-cars-stage-vi.

Japanese Ministry of Economy, Trade, and Industry. 2020. "Fuel Efficiency Standards for Passenger Vehicles in FY2030 Formulated." Ministry of Economy, Trade, and Industry. March 31. https://www.meti.go.jp/english/press/2020/0331_009.html.

Jenn, A., I.M.L. Azevedo, and J.J. Michalek. 2016. Alternative fuel vehicle adoption increases fleet gasoline consumption and greenhouse gas emissions under United States corporate average fuel economy policy and greenhouse gas emissions standards. *Environmental Science and Technology* 50(5):2165–2174. https://doi.org/10.1021/acs.est.5b02842.

Kim, H.-J., C. McMillan, G.A. Keoleian, and S.J. Skerlos. 2010. Greenhouse gas emissions payback for lightweighted vehicles using aluminum and high-strength steel. *Journal of Industrial Ecology* 14(6):929–946. https://doi.org/10.1111/j.1530-9290.2010.00283.x.

Kiso, T. 2017. Evaluating new policy instruments of the corporate average fuel economy standards: Footprint, credit transferring, and credit trading. *Environmental and Resource Economics* 72(2):445–476. https://doi.org/10.1007/s10640-017-0200-1.

Kubota, Y. 2020. "China Auto Sales Slid 8.2% Last Year." Wall Street Journal. January 13, sec. Business. https://www.wsj.com/articles/china-auto-sales-slid-8-2-last-year-11578898123.

Kühlwein, J., J. German, and A. Bandivadekar. 2014. Development of test cycle conversion factors among worldwide light-duty vehicle co_2 emission standards. White paper. International Council on Clean Transportation. https://theicct.org/sites/default/files/publications/ICCT_LDV-test-cycle-conversion-factors_sept2014.pdf.

Laing, K. 2018. "Carmakers Face Higher MPG Fines." Detroit News. April 24. https://www.detroitnews.com/story/business/autos/2018/04/24/higher-MPG-fines/34214385/.

Lattanzio, R.K., L. Tsang, and B. Canis. 2020. "Vehicle Fuel Economy and Greenhouse Gas Standards." Congressional Research Service. https://fas.org/sgp/crs/misc/IF10871.pdf.

Lave, L.B., C.T. Hendrickson, and F.C. McMichael. 1995. Environmental implications of electric cars. *Science* 268(5213):993–995.

Leard, B., and V. McConnell. 2017. *New Markets for Credit Trading Under US Automobile Greenhouse Gas and Fuel Economy Standards*. RFF Report, Resources for the Future, Washington, DC. August.

Leard, B., V. McConnell, and Y.C. Zhou. 2019. The effect of fuel price changes on fleet demand for new vehicle fuel economy. *Journal of Industrial Economics* 67(1):127–159. https://doi.org/10.1111/joie.12198.

Lee D., A. Elgowainy, and R. Vijayagopal. 2019. Well-to-wheel environmental implications of fuel economy targets for hydrogen fuel cell electric buses in the United States. *Energy Policy* 128:565–583. May.

Lewis, G.M., C.A. Buchanan, K.D. Jhaveri, J.L. Sullivan, J.C. Kelly, S. Das, A.I. Taub, and G.A. Keoleian. 2019. Green principles for vehicle lightweighting. *Environmental Science & Technology* 53(8):4063–4077. https://doi.org/10.1021/acs.est.8b05897.

Li, Y., L. Miao, Y. Chen, and Y. Hu. 2019. Exploration of sustainable urban transportation development in China through the forecast of private vehicle ownership. *Sustainability* 11(16):1–18.

Lin, Z., and S. Ou. 2019. "Dual-Credit Policy: Impact on PEV Sales and Industry Profits in China." University of Michigan Energy Institute. October 18. https://energy.umich.edu/te3/wp-content/uploads/sites/2/2019/10/Lin-slides-TE3-2019.pdf.

Linn, J., and V. McConnell. 2019. Interactions between federal and state policies for reducing vehicle emissions. *Energy Policy* 126(March 1):507–517. https://doi.org/10.1016/j.enpol.2018.10.052.

Liu, C., D.L. Greene, and D.S. Bunch. 2014. Vehicle manufacturer technology adoption and pricing strategies under fuel economy/emissions standards and feebates. *Energy Journal* 35(3):71–89.

Lutsey, N. 2018. "Modernizing Vehicle Regulations for Electrification." International Council on Clean Transportation. October. https://theicct.org/sites/default/files/publications/ZEV_Regulation_Briefing_20181017.pdf.

Macauley, N., R.W. Lujan, D. Spernjak, D.S. Hussey, D.L. Jacobson, K. More, R.L. Borup, et al. 2016. Durability of polymer electrolyte membrane fuel cells operated at subfreezing temperatures. *Journal of the Electrochemical Society* 163(13):F1317–F1329. https://www.osti.gov/servlets/purl/1374332.

MacDuffie, P., and E. Orts. 2019. "Why Automakers Are Driving for Uniform Fuel Efficiency Standards." Knowledge@ Wharton. July 14. https://knowledge.wharton.upenn.edu/article/end-california-emissions-standards/.

Matthews, H.S., C.T. Hendrickson, and D. Matthews. 2014. *Life Cycle Assessment: Quantitative Approaches for Decisions That Matter*. Open access textbook. https://www.lcatextbook.com/.

McNutt, B.D., D. Pirkey, R. Dulla, and C. Miller. 1978. "A Comparison of Fuel Economy Results from EPA Tests and Actual In-Use Experience 1974–1977 Model Year Cars." SAE Technical Paper 780037. Warrendale, PA: Society of Automotive Engineers.

Ministry of Industry and Information Technology. 2019. **工业和信息化部关于《乘用车企业平均燃料消耗量与新能源汽车积分并行管理办法》修正案（征求意见稿）公开征求意见的通知.** [Notice of the Ministry of Industry and Information Technology on the Public Consultation on the Amendments to the "Dual Regulation for the Average Fuel Consumption of Passenger Vehicle Enterprises and New Energy Vehicles" (Draft for Comment).] http://www.miit.gov.cn/n1146285/n1146352/n3054355/n3057585/n3057592/c7027601/content.html.

Ministry of Industry and Information Technology of the People's Republic of China. 2020a. Measures for the Parallel Administration of the Average Fuel Consumption and New Energy Vehicle Credits of Passenger Vehicle Enterprises. http://www.miit.gov.cn/n1146295/n1652858/n7280902/c7988463/content.html.

Ministry of Industry and Information Technology of the People's Republic of China. 2020b. "**中华人民共和国工业和信息化部 商务部 海关总署 市场监管总局公告.**" http://miit.gov.cn/n1146295/n1652858/n1652930/n4509607/c7991697/content.html.

Mock, P. 2019. CO_2 emission standards for passenger cars and light-commercial vehicles in the European Union. International Council on Clean Transportation. https://theicct.org/publications/ldv-co2-stds-eu-2030-update-jan2019.

NHTSA (National Highway Traffic Safety Administration). 2019. *CAFE Public Information Center: Manufacturer Performance Report, Credit Status Report*. https://one.nhtsa.gov/cafe_pic/CAFE_PIC_Mfr_LIVE.html and https://one.nhtsa.gov/cafe_pic/CAFE_PIC_Credit_LIVE.html.

NHTSA. 2020. "Fact Sheet: SAFE Vehicles Rule." March 31. https://www.nhtsa.gov/corporate-average-fuel-economy/safe-fact-sheet.

NHTSA/EPA (National Highway Traffic Safety Administration and U.S Environmental Protection Agency). 2020. The Safer Affordable Fuel-Efficient (SAFE) Vehicles Rule for Model Year 2012-2026 Passenger Cars and Light Trucks: Final Environmental Impact Statement. Docket No. NHTSA-2017 0069. https://www.nhtsa.gov/sites/nhtsa.dot.gov/files/documents/safe_vehicles_rule_feis.pdf.

Nordelöf, A., M. Messagie, A. Tillman, M.L. Söderman, and J. Van Mierlo. 2014. Environmental impacts of hybrid, plug-in hybrid, and battery electric vehicles—What can we learn from life cycle assessment? *International Journal of Life Cycle Assessment* 19(11):1866–1890. https://doi.org/10.1007/s11367-014-0788-0.

NSTC/DOT (National Science and Technology Council and U.S. Department of Transportation). 2020. "Ensuring American Leadership in Automated Vehicle Technologies: Automated Vehicles 4.0." https://www.transportation.gov/sites/dot.gov/files/2020-02/EnsuringAmericanLeadershipAVTech4.pdf.

Pike, E. 2012. "Calculating Electric Drive Vehicle Greenhouse Gas Emissions." International Council on Clean Transportation. https://theicct.org/sites/default/files/publications/ICCT_CalculatingEdriveGHG_082012.pdf.

Posada, F., and J. German. 2013. Measuring in-use fuel economy in Europe and the US: Summary of pilot studies. Working Paper. International Council on Clean Transportation. https://theicct.org/sites/default/files/publications/ICCT_FuelEcon_pilotstudies_20131125.pdf.

Pyper, J. 2020. "Automaker Electric Vehicle Plans 'Progressing at a Rapid Pace' Despite Pandemic, Economic Downturn." Atlantic Council (blog). June 24. https://www.atlanticcouncil.org/blogs/energysource/automaker-electric-vehicle-plans-progressing-at-a-rapid-pace-despite-pandemic-economic-downturn/.

Rabson, M. 2019. "Sales of Electric Cars in Canada Fall Well Short of Goal Set in 2009." Globe and Mail. January 7. https://www.theglobeandmail.com/business/article-sales-of-electric-cars-are-soaring-in-canada-but-still-fall-short-of/.

Randall, C. 2019. "France Increases EV Subsidy System." electrive.com. December 19. https://www.electrive.com/2019/12/19/france-revises-their-ev-subsidy-system/.

Randall, C. 2020. "Germany doubles EV subsidies, no more diesel support." electrive.com. June 4. https://www.electrive.com/2020/06/04/germany-doubles-ev-subsidies-no-more-diesel-support/.

Rubin, J., and P. Leiby. 2000. An analysis of alternative fuel credit provisions of US automotive fuel economy standards. *Energy Policy* 13:589–601.

Rubin, J., P.N. Leiby, and D.L. Greene. 2009. Tradable fuel economy credits: Competition and oligopoly. *Journal of Environmental Economics and Management* 58:315–328.

State of Minnesota. 2019. "Governor Tim Walz Announces Clean Car Standards in Minnesota." https://mn.gov/governor/news/?id=1055-403887.

State of New Mexico. 2019. Executive order 2019-003: Executive order on addressing climate change and energy waste prevention. https://www.governor.state.nm.us/wp-content/uploads/2019/01/EO_2019-003.pdf.

Temple, J. 2019. "Automakers Have Agreed with California to Make More Efficient Cars, in a Rebuke to Trump." MIT Technology Review. July 25. https://www.technologyreview.com/2019/07/25/134052/automakers-have-agreed-with-california-to-make-more-efficient-cars-in-a-rebuke-to-trump/.

U.K. Department for Transport. 2018. "The Road to Zero: Next Steps Towards Cleaner Road Transport and Delivering Our Industrial Strategy." London. https://assets.publishing.service.gov.uk/government/uploads/system/uploads/attachment_data/file/739460/road-to-zero.pdf.

United States Code. 1990. Clean Air Act Amendments, Pub. L. No. 101–549 (1990). https://www.govinfo.gov/content/pkg/STATUTE-104/pdf/STATUTE-104-Pg2399.pdf.

Vishvak, R., and R. V V. 2019. Quest for control of fuel consumption and CO_2 emission in Indian passenger cars—an analysis of the progress and prognosis. *International Journal of Mechanical Engineering and Technology* 10(3):1094–1099.

Wali, B., D.L. Greene, A.J. Khattak, and J. Liu. 2018. Analyzing within garage fuel economy gaps to support vehicle purchasing decisions–A copula-based modeling and forecasting approach. *Transportation Research D* 63:186–208.

Wan, Z., H. Chang, S. Shu, Y. Wang, and H. Tang. 2014. A review on cold start of proton exchange membrane fuel cells. *Energies* 7:3179–3203. doi:10.3390/en7053179.

Washington State Legislature. 2020. An act relating to reducing emissions by making changes to the clean car standards and clean car program. http://search.leg.wa.gov/search.aspx#document.

Yang, Z., and D. Rutherford. 2019. "Japan 2030 Fuel Economy Standards." Policy update. International Council on Clean Transportation. September 27. https://theicct.org/publications/japan-2030-fuel-economy-standards.

13

Emergent Findings, Recommendations, and Future Policy Scenarios for Continued Reduction in Energy Use and Emissions of Light-Duty Vehicles

The period that this report focuses on, from 2025 to 2035, could bring the most fundamental disruption in the 100-plus year history of the automobile. Light-duty vehicle (LDV) manufacturers are planning major introductions and penetrations into their product mix of electrified and connected and automated vehicles in this time period. These technologies will fundamentally change how consumers interact with and use their vehicles, contributing to ongoing fundamental changes in vehicle ownership models, ride sharing practices, traffic planning and management, urban design, and refueling infrastructure. As General Motors' chief executive officer Mary Barra recently stated, "I have no doubt that the automotive industry will change more in the next 5 to 10 years than it has in the past 50. The convergence of connectivity, vehicle electrification, and evolving customer needs demand new solutions" (GM Chevrolet Pressroom, 2016).

Other major factors will also influence and complicate projections of motor vehicle fleets and technologies in the 2025–2035 period. In addition to electrified and autonomous vehicles, other advances in vehicle technology are expected, including further developments in internal combustion engine vehicles, hybrid vehicles, and fuel cell vehicles, and new or increased reliance on alternative fuels such as biofuels and other low-carbon fuels. Political, economic, and regulatory pressures to decrease vehicle fuel consumption and reduce greenhouse gas (GHG) emissions may grow not only at the federal level but also at the state and local level. Increasing demand for travel is leading to increasing congestion, and automakers are facing limits on petroleum-fueled or even all LDVs in certain urban areas. International developments and regulatory pressures will also be critical, as the automobile industry becomes more globalized over this period. New modes of mobility, from ride sharing vehicles, to scooters, to new types of public transportation, and perhaps even flying cars, will all expand and diversify in the growth of mobility as a service.

These disruptive changes in the automotive industry and in transportation will have impacts, direct and indirect, on fuel consumption and GHG emissions. The development of new engine and vehicle technologies will be critical to improving fuel efficiency and environmental performance, but technology alone will not determine the performance of the vehicle fleet in the 2025–2035 period. Government regulation will continue to play a critical role in the development and implementation of more fuel-efficient vehicles and vehicle technology. Because of the fundamental changes in vehicle types and operation that may be possible, consumer acceptance, infrastructure development, transportation planning, and other off-vehicle factors will play key roles in determining the fuel consumption and GHG emissions of the vehicle of the future. These goals are pursued in a global market in which automotive manufacturers and suppliers comply with varying fuel economy and emissions standards in different countries around the world. The trend toward globalized markets and vehicle platforms adds complexity in projecting future fuel economy improvements in the U.S. market.

Section 13.1 highlights the findings and recommendations that emerge from the preceding chapters of this report. Section 13.2 identifies broader concerns and provides recommendations that go beyond the existing statutory authority, and are primarily directed to Congress for updating and refocusing the corporate average fuel economy (CAFE) program in light of evolving legal, scientific, policy, technological, and economic factors.

13.1 EMERGENT FINDINGS AND RECOMMENDATIONS

The findings and recommendations in the following section emerged from synthesis of learning across the technology, consumer, market, and regulatory sections of the report. The findings and recommendations in this section are premised on continued increase in requirements for energy efficiency and reduced GHG emissions of vehicles but do not assume any major changes in policy, such as requirements that all vehicles be zero-emission vehicles at the tailpipe. It considers current statutory authority or straightforward changes to statutory authority. *Section 13.2 discusses transformative ideas for fuel economy, vehicle efficiency, and GHG emissions that go* beyond the current statutory authority.

13.1.1 Synthesized Report Findings and Recommendations

13.1.1.1 Zero-Emission Vehicle (ZEV) Penetration

SUMMARY FINDING 1. ***ZEV Transition*:** The greatest opportunity and uncertainty for light-duty vehicle energy efficiency in 2025–2035 will be the increasing penetration of zero-emission vehicles (ZEVs). The price of the vehicles, fueling infrastructure, performance attributes, and consumer interest in and comfort with the technology will be major determinants in their uptake. For the mass-market consumer, electric and fuel cell vehicles represent a different energy source and different fueling behavior than is the norm today. They have lower operating and repair costs, and may have better vehicle performance. A transition to ZEVs is more significant and disruptive than other vehicle technologies that do not impact the fuel/energy source or refueling behavior.

SUMMARY FINDING 2. ***Global and U.S. ZEV Penetration*:** Regulations, incentives, and the interconnected international markets for zero-emission vehicles (ZEVs) will also affect their sales and technological development. Automakers are predicting deployment of tens of millions of ZEVs globally during the period of 2025–2035, aiming to achieve at least 50%–100% ZEV sales by 2030–2035 in leading jurisdictions (e.g., California, China, Europe). High penetration of ZEVs will involve profound changes to the vehicle fleet, charging/fueling infrastructure, business models for dealers, driver behaviors, repairs, emergency responders, materials, and battery recycling/second life. These changes will impact consumers, automakers, suppliers, dealers, fleet owners, and many others in the light-duty vehicle transportation system.

SUMMARY FINDING 3. ***Fuel Cell Electric Vehicle Deployment*:** Availability of commercial fuel cell electric vehicles (FCEVs) in the United States is limited to California and Hawaii. Introduction of FCEVs in the Northeast United States has been delayed, largely owing to the ban on hydrogen vehicles in tunnels and on the lower deck of two-tier bridges in the region. Recent studies have examined the behavior of hydrogen in tunnels, providing results of risk and scenario analyses to enable informed decisions on FCEV use in tunnels moving forward.

SUMMARY RECOMMENDATION 1. ***Growing Role of ZEVs*:** The agencies should use all their delegated authority to drive the development and deployment of zero-emission vehicles (ZEVs), because they represent the long-term future of energy efficiency, petroleum reduction, and greenhouse gas emissions reduction in the light-duty vehicle fleet. Vehicle efficiency standards for 2035 should be set at a level consistent with market dominance of ZEVs at that time, unless consumer acceptance presents a barrier that cannot be overcome by public policy and private sector investment. At the same time, maximum feasible fuel economy of petroleum-fueled vehicles should be pursued, under the National Highway Traffic Safety Administration's interpretation of its existing authority, and as a portion of the U.S. Environmental Protection Agency's combined stringency assessment. The pathway to zero emissions should be pursued in a technology-neutral manner.

SUMMARY RECOMMENDATION 2. *Purchase Subsidies*: The U.S. federal battery electric vehicle, plug-in hybrid electric vehicle, and fuel cell electric vehicle purchase subsidies should be continued until financial and psychological consumer barriers to purchasing such vehicles have been overcome. However, it should be changed to point-of-sale rebates to increase effectiveness and lower fiscal burdens. Income eligibility should be considered for both policy equity and effectiveness. Research organizations in partnership with federal agencies should conduct studies to optimize which type of vehicles and electric ranges should receive more or less subsidy, with considerations of equity and policy effectiveness in promoting zero-emission vehicle sales and/or electric vehicle miles traveled share.

SUMMARY FINDING 4. *Battery Electric Vehicle Charging*: Battery electric vehicle (BEV) charging is a paradigm shift from gasoline refueling. In a BEV charging ecosystem, charging opportunities are ubiquitous and frequent and part of normal vehicle parking, rather than a separate activity, like going to a gas station. The convenience of BEVs centers on at-home, overnight charging. Besides private home charging, semipublic infrastructure like multifamily dwelling and workplace charging are most important for increasing electric miles for daily trips. Public, fast charging on major corridors is most important for increasing electric miles for longer trips and providing assurance of urgent range extension.

SUMMARY FINDING 5. *Fuel Cell Electric Vehicle Fueling Infrastructure*: Hydrogen infrastructure build-out is the most significant challenge for fuel cell electric vehicle (FCEV) deployment. FCEV fueling infrastructure deployment would benefit from broader regional and national strategies and increased engagement with policy makers and local jurisdictions. Hydrogen infrastructure development for industrial and utility applications as well as for medium- and heavy-duty FCEVs have potential to accelerate build-out of the refueling network and reduce the cost of hydrogen.

SUMMARY RECOMMENDATION 3. *Charging Infrastructure*: The U.S. Department of Transportation, the U.S. Environmental Protection Agency, and the U.S. Department of Energy should coordinate to facilitate electric charging and hydrogen refueling infrastructure deployment with relevant stakeholders, including state and local government agencies, business associations, and entities. Congress should appropriate funds for and the agencies should create a national public-private partnership to lead this coordinating effort. For plug-in electric vehicle (PEV) charging, this coordinated effort should explicitly incorporate corridor fast charging, public charging at public parking spaces, PEV readiness of new and renovated homes and communities, and PEV readiness of workplace parking. For fuel cell electric vehicles, this coordinated effort should include support of hydrogen fuel infrastructure for light-duty vehicle (LDV) users in conjunction with medium- and heavy-duty vehicles and industry users, and deployment of LDV hydrogen refueling stations.

13.1.1.2 Agency Coordination

SUMMARY FINDING 6. *Agency Coordination of Different Authorities*: The National Highway Traffic Safety Administration and the U.S. Environmental Protection Agency (EPA) regulate under different statutory authorities, and for as long as these statutes remain as currently written, each agency is required to continue to adopt its own standards pursuant to its statutory criteria. Since 2010, the two agencies have coordinated their light-duty vehicle (LDV) regulations under the corporate average fuel economy (CAFE) and greenhouse gas (GHG) programs, which have minimized to the extent possible conflicting and duplicative requirements for industry. However, as the costs, performance, and sales of dedicated alternative fuel vehicles (AFVs), especially battery electric vehicles (BEVs) and fuel cell electric vehicles (FCEVs), improve substantially over the next decade, the two programs may increasingly diverge. EPA is permitted to consider the availability and feasibility of zero-emission vehicles (ZEVs) in setting its standards, and so as ZEVs become a growing portion of new LDV sales, the EPA standards will become progressively more stringent by incorporating the associated reduced GHG emissions. As currently implemented, however, the CAFE program does not include dedicated AFVs such as BEVs and FCEVs in its stringency feasibility analysis. If this practice continues, the CAFE standards will grow increasingly less stringent than the corresponding EPA standards over time, and the CAFE program will become less relevant to meeting its efficiency and petroleum reduction goals.

SUMMARY RECOMMENDATION 4. *Agency Coordination of Different Authorities*: The efforts of the National Highway Traffic Safety Administration (NHTSA) and the U.S. Environmental Protection Agency (EPA) to coordinate their fuel economy and greenhouse gas (GHG) emission standards since 2010 have been beneficial and should be continued to the extent feasible. However, the separate agency standards may now diverge because of the growing availability and benefits from zero-emission vehicles (ZEVs) and the agencies' different statutory authorities. The EPA can and must consider the availability and benefits of ZEVs and more efficient petroleum-fueled vehicles in setting the most stringent feasible GHG emission standards. In order to remain binding and relevant, NHTSA's program must consider the fuel economy or energy efficiency benefits provided by alternative fuel vehicles such as battery electric vehicles and fuel cell electric vehicles in setting the stringency of its corporate average fuel economy standards, either by NHTSA's interpretation of existing statute or by Congress passing a new or amended statute.

SUMMARY RECOMMENDATION 5. *NHTSA ZEV Authority*: To fulfill its statutory mandate of obtaining the maximum feasible improvements in fuel economy, the National Highway Traffic Safety Administration should consider the fuel economy benefits of zero-emission vehicles (ZEVs) in setting future corporate average fuel economy (CAFE) standards. The simplest way to accomplish that would be for Congress to amend the statute to delete the prohibition (42 U.S.C. § 32902[h][1]) on considering the fuel economy of dedicated alternative-fueled vehicles in setting CAFE standards. If Congress does not act, the Secretary of Transportation should consider ZEVs in setting the CAFE standards by using the broad authority under the statute to set the standards as a function of one or more vehicle attributes related to fuel economy, and define the form of the mathematical function. For example, recognizing that the maximum feasible average fuel economy depends on the market share of gasoline and diesel vehicles relative to ZEVs, the Secretary could consider redefining the function used for setting the standards to account for the expected decreasing share of gasoline and diesel vehicles relative to ZEVs. One possible mechanism to do this could be setting the standard as a function of a second attribute in addition to footprint—for example, the expected market share of ZEVs in the total U.S. fleet of new light-duty vehicles—such that the standards increase as the share of ZEVs in the total U.S. fleet increases.

SUMMARY RECOMMENDATION 6. *Fulfilling EPA Mandate*: If the National Highway Traffic Safety Administration is unable to consider alternative-fuel vehicles, and in particular zero-emission vehicles (ZEVs), in its stringency analysis, then the U.S. Environmental Protection Agency should continue under its mandate with divergent, more stringent standards, based on the advancements in ZEVs.

13.1.1.3 Life Cycle Emissions and Energy Use

SUMMARY FINDING 7. *Life Cycle LDV Transportation Emissions, Fuel, and Energy Use*: In current practice, vehicle manufacturers are responsible for meeting onboard per-mile fuel efficiency and emissions requirements. A full-fuel-cycle assessment more fully captures the total light-duty vehicle (LDV) system greenhouse gas emissions and energy consumption than an onboard, in-use consumption or emissions metric, and more evenly compares vehicles using different fuels. A full vehicle life cycle assessment including vehicle manufacture and disposal is an even more comprehensive measure of LDV system emissions or fuel and energy consumption.

SUMMARY FINDING 8. *ZEV Upstream Emissions Accounting*: Zero-emission vehicles (ZEVs) have zero greenhouse gas (GHG) emissions at the tailpipe, but have upstream emissions and energy use associated with processes to generate electricity, hydrogen, or other zero-emission fuels. Internal combustion engine vehicles including hybrid electric vehicles have both tailpipe and upstream emissions and energy use. Currently, neither the National Highway Traffic Safety Administration (NHTSA) nor the U.S. Environmental Protection Agency (EPA) account for the full fuel cycle including upstream emissions in regulatory compliance treatment in order to incentivize ZEVs. EPA currently assumes a 0 g/mi upstream carbon dioxide emission factor for ZEVs, and NHTSA uses a petroleum equivalency factor for upstream emissions that applies to only 15% of energy utilized by ZEVs to incentivize such vehicles (consistent with other alternative fueled vehicles). However, the full-fuel-cycle treatment is used in the benefit cost assessment of the regulations. The exclusion of upstream emissions in the corporate average fuel economy and GHG regulatory compliance metrics provides an incentive to produce ZEVs, relative to internal combustion engine only and hybrid electric vehicles, in states where there is no binding ZEV mandate, and could help to overcome significant market barriers that ZEVs face during a transition in the market toward the long-term goal of zero tailpipe emissions.

SUMMARY RECOMMENDATION 7. *Life Cycle Emissions*: Congress should define long-term energy and emissions goals for the corporate average fuel economy (CAFE) and greenhouse gas (GHG) programs, and the National Highway Traffic Safety Administration (NHTSA) and the U.S. Environmental Protection Agency (EPA) should set regulations that put the United States on a path to meet those goals. Considering other regulatory systems that may be implemented as part of a national program to reduce energy use and emissions in the fuel, electricity, and manufacturing sectors, the light-duty vehicle CAFE and GHG programs may or may not need to address the full vehicle and fuel life cycle emissions and energy consumption. Any vehicle or fuel life cycle requirements within the NHTSA or EPA programs should be set with appropriate lead-time for manufacturers to revise their upcoming product plans.

SUMMARY RECOMMENDATION 8. *ZEV Upstream Emissions Accounting*: In the longer term, it makes sense to address the full-fuel-cycle emissions of all vehicles, including zero-emission vehicles (ZEVs), especially as ZEVs become a progressively larger portion of the light-duty vehicle fleet. The National Highway Traffic Safety Administration and the U.S. Environmental Protection Agency should undertake a study of how and when to implement a full-fuel-cycle approach, including consideration of the potential benefits and drawbacks of the current temporary exclusion of upstream emissions for compliance of ZEVs. Based on that study, the agencies should decide whether and when to adopt a different approach for accounting for upstream ZEV emissions for compliance.

13.1.1.4 Technology

SUMMARY FINDING 9. *Internal Combustion Engines*: Internal combustion engines (ICEs) will continue to play a significant role in the new vehicle fleet in MY 2025–2035 in ICE-only vehicles, as well as in hybrid electric vehicles (HEVs) from mild hybrids to plug-in hybrids, but will decrease in number with increasing battery electric vehicle (BEV) and fuel cell electric vehicle penetration. In this period, manufacturers will continue to develop and deploy technologies to further improve the efficiency of conventional powertrains, for ICE-only vehicles and as implemented in HEVs. Developments in the ICE for hybrids will advance toward engines optimized for a limited range of engine operating conditions, with associated efficiency benefits. Major automakers are on differing paths, with some focusing their research and development and advanced technology deployment more squarely on BEVs, and others more focused on advanced HEVs to maximize ICE efficiency.

SUMMARY FINDING 10. *Road Load Reduction*: Road load reduction leads to reduced fuel consumption and greenhouse gas emissions, and opportunities include mass reduction, improved aerodynamics, and reduced rolling resistance. In 2025–2035, mass reduction will be implemented for fuel consumption reduction and driveability for all vehicles, and also for increased driving range for battery electric vehicles and fuel cell electric vehicles. Improved aerodynamics will be challenged by the shift to taller vehicles with larger frontal area, and may be positively or negatively impacted by vehicle architectures responding to electrification. There will be incremental improvements available in tire rolling resistance. Mass and geometric disparity in the fleet may increase or decrease owing to electrified powertrains, new architectures, automated and connected vehicle technologies, and a shift from sedans to crossover vehicles, sport utility vehicles, and pickup trucks in 2025–2035.

SUMMARY RECOMMENDATION 9. ***Safety*****:** Improved crash compatibility will reduce the adverse effect of mass and geometric disparity on crash safety for passengers of all vehicles and vulnerable roads users, including pedestrians. The National Highway Traffic Safety Administration should study mass disparity in 2025–2035, improve federal motor vehicle safety standard testing protocols for crash compatibility, and further develop testing or computer-aided engineering fleet modeling to simulate real-world crash interactions between new vehicle designs and with vulnerable users at different impact speeds and impact configurations.

SUMMARY FINDING 11. ***Battery Technology*****:** Lithium-ion batteries will be the dominant battery technology for battery electric vehicles (BEVs) in 2025–2035. The chemistries within them will have incremental improvements in performance and cost during this time frame. There are opportunities for breakthroughs in battery technologies to go "beyond lithium"; however, such breakthroughs are not guaranteed. By 2035, there may be limited commercial sales of BEVs with "beyond lithium ion" technologies, most likely solid-state batteries. Engineering improvements at the module and pack level will contribute to further increases in energy density and cost reduction.

SUMMARY FINDING 12. ***Electric Vehicle Costs*****:** Battery electric vehicles (BEVs) with longer electric range (e.g., 300 miles) may reach first-cost parity with comparable internal combustion engine only vehicles by 2030, especially from high-volume BEV manufacturers. Shorter range BEVs may be favored by some consumers and would reach cost parity even sooner. When considering fuel and maintenance, BEVs have or will reach total cost of ownership parity earlier than first-cost parity. Reducing battery cost while meeting specifications for greater durability and rapid charging capabilities will widen their appeal. The BEV cost driver is the battery, which for high-volume battery production is expected to decrease to $90–$115/kWh by 2025 and $65–$80/kWh over 2030–2035 at the pack level.

SUMMARY FINDING 13. ***Electric Drive System Technologies*****:** While the majority of the automakers are still using insulated gate bipolar transistor power switching devices in their power electronic circuitry (inverter and converter), many are pursuing the use of wide-bandgap devices (silicon carbide, SiC, or gallium nitride, GaN) in their next-generation propulsion systems, owing to their size, weight, and efficiency benefits. Most automakers are expected to be using SiC in their vehicles by 2025 owing to its widespread availability. However, given the inherent cost advantage of GaN on silicon devices compared with SiC, if the problems with GaN device architecture can be resolved, it is expected that GaN on silicon devices will gradually replace SiC during the period 2025–2035.

SUMMARY FINDING 14. ***Fuel Cell Electric Vehicles*****:** Several automakers are releasing their second generation of fuel cell vehicles. A few major automakers are planning a strategy of high fuel cell vehicle deployment in the United States and elsewhere to take advantage of their long ranges and short refueling times relative to battery electric vehicles. Developers have identified pathways to reduce fuel cell powertrain and hydrogen tank cost through materials and manufacturing improvements and economies of scale. Fuel cell electric vehicles (FCEVs) could reach parity with internal combustion engine only vehicles in total cost of ownership in 2025–2035 if aggressive efficiency and cost targets are met. FCEVs are expected to be particularly valuable to operators and fleet owners that need constant operation and/or high daily vehicle miles traveled, and those that require continuous low-end torque, such as for towing.

SUMMARY FINDING 15. *Connected and Automated Vehicle Technologies*: Vehicle connectivity and automation technologies could improve the fuel efficiency of internal combustion engines by up to 9% in city driving and up to 5% on the highway by detecting upcoming conditions and adjusting acceleration and powertrain operation accordingly. In 2025–2035 new vehicles, there will be even more widespread implementation of automated vehicle technologies at the lower levels of automation for convenience and safety than exists in 2020, and growing but uncertain use of vehicle-to-vehicle and vehicle-to-infrastructure connectivity. Connected and automated vehicle technology efficiency benefits are not currently detectable in fuel economy certification testing but may be eligible for off-cycle credits when direct fuel savings can be demonstrated. However, fuel savings are not presently the primary driver of connected and automated vehicle technologies in the market, and the potential energy benefits of these technologies are unlikely to be realized absent incentives or other policies to ensure that they are implemented with fuel efficiency as a fundamental goal.

SUMMARY FINDING 16. *Autonomous Vehicles*: Fully capable, fully automated, Level 4 and 5 light-duty vehicles will be deployed in some ride hailing, delivery, and closed-campus fleets by 2025. More widespread adoption will require ensuring safety under all conditions, resolving cybersecurity issues, developing appropriate regulations, and gaining consumer acceptance of a radically different driving experience. Hence, autonomous vehicles' share of the market in 2035 is highly uncertain but likely to fall in the 0–40% range. Adoption of autonomous vehicles could greatly increase or reduce transportation energy use and the impact will be determined to a large degree by their effects on travel choices and vehicle ownership decisions, particularly vehicle miles traveled.

SUMMARY RECOMMENDATION 10. *Autonomous Vehicle Efficiency Regulation*: The agencies should consider regulating autonomous vehicles for fleet use differently from personally owned vehicles. Maximum feasible standards for these vehicles could be substantially more stringent than standards for personally owned vehicles; an all-electric requirement should be considered. To achieve the fuel-savings potential of autonomous driving and avoid its unintended consequences, the U.S. Department of Transportation should consider actions to guide the effects of autonomous driving on the U.S. transportation system and make recommendations accordingly to other agencies and to Congress.

13.1.1.5 Consumer Value of Fuel Economy

SUMMARY FINDING 17. *Consumers*: New vehicle purchasers select vehicles with a variety of factors in mind, including fuel economy. Manufacturers perceive that consumers expect higher fuel economy but will not pay for the full value of fuel saving technologies, while many academics think consumers almost fully value fuel savings. Some automakers are trying to engage their future consumers in new vehicle options, including changes in propulsion systems like battery electric vehicles and fuel cell electric vehicles. Many consumers initially resist new technologies that disrupt current practices and lifestyles, or create novel risks or uncertainties, even if the technology provides net benefits to them.

SUMMARY RECOMMENDATION 11. *Novel Technology Barriers*: Because consumer resistance to novel technology is a significant issue in market penetration and acceptance of new technologies, policy interventions beyond purchase subsidies may be needed to address these barriers. Such policies may include investment in charging and refueling infrastructure, or consumer education and exposure to the new technology and its benefits.

13.1.1.6 Test Cycles and Regulatory Structure

SUMMARY FINDING 18. *Test Cycles and New Vehicle Technologies*: Two test cycles for corporate average fuel economy compliance were established in 1975, a city and a highway cycle. In 2008, three additional cycles, originally developed for criteria pollutant measurement in high speed, air conditioning, and cold temperature operation, were incorporated into fuel economy testing to better reflect real-world fuel economy for vehicle labeling.

There have been modifications to the test procedure to accommodate alternative powertrains, including a test for electric vehicle range. The current test procedures are insufficient to test electric vehicle range and connected and automated vehicle operation, and they do not adequately reflect modern driving patterns of light-duty vehicles.

SUMMARY FINDING 19. *On-Road Fuel Economy*: There is no representatively sampled, empirical measure of on-road fuel consumption or greenhouse gas (GHG) emissions for the U.S. light-duty fleet. Using onboard diagnostics and customer data available, it is increasingly possible to assemble such a statistically valid and relevant data set. Such data could be used to monitor the fuel consumption and GHG emissions of the light-duty vehicle sector, the effectiveness of the corporate average fuel economy program, and the effectiveness of off-cycle technologies in reducing real-world emissions and fuel consumption.

SUMMARY RECOMMENDATION 12. *In-Use Performance*: The agencies should implement a program that measures fuel consumption and greenhouse gas emissions from the light-duty vehicle fleet in use. The purpose of the in-use program should be to evaluate and improve the effectiveness of the corporate average fuel economy program, not for year-by-year enforcement against individual manufacturers. New data sources and telematic technologies makes such in-use monitoring feasible, but safeguards must be established to minimize privacy risks for vehicle owners and operators.

SUMMARY RECOMMENDATION 13. *Driving Patterns and Emissions Certification*: The agencies (U.S. Department of Transportation, U.S. Environmental Protection Agency, U.S. Department of Energy) should conduct a study on how well current driving patterns and new vehicle technology impacts are reflected by current vehicle certification test cycles. The results of this study should then be used to propose new light-duty vehicle test cycles, or adoption of the current or a new weighting of the existing 5-cycle test. The study of driving patterns and emissions and resulting changes in the test cycle may make it possible to eliminate some off-cycle treatment of fuel efficiency technologies and evaluate the energy saving impacts of those that remain.

SUMMARY RECOMMENDATION 14. *Off-Cycle Technologies*: The U.S. Environmental Protection Agency (EPA) and the National Highway Traffic Safety Administration should consider off-cycle technologies in setting the stringency of the standards. The agencies should approve off-cycle credits on an annual cycle, require automakers to clearly and transparently document the test procedures and data analysis used to evaluate off-cycle technologies, and produce a compiled report on proposed credits that is available for public comment. The agencies should track the adoption of off-cycle credits in the vehicle fleet at the model level and report these data to the public, for example through the EPA Trends Report.

SUMMARY RECOMMENDATION 15. *CAV Efficiency Regulation*: In setting the level of the standards, the agencies should consider connected and automated vehicle (CAV) technologies that can save energy. Off-cycle credits should be available for CAV technologies only to the extent that they improve the fuel efficiency of the vehicle on which they are installed. Credits should be based on realistic assumptions, where needed, regarding technology adoption on other vehicles or infrastructure.

SUMMARY FINDING 20. *Passenger Car and Light Truck Standards*: Passenger cars and light trucks are regulated under separate standards. In some cases, light truck capability and use is very similar to counterpart vehicles classified as passenger cars, sometimes with the same make and model, only distinguished by two- or four-wheel drive. In other cases, light trucks have duty-cycle requirements of off-road capability, hauling, towing, and four-wheel drive that could justify a different efficiency or emissions standard than passenger cars not designed with those requirements.

SUMMARY RECOMMENDATION 16. *Car and Truck Standards*: The National Highway Traffic Safety Administration and the U.S. Environmental Protection Agency should commission an independent group to study the effectiveness and appropriateness of separate standards for passenger cars and light trucks.

13.2 BIG PICTURE: RETHINKING REGULATION OF FUEL ECONOMY IN 2025–2035 AND BEYOND

So far, this report has provided analysis, findings, and recommendations on the future of the CAFE program largely within the confines of the existing statutory authority. However, that existing statutory authority only prescribes fuel economy standards for specific years through 2030, so it is likely that the statute will be revised by Congress for the period of this study of 2025–2035. Given the increasing changes to vehicle technology, national goals for vehicle efficiency and emissions, and other changes to the LDV transportation system projected in 2025–2035, and the natural time to update the statute, the congressional amendments will likely consider other changes to the CAFE program. Because many possible changes to standard goals and structure will be relevant to this committee's assigned task in evaluating the CAFE program in the 2025–2035 period, the committee offers the following discussion and recommendations for Congress and policy makers with respect to the structure of the CAFE program from 2025 to 2035.[1]

The existing CAFE program is quickly becoming outdated and will need to be updated in the time period from 2025 to 2035 for legal, scientific, policy, technological, global leadership, and economic reasons.

- From a *legal* perspective, the existing CAFE statute was last amended in 2007 and only explicitly authorizes new standards through 2030, so it will need to be updated to provide relevant criteria for setting CAFE standards beyond 2030.
- From a *scientific* perspective, climate change has become an increasingly urgent problem for the nation and world since the last statutory update in 2007, and as discussed in greater detail below, the CAFE program needs to be updated to better focus on this urgent national need.
- From a *policy* perspective, the United States is likely to enact broader climate change legislation in the coming years, which will interact and overlap with the CAFE program, requiring alignment or synergy of the CAFE program with these broader efforts.
- From a *technological* perspective, vehicle manufacturers are planning large-scale deployment of advanced technology vehicles such as battery electric vehicles (BEVs), fuel cell vehicles, and connected and automated vehicles. The existing statute does not adequately address nor ensure the environmental benefits of these important new vehicle technologies that could become dominant in the next couple of decades.
- From a *global leadership* perspective, other nations are surging forward with new vehicle requirements that surpass those envisioned in the existing CAFE program, and the United States will need to upgrade its approach if it wishes to remain an international leader in clean vehicle technology.
- From an *economic* perspective, consumer behavior and automaker business plans are likely to change substantially as advanced vehicle technologies grow in the market.

FINDING 13.1: The current statutory authority for the CAFE program is becoming increasingly outdated as a result of legal, scientific, policy, technological, global leadership, and economic developments and trends.

RECOMMENDATION 13.1: Given the end of the latest legislative specification for corporate average fuel economy (CAFE) in 2030, Congress should extend the CAFE program and, as part of that reauthorization, evaluate and update the statutory goals of the CAFE program. With those goals in mind, Congress should consider changes to the program structure and design, and its interaction with other related policies and regulations.

13.2.1 Changes Within Existing CAFE Program

In considering statutory changes to the CAFE program, it is useful to think both within and outside the current CAFE structure. Looking within the existing program first, one statutory change would be to refocus the objectives

[1]In addition to the implications of the committee's statement of task, the committee held a public meeting with the U.S. Environmental Protection Agency (EPA) in June 2020 at which EPA senior staff explicitly asked the committee to advise them on "big ideas" of vehicle efficiency and GHG emissions regulations that fit in the changing world of LDVs expected in 2025–2035.

of the CAFE program. At the time the CAFE program was adopted, enhancing energy security by reducing reliance on petroleum imports was a primary objective of the CAFE program. Today, while energy security concerns remain relevant, their importance for the CAFE program have diminished as the United States has become a net energy-exporting nation. Energy conservation was another important goal of the original CAFE program, and that objective remains important today. A new objective, not present when the CAFE program was originally enacted, is addressing climate change, and this should now be expressly recognized as an important objective of the CAFE program.

RECOMMENDATION 13.2: The statutory authorization for the corporate average fuel economy (CAFE) program should be amended to expressly include climate change as a core objective of the program, along with existing objectives such as energy conservation. Specifically, the statutory considerations for setting CAFE standards in 49 U.S.C. § 32902(f) should be amended to include the goal of reducing greenhouse gas emissions.[2]

The emergence of reducing GHGs as a national goal, and the partial overlap with the CAFE program of the U.S. Environmental Protection Agency's (EPA's) role in directly addressing LDV GHG reduction, raises the question of whether it continues to make sense to have both a CAFE program administered by the National Highway Traffic Safety Administration (NHTSA) and a GHG reduction program administered by EPA. The existence of two partially overlapping programs does create some redundancy, which increases costs to the government and compliance burdens for manufacturers. However, these duplicated costs have been minimized by aligning the two programs as much as possible. Moreover, there are benefits from maintaining the two programs. While the EPA program focuses specifically on GHG emissions, the NHTSA CAFE program expressly considers other relevant factors such as energy conservation, energy imports, and vehicle safety. While again there is some overlap in the practical impact on standards of these factors and the GHG emissions considered by EPA, there is value in giving independent consideration to the CAFE-exclusive factors. In addition, NHTSA and EPA staff have different expertise, experience, and capabilities, and so the two agencies can provide a useful check on each other's analyses and estimates. Last, the two programs can provide a backstop to each other if one set of standards is delayed or rescinded by administrative or judicial actions.

FINDING 13.2: The continued existence of two partially overlapping programs, the CAFE program administered by NHTSA and the GHG emissions program administered by EPA, imposes some duplication and extra costs on government and industry, but these additional burdens can be mostly offset by coordinating the two programs. In addition, the continued existence of the two separate programs provides some benefits that outweigh the duplicated costs and burdens, including the consideration of different unique and relevant factors by each agency, and the benefits of having the two agencies check and backstop each other's activities.

RECOMMENDATION 13.3: Congress should reauthorize the continuation of the National Highway Traffic Safety Administration (NHTSA) corporate average fuel economy (CAFE) program, notwithstanding its practical overlap with the U.S. Environmental Protection Agency light-duty vehicle greenhouse gas program. Congress can minimize any disruption from having two programs by eliminating any obstacles to coordinating the two programs, such as by eliminating the current prohibition that prevents NHTSA from considering zero-emission vehicles and other dedicated alternative-fueled vehicles in setting CAFE standards.

Another useful change to the CAFE program would be for Congress to provide a longer-term target for vehicle manufacturers' planning. Vehicle manufacturers face many different options of where to invest their research and development (R&D) efforts, from further improving internal combustion engines and developing such vehicles

[2]In discussion with stakeholders after the prepublication version of this report was released, it became apparent that this recommendation could be interpreted in a way that would have unintended effects that were not examined by the committee. This footnote was added to this final publication of the report in order to clarify that in making this recommendation, the committee did not examine and did not intend for any changes to statute to disrupt the regulatory structure that provides for state GHG emissions standards, state ZEV requirements, and GHG emissions standards promulgated by EPA.

for low-carbon fuels, to developing and advancing hybrid technologies, to focusing on zero-emission technologies such as BEVs and fuel cell electric vehicles (FCEVs). Without long-term targets, there is a greater risk that vehicle manufacturers will choose a technology investment and development strategy that "locks in" technologies that represent wasteful dead-ends rather than stepping-stones to the longer-term goals (Coglianese and D'Ambrosio, 2008; Williams et al., 2015). Because NHTSA can only set CAFE standards for a maximum of 5-year periods, Congress should adopt a longer-term national goal to assist the future planning of both NHTSA and industry.

One of the critical goals for the countries of the world will be to deeply decarbonize their economies by 2050. Government jurisdictions, some major vehicle manufacturers, and academic and think tank analyses have converged on the concept that all new LDVs should achieve net-zero emissions by the 2035–2050 period (NASEM, 2021).[3] To provide a longer-term target for manufacturers to plan their ongoing and future R&D and product rollout, Congress should set an explicit goal that all new LDVs should achieve net-zero GHG emissions by a specified date.[4] This zero-emission requirement should be technology neutral, to allow each manufacturer to choose its own technology pathway. Summary Recommendation 8 recommends a transition to address full-fuel-cycle emissions of all vehicles, including ZEVs. It recommends a study of how and when to implement a full-fuel-cycle approach. The study should evaluate the cost and emissions effectiveness of incentivizing ZEVs by excluding upstream ZEV emissions, relative to other methods of incentivizing the transition to ZEVs. It should also address some of the complexities of considering upstream emissions including their uncertainties, their heterogeneity by region and fuel used for energy generation, the dynamic changing nature of upstream emissions over time, the metrics that should be used to measure upstream emissions, and the respective roles and approaches of NHTSA and EPA to account for upstream emissions. Last, the study should identify whether statutory changes are needed to best account for upstream emissions, and if so, what those changes might be. Such full accounting would ensure more informed and effective policy choices in setting the standards, as well as provide consumers full transparency in their vehicle choices.

While a zero-emissions goal should be the primary motivator behind both EPA and NHTSA's regulations, NHTSA's existing authorities related to energy efficiency, consumer fuel savings, and safety continue to be important. Given a hierarchy of goals, Congress and NHTSA might consider what metric (GHG emissions, fuel consumption, energy consumption, or some combination) is best to use for meeting the goals of its regulatory program. Congress should provide appropriate funding for consumer tax credits, refueling infrastructure, and other incentives to help enable this systemic transformation to a zero-emission light-duty fleet. To reach an economy-wide deep decarbonization goal, the transition in the vehicle fleet must occur alongside transformation in the full fuel cycle of all fuels to be net-zero emissions. The transformation to a zero-emission standard and ZEVs will not only help address climate change and other emission problems but will also help to ensure that U.S. vehicle manufacturers and suppliers remain at the forefront of new vehicle technologies and save consumers money.

FINDING 13.3: Many studies suggest that reaching an economy-wide deep decarbonization goal will require new vehicles to be zero-emissions. To comprehensively address climate change, a transition to ZEVs needs to be in concert with a full move to net-zero GHG fuels and electricity, and also net-zero vehicle manufacturing GHG emissions.

RECOMMENDATION 13.4: To provide vehicle manufacturers a longer-term target to assist planning their ongoing technology investments and pathways, Congress should set a goal that all new light-duty vehicles will have net-zero greenhouse gas emissions by a specific date that aligns with a national deep decarbonization goal, and includes interim goals. This target should be technology neutral, to allow each manufacturer to choose its compliance pathway and technology strategy.

---

[3]Example jurisdictions with goals of only ZEV sales, or zero internal combustion engine vehicle (ICEV) sales, have been California, 2035 (including PHEVs); UK, 2030 (only ICEVs, part of green economic recovery); France, 2040; Norway, 2025; Germany, Ireland, and the Netherlands, 2030.

[4]While this net-zero emissions goal would primarily be met by ZEVs, the standard should provide some flexibility for allowing some relatively small number of non-ZEV vehicles to be sold as necessary to meet extreme duty cycle, weather, or geographical needs, or because of the local deficiencies in ZEV infrastructure. Any emissions from these non-ZEVs could be offset by trades, offsets, credits, or other mechanisms to satisfy the net-zero emissions goal.

13.2.2 Changes Outside Existing CAFE Program

Given the fundamental changes expected in mobility and the transportation industry over the next couple decades, no single agency or program can develop an adequate regulatory framework in isolation. Rather, an integrated approach is required that considers the many facets of sustainable transportation:

- New vehicle technology;
- Fuel supply and infrastructure;
- Existing vehicle use and vehicle miles traveled (VMT);
- Connected and autonomous vehicles;
- Public transportation;
- New modes of transport including ride-sharing, scooters, and drones;
- Roads and other infrastructure;
- Smart cities and transportation planning;
- Congestion strategies including pricing and car-free zones;
- New vehicle ownership models;
- Consumer issues;
- Justice and fairness impacts;
- Safety;
- International developments and competitiveness; and
- Other society-wide strategies and policies for addressing climate change.

Many of these extrinsic factors will interact with and affect the CAFE program, so NHTSA will need to find new ways to coordinate and integrate the program with other related efforts.

One possible way to promote greater coordination would be to create a federal interagency task force on the changes that are occurring in transportation and mobility. NHTSA and other U.S. Department of Transportation (DOT) agencies and offices should participate, as well as several offices and programs from the EPA, the U.S. Department of Energy, the Department of Health and Human Services, the Department of Housing and Urban Development, and the White House. The goal of this task force should be to adopt and coordinate interagency efforts to move the nation toward a more sustainable system of transportation and mobility. The interagency task force could regularly report to Congress and the public on the progress and obstacles it identifies. The goals of the task force should be to expedite the transition to a cleaner, safer, and fairer transportation system and to promote U.S. leadership in developing the vehicle and mobility technologies of the 21st century.

RECOMMENDATION 13.5: The Executive Branch should create an interagency task force with the objective of coordinating and integrating government efforts to achieve a cleaner, safer, and fairer transportation and mobility system.

In addition to zero tailpipe emissions technologies such as BEVs and FCEVs, another important change in vehicle technology for the period 2025 and beyond will be connected and automated vehicles. Safety is likely to be the primary driver and determinant of when and how such vehicles will be deployed at the lower levels of automation, while the opportunity for cost savings will cause fleets to pursue fully autonomous vehicles. However, autonomous vehicles also could have substantial impacts on fuel consumption and GHG emissions, which could range from strongly beneficial to strongly detrimental, for the reasons discussed earlier in this report. The issue is complicated by the fact that a simple measure like VMT will not alone be a useful measure of environmental impact, as it will depend on the emissions the vehicles produce. If, for example, most autonomous vehicles in a region are BEVs or FCEVs, and have no carbon tailpipe emissions, then additional VMT may not substantially increase emissions, especially if renewable fuels are used to generate the electricity or hydrogen used to power those vehicles. Yet, they could add to congestion and safety issues, raising societal costs. New policies governing autonomous vehicles and their environmental and congestion impacts may be a relevant subject matter for the interagency federal committee discussed above but may also be the subject of state or local legislation or incentives on the occupancy of autonomous vehicles or the types of engine technologies used to power them in that jurisdiction.

> **RECOMMENDATION 13.6:** The federal interagency committee on new mobility, along with state and local policy makers, should consider rules or incentives to encourage future autonomous vehicles, especially in fleets, to use zero or near-zero emission technologies. Furthermore, the impact of any incentives should be evaluated for their ability to promote an overall reduction in vehicle miles traveled and increase in the use of transit and shared rides.

The most important large-scale and longer-term issue for the future is how the CAFE program, and GHG emissions from LDVs generally, fit within a broader national strategy or program to combat climate change. It is increasingly likely that the United States will and must adopt an economy-wide national program to reduce GHG emissions across all sectors before or during the 2025–2035 period of this study. That national program may include a national carbon tax or a national emissions-trading program, or another approach with a real or shadow price on carbon, and perhaps also a suite of facilitating policies to transition to a lower-emitting economy (NASEM, 2021). Regardless of the structure or approach of any nationwide climate change program, it will almost certainly intersect and interact with the CAFE program, given the large role the transportation sector plays in overall U.S. GHG emissions. Moreover, the transition to full-fuel-cycle ZEV technologies that operate on electricity and hydrogen will increasingly depend on the carbon intensity of the electricity generation and hydrogen production. Thus, carbon emissions must be addressed as a system, and one that is likely to be increasingly integrated.

A key issue going forward will therefore be how the CAFE program aligns with and contributes to the nationwide efforts to manage the carbon system. Will and should the CAFE program continue to operate under its old mandate, unaffected by these larger economy-wide programs? Or should the CAFE program be modified, or perhaps even eliminated, in response to a comprehensive, nationwide carbon regulatory system? The answers to these questions will depend in large part on how the nationwide carbon regulatory program is designed and implemented, and whether or how the CAFE program can be integrated into a coherent and effective national program to reduce GHGs.

****

CAFE has historically been the bedrock of U.S. vehicle energy efficiency and climate policy, eventually joined by EPA vehicle and other GHG programs. It is now entering a time of major change. New technologies are enabling a pathway to zero emissions, and the future of the LDV market is likely to have a diversity of energy sources and modes of mobility. The committee expects that CAFE will continue to play an important role in the future if the recommendations in this report are adopted, and serve as an example for other energy and climate policies administered by government agencies in the United States and around the world.

13.3 REFERENCES

Coglianese, C., and J. D'Ambrosio. 2008. Policymaking under pressure: The perils of incremental responses to climate change. *Connecticut Law Review* 40:1411–1429.

GM Chevrolet Pressroom. 2016. "GM Chairman and CEO Addresses CES." January 6. https://media.chevrolet.com/media/us/en/chevrolet/news.detail.html/content/Pages/news/us/en/2016/Jan/boltev/0106-barra-ces.html.

NASEM (National Academies of Sciences, Engineering, and Medicine). 2021. *Accelerating Decarbonization of the U.S. Energy System*. Washington, DC: The National Academies Press. https://doi.org/10.17226/25932.

Williams, J.H., B. Haley, F. Kahrl, J. Moore, A.D. Jones, M.S. Torn, and H. McJeon. 2015. *US 2050 Report, Volume 2: Policy Implications of Deep Decarbonization in the United States*. Energy and Environmental Economics, Inc.

Appendixes

A

Committee Biographical Information

GARY MARCHANT, *Chair*, is a Regent's Professor of Law and director of the Center for Law, Science, and Innovation at Arizona State University (ASU). Dr. Marchant's research interests include legal aspects of genomics and personalized medicine, the use of genetic information in environmental regulation, risk and the precautionary principle, and governance of emerging technologies such as nanotechnology, neuroscience, biotechnology, and artificial intelligence. He teaches courses in Law, Science, and Technology; Genetics and the Law; Biotechnology: Science, Law, and Policy; Health Technologies and Innovation; Privacy, Big Data, and Emerging Technologies; and Artificial Intelligence: Law and Ethics. Dr. Marchant was named a Regent's Professor in 2011 and also is a professor in the ASU School of Life Sciences, a Distinguished Sustainability Scientist in the ASU Julie Ann Wrigley Global Institute of Sustainability, and a Lincoln Professor of Emerging Technologies Law and Ethics with the Lincoln Center for Applied Ethics at ASU. Prior to joining ASU in 1999, Dr. Marchant was a partner at the Washington, D.C., office of Kirkland and Ellis, where his practice focused on environmental and administrative law. During law school, he was editor-in-chief of the *Harvard Journal of Law and Technology* and editor of the *Harvard Environmental Law Review*, and he was awarded the Fay Diploma (awarded to top graduating student at Harvard Law School). Dr. Marchant frequently lectures about the intersection of law and science at national and international conferences. He has authored more than 150 articles and book chapters on various issues relating to emerging technologies. Among other activities, Dr. Marchant has served on five previous National Academies committees (including the Committee on Assessment of Technologies and Approaches for Reducing the Fuel Consumption of Medium- and Heavy-Duty Vehicles, Phase 2, February 2013–2019; the Committee on Assessment of Solid-State Lighting, May 2011–February 2013; and the Committee on State Practices in Setting Mobile Source Emission Standards, 2004–2006), has been the principal investigator (PI) on several major grants, and has organized numerous academic conferences on law and science issues. He has a Ph.D. in genetics from the University of British Columbia.

CARLA BAILO is the president and chief executive officer of the Center for Automotive Research (CAR). Ms. Bailo is a leader in engineering and vehicle program management with 35 years of experience in the automotive industry. In addition to her role at CAR, Ms. Bailo is the 2016–2018 vice president of automotive for SAE International, a global association of more than 138,000 engineers and related technical experts in the aerospace, automotive, and commercial-vehicle industries. Prior to joining CAR, Ms. Bailo was the assistant vice president for mobility research and business development at Ohio State University (OSU). In that capacity, she assisted the university in accelerating sustainable mobility and transportation innovation, while integrating related research and education across OSU's academic units. She also helped coordinate OSU's involvement as the primary research partner for Smart Columbus, a $140 million program to transform central Ohio into a premier transportation innovation region. Ms. Bailo has 25 years of experience at Nissan North America, Inc., where in her most recent role at Nissan she served as senior vice president of research and development (R&D). She was responsible for vehicle engineering and development operations in Michigan, Arizona, Mexico, and Brazil, managing a $500 million budget and 2,500 employees. In this role, she improved the efficiency of Nissan's R&D functions. Ms. Bailo has an M.S. degree in mechanical engineering from the University of Michigan and a B.S. degree in mechanical engineering from Kettering University.

RODICA BARANESCU is retired as a professor in the Department of Mechanical Engineering at the University of Illinois, Chicago. Prior to that, Dr. Baranescu was manager of the fuels, lubricants, and engine group of the International Truck and Engine Corporation in Melrose Park, Illinois. She is an internationally sought-after public speaker on technical issues related to mobility technology, environmental control, fuels, and energy. Dr. Baranescu has extensive expertise in diesel engine technology and was elected to the National Academy of Engineering in 2001 for research leading to effective and environmentally sensitive diesel and alternative-fuel engines and leadership in automotive engineering. She is a fellow of SAE International and was its president in 2000. In 2003, she received the Internal Combustion Engine Award of the American Society of Mechanical Engineering (ASME). Dr. Baranescu received her M.S. and Ph.D. degrees in mechanical engineering in 1961 and 1970, respectively, from the Politehnica University in Bucharest, Romania, where she served as assistant professor (1964–1968), lecturer (1970–1974), and associate professor (1974–1978).

NADY BOULES is the president of NB Motors, LLC. Dr. Boules began his engineering and management consultancy practice with a focus on electrified, connected, and autonomous vehicles in September 2013 after 45 years of engineering experience (32 years with General Motors [GM] and Delphi), including 14 years as director of R&D and innovation. In his most recent capacity, as director of the GM R&D Electrical and Control Systems Research Lab, Dr. Boules was responsible for the development of advanced electrical systems and components for electrified, connected, and automated vehicles to enhance vehicle safety, comfort, and efficiency. He led all R&D activities in the areas of electronics and control software globally and coordinated research in this area—internally with research laboratories and engineering customers; and externally with collaborators at universities, national laboratories, suppliers, and other automotive original equipment manufacturers (including USCAR). Dr. Boules first joined General Motors Research Laboratories in 1982. He held several positions leading and managing research activities in automotive mechatronic and electric drive systems. In September 1999, he was named director of R&D for Delphi Steering Systems in Saginaw, Michigan. His responsibilities expanded to cover brakes and suspension when he was named director of the dynamics innovation center in 2002. In 2005, Dr. Boules was named director of the dynamics innovation center and materials engineering, and his responsibilities expanded to materials and to the energy and chassis division. From May 2006 until returning to GM, he held the position of director of innovation and technology leadership. Dr. Boules received his doctorate of engineering degree in 1978 from the Technical University of Braunschweig, Germany. He is the author of numerous patents and technical and invited papers, and he has received several awards from GM in recognition of his accomplishments, including the John Campbell Award (for scientific accomplishments), the Charles McCuen Award (recognizing contributions to the business success of GM), the Extraordinary Accomplishment Award, and the President's Council Honors Award. Dr. Boules has been a fellow of the Institute of Electrical and Electronics Engineers (IEEE) since 1991 and was named a life fellow in January 2015. He is also the recipient of the 2011 IEEE Nikola Tesla Award. He is a member of the Industry Applications Society and a past member of its executive board. Dr. Boules was also a member of the board of directors of the Intelligent Transportation Society of America and a member of the executive board of several university consortia, and served on the National Academies Committee on Review of the U.S. DRIVE Research Program.

DAVID L. GREENE is a senior fellow of the Howard H. Baker Jr. Center for Public Policy and a research professor in the Department of Civil and Environmental Engineering at the University of Tennessee. In 2013, Dr. Greene retired from Oak Ridge National Laboratory (ORNL) as a corporate fellow after a 36-year career. He holds a Ph.D. in geography and environmental engineering from Johns Hopkins University. Dr. Greene's research interests are focused on energy use in transportation and policies to reduce petroleum consumption and greenhouse gas emissions and achieve a transition to sustainable energy sources. He has published extensively on automotive fuel economy and the Corporate Average Fuel Economy (CAFE) standards, and he has served on all four National Academies committees that evaluated U.S. fuel economy policy and technologies for cars and light trucks. Dr. Greene was also a member of the Committee on Transitions to Alternative Vehicles and Fuels. A current focus of his research and modeling is how technology and policy can accomplish a cost-effective transition to sustainable energy for transportation. Dr. Greene is author of more than 275 professional publications on transportation and energy issues, including 100 articles in peer-reviewed journals and 12 National Academies reports. He is the 2012 recipient of the Transportation Research Board's (TRB's) Roy W. Crum Award, and he is an emeritus member of both the Energy and Alternative Fuels Committees of the TRB and a lifetime national associate of the National Academies. Dr. *Greene received the Society of Automotive Engineers 2007 Barry D. McNutt Award for Excellence in Automotive* Policy Analysis, U.S. Department of Energy (DOE) 2007 Hydrogen R&D Award, DOE 2011 Vehicle Technologies R&D Award, and DOE Distinguished Career Service Award. Dr. Greene was recognized by the Intergovernmental Panel on Climate Change (IPCC) for contributing to the IPCC's receipt of the 2007 Nobel Peace Prize.

DANIEL KAPP is the principal of D.R. Kapp Consulting, providing consulting services in the area of automotive powertrain product technology and strategy, following his retirement from Ford Motor Company in 2012. Mr. Kapp was with Ford since 1977, following his graduation from Michigan Technological University with a B.S.M.E. degree. He has spent his entire 35+ year career in the area of engine and powertrain product development. From the late 1980s through the mid-1990s, Mr. Kapp was involved in the design and development of the "Modular" V8 and V6 engines as Ford revamped its engine line-up to modern overhead cam designs. He was the program manager of the Triton V8 truck engines through their launch and then spent 3 years in the Truck Vehicle Center as the powertrain systems manager for full-size trucks and sport utility vehicles (SUVs). In 2001, Mr. Kapp was appointed to his first executive position as director of core and advanced powertrain engineering, responsible for powertrain controls, catalyst and emission systems, and calibration. One year later, he became executive director for powertrain operations, and for 5 years he led the product development of all engines and transmissions in North America, during which time he also acted as a global powertrain product development lead for the enterprise. In late 2006, Mr. Kapp moved to Ford's research and advanced activity and remained there until retiring in 2012. During that time, he led the development of advanced powertrain technologies such as EcoBoost. In that role, Mr. Kapp also led the development of Ford's technology roadmaps for future sustainability and emission reduction strategy. He served as an internal technical consultant in the field of powertrain technologies and did significant external interfacing as a spokesperson for Ford in this area.

ULRICH KRANZ currently directs R&D as chief technology officer at Canoo, an electric vehicle start-up company based in Los Angeles. Prior to his current role, Mr. Kranz spent more than 30 years working for BMW AG in the R&D division as an expert in suspension and chassis development. While at BMW, he led the development of highly innovative products and technologies, including the introduction of BMW's first SUV to the world market, while working as a project leader in South Carolina. Mr. Kranz also led the reinvention of the MINI brand and its successful introduction to the world market. He headed BMWi, the electric car division, and prepared BMW for the future of e-mobility, mobility services, car-sharing, charging, and lightweight materials such as carbon fiber and thermoplastics. Mr. Kranz has served as committee member for innovations for the State of Bavaria and as a member of BMW's supervisory board, representing BMW's upper management.

THERESE LANGER is currently a consultant on transportation sector energy efficiency and emissions reduction. Dr. Langer was transportation program director at the American Council for an Energy-Efficient Economy (ACEEE) from 2001 to 2020. Her current areas of activity include technologies and policies to improve light- and heavy-duty fuel economy, energy impacts of vehicle automation, applications of information and communications technology to improve freight transportation system efficiency, and transportation electrification at the state and local levels. Prior to joining ACEEE, Dr. Langer was staff scientist for the Rutgers University Environmental Law Clinic, working to make the transportation system in the greater New York metropolitan area more sustainable. Dr. Langer holds a Ph.D. in mathematics from the University of California, Berkeley. She served as a member of the National Academies Committee on Assessment of Technologies for Improving Fuel Economy of Light-Duty Vehicles, Phase 2.

ZHENHONG LIN is a senior R&D staff member at the Buildings and Transportation Science Division of ORNL. As PI and manager of the Transportation Energy Evolution Modeling program, Dr. Lin has authored more than 100 technical articles on technological cost-effectiveness, infrastructure optimization, behavior opportunities, and policy design of transitions to zero-emission transportation. He currently serves on the Alternative Fuels Committee of the TRB, the editorial board of *Transportation Research Part D*, and the National Academies Committee on Assessment of Technologies for Improving Fuel Economy of Light-Duty Vehicles—Phase 3. Dr. Lin received his Ph.D. in transportation engineering in 2008 and M.S. in transportation technology and policy from the University of California, Davis. Before that, he obtained his B.E./M.S. in automotive engineering from Tsinghua University in Beijing.

JOSHUA LINN is an associate professor at the University of Maryland and a senior fellow at Resources for the Future (RFF). Dr. Linn's research centers on the effect of environmental regulation and market incentives on technology, with particular focus on the electricity sector and markets for new vehicles. His work on the electricity sector has compared the effectiveness of cap and trade and alternative policy instruments in promoting new technology, including renewable electricity technologies. Studies on new vehicle markets investigate the effect of CAFE standards and fuel prices on new vehicle characteristics, technology, consumer well-being, and manufacturer profits. Dr. Linn has published in leading general-interest and field journals in environmental, energy, and health economics. He joined the University of Maryland in 2018, joined RFF in 2010, and was an assistant professor in the Economics Department at the University of Illinois, Chicago, and a research scientist at the Massachusetts Institute of Technology (MIT). Dr. Linn holds a Ph.D. in economics from MIT and a B.A. in astronomy and physics from Yale University.

NIC LUTSEY is the program director at the International Council on Clean Transportation (ICCT), where he directs its electric vehicle research and leads its U.S. activities. Dr. Lutsey manages the ICCT's role as the Secretariat for the International Zero-Emission Vehicle Alliance. He has co-authored 19 peer-reviewed journal articles and dozens of reports on technology potential, regulatory design, and policy cost-effectiveness. Dr. Lutsey has received awards from the U.S. Department of Transportation (DOT); the University of California, Davis; the TRB; and the California Air Resources Board for his research contributions. In 2015, he received the SAE International Barry D. McNutt Award for Excellence in Automotive Policy Analysis. Previously, with the California Air Resources Board, Dr. Lutsey participated in the regulatory development of the 2004 and 2012 greenhouse gas emission regulations for automobiles. He received a B.S. in agricultural and biological engineering from Cornell University and a Ph.D. in transportation technology and policy from the University of California, Davis.

JOANN MILLIKEN is currently self-employed as a senior energy consultant. Dr. Milliken has 34 years of federal program management experience, more than 20 of those with DOE, where she developed and directed clean-energy R&D portfolios having budgets of up to $200 million per year. She has a strong track record of success in advancing energy efficiency and renewable energy technologies, practices, and policy, working in collaboration with industry, universities, small businesses, and national laboratories. Dr. Milliken is a recognized expert in hydrogen and fuel cell systems, and she is experienced in leading federal programs in energy-efficient buildings and solar, wind, and geothermal energy. Prior to joining DOE in 1994, Dr. Milliken was a research chemist at the U.S. Naval Research Laboratory and a program manager at the Office of Naval Research, where she conducted and managed mission-related materials research. She earned a B.A. in chemistry from LaSalle University and a

Ph.D. in chemistry from the University of Pennsylvania, researching electronically conducting polymers under Nobel Laureate Professor Alan MacDiarmid. Dr. Milliken retired from DOE in 2015.

RANDA RADWAN is the director of the Highway Safety Research Center at the University of North Carolina, Chapel Hill. Dr. Radwan has more than 27 years of experience in transportation safety and vehicle crashworthiness research, including 17 years as a research program manager at the U.S. DOT National Highway Traffic Safety Administration (NHTSA). Dr. Radwan has forged alliances and successfully collaborated with the safety community at large, from government to industry on both national and international levels. She led a multidisciplinary research program from concept to the Notice for Proposed Rulemaking for NHTSA's 2007 FMVSS 214 upgrade forecast to save more than 300 lives and reduce 400 serious injuries per year. Dr. Radwan received multiple awards while at NHTSA, including the Secretary of Transportation Award and the NHTSA Administrator's Award, which she received four times. Dr. Radwan then spent 9 years as the director of advanced research and senior research scientist at the George Washington University (GWU) National Crash Analyses Center, where she engaged in and directed innovative analyses and methodologies in vehicle and transportation safety research. Dr. Radwan created strategy and modeling methodology to assess safety performance of new vehicle designs, resulting in the Vehicle Fleet Simulation methodology used for NHTSA's "Corporate Average Fuel Economy Standards (CAFE) and Midterm Evaluation for Light-Duty Vehicles, Model Years 2022–2025" safety studies. She also served as adjunct faculty in the School of Engineering and Applied Sciences at GWU (2009–2013). Dr. Radwan has authored 29 peer-reviewed professional publications on vehicle safety, including two reports to the U.S. Congress. She has a Ph.D. in transportation safety engineering from GWU and a master's degree and B.S. in electrical engineering from Rice University.

ANNA STEFANOPOULOU is the William Clay Ford Professor of Manufacturing at the University of Michigan. Dr. Stefanopoulou has been on the faculty of the Department of Mechanical Engineering since 2000. She obtained her Diploma (1991, National Technical University of Athens, Greece) in naval architecture and marine engineering and her Ph.D. (1996, University of Michigan) in electrical engineering and computer science. Dr. Stefanopoulou served as the director of the Automotive Research Center, a multi-university U.S. Army Center of Excellence in Modeling and Simulation of Ground Vehicles (2009–2018). She was an assistant professor (1998–2000) at the University of California, Santa Barbara, and a technical specialist (1996–1997) at Ford Motor Company, where she developed and implemented multivariable controllers for advanced engines and powertrains. Dr. Stefanopoulou has been recognized as a fellow of three societies: the ASME (2008), IEEE (2009), and SAE (2018). She is an elected member of the executive committee of the ASME Dynamics Systems and Control Division and the board of governors of the IEEE Control Systems Society. Dr. Stefanopoulou is the founding chair of the ASME Dynamic Systems and Control Division (DSCD) Energy Systems Technical Committee and was a member of the National Academies Committee on Assessment of Technologies for Improving Fuel Economy of Light-Duty Vehicles—Phase 2. She is a recipient of the 2018 Rackham Distinguished Graduate Mentor Award, the 2017 IEEE Control System Technology award, the 2012 College of Engineering Research Award, the 2009 ASME Gustus L. Larson Memorial Award, a 2008 University of Michigan Faculty Recognition award, the 2005 Outstanding Young Investigator by the ASME DSCD, a 2005 Henry Russel award, a 2002 Ralph Teetor SAE educational award, and a 1997 National Science Foundation CAREER award, and she was selected in 2002 as one of the world's most promising innovators from the *MIT Technology Review*. Dr. Stefanopoulou has co-authored a book, 20 U.S. patents, and more than 250 publications (5 of which have received awards) on estimation and control of internal combustion engines and electrochemical processes such as fuel cells and batteries.

DEIDRE STRAND is the chief scientific officer at Wildcat Discovery Technologies. Dr. Strand has more than 25 years of experience in materials research, development, and commercialization, primarily in the areas of energy storage (lithium-ion batteries) and electronic applications. Prior to joining Wildcat in 2012, Dr. Strand served as a research fellow at Dow Chemical, where she was the technical lead in Dow Energy Materials, as well as the PI on external research programs with universities and national laboratories on battery materials. Dr. Strand also has extensive experience in patent analysis and technical due diligence of new technologies. Dr. Strand completed her

Ph.D. in analytical chemistry at the University of Wisconsin, Madison, under the supervision of Professor John Schrag. Her Ph.D. research focused on rheology and birefringence of polymeric solutions. Dr. Strand also holds a master of science degree in chemistry from the California Institute of Technology and a bachelor of science degree in chemistry from North Dakota State University.

KATE WHITEFOOT is an assistant professor of mechanical engineering and engineering and public policy at Carnegie Mellon University. Dr. Whitefoot is a thrust leader of Technology Commercialization for the Next Manufacturing Center and a faculty affiliate at the Carnegie Mellon Scott Institute for Energy Innovation. Prior to her current position, she served as a senior program officer and the Robert A. Pritzker Fellow at the National Academies of Sciences, Engineering, and Medicine, where she directed the National Academies Manufacturing, Design, and Innovation Program. Dr. Whitefoot's research bridges engineering design theory and analysis with that of economics to study the design and manufacture of energy-efficient and low-carbon products and processes and their adoption in the marketplace. Her areas of expertise include vehicle fuel efficiency, consumer choice, design and adoption of green products, energy-efficient and productive manufacturing, and energy and environmental policies. Dr. Whitefoot has gained recognition nationally and internationally for her research and teaching. She served on the National Academies Committee on the Review of the National Institute of Standards and Technology Engineering Laboratory. Her research is featured in the *Washington Post*, *Popular Mechanics*, *Bloomberg Business*, and *Business Insider* and is referenced in the 2017–2025 CAFE rulemaking. Dr. Whitefoot has worked with several companies in the automotive, aerospace, and high-tech industries, and has been invited to present briefings at the White House, Capitol Hill, the U.S. Department of Commerce, and the U.S. Environmental Protection Agency. Dr. Whitefoot earned three degrees from the University of Michigan: a B.S. and M.S. in mechanical engineering and a Ph.D. in design science—a multidisciplinary program where she concentrated in mechanical engineering and economics, completing course sequences and having an advisory committee across both disciplines.

B

Disclosure of Conflicts of Interest

The conflict of interest policy of the National Academies of Sciences, Engineering, and Medicine (http://www.nationalacademies.org/coi) prohibits the appointment of an individual to a committee authoring a Consensus Study Report if the individual has a conflict of interest that is relevant to the task to be performed. An exception to this prohibition is permitted if the National Academies determines that the conflict is unavoidable and the conflict is publicly disclosed. When the committee that authored this report was established, a determination of whether there was a conflict of interest was made for each committee member given the individual's circumstances and the task being undertaken by the committee. A determination of a conflict of interest for an individual is not an assessment of that individual's actual behavior or character or ability to act objectively despite the conflicting interest.

Daniel Kapp has a conflict of interest in relation to his service on the Committee on Assessment of Technologies for Improving Fuel Economy of Light-Duty Vehicles—Phase 3, because he owns Ford Motor Company stocks, and because he has a consulting relationship with AVL Powertrain Engineering, a supplier of engineering services to automobile manufacturers.

Ulrich Kranz has a conflict of interest in relation to his service on the Committee on Assessment of Technologies for Improving Fuel Economy of Light-Duty Vehicles—Phase 3, because he is the chief technology officer at Canoo, an electric vehicle start-up company.

Deidre Strand has a conflict of interest in relation to her service on the Committee on Assessment of Technologies for Improving Fuel Economy of Light-Duty Vehicles—Phase 3, because she is the chief scientific officer of Wildcat Discovery Technologies, a battery material discovery firm.

The National Academies determined that the experience and expertise of the above individuals was needed for the committee to accomplish the task for which it was established. The National Academies could not find other available individuals with the equivalent experience and expertise who did not have a conflict of interest. Therefore, the National Academies concluded that the above conflicts were unavoidable and publicly disclosed them through the National Academies Projects and Activities Repository (NAPAR; http://webapp.nationalacademies.org/napar/).

C

Committee Activities

Committee Meeting 1: May 10–11, 2018, Keck Center, Washington, D.C.

Committee Meeting 2: July 16–17, 2018, Keck Center, Washington, D.C.

Driving the Future
Ann Wilson, Senior Vice President of Government Affairs
Motor and Equipment Manufacturers Association

Office of Transportation and Air Quality, U.S. Environmental Protection Agency
William Charmley, Director, Assessment and Standards Division

Policy Considerations for Reducing Fuel Use from Passenger Vehicles, 2025–2035
David Cooke, Senior Vehicles Analyst, Clean Vehicles Program, Union of Concerned Scientists

DOE's Research to Improve Transportation Energy Security and Affordability
Steven Chalk, Deputy Assistant Secretary for Transportation, Department of Energy

The Three Big Technology Trends
Andrew Higashi, Director, Strategy&, Part of the PwC Network

Alliance of Automobile Manufacturers
Mike Hartrick, Director of Fuel Economy and Climate, Alliance of Automobile Manufacturers
(Now Alliance for Automotive Innovation)

Committee Meeting 3: October 15–16, 2018, Center for Automotive Research, Ann Arbor, Michigan

Energy Saving Through Connected and Automated Vehicles—What We Learned at UM/Mcity
Huei Peng, University of Michigan

Ford Future Trends
Sheryl Connelly, Ford Motor Company

Nissan's Sustainability and Light-Duty FE Strategy 2025–2035
Chris Reed, Nissan North America, Inc.

Powertrain Technology 2025 and Beyond
John Juriga, Hyundai America Technical Center

Future Propulsion Systems
John E. Kirwan, Delphi Technologies

Enlighten Award 2018
Anthony Norton, Vehicle Design, Altair

Plastics in the Auto Industry, Today and into the Future
Matthew Marks, Plastics and Joining, SABIC

Committee Meeting 4: January 24–25, 2019, University of California, Davis, Institute of Transportation Studies, Davis, California

Presentation to the National Academies Committee on Assessment of Technologies for Improving Fuel Economy of Light-Duty Vehicles—Phase 3
Joshua Cunningham, California Air Resources Board

CPUC Transportation Electrification Activities
Carolyn Sisto, California Public Utilities Commission

A Shared Future of Mobility
Adam Gromis, Uber

CEC Investments in Alternative Transportation Fuels/Technology
Tim Olson, California Energy Commission

Life Cycle Carbon Intensity and Vehicle Trends
Alissa Kendall, University of California, Davis

Advanced Plug-In Electric Vehicle Travel and Charging Behavior
Gil Tal, University of California, Davis

Partially Automated Vehicles and Travel Behavior
Scott Hardman, University of California, Davis

Fuel Economy in the Future: Behavioral Considerations
David Rapson, University of California, Davis

Making the Transition to Light-Duty Electric-Drive Vehicles in the United States
Joan Ogden, University of California, Davis

(How) Do Car and Truck Buyers Think About Fuel Economy
Ken Kurani, University of California, Davis

Considerations for Improving Fuel Economy, 2025–2035
Alan Jenn, University of California, Davis

Electric Vehicle Charging Infrastructure Webinar: May 2, 2019, Remotely via Zoom

Charging Infrastructure for Shared and Autonomous EVs
John Smart, INL

UV EV Infrastructure: Analysis and Projections
Eric Wood, NREL

Cycle 2 National Outreach—Lessons Learned
Matthew Nelson, Electrify America

Tesla Vehicles and Charging Networks
Patrick Bean, Tesla

Norwegian EV Charging Infrastructure and User Experiences
Erik Figenbaum, Norwegian Institute of Transport Economics

Materials for Improved Fuel Economy Webinar: May 17, 2019, Remotely via Zoom

LANXESS High-Performance Materials Addressing the Trends in Automotive
Jose Chirino, LANXESS

Fuel Economy with Aluminum
Todd Summe, Novelis

Steel Developments for Automotive Lightweighting
George Coates, World Auto Steel

FCA Site Visit: June 5, 2019, Fiat Chrysler America, Troy, Michigan

Delphi Site Visit: June 6, 2019, Delphi Automotive, Troy, Michigan

Hydrogen Fueling Infrastructure Webinar: June 26, 2019, Remotely via Zoom

The California Fuel Cell Revolution: Activating the Commercial Market
Bill Elrick, CaFCP

Economic and Environmental Perspectives of Hydrogen Infrastructure Deployment Options
Amgad Elgowainy, ANL

H2 Energy at the Heart of the Energy Transition
Dave Edwards, Air Liquide

Hydrogen for Transport
Jason Munster, Shell

Global Hydrogen Mobility Applications
James Kast, Toyota

Safety Webinar: September 19, 2019, Remotely via Zoom

Relationships Between Mass, Footprint, and Societal Fatality Risk in Recent Light-Duty Vehicles
Tom Wenzel, LBNL

Fuel Economy and Highway Safety
Chuck Farmer, Insurance Institute of Highway Safety

CAE Methodology for Evaluation of Fleet Crash Protection of New Vehicle Designs
Randa Radwan, UNC Highway Safety Research Center

Safety Effects of 2025+ Fuel Economy Goals
Priya Prasad, Prasad Consulting, LLC

Munro Site Visit: September 24, 2019, Munro and Associates, Auburn Hills, Michigan

Bosch Site Visit: September 25, 2019, Bosch Group, Farmington Hills, Michigan

Committee Meeting 5: September 25–26, 2019, Engineering Society of Detroit, Southfield, Michigan

BMW Site Visit: October 14, 2019, BMW AG, Munich, Germany

Mercedes-Benz Daimler Site Visit: October 16, 2019, Mercedes-Benz Daimler, Sindelfingen, Germany

Volkswagen Site Visit: October 18, 2019, Volkswagen AG, Wolfsburg, Germany

Ford Site Visit: December 9, 2019, Ford Motor Company Headquarters, Dearborn, Michigan

Committee Meeting 6: December 9–11, 2019, Detroit, Michigan

Design Optimization Webinar: January 6, 2020, Remotely via Zoom

National Academies Design Optimization Webinar
Tim Skszek, Magna International

Simulation-Driven Lightweight Design for Automotive Structures
Richard Yen, Altair

Integrative Design of Automobiles
Amory Lovins, Rocky Mountain Institute

Tesla Site Visit: January 16, 2020, Tesla Factory, Fremont, California

General Motors Site Visit: January 30, 2020, General Motors Headquarters, Detroit, Michigan

Nissan Site Visit: February 4, 2020, Nissan Advanced Technology Center, Kanagawa, Japan

Toyota Site Visit: February 6, 2020, Toyota Motor Corporation Headquarters, Toyota City, Japan

Panasonic Site Visit: February 7, 2020, Panasonic Head Office, Osaka, Japan

LG Chem Site Visit: February 11, 2020, LG Twin Towers, Seoul, Korea

Hyundai Site Visit: February 12, 2020, Namyang R&D Center, Hwaseong, Korea

Committee Meeting 7: March 19–20, 2020, Remotely via Zoom

Committee Meeting 8: May 4–5, 2020, Remotely via Zoom

EPA *Information-Gathering Session:* June 16, 2020, Remotely via Microsoft Teams

Light-Duty Vehicle Powertrain Benchmarking and Technology Effectiveness Assessments
Dan Barba, National Center of Advanced Technology, and Andrew Moskalik, Ph.D. engineer

Electric Vehicle Technology Issues
Michael Safoutin, Ph.D. engineer

Economic and Consumer Issues
Gloria Helfand, Ph.D. economist; Christian Noyce, ORISE fellow; Asa Watten, ORISE fellow; and Dana Jackman, Ph.D. economist

ALPHA Model Development
Kevin Newman, M.Eng. engineer

OMEGA Model Development
Kevin Bolon, Ph.D. resource policy/B.S. engineer

Emerging Trends in Transportation
Karl Simon, Director, Transportation and Climate Division

Toyota Site Visit Follow-Up Closed Session: June 18, 2020, Remotely via Microsoft Teams

Committee Meeting 9: July 22–23, 2020, Remotely via Zoom

Honda Site Visit: September 10, 2020, Remotely via Microsoft Teams

Committee Meeting 10: September 29, 2020, Remotely via Zoom

Committee Meeting 10 Follow-Up: October 15, 2020, Remotely via Zoom

Committee Meeting 10 Second Follow-Up: October 27, 2020, Remotely via Zoom

Committee Meeting 11: January 20, 2021, Remotely via Zoom

D

Acronyms

A/C	air conditioning
AC	alternating current
ACC	adaptive cruise control
ACE	Affordable Clean Energy
ACEEE	American Council for an Energy-Efficient Economy
ADAS	advanced driver assistance system
AEM/AAEM	anion exchange membrane/alkaline anion exchange membrane
AEMFC	anion exchange membrane fuel cell
AFDC	Alternative Fuels Data Center
AFV	alternative fuel vehicle
AGS	active grill shutter
AHSS	advanced high-strength steel
AI	artificial intelligence
AMFA	Alternative Motor Fuels Act
ANL	Argonne National Laboratory
ARPA-E	Advanced Research Projects Agency—Energy
AV	autonomous vehicle
AVM	around-view monitor
AWD	all-wheel drive
BEV	battery electric vehicle
BiSG	belt integrated starter generator
BMEP	brake mean effective pressure
BMS	battery management system
BOL	beginning of life
BOP	balance-of-plant
BP	bipolar plates
BSFC	brake-specific fuel consumption

C-V2X	cellular vehicle to everything
CAA	Clean Air Act
CAE	computer-aided engineering
CAFC	corporate average fuel consumption
CaFCP	California Fuel Cell Partnership
CAFE	corporate average fuel economy
CAN bus	controller area network
CARB	California Air Resources Board
CAV	connected and automated vehicle
CCM	catalyst coated membrane
CCS	carbon capture and sequestration
CCUS	carbon capture, utilization, and sequestration
CDCS	charge depleting/charge sustaining
CEC	California Energy Commission
CEM	*compressor-expander-motor*
CF	carbon fiber
CFR	Code of Federal Regulations
CHS	Center for Hydrogen Safety
CMU	Carnegie Mellon University
CO	carbon monoxide
CO_2	carbon dioxide
COPV	composite overwrapped pressure vessel
CPP	Clean Power Plan
CR	compression ratio
CUV	crossover utility vehicle
CVT	continuously variable transmission
CVVD	continuously variable valve duration
DC	direct current
DCT	dual clutch transmission
DEAC	cylinder deactivation
DFMA	design for manufacture and assembly
DI	direct injection
DMC	direct manufacturing cost
DME	dimethyl ether
DOD	depth of discharge
DOE	U.S. Department of Energy
DOT	U.S. Department of Transportation
DSRC	dedicated short-range communication
EC	ethylene carbonate
ECMS	equivalent consumption minimization strategy
ECSA	electrochemically active surface area
EERE	Office of Energy Efficiency and Renewable Energy
EGR	exhaust gas recirculation
EIA	Energy Information Administration
EISA	Energy Independence and Security Act
EPA	U.S. Environmental Protection Agency
EPCA	Energy Policy and Conservation Act

EPS	electric power steering
EREV	extended range electric vehicle
EV	electric vehicle
eVMT	electric vehicle miles traveled
EVSE	electric vehicle supply equipment
FC	fuel consumption
FCA	Fiat Chrysler Automobiles
FCC	Federal Communications Commission
FCEV	fuel cell electric vehicle
FCHEA	Fuel Cell and Hydrogen Energy Association
FCH JU	Fuel Cells and Hydrogen Joint Undertaking
FE	fuel economy
FHWA	Federal Highway Administration
FMVSS	*Federal Motor Vehicle Safety Standards*
FOV	field-of-view
FTP	Federal Test Procedure
FWD	front-wheel drive
GaN	gallium nitride
GDI	gasoline direct fuel injection
GDL	gas diffusion layer
GDP	gross domestic product
GHG	greenhouse gas
GM	General Motors
GNSS	Global Navigation Satellite System
GPS	global positioning system
GTDI	gasoline turbocharged direct injection
GTL	gas-to-liquid
GTR	Global Technical Regulation
GWP	global warming potential
H_2	hydrogen
HCCI	homogeneous charge compression ignition
HD	high-definition
HEV	hybrid electric vehicle
HF	hydrogen fluoride
HOR	hydrogen oxidation reaction
HOV	high-occupancy vehicle
HP	horsepower
HyMARC	Hydrogen Materials Advanced Research Consortium
ICCT	International Council on Clean Transportation
ICE	internal combustion engine
ICEV	internal combustion engine vehicle
IEA	International Energy Agency
IGBT	insulated gate bipolar transistor
IIHS-HLDI	Insurance Institute for Highway Safety/Highway Loss Data Institute
IPCC	Intergovernmental Panel on Climate Change
ITS	Intelligent Transportation Systems

LBNL	Lawrence Berkeley National Laboratory
LCFS	low-carbon fuel standard
LDV	light-duty vehicle
LDW	lane departure warning
LFC	low-fuel-consumption
lidar	light detection and ranging
Li-ion	lithium-ion
LP-EGR	low-pressure loop cooled exhaust gas recirculation
LRR	long-range radar
LRR	low rolling resistance
LTE	Long-Term Evolution
MEA	membrane-electrode assembly
MHDVs	medium- and heavy-duty vehicles
MHEV	*mild hybrid electric vehicle*
MOF	metal-organic framework
MPC	model predictive control
MPG	miles per gallon
MPGe	miles per gallon of gasoline-equivalent
mph	miles per hour
MPL	microporous layer
MRR	medium-range radar
MSRP	manufacturer suggested retail price
MTG	methanol-to-gasoline
MY	model year
NA	naturally aspirated
NAAQS	National Ambient Air Quality Standards
NCA	nickel cobalt aluminum
NEDC	New European Driving Cycle
NETL	National Energy Technology Laboratory
NEV	new energy vehicle
NEXTCAR	NEXT-Generation Energy Technologies for Connected and Automated On-Road-Vehicles
NHTS	National Household Travel Survey
NHTSA	National Highway Traffic Safety Administration
NMC	nickel manganese cobalt
NO_x	nitrogen oxides
NRC	National Research Council
NREL	National Renewable Energy Laboratory
NSTC	National Science and Technology Council
NVH	noise, vibration, and harshness
O&M	operations and maintenance
OCV	on-chip variation
OEM	original equipment manufacturer
ORNL	Oak Ridge National Laboratory
ORR	oxygen reduction reaction
PAN	polyacrylonitrile
PBI	polybenzimidazole

PEM	proton exchange membrane
PEMFC	proton exchange membrane fuel cell
PEV	plug-in electric vehicle
PFCV	plug-in fuel cell vehicle
PFSA	perfluorosulfonic acid
PGM	platinum group metal
PHEV	plug-in hybrid electric vehicle
PM	permanent magnet
PMSM	permanent magnet synchronous motor
PS	powersplit
psi	pounds per square inch
PTFE	poly(tetrafluoroethylene)
PtM	platinum metal
R&D	research and development
radar	radio detection and ranging
RE	rare earth
RFS	Renewable Fuel Standard
RH	relative humidity
RON	research octane number
RPM	revolutions per minute
RRC	rolling resistance coefficient
RWD	rear-wheel drive
SAE	Society of Automotive Engineers
SAFE	Safer Affordable Fuel Efficient
SEI	solid electrolyte interphase
SiC	silicon carbide
SMR	steam-methane reforming
SOC	state of charge
SOH	state of health
SOP	state of power
SPaT	signal phase and timing
SPCCI	spark-controlled compression ignition
SPEEK	sulfonated polyether ether ketone
SRR	short-range radar
SSL	solid-state lidar
SUV	sport utility vehicle
TAR	technical assessment report
TCO	total cost of ownership
TFSI	turbo fuel stratified injection
TPO	thermoplastics polyolefin elastomers
TSI	turbo stratified injection
UF	utility factor
UHSS	ultra-high-strength steel
U.S. DRIVE	U.S. Driving Research and Innovation for Vehicle Efficiency and Energy Sustainability

V2B	vehicle to buildings
V2G	vehicle to grid
V2I	vehicle to infrastructure
V2V	vehicle to vehicle
V2X	vehicle to everything
VAT	value added tax
VCT	variable camshaft timing
VMT	vehicle miles traveled
VTG	variable turbine geometry
VTO	Vehicle Technologies Office
VVL	variable valve lift
VVT	variable valve timing
WBG	wide bandgap
WTW	*well-to-wheels*
xFC	extreme fast charging
ZEV	zero-emission vehicle
ZLEV	zero- or low-emissions vehicle

E

Center for Automotive Research Commissioned Study

Vehicle Mass Reduction Roadmap Study 2025-2035

CENTER FOR **AUTOMOTIVE RESEARCH**

Carla Bailo
Shashank Modi
Michael Schultz
Terni Fiorelli
Brett Smith
Nicklaus Snell

Table of Contents

List of Figures

List of Tables

Glossary

Steel Grades

Category	Acronym	Description	Ultimate Tensile Strength (MPa) Range
Low Strength (LSS)	Mild	Mild Steel	Less than 270
High Strength Steels (HSS)	IF	Intersitial Free	410-420
	BH	Bake Hardenable	340-400
	HSLA	High-Strength Low Alloy	450-780
Advanced High Strength Steels (AHSS)	DP	Dual Phase	440-1270
	FB	Ferritic-bainitic (SF - stretch flangeable)	450-600
	CP	Complex Phase	800-1470
	MS	Martensitic	1200-1500
Ultra High Strength Steels (UHSS)	TRIP	Transformation-induced plasticity	600-980
	HF	Hot-formed (boron)	480-1900
	TWIP	Twinning-induced plasticity	900-1200

Aluminum Grades

Category	Series	Acronym
Commercially Pure Aluminum	1xxx	Al
Heat-Treatable Alloys	2xxx , 6xxx , 7xxx	Al
Non Heat-Treatable Alloys	3xxx , 4xxx , 5xxx	Al

Acknowledgments

The authors of this report thank the industry experts from various companies that provided data for this research. Our special thanks to the National Academies of Sciences, Engineering, and Medicine (NASEM) Committee to entrust CAR to conduct this research.

Companies which contributed data to this study include:

- American Chemistry Council
- American Iron and Steel Institute
- ArcelorMittal
- Automotive Aluminum Advisors
- BASF
- Ducker Worldwide
- Fiat Chrysler Automobiles
- Ford Motor Company
- General Motors
- Hexion
- Nissan
- Novelis
- The Aluminum Association
- U.S. Steel

Sincerely

Bailo, Modi, Schultz, Fiorelli, Smith, Snell

For citations and reference to this publication, please use the following:

Bailo, C., Modi, S., Schultz, M., Fiorelli, T., Smith, B., & Snell, N. (2020). (publication). Vehicle Mass Reduction Roadmap Study 2025-2035. Ann Arbor, MI: Center for Automotive Research.

CENTER FOR AUTOMOTIVE RESEARCH

3005 Boardwalk, Suite 200
Ann Arbor, MI 48108
www.cargroup.org

CAR's mission is to conduct independent research and analysis to educate, inform and advise stakeholders, policymakers, and the general public on critical issues facing the automotive industry, and the industry's impact on the U.S. economy and society.

Background and Objective

The National Academies of Sciences, Engineering, and Medicine (NASEM) Committee on Assessment of Technologies for Improving Fuel Economy of Light-Duty Vehicles, Phase 3 is tasked by The National Highway Traffic Safety Administration (NHTSA) with providing estimates of the potential cost, fuel economy improvements, and barriers to deployment of technologies for improving fuel economy in 2025-2035 light-duty vehicles. The National Academies Committee is currently investigating the state of vehicle mass reduction technology readiness and the impact of mass reduction on fuel economy while maintaining vehicle performance and safety requirements.

The reduction of in-use GHG emissions from light-duty passenger vehicles presents a vital opportunity to minimize motor vehicles' environmental impact. Vehicle mass reduction is one pathway to reduce vehicle emissions while improving performance. Considerable resources have been expended by the regulators trying to estimate the lowest cost feasible for the mass reduction of light-duty vehicles. Detailed teardown and cost studies performed by reputable engineering firms have aggressively approached lightweighting on a handful of vehicles, producing several innovative ideas. However, automakers respond by pointing out that there are risks, business constraints, and customer requirements that these studies do not address. Further, extrapolating the results from one or a few teardown studies to over 1,000 vehicle models for sale in the U.S. market is inappropriate. The task of assessing mass reduction trends is critical and challenging.

The Center for Automotive Research (CAR) was hired under contract by NASEM to study vehicle mass reduction for model years 2025-2035. Over the past decade, CAR has been a leader in light-duty vehicle mass reduction research. CAR has done work on assessing the real-world barriers to implementing mass reduction technologies (J. Baron, 2016). CAR also worked with nine global vehicle manufacturers to examine material trends over the next decade (Baron & Modi, 2016). The project still stands as one of the most cooperative and thorough analyses done to date. CAR collected data on 42 vehicles from 4 segments representing 50 percent of the U.S. light-duty fleet. Most recently, CAR has published a Materials and Manufacturing Technology Roadmap (Modi & Vadhavkar, 2019) that examined the likely penetration of materials for the automotive body-in-white.

The objective of this Study is as below:

- Describe the available opportunities for shifts in vehicle body-in-white and closure material application in vehicle model years (MY) 2025-2035.
- Estimate mass-reduction opportunity by each segment in MY2025, MY2030, MY2035, keeping in mind technology readiness, automaker development plans, SME opinion, and CAFE/GHG regulations.
- Based on mass reduction opportunities, create scenarios for possible material penetration by each vehicle type.
- Use the scenarios to estimate the incremental costs for mass reduction opportunities.

Research Approach

The project's research relies on CAR material and manufacturing roadmaps, publications by other organizations, and information collected through several automaker and supplier interviews. Data sources include, but are not limited to:

- Vehicle repair manuals
- Conference presentations
- NHTSA and EPA sponsored vehicle lighweighting studies
- SAFE Rule: Final Regulatory Impact Analysis
- Automaker and supplier interviews

The research focuses on vehicle body-in-white and closures for primary mass reduction opportunities. Powertrain mass-reduction is considered only for calculating mass decompounding opportunities. The baseline for this project is the U.S. MY2020 light-duty vehicle fleet.

Estimating Baseline Material Technology

The baseline model year for this research is 2020. There are hundreds of different vehicles (nameplates) in the U.S. fleet. Studying the material technology of every vehicle is not practical. Moreover, material information is not readily available in public databases. Therefore, CAR researchers decided to investigate the top-selling vehicles in the U.S. fleet. Table 1 lists the vehicles studied to establish the baseline material technology. The segments are the same as defined by NHTSA.[1] These 33 vehicles represent greater than 50 percent of the U.S. vehicle sales in 2019. This reports refer to these vehicles as the baseline fleet.

Table 1: Vehicles studied for baseline material analysis

Small Car	Mid-Size Car	Small SUV	Mid-Size SUV	Pickup
Honda Civic	Ford Fusion	Jeep Wrangler	Traverse	Silverado
Honda Accord	Tesla Model 3	Chevy Equinox	Pacifica	Ram Pickup
Hyundai Elantra		Ford Escape	Explorer	F Series
Nissan Altima		Edge	Pilot	Sierra
Nissan Sentra		CR-V	Grand Cherokee	Tacoma
Toyota Camry		Tucson	Highlander	
Toyota Corolla		Cherokee		
		Compass		
		CX-5		
		Rogue		
		Outback		
		Forester		
		Rav4		

[1] NHTSA further categorizes vehicles in mass market and performance for each segment.

Analysis of the baseline fleet (the 33 vehicles) revealed that 73 percent of the vehicles were redesigned (generation change) after MY2016. According to vehicle production forecasts by IHS Markit Inc., 40 percent of the baseline fleet vehicles are expected to be redesigned by MY2025.

CAR researchers referred to the vehicle repair manuals to study the materials used in the baseline fleet. The repair manuals are published for the collision repair shops to inform them about the vehicles' materials used in the critical structural components. Most repair manuals contain information on safety-critical body-in-white (BIW) and closure parts. In a 2016 CAR study (Baron & Modi, 2016), the researchers identified key components essential to understand mass-reduction (M.R.) initiatives. The research team studied the materials used in these components for the baseline fleet vehicles (using repair manuals). Table 2 shows the materials used in the MY2020 baseline fleet versus the materials identified for the same components for the five percent mass-reduction level in the 2016 CAR study.[2]

Table 2: Materials in MY2020 versus MY2016

Components	MY2020 baseline fleet	2016 study: 5% M.R. level
Fender	BH Steel and Aluminum (50:50)	HSS/BH Steel
A-pillar	UHSS 1500 Hot Formed	UHSS HF Steel
Floor	HSS 440-590 with UHSS Reinforce.	Mild with AHSS
Front Bumper Structure	Mostly Aluminum with some Steel	Aluminum
Roof Panel	Mild/B.H. Steel	Mild Steel
Door Outer	LSS And Aluminum	B.H. Steel and Aluminum
Hood	95% Aluminum	Aluminum
Decklid	LSS, Al, Mag, Comp.	B.H. Steel and Aluminum
Engine Cradle/Front frame	LSS 400-600	HSLA
Steering Knuckle	HSS 400-500 And Aluminum	Aluminum
IP Beam	HSS And Two Magnesium	AHSS

Source: CAR Research, Vehicle repair manuals

Material analysis suggests that the MY2020 fleet has advanced material technologies that should result in around five percent lighter curb weight than the MY2016 baseline. However, this does not tell the entire story. CAR researchers compared the actual curb weights and footprints of the MY2020 to the MY2016 vehicles (see Appendix A). As shown in Table 3, the finding confirms that the real world curb weight reduction is much lower than five percent. Also, the average footprint has increased for all vehicle segments.

[2] The 2016 study asked automakers the material roadmap (for key components) for 5%, 10%, and 15% target curb weight reduction over the 2016 fleet.

Table 3: Curb weight and Footprint Comparison: current model year 2020 vs previous model year 2016[3]

	SmallCar	SmallSUV	MidSUV	Pickup
Avg. Curb Weight Decrease	0%	2%	1%	4%
Avg. Footprint Increase	2%	1%	4%	6%

	Unibody	Body on Frame
Avg. Curb Weight Decrease	1%	3%
Avg. Footprint Increase	2%	6%

Source: CAR Research

The real-world mass reduction achievements do NOT match the mass reduction potential of the material technologies already implemented in the baseline MY2020 fleet. CAR research found two primary reasons for this discrepancy:

Q1. **Mass Add-Back**: automakers often need to add weight to improve vehicles' safety, performance, and customer expectations. Mass for safety is required for crashworthiness and electronic devices such as cameras, sensors, computers, etc. Mass for performance might be added for attributes such as improvements in stiffness, the quietness of the ride, lowering the center of gravity, equalizing the load distribution, unsprung mass reduction, etc. Automakers also add mass to satisfy customer demand for better comfort. In the 2016 CAR study, automakers indicated that the total mass add-back (safety plus performance) expected for cars today averaged 4.9 percent for cars and 4.6 percent for light-duty trucks.

Q2. **Footprint Increase**: vehicles, in general, have increased in size because of customer demands. The baseline fleet has a 2-6 percent larger footprint than the MY2016 fleet. This increase takes away the real-world curb weight reduction.

Therefore, we conclude that the MY2020 baseline fleet has advanced material technology but is not significantly lighter than the MY2016 fleet.

---

[3] Analysis includes only the top selling baseline fleet vehicles (33 nameplates). Analysis of the entire US fleet might produce different results. MidCar segment has only two vehicles in top selling 1)Ford Fusion 2)Tesla Model 3. Ford Fusion footprint increased by 6% and mass increased by 4%. Tesla Model 3 has no previous generation for comparison.
To avoid model year and production year confusion, the comparisons were made for two model years before and after the latest generation change model year. For example,

Ford F150 redesign in MY2015
MR = (MY2013 CW) – (MY2017 CW),
MR is positive if the vehicle got lighter.

Jeep Cherokee redesign in MY2020
Delta footprint = (MY2020 CW) – (MY2018 CW) [since 2022 does not exist)
Footprint is positive if the vehicle got bigger.

Automaker Interview Summary

CAR researchers interviewed several automakers to understand future mass-reduction opportunities. Below are the questions and key points from the conversations.

Q1. Is lightweighting an important decision factor for material selection in MY2025-2035 vehicles? What factors will affect its increase or decrease in importance?

Almost all interviewees said lightweighting would be an essential factor for material selection in MY2025-2035 vehicles. However, the importance will depend on multiple variables such as cost, system integration, supply base/chain, sustainability, vehicle program targets and timing, and production volume. Vehicle performance requirements must be met regardless of lightweighting importance. Table 4 lists positive, negative, and uncertain drivers that will govern lightweighting's importance in the 2025-2035 timeframe.

Table 4: Drivers that will govern the importance of ligthweighting in the 2025-2035 timeframe

Positive Drivers	Negative Drivers	Uncertain
•Increase in on-board technology (customer facing, sensors, control modules, etc.) •Increase of vehicle electrification •Vehicle redesign cycles •Automaker sustainability targets •Improvements in dissimilar material joining	•Significant increase in battery energy density and reduction in battery cost •Significant improvements in powertrain technology for conventional ICE •Investments shifting towards ADAS	•CAFE and GHG regulations •A potential shift in the regulatory narrative towards full life vehicle emissions

Source: CAR Research

Q2. Are the criteria for lightweighting different for battery electric vehicles than internal combustion vehicles? Will increases in energy density and battery cost reduction alter the selection criterion for lightweighting of the electric vehicles?

Most respondents said that the primary purpose of mass-reduction for fossil fuel vehicles is controlling fuel economy and greenhouse gas emissions. For electric vehicles, the primary goal is to increase the range. However, the material strategies for lightweighting may not differ much for BEVs versus ICE once batteries are lightweight and inexpensive.

Both battery energy density and battery costs are critical decision factors in the mass-reduction targets and materials qualification criterias to achieve the targets. Energy density, if it drives a lighter battery pack (as opposed to, for example, driving up consumer range expectations), will lower the lightweighting targets. Also, as the cost of battery storage decreases, the value of lightweighting decreases because the automakers can put a less expensive, bigger battery to match the range, compared to more expensive lightweighting. An opposing argument is that lower battery costs may free up material cost budgets to spend on lightweighting. However, respondents indicate that the former scenario is more likely than the latter. Thus, CAR researchers assume that higher battery energy density and lower battery cost will lower lightweighting targets and automakers' appetite to pay for material technology for this analysis.

Q3. What are the critical material and manufacturing technologies on which the committee should focus its attention for MY2025-2035? Comment on cost, supply chain, design optimization, capital investment, after-sales service, etc.

An overwhelming response to this question was that the committee should focus on low cost, high volume material, and manufacturing technologies. The primary reason for such a response is that the automotive industry is very cost-sensitive and produces large volumes of vehicles. As automakers are stressed to develop many new technologies based on global demand, there are limited resources to develop unless the production volume is large. New-age technologies frequently mentioned in the interviews include:

Materials Systems: novel low-cost, high-performance composite sandwich construction with *honeycomb cores, next-generation low-cost carbon fibers, 7xxx series aluminum, gen-3 steels*, graphene, and nano-based composites.

Manufacturing: significantly lower-cost high-volume fully-automated polymeric composite manufacturing methods (e.g., HP-RTM, Spray Transfer Molding (STM), straight and curved pultrusion, toolless manufacturing, and high-volume additive manufacturing.

Enablers: multi-material (dissimilar material) joining for assembly and disassembly, and predictive computational and design optimization tools.

CAR research found ongoing research and development efforts in the technologies mentioned above. The timeline for high-volume commercial production varies significantly by technology, and the uncertainty is high.

Since most automakers are global companies with standard vehicle platforms that share parts, the supply chain is critical in the material selection decision process. Experts admitted that the supply base is limited for carbon fiber and natural fibers.

Q4. How important is sustainability in material qualification? What role does life cycle assessment (LCA) play in material qualification?

All automakers and suppliers said that sustainability is becoming increasingly important in their companies. Many automakers have specific sustainability goals, the progress of which they report annually to the investors. Suppliers are getting a greater push from the automakers to have a more sustainable supply chain. Life cycle analysis (LCA) was also frequently mentioned in the interviews. LCA is an important tool for selecting sustainable materials; However, quantifiable metrics and long-term targets are not well defined. Also, there is a lack of standardization for LCA in the automotive industry.

Q5. With regard to lightweighting, what is fundamentally important for the NAS committee to understand and consider when writing the report?

Automakers said that the market forces value per kilogram, as driven by the consumers (price point) and regulatory requirements (additional heavy content). It is paramount for the committee to understand the dollar value consumers put on higher performance, higher range, and more features. Furthermore, the effects of the market forces on the amount of lightweighting automakers can do.

Another factor is manufacturing costs, especially for complex designs (geometry and/or materials). Manufacturing costs can be higher than material costs. In such a situation, focusing on low-cost raw materials does not significantly impact the overall cost (or piece cost). Therefore, because of the manufacturing process (not material costs), a lightweighting concept or lightweighting innovation may not be considered until the industry develops a lower-cost manufacturing method.

Q6. What vehicle systems do you target for lightweighting?

CAR researchers found that closures are the top priority because they are mostly large, flat panels which can be a bolt-on to the body-in-white. The body-in-white (BIW) is the second priority as it adds significant weight to the vehicle. Next in line is unsprung mass, interior components, and powertrain.

Vehicle System
Closures
BIW
Unsprung Mass
Interiors
Powertrain

Decreasing priority

Opportunities for Steel

Steel is a known commodity and has been utilized and continues to be optimized. Therefore, automakers are familiar with the material for high-volume production. With economies of scale and a millennium of efficiency improvements, steel costs are lower than other materials. The steel industry is continuously improving and launching better grades of steel. Push for higher lightweighting targets will take few vehicle parts away from steel, but steel will remain a dominant material for BIW at least till 2035 for mass-market vehicles.

Current Advantage

- Broad UTS range: 200 Mpa to 2000 Mpa in the last 20 years
- Low Cost
- Reasonable weight savings using UHSS, Gen-3
- Familiarity
- Global supply chain

Future Opportunities

- Steel industry is focusing on improving formability while increasing strength
- All steel makers are actively updating infrastructure to lower their carbon footprint. Example: Nucor, Big River and others moving heavily to EAF (electric arc furnaces)
- Significant increase in battery energy density and reduction in battery cost might drive automakers to reassess their priorities for lightweighting for shifting investments to other technologies like ADAS.
- Expected to be 50%-55% of the BIW+closures

Negatives

- Aluminum and polymer composites provide 30%-60% MR and actively investing in cost reduction.
- Mass add-back due to autonomous technology can force automakers to lightweight further by switching to other materials.

Opportunities for Aluminum

Aluminum is the prime competitor of steel. It provides a 35-40 percent mass reduction over lower grade steels. Like steel, aluminum is a known commodity with a well-established global supply chain. CAR research found aluminum to become a prime material for use in closures and outer body panels. Aluminum will increase from the current 10-13 percent to 20-22 percent of the BIW and closures subsystem.

Current Advantage

- Aluminum provides 35-40% mass reduction over mild steel.
- Aluminum content has increased from 200/lbs per vehicles in 2000 to over 400 lbs/vehicles in 2020. This growth is driven by use of aluminum in hoods and other closures
- Aluminum claims to be the most recycled material in the world, 70-80%
- Global supply chain

Future Opportunities

- Doors and other bolt-on components will continue to be an opportunity for Aluminum.
- Industry is working on new 6xxx and 7xxx grades (gen-2 and gen-3)
- Continuous casting has potential to reduce conversion cost of aluminum sheet products.
- Few premium brands such as Audi, Jaguar prefer aluminum
- Expected to be 20-25% of the BIW+closures

Negatives

- The manufacturing plant is totally disrupted by AL alloys and the usage of adhesives. This really explodes the cost, so high volume, totally new platform vehicles are the best applications.
- BIW shift to aluminum has not picked up after Ford F-series trucks and Tesla Model S. In-fact, other automakers have chosen a AHSS strategy for high-volume vehicles (for example, Chevy Silverado, Tesla Model 3).
- Significant increase in battery density and shift of investment to ADAS can limit lightweighting initiatives

Opportunities for Polymer Composites

Polymer composites offer significant lightweighting over steel and aluminum. They also simplify the design and reduce the number of parts. Polymers will play a significant role in achieving higher lightweighting requirements. For premium, performance-driven vehicles, they might be a dominant material. Since polymer composites are application-specific and not a commodity product, the cost will remain a challenge for high-volume production. With the increasing push towards sustainability, recyclability and reusability of polymers also remain a concern. Interviews with experts reveal that suppliers are diligently working to find sustainable plastics and polymer composites.

Current Advantage

- Polymer composites like CFRP, GFRP, SMC offer potential for 60-70% lightweighting
- Part consolidation and low tooling cost lowers overall cost.

Future Opportunities

- Opportunities in liftgate, door inner, fender, roof panel, front bulkhead, floor reinforcement, A/B pillar reinforcement, truck bed, seats
- Expected to be 8-12% of the BIW+Closures

Negatives

- In terms of $/lbs, polymer composites will remain expensive over metals.
- Major barriers
  - High raw material cost
  - Different tooling than metals
  - Paint shop for BIW applications
  - Joining
  - Design

Opportunities for Magnesium

Magnesium is at par with polymer composites in mass reduction opportunities. However, magnesium suffers from galvanic corrosion issues when used with other materials, and the supply chain is not well established. Also, there are issues with forming and brittleness in magnesium parts. Therefore, magnesium will have limited use in the non-exposed parts of BIW and closures.

Current Advantage

- Magnesium provides 60-70% mass reduction over mild steel.

Future Opportunities

- Limited opportunities in vehicle front end components and powertrain castings
- Expected to be 3-6% of the BIW+closures

Negatives

- Use of magnesium remains low (1%) and is limited to inner body parts. Examples, liftgate inner, IP beam, seats
- Major barriers:
  - High cost and limited supply chain
  - Low formability
  - Corrosion issues
  - Joining

Material Mass Reduction Percentages used in this Study

The weight reduction opportunity for materials depends on the design of the part and the manufacturing processes. However, there is a general understanding of the expected range of mass reduction each material can provide. Table 5 from the U.S. Department of Energy lists the range of mass-reduction estimates for each material.

Table 5: Mass Reduction Potential of Materials

Lightweight Material	Mass Reduction Opportunity
Magnesium	30-70%
Carbon fiber composites	50-70%
Aluminum and Al matrix composites	30-60%
Titanium	40-55%
Glass fiber composites	25-35%
Advanced high strength steel	15-25%
High strength steel	10-28%

Source: U.S. Department of Energy

Based on the interviews conducted for this project and the literature survey, the authors decided to use the mass reduction percentages listed in Table 6.

Table 6: Mass Reduction percentages of each material used for this project

Material	**MR%** (relative to mild steel)
AHSS	10%
UHSS	25%
Aluminum	45%
Magnesium	50%
Polymer Composite	60%

Source: CAR Research

Secondary Mass Reduction

The 2015 NASEM report suggested that a powertrain downsizing opportunity exists when the glider mass is lightweighted by at least 10 percent (*Cost, Effectiveness, and Deployment of Fuel Economy Technologies for Light-Duty Vehicles*, 2015). The committee found that a reduced mass vehicle would allow an additional 40 percent of the primary mass removed from cars (unibody) and an additional 25 percent of the primary mass removed for trucks (body-on-frame).

Automakers' interviews confirmed the NASEM committee findings. Therefore, we have used the same approach to calculate secondary mass reduction opportunities in this report.

Understanding the Cost of Lightweighting

The cost of lightweighting may seem like a simple material replacement cost. However, many factors influence the real-world lightweighting cost, including part complexity, part size, manufacturing process, annual volume, capital investment, additional labor, factory location, and reengineering effort. Moreover, mass add-back takes away a large portion of the mass-reduction achieved. Figure 1 shows a generic mass-reduction cost curve for an automaker. Please note:

- The model year 2020 baseline vehicle has advanced material technology and, therefore, already ahead in the curve.
- There is a distribution of material technology in the U.S. fleet with vehicles more and less advanced than the average vehicle.
- Mass reduction becomes expensive with real-world constraints (curve shifts upwards).
- Mass add-back reduces the actual curb weight reduction even after implementing lightweight material technology.

Figure 1: Generic Mass-Reduction Cost Curve

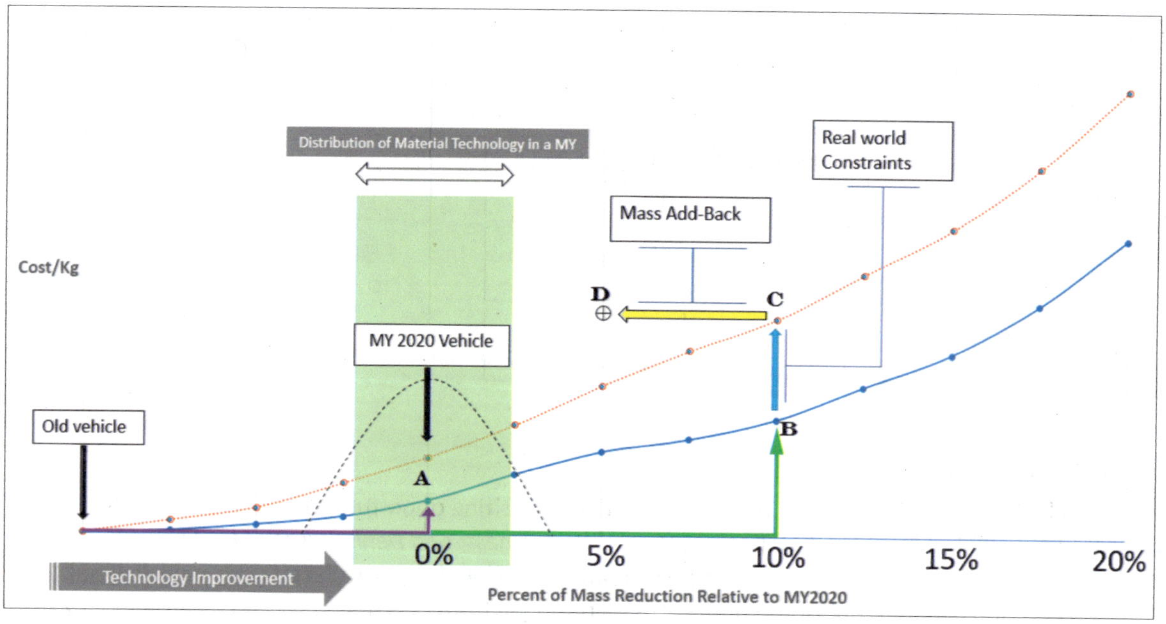

Source: CAR Research

Cost of Materials and Manufacturing

For understanding the cost of lightweighting, it is vital to understand the raw material and manufacturing costs. Vehicle teardown studies sponsored by the NHTSA and EPA have used detailed cost models to account for raw material, equipment, labor, energy, scrap, engineering, overhead, etc. Creating such a model is out-of-scope of this project. To arrive at a reasonable range of estimates for different levels of mass-reduction, CAR researchers performed an extensive literature survey and interviewed material suppliers to understand the raw material and manufacturing costs.

Raw Material Prices

Below is the discussion on individual raw material prices. The information presented below refers to multiple sources. Please refer to the bibliography for citations.

Statistical forecasting models' performance degrades rapidly as the forecast horizon expands. A review of the commodity price forecasting literature indicates that forecasts are, at best, accurate to a time horizon of three months. Given a fifteen-year forecast horizon, statistical methods are inappropriate. Thus, we rely on long-term historical patterns in the data for steel and aluminum prices.

Steel Prices

For steels, long-run inflation-adjusted (using the implicit GDP deflator, base year 2019) prices are relatively constant, with deviations driven by demand shocks and geopolitical events (collapse of the USSR, the industrialization of China). Given this stability of real prices, we assume steel will remain at roughly its long-term real price throughout the forecast period. The United States Geological Survey (USGS) historical data on steel prices, which uses hot-rolled special bar quality carbon steel as its benchmark, has averaged $0.85 per kg since the late 1950s. At current cost deltas across steel grades, this implies a long-run, inflation-adjusted average price between $0.70 and $0.75 per kg for cold-rolled mild steel.

Discussion with industry participants and available data indicate that cold-rolled steel is almost always more expensive than hot-rolled special bar quality (SBQ). The $0.70 per kg figure is definitively a lower bound and may result from idiosyncrasies in the limited data available for translating the USGS SBQ data to cold-rolled prices. To account for uncertainties, we use a range of $0.70-$1.00 per kg for cold-rolled mild steel.

Figure 2: Hot-Rolled Steel Price Trend

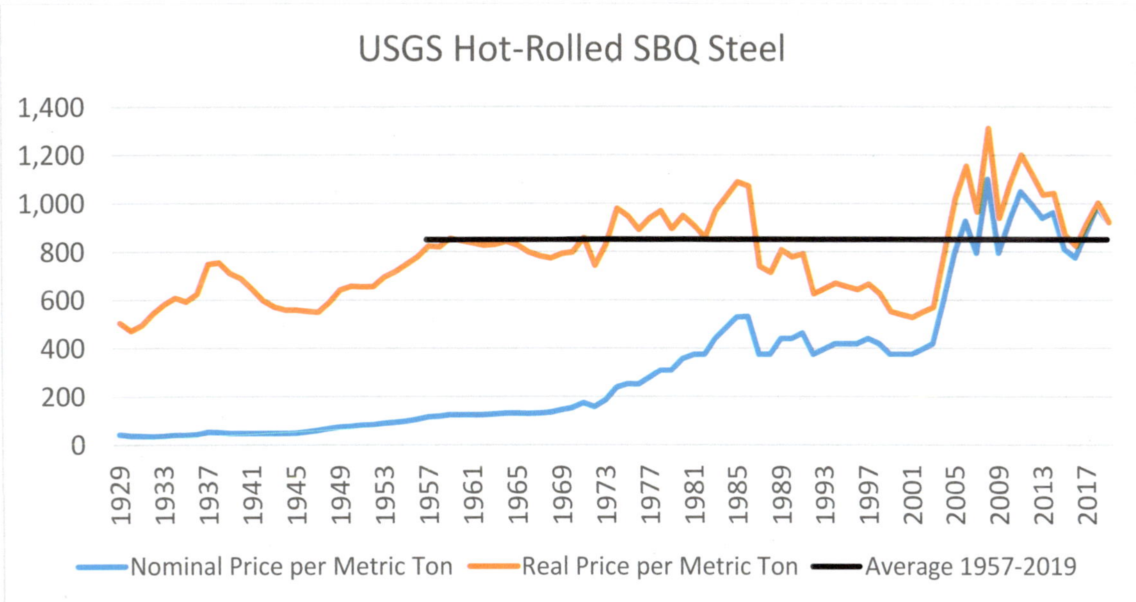

Source: United States Geological Survey

Aluminum Prices

Inflation-adjusted aluminum prices are less stable than steel prices, with a pronounced and rapid decline in the pre-WWII period. Since 1990, real prices for aluminum ingot have averaged and remained relatively steady near $2.44 per kg. To reflect uncertainty, we report a high and low around this figure, based upon the price volatility since 1990. Our low-cost aluminum scenario sees the real price per kg at $2.135, while the high real price scenario indicates prices at $2.745 per kg.

Figure 3: Aluminum Price Trend

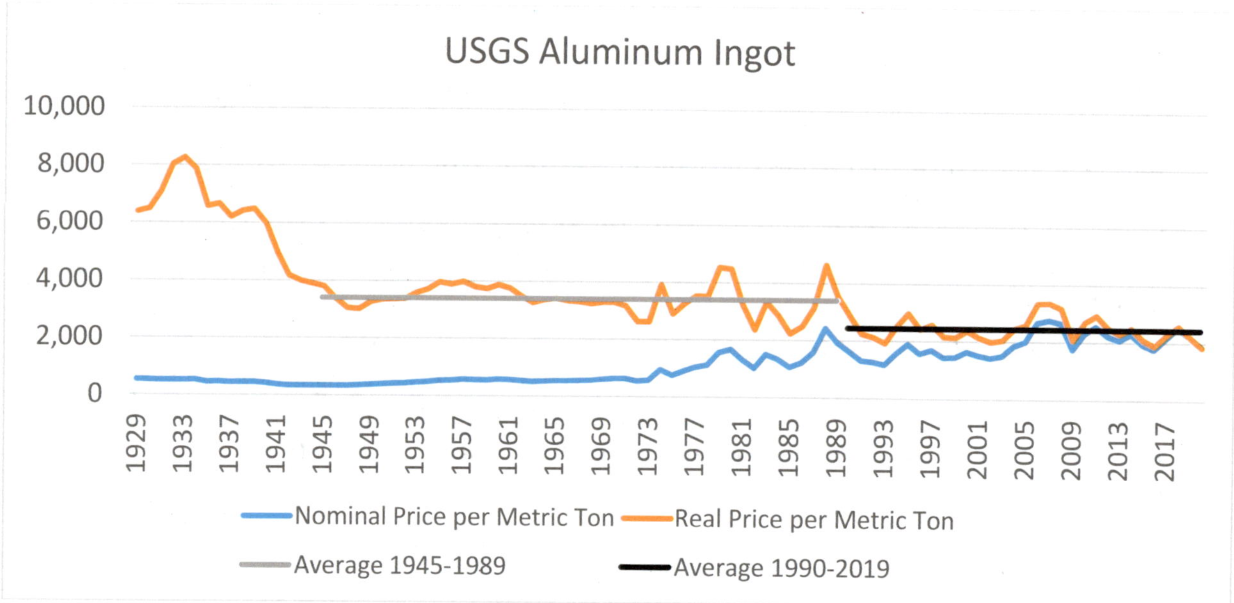

Source: United States Geological Survey

Carbon Fiber Prices

Little information is available for the cost of carbon fiber (C.F.), whether C.F. itself or for carbon fiber composites (CFC). Those documents and reports that are available are inconsistent as to materials discussed and often do not clearly specify what is being referenced.[4] As baselines are inconsistent and seldom fully defined, we focus on each document's anticipated percent change from its own baseline. Overall, documents discussing cost-reduction pathways for C.F. and CFC expect that the development of alternative precursors will lower the cost of approximately 35 percent for CFC. Process improvements and fully-scaled production are anticipated to each provide a further 20 percent cost reduction. These cost reductions are multiplicative, not additive. The expected combined effect of full-scale production wholly implemented process improvements, and an alternative precursor is approximately a 50 percent cost reduction for CFC.

Figure 4: Factors affecting CF and CFRP Cost or Price

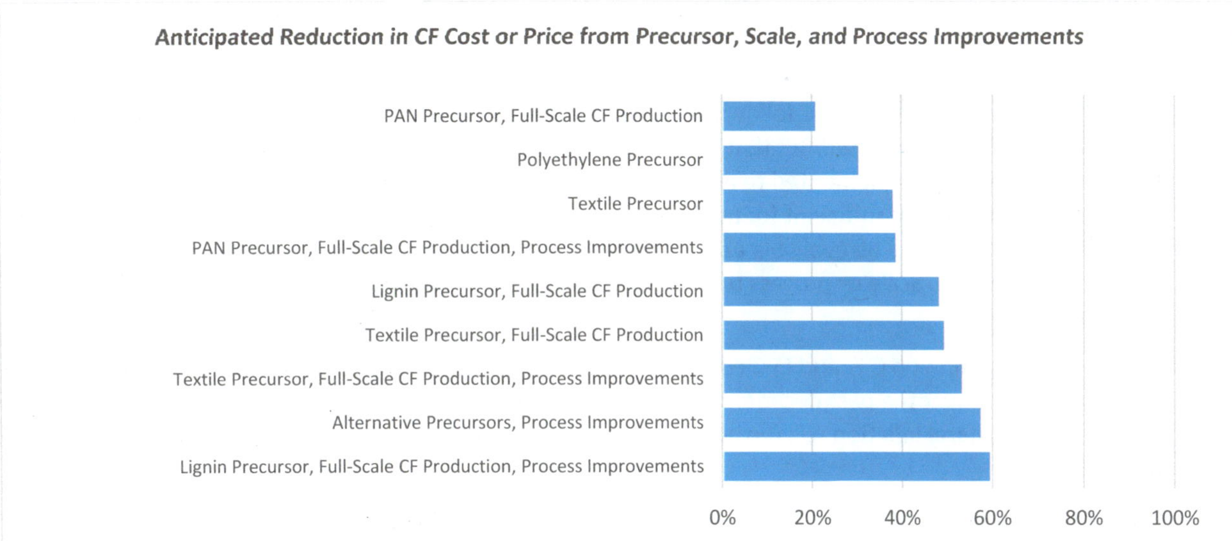

[4] Documents and reports are inconsistent with the materials discussed and often do not specify what is being referenced. Critical information missing includes:

- Differing tow sizes (large tow size CF is significantly lower cost than low tow size CF)
- Inconsistent performance characteristics/material qualities
- Does "carbon fiber" refer to...
  - Carbon Fiber (CF),
  - Carbon Fiber Reinforced Polymer / Carbon Fiber Composite (CFRP/CFC), or
  - A product manufactured from CFRP (e.g., CFRP-based door inner)?
- Are figures the direct cost of material manufacture or the price paid by a purchaser of the material? CF price roughly = CF Production Cost x 1.45*
- As a result, even reports citing similar data and time periods provide meaningfully different cost figures for CF, CFRP, and parts made from CFRP
- Quoted baseline costs/prices vary dramatically across sources, even for consistent materials, e.g. 50k tow CF (roughly $12 - $24 / kg)

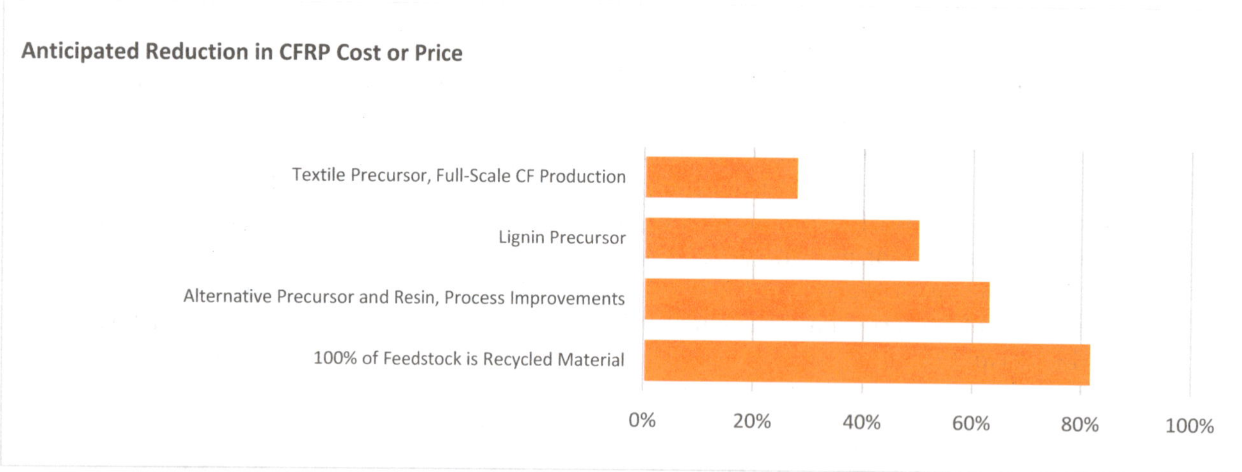

Source: Warren, 2011; Ennis et al., 2019; Bregar, 2014; Vehicles Technology Office, 2013; Heuss et al., 2012

Our future cost scenarios for CFC follow the same technology pathway, but at different time horizons. The low-cost CFC scenario assumes that, by 2025, sufficient scale and/or sufficient process improvements will be implemented to achieve a 20 percent reduction in CFC cost. Further improvements will occur sufficient for an additional 20 percent cost reduction by 2030, and an alternative precursor will be available and in-use by 2035. In the high-cost scenario, we did not consider any improvements to scale or process until 2030, and the experts do not expect an alternate precursor until after 2035.

Available information on C.F. and CFC costs and prices suggest that per kg nominal costs have remained roughly constant over the past decade; thus, real costs have fallen. We assume this will continue, and the effects of general inflation will further reduce CFC's real cost throughout the 15-year forecast horizon.

Given the full range of baseline cost estimates and anticipated cost reduction pathways, the range of costs per kg for C.F. and CFRP achievable after full implementation of fully-scaled production, process improvements, and alternate precursors are:

- CF: $5 – 11 per kg (typical $7 per kg)
- CFC: $9 – 15 per kg (typical $12 per kg)

These figures do not incorporate general inflation. Incorporating inflation and assuming all cost reduction possibilities are implemented by 2035, CAR estimates CFRP prices could be $8.74 in 2035.

Magnesium Prices

Magnesium is the eighth-most abundant element in the earth's crust, but raw magnesium's price instability inhibits broader use. The magnesium price is impacted by the demand for aluminum, titanium, and steel because magnesium is used to make these metals. The Platts Metals Week U.S. spot Western magnesium price range was $2.10 to $2.20 per pound ($4.62 to $4.85 per kg) throughout the entire year for an annual average price of $2.15 per pound in 2017, unchanged from the average price since the beginning of 2014. Global consumption of magnesium is expected to increase by a compound annual growth rate of about five percent per year from 2017 through 2027 (*2017 Minerals Yearbook*, 2017). The price of magnesium is unstable and depends on geopolitical factors. For this research, we have assumed magnesium prices to reduce going forward.

Material Costs used for this research

For this analysis, CAR researchers created high and low-cost scenarios for 2025-2035. Based on the research described above, steel prices are kept constant. Aluminum, CFRP, and Magnesium costs are assumed to decrease over the years. Table 7 shows the cost in dollars per kilogram for each material.

Table 7: High and Low-Cost Scenarios for Materials 2025-2035

LOW Material Cost $/kg range				
Material	**2020**	**2025**	**2030**	**2035**
Mild	0.70	0.70	0.70	0.70
HSS	0.83	0.83	0.83	0.83
AHSS	1.10	1.10	1.10	1.10
UHSS	1.15	1.15	1.15	1.15
Al	2.13	2.13	2.13	2.13
Mag	4.50	4.00	3.50	3.00
CFRP	24.00	19.71	16.41	8.74

HIGH Material Cost $/kg range				
Material	**2020**	**2025**	**2030**	**2035**
Mild	1.00	1.00	1.00	1.00
HSS	1.13	1.13	1.13	1.13
AHSS	1.40	1.40	1.40	1.40
UHSS	1.45	1.45	1.45	1.45
Al	2.74	2.74	2.74	2.74
Mag	4.80	4.30	4.00	3.50
CFRP	24.00	24.00	19.71	16.41

Source: CAR Research

Manufacturing Cost

Manufacturing cost is more difficult to estimate since it depends on many factors such as part complexity, part size, manufacturing process, the volume of production, and manufacturer's "tribal" knowledge. CAR researchers interviewed material suppliers and consortiums to understand the manufacturing cost part of the total part cost. Most respondents gave CAR a range of estimates. To control the complexity of manufacturing cost, we used averages of the estimates provided to CAR.

In a 2016 CAR study (Baron & Modi, 2016), the authors estimated manufacturing cost reduction due to time and volume-based learning for different materials[5]. These learning percentages are used in this analysis to estimate manufacturing cost reduction through 2035. See Table 8 and Table 9.

[5] There are two different types of learning curves:
Time-Based – Over time, plant operations adapt to using new material and new techniques for processing and handling of these materials. This enables the procedure to increase process speed and efficiency, resulting in lower overall cost.
Volume-Based – Higher volume bring opportunities for economies of scale. It can lead to a reduction in the cost of production per part. Higher volume demands drive more rigorous processing, which can create additional opportunities to improve overall process times.

Table 8: Manufacturing cost and learning

Material	Manufacturing cost as part of total cost %	Manufacturing cost % Reduction/ year (learning)
Mild	47%	0.00%
HSS	41%	0.00%
AHSS	41%	0.56%
UHSS (HF)	41%	0.56%
AL	40%	1.26%
Mag	24%	0.88%
Comp	58%	2.36%

Source: CAR Research

Table 9: Manufacturing Cost with Learning ($/kg) for key years

Material	2020	2025	2030	2035
Mild	0.75	0.75	0.75	0.75
HSS	0.68	0.68	0.68	0.68
AHSS	0.87	0.84	0.82	0.80
UHSS	0.90	0.88	0.85	0.83
Al	1.63	1.53	1.43	1.34
Mag	1.47	1.40	1.34	1.29
Comp	33.14	29.41	26.10	23.16

Source: CAR Research

Total Cost

To calculate the total cost for material changes, the team added the low and high material costs and each material's manufacturing cost. This exercise resulted in Table 10 and Table 11.

Table 10: LOW Total Cost $/kg range (material + processing)

Material	2020	2025	2030	2035
Mild	1.45	1.45	1.45	1.45
HSS	1.51	1.51	1.51	1.51
AHSS	1.97	1.94	1.92	1.90
UHSS	2.05	2.03	2.00	1.98
Al	3.76	3.66	3.57	3.48
Mag	5.97	5.40	4.84	4.29
Comp	57.14	49.13	42.51	31.90

Source: CAR Research

Table 11: HIGH Total Cost $/kg range (material + processing)

Material	2020	2025	2030	2035
Mild	1.75	1.75	1.75	1.75
HSS	1.81	1.81	1.81	1.81
AHSS	2.27	2.24	2.22	2.20
UHSS	2.35	2.33	2.30	2.28
Al	4.37	4.27	4.18	4.09
Mag	6.27	5.70	5.34	4.79
Comp	57.14	53.41	45.81	39.57

Source: CAR Research

Factors Affecting Automaker Lightweighting Targets

Automaker interviews revealed that lightweighting targets would depend on four primary factors 1) fuel economy and GHG regulations, 2) electrification volume, 3) battery cell energy density (weight of the battery pack) and, 4) battery pack cost.

Fuel Economy and GHG regulations – the government regulations on fuel economy and greenhouse gases affect automakers' mass-reduction target the most. *Since most automakers sell in multiple countries and have common vehicle platforms, worldwide government regulations affect material strategies, not just the U.S. regulations.*

An industry estimate is that a 10 percent reduction in vehicle's mass will produce approximately six to seven percent reduction in fuel consumption for passenger cars and four to five percent reduction for light-duty trucks (*Cost, Effectiveness, and Deployment of Fuel Economy Technologies for Light-Duty Vehicles*, 2015). Therefore, lightweighting is an essential tool for automakers to achieve fleet fuel economy targets.

Fuel economy and GHG regulations also affect electrification volume in the fleet. Since electric vehicles' fuel economy (miles per gallon equivalent) is higher than internal combustion engines, automakers can positively impact their CAFE targets by adding electric vehicles to the fleet. For this research, we have assumed electrification volume as a proxy variable for fuel economy and GHG regulations.

Battery Cell Energy Density – Assuming similar performance, a battery-electric vehicle usually has a higher curb weight than an internal combustion engine vehicle. The primary reason for the weight difference is because the current batteries have lower energy density than gasoline. A vehicle with 10 gallons of onboard fuel (337 kWh energy) weighs an additional 27-30 kg, and the vehicle gradually drops that weight as the fuel is combusted. On the other hand, a battery-electric vehicle's (BEV) battery pack may contain 100 kWh of energy but weigh 385-544 kg. Automakers need a higher battery capacity to meet range targets, which means adding significant weight to the vehicle with the current battery density. Therefore, battery weight is an essential factor affecting lightweighting targets.

Battery Pack Cost – Achieving long-range is paramount for selling electric vehicles in high volume. The range can be increased by adding battery capacity or reducing the weight of the vehicle. Since batteries are expensive, adding battery capacity adds a high cost to the vehicle. However, if the battery pack's cost decreases significantly, adding battery capacity becomes a cheaper option than lightweighting.

Automaker interview reveals that decreasing battery pack cost might negatively impact lightweighting targets.

For each of the above variables, high and low values were defined (see Table 12). Table 13 shows the expected values of the selected variables for the study years.

Table 12: High and Low Definitions for Selected Variables

	High	Low
Electrification Volume (CAFE/GHG proxy)	>25% BEV, 30-50% Hybrids	<15% BEV, 20-25% Hybrids
Battery Pack cost	$145-$160 per kWh	<$100 per kWh (2030+ projected)
Battery Cell Energy Density	900 Wh/Liter	700 Wh/Liter

Source: CAR Research

Table 13: Expected Variable Value for Select Years

Year/Variable	Electrification Volume	Battery Pack Cost	Battery Cell Energy Density
2020-2025	Low	High	Low
2025-2030	Mass Market: Low Premium - High	High	Low
2030-2035	High	Low	Low

Source: CAR Research

Method for Estimating Material Distribution

Generalizing material penetration in the U.S. fleet is challenging because of the wide distribution of technology and vehicle-specific lightweighting targets. Furthermore, the data on vehicles' material, weight, and manufacturing technology is not public information.

CAR research shows that material choices heavily depend on the vehicle's price than the vehicle's segment. Therefore, material distribution in future vehicles will differ more for premium versus mass-market vehicles. However, there are structural differences between unibody and body-on-frame vehicles, which affects material choices. Body-on-frame vehicles are also, in general, more expensive than unibody vehicles. After considering these points and in the spirit of using publicly available information, CAR researchers decided to use the vehicles studied by the regulating agencies – the 2011 Honda Accord as a reference for unibody vehicles, and the 2014 Chevy Silverado for body-on-frame vehicles.

The two vehicles were updated to represent a generic mass-produced unibody and body-on-frame vehicles based on the data collected from 2020 repair manuals, the material forecast information from the 2016 CAR study, and engineering judgment.[6] Figure 5 and Figure 6 show the material distribution in the original and updated vehicles.

[6] The analysis considers the two vehicles updated to MY2020 materials to represent the unibody and body-on-frame vehicles in the current US fleet. In reality, vehicle material technology has a wide distribution in a given model year.

Figure 5: 2020 Baseline vehicle material distribution- Unibody

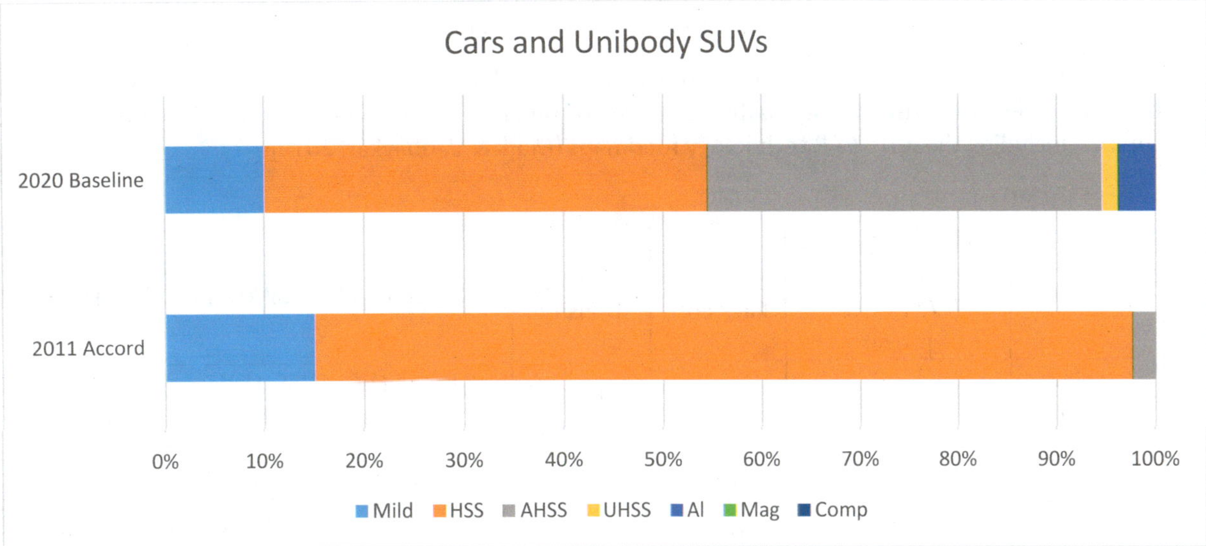

Figure 6: 2020 Baseline vehicle material distribution - Body-on-Frame

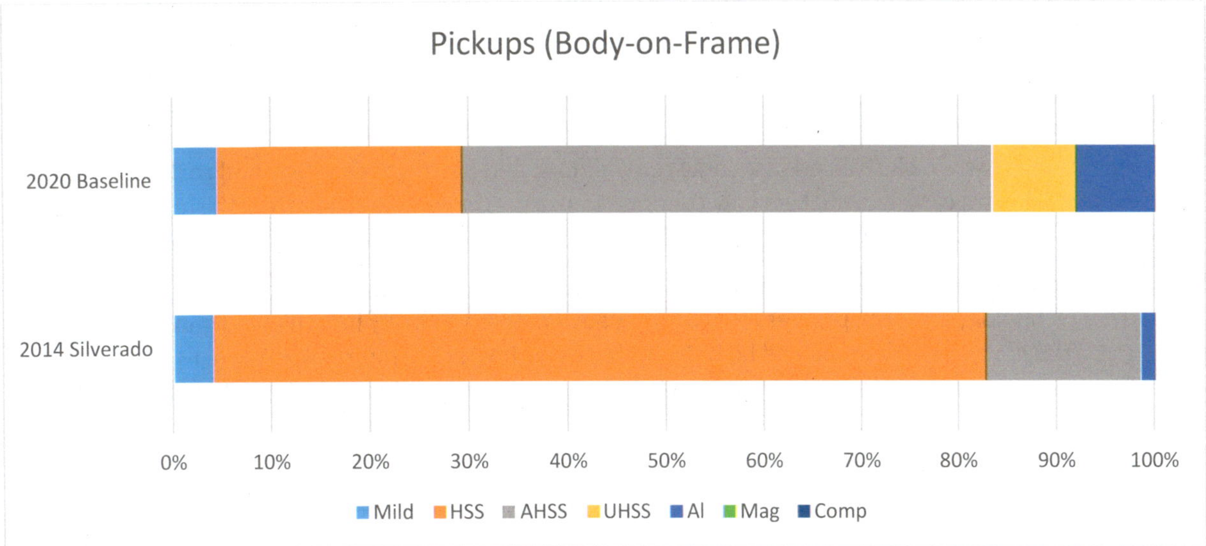

Source: CAR Research

As discussed in the previous section, future vehicle material distribution will depend in part on electrification volume, battery pack cost, and battery cell energy density. CAR researchers created scenarios using the three variables for the study years 2020-2025, 2025-2030, and 2030-2035. For each scenario, we created a material roadmap for unibody and body-on-frame vehicles. The following sections will discuss these scenarios and corresponding material trends.

Material Roadmap and Incremental Cost ($/kg) – Unibody Vehicles

The updated Honda Accord was used to study material distribution scenarios for unibody vehicles.

Table 14 lists the unibody scenarios, expected material trends, and the expected year range. We have based these estimates on the information collected through automaker interviews and literature surveys, including the 2019 CAR technology roadmaps (Modi & Vadhavkar, 2019).

Table 14: Scenarios for Unibody Vehicles

Scenario	Electrification Volume (CAFE/GHG proxy)	Battery Pack cost	Battery Density	Expected Material Trend	Expected Year
Baseline 2020	Low	High	Low	Body: HSS, AHSS, UHSS Closures: HSS, low Al	NA
Scenario One	Low	High	Low	Body: HSS, AHSS, UHSS Closures: HSS, Al	Mass Market 2025-2030
Scenario Two	High	High	Low	Body: Aluminum, AHSS, UHSS Closures: Al, comp, Mag	Premium Vehicles 2025-2030
Scenario Three	High	Low	Low	Body: AHSS, UHSS, low Al Closures: Al	Mass Market 2030-35
Scenario Four	High	Low	High	AHSS intensive	Low

Source: CAR Research

Scenario one is where electrification volume remains low, and battery technology and cost remain at 2020 levels. In this case, automakers may continue to use predominantly high strength steel BIW and a mix of steel and aluminum in the closures. This scenario may apply for the 2025-2030 timeframe.

In scenario two, electrification volume becomes high for performance and customer demand, but the battery technology does not progress. In this case, automakers will likely have higher lightweighting targets. They may use aluminum and high strength steel in BIW and a mix of aluminum, magnesium, and polymer composites in the closures. Scenario two may apply for premium vehicles in the 2025-2030 timeframe.

In scenario three, the electrification volume is high, and the battery pack cost has come down significantly to support mass-market electric vehicles. In this case, there is medium pressure for lightweighting since automakers can increase vehicle range at lower costs by adding batteries. Thus, in scenario three, automakers will likely use high strength steel and aluminum in BIW, and aluminum for closures.

Scenario four is unlikely in the timeframe because experts do not expect battery density to fall into the high-value range before 2035.

Based on the material trends for unibody vehicles listed in Table 14, CAR researchers upgraded the materials for each of the BIW and closure component in a spreadsheet. The team upgraded each component's material based on CAR's internal material databases and the research team's collective engineering judgment. The spreadsheet automatically calculates the components' new weights using the data from Table 6 when the materials are changed, thereby giving the expected mass reduction for each scenario. The team repeated the exercise for the three unibody scenarios from Table 14.

Once the materials and weights of BIW and closure components were known, it became possible to create a graph of each scenario's material distribution (see Figure 7).

Figure 7: Material Penetration for each Scenario - Unibody Vehicles

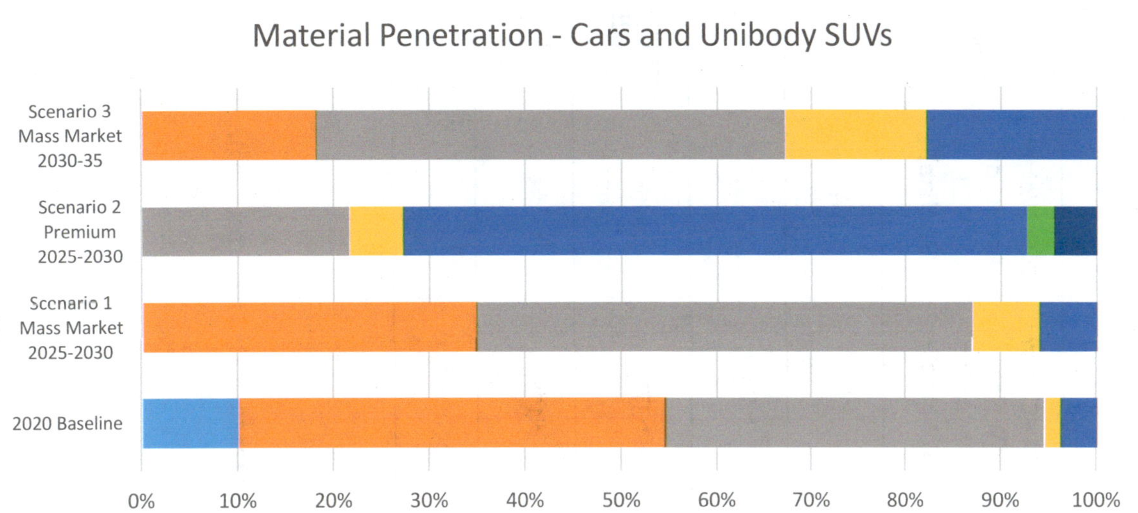

	2020 Baseline	Scenario 1 Mass Market 2025-2030	Scenario 2 Premium 2025-2030	Scenario 3 Mass Market 2030-35
Mild	10%	0%	0%	0%
HSS	44%	35%	0%	18%
AHSS	40%	52%	22%	49%
UHSS	2%	7%	6%	15%
Al	4%	6%	65%	18%
Mag	0%	0%	3%	0%
Comp	0%	0%	4%	0%

Source: CAR Research

To understand the cost penalty for ligthweighting, the team multiplied each component mass with the high and low cost of material change from Table 10 and Table 11 to understand the incremental cost range. We then divided these costs by each scenario's mass reduction (in kilograms) to get to the dollars per kilogram figure. Wherever the cost penalty range was too broad, we consulted with subject-matter-experts to narrow down the range. Table 15 shows the cost penalty and expected mass reduction for each of the unibody scenarios. Figure 8 is the expected material roadmap for unibody vehicles with cost penalties.

Table 15: Cost and Mass Reduction Analysis for Unibody Vehicle Scenarios

Scenario	**Electrification Volume (CAFE/GHG proxy)**	**Battery Pack cost**	**Battery Density**	**Cost penalty per kg of weight saved**	**Expected Material Trend**	**BIW+Closures Weight Reduction (2020 baseline)**	**Curb Weight Reduction (2020 baseline)**	**Expected Year**
Baseline	Low	High	Low	NA	Body: HSS, AHSS, UHSS Closures: HSS, low Al	NA	NA	NA
Scenario One	Low	High	Low	\$0.5- \$1.5 CY: 2030	Body: HSS, AHSS, UHSS Closures: HSS, Al	~4%	1.0% - 1.5%	Mass Market 2025-2030
Scenario Two	High	High	Low	\$4.0 -\$6.0 C.Y.: 2030	Body: Aluminum, AHSS, UHSS Closures: Al, CFRP, Mag	~37%	12 - 14% (with secondary)	Premium Vehicles 2025-2030
Scenario Three	High	Low	Low	\$1.5 - \$3.5 CY: 2035	Body: AHSS, UHSS, low Al Closures: Al	~12%	4 - 6%	Mass Market 2030-35
Scenario Four	High	Low	High*	-	AHSS intensive	not in scope		Low

Source: CAR Research

Real-world mass reduction might be less due to mass add-back

*High battery density is unlikely in the timeframe

2020 baseline costs used for incremental cost calculations are shown in Table 10: LOW Total Cost $/kg range (material + processing).

Figure 8: Technology Pathway and Relative Cost for Unibody Cars and SUVs

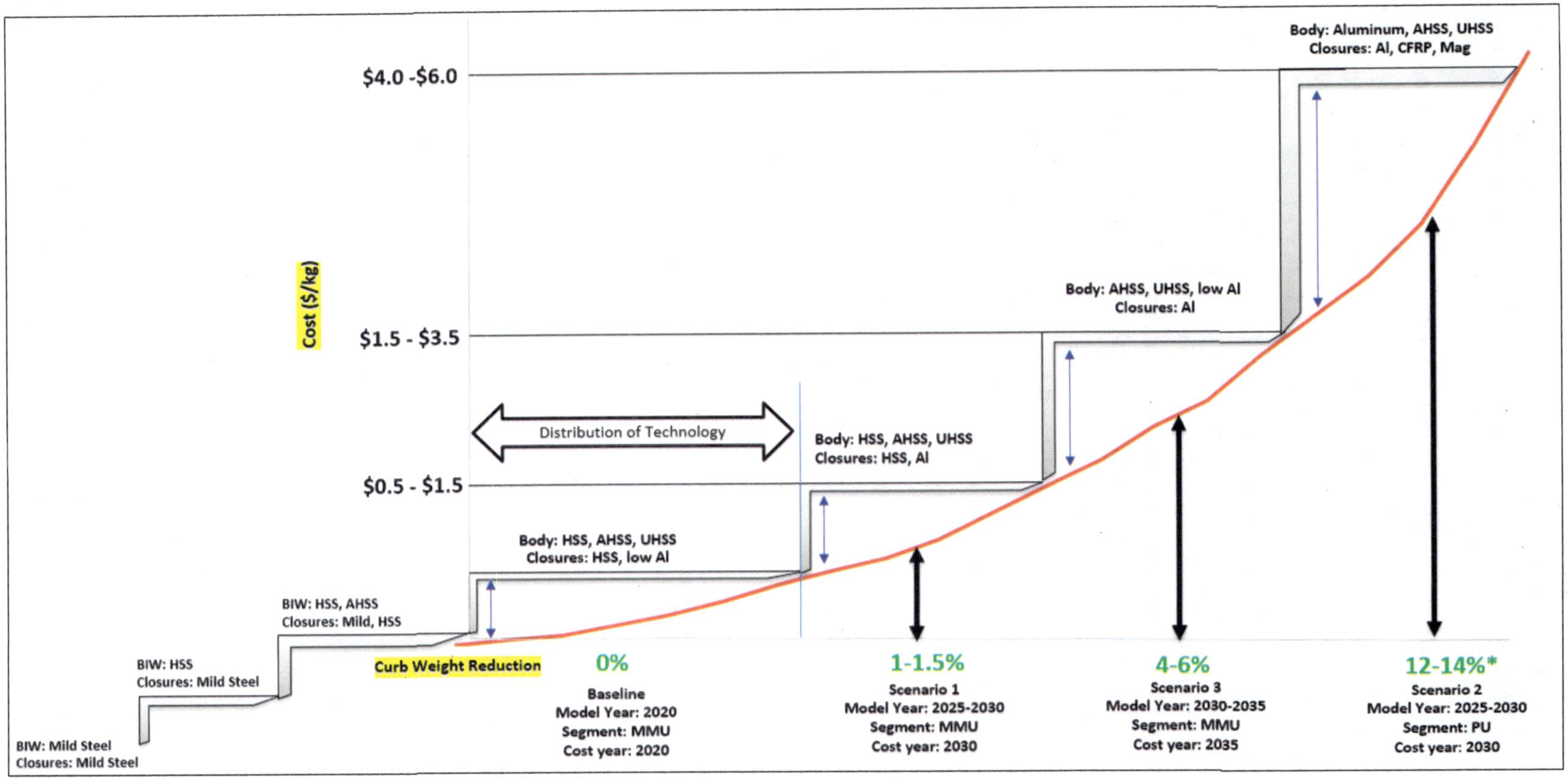

Source: CAR Research

MMU – Mass Market Unibody

PU – Premium Unibody

Cost Year – projected material and manufacturing cost for analysis in the year

*with secondary mass reduction
Real world cost curve will be higher than the shown curve due to real-world constraints and mass add-back.

Unibody Cost Curve Discussion

Most of the unibody vehicles in the MY2020 fleet use advanced steels for the BIW and a mix of steel and aluminum for the closures. Examples of aluminum use in closures include hoods, door inner, liftgate, and roof panel. Polymer composites find some use in the BIW, mostly in higher priced vehicles such as the Audi A8, Cadillac CT6, and BMW 7 series. However, polymer composites can be found in closure applications for mass-market vehicles, for example, the Nissan Rogue liftgate.

Automaker interviews reveal a strategy to use mixed-material solutions for lightweighting. Future vehicle structures might use a mix of materials, including high-strength steel, high-strength aluminum, magnesium, plastics, and polymer composites. There will be an increase in the number of steel grades with new generation steels replacing lower grade steels. Aluminum is the primary candidate to replace mild steel closures (doors, hoods, liftgate, and fenders), roof panels, and bodysides.

The cost penalty increases exponentially for achieving higher levels of mass reduction. In general, *carbon fiber composites' use makes lightweighting very expensive because of the high raw material and* manufacturing cost. Therefore, new generation steels and aluminum will remain the material of choice for lightweighting of unibody vehicles until polymer composites become affordable for high-volume vehicles. However, for premium performance-oriented vehicles, like the Chevy Corvette, polymer composites could remain the material of choice.

As shown in Figure 1, mass reduction's real cost is much higher than a simple material replacement. The roadmap showed in Figure 8 is the general trend identified through CAR research. In the real world, the lightweighting targets, choice of materials, and cost penalty depends on the specific vehicle program and the advancements in materials, battery technology, and government regulations.

Material Roadmap and Incremental Cost ($/kg) – Body-on-Frame Vehicles

For studying body-on-frame (BoF) vehicles, the team updated the 2014 Chevy Silverado to represent a generic MY2020 body-on-frame vehicle. Silverado is a good representation of BoF vehicles since most BoF vehicles in the U.S. fleet are pickups. In general, SUVs in the 2020 fleet have unibody construction, except for a few like the Jeep Wrangler. Table 16 lists the scenarios, material trends, and the year range for BoF vehicles. Similar to the unibody scenarios, the estimates for BoF vehicles are based on automaker interviews and literature surveys.

Table 16: Scenarios for Body-on-Frame vehicles

Scenario	Electrification Volume (CAFE/GHG proxy)	Battery Pack cost	Battery Density	Expected Material Trend	Expected Year
Baseline 2020	Low	High	Low	Body: AHSS, UHSS Frame: AHSS, UHSS Closures: HSS, Al	NA
Scenario One	Low	High	Low	Body: AHSS, UHSS, Al Frame: AHSS, UHSS Closures: Al	Mass Market 2025-2030
Scenario Two	High	High	Low	Body: Aluminum Frame: AHSS, UHSS Closures: Al *OR* Body: Aluminum, CFRP Frame: AHSS, UHSS Closures: Al, CFRP, Mag	Premium Vehicles 2025-2030
Scenario Three	High	Low	Low	Body: AHSS, UHSS, Al Frame: AHSS, UHSS Closures: Al	Mass Market 2030-35

Source: CAR Research

The baseline MY2020 BoF vehicles use more advanced materials than mass-market unibody vehicles. The primary reason for this difference is the higher price of pickups and performance loving customers, which allow automakers to pay a higher dollar per kg penalties for lightweighting of pickups. In the past, automakers have used lightweighting primarily to improve the pickups' performance and increase the footprint to satisfy customer demand. However, pickups are very high volume vehicles that affect automakers' fleet average CAFE and GHG significantly. Therefore, small improvements in the pickup fuel economy can have an enormous impact on the fleet average. The paragraphs below discuss future material scenarios for BoF vehicles.

In scenario one, electrification volume remains low, and battery technology and cost remain at 2020 levels. In this case, automakers might continue using advanced steels for BIW and frame with higher aluminum use in the closures due to lower lightweighting targets. Scenario one may apply to mass-market BoF vehicles in the 2025-2030 timeframe.

In scenario two, where electrification is high, but battery technology and cost are not low enough, interesting material decisions may happen. Due to range anxiety in electric vehicles and heavy battery, automakers will need to increase their lightweighting targets. To achieve these, vehicle manufacturers may use aluminum for the BIW, advanced steels for the frame, and aluminum for the closures. Another option is to use carbon fiber composites to achieve higher mass-reduction; however, this will substantially increase the cost penalty. Scenario two may apply to premium vehicles in the 2025-2030 timeframe.

Scenario three, where electrification is high, but battery costs have come down below the $100 per kWh, mass-reduction may not be a priority for automakers because they can increase the range by adding low-cost batteries. Thus, material choices for scenario three will be similar to scenario one. Scenario three may apply to the mass-market vehicles in the 2030-2035 timeframe.

CAR researchers upgraded the materials for each BIW, frame, and closure component as per the identified material trends. We divided the costs of each scenario to its mass reduction opportunity (in kilograms) to get to the dollars per kilogram figure. Wherever the cost penalty range was too broad, we consulted with subject-matter-experts to narrow down the range. Table 17 shows the cost penalty range and mass reduction for each of the BoF scenarios. Figure 9 is the material roadmap for BoF vehicles with cost penalties.

Table 17: Cost and Mass Reduction Analysis for Body-on-Frame Vehicle Scenarios

Scenario	Electrification Volume (CAFE/GHG proxy)	Battery Pack cost	Battery Density	Cost penalty per kg of weight saved	Expected Material Trend	Body+Frame Weight Reduction (2020 baseline)	Curb Weight Reduction (2020 baseline)
Baseline	Low	High	Low	NA	Body: AHSS, UHSS Frame: AHSS, UHSS Closures: HSS, Al	NA	NA
Scenario One	Low	High	Low	\$0.5- \$1.5 CY: 2030	Body: AHSS, UHSS, Al Frame: AHSS, UHSS Closures: Al	5%	2-3%
Scenario Two	High	High	Low	\$1.5 - \$2.5 C.Y.: 2030	Body: Aluminum Frame: AHSS, UHSS Closures: Al	21%	8-10% (with secondary)
Scenario Two (alternative)	High	High	Low	\$6.0 - \$8.0 CY: 2035	Body: Aluminum, CFRP Frame: AHSS, UHSS Closures: Al, CFRP, Mag	23%	10-12% (with secondary)
Scenario Three	High	Low	Low	\$0.0 - \$1.0 CY: 2035	Body: AHSS, UHSS, Al Frame: AHSS, UHSS Closures: Al	5%	2-3%

Source: CAR Research

Real world mass reduction might be less due to mass add-back

For scenario two alternative, CFRP is used for Fender, Pickup box floor, and tailgate. Magnesium for I.P. beam and Radiator

Scenario one and three have similar materials but have different cost penalties due to different cost years.

Figure 9: Technology Pathway and Relative Cost for Body-on-Frame Vehicles

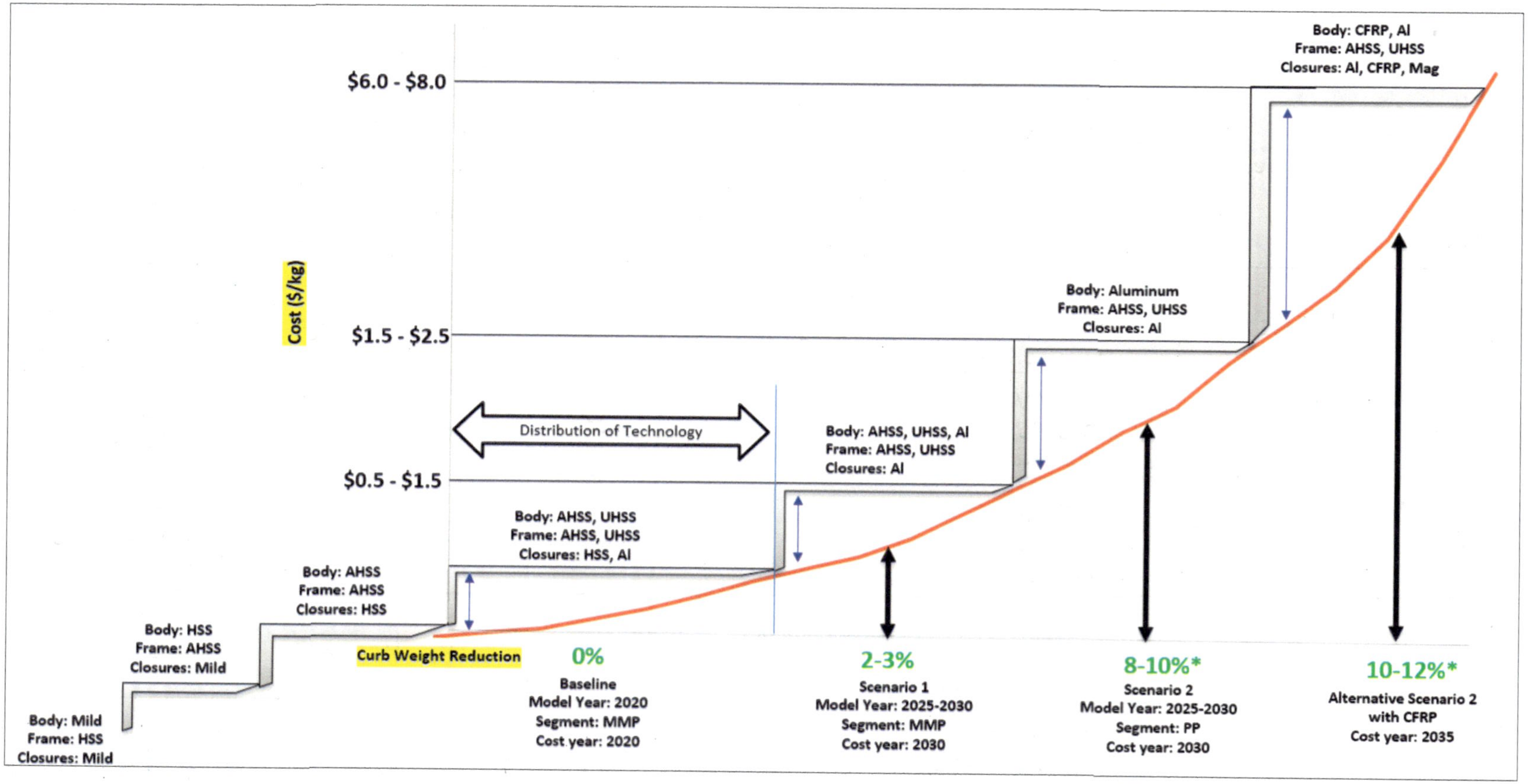

Source: CAR Research

MMP – Mass Market Pickup

P.P. – Premium Pickups, for example GMC Sierra Denali

Cost Year – projected material and manufacturing cost year for analysis

*with secondary mass-reduction
Real world cost curve will be higher than the shown curve due to real-world constraints and mass add-back.

Body-on-Frame Cost Curve Discussion

Body-on-frame vehicles, which mostly comprise of pickups, currently use advanced material and manufacturing technologies. Two good examples are the Chevy Silverado, which primarily uses advanced steel grades to achieve lightweighting and performance targets, and the Ford F-150, which has an all-aluminum body with a high strength steel frame. Polymer composites are also currently used for applications such as pickup beds. Pickups in the U.S. are high volume, high price, performance-driven vehicles, which allow automakers to use advanced lightweight materials.

Similar to unibody vehicles, automakers may use a mixed-material strategy for BoF lightweighting. BIW likely be a mix of steel and aluminum. The frame will continue to use advanced steels with further design optimization. Aluminum will also play a key role in mass-reduction of closures. Polymer composites will see limited use in truck beds, tailgates, and inner closure components.

The cost penalty for using advanced materials increases when switching parts from steel or aluminum, but it increases exponentially when carbon fiber composites are introduced. Since the industry is very cost-sensitive, automakers will likely choose steel or aluminum as primary materials for high-volume pickups.

The roadmap showed in Figure 9 is the general trend identified through CAR research. The lightweighting targets, choice of materials, and cost penalty will depend on the specific vehicle program and the advancements in materials, battery technology, and government regulations.

Real-World Challenges

There are several real-world constraints and manufacturing challenges that automakers face when looking to include a new material into production. Examples of some of these challenges include the following (J. Baron, 2016):

- **Global Platforms –** Automakers develop common parts for use across global platforms to share engineering costs. The part sharing inhibits the individual design and mass optimization of vehicle models due to the use of these parts across various locations worldwide.
- **Material availability –** Vehicle manufacturing requires assembling thousands of parts built around the world. The sales volumes are very high. Therefore, automakers look for materials with a robust supply chain. Preference is given to materials with at least two suppliers to manage risks. Supply chains for new materials are not well established, making them difficult to use in high-volume vehicles.
- **Material Qualification –** When automakers redesign parts to optimize mass, they test them virtually and physically to ensure performance. Every automaker designs cars and trucks using internal design standards and best practices. The time and development process to qualify new materials and derive appropriate product specifications is extensive. The process can take as long as 3 to 4 years (Modi, n.d.).
- **Manufacturing Technologies –** manufacturing technology is different for almost every new material used in the automotive industry. Different materials also require different equipment and tooling. The assembly and handling processes also differ in few cases; for example, magnesium requires special handling and care to avoid chipping and the special care taken to avoid sanding and grinding, which produce potentially flammable dust. Mixed material vehicles can require mechanical fasteners, adhesives, and welding, all in the same assembly. Mixing materials involves isolation of the various materials to avoid corrosion.

- **Cost versus Benefit** - A material change may improve the product's performance, but the cost may far exceed the real-life benefits. The car companies are constrained by cost and are forced to make practical decisions on material choices to make sure that the vehicle's cost does not increase by an unprecedented amount. For example, carbon fiber composite materials provide strength to the vehicles' structure while reducing weight by almost 50-60 percent compared to mild steel but are very expensive. On the other hand, aluminum can provide 20-30 percent mass-reduction at less cost than carbon fiber composites. Automakers balance cost versus benefit depending on vehicle programs.
- **Stranded Capital** – The established infrastructure with its sunk costs limits the speed of introducing new materials. For example, equipment for metal parts production is very different from equipment for polymer composites. Automakers often invest and commit to technologies for at least two redesign cycles before investing in equipment.
- **Risk of part failure**: New materials used in new ways can behave in unexpected ways. Part failure is constant risk companies have when new materials are introduced into the vehicles. These disruptions can drive recall costs or increase incidents of rework.
- **Consumer Demand** - Ride and handling quality are highly competitive differentiators in the market. While lighter cars generally handle better, other tradeoffs arise from lightweighting, such as transmitted noise and vibrations introduced by lightweight materials. Solving these issues adds weight back, taking away some of the lightweighting achieved. As discussed in the previous sections, automakers strive to continuously improve performance and safety driven by regulations and the competitive market. These improvements often add mass to the vehicles.
- **Sustainability** – Sustainability is becoming a crucial factor in the automotive industry. Many automakers and suppliers have begun to set high-level sustainability targets over the next decade and beyond. With the gamut of environmental laws and ethical reasons, the industry needs to ensure that the materials are not detrimental to the environment or general public health.
- **Noise, vibration, and harshness (NVH)** - As the customer has expectations to utilize their phone, laptops, etc. in their vehicles, the NVH demands have grown exponentially. Electric vehicles will further this need for sound deadening. Thin sheets of lightweight materials reduce mass but aggravate NVH. Solutions to counter the effect typically means a mass increase. Automakers struggle to balance lightweighting versus NVH requirements.

Bibliography

Automotive Plastics & Polymer Composites: A Roadmap for Future Mobility—Automotive Plastics. (n.d.). Automotive Plastics. Retrieved September 29, 2020, from https://www.automotiveplastics.com/roadmap-for-future-mobility/

Baron, J. (2016). Identifying Real World Barriers to Implementing Lightweighting Technologies and Challenges in Estimating the Increase in Costs. https://www.cargroup.org/publication/identifying-real-world-barriers-to-implementing-lightweighting-technologies-and-challenges-in-estimating-the-increase-in-costs/

Baron, J. S., & Modi, S. (2016). Assessing the Fleet-wide Material Technology and Costs to Lightweight Vehicles. http://www.cargroup.org/wp-content/uploads/2017/02/Assessing-the-Fleet-wide.pdf

Bernard, J., Khalaf, L., Kichian, M., & Mcmahon, S. (2008). Forecasting commodity prices: GARCH, jumps, and mean reversion. Journal of Forecasting, 27(4), 279–291. https://doi.org/10.1002/for.1061

Bregar, B. (2014, August 5). Price keeping carbon fiber from mass adoption. Plastic News. https://www.plasticsnews.com/article/20140805/NEWS/140809971/price-keeping-carbon-fiber-from-mass-adoption

Breitung, J., & Knuppel, M. (2018). How far can we forecast? Statistical tests of the predictive content. Deutsche Bundesbank Discussion Paper. https://www.econstor.eu/bitstream/10419/178500/1/1020157062.pdf

Brosius, D., & Das, S. (2017). IACMI Baseline Cost and Energy Metrics. IACMI. https://iacmi.org/wp-content/uploads/2017/12/IACMI-Baseline-Cost-and-Energy-Metrics-March-2017.pdf

Center for Automotive Research. (2020). Communications with industry stakeholders, companies and trade associations.

Chen, S.-L., Jackson, J. D., Kim, H., & Resiandini, P. (2014). What Drives Commodity Prices? American Journal of Agricultural Economics, 96(5), 1455–1468. https://doi.org/10.1093/ajae/aau014

Cook, J. J., & Booth, S. (2017a). Carbon Fiber Manufacturing Facility Siting and Policy Considerations: International Comparison. Clean Energy Manufacturing Analysis Center. https://www.nrel.gov/docs/fy17osti/66875.pdf

Cook, J. J., & Booth, S. (2017b). Carbon Fiber Manufacturing Facility Siting and Policy Considerations: International Comparison (NREL/TP--6A20-66875, 1365711; p. NREL/TP--6A20-66875, 1365711). https://doi.org/10.2172/1365711

Cost, Effectiveness, and Deployment of Fuel Economy Technologies for Light-Duty Vehicles (p. 21744). (2015). National Academies Press. https://doi.org/10.17226/21744

Das, S., Warren, J., West, D., & Laboratory, O. R. N. (2016). Global Carbon Fiber Composites Supply Chain Competitiveness Analysis. Technical Report, 116.

Das, S., Warren, J., West, D., & Schexnayder, S. M. (2016). Global Carbon Fiber Composites Supply Chain Competitiveness Analysis. Clean Energy Manufacturing Analysis Center. https://www.nrel.gov/docs/fy16osti/66071.pdf

DuckerFrontier North American Light Vehicle Aluminum Content and Outlook (August 2020). (n.d.). Retrieved September 29, 2020, from https://www.drivealuminum.org/research-resources/duckerfrontier-north-american-light-vehicle-aluminum-content-and-outlook-august-2020/

Ennis, B. L., Kelley, C. L., Naughton, B. T., Norris, R. E., Das, S., Lee, D., & Miller, D. A. (2019). Optimized Carbon Fiber Composites in Wind Turbine Blade Design. Sandia National Laboratories. https://www.energy.gov/sites/prod/files/2019/12/f69/SAND2019-14173-Optimized.pdf

Fuchs, E. R. H., Field, F. R., Roth, R., & Kirchain, R. E. (2008). Strategic materials selection in the automobile body: Economic opportunities for polymer composite design. Composites Science and Technology, 68(9), 1989–2002. https://doi.org/10.1016/j.compscitech.2008.01.015

Gargano, A., & Timmermann, A. (2014). Forecasting commodity price indexes using macroeconomic and financial predictors. International Journal of Forecasting, 30(3), 825–843. https://doi.org/10.1016/j.ijforecast.2013.09.003

Groen, J. J. J., & Pesenti, P. (2011, June 27). How Easy Is It to Forecast Commodity Prices? -Liberty Street Economics. Federal Reserve Bank of New York, Liberty Street Economics. https://libertystreeteconomics.newyorkfed.org/2011/06/how-easy-is-it-to-forecast-commodity-prices.html

Groen, J. J. J., & Pesenti, P. A. (2010). Commodity prices, commodity currencies, and global economic developments (Working Paper No. 15743; Working Paper Series). National Bureau of Economic Research. https://doi.org/10.3386/w15743

Harper International. (2020, April 13). How to Turn Pitch into Carbon Fiber for Automotive Applications. AZO Materials. https://www.azom.com/article.aspx?ArticleID=19200

Heuss, R., Müller, N., van Sintern, W., Starke, A., & Tschiesner, A. (2012). Lightweight, heavy impact. McKinsey & Company. https://www.mckinsey.com/~/media/mckinsey/dotcom/client_service/automotive%20and%20assembly/pdfs/lightweight_heavy_impact.ashx

Husain, A. M., & Bowman, C. (2004). Forecasting Commodity Prices: Futures Versus Judgment. IMF Working Papers, 04(41), 1. https://doi.org/10.5089/9781451846133.001

Issler, J. V., Rodrigues, C. F., & Burjack, R. (2013). Using common features to understand the behavior of metal-commodity prices and forecast them at different horizons [Working Paper]. Fundação Getulio Vargas. Escola de Pós-graduação em Economia. http://bibliotecadigital.fgv.br/dspace/handle/10438/10351

Jacks, D. S., & Stuermer, M. (2018). What Drives Commodity Price Booms and Busts? (SSRN Scholarly Paper ID 3265115). Social Science Research Network. https://doi.org/10.2139/ssrn.3265115

Kelly, T. D., & Matos, G. R. (2015). Historical Statistics for Mineral and Material Commodities in the United States. U.S. Geological Survey. https://www.usgs.gov/centers/nmic/historical-statistics-mineral-and-material-commodities-united-states

Krupitzer, R. (2012). A Steel Industry Perspective on Advanced High-Strength Steels. Steel Market Development Institute. https://www.nist.gov/system/files/documents/mml/acmd/structural_materials/2012_FEB-9-10-Krupitzer_AHSS-Workshop-FINALFINAL.pdf

Labys, W. C. (2003). New Directions in the Modeling and Forecasting of Commodity Markets. Mondes en developpement, no 122(2), 3–19. https://www.cairn.info/revue-mondes-en-developpement-2003-2-page-3.htm

Matos, G. R. (2015). Historical Global Statistics for Mineral and Material Commodities. U.S. Geological Survey. https://www.usgs.gov/centers/nmic/historical-global-statistics-mineral-and-material-commodities

Mazumdar, S. (2016). How to Succeed in the Automotive Market with Composites. Lucintel presentation to Composites Europe. https://www.composites-europe.com/dl/71/download/pdf/CE2016-Vortrag_01-12-2016_Lucintel.pdf

Modi, S. (2016). Material Qualification in the Automotive Industry. https://www.cargroup.org/publication/material-qualification-in-the-automotive-industry-2/

Modi, S., & Vadhavkar, A. (2019). Technology Roadmap: Materials and Manufacturing. Center for Automotive Research. https://www.cargroup.org/wp-content/uploads/2019/10/Technology-Roadmap_Materials-and-Manufacturing.pdf

Nunna, S., Blanchard, P., Buckmaster, D., Davis, S., & Naebe, M. (2019). Development of a cost model for the production of carbon fibres. Heliyon, 5(10). https://doi.org/10.1016/j.heliyon.2019.e02698

Papp, J. F., Bray, E. L., Edelstein, D. L., Fenton, M. D., Guberman, D. E., Hedrick, J. B., Jorgenson, J. D., Kuck, P. H., Shedd, K. B., & Tolcin, A. C. (2008). Factors that influence the price of Al, Cd, Co, Cu, Fe, Ni, Pb, Rare Earth Elements, and Zn. https://pubs.usgs.gov/of/2008/1356/

Pincheira, P., & Hardy, N. (n.d.). Forecasting Aluminum Prices with Commodity Currencies. November 2019. https://doi.org/10.13140/RG.2.2.32843.95525

Reinhart, C., & Borensztein, E. (1994, June). The Macroeconomic Determinants of Commodity Prices [MPRA Paper]. https://mpra.ub.uni-muenchen.de/6979/

Rocky Mountain Institute. (2013). Kickstarting the Widespread Adoption of Automotive Carbon Fiber Composites. Rocky Mountain Institute. https://rmi.org/wp-content/uploads/2017/05/RMI_Document_Repository_Public-Reprts_2013-01_AutocompositesFinalReport.pdf

Singh, H. (2016). Draft Final Report Mass Reduction for Light-Duty Vehicles for Model Years 2017-2025. EDAG, Inc. https://www.nhtsa.gov/staticfiles/rulemaking/pdf/cafe/Mass-Reduction-Feasibility-2014-Silverado.zip

Sloan, J. (2017). Autocomposites and the myth of $5/lb carbon fiber. CompositesWorld. https://www.compositesworld.com/articles/autocomposites-and-the-myth-of-5lb-carbon-fiber

United Nations. (2020, August). U.N. Comtrade Database. https://comtrade.un.org/data

U.S. Bureau of Economic Analysis. (2020a). Gross domestic product (implicit price deflator) [A191RD3A086NBEA] (Annual, 1929-2019). retrieved from FRED, Federal Reserve Bank of St. Louis. https://fred.stlouisfed.org/series/A191RD3A086NBEA

U.S. Bureau of Economic Analysis. (2020b). Gross Domestic Product: Implicit Price Deflator [GDPDEF] (Quarterly, 1947-2020:Q2). retrieved from FRED, Federal Reserve Bank of St. Louis. https://fred.stlouisfed.org/series/GDPDEF

U.S. Bureau of Labor Statistics. (2020). Producer Price Index by Commodity for Metals and Metal Products: Iron and Steel [WPU101]. retrieved from FRED, Federal Reserve Bank of St. Louis. https://fred.stlouisfed.org/series/WPU101

U.S. Geological Survey. (2012). Mineral Commodity Summaries: Aluminum. U.S. Geological Survey. https://s3-us-west-2.amazonaws.com/prd-wret/assets/palladium/production/mineral-pubs/aluminum/mcs-2012-alumi.pdf

U.S. Geological Survey. (2013). Metal Prices in the United States Through 2010. U.S. Geological Survey Scientific Investigations Report 2012–5188. https://pubs.usgs.gov/sir/2012/5188/

U.S. Geological Survey. (2016). Mineral Commodity Summaries: Aluminum. U.S. Geological Survey. https://s3-us-west-2.amazonaws.com/prd-wret/assets/palladium/production/mineral-pubs/aluminum/mcs-2016-alumi.pdf

U.S. Geological Survey. (2020). Mineral Commodity Summaries: Aluminum. U.S. Geological Survey. https://pubs.usgs.gov/periodicals/mcs2020/mcs2020-aluminum.pdf

Vehicle Technologies Office. (2013). WORKSHOP REPORT: Trucks and Heavy-Duty Vehicles Technical Requirements and Gaps for Lightweight and Propulsion Materials. Department of Energy, Energy Efficiency & Renewable Energy. https://www.energy.gov/sites/prod/files/2014/03/f13/wr_trucks_hdvehicles.pdf

Vehicle Technologies Office. (2019). Vehicle Lightweighting Program Progress Status – 2019. U.S. Department of Energy. https://www.energy.gov/sites/prod/files/2020/06/f75/DOR_Vehicle_Lightweighting_FINAL_DRAFT.pdf

Vehicles Technology Office. (2013). WORKSHOP REPORT: Light-Duty Vehicles Technical Requirements and Gaps for Lightweight and Propulsion Materials. Department of Energy, Energy Efficiency & Renewable Energy. https://www.energy.gov/sites/prod/files/2014/03/f13/wr_ldvehicles.pdf

Warren, C. D. (2011). Lower Cost, Higher Performance Carbon Fiber. Oak Ridge National Laboratory. https://www.energy.gov/sites/prod/files/2014/03/f11/compressed_hydrogen2011_4_warren.pdf

Winter, D. (n.d.). Lightweighting vs. Bigger Batteries in Electric Vehicles. 23.

WorldAutoSteel. (n.d.). Advanced High-Strength Steel (AHSS) Definitions. WorldAutoSteel. https://www.worldautosteel.org/steel-basics/automotive-advanced-high-strength-steel-ahss-definitions/

Appendix A: Curb Weight and Footprint Comparison

The charts below compare the baseline fleet's (33 top-selling vehicles in the MY2020 fleet) curb weight and footprints of vehicles with their previous generation.

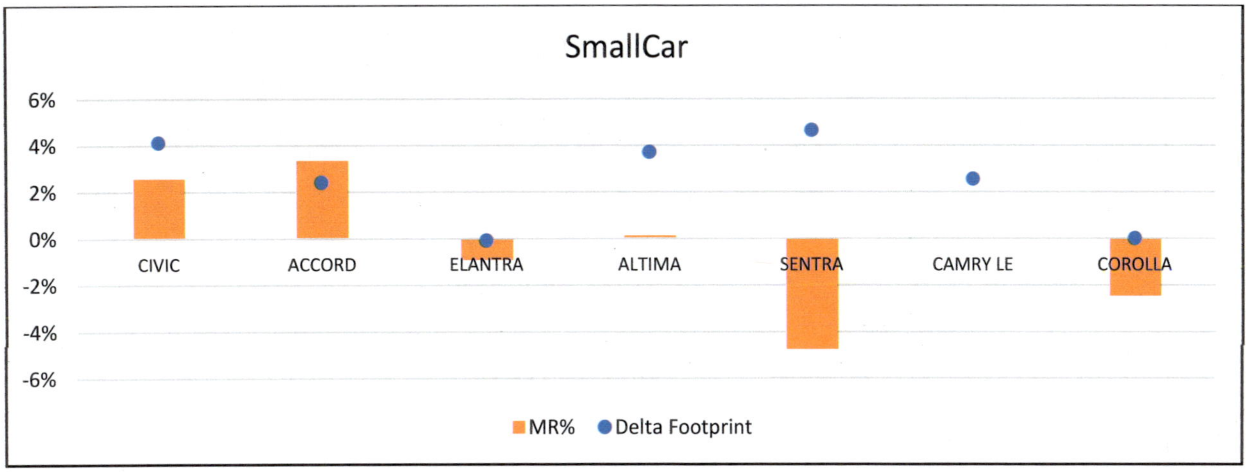

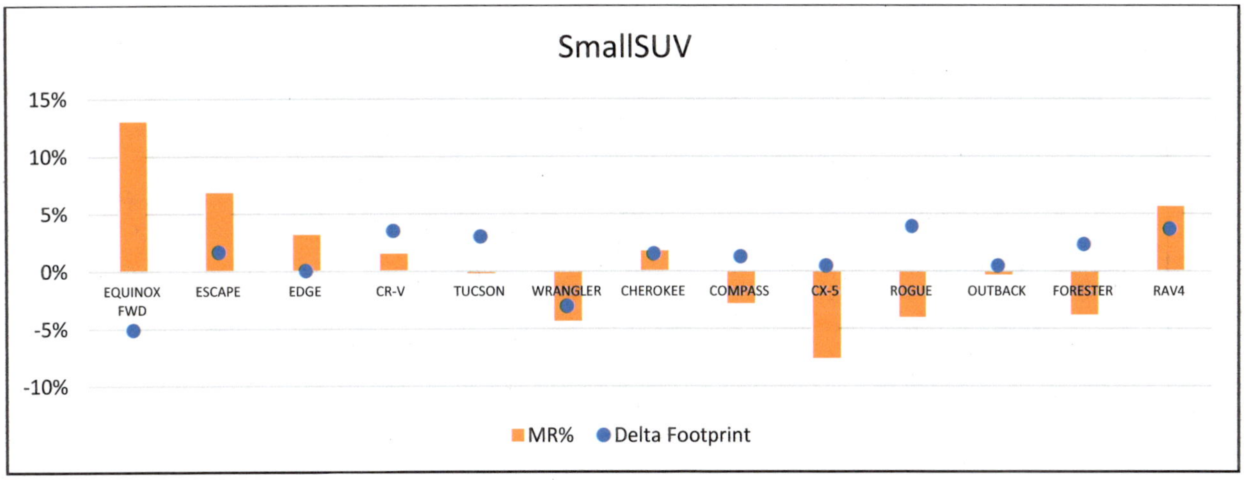

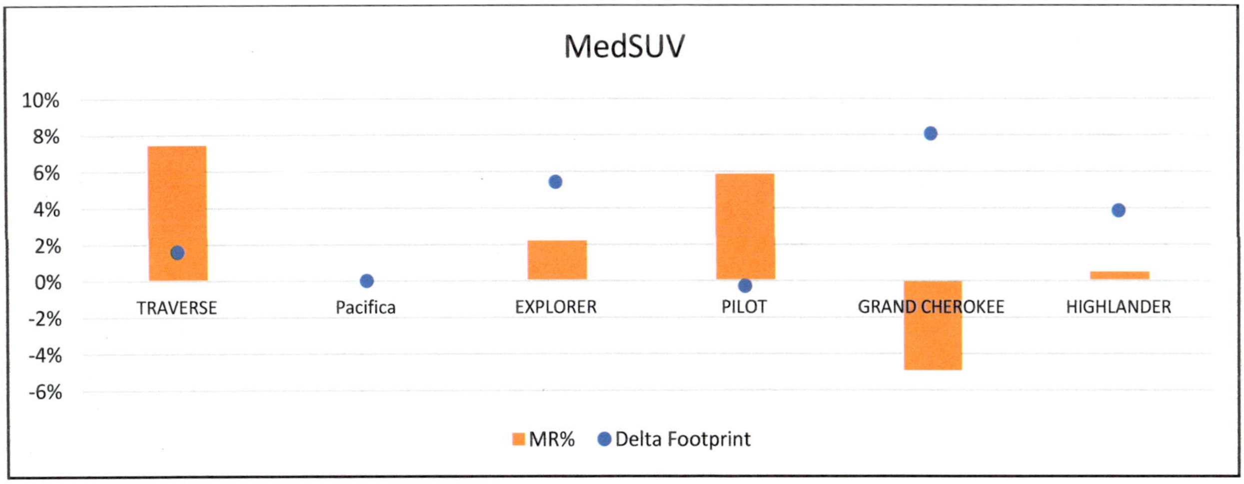

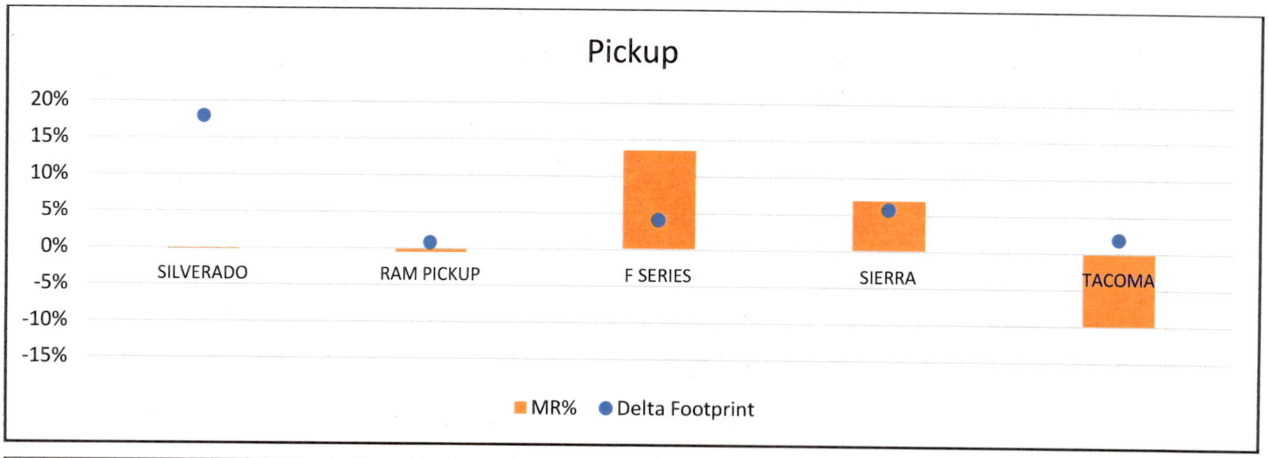
Pickup
20%
15%
10%
5%
0%
-5%
-10%
-15%
SILVERADO
RAM PICKUP
F SERIES
SIERRA
TACOMA
MR%
Delta Footprint

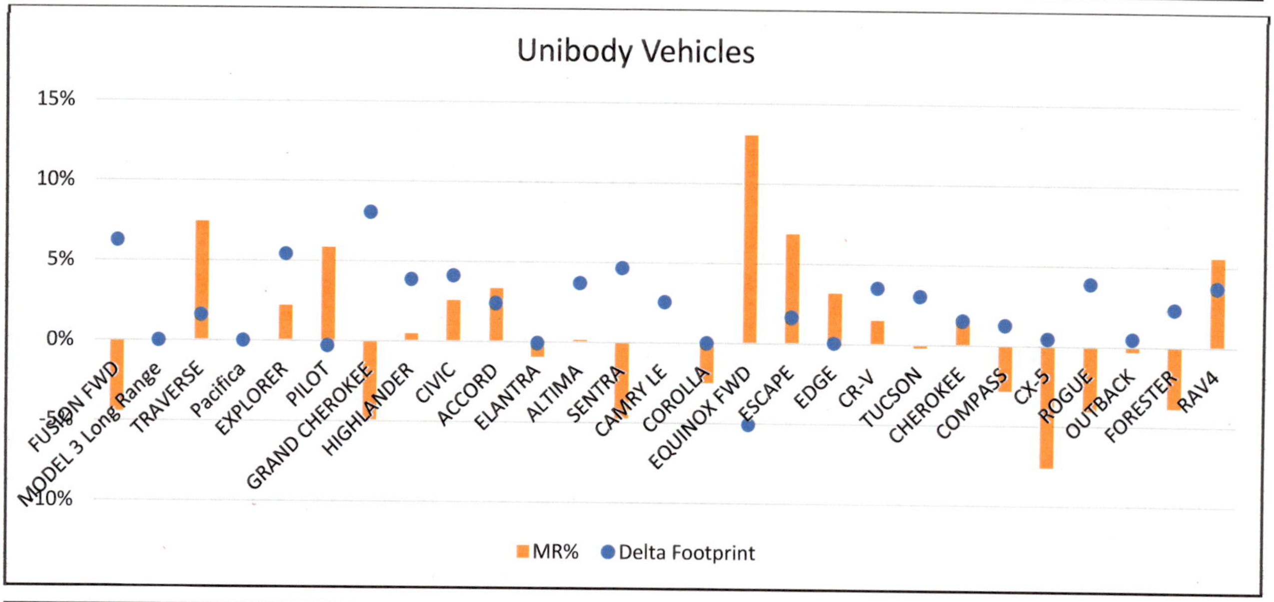
Unibody Vehicles
15%
10%
5%
0%
-10%
FUSION FWD
MODEL 3 Long Range
TRAVERSE
Pacifica
EXPLORER
PILOT
GRAND CHEROKEE
HIGHLANDER
CIVIC
ACCORD
ELANTRA
ALTIMA
SENTRA
CAMRY LE
COROLLA
EQUINOX FWD
ESCAPE
EDGE
CR-V
TUCSON
CHEROKEE
COMPASS
CX-5
ROGUE
OUTBACK
FORESTER
RAV4
MR%
Delta Footprint

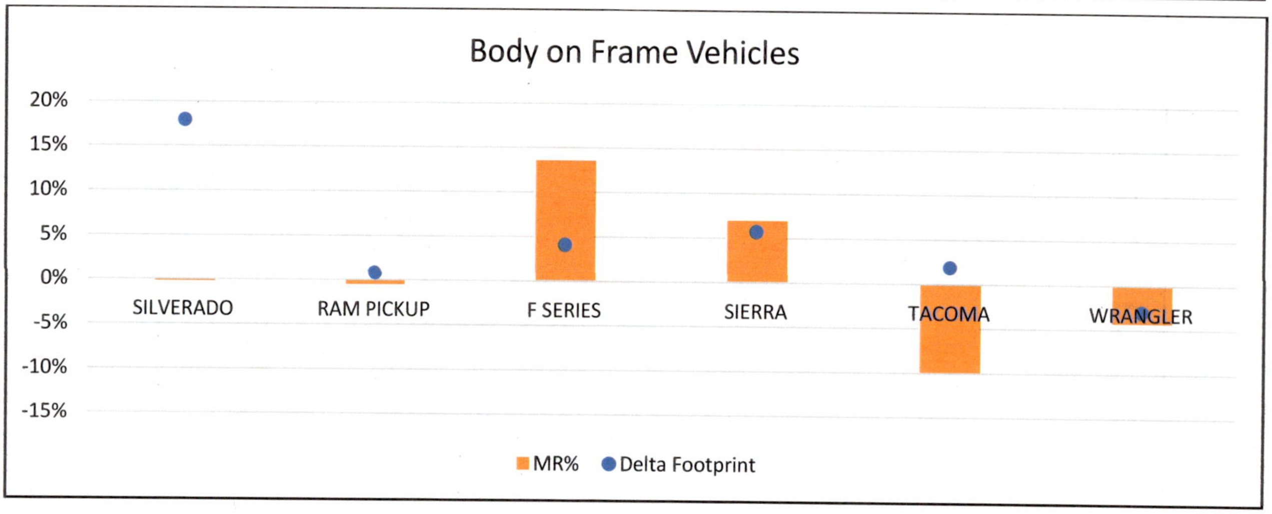
Body on Frame Vehicles
20%
15%
10%
5%
0%
-5%
-10%
-15%
SILVERADO
RAM PICKUP
F SERIES
SIERRA
TACOMA
WRANGLER
MR%
Delta Footprint